Biology

Laboratory Manual

seventh edition

Biology

Laboratory Manual

Darrell S. Vodopich
Baylor University

Randy Moore
University of Minnesota

Boston Burr Ridge, IL Dubuque, IA Madison, WI New York San Francisco St. Louis
Bangkok Bogotá Caracas Kuala Lumpur Lisbon London Madrid Mexico City
Milan Montreal New Delhi Santiago Seoul Singapore Sydney Taipei Toronto

BIOLOGY LABORATORY MANUAL, SEVENTH EDITION

Published by McGraw-Hill, a business unit of The McGraw-Hill Companies, Inc., 1221 Avenue of the Americas, New York, NY 10020.

Some ancillaries, including electronic and print components, may not be available to customers outside the United States.

 This book is printed on recycled, acid-free paper containing 10% postconsumer waste.

1 2 3 4 5 6 7 8 9 0 QPD/QPD 0 9 8 7 6 5 4

ISBN 0-07-255287-5

Publisher: *Martin J. Lange*
Senior sponsoring editor: *Patrick E. Reidy*
Developmental editor: *Anne L. Winch*
Marketing manager: *Tami Petsche*
Senior project manager: *Kay J. Brimeyer*
Production supervisor: *Kara Kudronowicz*
Senior media project manager: *Jodi K. Banowetz*
Senior media technology producer: *John J. Theobald*
Designer: *Rick D. Noel*
Cover designer: *Rokusek Design*
Cover photos courtesy of the author
Senior photo research coordinator: *Lori Hancock*
Photo research: *Mary Reeg*
Compositor: *GAC--Indianapolis*
Typeface: *10/12 Goudy*
Printer: *Quebecor World Dubuque, IA*

The credits section for this book begins on page 553 and is considered an extension of the copyright page.

Some of the laboratory experiments included in this text may be hazardous if materials are handled improperly or if procedures are conducted incorrectly. Safety precautions are necessary when you are working with chemicals, glass test tubes, hot water baths, sharp instruments, and the like, or for any procedures that generally require caution. Your school may have set regulations regarding safety procedures that your instructor will explain to you. Should you have any problems with materials or procedures, please ask your instructor for help.

www.mhhe.com

Contents

Preface vii
Welcome to the Biology Laboratory viii

Exercise 1
Measurements in Biology: The Metric System and Data Analysis 1

Exercise 2
The Microscope: Basic Skills of Light Microscopy 9

Exercise 3
The Cell: Structure and Function 21

Exercise 4
Solutions, Acids, and Bases: The pH Scale 37

Exercise 5
Biologically Important Molecules: Carbohydrates, Proteins, Lipids, and Nucleic Acids 45

Exercise 6
Separating Organic Compounds: Column Chromatography, Paper Chromatography, and Gel Electrophoresis 57

Exercise 7
Spectrophotometry: Identifying Solutes and Determining Their Concentration 67

Exercise 8
Diffusion and Osmosis: Passive Movement of Molecules in Biological Systems 79

Exercise 9
Cellular Membranes: Effects of Physical and Chemical Stress 93

Exercise 10
Enzymes: Factors Affecting the Rate of Activity 101

Exercise 11
Respiration: Aerobic and Anaerobic Oxidation of Organic Molecules 113

Exercise 12
Photosynthesis: Pigment Separation, Starch Production, and CO_2 Uptake 125

Exercise 13
Mitosis: Replication of Eukaryotic Cells 137

Exercise 14
Meiosis: Reduction Division and Gametogenesis 147

Exercise 15
Molecular Biology and Biotechnology: DNA Isolation and Bacterial Transformation 157

Exercise 16
Genetics: The Principles of Mendel 165

Exercise 17
Evolution: Natural Selection and Morphological Change in Green Algae 175

Exercise 18
Human Evolution: Skull Examination 189

Exercise 19
Ecology: Diversity and Interaction in Plant Communities 197

Exercise 20
Community Succession 207

Exercise 21
Population Growth: Limitations of the Environment 213

Exercise 22
Pollution: The Effects of Chemical, Thermal, and Acid Pollution 221

Exercise 23
Survey of Bacteria: Kingdoms Archaebacteria and Bacteria 231

Exercise 24
Survey of the Kingdom Protista: The Algae 247

Exercise 25
Survey of the Kingdom Protista: Protozoa and Slime Molds 261

Exercise 26
Survey of the Kingdom Fungi: Molds, Sac Fungi, Mushrooms, and Lichens 271

Exercise 27
Survey of the Plant Kingdom: Liverworts, Mosses, and Hornworts of Phyla Hepaticophyta, Bryophyta, and Anthocerophyta 283

Exercise 28
Survey of the Plant Kingdom: Seedless Vascular Plants of Phyla Pterophyta, Lycophyta, Psilophyta, and Sphenophyta 293

Exercise 29
Survey of the Plant Kingdom: Gymnosperms of Phyla Cycadophyta, Ginkgophyta, Coniferophyta, and Gnetophyta 303

Exercise 30
Survey of the Plant Kingdom: Angiosperms 313

Exercise 31
Plant Anatomy: Vegetative Structure of Vascular Plants 327

Exercise 32
Plant Physiology: Transpiration 341

Exercise 33
Plant Physiology: Tropisms, Nutrition, and Growth Regulators 347

Exercise 34
Bioassay: Measuring Physiologically Active Substances 359

Exercise 35
Survey of the Animal Kingdom: Phyla Porifera and Cnidaria 365

Exercise 36
Survey of the Animal Kingdom: Phyla Platyhelminthes and Nematoda 377

Exercise 37
Survey of the Animal Kingdom: Phyla Mollusca and Annelida 389

Exercise 38
Survey of the Animal Kingdom: Phylum Arthropoda 401

Exercise 39
Survey of the Animal Kingdom: Phyla Echinodermata, Hemichordata, and Chordata 415

Exercise 40
Vertebrate Animal Tissues: Epithelial, Connective, Muscular, and Nervous Tissues 435

Exercise 41
Human Biology: The Human Skeletal System 451

Exercise 42
Human Biology: Muscles and Muscle Contraction 457

Exercise 43
Human Biology: Breathing 465

Exercise 44
Human Biology: Circulation and Blood Pressure 473

Exercise 45
Human Biology: Sensory Perception 485

Exercise 46
Vertebrate Anatomy: External Features and Skeletal System of the Rat 495

Exercise 47
Vertebrate Anatomy: Muscles and Internal Organs of the Rat 503

Exercise 48
Vertebrate Anatomy: Urogenital and Circulatory Systems of the Rat 511

Exercise 49
Embryology: Comparative Morphologies and Strategies of Development 521

Exercise 50
Animal Behavior: Taxis, Kinesis, and Agonistic Behavior 531

Appendix I
Dissection of a Fetal Pig 539

Appendix II
How to Write a Scientific Paper or Laboratory Report 547

Appendix III
Conversion of Metric Units to English Units 551

Credits 553

Preface

We designed this laboratory manual for an introductory biology course with a broad survey of basic laboratory techniques. The experiments and procedures are simple, safe, easy to perform, and especially appropriate for large classes. Few experiments require more than one class meeting to complete the procedure. Each exercise includes many photographs, traditional topics, and experiments that help students learn about life. Procedures within each exercise are numerous and discrete so that an exercise can be tailored to the needs of the students, the style of the instructor, and the facilities available.

TO THE STUDENT

We hope this manual is an interesting guide to many areas of biology. As you read about these areas, you'll probably spend equal amounts of time observing and experimenting. Don't hesitate to go beyond the observations that we've outlined—your future success as a scientist depends on your ability to seek and notice things that others may overlook. Now is the time to develop this ability with a mixture of hard work and relaxed observation. Have fun, and learning will come easily. Also, remember that this manual is designed with your instructors in mind as well. Go to them often with questions—their experience is a valuable tool that you should use as you work.

TO THE INSTRUCTOR

This manual's straightforward approach emphasizes experiments and activities that optimize students' investment of time and your investment of supplies, equipment, and preparation. Simple, safe, and straightforward experiments are most effective if you interpret the work in depth. Most experiments can be done easily by a student in two to three hours. Terminology, structures, photographs, and concepts are limited to those the student can readily observe and understand. In each exercise we have included a few activities requiring a greater investment of effort if resources are available, but omitting them will not detract from the objectives.

This manual functions best with an instructor's guidance, and is not an autotutorial system. We've tried to guide students from observations to conclusions, to help students make their own discoveries, and to make the transition from observation to biological principles. But discussions and interactions between student and instructor are major components of a successful laboratory experience. Be sure to examine the "Questions for Further Thought and Study" in each exercise. We hope they will help you expand students' perceptions that each exercise has broad applications to their world.

THE SEVENTH EDITION

All exercises in this edition were critiqued by a review panel of current users, and their suggested revisions were carefully considered and incorporated. Classifications schemes presented in the diversity exercises have been updated to reflect the onslaught of molecular data being used in modern systematics. Introductory material has been reviewed and revised for completeness. The number of tables and figures has been extended with more than 70 figures being either revised or replaced.

Darrell S. Vodopich
Randy Moore

Reviewers

We thank the following reviewers for their helpful comments and suggestions during the preparation of this new edition.

Ralph G. Benedetto, Jr.	Wayne Community College
Lynn C. Burgess	Dickinson State University
Peter Eden	Marywood University
M. C. Hart	Minnesota State University–Mankato
Ramsey M. Kafoury	Jackson State University
Tim H. Lindblom	Lyon College
Craig Longtine	North Hennepin Community College
Kelly M. Major	University of South Alabama
Mary Catharine McElwain	Loyola Marymount University
Gary B. Peterson	South Dakota State University
Barbra A. Roller	Florida International University
Patricia Rugaber	Coastal Georgia Community College
Sarah H. Swain	Middle Tennessee State University
Franklyn Tan Te	Florida International University
David J. Thomas	Lyon College
Rukmani Viswanath	Laredo Community College
Carol Wake	South Dakota State University

Welcome to the Biology Laboratory

Welcome to the biology laboratory. Although reading your textbook and attending lectures are important ways of learning about biology, nothing can replace the importance of the laboratory. Indeed, in lab you'll get hands-on experience with what you've heard and read about biology—for example, you'll observe organisms, do experiments, test ideas, collect data, and make conclusions about what you've learned. That is, you'll *do* biology.

You'll enjoy the exercises in this manual—they're interesting, informative, and can be completed within the time limits of your laboratory period. We've provided questions to test your understanding of what you've done; in some of the exercises, we've also asked you to devise your own experiments to answer questions that you've posed. To make these exercises most useful and enjoyable, follow the guidelines discussed below:

THE IMPORTANCE OF COMING TO CLASS

Biology labs are designed to help you experience biology first-hand. To do that, you must attend class. Indeed, if you want to do well in your biology course, attend class and pay attention. To appreciate the importance of class attendance for making a good grade in your biology course, examine the following graph of how students' grades in an introductory biology course relate to their rates of class attendance. Data are from a General Biology class, University of Minnesota, 2003.

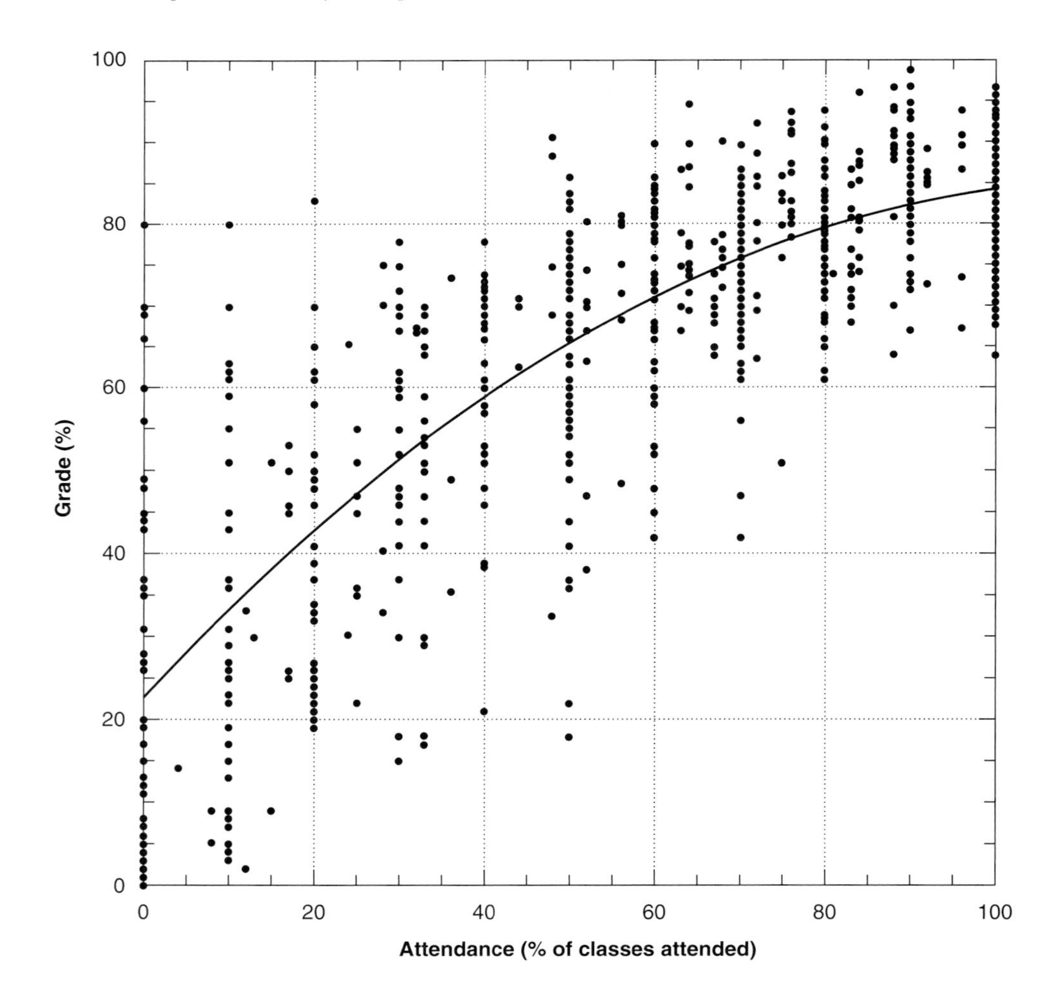

In the space below, write your interpretation of these data.

BEFORE COMING TO LAB

Read the exercise before coming to lab. This will give you a general idea about what you're going to do, as well as why you're going to do it. Knowing this will not only save time, it will also help you finish the experiments.

WHEN IN LAB

1. Don't start the exercise until you've discussed the exercise with your laboratory instructor. She/he will give you specific instructions about the lab and tell you how the exercise may have been modified.
2. Stay focused as you work. You'll be able to finish each exercise within the allotted time if you stay busy. You'll not be able to finish the exercise if you spend your time talking about this weekend's party or last week's big game.
3. Discuss your observations, results, and conclusions with your instructor and lab partners. Perhaps their comments and ideas will help you better understand what you've observed.
4. Always follow instructions and follow safety guidelines presented by your instructor.
5. If you have questions, ask your instructor.

SAFETY IN THE LABORATORY

The exercises in this manual were designed with safety as a top priority. Always follow these safety precautions:

1. Do not eat, drink, smoke, or apply cosmetics when in the lab.
2. Use the equipment properly. If you have any questions or problems, contact your instructor.
3. Clean up spills immediately.
4. Report all injuries—no matter how minor—immediately to your instructor.
5. If you have long hair, tie it back. If you use open flames, roll up loose sleeves.
6. Never taste any substance or solution. Do not put anything in the lab into your mouth.
7. Treat all live animals gently and with respect.
8. The locations of lab safety equipment will be pointed out to you during the first lab. Familiarize yourself with the location and operation of this equipment.
9. At the end of each lab, clean your work area, wash your hands thoroughly with soap, and return all equipment and supplies to their original locations.

AFTER EACH LABORATORY

Soon after each lab, review what you did. What questions did you answer? What data did you gather? What conclusions did you make?

Also note any questions that remain. Try to answer these questions by using your textbook or visiting the library. If you can't answer the questions, discuss them with your instructor.

Welcome to the laboratory!

1

Measurements in Biology
The Metric System and Data Analysis

Objectives

By the end of this exercise you should be able to:

1. Identify the metric units used to measure length, volume, mass, and temperature.
2. Measure length, volume, mass, and temperature in metric units.
3. Convert one metric unit to another (e.g., grams to kilograms).
4. Use measures of volume and mass to calculate density.
5. Practice the use of simple statistical calculations such as mean, median, range, and standard deviation.
6. Analyze sample data using statistical tools.

Every day we're bombarded with numbers and measurements. They come at us from all directions, including while we're at the supermarket, gas station, golf course, and pharmacy, as well as while we're in our classrooms and kitchens. Virtually every package that we touch is described by a measurement.

Scientists use a standard method to collect data as well as use mathematics to analyze those data. Measuring things is a must before we can objectively describe what we are observing, before we can experiment with biological processes, and before we can predict how organisms respond, adjust to, and modify their world. Once we have made our measurements, we can analyze our data and look for variation and the sources of that variation. Then we can infer the causes and effects of the biological processes that interest us.

THE METRIC SYSTEM

Scientists throughout the world use the **metric system** to make measurements. The metric system is also used in everyday life virtually everywhere except the United States. With few exceptions (e.g., liter bottles of soda, 35-mm film), most measurements in the United States use the antiquated English system of pounds, inches, feet, and so on. Check with your instructor about bringing to class common grocery store items with volumes and weights in metric units, or examining those items on display.

Scientists make all of their measurements in the metric system; they do not routinely convert from one system to another. Thus, this exercise will not involve conversions from the English to metric systems (if you want to know those conversions, see Appendix III). Rather, this exercise will introduce you to making metric measurements of length, mass, volume, and temperature. During this lab, you should spend your time making measurements, not reading background information. Therefore, *before lab, read this exercise carefully to familiarize yourself with the basic units of the metric system*.

Metric units commonly used in biology include:

meter (m)—the basic unit of length

liter (L)—the basic unit of volume

kilogram (kg)—the basic unit of mass

degree Celsius (°C)—the basic unit of temperature

Unlike the English system with which you are already familiar, the metric system is based on units of ten. This simplifies conversions from one metric unit to another (e.g., from kilometers to meters). This base-ten system is similar to our monetary system, in which 10 cents equals a dime, 10 dimes equals a dollar, and so forth. Units of ten in the metric system are indicated by Latin and Greek prefixes placed before the base units:

Prefix (Latin)		*Division of Metric Unit*	
deci	(d)	0.1	10^{-1}
centi	(c)	0.01	10^{-2}
milli	(m)	0.001	10^{-3}
micro	(μ)	0.000001	10^{-6}
nano	(n)	0.000000001	10^{-9}
pico	(p)	0.000000000001	10^{-12}

Prefix (Greek)		*Multiple of Metric Unit*	
deka	(da)	10	10^{1}
hecto	(h)	100	10^{2}
kilo	(k)	1000	10^{3}
mega	(M)	1000000	10^{6}
giga	(G)	1000000000	10^{9}

Thus, multiply by:

0.01 to convert centimeters to meters

0.001 to convert millimeters to meters

1000 to convert kilometers to meters

0.1 to convert millimeters to centimeters

For example, 620 meters = 0.620 kilometers = 620,000 millimeters = 62,000 centimeters.

Question 1

Make the following metric conversions:

1 meter = ___ centimeters = ___ millimeters

92.4 millimeters = ___ meters = ___ centimeters

10 kilometers = ___ meters = ___ decimeters

Length and Area

The **meter** (m) is the basic unit of length. Units of area are squared units (i.e., two-dimensional) of length.

1 m = 100 cm = 1000 mm = 0.001 km = 1×10^{-3} km

1 km = 1000 m = 10^3 m

1 cm = 0.01 m = 10^{-2} m = 10 mm

470 m = 0.470 km

1 cm^2 = 100 mm^2 (i.e., 10 mm × 10 mm = 100 mm^2)

To help you appreciate the magnitudes of these units, here are the lengths and areas of some familiar objects:

Length	
Housefly	0.5 cm
Mt. Everest	8848 m
Diameter of penny	1.9 cm
Toyota Camry	4.7 m

Area	
Total skin area of adult human male	1.8 m^2
Football field (goal line to goal line)	4459 m^2
Surface area of human lungs	80 m^2
Central Park (New York City)	3.4 km^2
Ping-Pong table	4.18 m^2
Credit card	46 cm^2

Procedure 1.1

Make metric measurements of length and area

Most biologists measure lengths with metric rulers or metersticks.

1. Examine intervals marked on the metric rulers and metersticks available in the lab.
2. Make the following measurements. Be sure to include units for each measurement.

Length of this page ________

Width of this page ________

Area of this page (Area = Length × Width) ________

Your height ________

Thickness of this manual ________

Height of a 200-ml beaker ________

Height of ceiling ________

Question 2

What are some potential sources of error in your measurements?

Volume

Volume is the space occupied by an object. Units of volume are cubed (i.e., three-dimensional) units of length. The liter (L) is the basic unit of volume.

$$1 \text{ L} = 1000 \text{ cm}^3 = 1000 \text{ mL}$$
$$1 \text{ L} = 0.1 \text{ m} \times 0.1 \text{ m} \times 0.1 \text{m}$$
$$1 \text{ cm}^3 = 0.000001 \text{ m}^3$$

To help you appreciate the magnitudes of these units, here are the volumes of some familiar objects:

Chicken egg	60 mL
One breath of air	500 cm^3
Coke can	355 mL

Scientists often measure volumes with pipets and graduated cylinders. Pipets are used to measure small volumes, typically 25 mL or less. Liquid is drawn into a pipet using a bulb or pipet pump (fig. 1.1). Never pipet by mouth.

Graduated cylinders are used to measure larger volumes. To appreciate how to make a measurement accurately, pour 40–50 mL of water into a 100-mL graduated cylinder

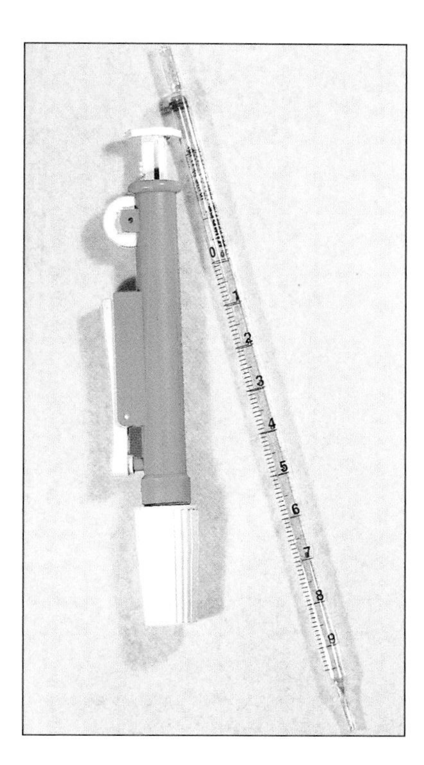

Figure 1.1

A pipet is used to extract and dispense volumes of liquid. A suction bulb (shown in green on the left) draws fluid into the pipet, and graduated markings on the pipet allow precise measurement of a fluid's volume. Never use your mouth to suck fluid into a pipet.

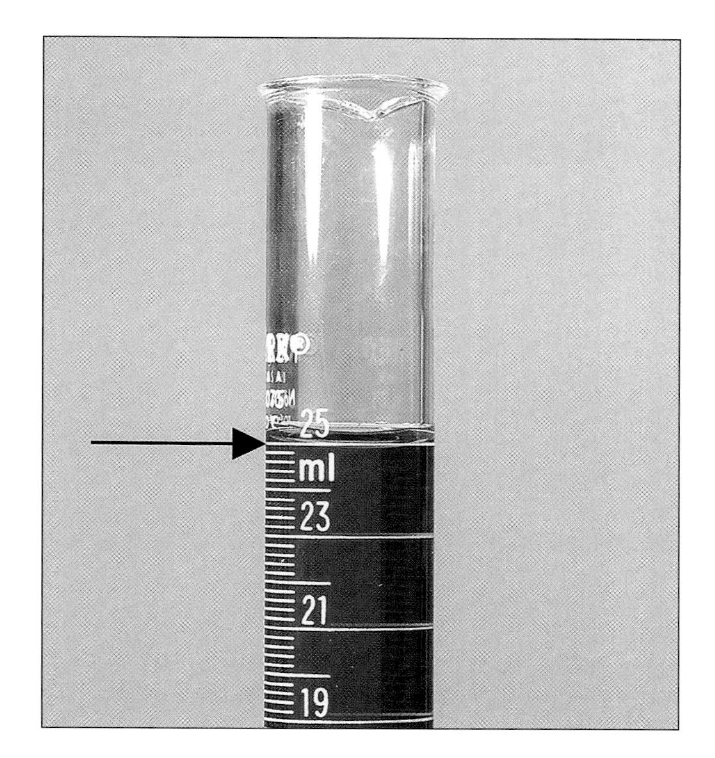

Figure 1.2

When measuring the volume of liquid in a graduated cylinder, always measure at the bottom of the meniscus. The bottom of the meniscus in this photograph is indicated by the arrow. The correct volume is 25 mL.

and observe the interface between the water and air. This interface, which is called the **meniscus,** is curved because of surface tension and the adhesion of water to the sides of the cylinder. When measuring the liquid in a cylinder such as a graduated cylinder, always position your eyes level with the meniscus and read the volume at the lowest level (fig. 1.2).

Procedure 1.2

Make metric measurements of volume

1. Biologists often use graduated cylinders to measure volumes. Locate the graduated cylinders available in the lab to make the following measurements. Determine what measurements the markings on the graduated cylinder represent. Be sure to include units for each measurement.
2. Measure the milliliters needed to fill a cup (provided in the lab). ________
3. Measure the liters in a gallon. ________

Procedure 1.3

Measure the volume of a solid object by water displacement

1. Obtain a 100-mL graduated cylinder, a thumb-sized rock, and a glass marble.
2. Fill the graduated cylinder with 70 mL of water.
3. Submerge the rock in the graduated cylinder and notice that the volume of the contents rises.
4. Carefully observe the meniscus of the fluid and record its volume.
5. Calculate and record the volume of the rock by subtracting the original volume (70 mL) from the new volume.

   Rock volume ________
6. Repeat steps 2–5 to measure and record the volume of the marble.

   Marble volume ________

Biologists use pipets to measure and transfer small volumes of liquid from one container to another. The following procedure will help you appreciate the usefulness of pipets.

Procedure 1.4

Learn to use a pipet

1. Add approximately 100 mL of water to a 100-mL beaker.
2. Use a 5-mL pipet with a bulb or another filling device provided by your instructor to remove some water from the beaker.
3. Fill the pipet to the zero mark.

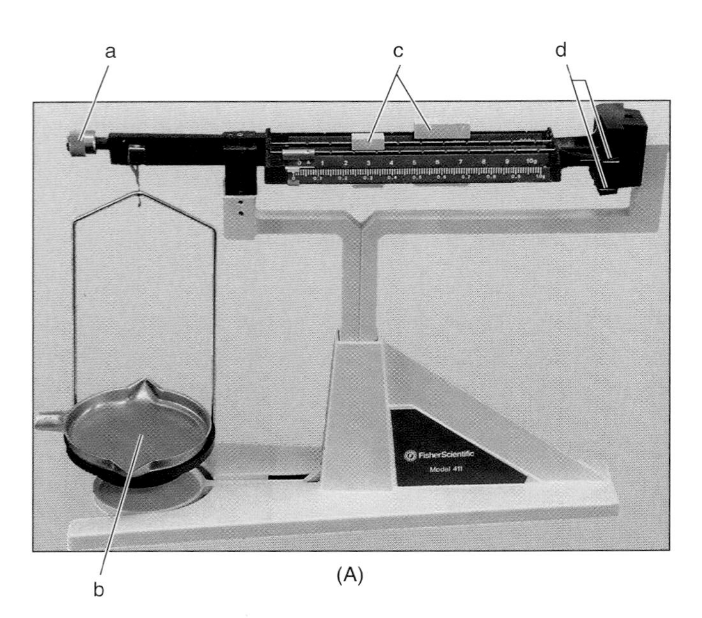

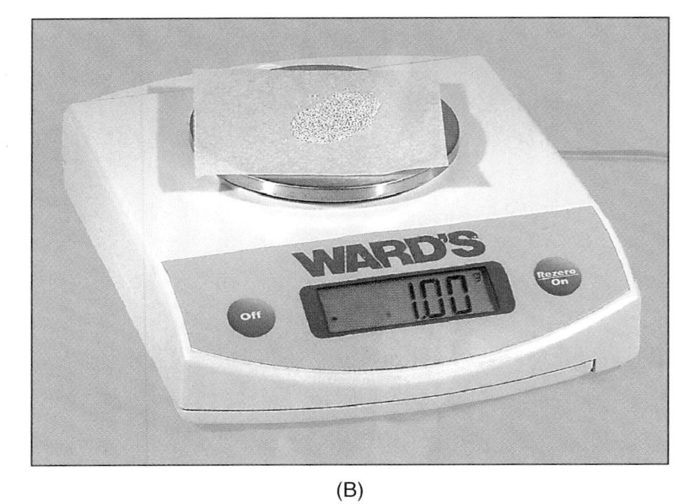

Figure 1.3
Biologists use balances to measure mass. (A) The parts of a triple-beam balance include (*a*) the zero-adjustment knob, (*b*) measuring pan, (*c*) movable masses on horizontal beams, and (*d*) balance marks. (B) The top loading balance has a measuring pan, a power switch, and a zero calibration button.

4. To read the liquid level correctly, your eye must be directly in line with the bottom of the meniscus.
5. Release the liquid into another container.

Question 3
What volume of liquid did you measure?

Mass

The **kilogram** (kg) is the basic unit of mass.[1] A kilogram equals the mass of one thousand cubic centimeters (cm^3) of water at 4°C. Similarly,

$$1 \text{ kg} = 1000 \text{ g} = 10^3 \text{ g}$$
$$1 \text{ mg} = 0.001 \text{ g} = 10^{-3} \text{ g}$$

Here are the masses of some familiar objects:

9V battery	40 g
Ping-Pong ball	2.45 g
Basketball	0.62 kg
Quarter	6.25 g

Biologists often measure mass with a triple-beam balance (fig 1.3), which gets its name from its three horizontal beams. Suspended from each of the three beams are movable masses. Each of the three beams of the balance are marked with graduations: the closest beam has 0.1-g graduations, the middle beam has 100-g graduations, and the farthest beam has 10-g graduations.

[1] *Remember that mass is not necessarily synonymous with weight. Mass measures an object's potential to interact with gravity, whereas weight is the force exerted by gravity on an object. Thus, a weightless object in outer space has the same mass as it has on earth.*

Before making any measurements, clean the weighing pan and move all of the suspended weights to the far left. The balance marks should line up to indicate zero grams; if they do not, turn the adjustment knob until they do. Measure the mass of an object by placing it in the center of the weighing pan and moving the suspended masses until the beams balance. The mass of the object is the sum of the masses indicated by the weights on the three beams.

Procedure 1.5
Make metric measurements of mass

1. Biologists often use a triple-beam balance or a top loading scale to measure mass. Locate the triple-beam balances or scales in the lab.
2. Measure the masses of the following items. Be sure to include units for each measurement.

   Nickel ________
   Paper clip ________
   Pencil ________
   Rock (used in procedure 1.3) ________
   100-mL beaker (empty) ________
   100-mL beaker containing 50 mL of water ________

Question 4
a. **Density** is mass per unit volume. Use data that you've gathered to determine the density of water at room temperature.

Density of water = (mass/volume) = ________

b. What is the density of the wooden pencil? Does it float? Why?

c. What is the density of the rock? Does it sink? Why?

Temperature

Temperature is the measure of the kinetic energy of molecules—that is, the amount of heat in a system. Biologists measure temperature with a thermometer calibrated in degrees Celsius (°C). The Celsius scale is based on water freezing at 0°C and boiling at 100°C. You can interconvert °C and degrees Fahrenheit (°F) using the formula 5(°F) = 9(°C) + 160. Here are some typical temperatures:

40°C	a very hot summer day
30.6°C	butter melts
75°C	hot coffee
−20°C	temperature in a freezer
37°C	human body temperature

Procedure 1.6

Make metric measurements of temperature

1. Obtain a thermometer in the lab. Handle a thermometer with care. If it breaks, notify your instructor immediately.
2. Determine the range of the temperatures that can be measured with your thermometer by examining the scale imprinted along the barrel of the thermometer.
3. Measure the following temperatures:

   Room temperature ________ °C

   Cold tap water ________ °C

   Hot tap water ________ °C

   Inside refrigerator ________ °C

Hints for Using the Metric System

1. Express measurements in units requiring only a few decimal places. For example, 0.3 m is more easily manipulated and understood than 300000000 nm.
2. When measuring pure water, the metric system offers an easy and common conversion from volume measured in liters to volume measured in cubic meters to mass measured in grams: 1 mL = 1 cm^3 = 1 g.
3. The metric system uses symbols rather than abbreviations. Therefore, do not place a period after metric symbols (e.g., 1 g, not 1 g.). Use a period after a symbol only at the end of a sentence.
4. Do not mix units or symbols (e.g., 9.2 m, not 9 m 200 mm).
5. Metric symbols are always singular (e.g., 10 km, not 10 kms).
6. Except for degree Celsius, always leave a space between a number and a metric symbol (e.g., 20 mm, not 20mm; 10°C, not 10° C).
7. Use a zero before a decimal point when the number is less than one (e.g., 0.42 m, not .42 m).

UNDERSTANDING NUMERICAL DATA

Statistics offer a way to organize, summarize, and describe data—the data are usually samples of information from a much larger population of values. Statistics and statistical tests allow us to analyze the sample and draw inferences about the entire population. Consequently, the use of statistics enables us to make decisions even though we have incomplete data about a population. Although this may seem unscientific, we do it all the time; for example, we diagnose diseases with a drop of blood. Decisions are based on statistics when it is impossible or unrealistic to analyze an entire population.

Let's say that you want to know the mass of a typical apple in your orchard. To obtain this information, you could analyze one apple, but how would you know that you'd picked a "typical" sample? After all, the batch from which you chose the apple may contain many others, each a little bit different. You'd get a better estimate of "typical" if you increased your sample size to a few hundred apples, or even to 10,000. Or, better yet, to 1,000,000.

The only way to be certain of your conclusions would be to measure all the apples in your orchard. Since this is clearly impossible, you must choose apples that *represent* all of the other apples—that is, you must be working with a *representative sample*. A statistical analysis of those sample apples reduces the sample-values to a few characteristic measurements (e.g., mean mass). As you increase the size of the sample, these characteristic measurements provide an ever-improving estimation of what is "typical."

There are a variety of software programs that perform statistical analyses of data; all you have to do is enter your data into a spreadsheet, select the data that you want to analyze, and perform the analysis. Although these software packages save time and can increase accuracy, you still need to understand a few of the basic variables that you'll use to understand your numerical data. We'll start with the mean and median:

The **mean** is the arithmetic average of a group of measurements. Chance errors in measurements tend to cancel themselves when means are calculated for samples that are relatively large; a value that is too high because of random error is often balanced by a value that is too low for the same reason.

The **median** is the middle value of a group of measurements.

The median is less sensitive to extreme values than is the mean. To appreciate this, consider a sample consisting of 14 leaves having the following lengths (all in mm):

80 69 62 74 69 51 45 40 9 64 65 64 61 67

The mean length is 58.6 mm. However, none of the leaves are that length, and most of the leaves are longer than 60 mm.

Question 5

a. Does the mean always describe the "typical" measurement? Why or why not?

b. What information about a sample does a mean *not* provide?

Determine the median by arranging the measurements in numerical order:

9 40 45 51 61 63 64 64 65 67 69 69 73 80

The median is between the seventh and eighth measurement: 64 mm. Note that in this sample, the mean differs from the median.

Question 6

a. What is responsible for this difference between the mean and median?

b. How would the median change if the 9-mm-long leaf was not in the sample?

c. How would the mean change if the 9-mm-long leaf was not in the sample?

d. Consider these samples:

Sample 1: 25 35 32 28

Sample 2: 15 75 10 20

What is the mean for Sample 1? ________

What is the mean for Sample 2? ________

In most of the exercises in this manual, you'll have time to make only one or two measurements of a biological structure or phenomenon. In these instances, a mean may be the only descriptor of the sample. However, if your class combines its data so that there are many measurements, you'll need to know how to do a couple of other calculations so that you understand the variation within your sample.

Variability

As you can see, the samples in Question 6d are different, but their means are the same. Thus, the mean does not reveal all there is to know about these samples. To understand how these samples are different, you need other statistics: the range and standard deviation.

The **range** is the difference between the extreme measurements (i.e., smallest and largest) of the sample. In Sample 1, the range is 35 − 25 = 10; in Sample 2 the range is 75 − 10 = 65. The range provides a sense of the variation of the sample, but the range can be artificially inflated by one or two extreme values. Notice the extreme values in the sample of leaf measurements previously discussed. Moreover, ranges do not tell us anything about the measurements between the extremes.

Question 7

a. Could two samples have the same mean but different ranges? Explain.

b. Could two samples have the same range but different means? Explain.

The **standard deviation** indicates how measurements vary about the mean. The standard deviation is easy to calculate. Begin by calculating the mean, measuring the deviation of each sample from the mean, squaring each deviation, and then summing the deviations. This summation results in the **sum of squared deviations.** For example, consider a group of shrimp that are 22, 19, 18, and 21 cm long. The mean length of these shrimp is 20 cm.

Sample Value	*Mean*	*Deviation*	*(Deviation)^2*
22	20	2	4
19	20	−1	1
21	20	1	1
18	20	−2	4

Sum of Squared Deviations = 10

The summary equation for the sum of squared deviations is:

$$\textbf{Sum of squared deviations} = \sum_{i=1}^{N} (x_i - \bar{x})^2$$

1–6

where

N = total number of samples

$\bar{x}$ = the sample mean

x_i = measurement of an individual sample

This formula is really quite simple. The summation sign ($\sum_{i=1}^{N}$) simply means to add up all the squared deviations from the first one ($i = 1$) to the last one ($i = N$). The sum of squared deviations (10) divided by the number of samples minus one ($4 - 1 = 3$) produces a value of $10/3 = 3.3$ cm^2 (note that the units are centimeters squared). This is the **variance:**

$$\text{Variance} = \frac{\text{sum of squared deviations}}{N - 1}$$

The square root of the variance, 1.8 cm, equals the **standard deviation (SD):**

$$\text{SD} = \sqrt{\text{Variance}} = \sqrt{3.3} = 1.8$$

The standard deviation is usually reported with the mean in statements such as, "The mean length of the leaf was 20 ± 1.8 cm."

The standard deviation helps us understand the spread or variation of a sample. For many distributions of measurements, the mean ± 1 SD includes 68% of the measurements, whereas the mean ± 2 SD includes 95% of the measurements.

Procedure 1.7

Gather and analyze data statistically

1. Use a meterstick or tape measure to measure your height in centimeters. Record your height here: ________ cm
2. Record your height and gender (male or female) on the board in the lab.
3. After all of your classmates have reported their heights, calculate the following:

Size of sample

All classmates ________

Male classmates ________

Female classmates ________

Mean height

All classmates ________

Male classmates ________

Female classmates ________

Median height

All classmates ________

Male classmates ________

Female classmates ________

Range

All classmates ________ to ________

Male classmates ________ to ________

Female classmates ________ to ________

Standard deviation

All classmates ± ________

Male classmates ± ________

Female classmates ± ________

Question 8

What can you conclude from these statistics?

Your instructor may ask you to do other statistical tests, such as Student's t, chi-square, and analysis of variance (ANOVA). The type of test you'll do will depend on the amount and type of data you analyze, as well as the hypotheses you are trying to test.

Questions for Further Thought and Study

1. What are the advantages and disadvantages of using the metric system of measurements?

2. Why is it important for all scientists to use a standard system of measures rather than the system that may be most popular in their home country or region?

3. Do you lose or gain information when you use statistics to reduce a population to a few characteristic numbers? Explain your answer.

4. Suppose that you made repeated measurements of your height. If you used good technique, would you expect the range to be large or small? Explain your answer.

5. Suppose that a biologist states that the average height of undergraduate students at your university is 205 cm plus or minus a standard deviation of 17 cm. What does this mean?

6. What does a small standard deviation signify? What does a large standard deviation signify?

2

The Microscope
Basic Skills of Light Microscopy

Objectives

By the end of this exercise you should be able to:

1. Identify and explain the functions of the primary parts of a compound microscope and dissecting (stereoscopic) microscope.
2. Practice carrying and focusing a microscope properly.
3. Use a compound microscope and dissecting microscope to examine biological specimens.
4. Prepare a wet mount, determine the magnification and size of the field of view, and determine the depth of field.

Many organisms and biological structures are too small to be seen with the unaided eye (fig. 2.1). Biologists often use a light microscope to observe such specimens. A **light microscope** is a coordinated system of lenses arranged to produce an enlarged, focusable image of a specimen. A light microscope **magnifies** a specimen, meaning that it increases its apparent size. Magnification with a light microscope is usually accompanied by improved **resolution,** which is the ability to distinguish two points as separate points. Thus, the better the resolution, the sharper or crisper the image appears. The resolving power of the unaided eye is approximately 0.1 mm (1 in = 25.4 mm), meaning that our eyes can distinguish two points that are 0.1 mm apart. A light microscope, used properly, can improve resolution as much as 1000-fold (i.e., to 0.1 μm).

The ability to discern detail also depends on **contrast,** which is the amount of difference between the lightest and darkest parts of an image. Therefore, many specimens examined with a light microscope are stained with artificial dyes that increase contrast and make the specimen more visible.

The invention of the light microscope was profoundly important to biology, because it was used to formulate the cell theory and study biological structure at the cellular level. Light microscopy has revealed a vast new world to the human eye and mind (fig. 2.2). Today, the light microscope is the most fundamental tool of many biologists.

THE COMPOUND LIGHT MICROSCOPE

Study and learn the parts of the typical compound light microscope shown in figure 2.3. A light microscope has two, sometimes three, systems: an illuminating system, an imaging system, and possibly a viewing and recording system.

Illuminating System

The illuminating system, which concentrates light on the specimen, usually consists of a light source, condenser lens, and iris diaphragm. The **light source** is a lightbulb located at the base of the microscope. The light source illuminates the specimen by passing light through a thin, almost transparent part of the specimen. The **condenser lens,** located immediately below the specimen, focuses light from the light source onto the specimen. Just below the condenser is the **condenser iris diaphragm,** which is a knurled ring or lever that can be opened and closed to regulate the amount of light reaching the specimen. When the condenser iris diaphragm is open, the image will be bright; when closed, the image will be dim.

Imaging System

The imaging system improves resolution and magnifies the image. It consists of the objective and ocular (eyepiece) lenses and a body tube. The **objectives** are three or four lenses mounted on a revolving nosepiece. Each objective is actually a series of several lenses that magnify the image, improve resolution, and correct aberrations in the image. The most common configuration for student microscopes includes four objectives: low magnification (4×), medium magnification (10×), high magnification (40×), and oil immersion (100×). Using the oil immersion objective requires special instructions, as explained in Exercise 23 to study bacteria. To avoid damaging your microscope do not use the oil immersion objective during this exercise.

The magnifying power of each objective is etched on the side of the lens (e.g., 4×). The **ocular** is the lens that

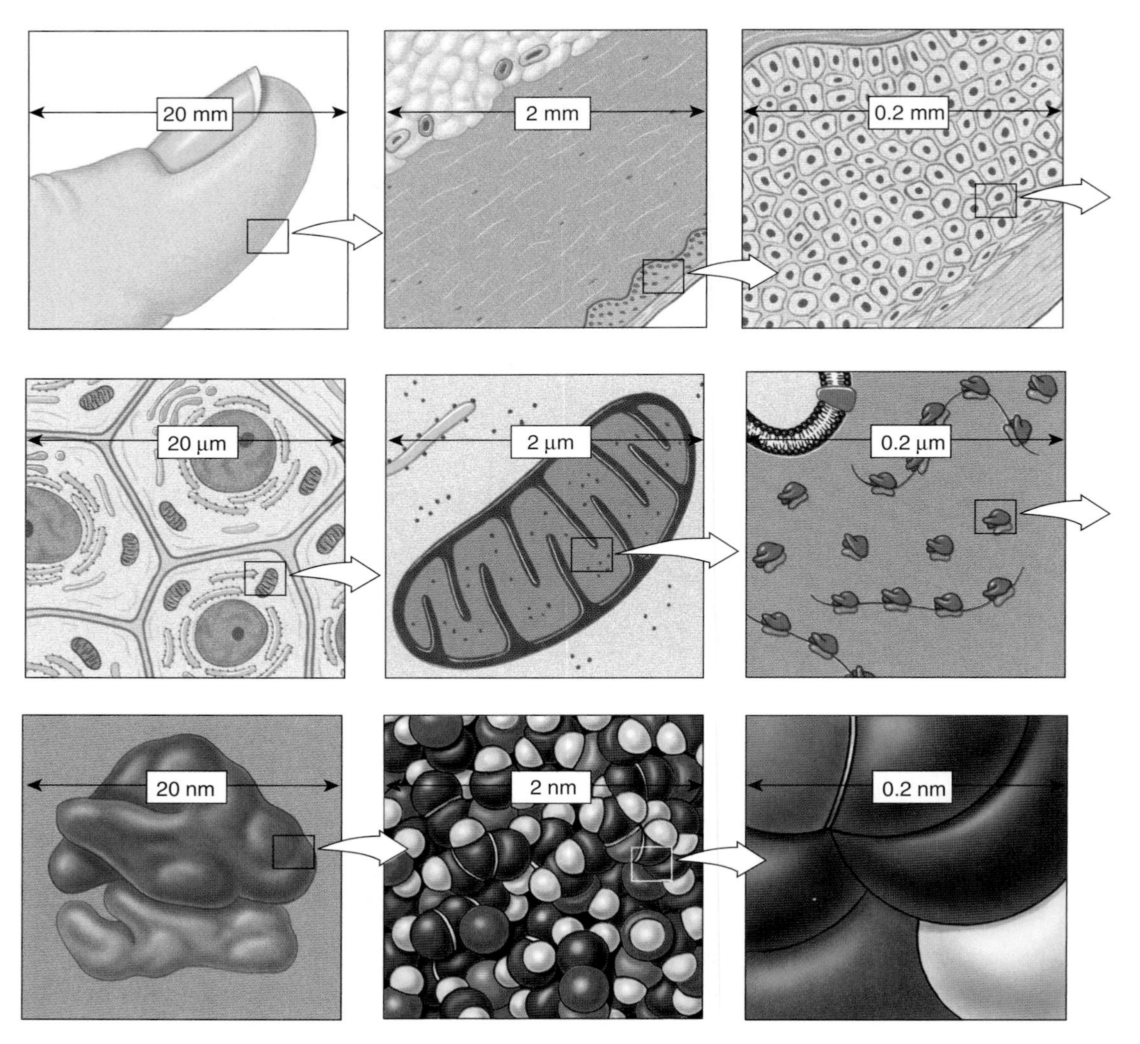

Figure 2.1

The size of cells and their contents. This diagram shows the size of human skin cells, organelles, and molecules. In general, the diameter of a human skin cell is about 20 micrometers (µm), of a mitochondrion is 2 µm, of a ribosome is 20 nanometers (nm), of a protein molecule is 2 nm, and of an atom is 0.2 nm.

you look through. Microscopes with one ocular are **monocular** microscopes, and those with two are **binocular** microscopes. Oculars usually magnify the image ten times. The **body tube** is a metal casing through which light passes to the oculars. In microscopes with bent body tubes and inclined oculars, the body tube contains mirrors and a prism that redirects light to the oculars. The **stage** secures the glass slide on which the specimen is mounted.

Viewing and Recording System

The viewing and recording system, if present, converts radiation to a viewable and/or permanent image. The viewing and recording system usually consists of a camera or video screen. Most student microscopes do not have viewing and recording systems.

USING A COMPOUND MICROSCOPE

Although the maximum magnification of light microscopes has not increased significantly during the last century, the construction and design of light microscopes has improved the resolution of newer models. For example, built-in light sources have replaced adjustable mirrors in the illuminating system, and lenses are made of better glass than they were in the past.

Your lab instructor will review with you the parts of the microscopes (and their functions) you will use in the lab. After familiarizing yourself with the parts of a microscope, you're now ready for some hands-on experience with the instrument.

Figure 2.2
"Egad, I thought it was tea, but I see I've been drinking a blooming micro-zoo!" says this horrified, proper nineteenth-century London woman when she used a microscope to examine her tea. People were shocked to learn that there is an active, living world too small for us to see.

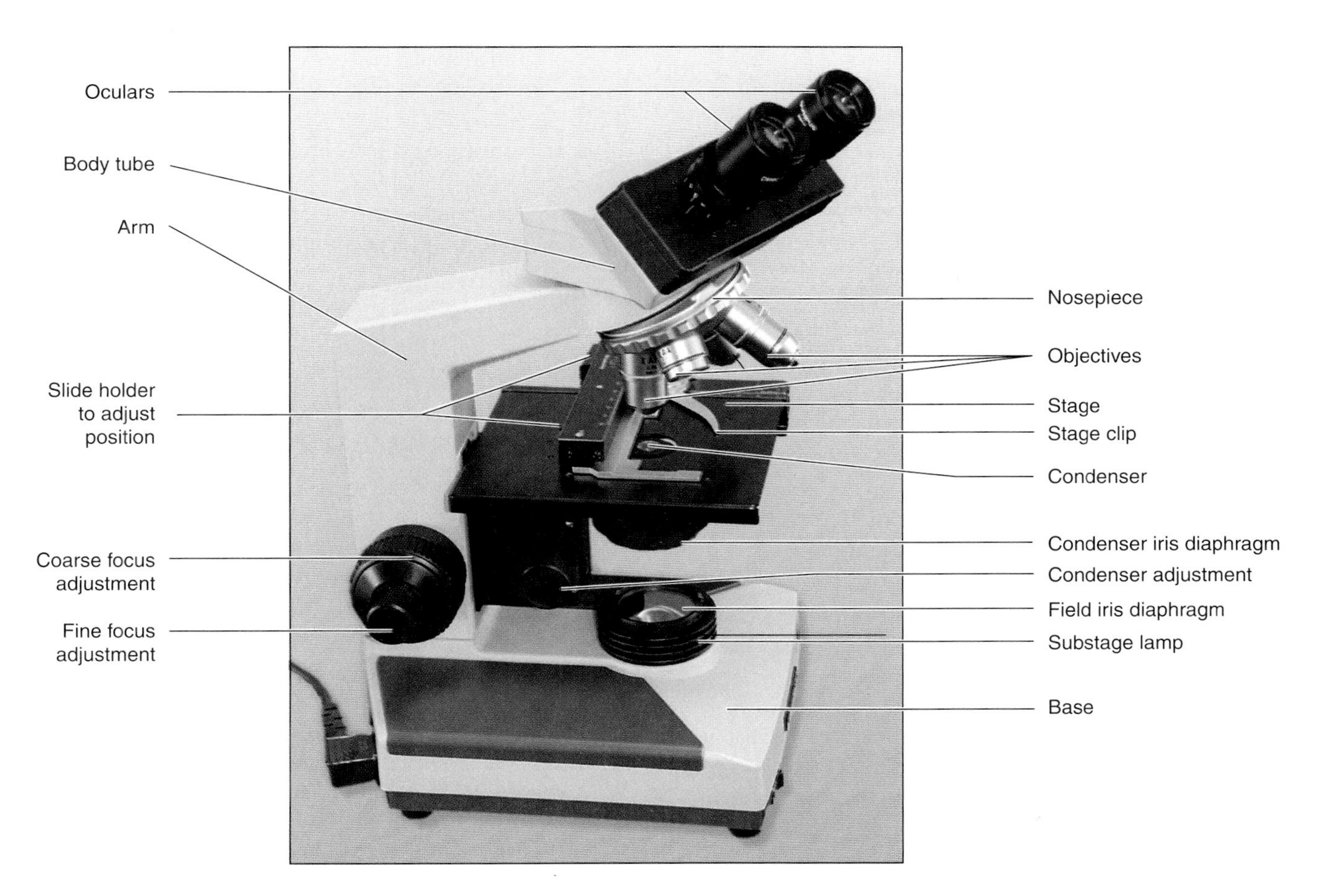

Figure 2.3
Major parts of a compound light microscope.

Procedure 2.1

Use a compound microscope

1. Remove the microscope from its cabinet and carry it upright with one hand grasping the arm and your other hand supporting the microscope below its base. Place your microscope on the table in front of you.

CAUTION

Do not use paper towels or Kimwipes to clean the lenses of your microscope; they can scratch the lenses.

2. Plug in the microscope and turn on the light source.
3. If it isn't already in position, rotate the nosepiece until the low-power (4×) objective is in line with the body tube.[1] You'll feel the objective click into place when it is positioned properly. *Always begin examining slides with the low-power objective*.
4. Locate the coarse adjustment knob on the side of the microscope. Depending on the type of microscope that you're using, the coarse adjustment knob moves either the nosepiece (with its objectives) or the stage to focus the lenses on the specimen. Only a partial turn of the coarse adjustment knob moves the stage or nosepiece a relatively large distance. *The coarse adjustment should only be used when you're viewing a specimen with the 4× or 10× objective lens*.
5. If your microscope is binocular, adjust the distance between the oculars to match the distance between your pupils. If your microscope is monocular, keep both eyes open when using the microscope. After a little practice you will ignore the image received by the eye not looking through the ocular.
6. Focus a specimen by using the following steps:
   a. Place a microscope slide of newsprint of the letter *e* on the horizontal stage so that the *e* is directly below the low-power objective lens and is right side up. It should be centered over the hole in the stage.
   b. Rotate the coarse adjustment knob clockwise to move the objective within 1 cm of the stage (1 cm = 0.4 in).
   c. Look through the oculars with both eyes open.
   d. Rotate the coarse adjustment knob counterclockwise (i.e., raising the objective lens or lowering the stage) until the *e* comes into focus. If you don't see an image, the *e* is probably off center. Be sure that the *e* is directly below the objective lens and that you can see a spot of light surrounding the *e*.
   e. Focus up and down to achieve the crispest image.
   f. Adjust the condenser iris diaphragm so that the brightness of the transmitted light provides the best view.
   g. Observe the letter, then rotate the nosepiece to align the 10× objective to finish your observation. Do not use the oil immersion objective.

Question 1

a. As you view the letter *e*, how is it oriented? Upside down or right side up?

b. How does the image move when the slide is moved to the right or left? Toward you or away from you?

c. What happens to the brightness of the view when you go from 4× to 10×?

Magnification

Procedure 2.2

Determine magnification

1. Estimate the magnification of the *e* by looking at the magnified image on lowest magnification (4×), and then at the *e* without using the microscope.
2. Examine each objective and record the magnifications of the objectives and oculars of your microscope in table 2.1.
3. Calculate and record in table 2.1 the total magnification for each objective following this formula:

$$\mathbf{Mag_{Tot} = Mag_{Obj} \times Mag_{Ocu}}$$

   where

   $\mathbf{Mag_{Tot}}$ = **total magnification of the image**

   $\mathbf{Mag_{Obj}}$ = **magnification of the objective lens**

   $\mathbf{Mag_{Ocu}}$ = **magnification of the ocular lens**

   For example, if you're viewing the specimen with a 4× objective lens and a 10× ocular, the total magnification of the image is 4 × 10 = 40×. That is, the specimen appears 40 times larger than it is.
4. Slowly rotate the high-power (i.e., 40×) objective into place. *Be sure that the objective does not touch the slide!* If the objective does not rotate into place without touching the slide, do not force it; ask your lab instructor to help you. After the 40× objective is in place, you should notice that the image remains near focus. Most light microscopes are **parfocal,** meaning that the image will remain nearly focused after the 40× objective lens is moved into place. Most light microscopes are also **parcentered,** meaning that the image will remain centered in the field of view after the 40× objective lens is in place.

---

[1] *The 4× objective is sometimes called the "scanning objective" because it enables users to scan large areas of a specimen.*

TABLE 2.1

TOTAL MAGNIFICATIONS AND AREAS OF FIELD OF VIEW (FOV) FOR THREE OBJECTIVES

Objective Power	Objective Magnification	×	Ocular Magnification	=	Total Magnification	FOV Diameter (mm)	FOV Area (mm²)	Measurement (mm) for 1 Ocular Space
4×	________	×	________	=	________	________	________	________
10×	________	×	________	=	________	________	________	________
40×	________	×	________	=	________	________	________	________

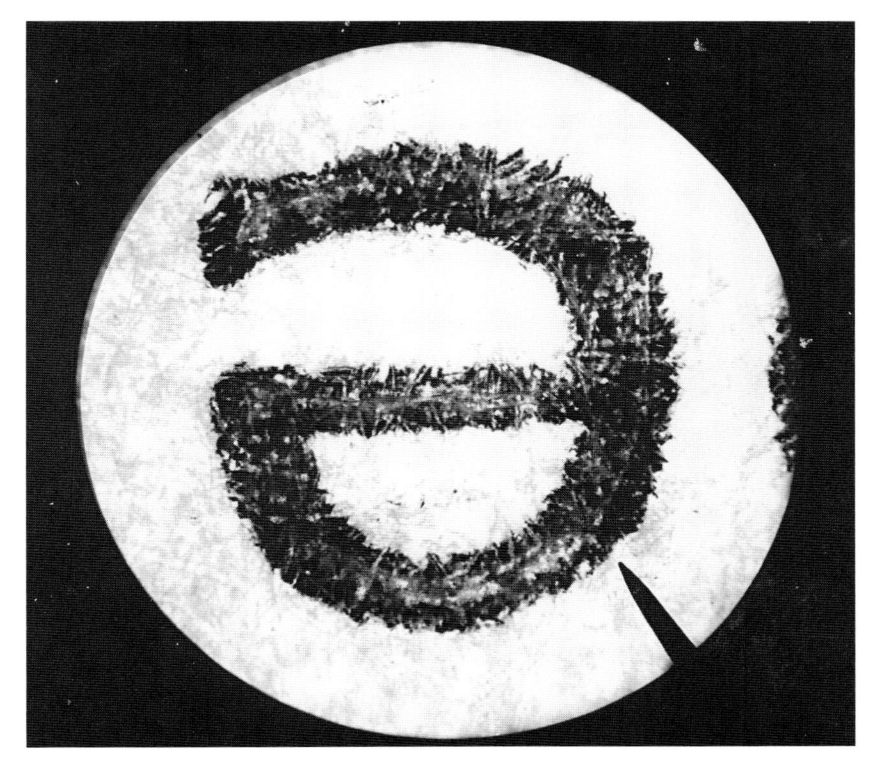

Figure 2.4
The circular, illuminated field of view of a compound light microscope. Shown here is the letter *e* from newsprint that is magnified 40 times.

5. You may need to readjust the iris diaphragm because the high-magnification objective allows less light to pass through to the ocular.
6. To fine-focus the image, locate the **fine adjustment knob** on the side of the microscope. Turning this knob changes the specimen-to-objective distance slightly and therefore makes it easy to fine-focus the image.

CAUTION

Never use the coarse adjustment knob to fine-focus an image on high power.

7. Compare the size of the image under high magnification with the image under low magnification.

Question 2

a. How many times is the image of the *e* magnified when viewed through the high-power objective?

b. If you didn't already know what you were looking at, could you determine at this magnification that you were looking at a letter *e*? How?

Determining the Size of the Field of View

The **field of view** is the area that you can see through the ocular and objective (fig. 2.4). Knowing the size of the field of view is important because you can use it to determine the

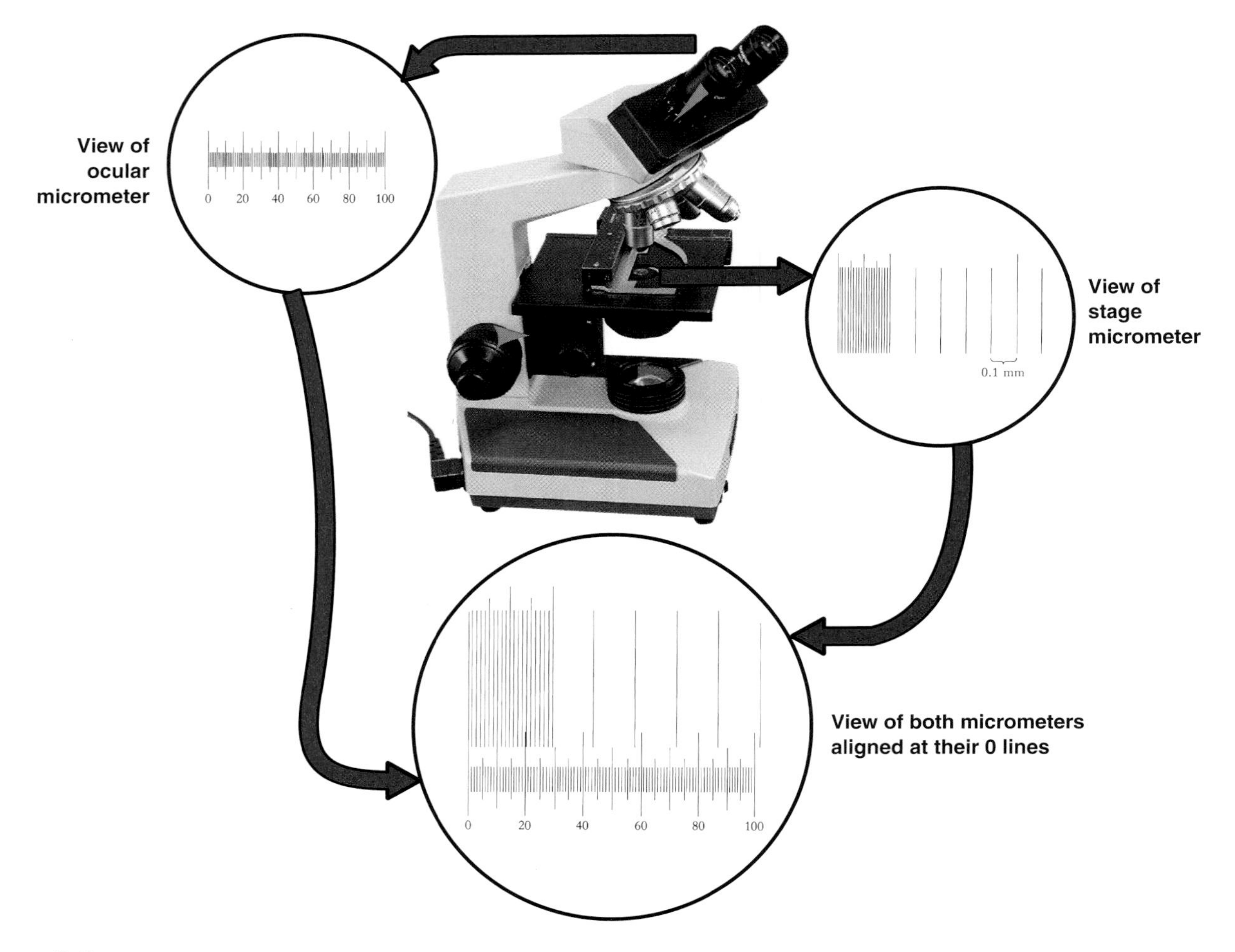

Figure 2.5

Stage and ocular micrometers. Micrometers are used to calibrate microscopes and measure the size of specimens.

approximate size of an object you are examining. The field of view can be measured with ruled **micrometers** (fig. 2.5). An **ocular micrometer** is a small glass disk with thin lines numbered and etched in a row. It was put into an ocular on your microscope so that the lines superimpose on the image and allow you to measure the specimen. Before you can use the micrometer you must determine for each magnification the apparent distance between the lines on the ocular micrometer. This means that you must calibrate the ocular micrometer by comparing its lines to those lines on a standard ruler called a **stage micrometer.** A stage micrometer is a glass slide having precisely spaced lines etched at known intervals.

Procedure 2.3

Use a stage micrometer to calibrate the ocular micrometer and determine the size of the field of view

1. Rotate the ocular until the lines of the ocular micrometer parallel those of the stage micrometer (fig. 2.5).
2. Align lines at the left edges (0 lines) of the two micrometers by moving the stage micrometer (fig. 2.5).
3. Count how many spaces on the stage micrometer fit precisely in a given number of spaces on the ocular micrometer. Record the values below.

   y ocular spaces = x stage spaces

   y = ________

   x = ________

   Since the smallest space on a stage micrometer = 0.01 mm, then

   y ocular spaces (mm) = x stage spaces × 0.01

   1 ocular space (mm) = $(x/y) \times 0.01$

4. Calculate the distance in millimeters between lines of the ocular micrometer. For example, if the length of ten spaces on the ocular micrometer equals the length of seven spaces on the stage micrometer, then

$$y = 10$$
$$x = 7$$
$$10 \text{ ocular spaces (mm)} = 7 \text{ stage spaces} \times 0.01 \text{ mm}$$
$$1 \text{ ocular space (mm)} = (7 \times 0.01 \text{ mm})/10$$
$$1 \text{ ocular space (mm)} = 0.007 \text{ mm}$$
$$1 \text{ ocular space} = 7\ \mu\text{m}$$

Therefore, if a specimen spans eight spaces on your ocular micrometer with that objective in place, that specimen is 56 μm long.

5. Calibrate the ocular micrometer for each objective on your microscope. Record in table 2.1 the diameter of the field of view (FOV) for each objective. Also record for each objective lens in table 2.1 the measurement (mm) for 1 ocular space. You can use this information in future labs as you measure the sizes of organisms and their parts.
6. Calculate the radius which is half the diameter.
7. Use this information to determine the area of the circular field of view with the following formula:

$$\text{Area of circle} = \pi \times \text{radius}^2$$
$$(\pi = 3.14)$$

8. Record your calculated FOV areas in table 2.1

Alternate Procedure 2.3

Use a transparent ruler to determine the size of the field of view

1. Obtain a clear plastic ruler with a metric scale.
2. Place the ruler on the stage and under the stage clips of your microscope. If your microscope has a mechanical stage, ask your instructor how to place the ruler to avoid damage. Carefully rotate the nosepiece to the objective of lowest magnification.
3. Slowly focus with the coarse adjustment and then the fine adjustment until the metric markings on the ruler are clear.
4. Align the ruler to measure the diameter of the circular field of view. The space between each line on the ruler should represent a 1-mm interval.
5. Record in table 2.1 the diameter of this low-magnification field of view. Also calculate the radius, which is half the diameter.
6. The ruler cannot be used to measure the diameters of the field of view at medium and high magnifications because the markings are too far apart. Therefore, these diameters must be calculated using the following formula:

$$FOV_{low} \times Mag_{low} = FOV_{hi} \times Mag_{hi}$$

where

FOV_{low} = diameter of the field of view of the low-power objective

Mag_{low} = magnification of the low-power objective (Be consistent and use the magnification of the objective, not total magnification.)

FOV_{hi} = diameter of the field of view of the high-power objective

Mag_{hi} = magnification of the high-power objective

For example, if 3.0 mm is the diameter of the field of view for a 4× low-power objective, then what is the diameter of the field of view of the 40× high-power objective?

$$3.0 \text{ mm} \times 4 = FOV_{hi} \times 40$$
$$0.30 \text{ mm} = FOV_{hi}$$

7. Calculate and record in table 2.1 the diameters of the field of view for the 10× and 40× magnifications.
8. Calculate and record in table 2.1 the circular area of the field of view for the three magnifications by using the following formula.

$$\text{Area of circle} = \pi \times \text{radius}^2$$
$$(\pi = 3.14)$$

Question 3

a. Which provides the largest field of view, the 10× or 40× objective?

b. How much more area can you see with the 4× objective than with the 40× objective?

c. Why is it more difficult to locate an object starting with the high-power objective than with the low-power objective?

d. Which objective should you use to initially locate the specimen? Why?

Depth of Field

Depth of field is the thickness of the object that is in sharp focus (fig. 2.6). Depth of field varies with different objectives and magnifications.

Procedure 2.4

Determine the depth of the field of view

1. Using the low-power objective, examine a prepared slide of three colored threads mounted on top of each other.
2. Focus up and down and try to determine the order of the threads from top to bottom. The order of the threads will not be the same on all slides.
3. Re-examine the threads using the high-power objective lens.

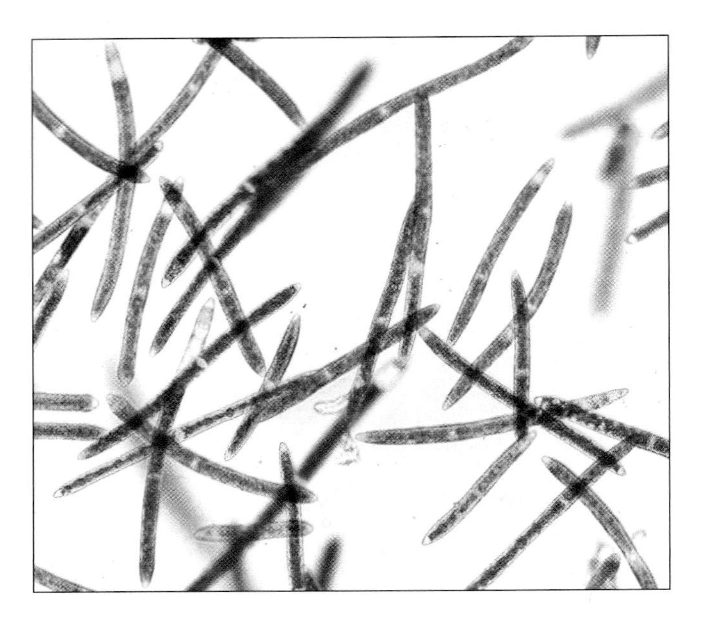

Figure 2.6
A thin depth of field is apparent in this 100× image of cells of *Closterium*, a green alga. The upper and lower layers of cells are out of focus, while the midlayer of cells is within the thin depth of field and is clearly focused.

Question 4

a. Are all three colored threads in focus at low power?

b. Can all three threads be in focus at the same time using the high-power objective?

c. Which objective, high- or low-power, provides the greatest depth of field?

Preparing a Wet Mount of a Biological Specimen

Procedure 2.5

Prepare a wet mount of a biological specimen

1. Place a drop of water containing algal cells from a culture labeled "algae" on a clean microscope slide.
2. Place the edge of a clean coverslip at an edge of the drop at a 45° angle; then slowly lower the coverslip onto the drop so that no air bubbles are trapped (fig. 2.7). (Your instructor will demonstrate this technique.) This fresh preparation is called a **wet mount** and can be viewed with your microscope.
3. Experiment with various intensities of illumination. To do this, rotate the 4× objective into place and adjust the condenser iris diaphragm to produce the least illumination. Observe the image; note its clarity, contrast, and color. Repeat these observations with at least four different levels of illumination. The fourth level should have the diaphragm completely open.
4. Repeat step 3 for the 10× and 40× objectives.

Question 5

a. Is the image always best with highest illumination?

b. Is the same level of illumination best for all magnifications?

c. Which magnifications require the most illumination for best clarity and contrast?

5. Examine your preparation of algae, and sketch in the space below the organisms that you see. Don't mistake air bubbles for organisms! Air bubbles appear as uniformly round structures with dark, thick borders.

6. Prepare a wet mount of some newly hatched brine shrimp (*Artemia*) and their eggs. Use your calculations for the diameter of the field of view to estimate the length of the shrimp.

Question 6

a. Why is it important to put a coverslip over the drop of water when you prepare a wet mount?

b. Approximately how long and wide is a brine shrimp?

Practice

For practice using your microscope, prepare some wet mounts of pond water or a hay infusion to view the diversity of protozoa and algae (fig. 2.8). If the protozoa are moving too fast for you to examine carefully, add a drop of methylcellulose (often sold commercially as Proto-Slo) to your sample. (The methylcellulose will slow the movement of the protozoa.) Also examine the prepared slides available in

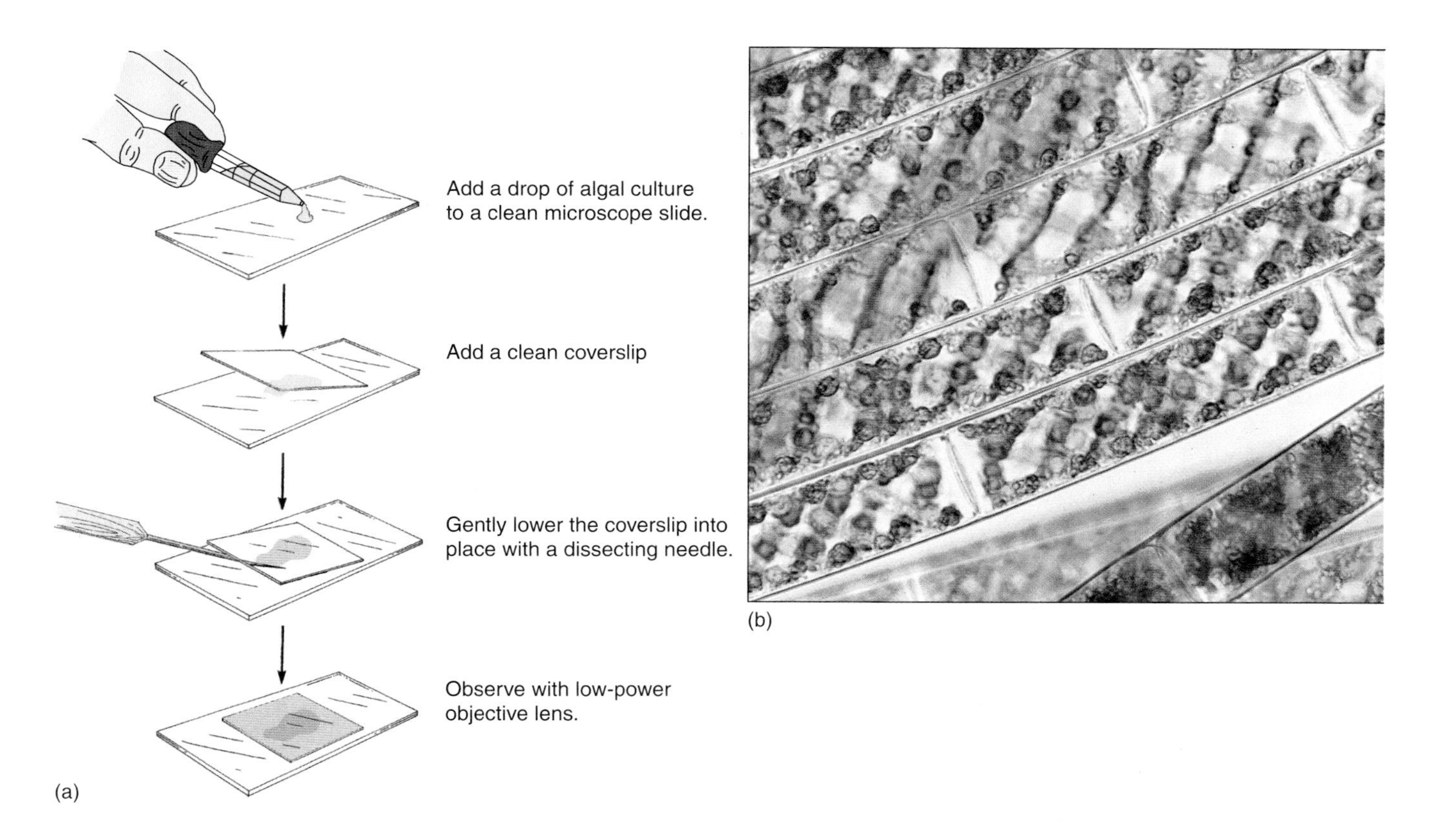

Figure 2.7

(*a*) Preparing a wet mount of a biological specimen. (*b*) A wet mount might include the common alga *Spirogyra*, 800×. See also figures 2.6, 23.9, 24.1–24.4.

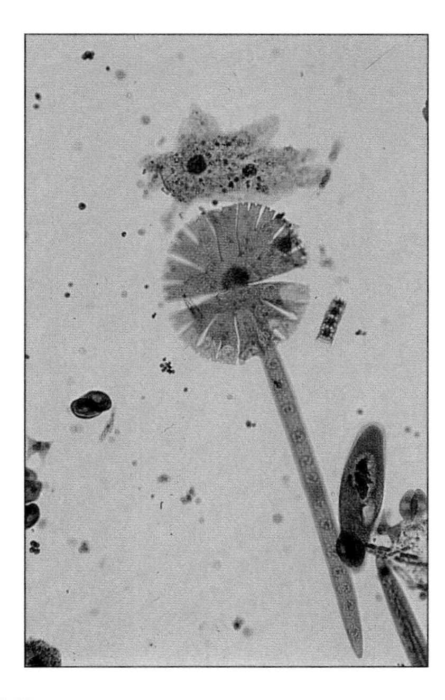

Figure 2.8

The diversity of organisms in pond water.

the lab. You'll examine these slides in more detail in the coming weeks, so don't worry about their contents. Rather, use this exercise to familiarize yourself with the microscope. Also prepare wet mounts of the cultures available in the lab and sketch the organisms that you see. When you've finished, turn off the light source, cover your microscope, and store the microscope in its cabinet.

THE DISSECTING (STEREOSCOPIC) MICROSCOPE

A **dissecting (stereoscopic) microscope** offers some advantages over a compound microscope. Although a compound microscope can produce high magnifications and excellent resolution, it has a small **working distance,** which is the distance between the objective lens and specimen. Therefore, it is difficult to manipulate a specimen while observing it with a compound microscope. Specimens that can be observed with a compound microscope are limited to those thin enough for light to pass through them. In contrast, a dissecting microscope is used to view objects that are opaque or too large to see with a compound microscope.

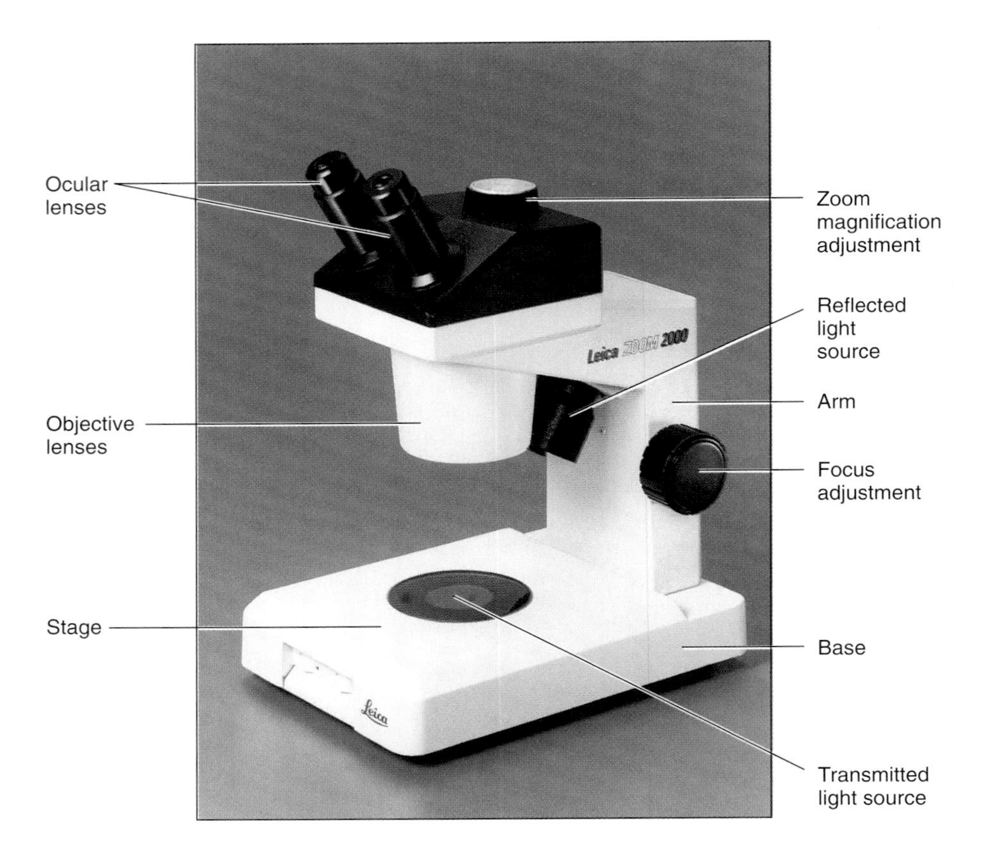

Figure 2.9
Dissecting (stereoscopic) microscope.

A dissecting microscope provides a much larger working distance than does a compound microscope. This distance is usually several centimeters (compared to a centimeter or less for a compound microscope), making it possible to dissect and manipulate most specimens. Also, most specimens for dissection are too thick to observe with transmitted light from a light source below the specimen. Therefore, many dissecting microscopes use a light source above the specimen; the image is formed from reflected light.

Dissecting microscopes are always binocular (fig. 2.9). Each ocular views the specimen at different angles through one or more objective lenses. This arrangement provides a three-dimensional image with a large depth of field. This is in contrast to the image in a compound microscope, which is basically two-dimensional. However, the advantages of a stereoscopic microscope are often offset by lower resolution and magnification than a compound microscope. Most dissecting microscopes have magnifications of 4× to 50×.

Procedure 2.6

Use a dissecting microscope

1. Carry the dissecting microscope to your desk.
2. Use figure 2.9 to familiarize yourself with the parts of your microscope.
3. Use your dissecting microscope to examine the organisms available in lab. Sketch some of these organisms.
4. Use a ruler to measure the diameter of the field of view with your dissecting microscope at several levels of magnification.

Question 7

a. What is the area of the field of view when you use the lowest magnification of your dissecting microscope?

b. What is the area when you use the highest magnification?

c. Place a microscope slide of the letter *e* on the stage. As you view the letter *e* how is it oriented?

d. How does the image through a dissecting microscope move when the specimen is moved to the right or left? Toward you or away from you?

e. How does the direction of illumination differ in dissecting as opposed to compound microscopes?

A COMPARISON OF COMPOUND AND DISSECTING MICROSCOPES

Complete table 2.2 comparing magnification, depth of field, size of the field of view, and resolution of a dissecting microscope and a compound microscope. Use the terms *high*, *low*, or *same* to describe your comparisons.

Question 8
What other differences are there between compound and dissecting microscopes?

TABLE 2.2
A COMPARISON OF DISSECTING AND COMPOUND MICROSCOPES

Characteristic	Dissecting Microscope	Compound Microscope
Magnification	______________	______________
Resolution	______________	______________
Size of field of view	______________	______________
Depth of field	______________	______________

Questions for Further Thought and Study

1. What are the advantages of knowing the diameter of the field of view at a given magnification?

2. Why must specimens viewed with a compound microscope be thin? Why are they sometimes stained with dyes?

3. Why is depth of field important in studying biological structures? How can it affect your ability to find and examine a specimen?

4. What is the importance of adjusting the light intensity when viewing specimens with a compound microscope?

5. What is the function of each major part of a compound and dissecting microscope?

WRITING TO LEARN BIOLOGY

The smallest structures of cells are best seen with a transmission electron microscope. Refer to your textbook or other book and describe how an electron microscope can resolve such small structures. Write a short essay about the advantages and limitations of a **transmission electron microscope.**

2–12

3

The Cell
Structure and Function

Objectives

By the end of this exercise you should be able to:

1. Understand the differences between prokaryotes and eukaryotes and identify structures characteristic of each.
2. Prepare a wet mount to view cells with a compound microscope.
3. Understand the function of organelles visible with a light microscope.
4. Examine a cell's structure and determine whether it is from a plant, animal, or protist.

Cells are considered the basic unit of living organisms because they perform all of the processes we collectively call "life." All organisms are made of cells. Although most individual cells are visible only with the aid of a microscope, some may be a meter long (e.g., nerve cells) or as large as a small orange (e.g., the yolk of an ostrich egg). Despite these differences, all cells are designed similarly and share fundamental features.

Cytology is the study of cellular structure and function. The major tools of cytologists are light microscopy, electron microscopy, and cell chemistry. By studying the anatomy of a cell, we can find clues to how the cell works.

In today's lab, you will study some of the features and variations among living cells to understand the life processes of organisms. Prior to this exercise, review in your textbook the general features of cellular structure and function.

PROKARYOTIC CELLS

Bacteria and cyanobacteria are **prokaryotes** (fig. 3.1), and thus do not contain a membrane-bound nucleus or any other membrane-bound **organelles.** Organelles are organized structures of macromolecules having a specialized function and are suspended in the **cytoplasm.** The cytoplasm of prokaryotes is enclosed in a **plasma membrane** (cellular membrane) and is surrounded by a supporting **cell wall** covered by a gelatinous **capsule. Flagella** and hairlike outgrowths called **pili** are common in prokaryotes; flagella are used for movement, and pili are used to attach some types of bacteria to surfaces or to exchange genetic material with other bacteria. Within the cytoplasm of prokaryotes are **ribosomes** (small particles involved in protein synthesis), **mesosomes** (internal extensions of the plasma membrane), and **chromatin bodies** (concentrations of DNA). Prokaryotes do not reproduce sexually, but they have mechanisms for genetic recombination (see Exercise 15).

Cyanobacteria

The largest prokaryotes are **cyanobacteria,** also called blue-green algae. They contain chlorophyll *a* and accessory pigments for photosynthesis, but these pigments are not contained in membrane-bound chloroplasts. Instead, the pigments are held in photosynthetic membranes called **thylakoids** (fig. 3.2). Cyanobacteria are often surrounded by a **mucilaginous sheath.** Their ability to photosynthesize made them the primary contributors to the early oxygenation of the ancient earth's atmosphere.

Procedure 3.1

Examine cyanobacteria

1. Prepare a wet mount of *Oscillatoria,* a filament of cells, and one of *Gloeocapsa,* a loosely arranged colony (fig. 3.3). Review procedure 2.5 in Exercise 2 for preparing a wet mount.
2. Focus with the low-power objective.
3. Rotate the high-power objective into place to see filaments and masses of cells.
4. Observe the cellular structures and draw the cellular shapes and relative sizes of *Oscillatoria* and *Gloeocapsa* in the space below. Use an ocular micrometer to measure their dimensions.

Oscillatoria *Gloeocapsa*

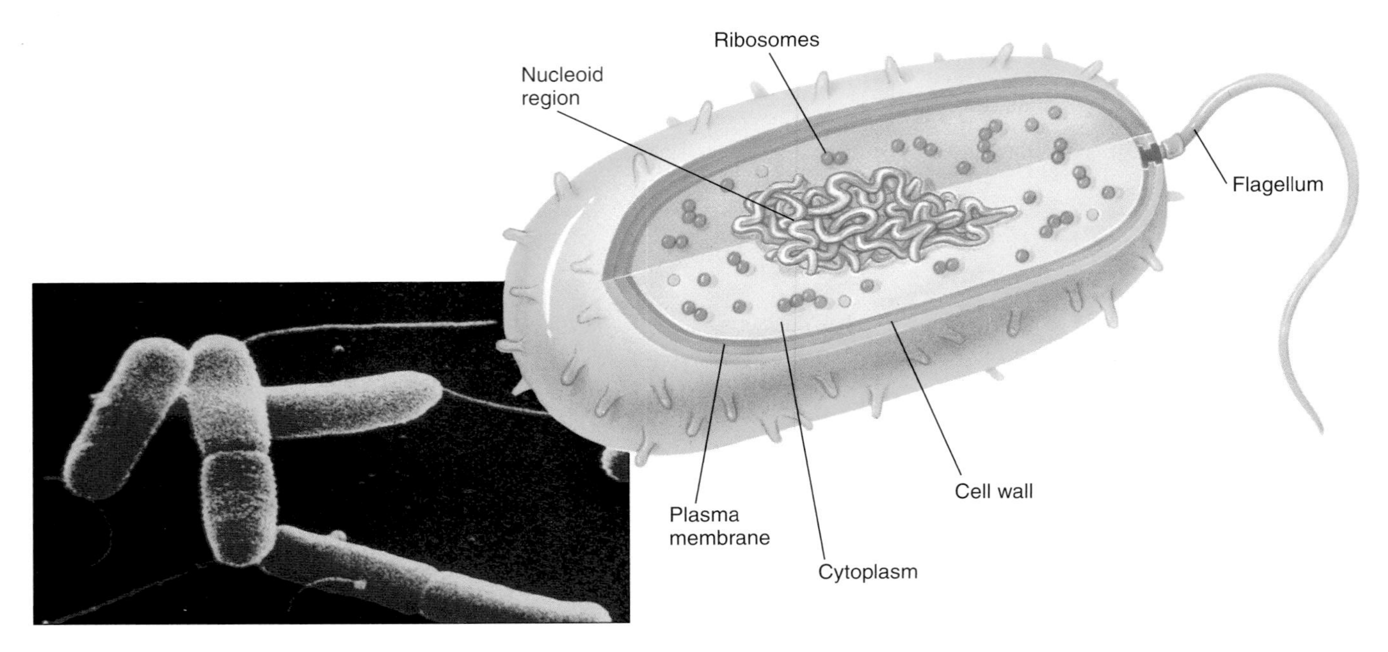

Figure 3.1

The structure of a bacterial cell. Bacteria lack a nuclear membrane. All prokaryotic (bacterial) cells have a nucleoid, ribosomes, plasma membrane, cytoplasm, and cell wall, but not all have flagella.

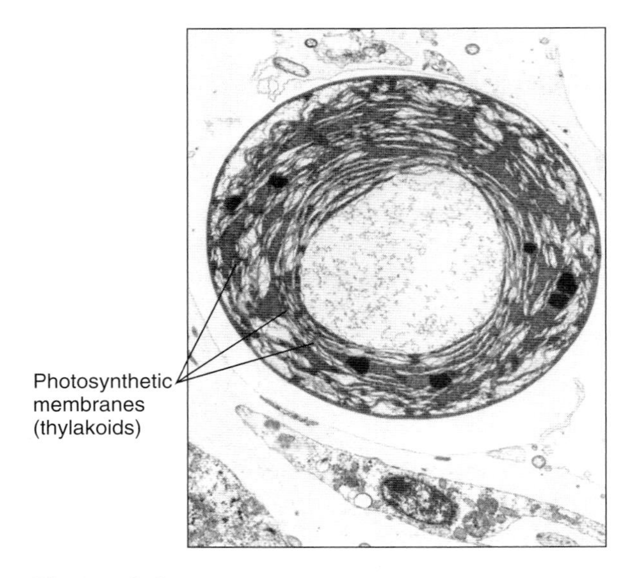

Figure 3.2

Electron micrograph of a photosynthetic bacterial cell, *Prochloron*, showing extensively folded photosynthetic membranes. The DNA is in the clear area in the central region of the cell; it is not membrane-bound, 5,200×.

Question 1

a. Where are the pigments located in these cyanobacteria?

b. Are nuclei visible in cyanobacterial cells?

c. Which of these two genera has the most prominent mucilaginous sheath?

d. How many cells are held within one sheath of *Gloeocapsa?*

Bacteria

Most bacteria are much smaller than cyanobacteria and do not contain chlorophyll. Yogurt is a nutrient-rich culture of bacteria. The bacterial cells composing most of the yogurt are *Lactobacillus*, a bacterium adapted to live on milk sugar (lactose). *Lactobacillus* converts milk to yogurt. Yogurt is acidic and keeps longer than milk. Historically, *Lactobacillus* has been used in many parts of the world by peoples deficient in lactase, an enzyme that breaks down lactose. Many Middle Eastern and African cultures use the more digestible yogurt in their diets instead of milk.

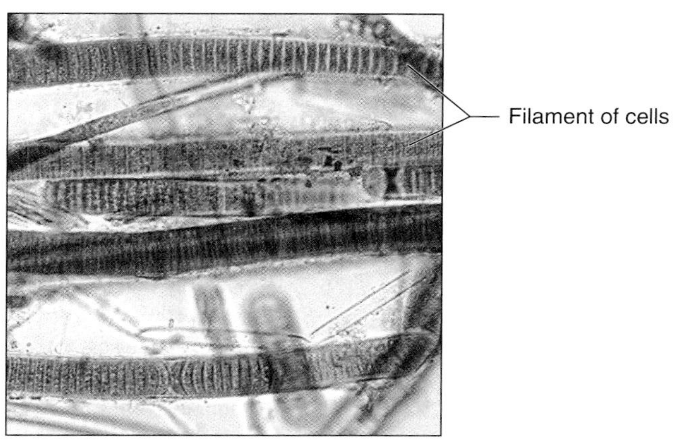

(a)

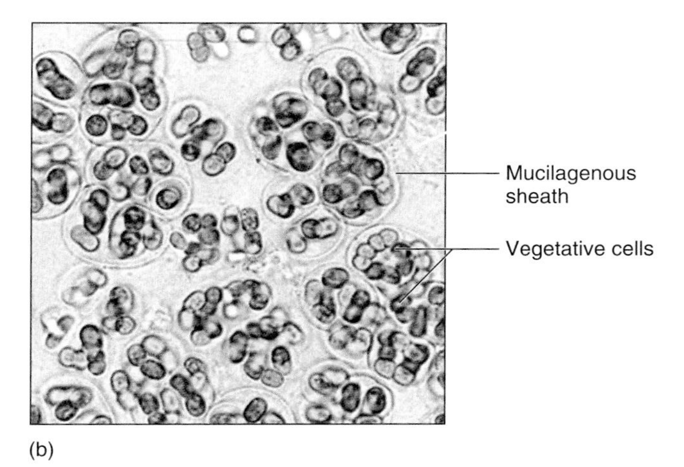

(b)

Figure 3.3
Common cyanobacteria. (*a*) *Oscillatoria*, 40×. (*b*) *Gloeocapsa*, 400×.

Procedure 3.2

Examine bacteria

1. Place a tiny dab of yogurt on a microscope slide.
2. Mix this small amount of yogurt in a drop of water, add a coverslip, and examine the yogurt with a compound microscope.
3. Focus with the low-power objective.
4. Rotate the high-power objective (40×) into place to see masses of rod-shaped cells.
5. Observe the simple, external structure of the bacteria and draw their cellular shapes in the space below:

Question 2

How does the size of *Lactobacillus* compare with that of *Oscillatoria* and *Gloeocapsa?*

EUKARYOTIC CELLS

Eukaryotic cells contain membrane-bound **nuclei** and other organelles (figs. 3.4, 3.5). Nuclei contain genetic material of a cell and control metabolism. **Cytoplasm** forms the matrix of the cell and is contained by the plasma membrane. Within the cytoplasm are a variety of organelles. **Chloroplasts** are elliptical green organelles in plant cells. Chloroplasts are the site of photosynthesis in plant cells and are green because they contain chlorophyll, a photosynthetic pigment capable of capturing light energy. **Mitochondria** are organelles found in plant and animal cells. These organelles are where aerobic respiration occurs. When viewed with a conventional light microscope, mitochondria are small, dark, and often difficult to see. All of the material and organelles contained by the plasma membrane are collectively called the **protoplast.**

Eukaryotic cells are structurally more complex than prokaryotic cells. Although some features of prokaryotic cells are in eukaryotic cells (e.g., ribosomes, cell membrane), eukaryotic cells also contain several organells not found in prokaryotic cells (table 3.1). In the following exercise you will investigate some of these organelles.

PLANT CELLS

Procedure 3.3

Examine living *Elodea* cells and chloroplasts

1. Remove a young leaf from the tip of a sprig of *Elodea*. *Elodea* is a common pond weed used frequently in studies of photosynthesis, cellular structure, and cytoplasmic streaming.
2. Place this leaf in a drop of water on a microscope slide with the top surface facing up. The cells on the upper surface are larger and more easily examined. Add a coverslip, but do not let the leaf dry. Add another drop of water if necessary.
3. Examine the leaf with your microscope. First use low, then high, magnification to bring the upper layer of cells into focus (fig. 3.6). Each of the small, regularly shaped units you see are cells surrounded by cell walls made primarily of **cellulose** (fig. 3.7). Cellulose is a complex carbohydrate made of glucose molecules attached end-to-end. The plasma membrane lies just inside the cell wall. Sketch what you see.

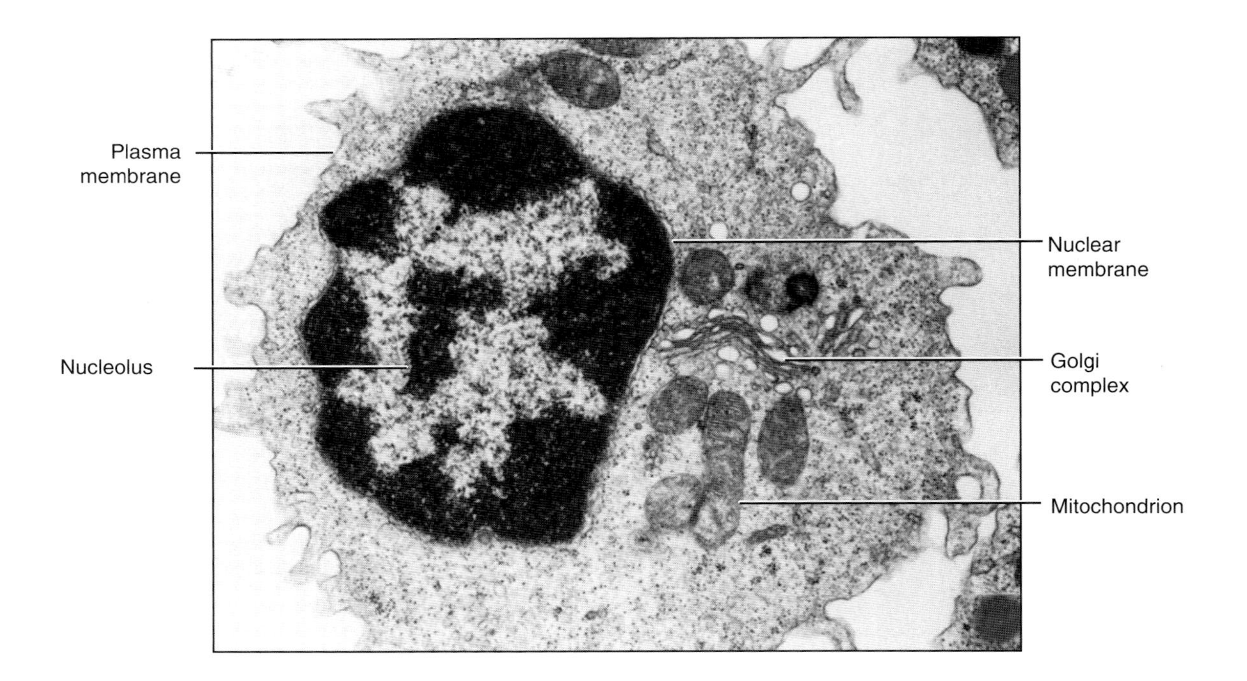

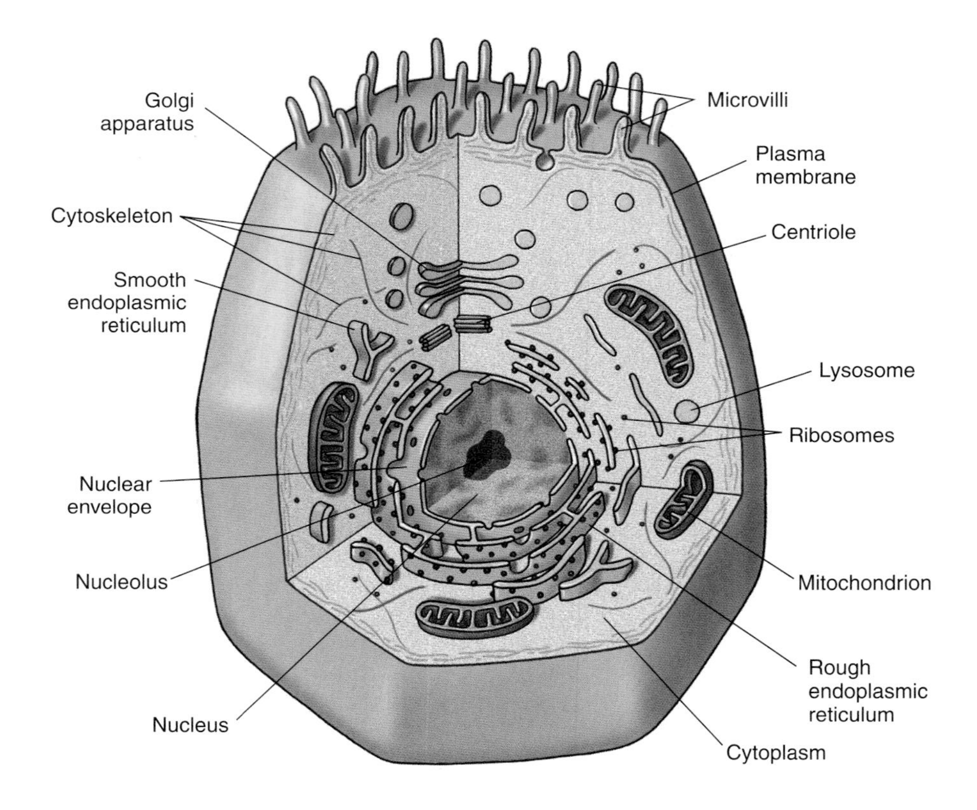

Figure 3.4

Structure of animal cells. Most organelles of animal cells are not visible with common light microscopes—therefore, our understanding of cellular structures is based mainly on research using electron micrographs, as shown with the upper photo and diagram. Cells are surrounded by a bilayered plasma membrane containing phospholipids and proteins. The nucleus houses chromosomal DNA and is surrounded by a double-membraned nuclear envelope. Centrioles organize spindle fibers during cell division. Endoplasmic reticulum (ER) is a system of membranes inside the cell. Rough ER has many ribosomes, and smooth ER has fewer ribosomes. Mitochrondria are sites of oxidative respiration and ATP synthesis. Microvilli are cytoplasmic projections that increase the surface area of some specialized animal cells. Golgi complexes are flat sacs and vesicles that collect and package substances made in the cell. Ribosomes are aggregations of proteins that conduct protein synthesis. Lysosomes contain enzymes important in recycling worn-out cellular debris, 17,500×.

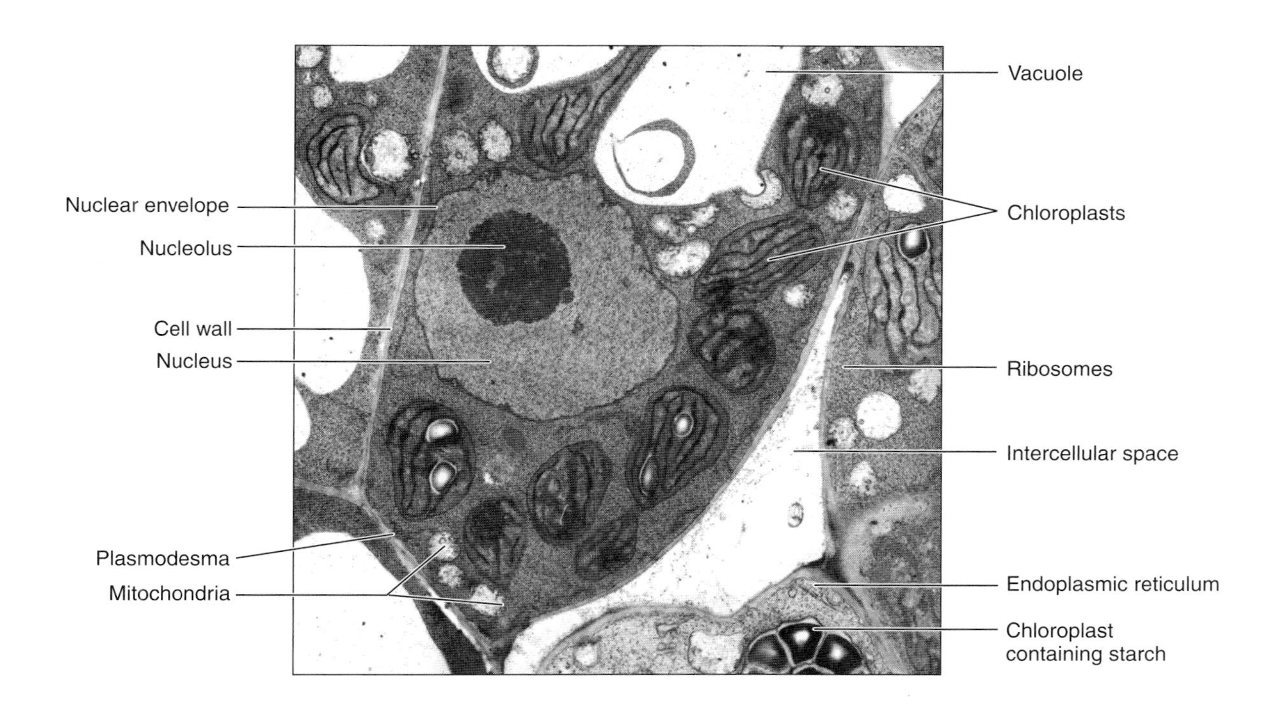

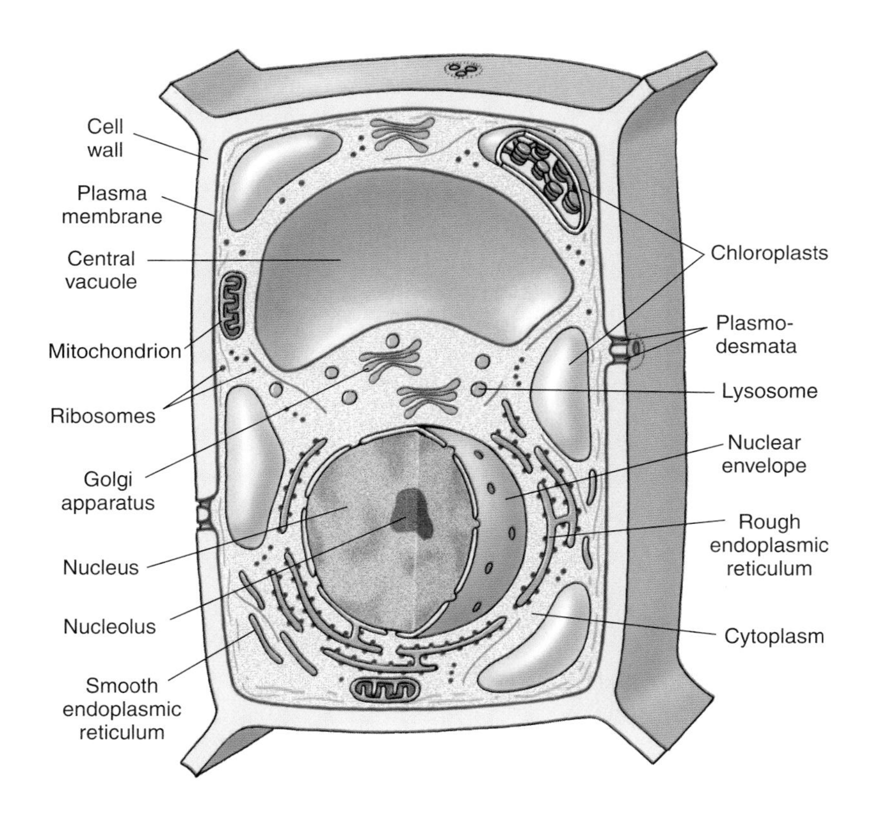

Figure 3.5

Structure of plant cells. This illustration shows relative proportions of the different parts of a plant cell. Most mature plant cells contain large central vacuoles, which occupy most of the volume of the cell. Cytoplasm is often a thin layer between the vacuole and the plasma membrane. Cytoplasm contains the cell's organelles.

Table 3.1

Some of the Major Differences between Prokaryotic and Eukaryotic Cells, and between Plant and Animal Cells

	Prokaryote	Eukaryote: Animal	Eukaryote: Plant
Exterior Structures			
Cell wall	Present (protein-polysaccharide)	Absent	Present (cellulose)
Cell membrane	Present	Present	Present
Flagella	May be present (single strand)	May be present	Absent except in sperm of a few species
Interior Structures			
ER	Absent	Usually present	Usually present
Ribosomes	Present	Present	Present
Microtubules	Absent	Present	Present
Centrioles	Absent	Present	Absent
Golgi apparatus	Absent	Present	Present
Other Organelles			
Nucleus	Absent	Present	Present
Mitochondria	Absent	Present	Present
Chloroplasts	Absent	Absent	Present
Chromosomes	A single circle of naked DNA	Multiple; DNA-protein complex	Multiple; DNA-protein complex
Vacuoles	Absent	Absent or small	Usually a large single vacuole

Question 3

a. What three-dimensional shape are *Elodea* cells?

b. Examine various layers of cells by focusing up and down through the layers. About how many cells thick is the leaf that you are observing?

c. What are the functions of the cell wall?

d. Use an ocular micrometer or refer to the dimensions of the field of view calculated in Exercise 2 to measure the dimensions of an *Elodea* cell. What are the cell's approximate dimensions?

4. Chloroplasts appear as moderately-sized green spheres within the cells (figs. 3.6, 3.8). Locate and sketch cells having many chloroplasts; estimate the number of chloroplasts in a healthy cell. Remember that a cell is three-dimensional, and some chloroplasts may obscure others.

Question 4

a. What shape are the chloroplasts? What is their function?

b. Where are the chloroplasts located within the *Elodea* cell—towards the perimeter, or centrally located?

5. Determine the spatial distribution of chloroplasts within a cell. They may be pushed against the margins of the cell by the large **central vacuole** containing mostly water and bounded by a **vacuolar membrane.** The vacuole occupies about 90% of the volume of a mature cell. Its many functions include storage of organic and inorganic molecules, ions, water, enzymes, and waste products.
6. Search for a **nucleus;** it may or may not be readily visible. Nuclei usually are appressed to the cell wall as a faint gray sphere the size of a chloroplast or larger. Staining the cells with a drop of iodine may enhance the nucleus. If your preparation is particularly good, a **nucleolus** may be visible as a dense spot in the nucleus.
7. Search for some cells that may appear pink due to water-soluble pigments called anthocyanins. These pigments give many flowers and fruits their bright reddish color.

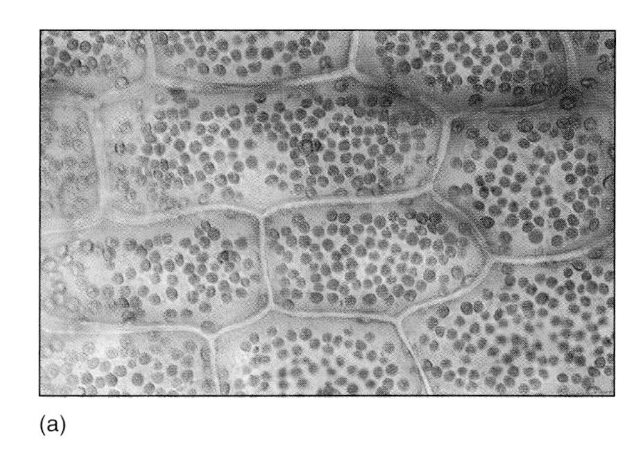

(a)

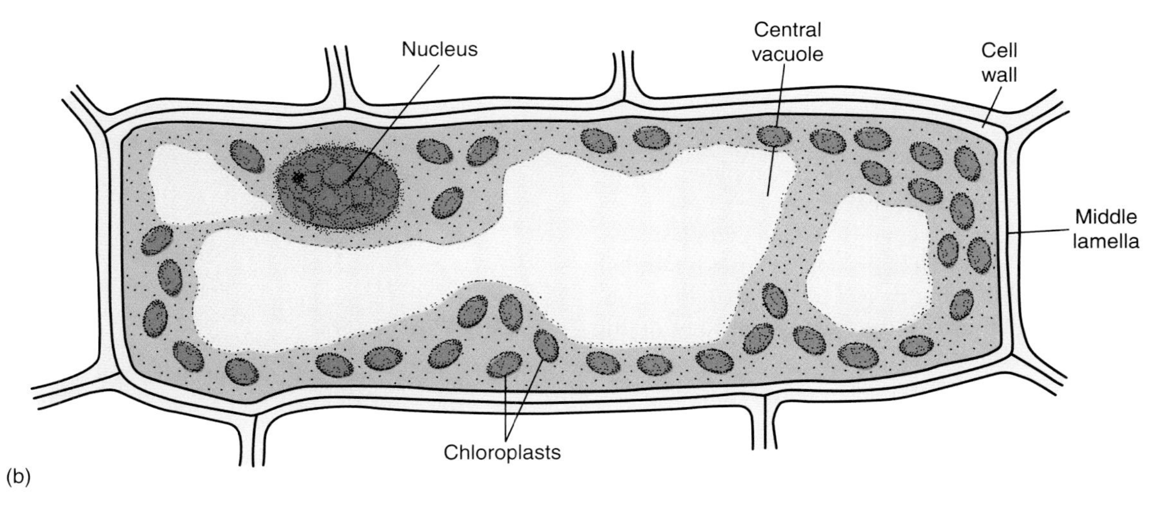

(b)

Figure 3.6
(*a*) *Elodea* cells showing abundant chloroplasts, 400×. (*b*) The cellular structure of *Elodea*, 150×.

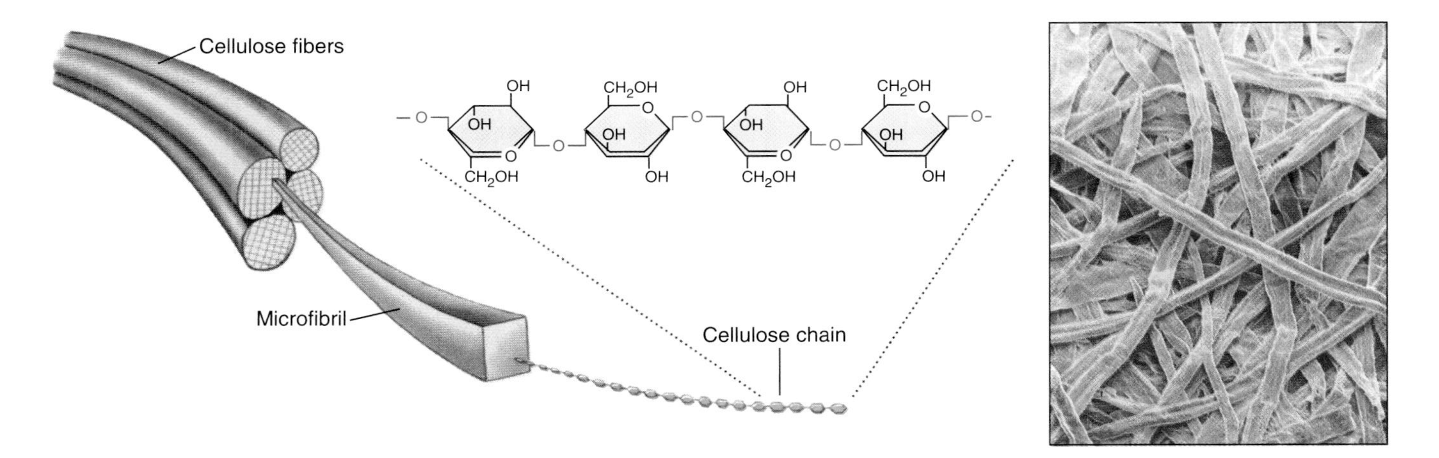

Figure 3.7
Cellulose is the most abundant organic compound on earth and is a polymer of glucose molecules. Free hydroxyl (OH^-) groups of the glucose molecules form hydrogen bonds between adjacent cellulose molecules to form cohesive microfibrils. Microfibrils align to form strong cellulose fibers that resist metabolic breakdown. Because humans cannot hydrolyze the bonds between glucose molecules of cellulose, cellulose is indigestible and its energy is unavailable. Cellulose passes through the human digestive tract as bulk fiber. 20×.

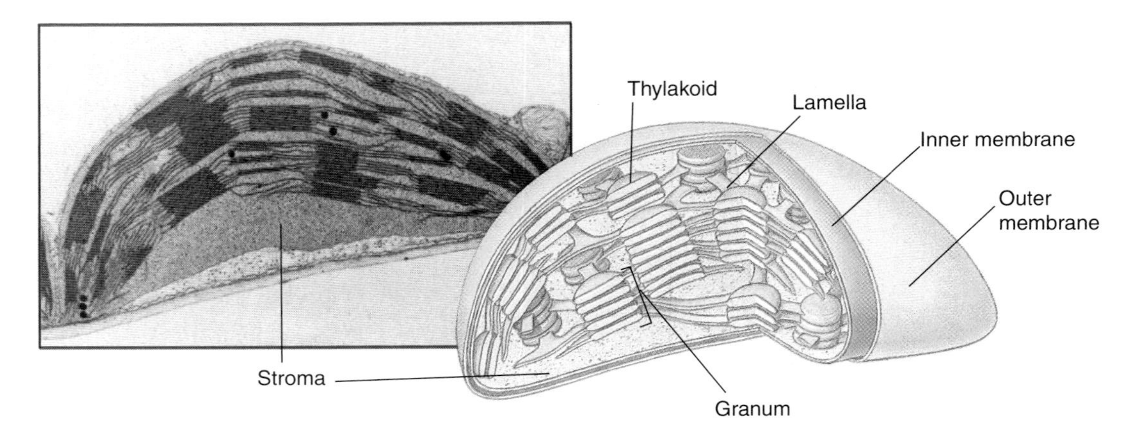

Figure 3.8
Chloroplast structure. The inner membrane of a chloroplast is fused to form stacks of closed vesicles called thylakoids. Photosynthesis occurs within these thylakoids. Thylakoids are typically stacked one on top of the other in columns called grana.

8. Warm the slide with intense light for about 10 min and search for movement of the chloroplasts. You may need to search many cells or make a new preparation. This movement is called **cytoplasmic streaming,** or **cyclosis.** Chloroplasts are not motile; instead, they are being moved by the activity of the cytoplasm. Add water if the cells appear to be drying out.
9. In the space below sketch a few cells of *Elodea;* compare the cells with that shown in figure 3.6.

Question 5

a. Can you see nuclei in *Elodea* cells?

b. What are the functions of nuclei?

c. Which are larger, chloroplasts or nuclei?

d. What is the approximate size of a nucleus?

e. Why is the granular-appearing cytoplasm more apparent at the sides of a cell rather than in the middle?

Question 6

a. Are all cellular components moving in the same direction and rate during cytoplasmic streaming?

b. What do you conclude about the uniformity of cytoplasmic streaming?

Cell Walls

Cell walls include an outer **primary cell wall** deposited during growth of the cell and a **middle lamella,** which is the substance holding walls of two adjacent cells together. The protoplasm of adjacent cells is connected by cytoplasmic strands called **plasmodesmata** that penetrate the cell walls (fig. 3.9).

Procedure 3.4

Examine cell walls and plasmodesmata

1. Prepare a wet mount of *Elodea* and examine the cell walls at high magnification. The middle lamella may be visible as a faint line between cells.
2. Obtain a prepared slide of tissue showing plasmodesmata. This tissue may be persimmon (*Diospyros*) endosperm, which has highly thickened primary walls. Sketch what you see.
3. Locate the middle lamella as a faint line between cell walls.
4. Locate the plasmodesmata appearing as darkened lines perpendicular to the middle lamella and connecting the protoplasts of adjacent cells (fig. 3.9).

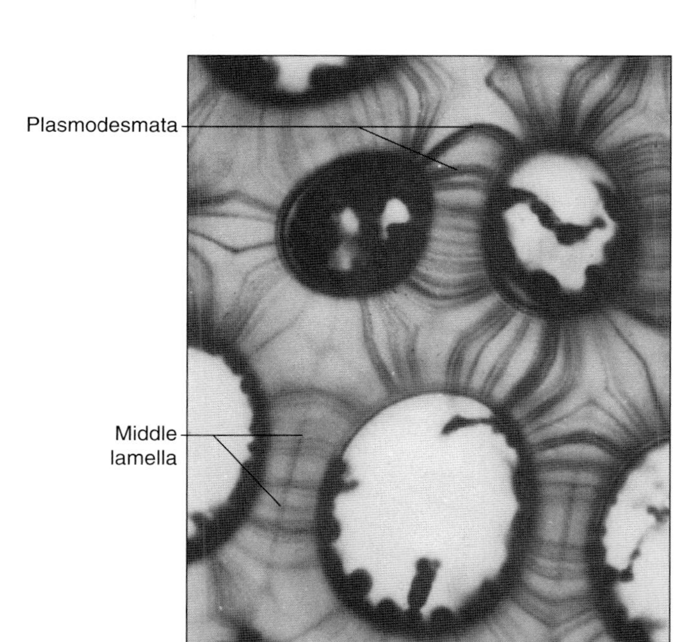

Figure 3.9

The thickened primary cell walls of persimmon endosperm show plasmodesmata connecting adjacent cells. Middle lamella appear as faint lines parallel to the cell surface. 950×.

Question 7

a. What are the functions of plasmodesmata?

b. Why do you suspect that there are so many plasmodesmata connecting the cells in this fruit?

Onion Cells

Staining often reveals the structure of cells and cell organelles more clearly. A specimen is **stained** by adding a dye that preferentially colors some parts of the specimen but not others. Neutral red is a common stain that accumulates in the cytoplasm of the cell, leaving the cell walls clear. Nuclei appear as dense bodies in the translucent cytoplasm of the cells.

Procedure 3.5

Examine stained onion cells

1. Cut a red onion into eighths and remove a fleshy leaf.
2. Snap the leaf backward and remove the thin piece of the inner epidermis formed at the break point (fig. 3.10), as demonstrated by your lab instructor.
3. Place this epidermal tissue in a drop of water on a microscope slide, add a coverslip, and examine the tissue. This preparation should be one cell thick.
4. Stain the onion cells by placing a small drop of 0.1% neutral red at the edge of the coverslip. Draw the neutral red across the specimen by wicking. To wick the solution, hold the edge of a small piece of paper towel at the opposite edge of the coverslip and it will withdraw some fluid. This will cause the neutral red to diffuse over the onion.
5. Stain the tissue for 5–10 min.
6. Carefully focus to distinguish the vacuole surrounded by the stained cytoplasm.
7. Search for the nucleus of a cell (fig. 3.11). The nucleus may appear circular in the central part of the cell. In other cells it may appear flattened.

Question 8

How do you explain the differences in the apparent shapes and positions of the nuclei in different cells?

8. Repeat steps 1–7 and stain a new preparation of onion cells with other stains that are available such as methylene blue.
9. In the space below sketch a few of the stained onion cells.

Question 9

a. What cellular structures of onion are more easily seen in stained as compared to unstained preparations?

b. Which of the available stains enhanced your observations the most?

c. Do onion cells have chloroplasts? Explain.

d. Use an ocular micrometer or the dimensions of the FOV calculated in Exercise 2 to measure the dimensions of an onion epidermal cell. Are these cells larger or smaller than the *Elodea* cells you examined in procedure 3.3?

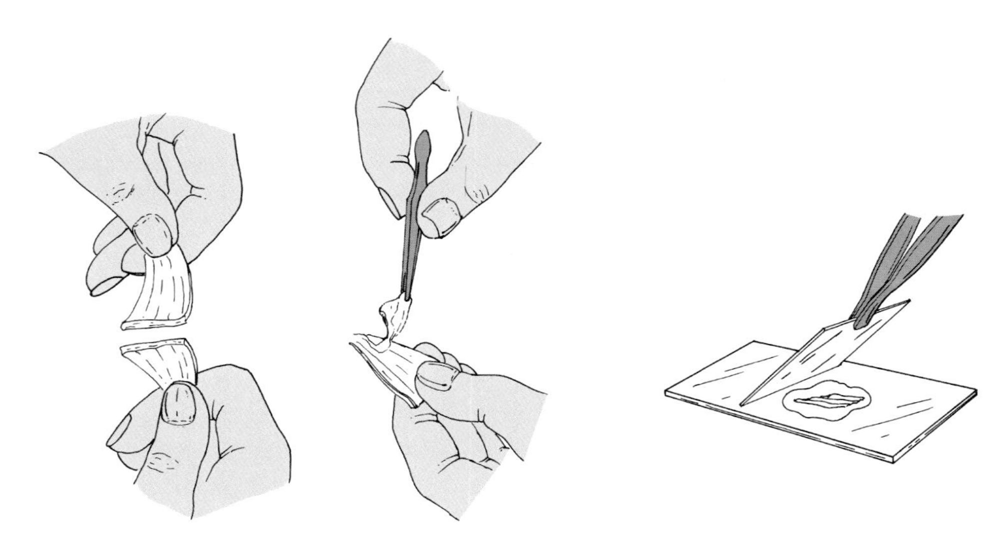

Figure 3.10
Preparing a wet mount of an onion epidermis.

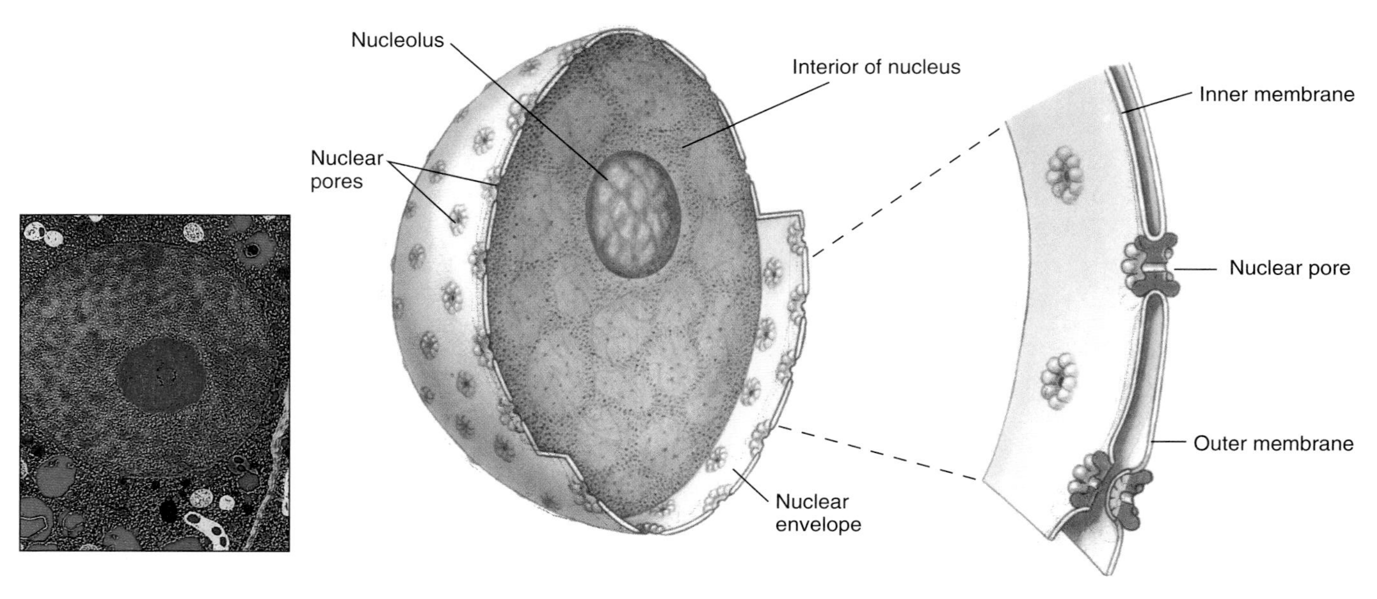

Figure 3.11
The nucleus. The nucleus consists of a double membrane, called a nuclear envelope, enclosing a fluid-filled interior containing the DNA. In cross section, the individual nuclear pores extend through the two membrane layers of the envelope; the material within the pore is protein, which controls access through the pore. 1765×.

Mitochondria

Mitochondria are surrounded by two membranes (fig 3.12). The inner membrane folds inward to form **cristae,** which hold respiratory enzymes and other large respiratory molecules in place. Some DNA also occurs in mitochondria. Chloroplasts also are double-membraned and contain DNA.

Procedure 3.6

Examine mitochondria in onion cells

1. On a clean glass slide mix two or three drops of the stain Janus Green B with one drop of 7% sucrose.
2. Prepare a thin piece of onion epidermis (as instructed in procedure 3.5) and mount it in the staining solution. The preparation should be 1-cell thick. For mitochondria to stain well, the onion cells must be healthy and metabolically active. Add a coverslip.
3. Search the periphery of cells to locate stained mitochondria. They are small blue spheres about 1 μm in diameter. The color will fade in 5–10 min, so examine your sample quickly and make a new preparation if needed.

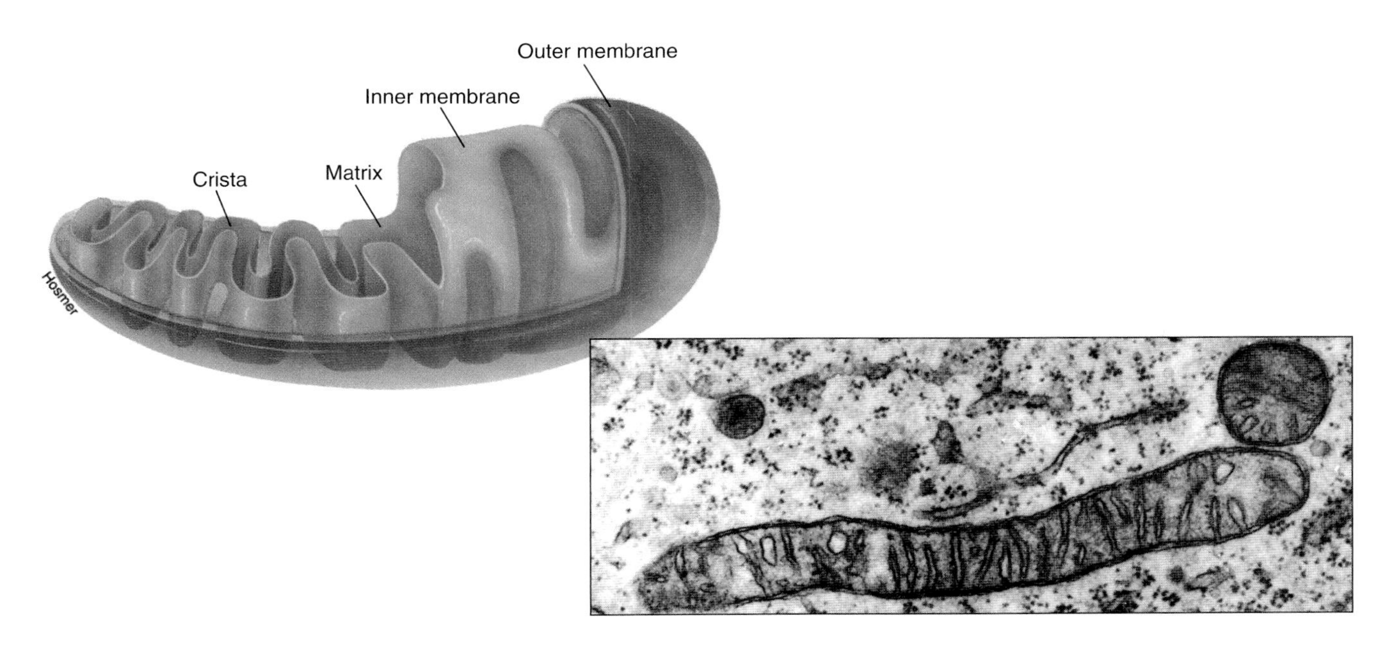

Figure 3.12
Mitochondrion in cross section. Mitochondria evolved from bacteria that long ago took up residence within the ancestors of present-day eukaryotes. 80,000×.

Plastids

Plastids are organelles where food is made and stored. You have already examined chloroplasts, a type of plastid in which photosynthesis occurs. Other plastids have different functions. We will examine **amyloplasts,** which are plastids that store starch and therefore will stain darkly with iodine.

Procedure 3.7

Examine amyloplasts

1. Use a razor blade to make a thin section of a potato tuber. Make the section as thin as you can.
2. Place the section in a drop of water on a microscope slide and add a coverslip. Add another drop of water to the edge if needed.
3. Locate the small, clam-shaped amyloplasts within the cells. High magnification may reveal the eccentric lines distinguishing layers of deposited starch on the grains.
4. Stain the section by adding a drop of iodine to the edge of the coverslip. Iodine is a stain specific for starch (see Exercise 5, "Biologically Important Molecules"). If necessary, pull the stain under the coverslip by touching a paper towel to the water at the opposite edge.

Question 10

a. Are any cellular structures other than amyloplasts stained intensely by iodine?

b. What can you conclude about the location of starch in storage cells of potato?

c. What are the functions of amyloplasts in potatoes?

d. Why are potatoes a good source of carbohydrates?

ANIMAL CELLS

Animals, like plants, are eukaryotes. They share many similarities, and also have several differences (table 3.1).

Human Epithelial Cells

Human epithelial cells are sloughed from the inner surface of your mouth. They are flat cells with a readily visible nucleus.

Procedure 3.8

Examine human epithelial cells

1. Gently scrape the inside of your cheek with the broad end of a clean toothpick.

2. Stir the scrapings into a drop of water on a microscope slide, add a coverslip, and examine with your compound microscope. Dispose of used toothpicks in a container designated by your instructor.
3. Stain the cells by placing a small drop of methylene blue at one edge of the coverslip and drawing it under the coverslip with a piece of absorbent paper towel placed at the opposite side of the coverslip.
4. Prepare another slide and stain the cells with Janus Green B. Observe the mitochondria.
5. Use an ocular micrometer or the dimensions of the FOV calculated in Exercise 2 to measure the dimensions of a human epithelial cell.

Question 11

a. What structures visible in the stained preparation were invisible in the unstained preparation?

b. Were mitochondria as abundant in human epithelial cells as in onion epidermal cells (procedure 3.6)? Explain.

c. What similarities and differences are there between plant and animal cells?

d. How do the size and shape of a human epithelial cell differ from those of the *Elodea* and onion cells that you examined earlier?

e. Why do *Elodea* and onion cells have more consistent shapes than human epithelial cells?

6. After viewing the preparation, put the slides and coverslips in a container of 10% bleach.

PROTISTS

Amoeba and *Paramecium* are members of the kingdom Protista, a large group of eukaryotic organisms that includes algae and sporozoans. You will learn more about protists in Exercises 23 and 24. In today's exercise, you'll examine *Amoeba* and *Paramecium*. Protists are single-celled organisms, although some are colonial.

Amoeba

Amoeba is an irregularly shaped protist with many internal organelles (fig. 3.13). *Amoeba* move via amoeboid movement. **Amoeboid movement** occurs by means of **pseudopodia,** which are temporary protrusions of the cell. Pseudopodia also surround food particles and create food vacuoles, where food is digested. Another important structure in *Amoeba* is the **contractile vacuole** that accumulates and expels water and waste products.

Procedure 3.9

Examine *Amoeba*

1. Use an eyedropper to obtain a few drops from the bottom of an *Amoeba* culture. Examining the culture with a dissecting microscope may help you locate some organisms.
2. Place the organisms on a microscope slide.
3. Add a coverslip and use a compound microscope to locate a living *Amoeba*. Your instructor may allow you to view the *Amoeba* without using a coverslip, but view them *only* on 4× or 10× magnification.
4. Decrease the light intensity and observe an *Amoeba* for a few minutes.
5. Locate the structures shown in figure 3.13.
6. Examine a prepared slide of stained *Amoeba*; then observe a demonstration of *Amoeba* on a dark-field microscope if one is available.
7. Sketch an *Amoeba* in the space below.

Question 12

a. List the organelles found in plant cells, in *Amoeba*, and common to both.

b. Does *Amoeba* have a cell wall? How can you tell?

c. How do the appearances of *Amoeba* differ in live cells and preserved cells?

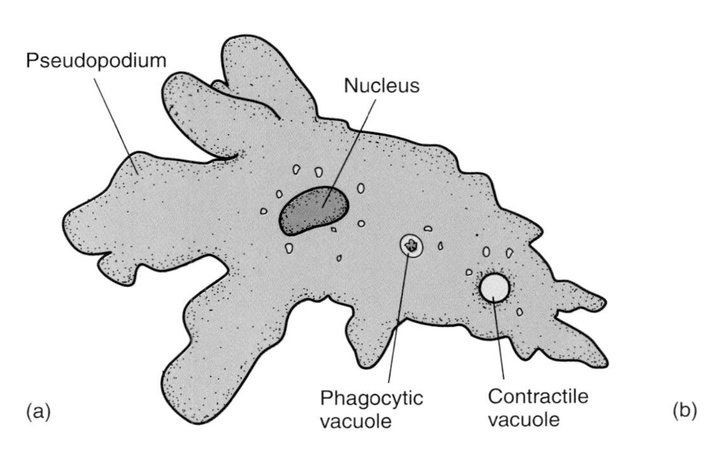

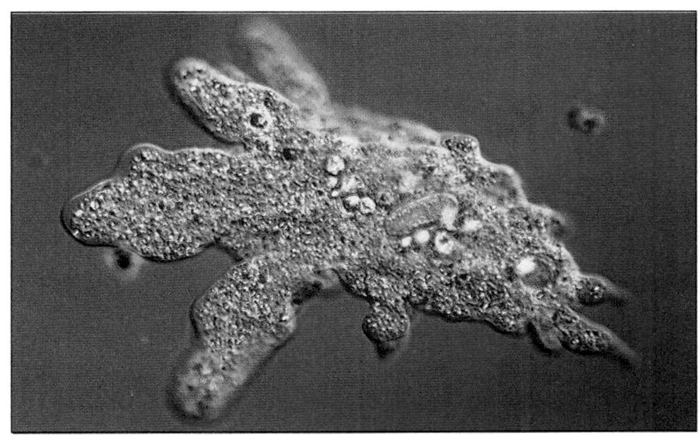

Figure 3.13
(*a*) Diagram of *Amoeba*. (*b*) Light micrograph of a living *Amoeba*. 160×.

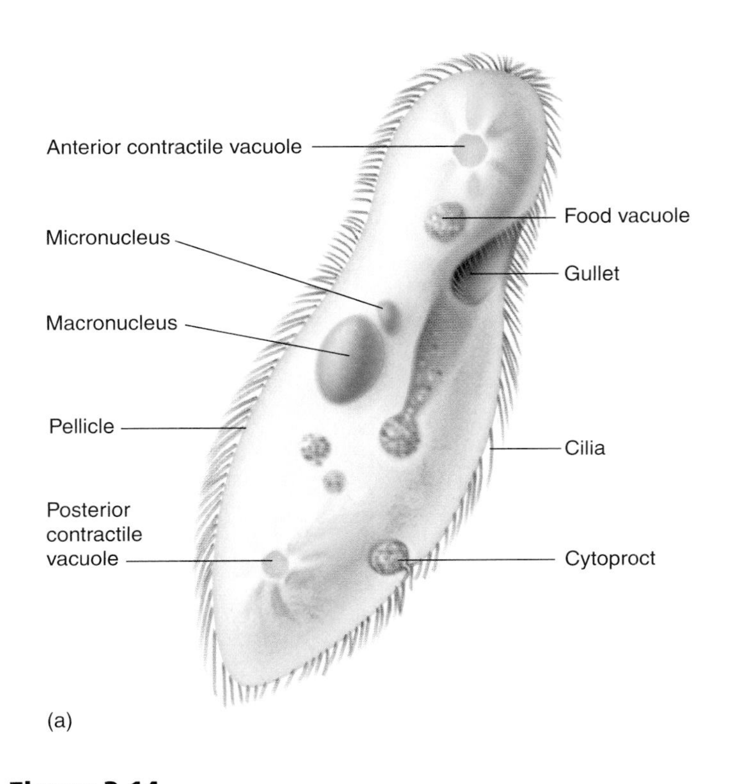

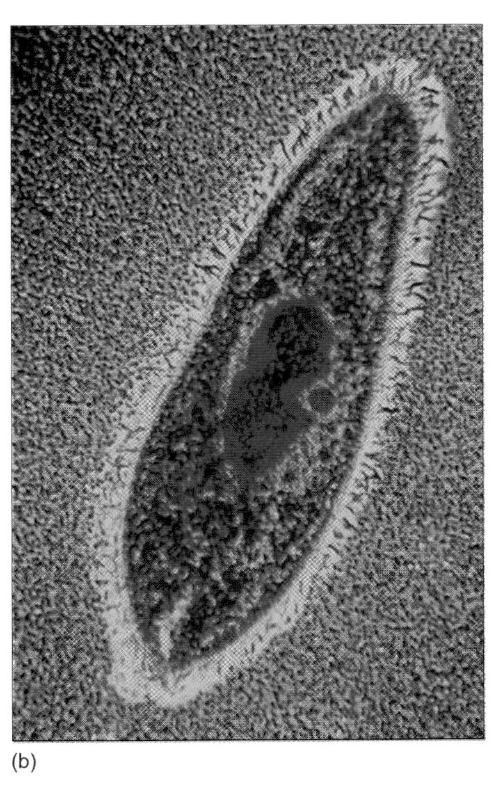

Figure 3.14
(*a*) Diagram of *Paramecium*, 150×. (*b*) Light micrograph of a living *Paramecium*. Note the abundant cilia, 150×.

Paramecium

Like *Amoeba*, *Paramecium* is also a single-celled organism (fig. 3.14).

Procedure 3.10
Examine *Paramecium*

1. Place a small ring of methylcellulose on a microscope slide to slow the *Paramecium*.
2. Place a drop from a culture containing *Paramecium* inside the methylcellulose ring.
3. Use a toothpick to mix the methylcellulose with the drop of water from the culture of *Paramecium*.
4. Add a coverslip and examine *Paramecium* with your compound microscope. On the surface of *Paramecium* are cilia, which are short hairlike structures used for locomotion.
5. Examine a prepared slide of stained *Paramecium*.

6. Sketch a *Paramecium* below.

Question 13

a. How does movement of *Paramecium* compare to that of *Amoeba?*

b. How does shape and body consistency differ between *Amoeba* and *Paramecium?*

c. What structures in *Amoeba* and *Paramecium* also occur in plant cells? What structures in *Amoeba* and *Paramecium* do not occur in plant cells?

Procedure 3.11

You will be given a slide of an unknown organism. Use what you've learned in today's lab to identify the cells as prokaryotic or eukaryotic; if eukaryotic, identify the cells as plant, animal, or protist. Complete table 3.2 before leaving the lab. If instructed to do so, turn in table 3.2 before leaving the lab.

INVESTIGATION

Response of Single-Celled Organisms to Environmental Stimuli

Like all organisms, *Amoeba* and *Paramecium* are sensitive to environmental stimuli.

a. Choose a stimulus you would like to test (e.g., temperature, salinity, acidity, etc.).

b. Form a hypothesis about the result you expect. Your instructor will advise you about how to write a testable hypothesis. Write your hypothesis here:

c. Decide how you will test your hypothesis. Describe your experimental design here:

d. Do your experiment. What did you conclude? Do your data support your hypothesis?

Questions for Further Thought and Study

1. What is a cell?

2. Why would some biologists refer to single-celled organisms such as *Amoeba* and *Paramecium* as "acellular" (i.e., without cells) rather than "unicellular" (i.e., one cell)?

3. Describe the structure and function of each component of the cells that you observed with a light microscope.

4. What are the advantages and disadvantages to an individual cell of being part of a multicellular organism?

5. What is the purpose of using a biological stain when microscopically examining cellular components?

DOING BIOLOGY YOURSELF

Determine the total surface area of the chloroplasts in a typical *Elodea* cell. Assume that each chloroplast is a sphere of 5 μm diameter. (The surface area of a sphere = πd^2.) What is the significance of this surface area? Would it be advantageous for a cell to be filled with chloroplasts? Why or why not?

WRITING TO LEARN BIOLOGY

What criteria might you use to distinguish colonial organisms such as many cyanobacteria from truly multicellular organisms?

Table 3.2

Using Distinguishing Features to Identify an Unknown Organism

Overall description of specimen:

Name ______________________

Unknown No: ______________

Lab Section ______________

Based on the above, my unknown organism is a: (Circle One)	Prokaryote	Eukaryote	
If the specimen is a eukaryote, it is a(n): (Circle One)	Plant	Animal	Protist

4

Solutions, Acids, and Bases
The pH Scale

Objectives

By the end of this exercise you should be able to:
1. Apply the concepts of mole and molarity to prepare solutions.
2. Measure the pH of various liquids.
3. Demonstrate that buffers stabilize the pH of a liquid.
4. Measure the ability of commercial antacids to buffer the pH of a liquid.

Chemicals in living systems are in solution. Biologists experiment with solutions because dissolved chemicals are more readily available to react than are solid, crystaline chemicals. A **solution** consists of a **solute**(s) dissolved in a **solvent.** For example, salt water is a solution in which salt (i.e., the solute) is dissolved in water (i.e., the solvent).

The concentration of a solute is often expressed as a percentage of the total solution (e.g., weight/volume or grams solute/100 mL solution). For example, a 3% (weight/volume) solution of sucrose is prepared by dissolving 30 g of sucrose in water for a total solution of 1 L (or 3 g of sucrose in water for a total volume of 100 mL).

Question 1

a. How many grams of sucrose would you dissolve in water for a total volume of 500 mL to make a 5% (weight/volume) solution?

b. How many grams of calcium chloride would you add to water for a total volume of 500 mL to make a 5% (weight/volume) solution?

c. How many grams of calcium chloride would you add to water for a total volume of 100 mL to make a 5% (weight/volume) solution?

Molarity is the most common measure of concentration. To understand how to prepare a molar solution you must first understand what is meant by a **mole** of a chemical. A mole is a standard measure of the amount of a chemical—one mole of any substance has 6.02×10^{23} molecules (Avogadro's number). One mole of NaCl and one mole of sucrose contain the same number of molecules. However, a mole of NaCl and a mole of sucrose weigh different amounts. This is because each chemical has a different **molecular weight,** and the weight of 1 mole of a chemical equals that chemical's molecular weight in grams. For example the molecular weight of water (H_2O) is 18 g ($2H = 2 \times 1 = 2$; $O = 16$; $16 + 2 = 18$). A mole of water weighs 18 g. A mole of NaCl weighs 58.5 g (fig. 4.1). A chemical's molecular weight is the sum of the atomic weights of its component elements.

To further understand why biologists usually prepare solutions in molar concentrations rather than as percents you must remember that chemicals react on a molecule by molecule basis—the number of molecules is more critical than the weight. It follows that expressing a solution's concentration in moles is a better measure of how much chemical is available to react. A solution that contains one mole of a chemical in 1 liter of solution has 6.02×10^{23} molecules available and is a 1-molar (1 M) solution. For example, a liter of solution containing 58.5 g of NaCl is a 1 M solution of NaCl (fig. 4.2).

Question 2

a. How many grams of NaCl (molecular weight = 58.5 g $mole^{-1}$) would you dissolve in water to make a 0.5 M NaCl solution with 500 mL final volume?

_________ g

b. How many grams of NaCl (molecular weight = 58.5 g $mole^{-1}$) would you dissolve in water to make a 50 mM NaCl solution with 500 mL final volume?

_________ g

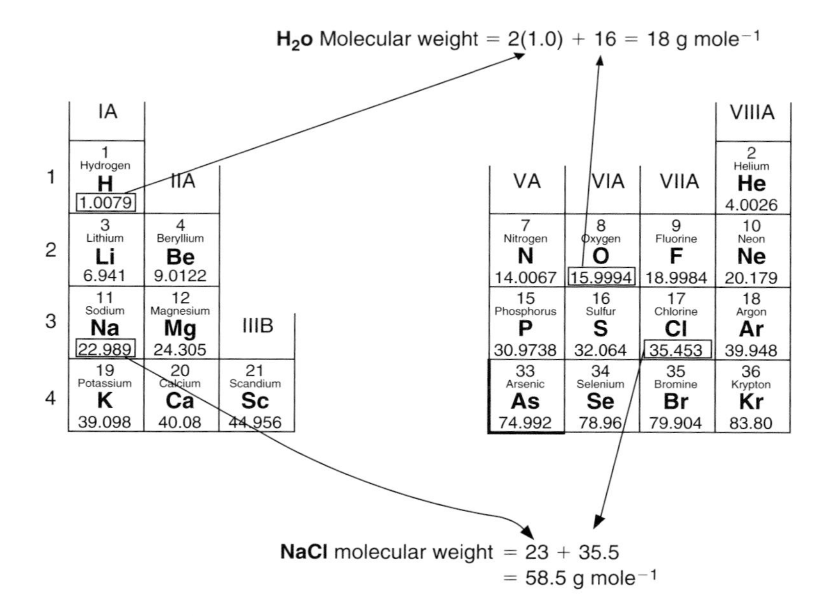

Figure 4.1

The atomic weights of elements are listed in the periodic table. Shown here are the portions of the periodic table that would be used to calculate the molecular weights of water (H_2O) and table salt (sodium chloride, NaCl).

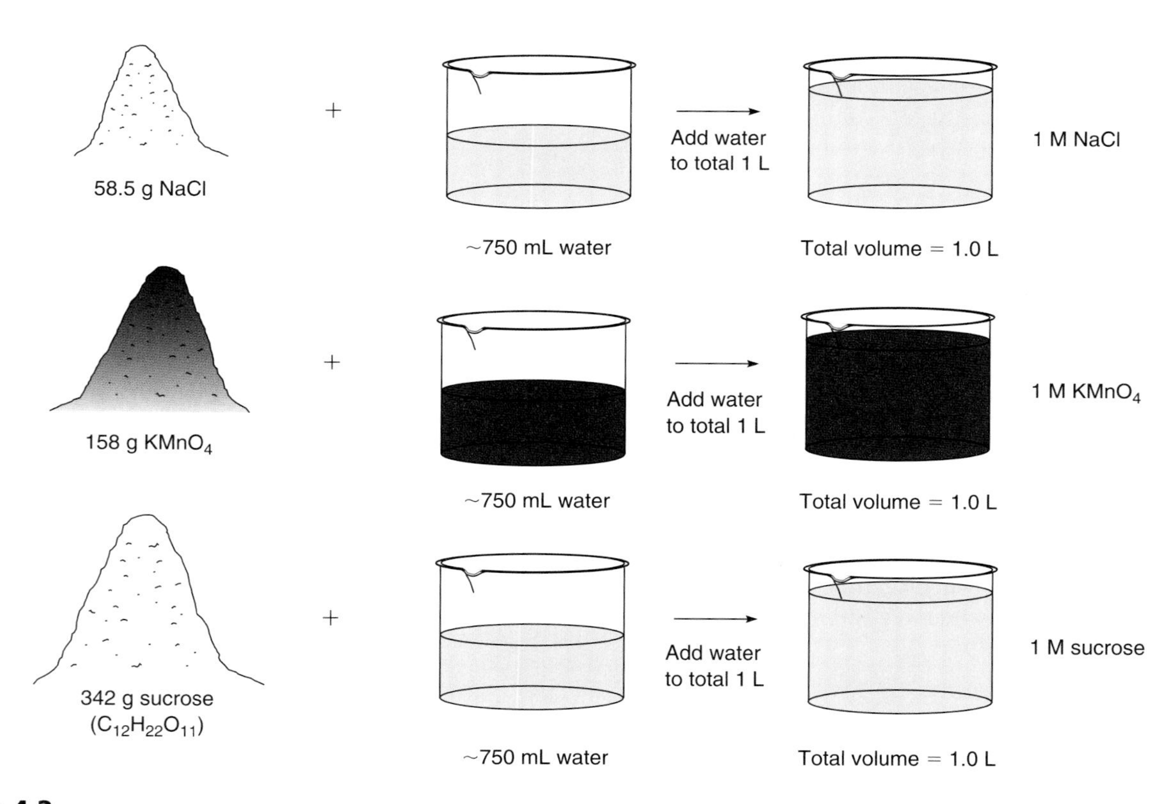

Figure 4.2

Preparing 1.0 M solutions of sodium chloride (NaCl; molecular weight = 58.5 g mole^{-1}), potassium permanganate ($KMnO_4$; molecular weight = 158 g mole^{-1}), and sucrose ($C_{12}H_{22}O_{11}$; molecular weight = 342 g mole^{-1}). Each of these solutions contains the same number of units of solutes (i.e., 6.02×10^{23} molecules).

c. How many grams of sucrose (molecular weight = 342 g $mole^{-1}$) would you dissolve in water to make a 0.22 M sucrose solution with 1 L final volume?

_________ g

d. How many grams of sucrose (molecular weight = 342 g $mole^{-1}$) would you dissolve in water to make a 0.22 mM sucrose solution with 100 mL final volume?

_________ g

e. How many grams of calcium chloride ($CaCl_2$; molecular weight = 111 g $mole^{-1}$) would you dissolve in water to make a total 0.111 M $CaCl_2$ solution with 1 L final volume?

_________ g

f. How many grams of calcium chloride ($CaCl_2$; molecular weight = 111 g $mole^{-1}$) would you dissolve in water to make a 0.2 M $CaCl_2$ solution with 200 mL final volume?

_________ g

g. If you were presented with 2 L of a 2 M sucrose stock solution, how many grams of sugar would be in a 100 mL aliquot?

_________ g

h. To prepare the 5% sucrose solution called for in Question 1a how many moles of sugar did you add? What was the molarity of that solution?

i. To prepare the 5% calcium chloride solution called for in Question 1b how many moles of calcium chloride did you add? What was the molarity of that solution?

j. How many milliliters of a 2 M sucrose solution would contain 1 mole of sucrose?

Dilutions

To save time and space, biologists often prepare commonly used solutions in concentrated forms called **stock solutions.** These stock solutions are then diluted with water to make new solutions having a desired molarity. This process is called **dilution.**

Dilution involves spreading a given amount of solute throughout a larger solution. That is, the number of moles of solute doesn't change when a solution is diluted but the volume of solution containing those moles increases. This means that the product of the initial volume (V_i) and initial molarity (M_i) must equal the product of the final volume (V_f) and final molarity (M_f):

$$V_iM_i = V_fM_f$$

where

$$V_i = \text{initial volume}$$
$$M_i = \text{initial molarity}$$
$$V_f = \text{final volume}$$
$$M_f = \text{final molarity}$$

Let's now use this simple equation to solve a dilution problem. Suppose we want to know how much water to add to 25 mL of a 0.50 M KOH solution to produce a solution having a KOH concentration of 0.35 M. In this case,

$$M_i = 0.5 \text{ M}$$
$$V_i = 25 \text{ mL}$$
$$M_f = 0.35 \text{ M}$$
$$V_f = ?$$

We can now solve the problem:

$$V_iM_i = V_fM_f$$
$$(25 \text{ mL})(0.5 \text{ M}) = (V_f)(0.35 \text{ M})$$
$$V_f = 35.7 \text{ mL}$$

Since the initial volume (V_i) was 25 mL, we must subtract 25 mL from 35.7 mL to get our answer: 35.7 mL − 25 mL = 10.7 mL of water to produce a KOH solution having a concentration of 0.35 M.

Question 3

a. How many mL of concentrated (18 M) sulfuric acid (H_2SO_4) are required to prepare 750 mL of 3 M sulfuric acid?

b. How would you prepare 100 mL of 0.4 M $MgSO_4$ from a stock solution of 2 M $MgSO_4$?

c. How many milliliters of water would you add to 100 mL of 1.0 M HCl to prepare a final solution of 0.25 M HCl?

ACIDS AND BASES

One of the most important applications of molarity involves the concentration of hydrogen ions (H^+) in a solution. Pure water is the standard by which all other solutions are compared, because water is an ionically neutral solution. This neutrality is not due to the absence of ions, but rather to the equal concentrations of positive and negative ions. When the oxygen of water pulls hard enough on an electron from one of its hydrogens, two ions form:

$$H_2O \longleftrightarrow H^+ + OH^-$$

This dissociation of water is rare and reversible, but it happens often enough for the concentration of H^+ in pure water to be 10^{-7} M. The solution is neutral because the concentration of OH^- is also 10^{-7} M. The sum of H^+ and OH^- ions will always equal 10^{-14}.

Any substance that increases the concentration of H^+ is an **acid;** any substance that decreases the concentration of H^+ is a **base.** When the concentration of one ion increases, the concentration of the other becomes proportionately less. For example, hydrocholoric acid (HCl) quickly ionizes in water and increases the concentration of H^+; therefore, HCl is an acid. In contrast, sodium hydroxide (NaOH) is a base because it ionizes and increases the concentration of OH^-, thereby lowering the relative proportion of H^+. Thus, if enough acid is added to water to raise the H^+ concentration to 10^{-6} M, the OH^- concentration would decrease to 10^{-8} M.

By general agreement, the scale we use to measure acidity is the **pH scale** (*pH* stands for the *p*otential of *H*ydrogen ions). The pH is the negative logarithm of the concentration of H^+; that is,

$$\mathbf{pH = -\log [H^+]}$$

The brackets indicate concentration of hydrogen ions. The pH scale ranges from 0 ($-\log 10\infty$; most acidic) to 14 ($-\log 10^{-14}$; most basic). On this scale, pure water has a pH of 7 ($-\log 10^{-7}$); pH values less than 7 are acidic, whereas those above 7 are basic (fig. 4.3).

Figure 4.3 shows some pHs of some common (and a few not-so-common) substances. Note that the pH scale is a logarithmic scale; each unit represents a change of tenfold. Thus, a lime with a pH of 2 is ten times more acidic than an apple with a pH of 3, and 100 times more acidic than a tomato having a pH of 4. Each decrease of 1.0 pH unit represents a tenfold increase in acidity. Each increase of 1.0 pH unit represents a tenfold decrease in acidity.

Measuring pH

A convenient way of measuring the pH of a solution is with **pH paper.** pH paper is treated with a chemical indicator that changes colors depending on the concentration of H^+ in the solution that it has contacted. Here are some examples of pH indicators:

Indicator	*Range*	*Color Change*
Methyl violet	0.2–3.0	yellow to blue-violet
Bromphenol blue	3.0–4.6	yellow to blue
Methyl red	4.4–6.2	red to yellow
Litmus	4.5–8.3	red to blue
Bromcresol purple	5.2–6.8	yellow to purple
Phenol red	6.8–8.0	yellow to red
Thymol blue	8.0–9.6	yellow to blue
Phenolphthalein	8.3–10.0	colorless to red

The color chart on the container of pH paper relates the color of the pH paper to the pH of the solution.

Procedure 4.1

Measure the pH of liquids

Use pH papers to measure the pH of the following liquids. Be as accurate as possible and use a fresh piece of pH paper or pH dipstick for each test.

Vinegar	________
Skim milk	________
Apple juice	________
Grapefruit juice	________
Buttermilk	________
Black coffee	________
Sprite	________
Solution of baking soda	________
Mixture of Sprite and baking soda	________
10 mM hydrochloric acid	________
1.0 mM hydrochloric acid	________
0.01 mM hydrochloric acid	________
Distilled water	________
Tap water	________

Check your measurements of the hydrochloric acid solutions by comparing them with calculations using the formula below. For example,

$$\mathbf{pH = -\log[H^+]}$$
$$\mathbf{10\ mM\ HCl = 10^{-2}\ M\ HCl}$$
$$\mathbf{pH = -\log[10^{-2}] = 2}$$

Question 4

Are your measured pH values similar to the calculated pH values? What are possible sources of error?

Buffers

In most organisms, the pH is kept relatively constant by **buffers,** which are chemicals that absorb excess H^+ as the pH decreases or release H^+ as the pH increases. Buffers minimize changes in pH (fig. 4.4). The addition of a small amount of acid to a buffered solution produces a small change in pH, whereas adding the same amount of acid to an unbuffered solution changes the pH drastically. Most biological fluids (e.g., milk, blood) contain buffers. For example, human blood contains buffers that maintain a pH of 7.3–7.5; slight changes in the pH of blood can cause illness and even death.

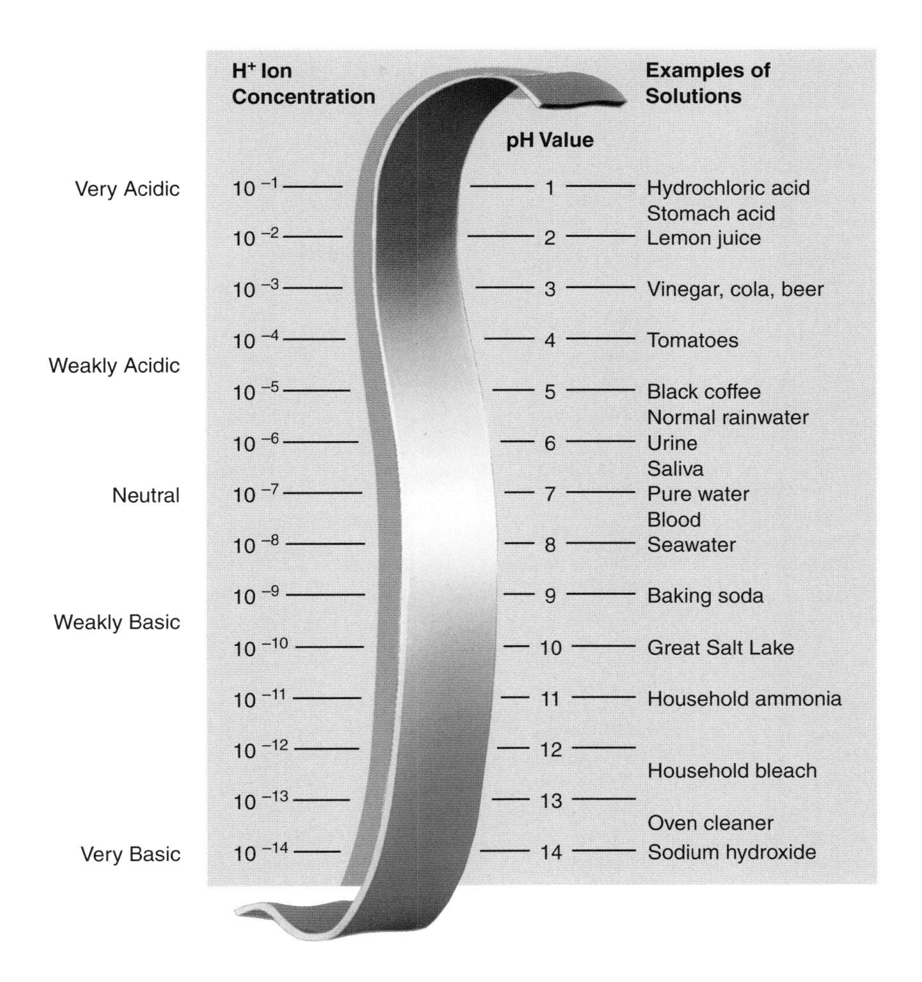

Figure 4.3

The pH scale. The pH value of a solution indicates its concentration of hydrogen ions. Solutions with a pH less than 7 are acidic, whereas those with a pH greater than 7 are basic. The pH scale is logarithmic: a pH change of 1 means a tenfold change in the concentration of hydrogen ions. Thus, lemon juice is 100 times more acidic than tomato juice, and seawater is 10 times more basic than pure water, which has a pH of 7.

TABLE 4.1

TESTING THE BUFFERING CAPACITY OF VARIOUS SOLUTIONS

Procedure 4.2 Solution	Initial pH	pH after Adding Acid	Procedure 4.3 Solution	Drops of Acid
Water	________	________	Maalox	________
0.1 M NaCl	________	________	Alka-Seltzer	________
Skim milk	________	________	Rolaids	________
0.1 M phosphate buffer	________	________	Tums	________

Procedure 4.2

Test the ability of buffers to stabilize pH

1. Obtain and label four test tubes to receive the four solutions listed in table 4.1.
2. Place 5 mL of each solution into its appropriately labeled tube.
3. Measure the pH of each of the solutions in the tubes and record these initial values in table 4.1
4. Add 5 drops of acid (0.1 M HCl) to the first tube. Cover the tube with parafilm and invert the tube gently to mix the contents.
5. Measure the pH of the acidified solution and record it in table 4.1.
6. Repeat steps 4 and 5 for each of the remaining tubes. Record your results in table 4.1.

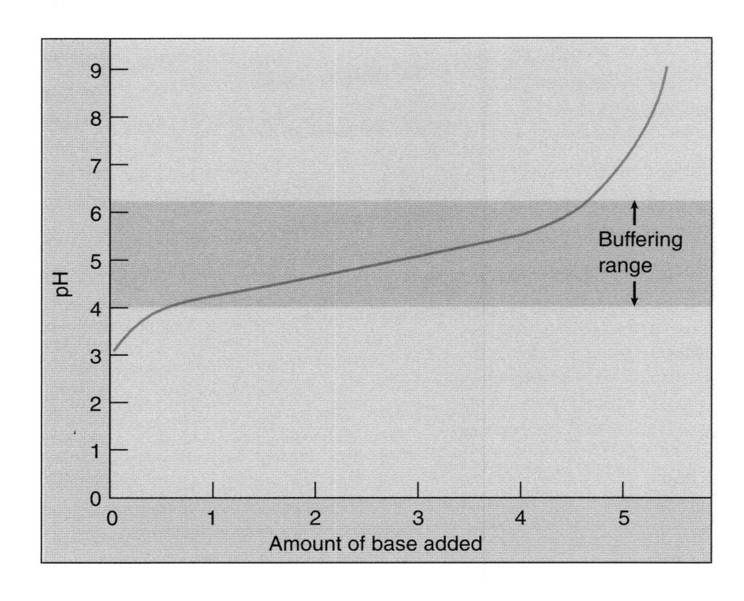

Figure 4.4

Buffers minimize changes in pH. Adding a base to a solution will raise the pH (neutralizes some of the acid present). Thus, as more and more base is added the pH continues to rise. However, a buffer makes the curve rise or fall very slowly over a portion of the pH scale, called the "buffering range" of that buffer.

Question 5

a. Compare the initial pH and the pH after acid addition for each sample. Which is the most effective buffer? Least effective?

b. What accounts for the different buffering capacities of these fluids?

c. What is the biological importance of what you observed?

Procedure 4.3

Test the effectiveness of commercial antacids

Commercial antacids such as Maalox, Alka-Seltzer, Rolaids, and Tums claim to "neutralize stomach acid" by absorbing excess H^+ (produced as hydrochloric acid by the stomach). To test the abilities of these products to absorb acids, do the following:

1. Use a mortar and pestle to pulverize the amount of antacid that is listed as one dose. Dissolve the crushed antacid in 100 mL of distilled water. Some of the products may require extensive stirring to get most or all of the powder to dissolve.
2. Using a pipet or 10-mL graduated cylinder add 5 mL of the antacid solution into a test tube. Add 4 drops of the indicator bromcresol purple to the tube. Cover the tube with parafilm and invert the tube to mix the contents.
3. Add 0.1 M hydrochloric acid (HCl) dropwise to the tube; mix after each drop. Continue this process until the solution turns yellow, indicating an acidic solution.
4. Record in table 4.1 the number of drops of acid needed to generate the change of color. This number of drops is an index to the amount of acid (H^+) that the solution neutralizes before the pH drops below the yellow end-point of bromcresol purple (pH 5.2).

Question 6

a. Which antacid neutralizes the most acid? Which neutralizes the least acid?

b. What is the effect of dose (size of tablets) on your results and conclusions?

Questions for Further Thought and Study

1. What do buffers do and why are they important in biological systems?

2. Your stomach secretes hydrochloric acid. How would antacids “settle an upset stomach”?

3. The soft drink Mr. Pibb contains (among other things) 39 g of sucrose in 355 mL of solution. What is the molarity of this sucrose solution? What is the percent (weight/volume) of sucrose in the solution?

5

Biologically Important Molecules
Carbohydrates, Proteins, Lipids, and Nucleic Acids

Objectives

By the end of this exercise you should be able to:
1. Perform tests to detect the presence of carbohydrates, lipids, proteins, and nucleic acids.
2. Explain the importance of a control in biochemical tests.
3. Use biochemical tests to identify an unknown compound.

Most organic compounds in living organisms are **carbohydrates, proteins, lipids,** or **nucleic acids.** Each of these macromolecules is made of smaller subunits. These subunits are linked by **dehydration synthesis,** an energy-requiring process in which a molecule of water is removed and the two subunits are bonded covalently (fig. 5.1). Similarly, breaking the bond between the subunits requires the addition of a water molecule. This energy-releasing process is called **hydrolysis.**

The subunits of macromolecules are held together by covalent bonds and have different structures and properties. For example, lipids (made of fatty acids) have many C—H bonds and relatively little oxygen, while proteins (made of amino acids) have amino groups (—NH_2) and carboxyl (—COOH) groups. These characteristic subunits and groups impart different chemical properties to macromolecules—for example, monosaccharides such as glucose are polar and soluble in water, whereas lipids are nonpolar and insoluble in water.

CONTROLLED EXPERIMENTS TO IDENTIFY ORGANIC COMPOUNDS

Scientists have devised several biochemical tests to identify the major types of organic compounds in living organisms. Each of these tests involves two or more treatments: (1) an **unknown solution** to be identified, and (2) **controls** to provide standards for comparison. As its name implies, an unknown solution may or may not contain the substance that the investigator is trying to detect. Only a carefully conducted experiment will reveal its contents. In contrast, controls are known solutions. We use controls to validate that our procedure is detecting what we expect it to detect and nothing more. During the experiment we compare the unknown solution's response to the experimental procedure with the control's response to that same procedure.

A **positive control** contains the variable for which you are testing; it reacts positively and demonstrates the test's ability to detect what you expect. For example, if you are testing for protein in unknown solutions, then an appropriate positive control is a solution known to contain protein. A positive reaction shows that your test reacts correctly; it also shows you what a positive test looks like.

A **negative control** does not contain the variable for which you are searching. It contains only the solvent (often distilled water with no solute) and does not react in the test. A negative control shows you what a negative result looks like.

Controls are important because they reveal the specificity of a particular test. For example, if water and a glucose solution react similarly in a particular test, the test cannot distinguish water from glucose. But if the glucose solution reacts differently from distilled water, the test can distinguish water from glucose. In this instance, the distilled water is a negative control for the test, and a known glucose solution is a positive control.

CARBOHYDRATES

Benedict's Test for Reducing Sugars

Carbohydrates are molecules made of C, H, and O in a ratio of 1:2:1 (e.g., the chemical formula for glucose is $C_6H_{12}O_6$). Carbohydrates are made of **monosaccharides,** or simple sugars (fig. 5.2). Paired monosaccharides form **disaccharides**—for example, sucrose is a disaccharide of glucose linked to fructose. Similarly, linking three or more monosaccharides forms a **polysaccharide** such as starch, glycogen, or cellulose (fig. 5.3).

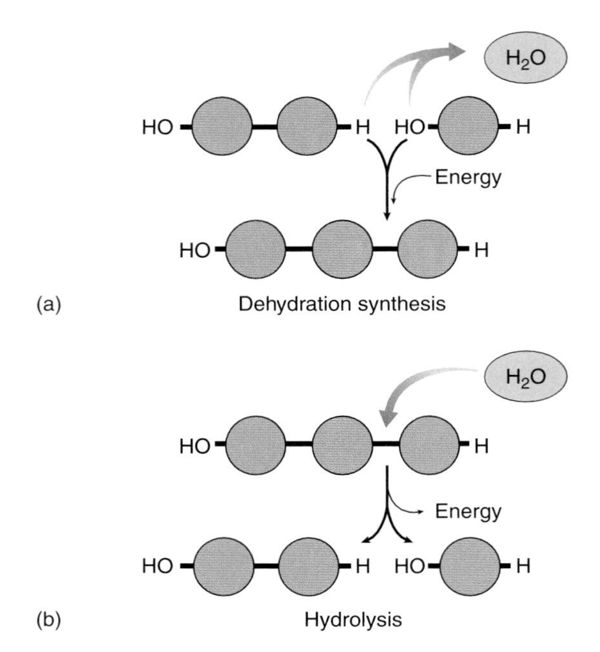

Figure 5.1

Making and breaking macromolecules. (*a*) **Dehydration synthesis.** Biological macromolecules are polymers formed by linking subunits together. The covalent bond between the subunits is formed by dehydration synthesis, an energy-requiring process that creates a water molecule for every bond formed. (*b*) **Hydrolysis.** Breaking the bond between subunits requires the returning of a water molecule with a subsequent release of energy, a process called hydrolysis.

Question 1

Examine figure 5.2. Which groups of a glucose molecule are involved in forming a polysaccharide? Shade the groups with a pencil.

As already mentioned, the linkage of subunits in carbohydrates, as well as other macromolecules, involves the removal of a water molecule (dehydration). Figure 5.4 depicts how dehydration synthesis is used to make maltose and sucrose, two common disaccharides.

Many monosaccharides such as glucose and fructose are **reducing sugars,** meaning that they possess free aldehyde (—CHO) or ketone (—C=O) groups that reduce weak oxidizing agents such as the copper in Benedict's reagent. **Benedict's reagent** contains cupric (copper) ion complexed with citrate in alkaline solution. Benedict's test identifies reducing sugars based on their ability to reduce the cupric (Cu^{2+}) ions to cuprous oxide at basic (high) pH. Cuprous oxide is green to reddish orange.

Oxidized Benedict's reagent (Cu^{2+}) + Reducing sugar (R-COH)
(blue)

Heat
High pH ↓

Reduced Benedict's reagent (Cu^{+}) + Oxidized sugar (R-COOH)
(green to reddish orange)

A green solution indicates a small amount of reducing sugars, and reddish orange indicates an abundance of reducing sugars. Nonreducing sugars such as sucrose produce no change in color (i.e., the solution remains blue).

Procedure 5.1

Perform the Benedict's test for reducing sugars

1. Obtain seven test tubes and number them 1–7.
2. Add to each tube the materials to be tested (table 5.1). Your instructor may ask you to test some additional materials. If so, include additional numbered test tubes. Add 2 mL of Benedict's solution to each tube.
3. Place all of the tubes in a gently boiling water-bath for 3 min and observe color-changes during this time.
4. After 3 min, remove the tubes from the water-bath and let them cool to room temperature. Record the color of their contents in table 5.1.

Question 2

a. Which of the solutions is a positive control? Negative control?

b. Which is a reducing sugar, sucrose or glucose? How do you know?

c. Which contains more reducing sugars, potato juice or onion juice? How do you know?

d. What does this tell you about how sugars are stored in onions and potatoes?

Iodine Test for Starch

Staining by iodine (iodine-potassium iodide, I_2KI) distinguishes starch from monosaccharides, disaccharides, and other polysaccharides. The basis for this test is that starch is a coiled polymer of glucose; iodine interacts with these coiled molecules and becomes bluish black. Iodine does not react with carbohydrates that are not coiled and remains yellowish brown. Therefore, a bluish-black color is a positive test for starch, and a yellowish-brown color (i.e., no color change) is a negative test for starch. Notably, glycogen, a common polysaccharide in animals, has a slightly different structure than does starch and produces only an intermediate color-reaction.

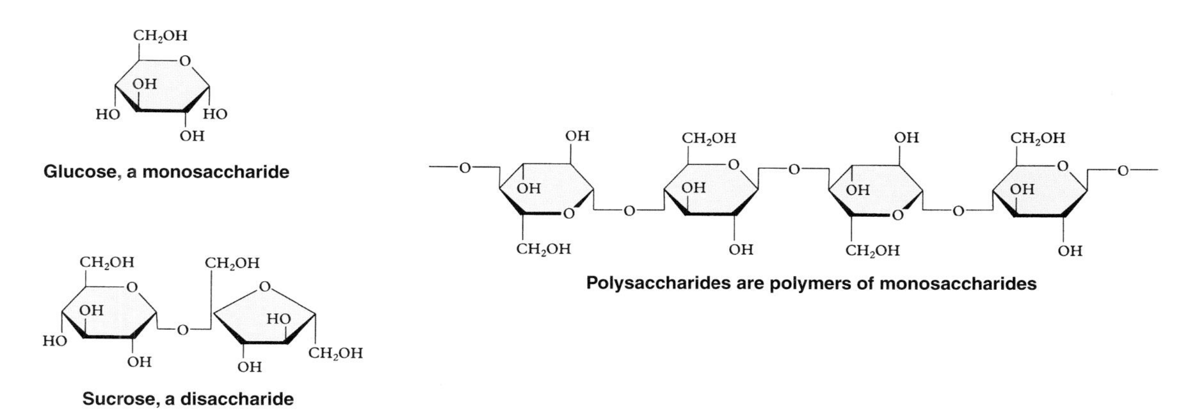

Figure 5.2

Carbohydrates consist of subunits of mono- or disaccharides. These subunits can be combined by dehydration synthesis (see fig. 5.4) to form polysaccharides.

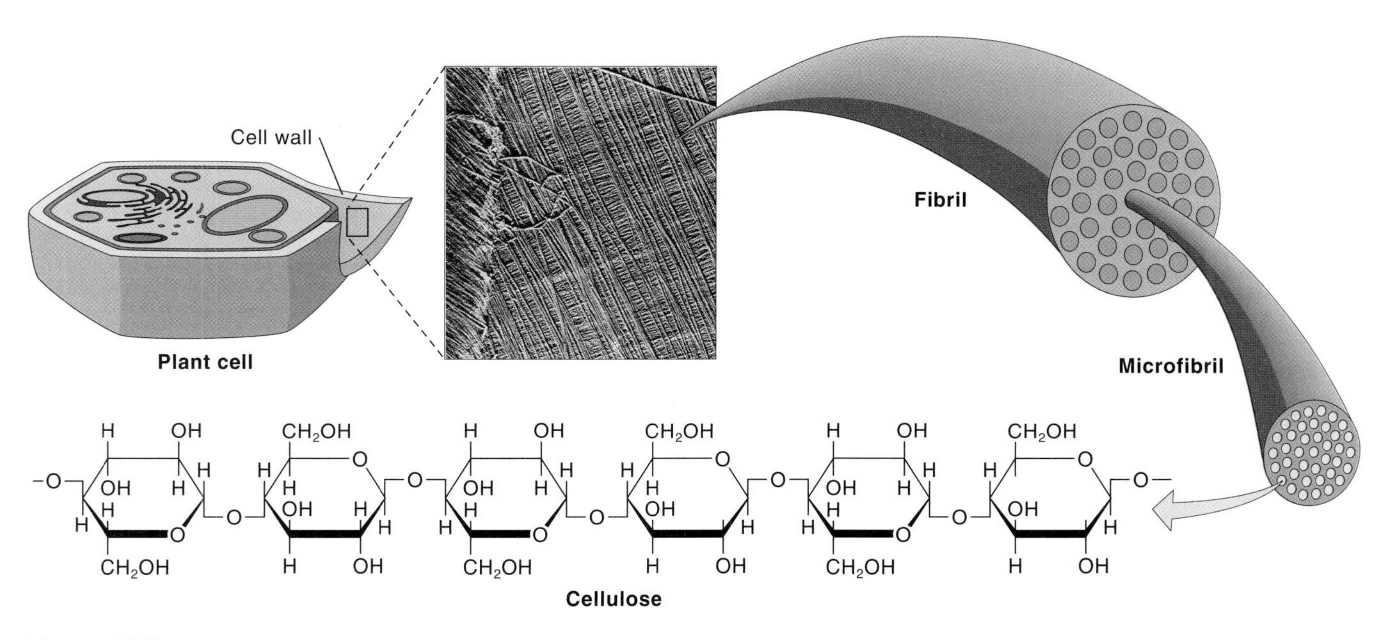

Figure 5.3

Plant cell walls are made of cellulose arranged in fibrils and microfibrils. The scanning electron micrograph shows the fibrils in a cell wall of the green alga *Chaetomorpha*, 30,000×.

Procedure 5.2

Perform the iodine test for starch

1. Obtain seven test tubes and number them 1–7.
2. Add to each tube the materials to be tested (table 5.1). Your instructor may ask you to test some additional materials. If so, include additional numbered test tubes.
3. Add seven to ten drops of iodine to each tube.
4. Record the color of the tubes' contents in table 5.1.

Question 3

a. Which of the solutions is a positive control? Which is a negative control?

b. Which colors more intensely, onion juice or potato juice?

c. What does this tell you about how these plants store carbohydrates?

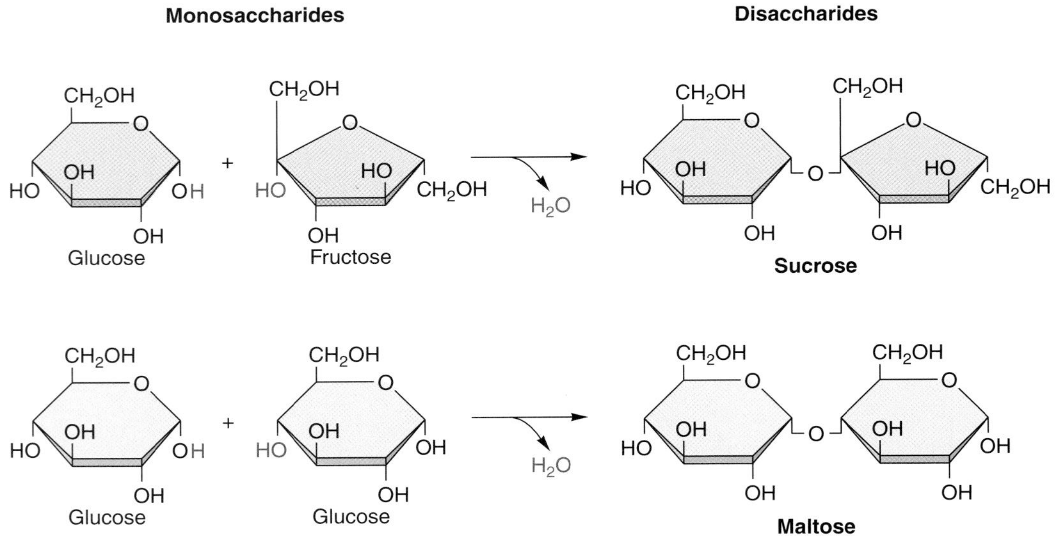

Figure 5.4
Dehydration synthesis is used to link monosaccharides (such as glucose and fructose) into disaccharides. The disaccharides shown here are maltose (malt sugar) and sucrose (table sugar).

TABLE 5.1

SOLUTIONS AND COLOR REACTIONS FOR (1) BENEDICT'S TEST FOR REDUCING SUGARS, AND (2) IODINE TEST FOR STARCH

Tube	Solution	Benedict's Color Reaction	Iodine Color Reaction
1	10 drops onion juice		
2	10 drops potato juice		
3	10 drops sucrose solution		
4	10 drops glucose solution		
5	10 drops distilled water		
6	10 drops reducing-sugar solution		
7	10 drops starch solution		
8			
9			

PROTEINS

Proteins are remarkably versatile structural molecules found in all life forms (fig. 5.5). Proteins are made of amino acids (fig. 5.6), each of which has an amino group (—NH_2), a carboxyl (acid) group (—COOH), and a variable side chain (R). A **peptide bond** (fig. 5.7) forms between the amino group of one amino acid and the carboxyl group of an adjacent amino acid and is identified by a **Biuret test.** Specifically, peptide bonds (C—N bonds) in proteins complex with Cu^{2+} in Biuret reagent and produce a violet color. A Cu^{2+} must complex with four to six peptide bonds to produce a color; therefore, individual amino acids do not react positively. Long-chain polypeptides (proteins) have many peptide bonds and produce a positive reaction.

Biuret reagent is a 1% solution of $CuSO_4$ (copper sulfate). A violet color is a positive test for the presence of protein; the intensity of color relates to the number of peptide bonds that react.

Figure 5.5

Common structural proteins. (*a*) Fibrin. This electron micrograph shows a red blood cell caught in threads of fibrin. Fibrin is important in the formation of blood clots. (*b*) Collagen. The so-called "cat-gut" strings of a tennis racket are made of collagen. (*c*) Keratin. This type of protein makes up bird feathers, such as this peacock feather. (*d*) Spider silk. The web spun by this agile spider is made of protein. (*e*) Hair. Hair is also a protein.

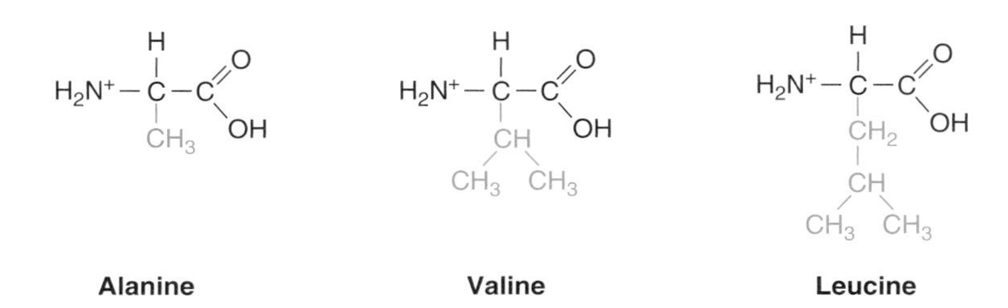

Figure 5.6

Structures of three amino acids that are common in proteins. Each amino acid has one carbon that is bonded to both an amine group (—NH_2) and a carboxyl group (—COOH). The side chains that make each amino acid unique are shown in red.

TABLE 5.2

SOLUTIONS AND COLOR REACTIONS FOR THE BIURET TEST FOR PROTEIN

Tube	Solution	Color
1	2 mL egg albumen	
2	2 mL honey	
3	2 mL amino acid solution	
4	2 mL distilled water	
5	2 mL protein solution	
6		
7		

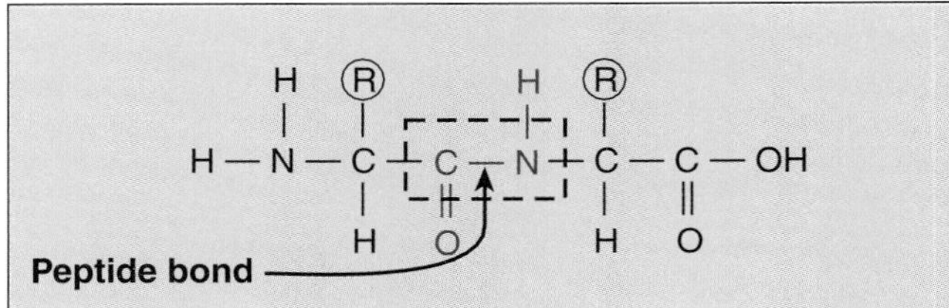

Figure 5.7

A peptide bond joins two amino acids, and peptide bonds link many amino acids to form polypeptides, or proteins. The formation of a peptide bond (i.e., between the carbon of one amino acid's carboxyl group and the nitrogen of another amino acid's amino group) liberates a water molecule. The R in these amino acids represents a variable side chain that characterizes each kind of amino acid.

Question 4

Examine figure 5.6. Shade with a pencil the reactive amino and carboxyl groups on the three common amino acids shown.

Procedure 5.3

Perform the Biuret test for protein

1. Obtain five test tubes and number them 1–5. Your instructor may ask you to test some additional materials. If so, include additional numbered test tubes.
2. Add the materials listed in table 5.2.
3. Add 2 mL of 2.5% sodium hydroxide (NaOH) to each tube.

CAUTION

Do not spill the NaOH—it is extremely caustic. Rinse your skin if it comes in contact with NaOH.

4. Add three drops of Biuret reagent to each tube and mix.
5. Record the color of the tubes' contents in table 5.2.

Question 5

a. Which of the solutions is a positive control? Which is a negative control?

b. Which contains more protein (C—N bonds), egg albumen or honey? How can you tell?

c. Do free amino acids have peptide bonds?

LIPIDS

Lipids include a variety of molecules that dissolve in nonpolar solvents such as ether, acetone, methanol, or ethanol, but not as well in polar solvents such as water. Triglycerides (fats) are abundant lipids that are made of glycerol and three fatty acids (fig. 5.8). Tests for lipids are based on a lipid's ability to selectively absorb pigments in fat-soluble dyes such as Sudan IV.

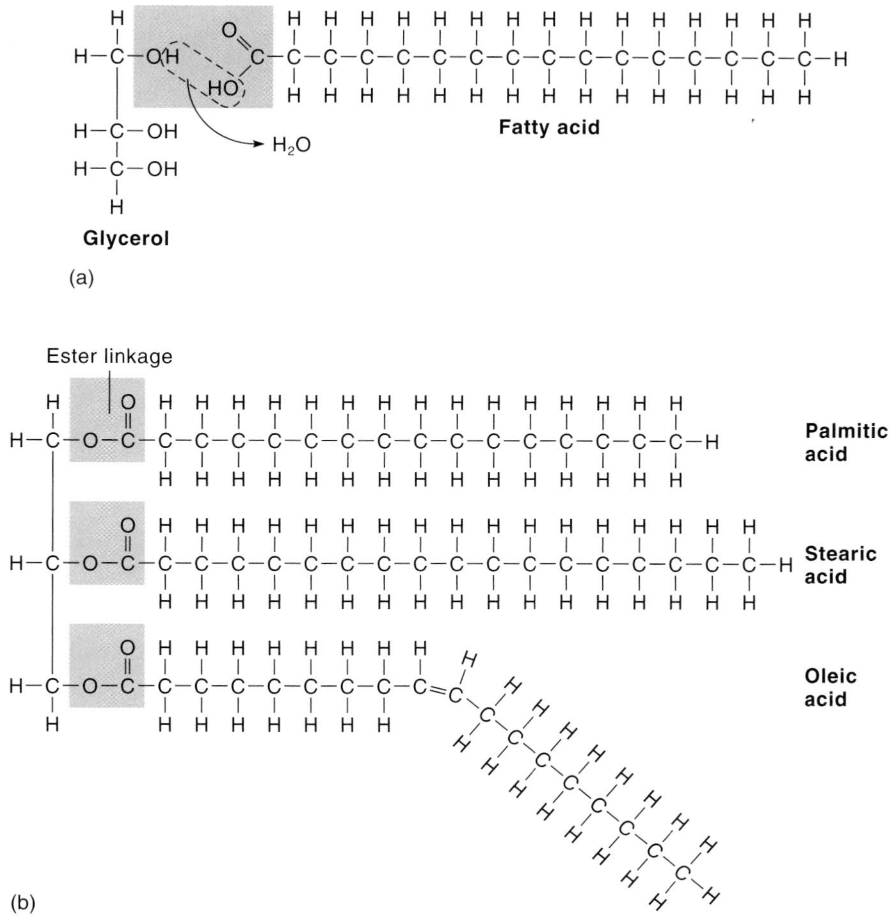

H_2O

Fatty acid

Glycerol

(a)

Ester linkage

Palmitic acid

Stearic acid

Oleic acid

(b)

Figure 5.8

The structure of a fat includes glycerol and fatty acids. (*a*) An ester linkage forms when the carboxyl group of a fatty acid links to the hydroxyl group of glycerol, with the removal of a water molecule. (*b*) Fats are triacylglycerides whose fatty acids vary in length and vary in the presence and location of carbon–carbon double bonds.

Question 6

Examine figure 5.8. What are the reactive groups of the fatty acids?

CAUTION

Handle acetone carefully; it is toxic.

Procedure 5.4

Solubility of lipids in polar and nonpolar solvents

1. Obtain two test tubes. To one of the tubes, add 5 mL of water. To the other tube, add 5 mL of acetone.
2. Add a few drops of vegetable oil to each tube.

Question 7

What do you conclude about the solubility of lipids in polar solvents such as water? In nonpolar solvents such as acetone?

TABLE 5.3

SOLUTIONS AND COLOR REACTIONS FOR THE SUDAN IV TEST FOR LIPIDS

Tube	Solution	Description of Reaction
1	1 mL salad oil + water	
2	1 mL salad oil + Sudan IV	
3	1 mL honey + Sudan IV	
4	1 mL distilled water + Sudan IV	
5	1 mL known lipid solution + Sudan IV	
6		
7		

Procedure 5.5

Perform the Sudan IV test for lipid

1. Obtain five test tubes and number them 1–5. Your instructor may ask you to test some additional materials. If so, include additional numbered test tubes.
2. Add the materials listed in table 5.3.
3. Add five drops of water to tube 1 and five drops of Sudan IV to each of the remaining tubes. Mix the contents of each tube. Record the color of the tubes' contents in table 5.3.

Question 8

a. Is salad oil soluble in water?

b. Compare tubes 1 and 2. What is the distribution of the dye with respect to the separated water and oil?

c. What observation indicates a positive test for lipid?

d. Does honey contain much lipid?

e. Lipids supply more than twice as many calories per gram as do carbohydrates. Based on your results, which contains more calories, oil or honey?

Grease-Spot Test for Lipids

A simpler test for lipids is based on their ability to produce translucent grease-marks on unglazed paper.

Procedure 5.6

Perform the grease-spot test for lipids

1. Obtain a piece of brown wrapping paper or brown paper bag from your lab instructor.
2. Use an eyedropper to add a drop of salad oil near a corner of the piece of paper.
3. Add a drop of water near the opposite corner of the paper.
4. Let the fluids evaporate.
5. Look at the paper as you hold it up to a light.
6. Test other food products and solutions available in the lab in a similar way and record your results in table 5.4.

Question 9

Which of the food products that you tested contain large amounts of lipid?

NUCLEIC ACIDS

DNA and RNA are nucleic acids made of nucleotide subunits (fig. 5.9). One major difference between DNA and RNA is their sugar: DNA contains deoxyribose, whereas RNA contains ribose. DNA can be identified chemically with the **Dische diphenylamine test.** Acidic conditions convert deoxyribose to a molecule that binds with diphenylamine to form a blue complex. The intensity of the blue color is proportional to the concentration of DNA.

Question 10

Examine figure 5.9*a*. Which groups on ribose and deoxyribose react when combining with the phosphate? Shade these groups. Also shade the reactive groups that combine with a nitrogenous base.

TABLE 5.4

MATERIALS AND GREASE-SPOT REACTION AS A TEST FOR LIPID CONTENT

	Food Product	Description of Grease-Spot Reaction
1		
2		
3		
4		
5		
6		

TABLE 5.5

SOLUTIONS AND COLOR REACTIONS FOR DISCHE DIPHENYLAMINE TEST FOR DNA

Tube	Solution	Color
1	2 mL DNA solution	
2	1 mL DNA solution, 1 mL water	
3	2 mL RNA solution	
4	2 mL distilled water	
5		
6		

Procedure 5.7

Perform the Dische diphenylamine test for DNA

1. Obtain four test tubes and number them 1–4. Your instructor may ask you to test some additional materials. If so, include additional numbered test tubes.
2. Add the materials listed in table 5.5.
3. Add 2 mL of the Dische diphenylamine reagent to each tube and mix thoroughly.

CAUTION

Handle the Dische diphenylamine reagent carefully; it is toxic. Wash your hands after the procedure.

4. Place the tubes in a gently boiling water-bath to speed the reaction.
5. After 10 min, transfer the tubes to an ice bath. Gently mix and observe the color of their contents as the tubes cool. Record your observations in table 5.5.

Question 11

a. How does the color compare between tubes 1 and 2? Why?

b. Do DNA and RNA react alike? Why or why not?

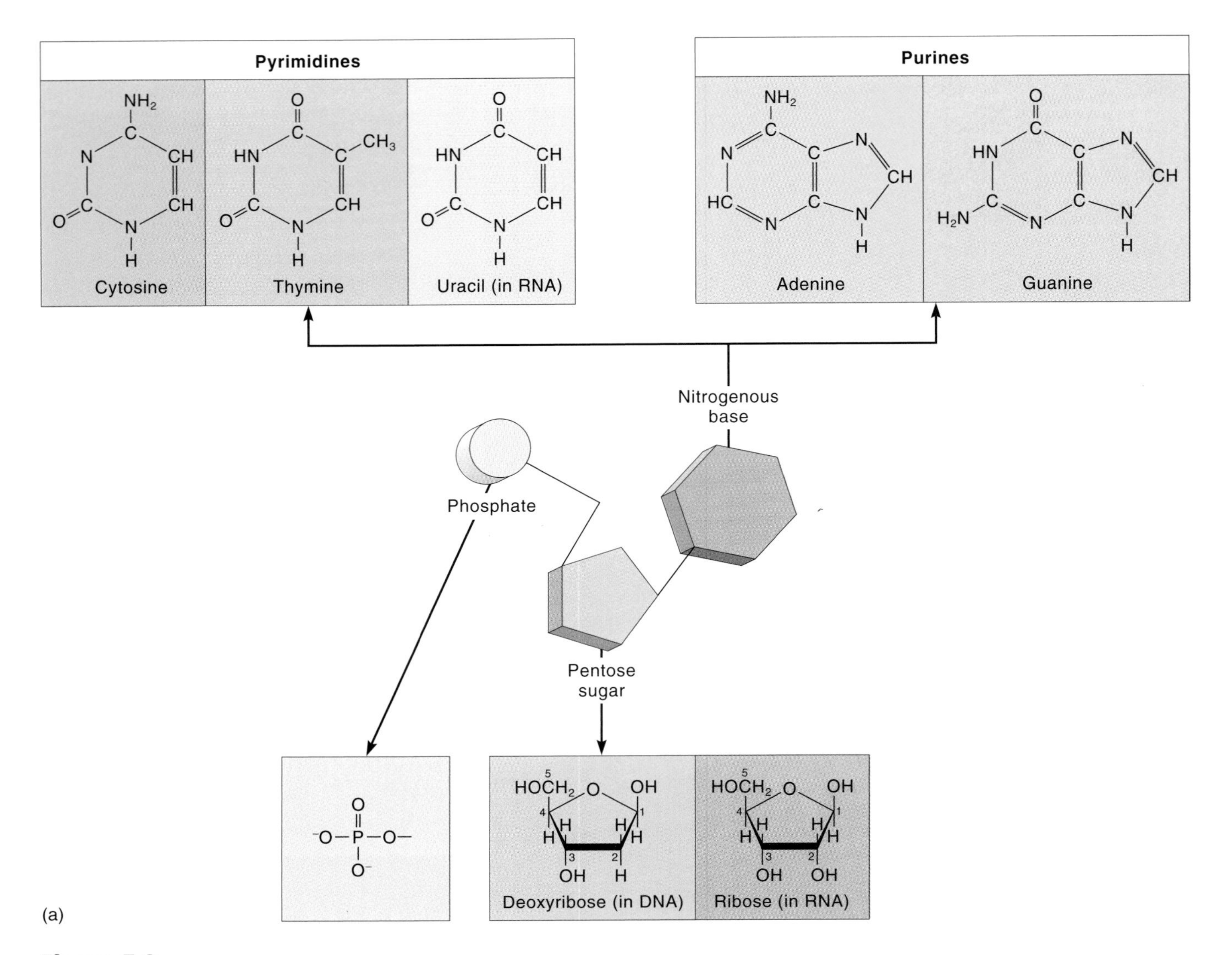

Figure 5.9

The structure of DNA and RNA. (*a*) Each nucleotide consists of three smaller building blocks: a nitrogenous base, a pentose sugar, and a phosphate group. (*b*) Nucleotides are bonded to each other by covalent bonds between the phosphate of one nucleotide and the sugar of the next nucleotide. (*c*) DNA is usually a double strand held together by hydrogen bonds between nitrogenous bases; A bonds only with T, and C bonds only with G. The double strand is twisted into a double helix.

INVESTIGATION

Identify Unknowns

Each of the previously described tests is relatively specific; that is, iodine produces a bluish-black color with starch but not with protein, lipid, nucleic acids, or other carbohydrates. This specificity can be used to identify the contents of an unknown solution.

a. Obtain an unknown solution from your laboratory instructor. Record its number in table 5.6.
b. Obtain 10 clean test tubes.
c. Number five tubes for the sample as S1–S5. Number the other five tubes as controls C1–C5.
d. Place 2 mL of your unknown solution into each of tubes S1–S5.
e. Place 2 mL of distilled water into each of tubes C1–C5.
f. Use procedures 5.1–5.5 to detect reducing sugars, starch, protein, lipid, and DNA in your unknown. Your unknown may contain one, none, or several of these macromolecules. Record your results in table 5.6. Show table 5.6 and the following report (page 56) to your instructor before you leave the lab.

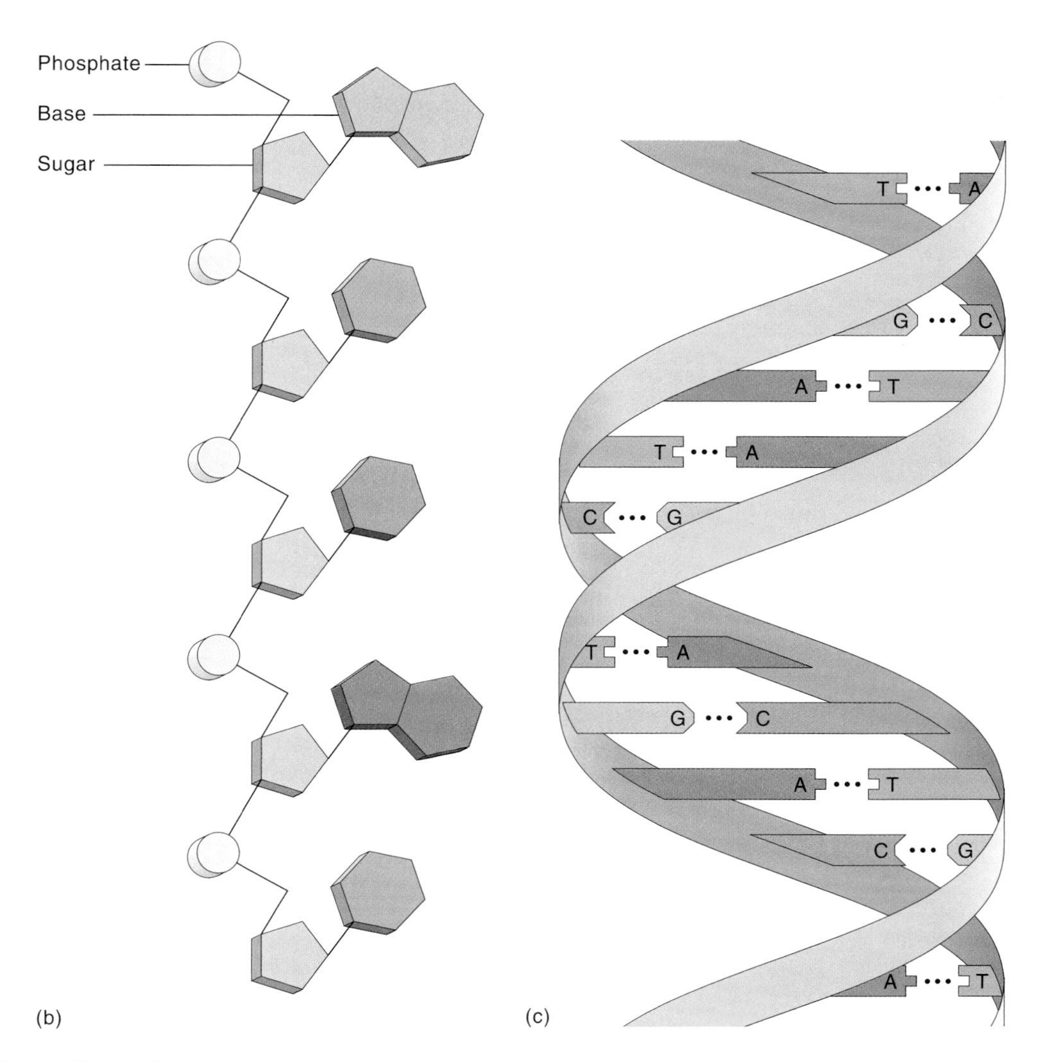

Figure 5.9 continued

Questions for Further Thought and Study

1. Why did you include controls in all of your tests?

2. Are controls always necessary? Why or why not?

3. What is a phospholipid? What functions do phospholipids have in cells?

WRITING TO LEARN BIOLOGY

What are the limitations of these common techniques in detecting the presence of a class of macromolecules? Do biologists who study plant cells commonly use the iodine test for starch? Why or why not?

DOING BIOLOGY YOURSELF

Design a procedure to indicate the amount of starch present in various plant tissue samples. How would you weigh your samples? How would you treat your samples? How would you quantify the iodine test?

Name ________________________ Lab Section ____________

Table 5.6

Chemical Testing to Identify an Unknown Unknown No: ____________

Biochemical Test	Color		Unknown Result
	Sample	Control	(+/−)
Benedict's test (reducing sugars)			
Iodine (starch)			
Biuret test (protein)			
Dische diphenylamine test (DNA)			
Sudan IV (lipid)			

Report: Identity of Unknown

Indicate which of the following are in your unknown:

Reducing sugars

Starch

Lipid

DNA

Protein

Comments:

6

Separating Organic Compounds
Column Chromatography, Paper Chromatography, and Gel Electrophoresis

Objectives

By the end of this exercise you should be able to:

1. Describe the basis for column chromatography, paper chromatography, and gel electrophoresis.
2. Use column chromatography, paper chromatography, and gel electrophoresis to separate organic compounds from mixtures.

Cells are a mixture of the kinds of organic compounds that you studied in Exercise 5: proteins, carbohydrates, lipids, and nucleic acids. Biologists characterize and study these compounds to understand how organisms function. This requires that biologists separate the compounds, such as amino acids and nucleotides, from mixtures. Three separation techniques that biologists use are column chromatography, paper chromatography, and gel electrophoresis.

In today's exercise you will use these common techniques to separate compounds from mixtures. The procedures are simple and model how these techniques are used by biologists in their research.

COLUMN CHROMATOGRAPHY

Column chromatography separates molecules according to their size and shape. The procedure is simple and involves placing a sample onto a column of beads having tiny pores. There are two ways that molecules can move through the column of beads: a fast route between the beads or a slower route through the tiny pores of the beads. Molecules too big to fit into the beads' pores move through the column quickly, whereas smaller molecules enter the beads' pores and move through the column more slowly (fig. 6.1). Movement of the molecules is analogous to going through or walking around a maze: It takes more time to walk through a maze than to walk around it.

The apparatus used for column chromatography is shown in figure 6.2 and consists of a chromatography column, a matrix, and a buffer.

- The **chromatography column** is a tube having a frit and a spout at its bottom. The frit is a membrane or porous disk that supports and keeps the matrix in the column but allows water and solutes to pass.
- The **matrix** is the material in the column that fractionates, or separates, the chemicals mixed in the sample. The matrix consists of beads having tiny pores and internal channels. The size of the beads' pores determines the matrix's **fractionation range,** which is the range of molecular weights the matrix can separate. These molecular weights are measured in units called daltons; 1 dalton $\approx$ 1 g mole^{-1}. Different kinds of matrices have different fractionation ranges. In today's exercise you'll use a matrix having a fractionation range of 1000 to 5000 daltons. As they move through the matrix, small molecules spend much time in the maze of channels and pores in the matrix. Large molecules do not.
- The **buffer** helps control the pH of the sample (see Exercise 4). A buffer is a solution with a known pH that resists changes in pH if other chemicals are added. The pH of a buffer remains relatively constant. This is important because the shapes of molecules such as proteins often vary according to their pH. The buffer carries the sample through the matrix, which separates the chemicals mixed in the sample.

Column chromatography can also separate compounds having the same molecular weight but different shapes. Compact, spherical molecules penetrate the pores and channels of the matrix more readily than do rod-shaped molecules. Thus, spherical molecules move through a column more slowly than do rod-shaped molecules.

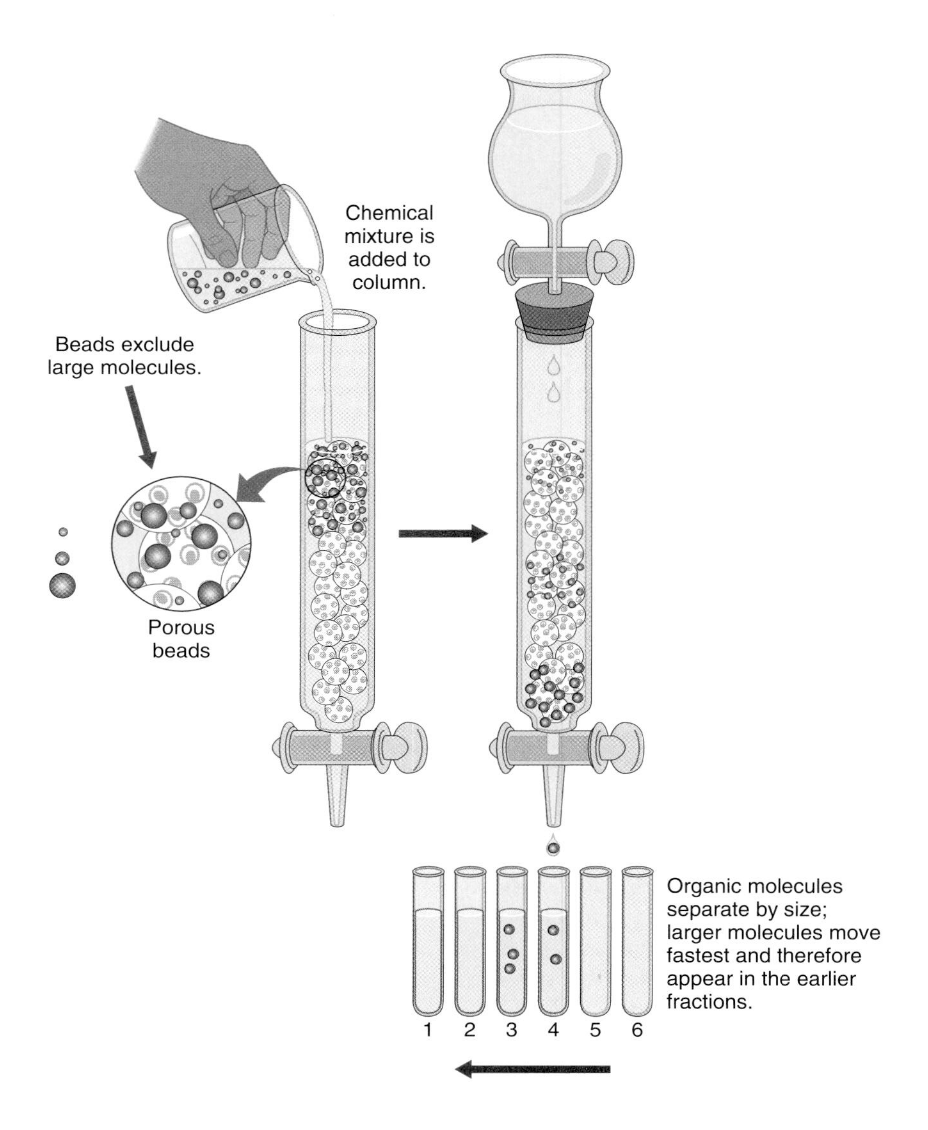

Figure 6.1

Separation of organic molecules by column chromatography. As the solution flows through the column, the smaller molecules are slowed down as they pass through the pores of the beads. Medium-sized molecules will pass through a bead with pores less frequently, and the largest molecules will quickly flow around all the beads. The exiting fluid is collected in fractions. The first fractions collected will contain the largest molecules.

During column chromatography, the buffer containing the sample mixture of chemicals moves through the column and is collected sequentially in test tubes from the bottom of the column. Biologists then assay the content of the tubes to determine which tubes contain the compounds they're interested in.

Question 1

In today's exercise you'll isolate colored compounds from mixtures. However, most biological samples are colorless. How would you determine the contents of the test tubes if all of the samples were transparent?

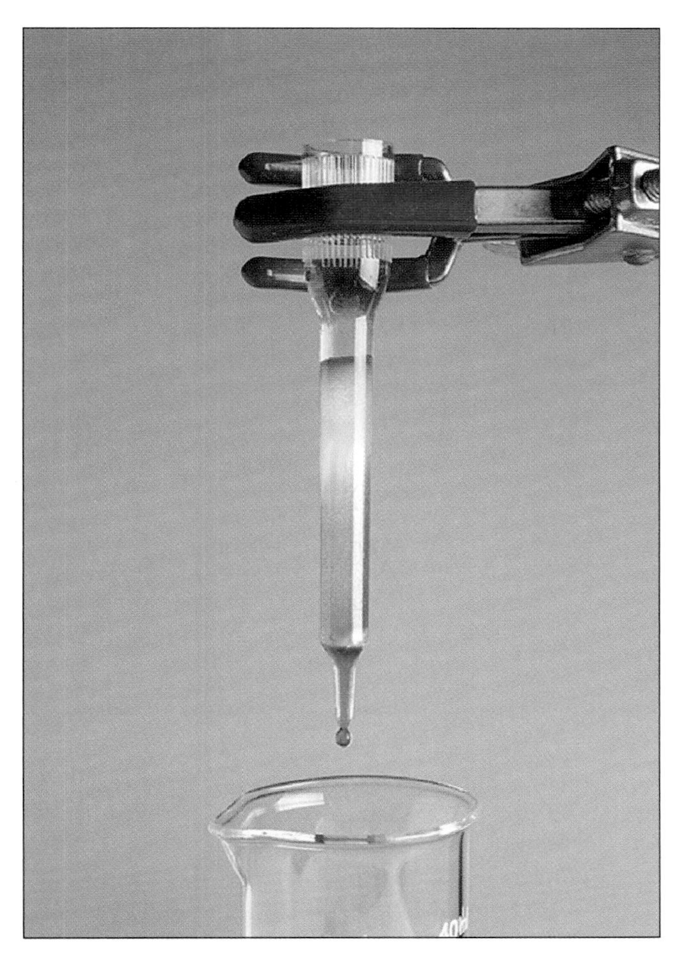

Figure 6.2
Apparatus for column chromatography. A fraction is being collected in the beaker.

Procedure 6.1

Separate compounds by column chromatography

1. Label nine microtubes 1–9.
2. Obtain an apparatus for column chromatography and carefully remove all of the buffer from above the bed with a transfer pipet. Do not remove any of the matrix.
3. Obtain a sample to be separated. The sample is a mixture of Orange G (molecular weight = 452 g mole^{-1}) and a rodlike polymer of glucose stained blue and having a molecular weight of about 2,000,000 g mole^{-1}.
4. Use a transfer pipet to slowly load 0.2 mL of the sample onto the top of the bed. Drip the sample down the inside walls of the column.
5. Place a beaker under the column.
6. Slowly open the valve. This will cause the sample to enter the bed. Close the valve after the sample has completely entered the bed (i.e., when the top of the bed is exposed to air).
7. Use a transfer pipet to slowly cover the bed with buffer. Add buffer until the reservoir is almost full.
8. Hold microtube 1 under the tube and open the valve until you've collected about 1.0 mL of liquid.
9. Repeat step 8 for tubes 2–9. The sample will separate in the column.
10. Identify the tubes containing (1) the most orange dye, and (2) the most blue dye that eluted from the column.
11. Refill the reservoir with buffer and cover the reservoir with Parafilm.

Question 2

a. Was the color separation distinctive? Would you expect a longer column to more clearly separate the compounds? Why or why not?

b. Suppose your sample had consisted of a mixture of compounds having molecular weights of 50,000, 100,000, and 1,000,000 g mole^{-1}. What kind of results would you predict? Explain your answer.

PAPER CHROMATOGRAPHY

Biologists often analyze the amino acid content of samples to determine protein sequences and enzyme structures. Amino acids can be separated by partitioning them between the stationary and mobile phases of paper chromatography. The **stationary phase** is the paper fibers, and the **mobile phase** is an organic solvent that moves along the paper.

Separation by paper chromatography begins by applying a liquid sample to a small spot on an origin line at one end of a piece of chromatography paper. The edge of the paper is then placed in a solvent. As the solvent moves up the paper, any sample molecules that are soluble in the solvent will move with the solvent. However, some molecules move faster than others based on their solubility in the mobile phase and their attraction to the stationary phase. These competing factors are different for different molecular structures, so each type of molecule moves at a different speed and occurs at a different position on the finished chromatogram.

Amino acids in solution have no color but react readily with molecules of ninhydrin to form a colored product. A completed chromatogram is sprayed with a ninhydrin solution and heated to detect the amino acids. The distance of these spots from the origin is measured and used to quantify

TABLE 6.1

CHROMATOGRAPHY DATA FOR DETERMINING AMINO ACID UNKNOWNS

Tick Mark Number	Amino Acid or Sample Number	Distance to Solvent Front	Distance Traveled by Sample	R_f	Identity of Unknown
1					
2					
3					
4					
5					

the movement of a sample. The resulting R_f value (retardation factor) characterizes a known molecule in a known solvent under known conditions, and is calculated as follows:

$$R_f = \frac{\text{Distance moved by sample}}{\text{Distance from origin to solvent front}}$$

Procedure 6.2

Separate amino acids and identify unknowns by paper chromatography

1. Obtain a piece of chromatography paper 15 cm square. Avoid touching the paper with your fingers. Use gloves, tissue, or some other means to handle the paper because oils from your skin will alter the migration of the molecules on the paper.
2. Lay the paper on a clean paper towel. Then use a pencil to draw a light line 2 cm from the bottom edge of the paper.
3. Draw five tick marks at 2.5 cm intervals from the left end of the line. Lightly label the marks 1–5 below the line.
4. Locate the five solutions available for the chromatography procedure. Three of the solutions are known amino acids. One solution is an unknown. The last solution is a plant extract or another unknown.
5. Use a wooden applicator stick to "spot" one of the solutions on mark #1. To do this, dip the stick in the solution and touch it to the paper to apply a small drop (2–3 mm in diameter). Let the spot dry; then make three to five more applications on the same spot. Dry between each application. Record in table 6.1 the name of the solution next to the appropriate mark number.
6. Repeat step 5 for each of the other solutions.
7. Staple or paper clip the edges of the paper to form a cylinder with the spots on the outside and at the bottom.
8. Obtain a quart jar containing the chromatography solvent. The solvent should be 1 cm or less deep. The solvent consists of butanol, acetic acid, and water (2:1:1).
9. Place the cylinder upright in the jar (fig. 6.3). *The solvent must be below the pencil line and marks*. Close the lid to seal the jar.
10. Keep the jar out of direct light and heat. Allow the solvent to move up the paper for 2 hours (h) but not all the way to the top.
11. Open the jar and remove the chromatogram. Unclip and flatten the paper. Dry it with a fan or hair dryer. Work under a hood if possible to avoid breathing the solvent vapors.
12. Spray the chromatogram with ninhydrin. Carefully dry the chromatogram with warm air.
13. Circle with a pencil each of the spots. Measure the distance each of the spots has traveled and calculate the R_f for each spot. Record the values in table 6.1.

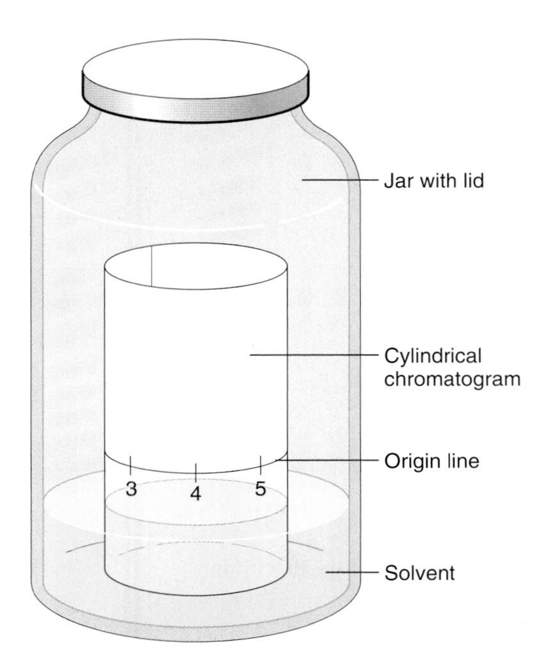

Figure 6.3
Apparatus for paper chromatography. Numbers on the chromatogram indicate the positions of multiple samples that are applied to the chromatogram. The samples will move up the chromatogram along with the solvent.

14. Determine the contents of the unknown solutions by comparing R_f values. Record the results in table 6.1.

GEL ELECTROPHORESIS

Gel electrophoresis separates molecules according to their charge, shape, and size (fig. 6.4). Buffered samples (mixtures of organic chemicals) are loaded into a Jello-like gel, after which an electrical current is placed across the gel. This current moves the charged molecules toward either the cathode or anode of the electrophoresis apparatus. The speed, direction, and distance that each molecule moves are related to its charge, shape, and size.

The apparatus for gel electrophoresis is shown in figure 6.5 and consists of an electrophoresis chamber, gel, buffer, samples, and a power supply.

- The gel is made by dissolving agarose powder (a derivative of agar) in hot buffer. When the solution cools, it solidifies into a gel having many pores that function as a molecular sieve. The gel is submerged in a buffer-filled chamber containing electrodes.

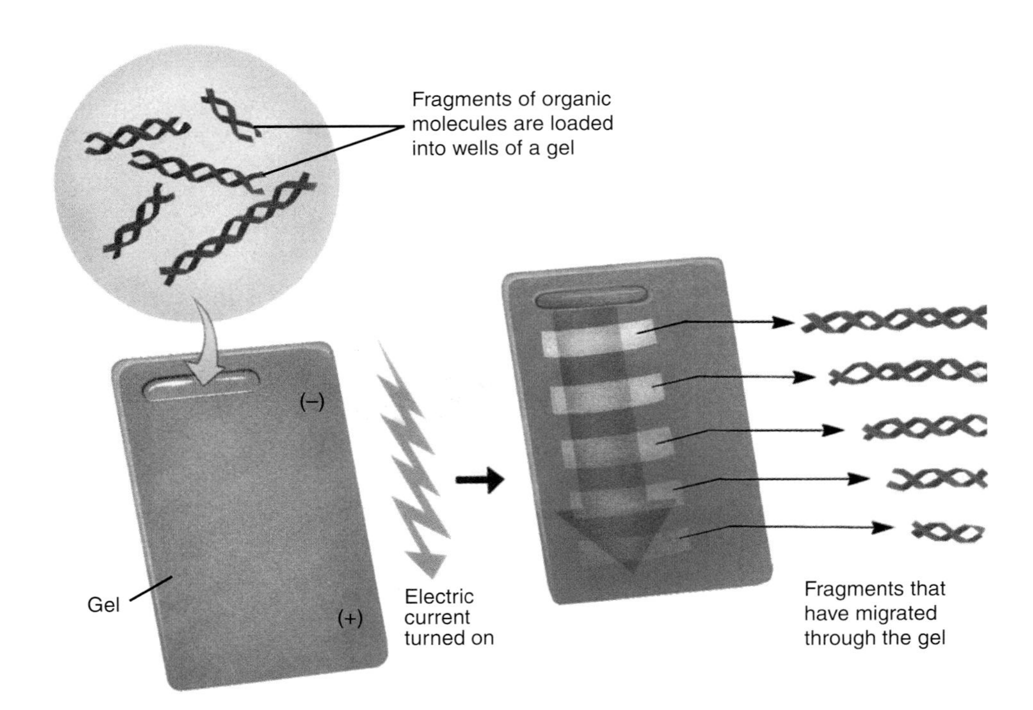

Figure 6.4
Gel electrophoresis. This process separates DNA fragments, protein fragments, and other organic compounds by causing them to move through an electrically charged gel. The fragments move according to their size, shape, and electrical charge; some fragments move slowly and some move quickly. When their migration is complete the fragments can be stained and visualized easily. In the example shown here, the DNA fragments were separated by size.

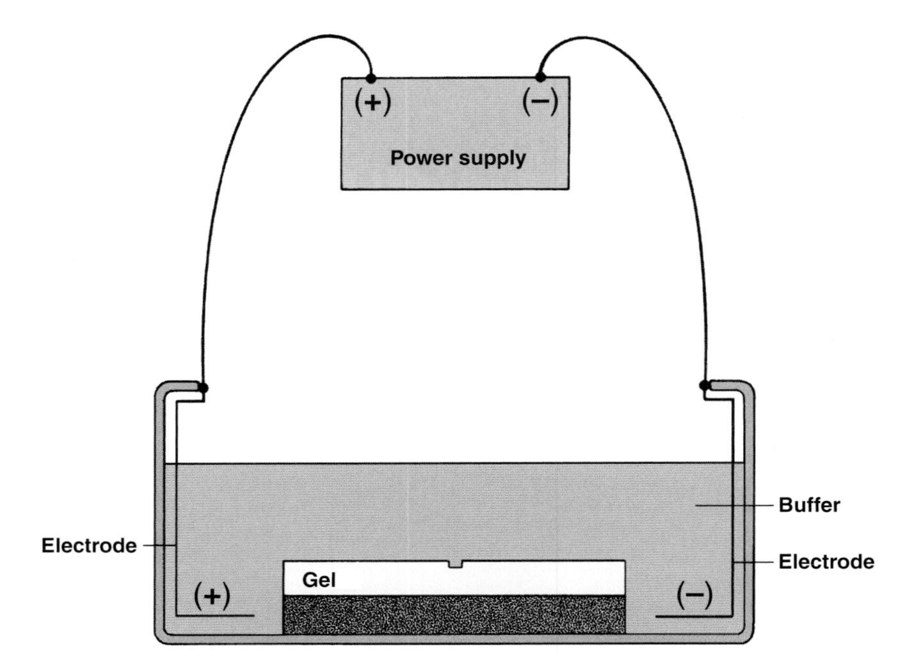

Figure 6.5
Apparatus for gel electrophoresis. The power supply produces an electrical gradient between the + and − poles and across the gel.

- The buffer conducts electricity and helps control the pH. The pH affects the stability and charge of the samples.
- The samples are mixtures of chemicals that are loaded into wells in the gel. These samples move in the gel during electrophoresis. Samples are often mixed with glycerol or sucrose to make them denser than the buffer so that they will not mix with the buffer.
- The power supply provides a direct current across the gel. Charged molecules respond to the current by moving from the sample wells into the gel. Negatively charged molecules move through the gel toward the positive electrode (anode), whereas positively charged molecules move through the gel toward the negative electrode (cathode). The greater the voltage, the faster the molecules move.

The sieve properties of the gel affect the rate of movement of a sample through the gel. Small molecules move easier through the pores than do larger molecules. Consequently, small, compact (e.g., spherical) molecules move faster than do large, rodlike molecules. If molecules have similar shapes and molecular weights, the particles having the greatest charge move fastest and, therefore, the farthest.

Procedure 6.3

Separate organic molecules by gel electrophoresis

1. Obtain an electrophoresis chamber. Cover the ends of the bed as shown in figure 6.6 and demonstrated by your instructor.
2. Place a six-tooth comb in or near the middle set of notches of the gel-cast bed. There should be a small space between the bottom of the teeth and the bed.
3. Mix a 0.8% (weight by volume) mixture of agarose powder in a sufficient volume of buffer to fill the gel chamber. Heat the mixture until the agarose dissolves.
4. When the hot agarose solution has cooled to 50°C, pour the agarose solution into the gel-cast bed (fig. 6.7).
5. After the gel has solidified, gently remove the comb by pulling it straight up (fig. 6.8). Use the sketch in figure 6.9 to label the wells formed in the gel by the comb.
6. Submerse the gel under the buffer in the electrophoresis chamber.

Figure 6.6
Cover the ends of the removable gel bed with rubber end-caps or tape.

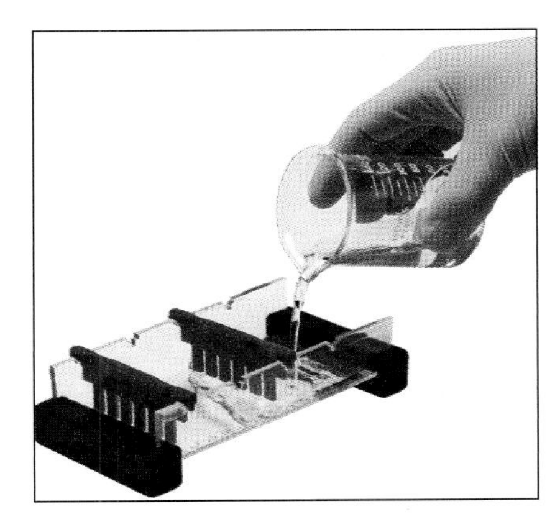

Figure 6.7
Place the comb near the center set of notches of the gel bed. Prepare the agarose solution and pour the gel.

7. You will study six samples:
   Sample 1: Bromophenol blue (molecular weight = 670 g $mole^{-1}$)
   Sample 2: Methylene blue (molecular weight = 320 g $mole^{-1}$)
   Sample 3: Orange G (molecular weight = 452 g $mole^{-1}$)
   Sample 4: Xylene cyanol (molecular weight = 555 g $mole^{-1}$)
   Samples 5 and 6: Unknowns

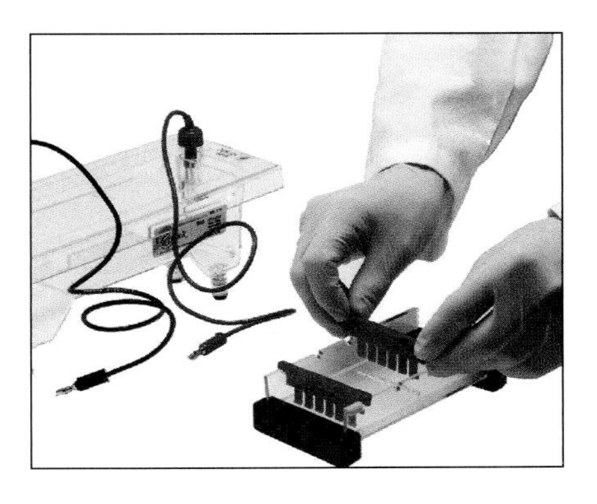

Figure 6.8
After the gel solidifies, gently remove the rubber end-caps (or tape) and pull the combs straight up from the gel.

Figure 6.9
Sketch of the wells formed in the gel by the comb as viewed from above.

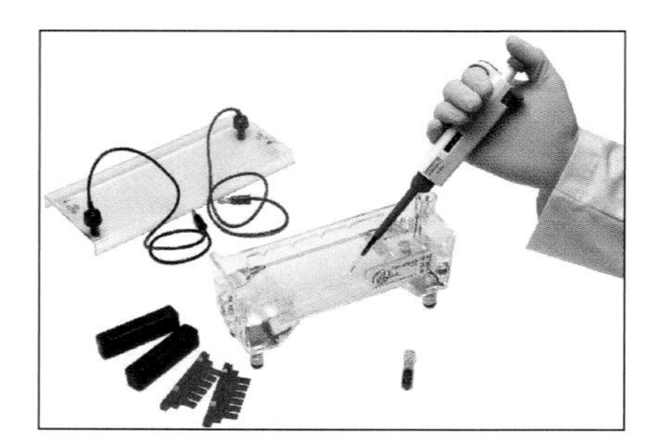

Figure 6.10
Submerge the gel in the buffer-filled electrophoresis chamber and load the samples into the wells of the gel.

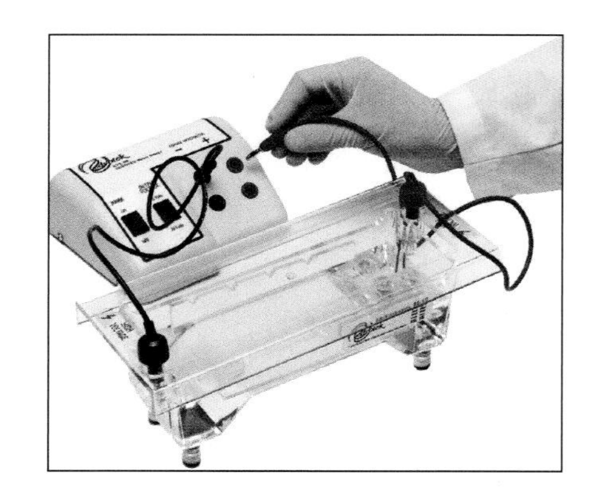

Figure 6.11
Attach the safety cover, connect the power source, and run the electrophoresis.

Use a micropipettor or a simple pipet and bulb to load the samples into the wells of the gel. If you use a micropipettor, your instructor will demonstrate its use. If you use a simple pipet and bulb, gently squeeze the pipet bulb to draw Sample 1 into the pipet. Be sure that the sample is in the lower part of the pipet. If the sample becomes lodged in the bulb, tap the pipet until the sample moves into the lower part.

8. To eliminate excess air hold the pipet above the sample tube and slowly squeeze the bulb until the sample is near the pipet's opening.
9. Place the pipet tip into the electrophoresis buffer so it is barely inside sample well 1 (fig. 6.10). Do not touch the bottom of the sample well. Maintain pressure on the pipet bulb to avoid pulling buffer into the pipet.
10. Slowly inject the sample into the sample well. Stop squeezing the pipet when the well is full. Do not release the pressure on the bulb. Remove the pipet from the well.
11. Thoroughly rinse the pipet with distilled water.
12. Load the remaining five samples into the gel by repeating steps 6–10 (fig. 6.10). Load Sample 2 into the second well, Sample 3 into the third well, etc.
13. Carefully snap on the cover of the electrophoresis chamber (fig. 6.11). The red plug in the cover should be placed on the terminal indicated by the red dot. The black plug in the cover should be placed on the terminal indicated by the black dot.
14. Insert the plug of the black wire into the black (negative) input of the power supply. Insert the plug of the red wire into the red (positive) input of the power supply.
15. Turn on the power and set the voltage at 90 V. You'll soon see bubbles forming on the electrodes. Examine the gel every 10 min.
16. After 30 min, turn off the power and disconnect the leads from the power source. Gently remove the cover from the chamber and sketch your results in figure 6.9.

Question 3

a. Orange G, bromophenol blue, and xylene cyanol each have a negative charge at neutral pH, whereas methylene blue has a positive charge at neutral pH. How does this information relate to your results?

b. Did Orange G, bromophenol blue, and xylene cyanol move the same distance in the gel? Why or why not?

c. What compounds do you suspect are in Samples 5 and 6? Explain your answer.

INTERPRETING A DNA-SEQUENCING GEL

Examine figure 6.12, which includes a photograph of a gel used to determine the order, or sequence, of nucleotides in a strand of DNA. To prepare the sample for electrophoresis, samples of the DNA being investigated were put into each of four tubes and induced to replicate. Also, into the first tube, an adenine-terminator was added in addition to all the other nucleotides. As the complementary strand was being constructed the terminators were occasionally incorporated wherever an adenine nucleotide was used. This random incorporation resulted in all possible lengths of DNA pieces

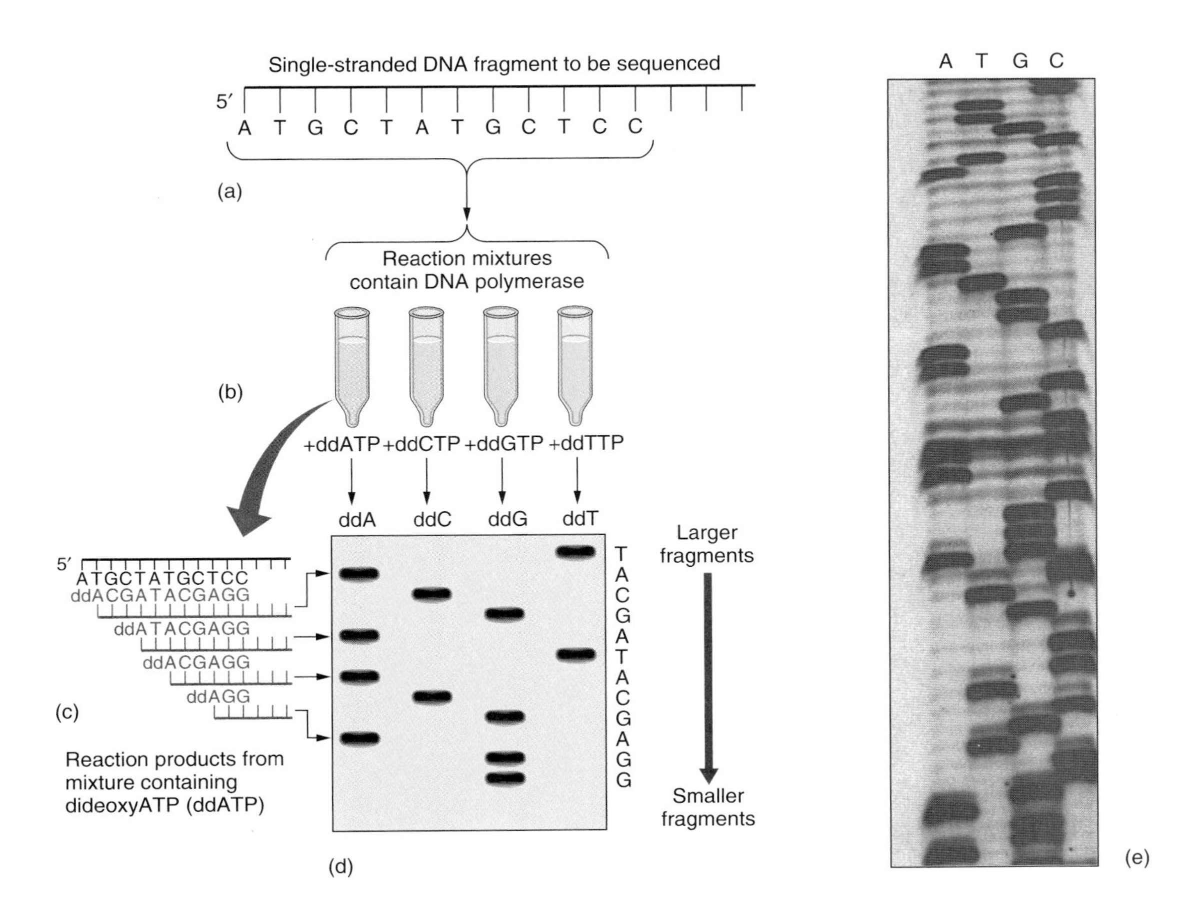

Figure 6.12

Determining the sequence of nucleotides in DNA. (*a*) Treating DNA with sodium hydroxide (NaOH) denatures double-stranded DNA into single-stranded DNA. One of the single strands of DNA to be sequenced is placed in each of four tubes. (*b*) The enzyme DNA polymerase is added to each tube along with a specific nucleotide-terminator. As polymerase replicates the DNA, the terminators are incorporated and will terminate various lengths of fragments of DNA. For example, the terminator ddATP will halt the reaction wherever adenosine occurs. The terminator ddATP (dideoxy adenosine triphosphate) will terminate a growing strand because it lacks a 3′ hydroxyl group and therefore cannot bond with the next deoxynucleotide. (*c*) Each tube will contain a sample of all possible replicated fragment lengths corresponding to the positions of that specific nucleotide. The sequences in red are the complement strands. (*d*) During electrophoresis, the fragments migrate at different rates according to their length. (*e*) The lanes of the resulting gel are labeled according to their base: A, adenine; T, thymine; G, guanine; and C, cytosine. This technique is usually referred to as "Sanger" sequencing in honor of Fred Sanger, a Nobel laureate who, in 1977, first sequenced a piece of DNA.

that had an adenine on the end. The same process was conducted in the other tubes with thymine-, guanine-, and cytosine-terminators; one treatment for each of the four lanes in the gel. Electrophoresis separated the replicated pieces of DNA by size. Staining the gel revealed which lengths of the complementary DNA were terminated by which nucleotide-terminators. Examine Figure 6.12d.

The gel consists of four "lanes," labeled A, T, G, and C, indicating either adenine-, thymine-, guanine-, or cytosine-terminated pieces of DNA. By "reading" down the gel, you can determine the sequence of nucleotides in the DNA. For example, the uppermost band of the gel is in the T (thymine) lane. Therefore, the first base of the piece of DNA is thymine. Similarly, the next bands are in the A, C, G, and A lanes. Thus, the first five bases of the complementary strand DNA are T-A-C-G-A. List the next 7 nucleotides of the DNA as indicated by the gel. Also list the sequence of the first 12 nucleotides in the original DNA being investigated.

Question 4

How did the sequence of nucleotides revealed on the gel differ from the sequence of the original strand of DNA?

INVESTIGATION

Refining the Paper Chromatography Procedure

Carefully planned and refined procedures are critical for laboratory techniques such as paper chromatography. The sensitivity of these techniques depends on a variety of factors, including the many parameters associated with timing, chemicals, measurements, and temperatures. In procedure 6.2 you were given a rather standardized protocol, but it can always be improved for specific experiments. For example, how would you modify the paper chromatography procedure to better resolve two amino acids having approximately the same R_f values? What parameter(s) of the experimental design might be tweaked to increase the technique's resolving power? We suggest that you begin your investigation in the following way:

a. List the parameters involved in paper chromatography. Think carefully; many factors are involved.

b. Choose one or two parameters that you can test for their impact on the chromatography results. Why did you choose these?

c. Choose two amino acids for experimentation. Why did you choose these two?

d. Choose your treatment levels for each parameter, and then do your experiment.

e. What did you conclude?

Questions for Further Thought and Study

1. How are gel chromatography, paper chromatography, and column chromatography different?

2. How could knowing the base sequence of a piece of DNA be important to a biologist?

3. How could knowing the base sequence of a piece of DNA be important to someone trying to solve a crime?

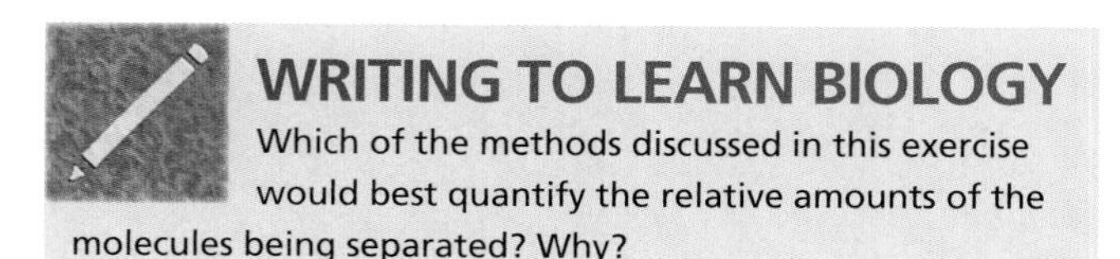

WRITING TO LEARN BIOLOGY

Which of the methods discussed in this exercise would best quantify the relative amounts of the molecules being separated? Why?

7

Spectrophotometry
Identifying Solutes and Determining Their Concentration

Objectives

By the end of this exercise you should be able to:

1. Operate a spectrophotometer.
2. Describe the parts of a spectrophotometer and the function of each.
3. Construct absorption spectra for cobalt chloride and chlorophyll.
4. Construct and use a standard curve to determine the unknown concentration of a dissolved chemical.

Absorption and reflection of different wavelengths of light are part of your everyday experience. Different materials absorb and reflect different wavelengths of light; therefore they have different colors. When you recognize things with color you are observing that they absorb and reflect different wavelengths of light. Light includes the visible wavelengths of the electromagnetic spectrum and is only a small part of the total spectrum. The entire electromagnetic spectrum includes radiation with wavelengths from less than 1 to more than 1 million nanometers. Visible light represents wavelengths between 380 and 700 nm (fig. 7.1). In this exercise we'll work within the visible portion of the electromagnetic spectrum.

Question 1

a. Chlorophyll reflects green light (540–560 nm) and absorbs other wavelengths. What biologically important molecules other than chlorophyll absorb and reflect certain colors?

b. Is the absorbance of light critical to these molecules' functions or just a consequence of their molecular structure?

Biologists routinely determine the presence and concentration of dissolved chemicals such as phosphates that may pollute lake water. **Spectrophotometry** is one of our most versatile and precise techniques for such chemical assays. Spectrophotometry is based on the principle that every different atom, molecule, or chemical bond absorbs a unique pattern of wavelengths of light. For example, the nitrogenous base cytosine absorbs a different pattern of light than does adenine, uracil, or any other molecule with a different structure. As part of this pattern, some wavelengths are absorbed and some are not. Conversely, a unique pattern of light is reflected as well as absorbed by each chemical. Each chemical has a unique pattern or "fingerprint" of various wavelengths that it absorbs and/or reflects. In this exercise you will (a) determine the unique pattern of absorption for two common molecules, and then (b) use spectrophotometry to measure the concentrations of one of these molecules.

SPECTROPHOTOMETER

We use an instrument called a **spectrophotometer** to measure the amount of light absorbed and transmitted by a dissolved chemical. For dissolved chemicals we usually refer to the nonabsorbed light as transmitted light rather than reflected light. With this measurement of absorbance or transmittance we can identify a chemical and its concentration.

A spectrophotometer is an instrument that separates white light into a spectrum of colors (wavelengths). It then directs a specific wavelength of light at a tube containing a solution that we are trying to measure. The light is either absorbed by the dissolved substance or transmitted through the solution and exits the sample tube. The spectrophotometer compares the amount of light exiting the tube (that is, the transmitted light) with the amount entering the tube and calculates transmittance; the more solute, the lower the transmittance. The spectrophotometer also calculates the amount of light absorbed; the more solute, the higher the absorbance. The basic parts of a spectrophotometer are shown in figure 7.2.

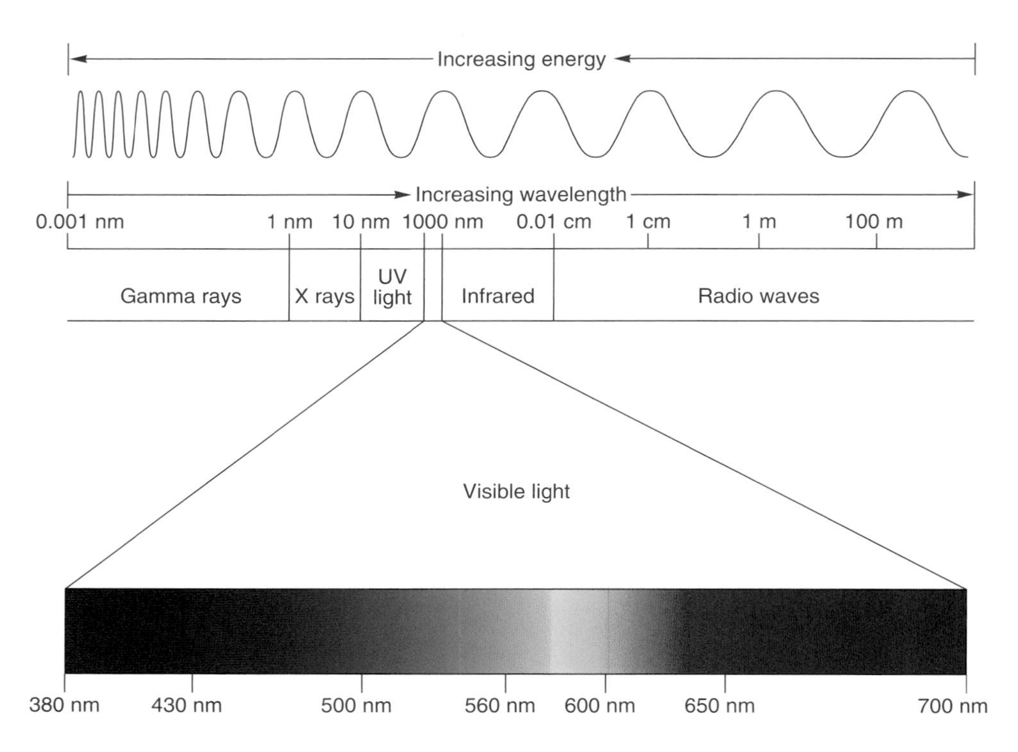

Figure 7.1

The electromagnetic spectrum. Light is a form of electromagnetic energy and is conveniently thought of as a wave. The shorter the wavelength of light, the greater the energy. Visible light, which represents only the small part of the electromagnetic spectrum, occurs between 380 and 700 nanometers (nm). (UV stands for ultraviolet light.) Also see figure 7.5.

The **light source** of a spectrophotometer produces white light, which is a combination of all visible wavelengths. A mixture of all colors of light is white. Spectrophotometry may involve wavelengths outside the visible range, such as ultraviolet and infrared, but the spectrophotometer must have special light sources to produce these wavelengths. In this exercise you will work only with visible light.

The **filter** is adjusted to select the wavelength that you wish to pass through the sample. The filter may be a prism that separates white light into a rainbow of colors and focuses the desired wavelength (color) on the sample. Or, the filter may be a series of colored glass plates that absorb all but the selected wavelength, which is focused on the sample.

The **sample** is a solution contained in a clear test tube or cuvette made of glass or quartz. Light of the specific wavelength determined by the filter passes into the sample, where it may be completely or partially absorbed or transmitted. The amount of light absorbed depends on the amount and kind of chemicals in the solution and the dimensions of the tube.

A **blank** is a test tube or cuvette containing only the solvent used to dissolve the chemical you are analyzing. A blank is used to calibrate the spectrophotometer for the solution used in your experiment. For most of your experiments, the solvent, and therefore the blank, is distilled water.

Any light transmitted through the sample exits the sample on the opposite side and is focused on a **photodetector** that converts light energy into electrical energy. The amount of electricity produced by the photodetector is proportional to the amount of transmitted light: The more light transmitted, the more electricity produced. The **meter** on the front of the spectrophotometer measures the electrical current produced by the photodetector and displays the results on a scale of absorbance or transmittance values. Note that if a chemical is in solution we usually refer to its transmittance rather than its reflectance, although both terms refer to nonabsorbed light.

The double scale on the spectrophotometer indicates that you can measure either absorbance or transmittance of radiation. **Absorbance** is the amount of radiation retained by the sample, and **transmittance** is the amount of radiation passing through the sample. In mathematical terms, transmittance is the intensity of light exiting the sample divided by the amount entering the sample. Transmittance is usually expressed as a percent:

$$\textbf{Percent transmittance} = (I_t/I_o) \times 100$$

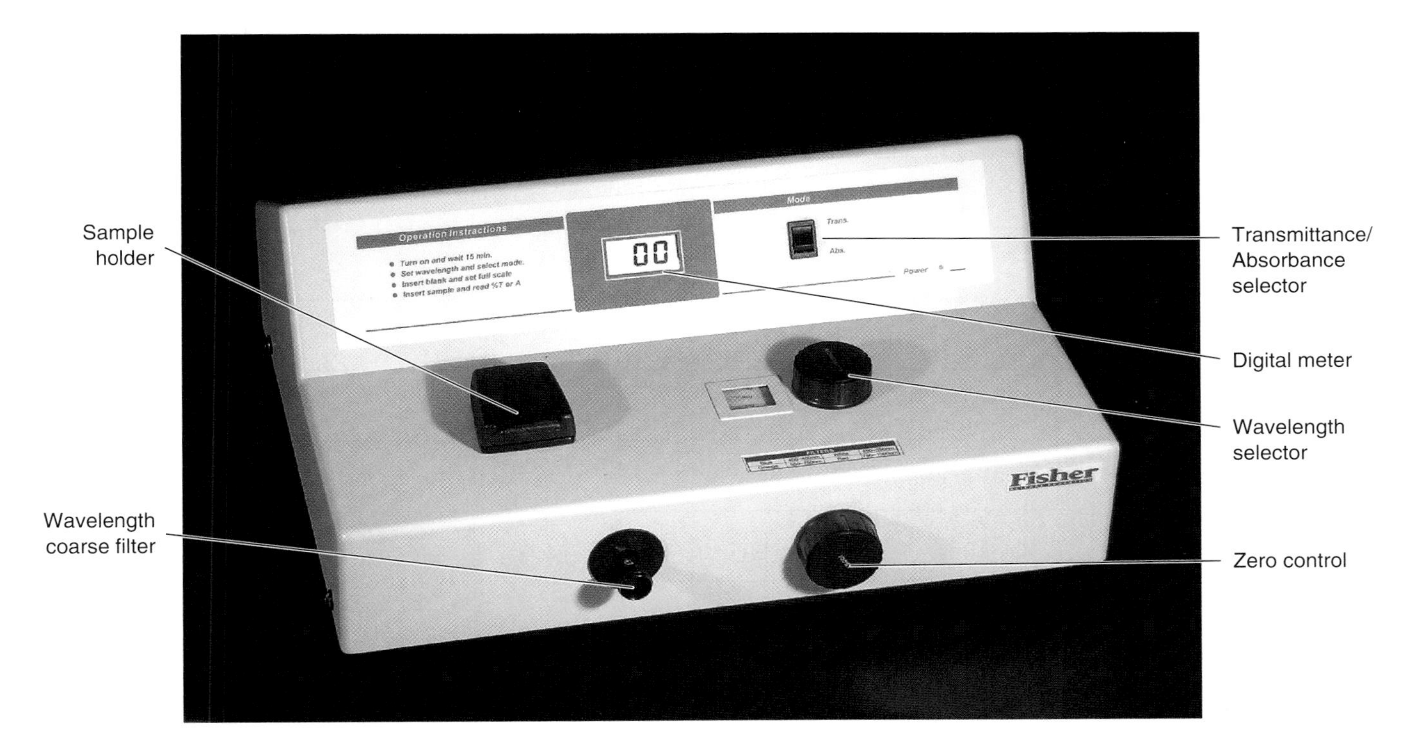

Figure 7.2
A spectrophotometer and its parts.

where

I_t = **transmitted (exiting) light intensity**
I_o = **original (entering) light intensity**

Absorbance is the logarithm of the reciprocal of transmittance and is expressed as a ratio with no units:

$$\text{Absorbance} = \log_{10}(I_o/I_t)$$

Notice that absorbance is not a percent and is not simply the opposite (reciprocal) of transmittance. Instead, it is a logarithmic function and has no units. This is done to make absorbance values directly proportional to the concentration of the substance in solution. Thus, a twofold increase in absorbance indicates a twofold increase in concentration. This convenient and direct relationship between concentration and absorbance helps scientists measure an unknown concentration of a chemical.

ABSORPTION SPECTRUM OF COBALT CHLORIDE

Your first task is to learn to operate a spectrophotometer while deriving the **absorption spectrum** of a common chemical, cobalt chloride ($CoCl_2$). The pattern of wavelengths absorbed by $CoCl_2$ is its "fingerprint" because it is unique to that chemical. This fingerprint is the **absorption spectrum** of the chemical and is represented as a graph relating absorbance to wavelength (fig. 7.3). Your instructor may choose to use red dye to simulate $CoCl_2$.

Procedure 7.1
Determine the absorption spectrum of $CoCl_2$

1. Turn on the spectrophotometer. Let it warm up for 10–15 min before you begin work.
2. Check with your instructor or manufacturer's directions for any special instructions for using the spectrophotometer in your lab.
3. Prepare seven solutions in spectrophotometer tubes (test tubes or cuvettes) with the mixtures of distilled water and stock solution of $CoCl_2$ (100 mg/mL) listed in table 7.1. Only tubes 0 and 6 are needed to determine the absorption spectrum. You will use the others later in the lab period.

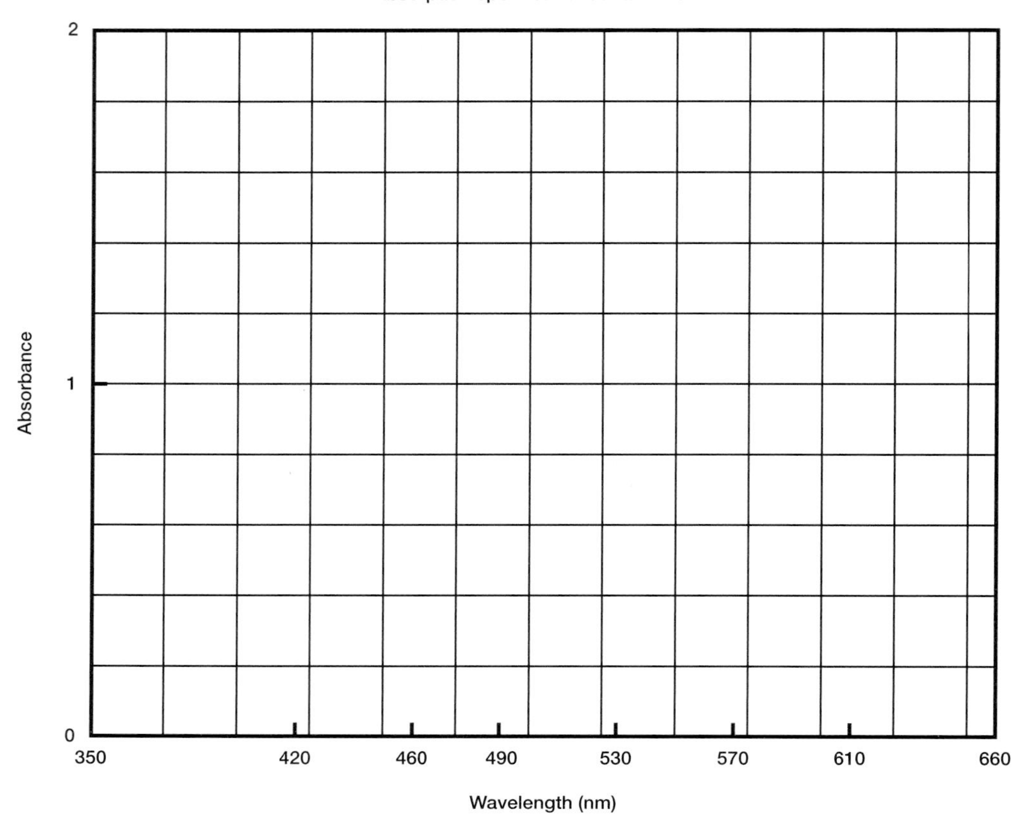

Figure 7.3
Absorption spectrum of cobalt chloride ($CoCl_2$).

TABLE 7.1

VOLUMES OF COBALT CHLORIDE STOCK SOLUTION (100 mg/mL) AND WATER USED TO PREPARE SEVEN KNOWN DILUTIONS

Tube Number	Concentration of $CoCl_2$ (mg/mL)	$CoCl_2$ Stock (mL)	Distilled H_2O (mL)
0	0	0	10.0
1	1	0.1	9.9
2	10	1.0	9.0
3	20	2.0	8.0
4	30	3.0	7.0
5	40	4.0	6.0
6	50	5.0	5.0

4. After you have prepared the dilutions, clean the outside of all the tubes with a cloth or paper towel.
5. Verify that the solutions in each of your tubes are free of particles (dust, chalk, etc.) that might scatter the light from the spectrophotometer and produce false absorbance values. If necessary, centrifuge the tubes at 2000 rpm for 5–10 min.

Question 2
Why is it important to clean the sample tubes?

TABLE 7.2

ABSORBANCE FOR $CoCl_2$ (50 mg/mL)

Wavelength	Absorbance	Wavelength	Absorbance
350 nm	______	530 nm	______
420 nm	______	570 nm	______
460 nm	______	610 nm	______
490 nm	______	660 nm	______

6. Cap the tubes and label the caps 0–6. If you label the tube rather than the caps, be sure that the labels don't interfere with light entering the tube while it is in the spectrophotometer.
7. Place the blank (tube 0) in the sample holder of the spectrophotometer.
8. Adjust the filter to the lowest wavelength (350 nm) and read the absorbance value indicated on the meter. The distilled water blank has no color and should not absorb any visible light.
9. If the absorbance is not zero, use the zero adjust knob to calibrate the meter to zero on the absorbance scale.
10. Remove the blank and replace it with tube 6 (50 mg/mL). This is the sample you will use to determine the absorption spectrum of cobalt chloride.
11. After the meter has stabilized (5–10 sec) read the absorbance value and record the wavelength and absorbance value in table 7.2.
12. Remove tube 6 and adjust the filter to 420 nm.
13. Put the blank back into the spectrophotometer and readjust for zero absorbance at the new wavelength. The spectrophotometer should be recalibrated with the blank often, especially when you change the wavelength.
14. Insert tube 6 and measure its absorbance at the new wavelength. Record the absorbance in table 7.2.
15. Complete table 7.2 for the other wavelengths by repeating steps 5–12 and measuring the absorbance of the contents of tube 6.

The absorbance values in table 7.2 represent the absorption spectrum for $CoCl_2$ and are expressed best with a graph. Plot on figure 7.3 the *Absorbance* vs. *Wavelength* values recorded in table 7.2. Connect the points with straight lines.

Question 3

a. What wavelength is the peak absorbance of $CoCl_2$?

b. Would you expect a curve of the same shape for another molecule such as chlorophyll? Why or why not?

THE STANDARD CURVE

A graph showing a chemical's concentration versus its absorbance of a wavelength of light is called a **standard curve,** and the relationship is a straight line. In this exercise you will construct a standard curve and then use it to determine some unknown concentrations of solutions of $CoCl_2$ prepared by your instructor. Use the six dilutions that you prepared previously to construct your standard curve for $CoCl_2$. These solutions are **standards** because their concentrations are known, and they are used to determine the concentration of an unknown solution. The absorbance of each standard is measured at the peak wavelength of the absorption spectrum for $CoCl_2$.

Procedure 7.2

Construct a standard curve for cobalt chloride

1. Refer to your data in table 7.2. Determine the wavelength of peak absorbance for $CoCl_2$ and set the filter of your spectrophotometer to this wavelength.
2. Insert the solvent blank (tube 0) and adjust the spectrophotometer for zero absorbance.
3. Replace the blank with tube 1 (1 mg $CoCl_2$/mL), measure its absorbance, and then record the absorbance value in table 7.3.
4. Repeat steps 2–3 for the other five tubes (standards). Be sure and check the zero-absorbance calibration with the blank before each standard measurement.
5. To construct your standard curve for $CoCl_2$, plot the data in table 7.3 with *Concentration (mg/mL)* on the horizontal axis and *Absorbance* on the vertical axis on the graph paper at the end of this exercise.

TABLE 7.3

ABSORBANCE VALUES FOR SIX SOLUTIONS OF KNOWN CONCENTRATION (STANDARDS) OF $CoCl_2$ AT THE PEAK ABSORBANCE WAVELENGTH

Concentration of Standards (mg $CoCl_2$/mL)	Absorbance	Peak Wavelength = ______
1	______	
10	______	
20	______	
30	______	
40	______	
50	______	

6. Because the relationship between concentration and absorbance is linear (directly proportional), you should draw a straight line that lies as close as possible to each data point. Do not merely connect the dots. Note that extremely high concentrations of a solute can produce a nonlinear segment of the standard curve. However, the concentrations used in this lab exercise are not high enough to produce such "saturation" effects.
7. If a computer and software are available, calculate and plot the line of best fit.

Question 4
Do the plotted data points of your standard curve lie on a straight line?

Using the Standard Curve to Measure the Unknown Concentration of a Solution

After you have created a standard curve, measuring the unknown concentration of a $CoCl_2$ solution is easy. Your instructor has prepared a series of numbered tubes containing unknown concentrations of $CoCl_2$.

Procedure 7.3

Determine unknown concentrations

1. Obtain a tube with an unknown solution and record the tube number in table 7.4.
2. Use the blank tube (tube 0) to zero the spectrophotometer at the wavelength of peak absorbance for $CoCl_2$.
3. Measure the absorbance of the unknown solution and record this value in table 7.4.
4. Find this absorbance value on the vertical axis of your standard curve and draw a line from this point parallel to the horizontal axis until the line intersects the standard curve (see the example in fig. 7.4).
5. Draw a line from the intersection straight down until it intersects the horizontal axis. This point on the horizontal axis marks the concentration of the unknown solution.
6. Record the concentration of the unknown solution in table 7.4.
7. Obtain three more tubes with unknown solutions, determine their absorbance and concentration, and record the values in table 7.4. Ask your instructor to check your results.

ABSORPTION SPECTRUM OF CHLOROPHYLL

To give you more experience with absorption spectra, your instructor has prepared a plant extract containing chlorophyll, a photosynthetic pigment (fig. 7.5). The extract was made by grinding leaves in acetone.

CAUTION

Acetone is flammable. Keep all solvents away from hotplates and flames at all times.

Procedure 7.4

Determine the absorption spectrum of chlorophyll

1. Obtain a tube of the extract and prepare a blank. Your instructor will provide the solvent used for the blank.

TABLE 7.4

MEASUREMENTS OF ABSORBANCE AND CONCENTRATION FOR FOUR UNKNOWN SOLUTIONS OF $CoCl_2$

Tube Number	Absorbance	Concentration (mg/mL)
Unknown ______	______	______
Unknown ______	______	______
Unknown ______	______	______
Unknown ______	______	______

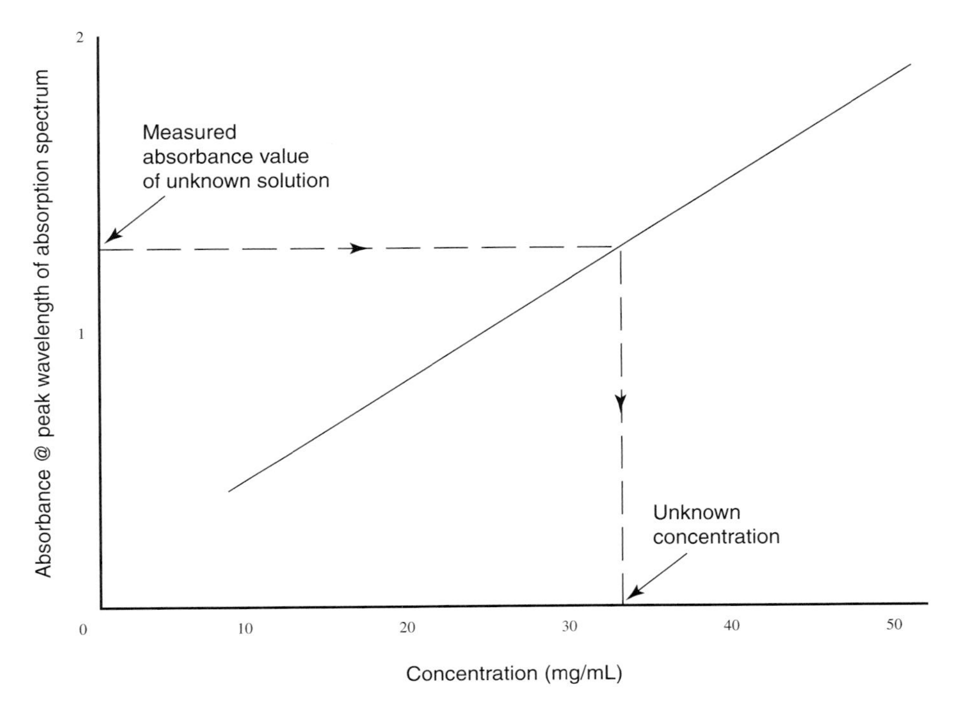

Figure 7.4

A standard curve showing the graphical determination of the $CoCl_2$ concentration of an unknown solution.

2. Using procedure 7.3 for determining an absorption spectrum, measure the absorbance of chlorophyll for at least eight wavelengths available on your spectrophotometer as listed in table 7.5.
3. Record your results in table 7.5.
4. Graph your results (in fig. 7.6) as you did for the absorption spectrum of $CoCl_2$.

Question 5

a. What is the proper blank for determining the absorption of chlorophyll in a plant extract?

b. Which wavelengths are least absorbed by chlorophyll?

c. Which wavelengths are most absorbed by chlorophyll?

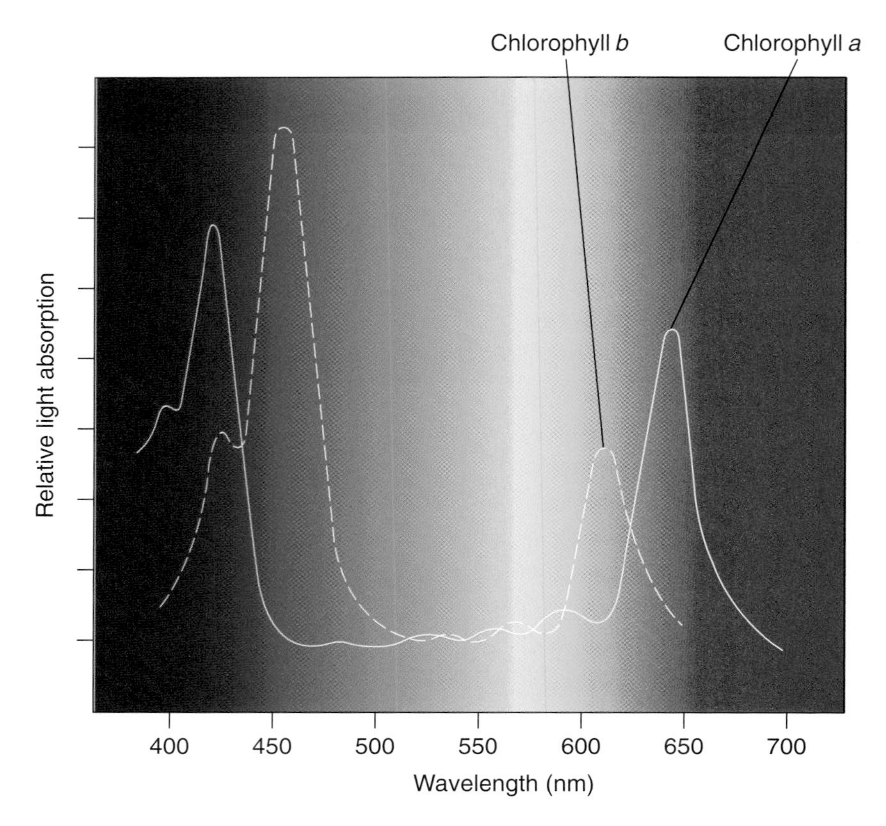

Figure 7.5
Absorption spectra for chlorophyll *a* and chlorophyll *b*. Chlorophyll *a*, along with other accessory pigments, occurs in all photosynthetic plants. Instructions for measuring the absorption spectrum of a pigmented plant extract are provided in the text of this exercise.

TABLE 7.5

ABSORBANCE VALUES FOR A PLANT EXTRACT CONTAINING CHLOROPHYLL

Wavelength	Absorbance	Wavelength	Absorbance
350 nm	______	530 nm	______
420 nm	______	570 nm	______
460 nm	______	610 nm	______
490 nm	______	660 nm	______

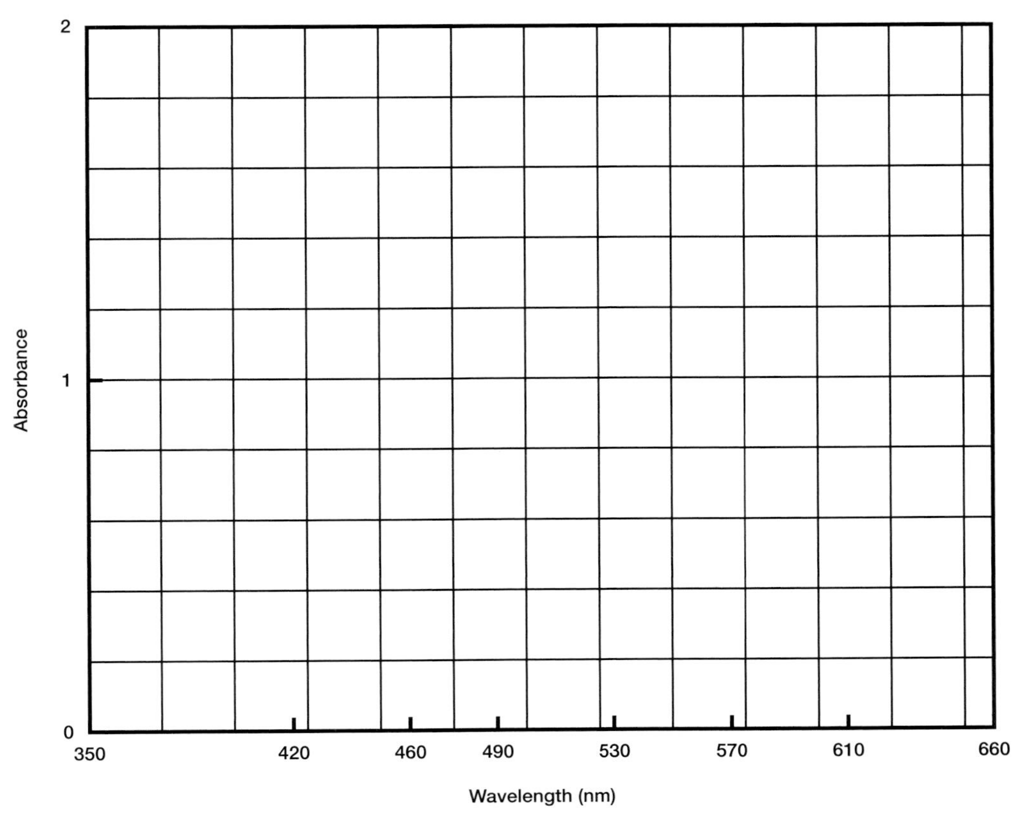

Figure 7.6
Absorption spectrum of chlorophyll.

7–10

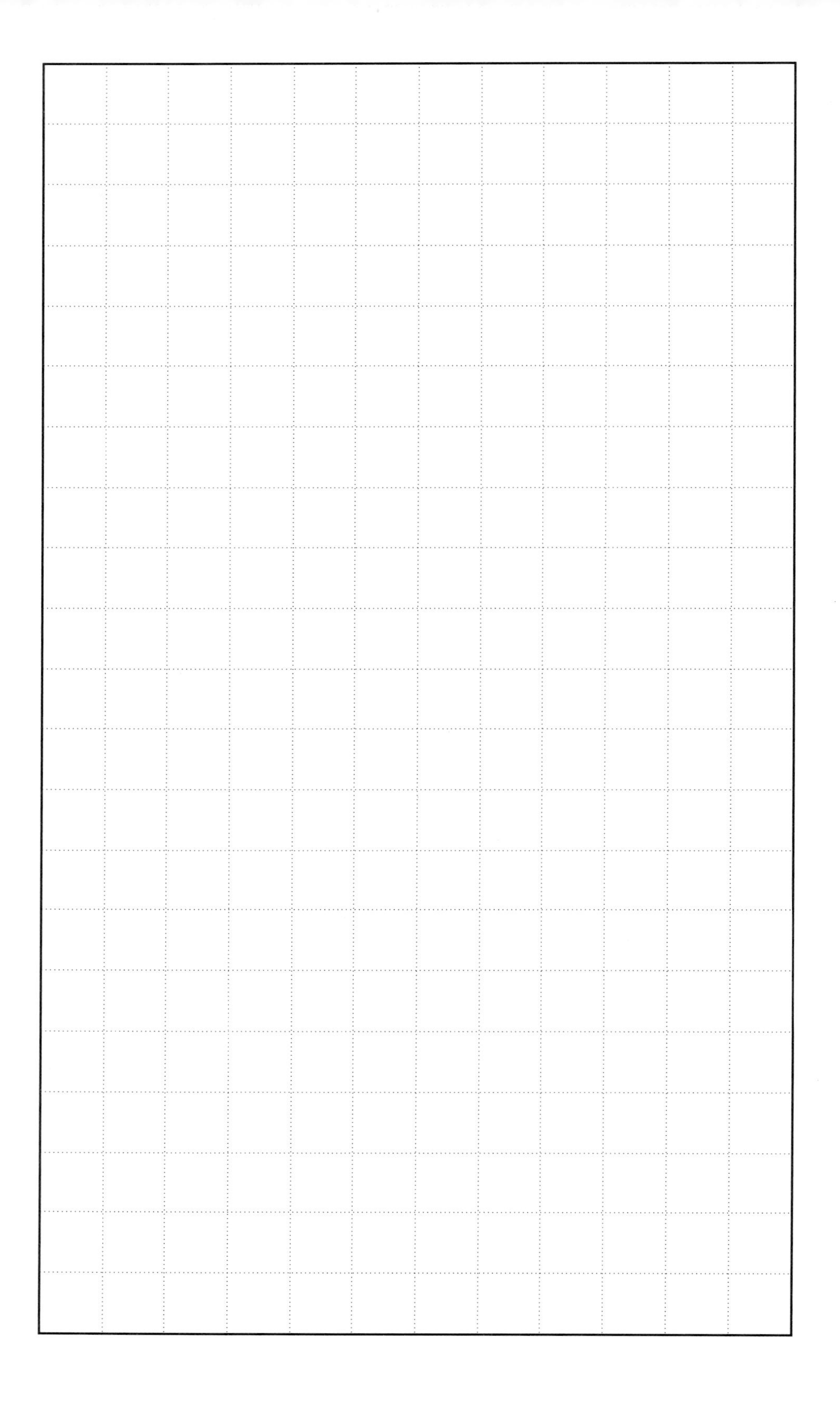

7–12

8

Diffusion and Osmosis
Passive Movement of Molecules in Biological Systems

Objectives

By the end of this exercise you should be able to:

1. Observe Brownian movement and understand its relationship to molecular movement.
2. Explain the factors controlling a substance's direction and rate of diffusion.
3. Determine the direction and relative rates of diffusion of molecules of different sizes.
4. Determine the direction and rate of osmosis into and out of simulated cells in hypotonic, hypertonic, and isotonic environments.
5. Describe how hypotonic, isotonic, and hypertonic solutions affect the volume and integrity of blood cells.
6. Describe how a hypertonic solution affects the volume and integrity of plant cells.

All molecules display random thermal motion, or kinetic energy; this is why a dissolved molecule tends to move around in a solution. Kinetic energy causes molecules to diffuse outward from regions of high concentration to regions of lower concentrations. This random movement is constant, but the net movement of molecules from high to low concentration continues until the distribution of molecules becomes homogenous throughout the solution. For example, when a dye has dissolved in a container of water, the dye disperses as the crystal dissolves. The rate of dispersal depends on the concentration of the dye, the size of the dye molecules, the temperature of the solution, and the medium through which diffusion occurs. Regardless of this rate, the dye will eventually become uniformly distributed throughout the solution. This phenomenon is easily illustrated by placing a drop of dye into a glass of water (fig. 8.1).

In this exercise you will study the diffusion of molecules in artificial and living systems.

Figure 8.1
Beakers of water before and after diffusion of a dye. Random movements of water and dye molecules drive diffusion, eventually resulting in a uniform distribution of the dye. Convection currents may also help distribute the dye in these solutions.

BROWNIAN MOVEMENT

Heat causes **random motion** of molecules and passively moves molecules in biological systems. Although we cannot directly see molecules move, we can see small particles move after collisions with moving molecules. This motion was originally described in 1827 by Robert Browning as he observed dead pollen grains in water and viewed them with a microscope. **Brownian movement** is visible using your microscope's high magnification. Carmine red dye mixed with soap produces a good suspension of small particles. The particles of red dye are small enough to vibrate when water molecules bump into them.

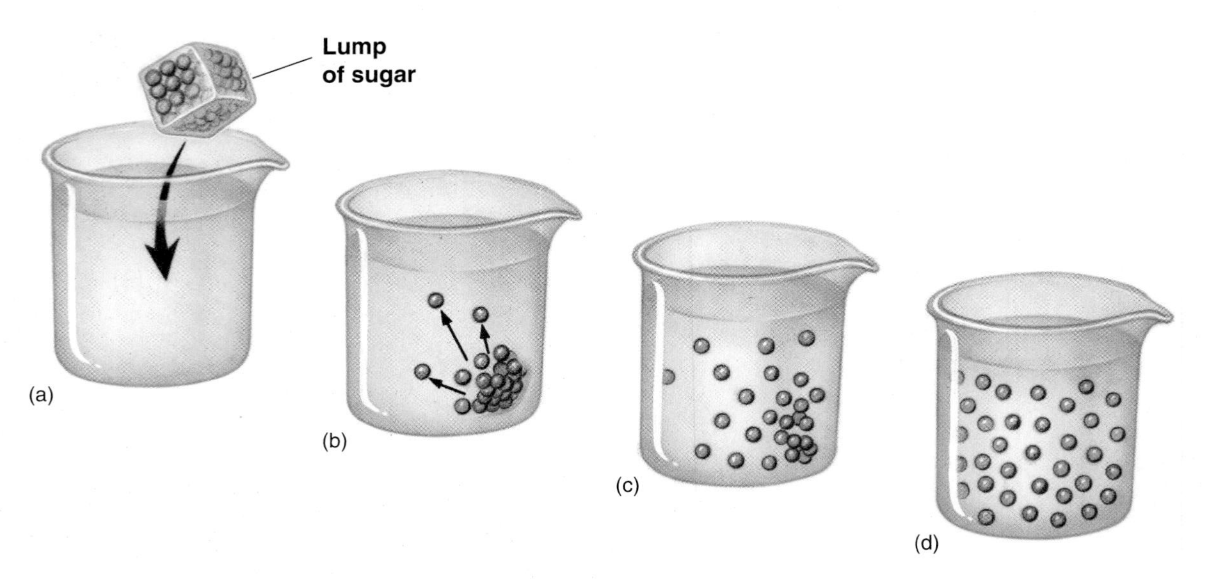

Figure 8.2
Diffusion. If a lump of sugar is dropped into a beaker of water, its molecules dissolve (*a*) and diffuse (*b* and *c*). Eventually, diffusion results in an even distribution of sugar molecules throughout the water (*d*).

Procedure 8.1

Observe Brownian movement

1. Place a small drop of a carmine red suspension on a microscope slide and cover the drop with a coverslip.
2. Focus first at low magnification; then rotate to higher power (40×). Be careful not to get dye on the objective lens.
3. Fine focus the image. At first the field of view will appear uniformly reddish gray. But with sharp focus, you will see thousands of small particles vibrating rapidly.
4. Check with your instructor to determine if your microscope has oil immersion magnification and if you need this to easily view the particles. If needed, follow their instructions for using this objective.
5. Leave the microscope light on. Observe any changes in motion with increased heat.

Question 1

a. Briefly describe your observation of the moving pigment particles.

b. Does the movement of particles change visibly with heat? How?

DIFFUSION

In biological systems, substances often move through solutions and across membranes in a predictable direction. This passive, directional movement of molecules is **diffusion** (fig. 8.2). The *direction* of diffusion depends on the presence of a gradient of concentration, heat, and pressure. Specifically, molecules diffuse from an area of high concentration, heat, and pressure to an area of low concentration, heat, and pressure. The *rate* of diffusion is determined by the steepness of the gradient and other characteristics of the specific molecule in question, such as its size, polarity, or solubility.

Even though temperature, pressure, and concentration all affect diffusion, temperature and pressure are relatively constant in most biological systems. Therefore, concentration is usually the best predictor of a substance's direction of diffusion. But remember that temperature and pressure gradients may also affect diffusion.

Diffusion and Molecular Weight

Before your class meeting your instructor inoculated some petri plates containing agar with either potassium permanganate (molecular weight = 158 g $mole^{-1}$), malachite green (molecular weight = 929 g $mole^{-1}$), or methylene blue (molecular weight = 374 g $mole^{-1}$).

Question 2
Which would you predict would diffuse faster: a substance having a high molecular weight or a substance having a low molecular weight? Why?

Procedure 8.2

Observe diffusion as affected by molecular weight

1. Examine one of the prepared agar plates and note the three halos of color. These halos indicate that the chemicals have diffused away from the two original spots and moved through the agar.
2. Measure the halos with a ruler.
3. Record within the outline of a petri dish your observations of the size of each halo.

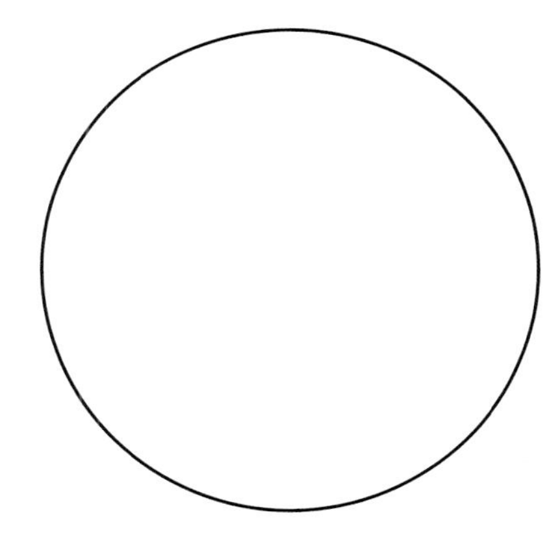

Question 3

a. Considering the different molecular weights of potassium permanganate, malachite green, and methylene blue, which should have the larger halo after the same amount of time? Why?

b. Do molecules stop moving when diffusion stops? Explain your answer.

DIFFUSION AND DIFFERENTIALLY PERMEABLE MEMBRANES

Membranes surround cells and organelles and organize an immense number of simultaneous reactions. However, the barrier imposed by a cellular membrane does not isolate a cell. Instead, it allows a cell to selectively communicate with its environment. Membranes are "alive" in the sense that they respond to their environment and allow some molecules to pass while retarding others. Thus, membranes are selective and **differentially permeable** (fig. 8.3). This selective permeability results from the basic structure of membranes. Membranes have a two-layered core of nonpolar lipid molecules that selects against molecules that are not readily soluble in lipids. You'll learn more about membrane structure in Exercise 9.

Two important characteristics of molecules govern their passive movement through a lipid membrane: polarity and size. **Polar molecules** have positively charged areas and negatively charged areas. **Nonpolar molecules** have no local areas of positive or negative charge. Small, nonpolar molecules pass through membranes most easily, whereas large polar molecules are retained.

In general, small molecules pass through a membrane more easily than do large molecules. We can demonstrate membrane selection for molecular size by using a bag made from **dialysis tubing** to model a differentially permeable membrane. **Dialysis** is the separation of dissolved substances by means of their unequal diffusion through a differentially permeable membrane. Dialysis membranes (or tubing) are good models of differentially permeable membranes because they have small pores that allow small molecules such as water molecules to pass but block large molecules such as glucose. However, remember that living cell membranes also discriminate among molecules based on charge and solubility whereas dialysis tubing does not. Dialysis tubing is only a physical model of a cell and its selectivity is based only on molecular size.

Examine some dialysis tubing. Although the dried material looks like a narrow sheet of cellophane it is actually a flattened, open-ended tube.

In procedure 8.3 you will use two indicators: **phenolphthalein** and **iodine.** Phenolphthalein is a pH indicator that turns red in basic solutions (see Exercise 4). Iodine is a starch indicator that changes from yellow to dark blue in the presence of starch (see Exercise 5, "Biologically Important Molecules").

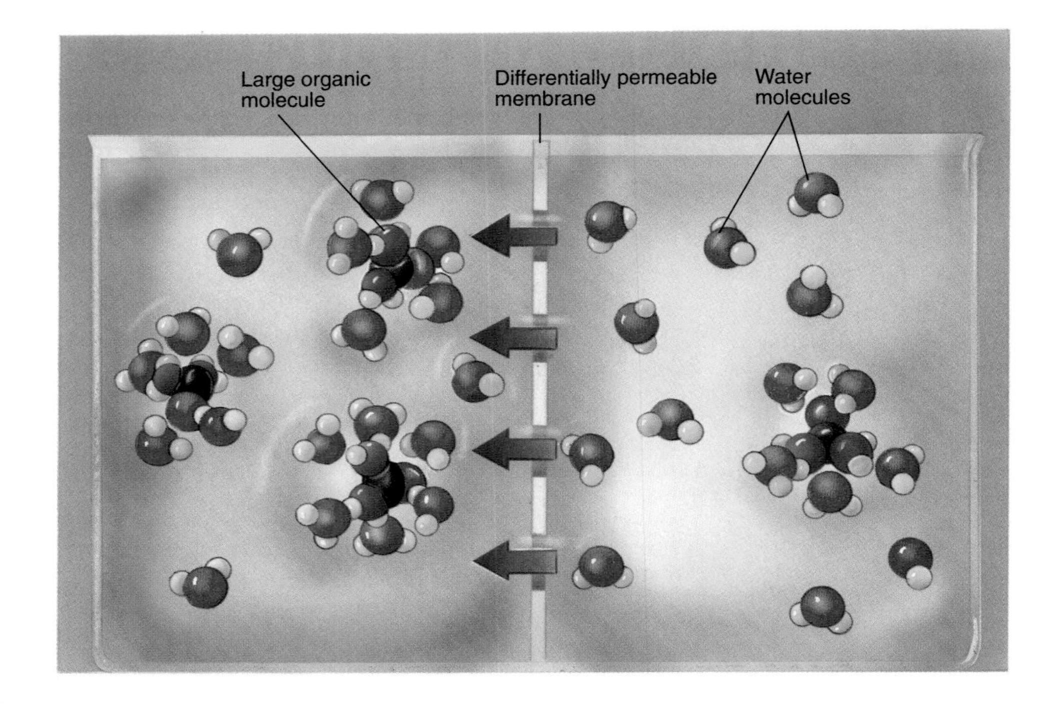

Figure 8.3

A differentially permeable membrane prevents the movements of some molecules but not others. Arrows indicate the movement of small molecules such as water from an area of high concentration to an area of lower concentration. The large molecules cannot pass through the membrane.

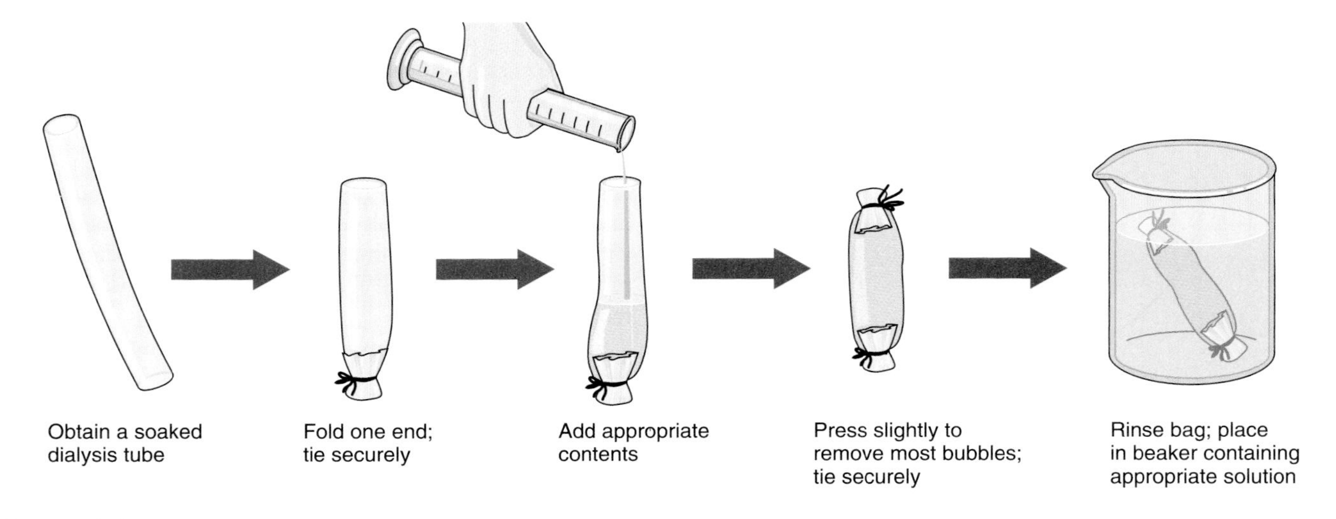

Figure 8.4

Preparation of dialysis tubing as a model of a cell surrounded by a differentially permeable membrane.

Procedure 8.3

Observe diffusion across a differentially permeable membrane

1. Obtain four pieces of string or dialysis clips and two pieces of water-soaked dialysis tubing approximately 15 cm long.
2. Seal one end of each bag by folding over 1–2 cm of the end. Then accordion-fold this end and tie it tightly with monofilament line or string (fig. 8.4). The ends of the tube must be sealed tightly to prevent leaks.
3. Roll the untied end of each tube between your thumb and finger to open it and form a bag.

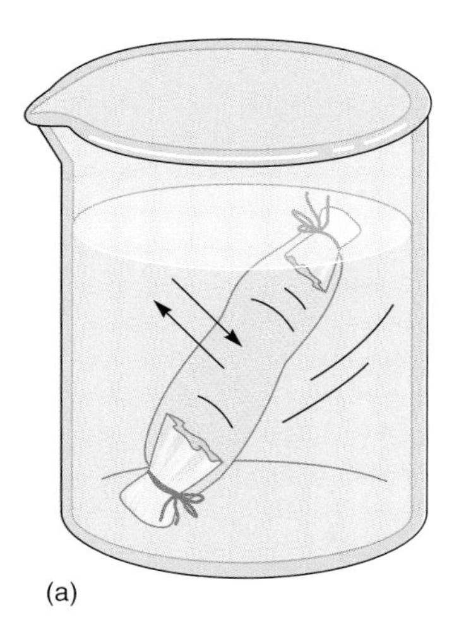

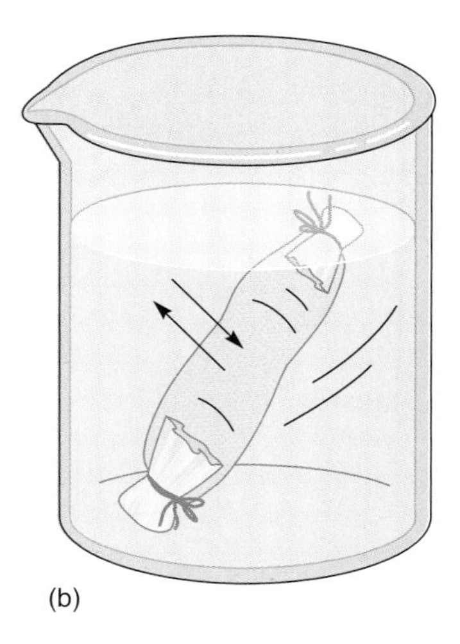

Figure 8.5

(*a*) Movements and reaction of sodium hydroxide and phenolphthalein through a differentially permeable membrane. (*b*) Movements and reaction of iodine and starch. Record the results of your experiment on this diagram.

4. Use either a graduated cylinder or pipet to fill one tube with 10 mL of water and add three drops of phenolphthalein. Seal the open end of the bag by folding the end and tying it securely.
5. Fill the other bag with 10 mL of starch suspension. Seal the open end of the bag by folding the end and tying it securely.
6. Gently rinse the outside of each bag in tap water.
7. Fill a beaker with 200 mL of tap water and add ten drops of 1 M sodium hydroxide (NaOH). Submerge the dialysis bag containing phenolphthalein in the beaker.

CAUTION

Do not spill the NaOH. It is extremely caustic.

8. Fill a beaker with 200 mL of tap water and add 20–40 drops of iodine. Submerge the dialysis bag containing starch in the beaker.
9. Observe color changes in the two bags' contents and the surrounding solutions.
10. In this experiment some of the solutes can move through the membrane and some cannot. Water can freely move through the membrane, but the movement of water is not of interest in this experiment.
11. Record in figure 8.5 the color inside and outside the bags. Label the contents inside and outside the bags.

Question 4

a. Describe color changes in the two bags and their surrounding solutions.

b. For which molecules and ions (phenolphthalein, iodine, starch, Na^+, OH^-) does your experiment give evidence for passage through the semipermeable membrane?

c. What characteristic distinguishes those molecules and ions passing through the membrane from those that do not pass through the membrane?

OSMOSIS AND THE RATE OF DIFFUSION ALONG A CONCENTRATION GRADIENT

The speed at which a substance diffuses from one area to another depends primarily on the concentration gradient between those areas. For example, if concentrations of a diffusing substance at the two areas differ greatly, then diffusion is rapid. Conversely, when the concentration of a substance at the two areas is equal, the diffusion rate is zero and there is no net movement of the substance.

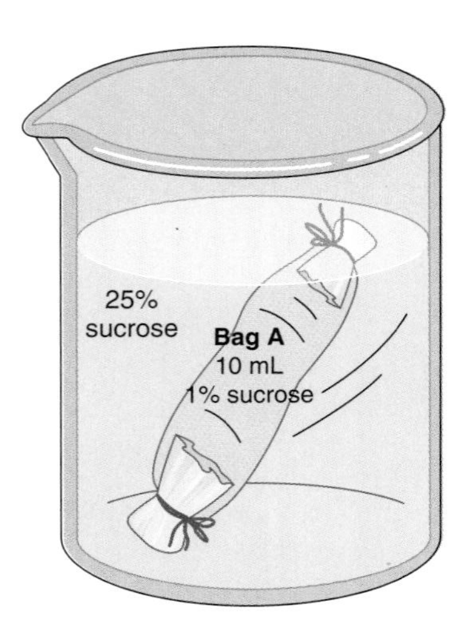

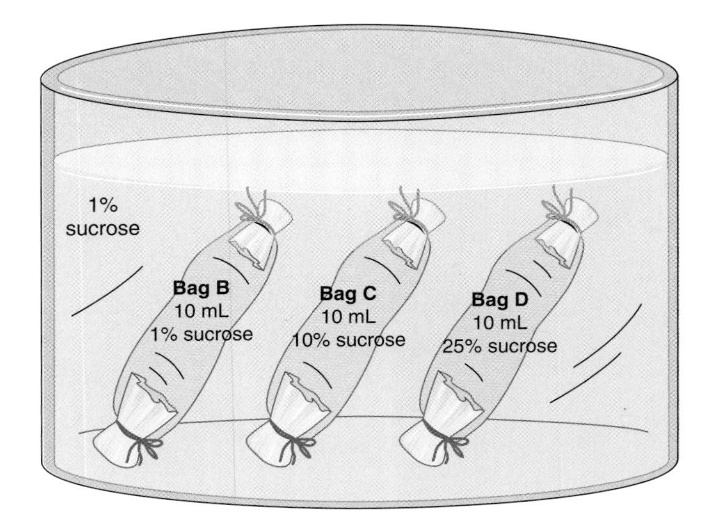

Figure 8.6
Experimental setup for four cellular models used to measure the rate of osmosis.

Osmosis is diffusion of water across a differentially permeable membrane. Osmosis follows the same laws as diffusion but always refers to water, the principal solvent in cells. A **solution** is a homogenous, liquid mixture of two or more kinds of molecules. A **solvent** is a fluid that dissolves substances, and a **solute** is a substance dissolved in a solution.

We can simulate osmosis by using dialysis bags to model cells under different conditions and measuring the direction and rate of osmosis. Each of the four dialysis bags in the following experiment is a model of a cell. Bag A simulates a cell with a solute concentration that is hypotonic relative to its environment. **Hypotonic** describes a solution with a lower concentration of solutes, especially those solutes that do not pass across the surrounding membrane. Water moves across semipermeable membranes out of hypotonic solutions. Conversely, the solution surrounding bag A is hypertonic relative to the cell. **Hypertonic** refers to a solution with a high concentration of solutes.

Bag B represents a cell whose solute concentration equals the concentration in the environment; that is, this cell (bag B) is isotonic to its environment. **Isotonic** refers to two solutions that have equal concentrations of solutes. Bags C and D are both hypertonic to their environment and have higher solute concentrations than the surrounding environment. Remember that the solute (sugar) does not pass through the membrane—only the water does.

Procedure 8.4

Observe osmosis across a concentration gradient

1. Obtain eight pieces of string and four pieces of water-soaked dialysis tubing 15 cm long. Seal one end of each tube by folding and tying it tightly.
2. Open the other end of the tube by rolling it between your thumb and finger.
3. Fill the bags with the contents shown in figure 8.6. To label each bag, insert a small piece of paper with the appropriate letter (A, B, C, or D written on it in pencil).
4. For each bag, loosely fold the open end and press on the sides to push the fluid up slightly and remove most of the air bubbles. Tie the folded ends securely, rinse the bags, and check for leaks.
5. Gently blot excess water from the outside of the bags and weigh each bag to the nearest 0.1 g.
6. Record these initial weights in table 8.1 in the first column.
7. Place bags B, C, and D in three individual beakers or one large bowl filled with 1% sucrose (fig. 8.6). Record the time.
8. Place bag A in a 250-mL beaker and fill the beaker with 150 mL of 25% sucrose. Record the time.
9. Remove the bags from the beakers at 15-min intervals for the next hour (or at intervals indicated by your instructor), gently blot them dry, and weigh them to the nearest 0.1 g. Handle the bags delicately to avoid leaks and quickly return the bags to their respective containers.

TABLE 8.1

CHANGES IN WEIGHT OF DIALYSIS BAGS USED AS CELLULAR MODELS*

	0 Min	15 Min		30 Min		45 Min		60 Min	
	Initial Weight	Total Weight	Change in Weight	Total Weight	Change in Weight	Total Weight	Change in Weight	Total Weight	Change in Weight
Bag A									
Bag B									
Bag C									
Bag D									

*Each change in weight is only for the previous 15-min interval.

10. During the 15-min intervals, use your knowledge of osmosis to make hypotheses about the direction of water-flow in each system (i.e., into or out of bag), and the extent of water flow in each system (i.e., in which system will osmosis be most rapid?).
11. For each 15-min interval record the total weight of each bag and its contents in table 8.1. Then calculate and record in table 8.1 the change in weight since the previous weighing.

Procedure 8.5

Graph osmosis

1. Use the graph paper at the end of this exercise to construct a graph with *Total Weight (g)* vs. *Time (min)*. *Time* is the variable that you established and actively controlled and, therefore, is the **independent variable.** The independent variable is always graphed on the horizontal axis. *Total Weight* changed in response to differences in the independent variable, so *Total Weight* is the **dependent variable.** The dependent variable is always graphed on the vertical axis.
2. Graphs must have a title (e.g., Relationship Between Time and Weight Gain), correctly labeled axes (e.g., *Total Weight, Time*), a label showing measurement units (e.g., g and min), and values along each axis (e.g., 0, 15, 30, 45, 60). Include these in your graph.
3. Plot the data for total weight at each time interval from table 8.1
4. Include the data for all four bags as four separate curves on the same graph.

Question 5

a. Did water move across the membrane in all bags containing solutions of sugar?

b. In which bags did osmosis occur?

c. A concentration gradient for water must be present in cells for osmosis to occur. Which bag represented the steepest concentration gradient relative to its surrounding environment?

d. The steepest gradient should result in the highest rate of diffusion. Examine the data in table 8.1 for Change in Weight during the 15- and 30-min intervals. Did the greatest changes in weight occur in cells with the steepest concentration gradients? Why or why not?

Question 6

a. Refer to your graph. How does the slope of a segment of a curve relate to the rate of diffusion?

b. What influence on diffusion (i.e., temperature, pressure, concentration) causes the curves for bags C and D eventually to become horizontal (i.e., have a slope = 0)?

TABLE 8.2

CHANGE IN LENGTH OF POTATO CYLINDERS SURROUNDED BY DIFFERENT SALT CONCENTRATIONS

Concentration of Salt Solution (%)	Initial Size of Cylinders (millimeters or grams)	Changes in Size of Three Sample Cylinders			Mean Change in Size
0	______	______	______	______	______
0.9	______	______	______	______	______
5	______	______	______	______	______
10	______	______	______	______	______
15	______	______	______	______	______

WATER POTENTIAL

Plants need to balance water uptake and loss as it moves from one part of a plant to another and in and out of cells by osmosis. However, the concentration gradient of water and solutes doesn't solely determine the direction and rate of water movement. Physical pressure influenced by cell walls and evaporation is also important. Plant physiologists refer to the combined effects of concentration and pressure such as that from cell walls as **water potential;** water will flow from an area of high water potential to an area of low potential. Both high water concentration (low solute concentration) and high pressure increase water potential. Similarly high solutes and low pressure decrease water potential. In simple terms, water flows through a plant from the higher water potentials of the root tissues toward the lower water potentials of leaves. These lower potentials in leaves are created by their loss of water to the atmosphere (see Exercise 32). In the following procedure you will measure the concentration of solutes in potato cells and relate this concentration to water potential.

Procedure 8.6

Determine the concentration of solutes in living plant cells

1. Locate the five beakers prepared by your instructor with five concentrations of salt (NaCl) solution.
2. The cylinders of potato that you see in the solutions were all originally the same size (i.e., the same length or weight). Check the beaker labels to determine which measure of size (length or weight) you will be using as your data.
3. Record the initial values in table 8.2.
4. Carefully remove three of the potato cylinders from each solution and measure their size.
5. Record your data in table 8.2
6. Calculate the mean change in size and record the data in table 8.2.
7. Your instructor may ask you to graph your data (see Question 7f). Follow their instructions.

Question 7

a. Which potato cylinders increased in size or weight? Why?

b. Which solution(s) contained a higher concentration of solutes and therefore a lower water potential than in the potato cells? Explain your answer.

c. Which salt solution best approximated the water potential in the potato cells?

d. For a growing potato plant what would you predict the water potential of the potato relative to the soil? Relative to the leaves?

e. What might be some sources of error in this experiment?

f. How could a graph of your data help you estimate the solute concentration of potato cells?

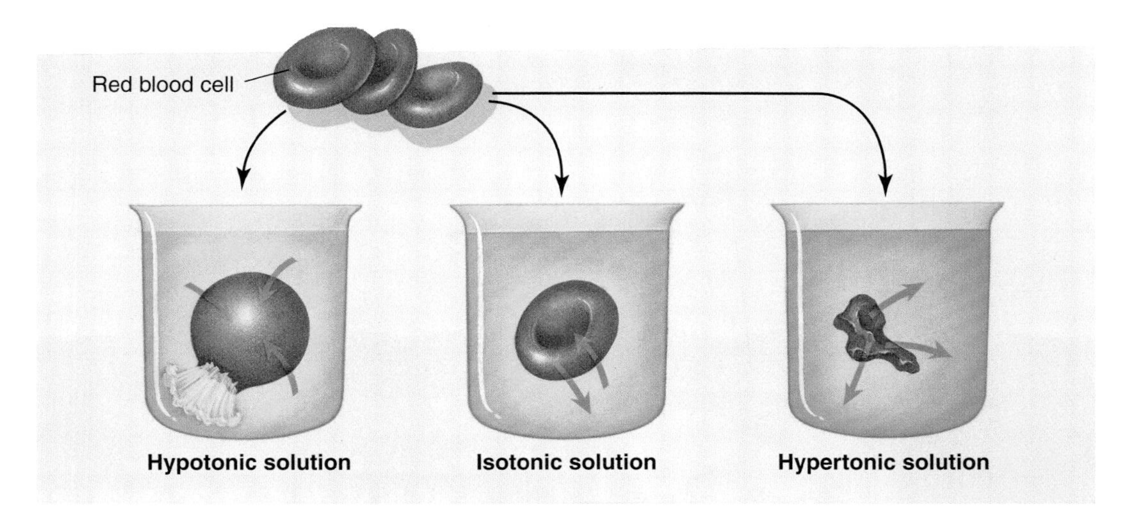

Figure 8.7
Osmosis of water surrounding animal cells. When the outer solution is hypotonic with respect to the cell, water will move into the cells and the cells will lyse; when it is hypertonic, water will move out of the cells and the cells will shrink (i.e., become crenate).

HEMOLYSIS OF BLOOD CELLS

Living red blood cells (erythrocytes) are good models for studying osmosis and diffusion in hypotonic, hypertonic, and isotonic solutions. Osmosis occurs when living cells are placed in a hypotonic or hypertonic environment and water diffuses into or out of the cell (fig. 8.7). For example, in the previous experiment water moved into cells toward the low concentration of water. However, osmosis into animal cells increases the hydrostatic (i.e., water) pressure and bursts the cells because they lack cell walls. This destruction of a cell by the influx of water (causing the cell to burst) is called **lysis.** Such destruction of a red blood cell is called **hemolysis.** If water flows out of a cell into a hypertonic solution, the cell will shrivel and become crenate.

Detect hemolysis and crenation in blood cells in three different solutions using the following procedure.

Procedure 8.7

Observe hemolysis

1. Obtain and label three test tubes and fill them with the solutions listed in table 8.3.
2. Add four drops of fresh sheep's blood to each tube.

CAUTION

Wash your hands thoroughly after working with blood products. Always handle sheep blood with caution and avoid skin contact.

3. Cover each tube with parafilm and invert the tubes to mix the contents.
4. Hold each tube in front of a printed page and determine if you can read the print through the solution. Record your results in table 8.3.
5. Obtain a microscope, slide, and coverslip.
6. Use an eyedropper or pipet to obtain one drop from each tube. Make a wet mount and examine the blood cells. Use low magnification first and then higher magnification.
7. Record in table 8.3 the cell's condition as crenate, normal, or lysed.

Question 8

a. Through which test tubes could you read the printed page? Why?

b. Which concentration of NaCl lysed the cells?

c. Which of the three solutions most closely approximates the solute concentration in a red blood cell? How do you know?

Table 8.3

Hemolysis of Red Blood Cells Exposed to Three Solutions with Different Solute Concentrations

Tube	Contents	Readable Print (yes/no)	Cell Condition (crenate/normal/lysed)
1	5 mL 10% NaCl	________	________
2	5 mL 0.9% NaCl	________	________
3	5 mL distilled water	________	________

Figure 8.8

Experimental setup for determining hemolysis. Hypertonic solutions will hemolyze cells.

PLASMOLYSIS OF PLANT CELLS

Plasmolysis is the shrinking of the cytoplasm of a plant cell in response to diffusion of water out of the cell and into a hypertonic solution (high salt concentration) surrounding the cell (fig. 8.9). During plasmolysis the cellular membrane pulls away from the cell wall (fig. 8.10).

In procedure 8.8 you will examine the effects of highly concentrated solutions on osmosis and cellular contents.

Procedure 8.8

Observe plasmolysis

1. Prepare a wet mount of a thin layer of onion epidermis or *Elodea* leaf. Examine the cells.
2. Add two or three drops of 30% NaCl to one edge of the coverslip.
3. Wick this salt solution under the coverslip by touching a piece of absorbent paper towel to the fluid at the opposite edge of the coverslip.
4. Examine the cells and note that the cytoplasm is no longer pressed against the cell wall. This shrinkage is **plasmolysis.**

Question 9

a. Why did the plant cells plasmolyze when immersed in a hypertonic solution?

b. What can you conclude about the permeability of the cell membrane (i.e., the membrane surrounding the cytoplasm) and vacuolar membrane (the membrane surrounding the vacuole) to water?

To observe the effects of cellular plasmolysis on a larger scale, compare petioles of celery that have been immersed overnight in distilled water or in a salt solution.

Question 10

What causes crispness in celery?

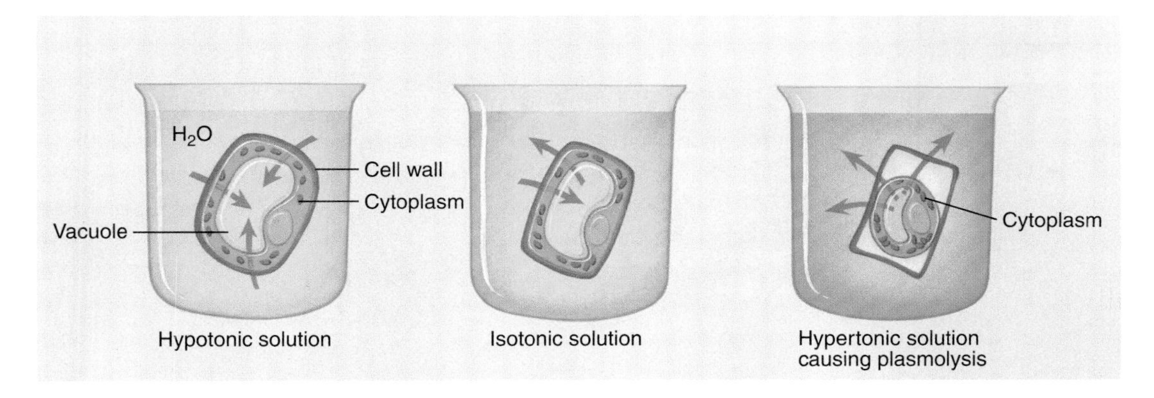

Figure 8.9

Osmosis of water into and out of plant cells. In most plant cells the large central vacuole contains a high concentration of solutes (i.e., the environment surrounding the cell is hypotonic to the cell), so water tends to diffuse into the cells, causing the cells to swell outward against their rigid cell walls. However, if a plant cell is immersed in a high-solute (hypertonic) solution, water will leave the cell, causing the cytoplasm to shrink and pull away from the cell wall.

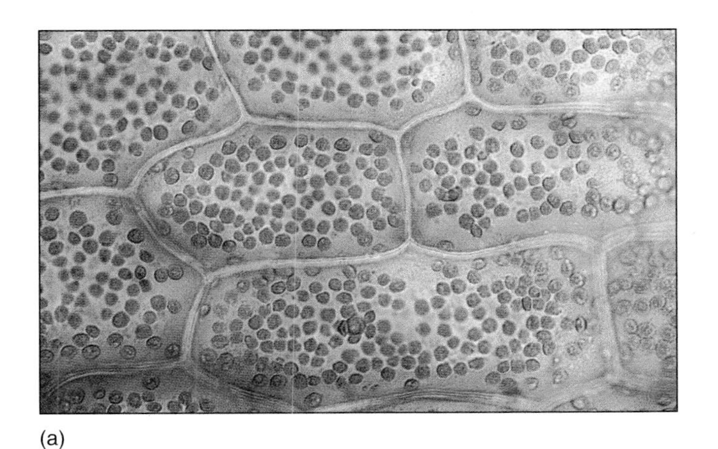

(a)

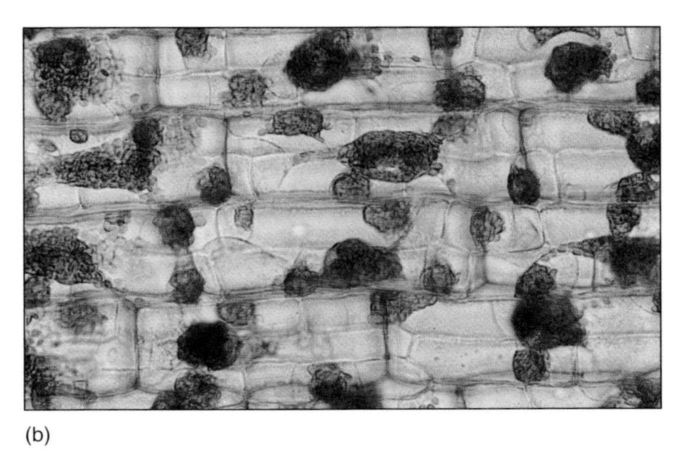

(b)

Figure 8.10

(*a*) Turgid *Elodea* cells. (*b*) Plasmolyzed *Elodea* cells showing the effects of exposure to a hypertonic solution.

INVESTIGATION

Determination of Unknown Solute Concentrations in Plant Tissue

Water moves in and out of cells along a concentration gradient. How would you design an experiment to determine the solute concentration within cells of a piece of apple?

a. Would you expect the solute concentrations to be high or low compared to other tissues of the plant? Why?

b. What materials would you need?

c. What would be the steps of your procedure?

d. Would you consider repeating your experiment? Why or why not?

Questions for Further Thought and Study

1. Why must particles be extremely small to demonstrate Brownian movement?

2. What is the difference between molecular motion and diffusion?

3. If you immerse your hand in distilled water for 15 min, will your cells lyse? Why or why not?

4. Your data for diffusion of water across a differentially permeable membrane in response to a sucrose gradient could be graphed with *Change in Weight* on the vertical axis rather than *Total Weight*. How would you interpret the slope of the curves that are produced when you do this?

5. How do cells such as algae and protists avoid lysis in fresh water?

WRITING TO LEARN BIOLOGY

Where in an animal might pressure affect diffusion of a substance?

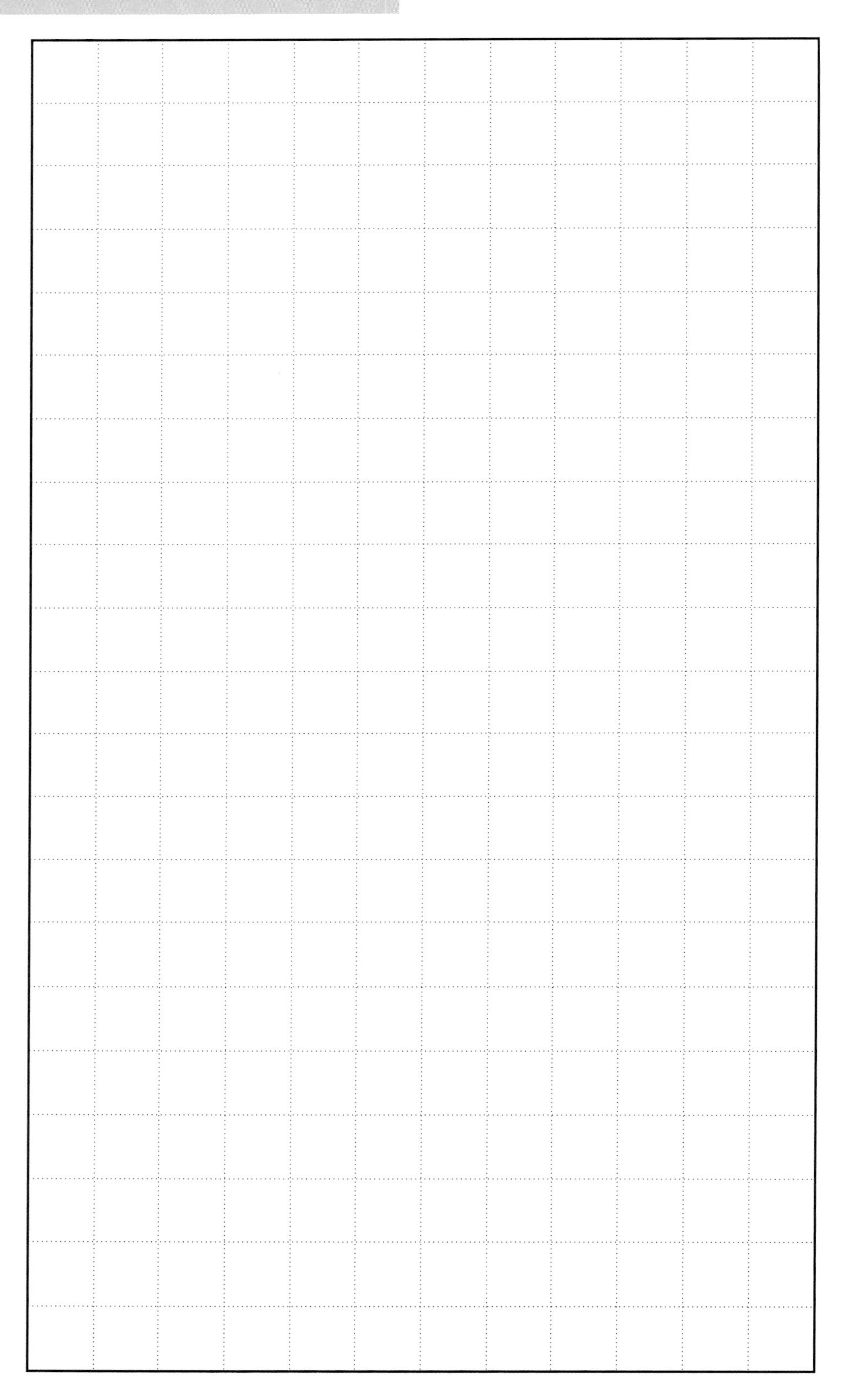

exercise nine

9

Cellular Membranes

Effects of Physical and Chemical Stress

Objectives

By the end of this exercise you should be able to:

1. Explain the function and structure of cellular membranes.
2. Predict the effect of common organic solvents and extreme temperatures on membrane integrity.
3. Relate the results of experiments with beet membranes to the general structure and function of membranes.

Membranes separate and organize the myriad of reactions within cells and allow communication with the surrounding environment. Although membranes are only a few molecules thick (6–10 nm), they (1) retard diffusion of certain molecules, (2) house receptor molecules that detect other cells or organelles, (3) provide sites for active and passive transport of selected molecules, (4) organize life processes, (5) maintain the integrity of cells, and (6) regulate cell function.

As with all other biological entities, the structure of a membrane reflects its function. Membranes consist of a phospholipid bilayer; attached to or embedded within this bilayer are thousands of proteins. A phospholipid molecule consists of a phosphate group and two fatty acids bonded to a three-carbon, glycerol chain (fig. 9.1). Phospholipids have an unevenly distributed charge; that is, they have charged (polar) and uncharged (nonpolar) areas. In phospholipids the phosphate group and glycerol are polar and **hydrophilic** ("water-loving"), whereas the fatty-acid chains are nonpolar and **hydrophobic** ("water-fearing"). Such molecules with hydrophobic and hydrophilic regions are **amphipathic,** and amphipathic phospholipids have a natural tendency to self-assemble into a double-layered sheet (fig. 9.2). In this double layer, the hydrophobic tails of lipids form the core of the membrane, and the hydrophilic phosphate groups line both surfaces. This elegant assembly is stable, self-repairing, and resists penetration by most hydrophilic molecules.

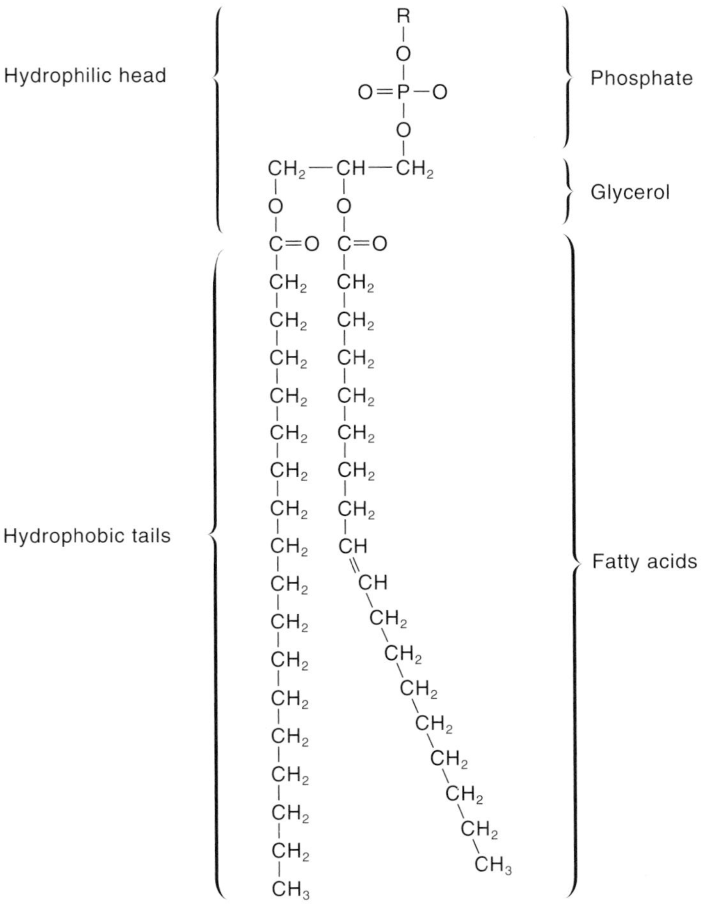

Hydrophilic head

R
O
O=P—O
O
CH_2—CH—CH_2
O O
C=O C=O

Phosphate

Glycerol

Hydrophobic tails

Fatty acids

CH_2–CH_2–CH_2–CH_2–CH_2–CH_2–CH_2–CH_2–CH_2–CH_2–CH_2–CH_2–CH_2–CH_2–CH_2–CH_2–CH_3

CH_2–CH_2–CH_2–CH_2–CH_2–CH_2–CH_2–CH=CH–CH_2–CH_2–CH_2–CH_2–CH_2–CH_2–CH_2–CH_3

Figure 9.1

The structure of a phospholipid. The backbone of a phospholipid is glycerol, a three-carbon alcohol. Glycerol is bonded to two fatty acids, which are both hydrophobic, and to one phosphate group, which is hydrophilic. Phospholipids vary by their fatty acids and by the side chains (R groups) that are attached to the phosphate. The R groups include glycerol, sugars, and amine-containing carbon chains.

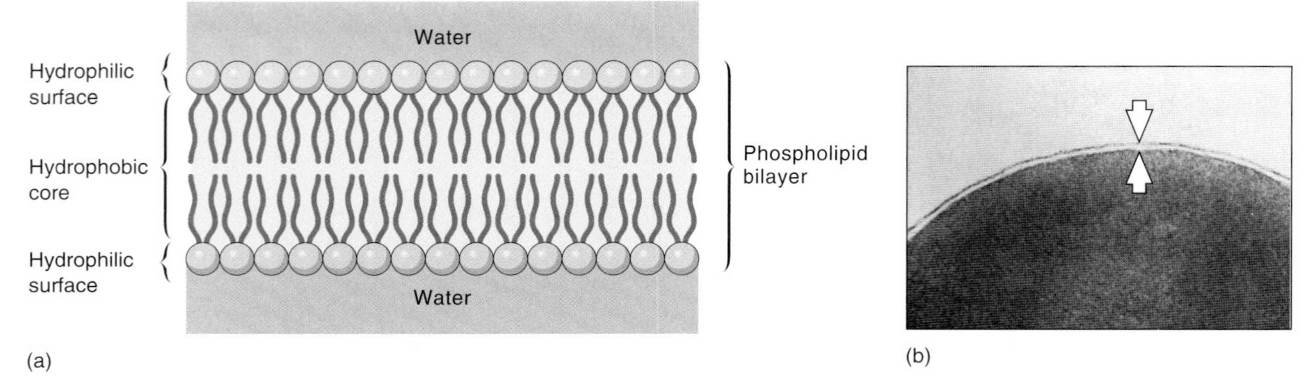

Figure 9.2

(*a*) Model of an artificial membrane of phospholipids. Phospholipids aggregate spontaneously into a bilayer. (*b*) Electron micrograph of a bilayer membrane. The arrowheads point to the bilayer.

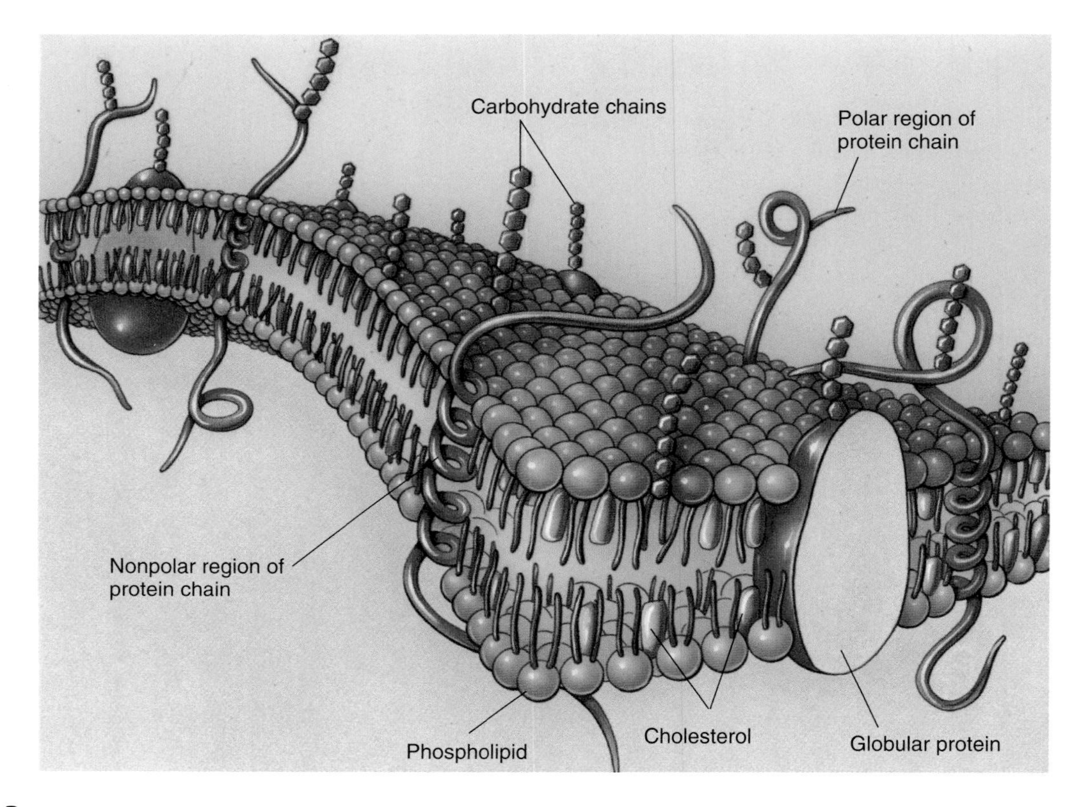

Figure 9.3

Fluid mosaic model of membrane structure. A variety of proteins protrude through the lipid membrane. Carbohydrate chains often bind to these proteins and are unique to particular types of cells.

Membranes also include proteins that are dispersed as a mosaic throughout the fluid bilayer of lipids (fig. 9.3). These proteins are not fixed in position; they move about freely and may be densely packed in some membranes and sparsely distributed in others. Carbohydrate chains (strings of sugar molecules) are often bound to these proteins and to lipids. These chains serve as distinctive identification tags, unique to particular types of cells. This elaborate combination forms the **fluid mosaic model** of membrane structure.

Membranes are selectively permeable. The proteins embedded in the phospholipid bilayer can selectively take up and expel molecules that otherwise could not penetrate the membrane. In doing so, these proteins function as pores, permitting and often facilitating passage of specific ions and polar molecules. In addition to forming pores and sites for active transport, membrane-bound proteins also function as enzymes and receptors that detect signals from the environment or from other cells.

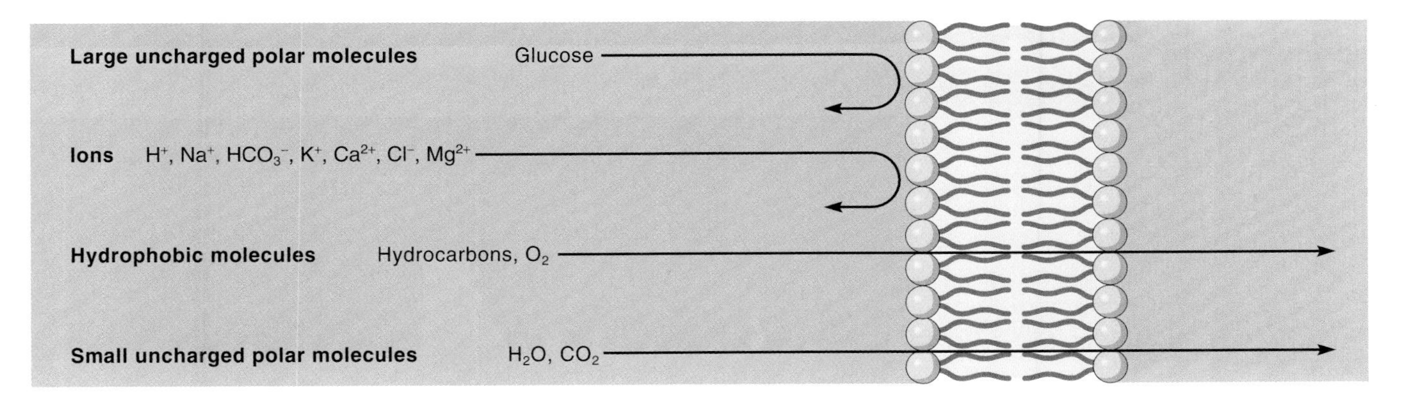

Figure 9.4
Phospholipid bilayers are selectively permeable. The bilayer is much more permeable to hydrophobic molecules and small uncharged polar molecules than it is to ions and large polar molecules.

The physical and chemical integrity of a membrane is crucial for the proper functioning of the cell or organelle that it surrounds. As a stable sheet of interlocking molecules, the membrane functions as a barrier to simple diffusion (fig. 9.4). Membrane permeability depends on the solubility and size of the molecules trying to penetrate the membrane. Molecules that are lipid-soluble pass easily through the lipid core of a membrane.

In this exercise you will physically and chemically stress the integrity of membranes.

Question 1

a. What ions must routinely move across cell membranes?

b. How could membranes promote the movement of ions out of or into cells? How could membranes restrict the movement of ions out of or into cells?

c. Could a cell survive without an intact cell membrane? Explain.

BEET CELLS AS AN EXPERIMENTAL SYSTEM

Beet tissue will be your model to investigate membrane integrity. Roots of beet (*Beta vulgaris*) contain large amounts of a reddish pigment called **betacyanin** that is localized almost entirely in the large central vacuoles of cells. Betacyanin in healthy cells remains inside the vacuoles, which are surrounded by a vacuolar membrane called the **tonoplast.** The entire cell (including vacuole, tonoplast, and cytoplasm) is surrounded by a cell membrane and cell wall.

In the two procedures you will subject beet cells to a range of temperatures and organic solvents, and determine which treatments stress and damage the membranes the most. If stress damages the membranes, betacyanin will leak through the tonoplast and plasma membrane. This leakage will produce a red color in the water surrounding the stressed beet. Thus, you can measure membrane damage by measuring the intensity of color resulting from a particular treatment.

THE EFFECT OF TEMPERATURE STRESS ON MEMBRANES

Membranes are sensitive to extreme temperatures. High temperatures cause violent molecular collisions that destroy a membrane as a physical barrier to diffusion. Conversely, freezing temperatures cause water to crystallize as ice and expand because of hydrogen bond alignment. This expansion and formation of ice often ruptures membranes.

Procedure 9.1

Observe the effect of temperature stress on cellular membranes

1. Cut six uniform cylinders of beet using a cork borer with a 5-mm inside diameter. Trim each cylinder to exactly 15 mm in length. All of the cylinders must be the same size.
2. Place these cylinders of beet tissue in a beaker and rinse them with tap water for 2 min to wash betacyanin from the injured cells on the surface. Be sure that all of the cylinders are washed in the same way. Discard the colored rinse water.

Table 9.1

The Color Intensity of Betacyanin Leaked from Damaged Cells Treated at Six Temperatures

Tube Number	Treatment (°C)	Color Intensity (0–10)	Absorbance (460 nm)
1	70		
2	55		
3	40		
4	20		
5	5		
6	−5		

3. Gently place one of the six beet sections into each of six dry test tubes. Do not crush, stab, or otherwise damage the cylinders when moving them to the test tubes.
4. Label the tubes 1–6 and write the temperature treatment on each tube as listed in table 9.1.
5. **FOR COLD TREATMENTS:**
   a. Place tube 5 in a refrigerator (5°C) and tube 6 in a freezer (−5°C). If a refrigerator and freezer are not available in your lab, give your labeled tubes to an assistant who will take them to another facility.
   b. Leave tubes 5 and 6 in the cold for 30 min. While waiting, proceed with hot treatments (step 6). However, watch your time and return to steps 5c and 5d after 30 min.
   c. After 30 min, remove the beets from the freezer and refrigerator and add 10.0 mL of distilled water at room temperature to each of the tubes.
   d. Let the cold-treated beets soak in distilled water for 20 min. Then remove and discard the beets from tubes 5 and 6.
6. **FOR HOT TREATMENTS:**
   a. Take the beet section out of tube 1 and immerse it in a beaker of hot water at 70°C for 1 min. If a 70°C water-bath is not available, hot tap water should be adequate, but carefully adjust the temperature to 70°C. Handle the beet gently with forceps, but don't squeeze it tightly because you may rupture the beet's cells.
   b. After 1 min at 70°C, return the beet to tube 1 and add 10.0 mL of distilled water at room temperature.
   c. If a 55°C water-bath is not available, slowly add ice chips or cold water to cool the beaker of hot water to 55°C. Then immerse the beet from tube 2 for 1 min. Return the beet to tube 2 and add 10.0 mL of distilled water at room temperature.
   d. If a 40°C water-bath is not available, cool the beaker of hot water to 40°C. Then immerse the beet from tube 3 for 1 min. Return the beet to tube 3 and add 10.0 mL of distilled water at room temperature.
   e. If a 20°C water-bath is not available, cool the beaker of hot water to 20°C. Then immerse the beet from tube 4 for 1 min. Return the beet to tube 4 and add 10.0 mL of distilled water at room temperature.
   f. Allow the treated beets in tubes 1–4 to soak in distilled water at room temperature for 20 min. Then remove and discard the beets and measure the extent of membrane injury according to the amount of betacyanin that diffused into the water.
7. **FOR ALL SIX TEMPERATURE TREATMENTS:**
   Quantify the relative color of each solution between 0 (colorless) and 10 (darkest red), and record your results in table 9.1. If color standards are available in the lab, examine them to determine relative values for the colors of your samples.

Use the graph paper at the end of this exercise to graph *Temperatures* vs. *Relative Color* according to a demonstration graph provided by your instructor. Your instructor may also ask you to quantify your results further using a spectrophotometer. If so, see Exercise 7 for instructions for using a spectrophotometer. Read the absorbance of the solutions at 460 nm and record your results in table 9.1. Then graph *Temperature* vs. *Absorbance*.

TABLE 9.2

THE COLOR INTENSITY OF BETACYANIN LEAKED FROM DAMAGED CELLS TREATED WITH VARIOUS CONCENTRATIONS OF TWO ORGANIC SOLVENTS

Tube Number	Treatment	Color Intensity (0–10)	Absorbance (460 nm)
1	1% acetone		
2	25% acetone		
3	50% acetone		
4	1% methanol		
5	25% methanol		
6	50% methanol		
7	Isotonic saline		

Question 2

a. Which temperature damaged membranes the most? Which the least? How do you know?

b. In general, which is more damaging to membranes, extreme heat or extreme cold? Why?

c. If the results of this experiment are easily observed with the unaided eye, why use a spectrophotometer?

d. The beets were subjected to cold temperatures longer than to hot temperatures to make sure that the beet sections were thoroughly treated. Why does the freezing treatment require more time?

THE EFFECT OF ORGANIC SOLVENT STRESS ON MEMBRANES

Organic solvents dissolve a membrane's lipids. Acetone and methanol are common solvents for various organic molecules, but acetone has the greater ability to dissolve lipids.

CAUTION

Organic solvents are flammable. Extinguish all open flames and heating elements before doing the following procedure. Do not pour organic solvents down the drain. Dispose of them properly.

Question 3

Which of the organic liquids (acetone or methanol) do you predict will damage membranes the most?

Procedure 9.2

Observe the effect of organic solvents on cellular membranes

1. Cut seven uniform cylinders of beet using a cork borer with a 5-mm inside diameter. Trim each cylinder to exactly 15 mm in length. All the cylinders must be the same size.
2. Place these cylinders of beet tissue in a beaker and rinse them with tap water for 2 min to wash betacyanin from the injured cells on the surface. Be sure that all of the cylinders are the same size. Discard the colored rinse water.
3. Place one of the seven beet sections into each of seven dry test tubes. Do not crush, stab, or otherwise damage the cylinders when moving them to the test tubes.
4. Label the tubes 1–7 and write the organic-solvent treatment on each tube as listed in table 9.2.
5. Add 10.0 mL of the appropriate solvent (see table 9.2) to each of the seven tubes.
6. Keep all beets at room temperature for 20 min and shake them occasionally. Then remove and discard the beet sections and measure the extent of membrane damage according to the amount of betacyanin that diffused into the water.

7. Quantify the relative color of each solution between 0 (colorless) and 10 (darkest red). Record your data in table 9.2. If color standards are available, examine them to determine the relative values for the colors of your samples.

CAUTION

Be sure to dispose of the organic solvents as directed by your instructor.

Graph *Concentration of Organic Solvent* vs. *Relative Color* according to a demonstration graph provided by your instructor. Your instructor may also ask you to further quantify your results using a spectrophotometer. If so, see Exercise 7 for instructions for using a spectrophotometer. Read the absorbance of the solutions at 460 nm and record your results in table 9.2. Then graph *Concentration of Organic Solvent* vs. *Absorbance*.

Question 4

a. Based on your results, are lipids soluble in both acetone and methanol?

b. Based on your results, which damages membranes more: 50% methanol or 25% acetone?

c. In which solvent are lipids most soluble?

d. The concentration of solvent affects its ability to dissolve lipids. Based on your results, did the highest concentration of both solvents cause the most damage?

e. What other solvents might be interesting to test in this experiment?

f. What was the purpose of tube 7?

INVESTIGATION

Effects of Environmental Stimuli on Membranes

This exercise has demonstrated that membranes are sensitive to a variety of environmental stimuli.

a. Form a hypothesis to determine how one of these stimuli affects membrane permeability. Write your hypothesis here.

b. Decide how you will test your hypothesis. Describe your experimental design here.

c. Do your experiment. What do you conclude? Do your data support your hypothesis?

Questions for Further Thought and Study

1. Are your conclusions about membrane structure and stress valid only for beet cells? Why or why not?

2. What characteristics of beets make them useful as experimental models for studying cellular membranes?

3. Explain why phospholipids have a natural tendency to self-assemble into a bilayer. Why is this biologically important?

4. Freezing temperatures are often used to preserve food. Considering the results of this experiment, which qualities of food are preserved and which are not?

5. Movement of water through membranes has long puzzled scientists. Why would you not expect water to move easily through a membrane?

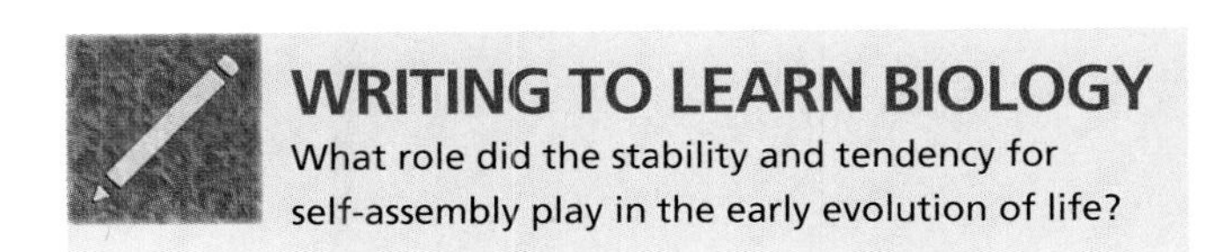

WRITING TO LEARN BIOLOGY

What role did the stability and tendency for self-assembly play in the early evolution of life?

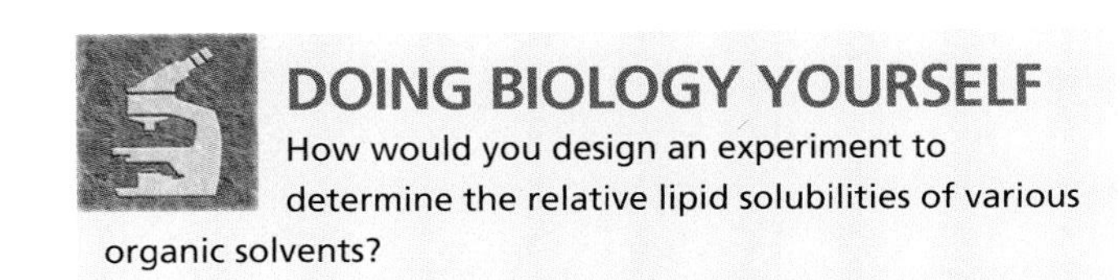

DOING BIOLOGY YOURSELF

How would you design an experiment to determine the relative lipid solubilities of various organic solvents?

9–8

10

Enzymes
Factors Affecting the Rate of Activity

Objectives

By the end of this exercise you should be able to:

1. Describe the structure and function of enzymes, and relate structure and function to active sites, modes of inhibition, and optimal conditions for enzymatic activity.
2. Predict how inhibitors and changes in temperature and pH affect enzymatic reaction rates.
3. Describe how some enzymatic reaction rates can be measured by color changes and gas liberation as products are formed.

Fortunately, not all chemical reactions within our cells occur spontaneously. If they did, our metabolism would be chaotic. Instead, most reactions in cells are controlled by proteins called **enzymes.** Enzymes are **biocatalysts,** meaning that they accelerate metabolic reactions to biologically useful rates. Specifically, enzymes catalyze (accelerate) reactions by lowering the activation energy needed for the reaction to occur (fig. 10.1).

Enzymes act by binding to reacting molecules, called the **substrate,** to form an **enzyme-substrate** complex. This complex stresses or distorts chemical bonds to form a **transition state** in which the substrate becomes more reactive and the metabolic reaction accelerates. The energy needed to form the transition state is called **energy of activation** and is supplied by the enzyme. The site of attachment and the surrounding parts of the enzyme that stress the substrate's bonds constitute the enzyme's **active site.**

Enzyme + Substrate → Enzyme-Substrate Complex → Enzyme + Product

The reaction is complete when the **product** forms and the enzyme is released in its original condition. The enzyme then repeats the process with other molecules of substrate (fig. 10.2).

Enzymes are proteins made of long chains of amino acids that form complex shapes. Although cells contain many enzymes, each kind of enzyme has a precise structure and function, and each enzyme catalyzes a specific reaction. This specificity results from an enzyme's unique structure and shape. For example, the shape of the active site on the enzyme's surface is complex and usually couples with only one kind of substrate. Any structural change in an enzyme may **denature** or destroy its effectiveness by altering the active site and slowing down the reaction rate. Therefore, the rate of an enzymatic reaction depends on conditions in the immediate environment. These conditions affect the shape of the enzyme and modify the active site and precise fit of an enzyme and its substrate.

The range of values for environmental factors such as temperature and pH at which an enzyme functions best represents that enzyme's **optimal conditions.** The optimal conditions for the enzymes of an organism may be specific for that species and usually are adaptive for the environment of the organism. Other factors such as the amount of substrate or concentration of enzyme also affect the reaction rate.

In this exercise you will learn that environmental factors such as temperature and pH affect enzymatic reactions (fig. 10.3). You will also investigate how inhibitors affect enzymatic activity.

TEMPERATURE AFFECTS THE ACTIVITY OF ENZYMES

Heat increases the rate of most chemical reactions. During enzymatic reactions, faster molecular motion caused by heat increases the probability that enzyme molecules will contact substrate molecules. The rate of chemical reactions generally doubles with a 10°C rise in temperature. However, higher temperatures do not always accelerate enzymatic reactions; enzymatic reactions have an optimal range of temperatures. Temperatures above or below this range decrease the reaction rate. Extreme temperatures often denature enzymes.

The effects of temperature on enzyme activity can be investigated with **catechol oxidase**, a plant enzyme that oxidizes catechol and converts it to benzoquinone. When

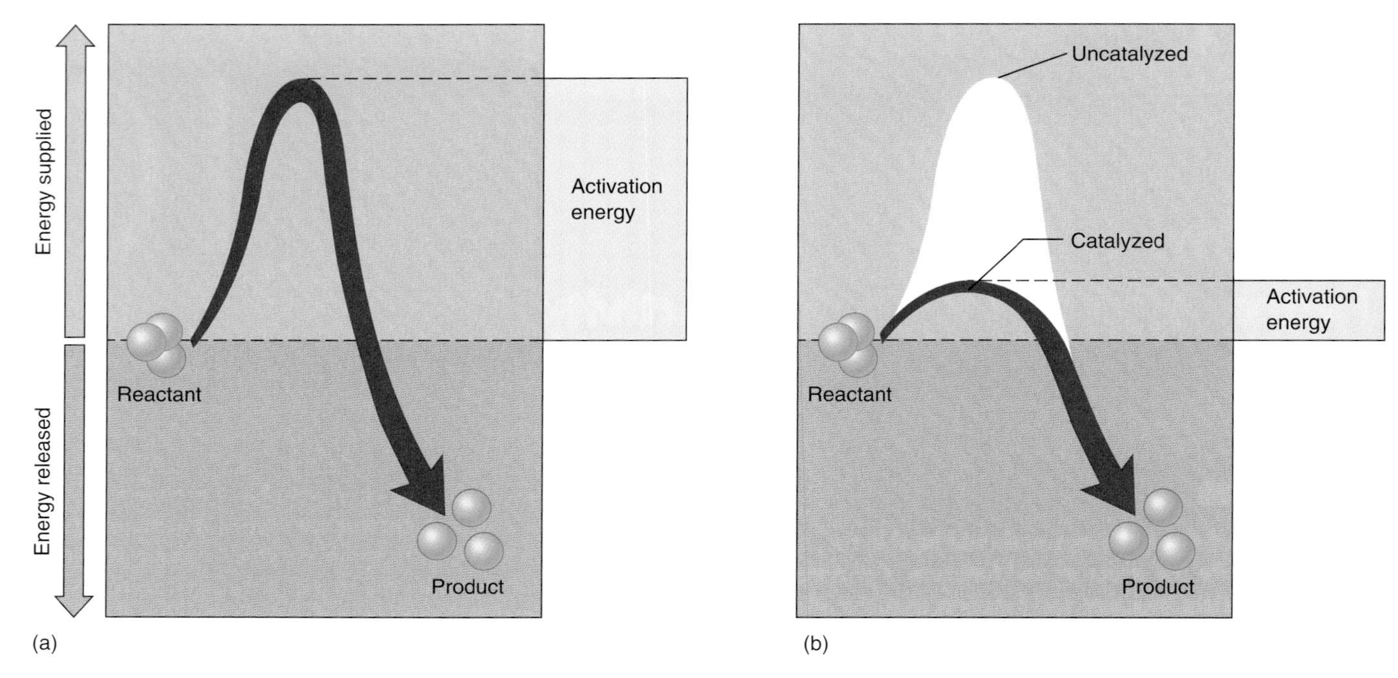

Figure 10.1
Activation energy and catalysis. (*a*) Exergonic reactions (those that release energy) do not necessarily proceed rapidly because energy must be supplied to destabilize existing chemical bonds. This extra energy is the activation energy for the reaction. (*b*) Catalysts accelerate particular reactions by lowering the amount of activation energy required to initiate the reaction.

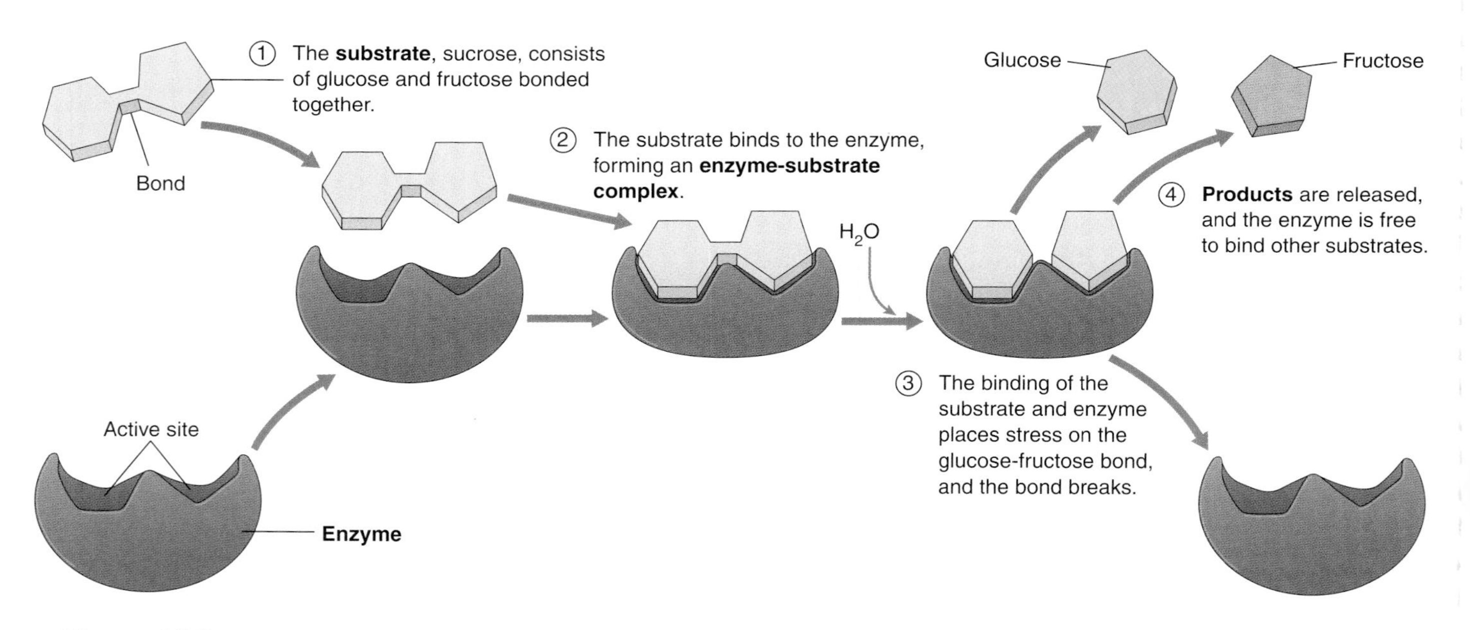

Figure 10.2
The catalytic cycle of an enzyme. Enzymes increase the speed of chemical reactions but are not themselves permanently altered by the process. Here, the enzyme splits the disaccharide sucrose (steps 1, 2, 3, and 4) into its two parts, the monosaccharides glucose and fructose. After the enzyme releases the glucose and fructose, it can bind another molecule of sucrose and begin the catalytic cycle again.

TABLE 10.1

EXPERIMENTAL CONDITIONS TO TEST THE EFFECT OF TEMPERATURE ON CATECHOL OXIDASE ACTIVITY

Tube	Distilled Water	pH 6 Buffer	Potato Extract (catechol oxidase)	1% Catechol	Temperature
1	2 mL	1 mL			22°C
2	1 mL	1 mL		1 mL	22°C
3	1 mL	1 mL	1 mL		22°C
4		1 mL	1 mL	1 mL	22°C
5		1 mL	1 mL	1 mL	4°C
6		1 mL	1 mL	1 mL	40°C
7		1 mL	1 mL	1 mL	80°C

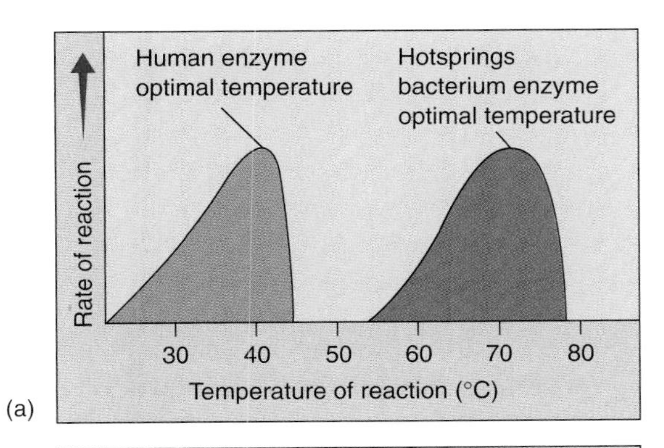

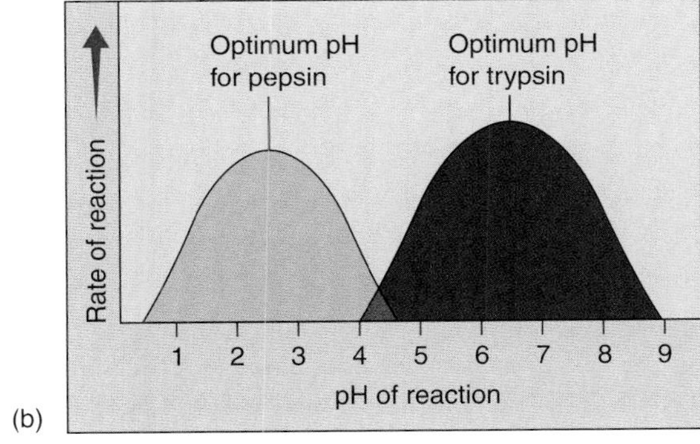

Figure 10.3

Enzymes are sensitive to their environment. The activity of an enzyme is influenced by both (a) temperature and (b) pH. Most enzymes in humans, such as the protein-degrading enzyme trypsin, work best at temperatures about 40°C and within a pH range of 6 to 8. As you can see, however, pepsin works best at a much lower pH than does trypsin.

fruit is bruised, injured cells release catechol and catechol oxidase, which react to form a brownish product, benzoquinone. Toxic to bacteria, benzoquinone prevents decay in damaged cells. Your source of catechol oxidase will be potato extract.

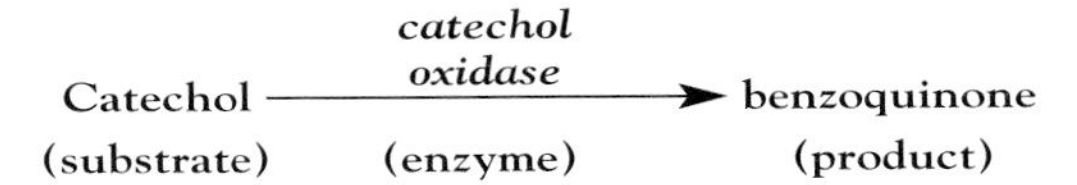

Procedure 10.1

Observe the effect of temperature on catechol oxidase activity

1. Prepare water-baths at 40°C and 100°C. Locate a refrigerator or ice bath at or below 4°C. Place a test tube rack in each bath and in the refrigerator.
2. Obtain seven test tubes and number them at the top 1–7.
3. Obtain a tube of potato extract and a tube of 1% catechol from your instructor.
4. Add distilled water, pH buffer, and potato extract to the tubes as listed in table 10.1. Shake or swirl to resuspend the potato extract.
5. Place the tubes in the appropriate bath or refrigerator. Allow each tube to stand undisturbed for 5 min at its respective temperature. Put tubes 1–4 in a test tube rack at room temperature (approximately 22°C).
6. Add 1% catechol solution to tubes 2 and 4–7 as listed in table 10.1. For each tube immediately record in table 10.2 any color changes for 0 min. Record qualitative color changes on a scale between 0–5 by using the color of tube 1 as 0 and the color of tube 6 as 5.

TABLE 10.2

QUANTITATIVE AND QUALITATIVE COLOR CHANGES AS CATECHOL OXIDASE ACTIVITY PRODUCES BROWN BENZOQUINONE

	Qualitative Color Change Results					Quantitative Absorbance Results				
Tube	0 min	5 min	10 min	15 min	20 min	0 min	5 min	10 min	15 min	20 min
1	0	0	0	0	0	0	0	0	0	0
2										
3										
4										
5										
6	5	5	5	5	5					
7										

CAUTION

Catechol is toxic. Wash well with soap and water after skin contact.

7. Every 5 min observe and note color changes in the seven tubes over the next 20 min.
8. If your instructor asks you to further quantify your data, then measure the absorbance of the solution in each tube using a spectrophotometer set to 470 nm with tube 1 as a blank. Refer to Exercise 7 for instructions on how to use the spectrophotometer.
9. Clean your work area and materials. Catechol must be disposed into waste containers, not down the sink drain.

Question 1

a. What were the enzyme, substrate, and product of the enzymatic reaction?

b. Why was each tube left undisturbed for 5 min in step 5 of procedure 10.1?

c. Explain the results observed for tubes 1–3. What was the purpose of these tubes?

d. Use your results for tubes 4–7 to construct a line graph of *Enzyme Activity* vs. *Time*. There will be four curves on the graph. Refer to the discussion of graphs in Appendix III.

e. Use your results to argue for or against the statement, "Catechol oxidase functions equally and efficiently at various temperatures."

f. Over what range of temperatures tested was catechol oxidase active? Should other temperatures be tested to more accurately determine the range of activity?

g. At which temperature was catechol oxidase activity greatest? Should more temperatures be tested to determine its optimum?

h. What temperature apparently denatured catechol oxidase? How do you know?

i. What is the effect of denaturing an enzyme?

TABLE 10.3

EXPERIMENTAL CONDITIONS TO TEST THE EFFECT OF pH ON CATALASE ACTIVITY

Tube	Distilled Water	Buffer	Hydrogen Peroxide	HCl	NaOH	pH	Catalase Solution
1	5 mL	1 mL, pH 7					
2	4 mL	1 mL, pH 7					1 mL
3	2 mL	1 mL, pH 7	3 mL				
4	1 mL		3 mL	1 mL			1 mL
5	1 mL	1 mL, pH 5	3 mL				1 mL
6	1 mL	1 mL, pH 7	3 mL				1 mL
7	1 mL	1 mL, pH 9	3 mL				1 mL
8	1 mL		3 mL		1 mL		1 mL
9	1 mL	1 mL, pH 7	3 mL	1 mL			
10	1 mL	1 mL, pH 7	3 mL		1 mL		

j. If an enzyme has a single optimal temperature, then an organism might have difficulty dealing with an environment with wide temperature variation. What adaptive advantage is there in having repetitive enzyme systems (i.e., more than one enzyme to catalyze the same reaction) that we know many organisms have?

pH AFFECTS THE ACTIVITY OF ENZYMES

Enzymatic activity is sensitive to pH because the surfaces and side groups of enzyme molecules are often charged. Acidic and basic solutions are rich in H^+ and OH^- ions (see Exercise 4), respectively, and they react with the side groups of the enzyme molecules. As the pH is lowered, side groups gain H^+ ions; as the pH is raised, side groups lose H^+ ions. In this way, solutions having an extreme pH can change an enzyme's conformation enough to alter its active site. Extreme pH can denature an enzyme just as drastically as can high temperatures. Many enzymes function optimally in the neutral pH range, while others (such as pepsin, an enzyme in your digestive tract) function optimally as low as pH 1.6.

The effects of pH can be investigated with **catalase**—an enzyme in plants and animals that speeds up the breakdown of hydrogen peroxide, which is toxic to cells. Hydrogen peroxide is broken down by catalase to water and oxygen.

$$2\ H_2O_2 \xrightarrow{\textit{catalase}} 2\ H_2O + O_2$$

Procedure 10.2

Observe the effects of pH on catalase activity

1. Prepare catalase solution.
    a. Use a mortar and pestle to macerate a marble-size portion of fresh, raw ground meat in 10 mL of distilled water.
    b. Filter the solution through cheesecloth into a test tube and add an equal volume of distilled water.
2. Obtain 10 test tubes and number them at the top 1–10.
3. Obtain stock solutions of distilled water, hydrogen peroxide, buffer pH 5, buffer pH 7, buffer pH 9, 1.0 mM HCl, and 1.0 mM NaOH.

CAUTION

HCl is a strong caustic acid, and NaOH is a strong caustic base. Follow your instructor's directions for handling, dispensing and disposing of these chemicals. Rinse immediately with water if you spill any acid or base on your skin.

4. Add distilled water and hydrogen peroxide to each tube as listed in table 10.3. If you are measuring by drops, then 1 mL equals about 20 medium-sized drops. Wait 2 min before proceeding to step 5.
5. Add 1 mL of HCl to tubes 4 and 9. Verify that the pH is approximately 3 or lower.
6. Add 1 mL of NaOH to tubes 8 and 10. Verify that the pH is approximately 11 or higher.
7. Add 1 mL of the buffer solutions as indicated in table 10.3.

TABLE 10.4

PRODUCTION OF OXYGEN BY CATALASE ACTIVITY. QUALITATIVE DATA ARE OBSERVATIONS OF INTENSITY OF OXYGEN EFFERVESCENCE RANGING FROM 1–5. QUANTITATIVE DATA ARE MILLILITERS OF OXYGEN PRODUCED.

Tube	Oxygen Production		Explanation
	Qualitative (0–5)	Quantitative (mL O_2)	
1			
2			
3			
4			
5			
6			
7			
8			
9			
10			

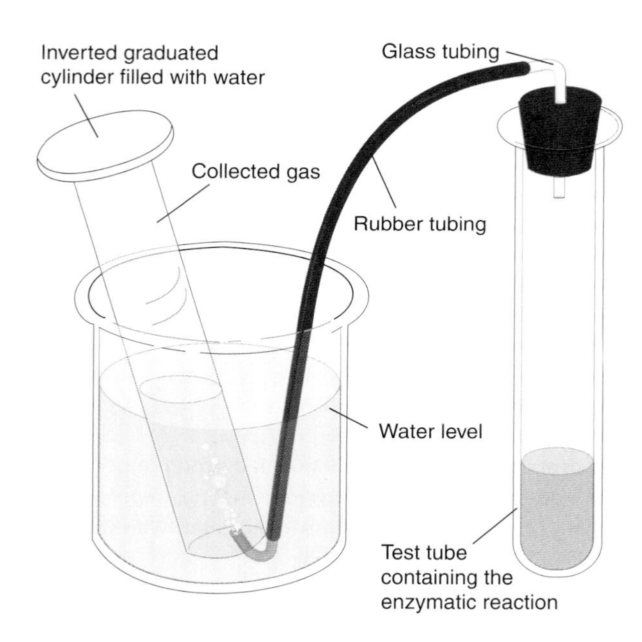

Figure 10.4
A method to capture oxygen released by catalase activity.

8. Your instructor may ask you to verify that the buffers produce the indicated pH. If so, use pH paper to measure the values for each solution and record them in table 10.3.
9. Beginning with tube 1, add the volume of catalase solution as indicated in table 10.3 (note that no catalase is added to tubes 1 or 3). Swirl gently and immediately record in table 10.4 qualitative changes in the bubbling intensity of oxygen production on a scale of 0 (no bubbling) to 5 (vigorous bubbling).
10. If your instructor asks you to more rigorously quantify your results, then immediately after adding the catalase place a stopper with tubing over each tube to collect and measure the volume of gases produced in a water-filled graduated cylinder inverted in a beaker of water (fig. 10.4). Be sure the cylinder is vertical when you measure volume. Record these results in table 10.4.
11. Repeat step 5 for each remaining solution.
12. After you have gathered your data for all 10 tubes, record in table 10.4 your explanation for the results of the catalase activity in each of the tubes.
13. Clean your work area and materials. Follow your instructor's directions concerning the disposal of waste solutions containing HCl and NaOH.

Question 2

a. What were the enzyme, substrate, and product of the enzymatic reaction?

b. What was the purpose of completing steps 1–8 for all tubes before adding the catalase in step 9?

c. What was the purpose of tubes 1, 2, 3, 9, and 10?

d. Use your data for tubes 4–8 to construct a line graph of *Enzyme Activity* vs. *pH*. Refer to the discussion of graphs in Appendix III.

e. Over what pH range was catalase active?

f. What pH levels denatured catalase? Specifically how do solutions of high or low pH change an enzyme's reactivity?

g. At which of the tested pH values did catalase react most rapidly? Should more values be tested to accurately determine its optimum?

h. After experimenting with the effects of pH on enzymes, would you suspect that human blood has a constant pH? Why? What would be the adaptive advantage of this?

INHIBITORS AFFECT THE ACTIVITY OF ENZYMES

Peroxidase is an enzyme in plants (such as turnips) and some bacteria that converts toxic hydrogen peroxide to H_2O and O_2 in a reaction similar to that of catalase. Peroxidase is a large protein with a reactive iron atom at its active site.

$$2\ H_2O_2 \xrightarrow{\text{peroxidase}} 2\ H_2O + O_2$$

Enzymes such as peroxidase can be inhibited by chemicals in various ways. One mechanism is **competitive inhibition.** Competitive inhibitors are molecules that are structurally similar to the substrate and compete for a position at the active site of an enzyme. This ties up the enzyme, thereby making it unavailable to bind with the substrate. For example, hydroxylamine ($HONH_2$) is structurally similar to hydrogen peroxide (H_2O_2) and binds to the iron atom at the active site of peroxidase. Thus, hydroxylamine competes with hydrogen peroxide for the active site on peroxidase, thereby preventing peroxidase from binding with hydrogen peroxide. This inhibits the reaction. A high enough concentration of enzyme with a constant concentration of inhibitor will reduce the inhibition.

The production of oxygen by peroxidase provides a method to measure the ongoing reaction rate. One method would be to capture liberated bubbles of oxygen and measure their total volume. But in the following procedure you will measure oxygen by combining it with a dye that changes color when it is oxidized. Guaiacol is a convenient dye that turns from colorless to brown as it is oxidized by oxygen. The amount of brown color in the final product is proportional to the amount of oxygen formed by the reaction. You can measure the color change qualitatively by rating the color of a reacting solution on an arbitrary scale from 0 to 5, or quantitatively with a spectrophotometer by measuring the solution's absorbance of 470 nm light. Review Exercise 7 for instructions on using a spectrophotometer.

$$O_2 + \underset{\text{(colorless)}}{\text{guaiacol}} \longrightarrow \underset{\text{(brown)}}{\text{oxidized guaiacol}}$$

Procedure 10.3

Observe the effects of an inhibitor on enzymatic activity

1. Prepare turnip extract.
   a. Thoroughly blend 6 g of the inner portion of a peeled turnip in a blender with 200 mL of cold water.
   b. Filter the turnip slurry through cheesecloth into a beaker.
   c. Pour about 7 mL of the extract into a test tube and determine its absorbance in a spectrophotometer. Refer to Exercise 7 for instructions on spectrophotometry. The absorbance for the turnip extract should be between 0.1 and 0.2 at 470 nm to give a reasonable concentration of enzyme.

10–7

TABLE 10.5

EXPERIMENTAL CONDITIONS TO TEST THE INHIBITION OF HYDROXYLAMINE ON PEROXIDASE ACTIVITY

Tube	Distilled Water	Guaiacol (25 mM)	Hydrogen Peroxide (3%)	Turnip Extract	Hydroxylamine (10%)
1	5.9 mL	0.1 mL			
2	5.8 mL		0.2 mL		
3	5.7 mL	0.1 mL	0.2 mL		
4	4.9 mL	0.1 mL		1.0 mL	
5	4.7 mL	0.1 mL	0.2 mL	1.0 mL	
6	4.2 mL	0.1 mL	0.2 mL	1.0 mL	0.5 mL
7	3.7 mL	0.1 mL	0.2 mL	1.5 mL	0.5 mL
8	3.2 mL	0.1 mL	0.2 mL	2.0 mL	0.5 mL
9	2.2 mL	0.1 mL	0.2 mL	3.0 mL	0.5 mL

TABLE 10.6

ABSORBANCE AT 470 NM OF PEROXIDE/PEROXIDASE SOLUTIONS

Tube	0.0 min	0.5 min	1.0 min	1.5 min	2.0 min	2.5 min	3.0 min	3.5 min	4.0 min	4.5 min	5.0 min
1											
2											
3											
4											
5											
6											
7											
8											
9											

d. Dilute or concentrate the suspension as necessary. Your instructor may provide directions for standardizing this enzyme solution more precisely.

2. Obtain nine test tubes and number them at the top 1–9.
3. Add distilled water to each tube as listed in table 10.5.
4. To tube 1, add 0.1 mL (2 drops) of guaiacol as listed in table 10.5, swirl the contents.
5. Immediately determine the solution's absorbance at 470 nm using a spectrophotometer. Record the absorbance value in table 10.6. Measure the absorbance every 30 sec for 5 min and record the value each time.
6. To tube 2, add 0.2 mL (4 drops) of hydrogen peroxide as listed in table 10.5. Swirl the contents. Repeat step 5 quickly.
7. To tube 3, add 0.1 mL (2 drops) of guaiacol and 0.2 mL of hydrogen peroxide as listed in table 10.5. Swirl the contents. Repeat step 5 quickly.
8. To tube 4, add 0.1 mL (2 drops) of guaiacol and 1.0 mL of turnip extract as listed in table 10.5. Swirl the contents. Repeat step 5 quickly.

9. Complete all of the measurements for steps 4–8 before proceeding to step 10.
10. For each of tubes 5–9, add 0.1 mL (2 drops) of guaiacol and 0.2 mL (4 drops) of hydrogen peroxide as listed in table 10.5.
11. For tube 5, add 1.0 mL of turnip extract and swirl the contents. Repeat step 5 quickly.
12. For tube 6, add 1.0 mL of turnip extract and 0.5 mL (10 drops) of hydroxylamine and swirl the contents. Repeat step 5 quickly.
13. For tube 7, add 1.5 mL of turnip extract and 0.5 mL (10 drops) of hydroxylamine and swirl the contents. Repeat step 5 quickly.
14. For tube 8, add 2.0 mL of turnip extract and 0.5 mL (10 drops) of hydroxylamine and swirl the contents. Repeat step 5 quickly.
15. For tube 9, add 3.0 mL of turnip extract and 0.5 mL (10 drops) of hydroxylamine and swirl the contents. Repeat step 5 quickly.
16. Clean your work area and materials.

Question 3

a. What were the enzyme, substrate, and product of this enzymatic reaction?

b. Explain the results you observed for tubes 1, 2, 3, and 4. What was the purpose of these tubes?

c. Use your data for tubes 5–9 to construct a line graph of *Enzyme Activity (Absorbance)* vs. *Time*. There will be five curves on the graph. Refer to the discussion of graphs in Appendix III. You will not graph the values for tubes 1–4.

d. For which tubes is peroxidase still active after 5 minutes?

e. How does hydroxylamine affect peroxidase activity?

f. Was it possible to detect peroxidase activity in the presence of the inhibitor by increasing enzyme concentration? Why or why not?

g. Inhibitors are common in biological systems. Why might some organisms release enzyme inhibitors into their surrounding environment?

Questions for Further Thought and Study

1. More substrate increases the probability that an enzyme will contact substrate and should increase the enzymatic reaction rate. How do you explain the increase in time to complete hydrolysis when more substrate was present?

2. What term describes the alteration of an enzyme's structure? What factors in addition to temperature influence a protein's structure?

DOING BIOLOGY YOURSELF

Review the structure of starch and the action of the enzyme amylase. Design an experiment that uses a spectrophotometer to detect the progress and completion of hydrolysis of starch by amylase.

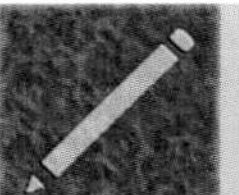

WRITING TO LEARN BIOLOGY

Propose a mechanism involving enzyme production by which a cell could counteract a sudden increase in the amount of substrate.

10–12

11

Respiration

Aerobic and Anaerobic Oxidation of Organic Molecules

Objectives

By the end of this exercise you should be able to:

1. Demonstrate carbon dioxide production during anaerobic respiration.
2. Understand the effects of inhibitors, intermediate compounds, and cofactors in anaerobic respiration.
3. Determine oxygen consumption during aerobic respiration.
4. Use a pH-indicator to measure the relative production of carbon dioxide by plants and animals.
5. Use a respirometer to determine the metabolic rate of an animal.
6. Demonstrate practical applications of anaerobic respiration such as making wine and kimchee.

All living organisms respire. Some need oxygen to do it, some don't, but they all respire because all organisms need usable chemical energy to fuel their life processes. Respiration is the chemistry that provides that energy. Usually the organic carbon molecules are the energy source and CO_2 and H_2O are released as waste. Humans release the waste as they exhale. Respiring yeast don't exhale, but they can "pump up" rising bread by liberating CO_2 as the yeast breaks down sugar (fig. 11.1).

Cellular respiration involves oxidation of organic molecules and a concomitant release of energy. Some of this energy is stored in chemical bonds of **adenosine triphosphate (ATP),** which is used later as a direct source of energy for cellular metabolism. Organisms use the energy stored in ATP to do work such as transport, synthesize new compounds, reproduce, contract muscles, and remove wastes.

Photosynthesis, the topic of Exercise 12, uses light energy to split H_2O and harvest high-energy electrons. These energetic electrons (and accompanying H^+) are passed to CO_2, thereby reducing CO_2 to energy-storing sugars. Respiration removes electrons from (i.e., oxidizes) glucose, captures some of the energy in ATP, and ultimately passes the electrons to oxygen to form H_2O.

In most cells, respiration begins with the oxidation of glucose to pyruvate via a set of chemical reactions called **glycolysis** (fig. 11.2). During glycolysis, some of the energy released from each glucose molecule is stored in ATP. Glycolysis occurs with or without oxygen. If oxygen is present, most organisms continue respiration by oxidizing pyruvate to CO_2 via chemical reactions of the **citric acid cycle** (also known as the **tricarboxylic acid cycle** and **Krebs cycle**). Organisms that use oxygen for respiration beyond glycolysis are called **aerobes.**

As aerobes oxidize pyruvate in the citric acid cycle, they store energy in electron carriers such as NAD (nicotinamide adenine dinucleotide). Specifically, aerobes store energy by reducing (adding high-energy electrons to) NAD and related compounds. These compounds later transfer their high-energy electrons to a series of compounds collectively called the electron transport chain. The **electron transport chain** generates proton gradients from energy stored in reduced NAD and related compounds to form approximately 18-times more ATP than that formed in glycolysis. Oxygen, which is the final electron-acceptor in the electron transport chain, is reduced to form H_2O. Without oxygen to accept electrons passed through the electron transport chain, the chain is not functional and an aerobic organism will quickly die.

Question 1

Why must aerobic organisms such as yourself inhale oxygen and exhale CO_2?

Other organisms called **anaerobes** live without oxygen and may even be killed by oxygen in the atmosphere. Some of these anaerobes are primitive bacteria that gather their energy with a pathway of anaerobic respiration that uses inorganic electron acceptors other than oxygen. For example, many bacteria use nitrate, sulfate, or other inorganic compounds as the electron acceptor instead of oxygen. Other anaerobes use glycolysis, but the pyruvate from glycolysis is reduced via anaerobic **fermentation** to either CO_2 and ethanol (in plants and some microbes such as yeast) or lactic acid (in other microbes and oxygen-stressed muscles of animals). We can summarize aerobic and anaerobic fermentation in the equations below:

Summary Equation for Aerobic Respiration

$$\underset{\text{Glucose}}{C_6H_{12}O_6} + \underset{\text{Oxygen}}{6\,O_2} \longrightarrow \underset{\text{Carbon Dioxide}}{6\,CO_2} + \underset{\text{Water}}{6\,H_2O} + \text{ATP} + \text{Heat}$$

Anaerobic Fermentation in Plants and Some Microbes

$$\underset{\text{Glucose}}{C_6H_{12}O_6} \longrightarrow \underset{\text{Ethanol}}{2\,C_2H_5OH} + \underset{\text{Carbon Dioxide}}{2\,CO_2} + \text{ATP} + \text{Heat}$$

Anaerobic Fermentation in Animals and Some Microbes

$$\underset{\text{Glucose}}{C_6H_{12}O_6} \longrightarrow \underset{\text{Lactic Acid}}{2\,CH_3CHOHCOOH} + \text{ATP} + \text{Heat}$$

Notice from these equations that plants (as well as microbial prokaryotes and eukaryotes such as yeasts) can temporarily conduct anaerobic fermentation that reduces pyruvate from glycolysis to ethanol and carbon dioxide. This occurs, for example, in roots that penetrate anaerobic soils and sediments.

Anaerobic fermentation does not involve or benefit from the additional ATP produced by the citric acid cycle or electron transport chain. Thus, the ability of an organism to live in the absence of oxygen comes at a price: Anaerobic fermentation produces 18-fold fewer ATP per glucose molecule than does aerobic respiration.

Question 2

What are the advantages and disadvantages of anaerobic fermentation?

Figure 11.1

Bread dough rises because respiring yeasts break down sugars to obtain their energy for growth and liberate CO_2, thereby forming small bubbles in the dough. The lower loaf has been rising 4 hr longer than the upper loaf.

In today's exercise, you will study the major features of cellular respiration. Let's begin with a type of anaerobic fermentation with which you are already familiar: alcoholic fermentation by yeast.

PRODUCTION OF CO_2 DURING ANAEROBIC FERMENTATION

Yeast are fungi used in baking and producing alcoholic beverages. They can respire in the absence of O_2 and can oxidize glucose to ethanol and CO_2. To demonstrate CO_2 production during anaerobic fermentation by yeast (fermentation), follow procedure 11.1. In this procedure you will observe the effects of these compounds on respiration:

- **Pyruvate**—a product of glycolysis; pyruvate is reduced to ethanol or lactic acid during anaerobic fermentation
- **Magnesium sulfate ($MgSO_4$)**—provides Mg^{2+}, a cofactor that activates some enzymes of glycolysis
- **Sodium fluoride (NaF)**—an inhibitor of some enzymes of glycolysis
- **Glucose**—a common organic molecule used as an energy source for respiration

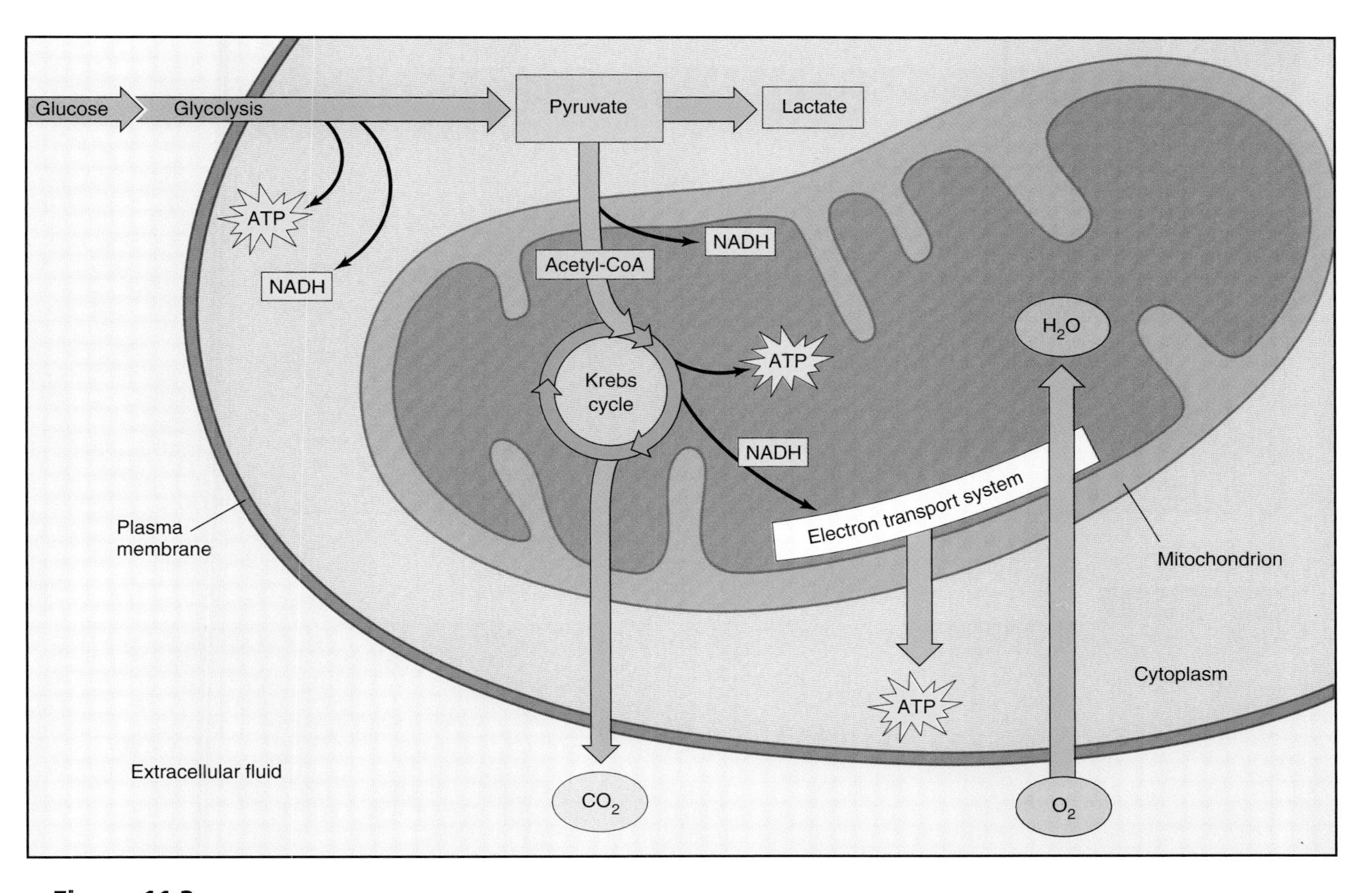

Figure 11.2

An overview of aerobic respiration. Glycolysis occurs in the cytoplasm, and the Krebs cycle and electron transport chain occur in mitochondria.

Procedure 11.1

Demonstrate CO_2 production during anaerobic fermentation

1. Label seven fermentation tubes and add the solutions listed in table 11.1.
2. Completely fill the remaining volume in fermentation tubes 1–6 with the yeast suspension provided. Fill the remaining volume in tube 7 completely with distilled water. Your instructor will demonstrate how to tip the fermentation tube to remove air bubbles.
3. Cover the outlet of each tube with a piece of plastic film or insert a foam plug.
4. Make a pinhole in the film to release pressure.
5. Incubate the tubes at 50°C for 40 min.
6. After 40 min measure the height (in millimeters) of the bubble of accumulated CO_2. Record your results in table 11.1.
7. The effects of pyruvate, $MgSO_4$, NaF, and glucose on CO_2 production are best determined by pair-wise comparisons of tubes rather than by ranking all of the treatment tubes. Determine which pairs of tubes best reveal the effects of each variable, and then complete table 11.2.

CAUTION

NaF is a poison. Handle it carefully.

Question 3

a. What was the purpose of tube 7?

b. How was the effect of concentration of inhibitor tested in this experiment? How did the concentration of NaF affect anaerobic fermentation in your experiment? Why?

TABLE 11.1

EXPERIMENTAL TREATMENTS AND CO_2 PRODUCTION DURING ANAEROBIC FERMENTATION

Tube	3 M Na Pyruvate (Activator)	0.1 M $MgSO_4$ (Activator)	0.1 M NaF (Inhibitor)	5.0% Glucose (Activator)	Water	Fill With	CO_2 Produced After 40 min (mm)
1	—	—	—	—	7.5 mL	Yeast suspension	______
2	—	—	—	2.5 mL	5.0 mL	Yeast suspension	______
3	—	5.0 mL	—	2.5 mL	—	Yeast suspension	______
4	—	—	0.5 mL	2.5 mL	4.5 mL	Yeast suspension	______
5	—	—	5.0 mL	2.5 mL	—	Yeast suspension	______
6	2.5 mL	—	2.5 mL	2.5 mL	—	Yeast suspension	______
7	—	—	—	2.5 mL	2.5 mL	Water	______

TABLE 11.2

EFFECTS OF FOUR CHEMICAL VARIABLES ON CO_2 PRODUCTION DURING ANAEROBIC FERMENTATION

Variable	Comparison Tubes	Effect of Variable on Respiration Rate	Mechanism for the Effect
Yeast			
Glucose			
NaF			
Na Pyruvate			
$MgSO_4$			

c. Which compounds in the list are intermediates in the respiratory pathway?

d. Why did tube 6 produce CO_2 even though an inhibitor of glycolysis was present?

e. Did magnesium (a cofactor that activates many enzymes) promote respiration? If not, what are some possible reasons?

f. Smell the contents of the tube containing the most CO_2. What compound do you smell?

g. What is the economic importance of fermentation by yeast?

h. What gas is responsible for the holes in baked bread?

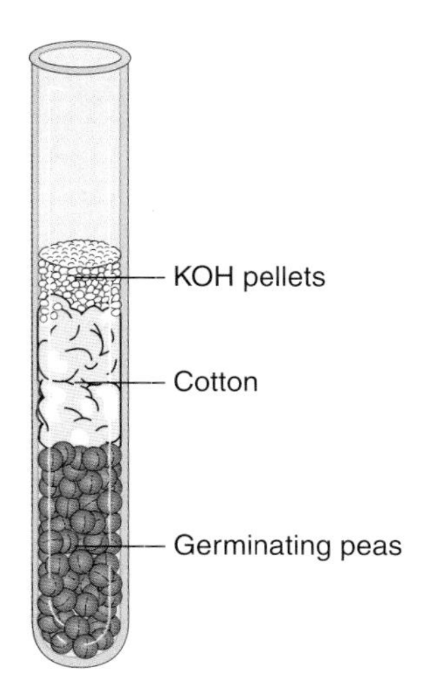

Figure 11.3
Test tube containing germinating peas, cotton, and KOH pellets.

If time and facilities are available, repeat procedure 11.1 and incubate the tubes at 4°C (refrigerator), 20°C (incubator), and/or 55°C (incubator). Use your data to explain the effect of temperature on fermentation by yeast.

OXYGEN CONSUMPTION DURING AEROBIC RESPIRATION

Aerobic respiration uses oxygen as the terminal electron-acceptor in the electron transport chain. Because this oxygen is reduced to water, you can measure aerobic respiration by measuring the consumption of oxygen. During respiration, CO_2 is produced while O_2 is consumed. In the following experiment, KOH is used to absorb the CO_2. Therefore, the net change in gas volume is a measure of oxygen consumption.

Procedure 11.2

Determine oxygen consumption during aerobic respiration (may be done as a demonstration)

1. Fill a test tube or flask half-full with germinating peas and another half-full with heat-killed peas. The germinating peas have been soaked in water in the dark for three to four days.
2. Cover the contents of each tube with a loose-fitting plug of cotton.
3. Cover the cotton with approximately 1 cm of loosely packed pellets of potassium hydroxide (KOH) (fig. 11.3).
4. Place a stopper containing a capillary tube or graduated pipet with an attached outlet tube into both tubes containing peas (fig. 11.4).
5. Cover the tube with foil to prevent light and photosynthesis.

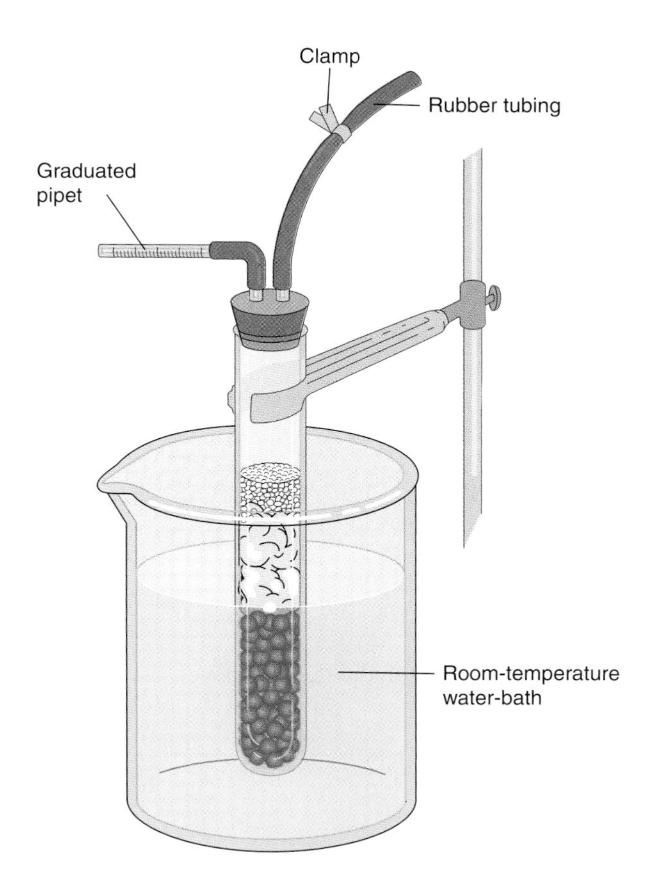

Figure 11.4
Test tube with stopper having capillary tubes attached. The tube will be covered in foil.

6. Vertically clamp the tubes to a ring stand so that the bottom of each tube is submerged in a room-temperature water-bath. The water-bath will minimize temperature fluctuations in the tube.
7. Use a Pasteur pipet to inject enough dye into each capillary tube so that approximately 1 cm of dye is drawn into each capillary tube.
8. After waiting 1 min for equilibration, attach a pinch clamp to the outlet tube and mark the position of the dye with a wax pencil.
9. Use a wax pencil to mark the position of the dye every 10 min for the next 30 min.
10. After each time interval measure the distance the fluid moved from its starting point; record your data in table 11.3.
11. Remove the pinch clamp from the outlet valve and return the dye to the end of the capillary tube by tilting the capillary tube.
12. Repeat steps 1–11 using tubes incubated in an ice bath and warm water-bath. Record your results in table 11.3.

TABLE 11.3

OXYGEN CONSUMPTION BY SEEDS AT THREE TEMPERATURES

		mL O2 Consumed					
		10 min		20 min		30 min	
Treatment	0 min	Alive	Heat-Killed	Alive	Heat-Killed	Alive	Heat-Killed
Room temperature	0	______	______	______	______	______	______
Ice bath	0	______	______	______	______	______	______
Warm water-bath	0	______	______	______	______	______	______

Question 4

a. What was the purpose of adding heat-killed peas to a tube?

b. In which direction did the dye move?

c. Why?

d. What does this experiment tell you about the influence of temperature on oxygen consumption during cellular respiration?

PRODUCTION OF CO_2 DURING AEROBIC RESPIRATION

CO_2 produced during cellular respiration can combine with water to form carbonic acid:

$$\underset{\text{Carbon dioxide}}{CO_2} + \underset{\text{Water}}{H_2O} \longleftrightarrow \underset{\text{Carbonic acid}}{H_2CO_3}$$

In this procedure (fig. 11.5) you will use phenolphthalein to detect changes in pH resulting from the production of CO_2 (and, therefore, carbonic acid) during cellular respiration. Phenolphthalein is red in basic solutions and colorless in acidic solutions. Thus, you can monitor cellular respiration by measuring acid production as change in pH. **pH** is a measure of the acidic or basic properties of a solution; pH 7 is neutral. Solutions having a pH < 7 are acidic, and solutions having a pH > 7 are basic (see Exercise 4).

In procedure 11.3 you will not directly measure the volume of CO_2 produced by a respiring organism. Instead, you will measure the volume of NaOH used to neutralize the carbonic acid produced by the CO_2, and thereby calculate a relative measure of respiration.

Question 5

The organisms you will study include a plant (*Elodea*) and an animal (snail). Which do you think will respire more? Write your hypothesis here:

Procedure 11.3

Measure relative CO_2 production by aerobic organisms

Experimental Setup

1. Obtain 225 mL of culture solution provided by your instructor. This solution has been dechlorinated and adjusted to be slightly acidic.
2. Place 75 mL of this solution in each of three labeled beakers.
3. Obtain the organisms listed in table 11.4 from your instructor and determine the volume of each organism by following steps 4–6.

Determine Volume by Water Displacement

4. Place exactly 25 mL of water in a 50-mL graduated cylinder.
5. Place the organism in the cylinder and note the increase in volume above the original 25 mL. This increase equals the volume of the organism.
6. Record the volumes in table 11.4. Gently place the organism in the appropriate beaker.

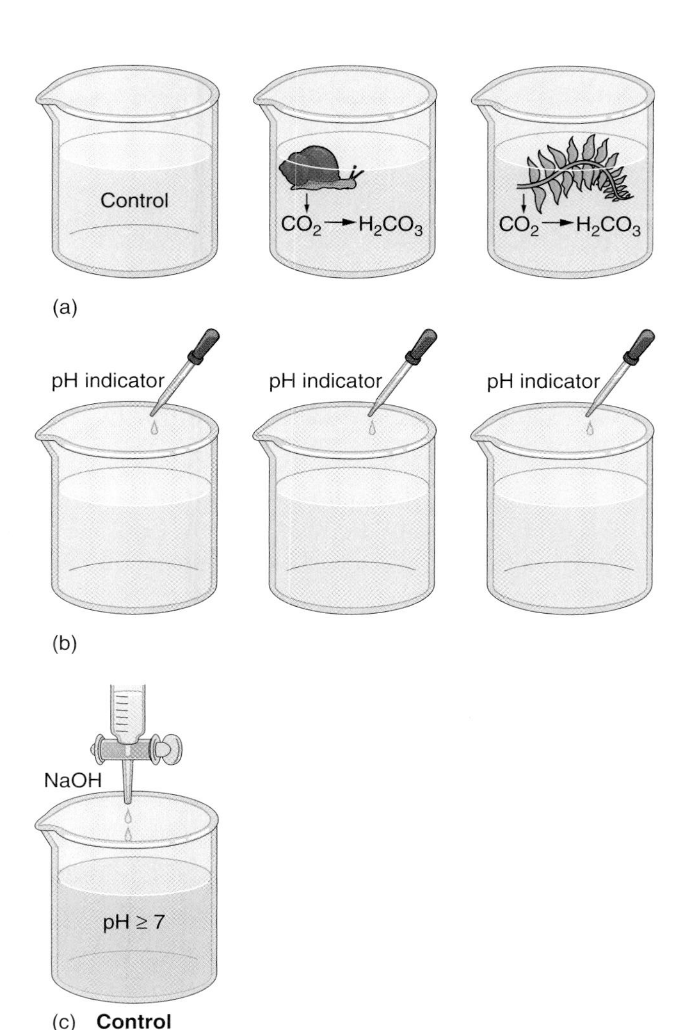

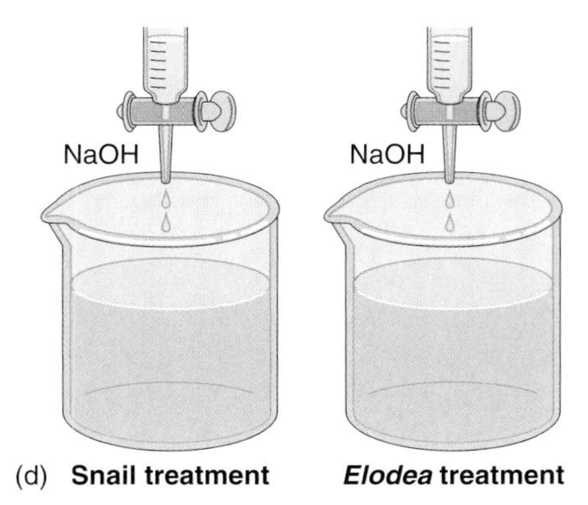

Figure 11.5

The procedure to determine the relative respiration rates of a plant and animal. (*a*) During respiration, organisms release CO_2 which combines with water to form carbonic acid (H_2CO_3). (*b*) The acidic solutions remain colorless after addition of phenolphthalein, a pH indicator. (*c*) Titration of the control with NaOH will make the solution basic and pink when the pH reaches the endpoint of phenolphthalein. (*d*) The treatment solutions are then titrated to the pink endpoint matching the control. The volume of NaOH needed to reach the endpoint indicates the relative amounts of dissolved CO_2 that were produced during respiration.

Incubate Experimental Treatments

7. Cover each beaker with a plastic film or petri dish top and set them aside on your lab bench. Place the beaker containing the *Elodea* in the dark by covering it with a coffee can or aluminum foil.
8. Allow the organisms to respire for 15 min.
9. Gently remove the organisms from the beakers and return them to their original culture bowls.

Titrate to Gather Your Raw Data

10. Add four drops of phenolphthalein to the contents of each beaker. The solutions should remain clear because the solutions are acidic.
11. Obtain a burette or dropper bottle to dispense NaOH (2.5 mM). Add NaOH drop by drop to the contents of the control beaker. Thoroughly mix the contents of the beaker after adding each drop. Record in table 11.4 the milliliters of NaOH required to reach the endpoint of phenolphthalein (i.e., make the solution pink).
12. Repeat step 11 for beaker 1; be sure to add NaOH only until the solution is the same shade of pink as the control beaker. Record the number of milliliters of NaOH added to beaker 1 in table 11.4.
13. Repeat step 11 for beaker 2.

Calculate Your Results

14. For beaker 1, determine the relative respiration rate for organisms by subtracting the milliliters NaOH added to the control beaker from the milliliters NaOH added to beaker 1. Record this value in table 11.4.
15. Repeat step 14 for beaker 2.
16. For beakers 1 and 2, determine the respiration rate per milliliter of organism by dividing the relative respiration rate for organisms by the volume of the organism(s). Record these values in table 11.4.

Question 6

a. In this exercise you measured the relative respiration rates of a plant and an animal. Why were you cautioned about having no algae in the control beaker?

b. Before you gathered your raw data, you formulated a hypothesis about the expected results. After considering your data, do you accept or reject your hypothesis? Why?

TABLE 11.4

DATA FOR MEASURING CO_2 PRODUCTION DURING RESPIRATION

Organisms	Total Volume of Organisms (mL)	Milliliters of NaOH to Reach Endpoint (mL NaOH)	Relative Respiration Rate of Organisms (mL NaOH)	Respiration Rate per Milliliter of Organism (mL NaOH/mL organisms)
Beaker 1: 4 snails	______	______	______	______
Beaker 2: *Elodea*	______	______	______	______
Control beaker	0	______	0	0

c. What is your major conclusion from the results of this procedure?

d. What features of the biology of the organisms that you used most likely contributed to the observed differences in respiration rate?

e. Do you feel justified in drawing conclusions from your work about all plants and animals? Or only about snails and *Elodea*? Why?

f. How would you expand this experiment to further test your conclusions about other plants and other animals?

g. What other organisms might you include in an expanded experiment? Why did you choose these organisms?

DEMONSTRATION: DETERMINATION OF METABOLIC RATE

The rate of O_2 uptake during cellular respiration indicates the metabolic rate of an organism. In procedure 11.4 you will measure O_2 uptake by measuring changes in air pressure as O_2 is removed from the air by a respiring mouse. Changes in air pressure can be attributed primarily to O_2 consumption (rather than CO_2 production or exhalation of water vapor) only if exhaled CO_2 and H_2O are removed from the air. This is accomplished by adding ascarite (which adsorbs CO_2) and drierite (which adsorbs H_2O) to the experimental setup (fig. 11.6). Use procedure 11.4 to estimate the metabolic rate of a mouse.

Procedure 11.4

Estimate the metabolic rate of a mouse

1. Weigh a mouse to the nearest 0.1 g. Record this weight in table 11.5 and place the mouse in the jar of a respirometer (fig. 11.6). Use a fan to circulate air in the jar and allow the mouse to get accustomed to the jar.
2. Attach a 10-mL syringe that is filled with air to the respirometer.
3. Seal the respirometer jar with a lid. Then close the air escape line with a clamp and record the position of the dye solution in the right column of the curved capillary tube. This tube is called a manometer.
4. Inject 10 mL of air into the respirometer. The level of dye in the right side of the manometer will rise because of the increased presence of air. Record the position of the dye solution in table 11.5.
5. Allow the mouse to respire. The air pressure in the respirometer will decrease as O_2 is consumed, and the dye level in the right column of the manometer will decrease.
6. Record in table 11.5 the elapsed time for the dye level to return to its original position. This is the time for the mouse to consume 10 mL of O_2.
7. Gently return the mouse to its cage.
8. Calculate and record in table 11.5 the mouse's metabolic rate in kcal/day, assuming that 4.8 kcal of energy are used for each liter of O_2 that is consumed.
9. Calculate and record the predicted metabolic rate obtained from the following general equation for metabolic rate of small mammals:

Predicted metabolic rate = 70 × (body weight in kg)$^{3/4}$

10. Compare your experimental value with the predicted value for metabolic rate.

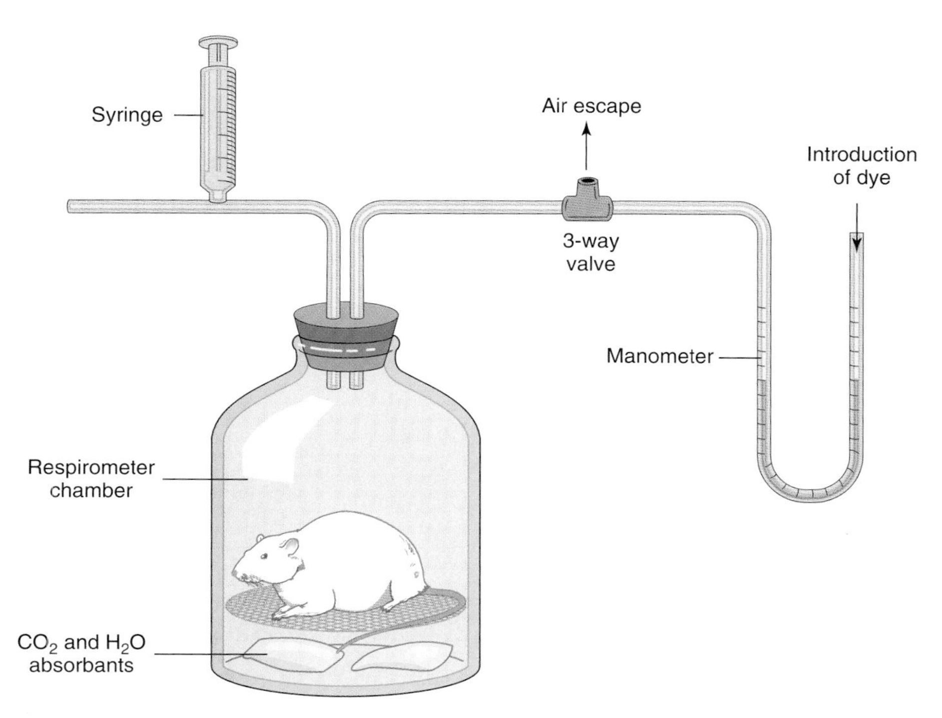

Figure 11.6
Respirometer with mouse.

TABLE 11.5

DATA FOR DETERMINATION OF METABOLIC RATE OF A RESPIRING MOUSE

Observations and Data

Weight of mouse: __________ grams

Initial position of dye solution: __________

Position of dye solution after injection of 10 mL of air: __________

Minutes for dye level to return to initial position: (minutes per 10 mL oxygen): __________ min

Calculations

A	Minutes to consume 1 liter of O_2 = (minutes per 10 mL oxygen) × 100 = __________ min
B	Liters of O_2 consumed per day = 1440 minutes per day ÷ A = __________ liters per day
C	Experimental metabolic rate as kcal per day = B × 4.8 kcal per liter O_2 = __________ kcal per day
D	Predicted metabolic rate = 70 × (weight of mouse)$^{3/4}$ = __________ kcal per day

Question 7

a. Is the predicted metabolic rate similar to that which you determined experimentally?

b. What could cause any differences in these values?

c. Determine the metabolic rate of other organisms available in the lab. How do their metabolic rates compare with that of a mouse?

INVESTIGATION

Effects of Environmental Stimuli on Respiration Rates

Respiration, like all biochemical processes, is sensitive to environmental stimuli. In the following investigation, use *Elodea* and/or snails to better understand how environmental stimuli affect respiration.

a. Choose a stimulus you would like to test (e.g., temperature, salinity, acidity, etc.).

b. Form a hypothesis about the result you expect. Your instructor will advise you about how to write a testable hypothesis. Write your hypothesis here:

c. Decide how you will test your hypothesis. Describe your experimental design here:

d. Do your experiment. What did you conclude? Do your data support your hypothesis?

APPLICATIONS OF ANAEROBIC RESPIRATION

Making Wine

In this exercise you've seen how easy it is to demonstrate alcoholic fermentation by yeast. Many biologists as well as nonbiologists use this reaction to make their own wine. If you're game for an introduction to home wine making, try the following procedure.

Procedure 11.5

Making wine

1. Thoroughly clean and sterilize all glassware.
2. Combine a cake of yeast with either bottled grape juice or cranberry juice. Mix the yeast and juice in a ratio of approximately 5 liters of juice to 1 gram of yeast.
3. Add approximately 650 mL of the juice-yeast mix to each of four 1-liter Erlenmeyer flasks (or use 1- to 2-liter recycled plastic pop bottles).
4. Dissolve the following amounts of sucrose in each flask:
   Flask 1: 75 g
   Flask 2: 150 g
   Flask 3: 300 g
   Flask 4: no sucrose
5. Set up the fermentation apparatus as shown in figure 11.7 for each flask.
6. Be sure to keep the procedure anaerobic by keeping the end of the exit tube under water in the adjacent flasks. This will prevent contamination by airborne bacteria and yeast.
7. Incubate the flasks at temperatures between 15°C and 22°C. Although fermentation will continue for a month or so, most fermentation will occur within the first 14 days. Fermentation is complete when bubbling stops.
8. To test your wine, remove the stopper and use a piece of tubing to siphon off the wine solution without disturbing the sediment in the bottom of the flasks. You may then want to filter the solution to remove any remaining yeast cells from the wine.
9. Taste your wine. What differences are there in wines produced with different amounts of sugar? If your wine has been contaminated by bacteria that produce acetic acid, vinegar may have been formed, so take your first sip cautiously.

If you're interested in the finer points of wine making, visit your local bookstore or library. There may also be a local society of amateur wine makers in your area who will be glad to give you some pointers on creating "a simply delightful bouquet."

Making Kimchee

Pickling is an ancient way of preserving food. Pickling involves the anaerobic fermentation of sugars to lactic acid; this acid lowers the pH of the medium, thereby creating an environment in which other food-spoiling organisms cannot grow. Common foods preserved with pickling include sauerkraut, yogurt, and dill pickles. The ancient Chinese cabbage product kimchee, still a major part of the Korean diet, is also made with pickling. Here's how to make kimchee.

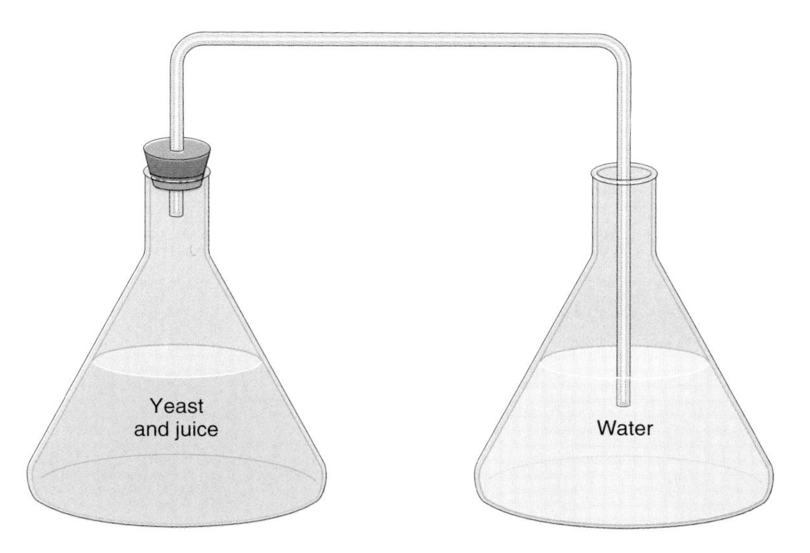

Figure 11.7
Experimental setup for home wine making.

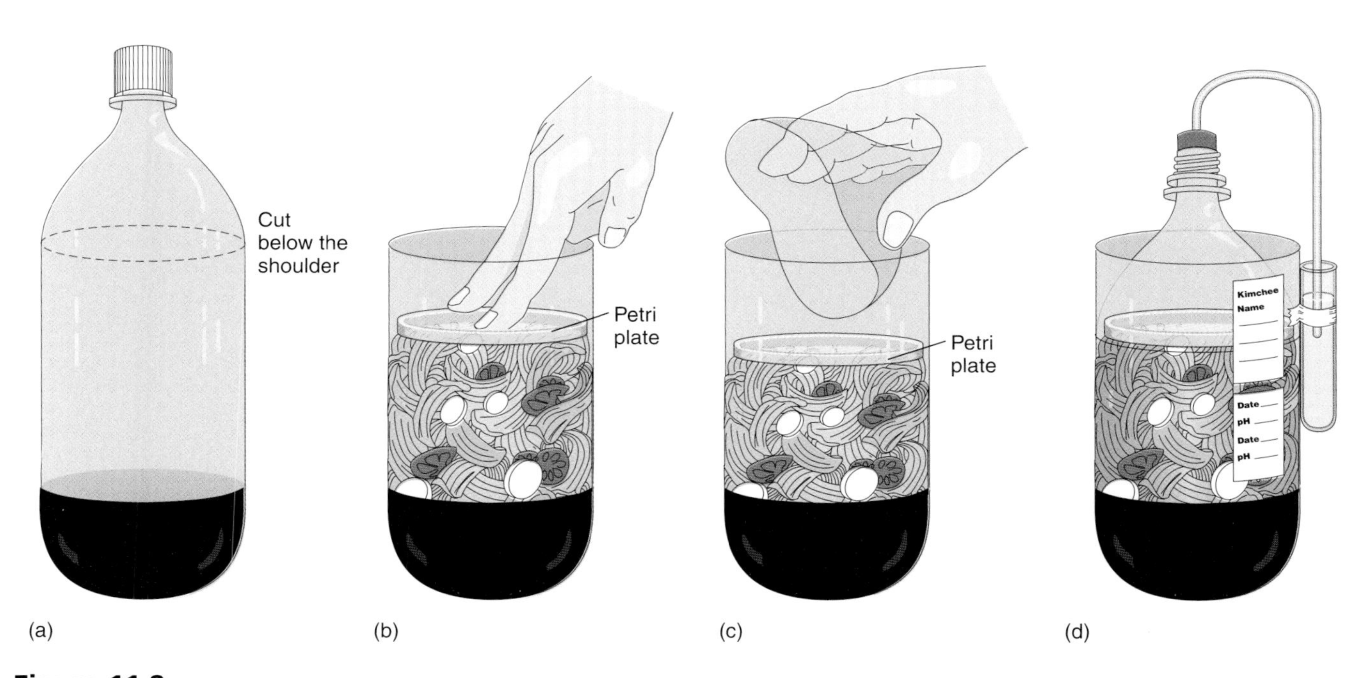

Figure 11.8
Experimental setup for making kimchee. See the text for the recipe and procedure.

Procedure 11.6
Making kimchee

1. Cut a 2-liter bottle just below the shoulder, as shown in figure 11.8*a*.
2. Add alternating layers of cabbage, garlic, pepper, and a sprinkling of salt in the bottle, pressing each layer down until the bottle is full. If you're using chilies or pepper, do not touch your eyes or mouth.
3. Place the lid, rim side up, atop the ingredients. Press down (fig. 11.8*b*). Within a few minutes, the salt will draw liquid from the cabbage; that liquid will begin to accumulate in the bottle.
4. For the next hour or so, continue to press the cabbage. You should then be able to fit the bottle top inside the bottle bottom, forming a sliding seal (fig. 11.8*c*). When you press with the sliding seal, cabbage juice will rise above the petri plate and air will bubble out around the edge of the plate.

5. The cabbage will pack half to two-thirds of the bottle's volume (fig. 11.8*d*). Every day, press on the sliding seal to keep the cabbage covered by a layer of juice. What happens when you press on the cabbage? How do you explain this?

6. Use pH-indicator paper to measure and record the pH of the juice each day (see Exercise 4).

7. After 4 to 7 days (depending on the temperature), the pH will have dropped from about 6.5 to about 3.5. Enjoy your kimchee!

Questions for Further Thought and Study

1. What is the difference between respiration and breathing?

2. Does cellular respiration occur simultaneously with photosynthesis in plants? How could you determine the relative rates of each?

3. What role does cellular respiration play in the metabolism of an organism?

4. What modifications of cellular respiration might you expect to find in dormant seeds?

DOING BIOLOGY YOURSELF

Repeat the procedure to measure relative CO_2 production by aerobic organisms and include in your design an animal such as a fish. Would you expect greater CO_2 production from a fish or a snail? Why?

DOING BIOLOGY YOURSELF

Repeat procedure 11.1 to measure CO_2 production in yeast and incubate the tubes at 4°C (refrigerator), 20°C (incubator), and/or 50°C (incubator). How does temperature affect the rate of fermentation by yeast?

12

Photosynthesis

Pigment Separation, Starch Production, and CO_2 Uptake

Objectives

By the end of this exercise you should be able to:

1. Relate each part of the summary equation for photosynthesis to the synthesis of sugar.
2. Describe the differences between the light-dependent and light-independent reactions involved in photosynthesis.
3. Separate the photosynthetic pigments using paper chromatography and calculate their R_f numbers.
4. Use a spectroscope to describe the absorption of visible light by chlorophyll.
5. Describe fluorescence and its role in photosynthesis.
6. Describe the process of electron transport in chloroplasts and its role in photosynthesis.
7. Describe the change of pH that occurs as plants take up CO_2 from their environment during photosynthesis.
8. Describe the distribution of starch in leaves resulting from photosynthesis relative to the amount of light they receive.

Figure 12.1

The energy that drives photosynthesis comes from the sun. Less than 1% of all the energy that reaches the earth from the sun is captured by photosynthesis. This 1% drives virtually all the activities of life on earth.

Photosynthesis is the most important series of chemical reactions that occurs on earth (fig. 12.1). Indeed, virtually all life depends on photosynthesis for food and oxygen. **Photosynthesis** is a complex chemical process that converts radiant energy (light) to chemical energy (sugar). A summary equation for photosynthesis is given here:

$$\underset{\text{Carbon Dioxide}}{6\,CO_2} + \underset{\text{Water}}{12\,H_2O} \xrightarrow[\textit{chlorophyll}]{\textit{light}} \underset{\text{Sugar}}{C_6H_{12}O_6} + \underset{\text{Water}}{6\,H_2O} + \underset{\text{Oxygen}}{6\,O_2}$$

Thus, photosynthesis is the light-dependent and chlorophyll-dependent conversion of carbon dioxide and water to sugar, water, and oxygen. Oxygen is released to the environment, and sugar is used to fuel growth or is stored as starch, a polysaccharide. Although water is present on both sides of the summary equation, these are not the same water molecules. The "reactant" water molecules (i.e., those on the left side of the equation) are split to release electrons during the photochemical (i.e., light-dependent) reactions. The "product" water molecules (i.e., those on the right side of the equation) are assembled from hydrogen and oxygen released during the photochemical and biochemical (i.e., light-independent) reactions. The photochemical reactions of photosynthesis are often referred to as the "light reactions." The biochemical reactions are often referred to as the "dark reactions" or the Calvin cycle, in honor of Melvin Calvin, the botanist who described the reactions.

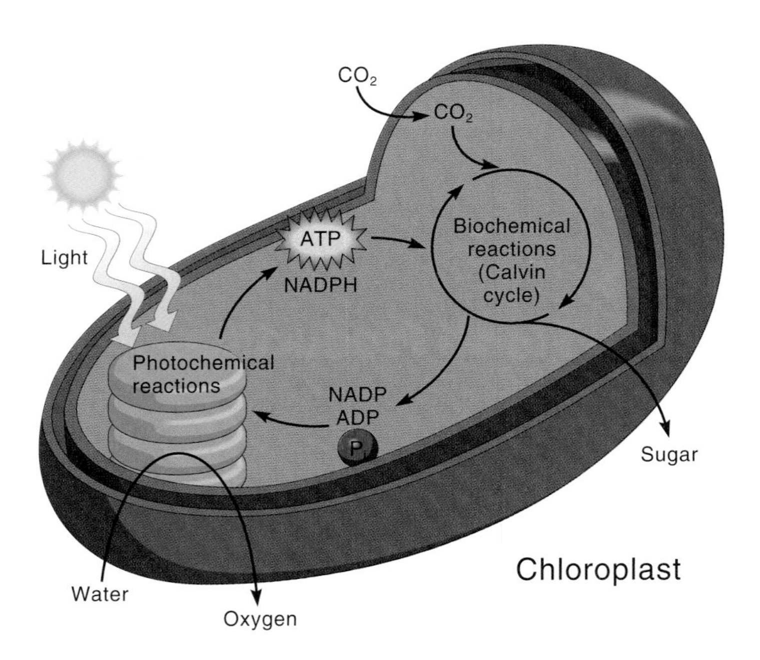

Figure 12.2

Photosynthesis consists of photochemical (the light-dependent "light reactions") and biochemical (the light-independent "dark reactions" including the Calvin cycle) reactions, both of which occur in chloroplasts. The photochemical (i.e., light) reactions convert light-energy to chemical energy captured in ATP and NADPH. The biochemical (i.e., dark) reactions use the ATP and NADPH produced by the photochemical reactions to reduce CO_2 to sugars. The photochemical reactions occur on thylakoid membranes, whereas the biochemical reactions occur in the stroma.

As already mentioned, photosynthesis can be divided into two sets of reactions (fig. 12.2). Some characteristics of these reactions are compared here:

Photochemical "Light" Reactions	*Biochemical "Dark" Reactions*
Fast (practically instantaneous)	Slower, but still extremely fast
Light-dependent	Light-independent
Splits water to release oxygen, electrons, and protons	Converts (fixes) carbon dioxide to sugar

In today's exercise, you'll investigate some of the major aspects of photosynthesis, beginning with the isolation and identification of photosynthetic pigments.

Before you begin studying photosynthesis, we should remind you that *all* organisms (including plants) carry out respiration in one form or another, but chlorophyll-containing organisms can *also* photosynthesize.

PAPER CHROMATOGRAPHY OF PHOTOSYNTHETIC PIGMENTS

Light must be absorbed before its energy can be used. A substance that absorbs light is a **pigment.** The primary photosynthetic pigments that absorb light for photosynthesis are **chlorophylls *a*** and ***b*.** However, chlorophylls are not the only photosynthetic pigment; **accessory pigments** such as **carotenes** and **xanthophylls** also absorb light and transfer energy to chlorophyll *a*.

Paper chromatography is a technique for separating dissolved compounds such as chlorophyll, carotene, and xanthophyll. When a solution of these pigments is applied to strips of paper, the pigments adsorb onto the fibers of the paper. When the tip of the paper is immersed in a solvent, the solvent is absorbed and moves up through the paper. As the solvent moves through the spot of applied pigments, the pigments dissolve in the moving solvent. However, the pigments do not always keep up with the moving solvent—some pigments move almost as fast as the solvent, whereas others move more slowly. This differential movement of pigments results from each pigment's solubility and characteristic tendency to stick (i.e., be adsorbed) to the cellulose fibers of the paper. A pigment's molecular size, polarity, and solubility determine the strength of this tendency; pigments adsorbed strongly move slowly, whereas those adsorbed weakly move fastest. Thus, each pigment has a characteristic rate of movement, and the pigments can be separated from each other. In procedure 12.1, four bands of color will appear on the strip—a yellow band of carotenes, a yellow-orange band of xanthophylls, a blue-green band of chlorophyll *a*, and a yellow-green band of chlorophyll *b*.

Figure 12.3
Application of pigment extract to a chromatography strip.

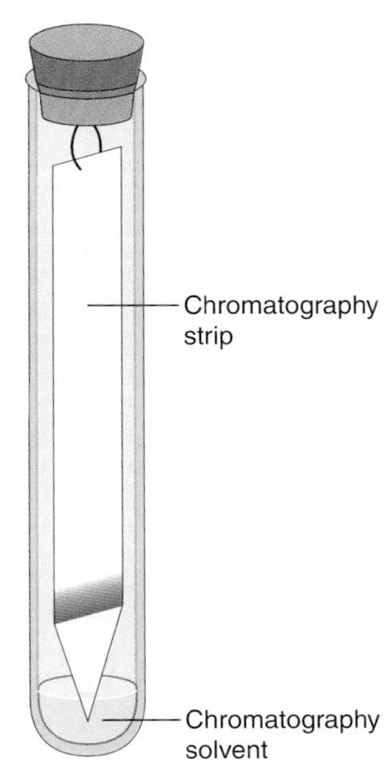

Figure 12.4
Chromatography setup.

The relationship of the distance moved by a pigment to the distance moved by the solvent front is specific for a given set of conditions. We call this relationship the $\mathbf{R_f}$ number and define it as follows:

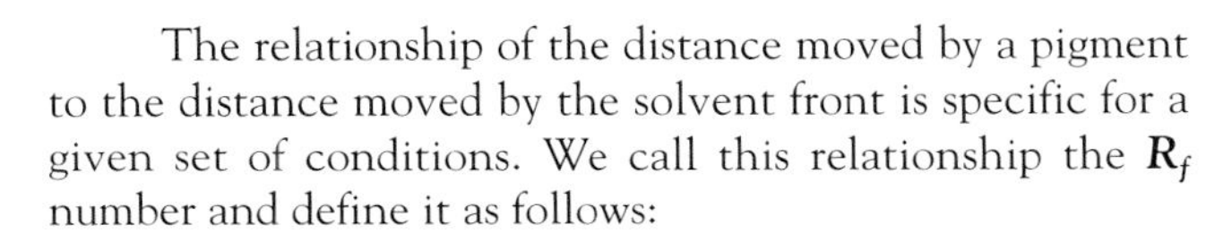

$$R_f = \frac{\text{Distance moved by pigment}}{\text{Distance from pigment origin to solvent front}}$$

Thus, paper chromatography can be used to identify each pigment by its characteristic R_f. This R_f is constant for a given pigment in a particular solvent-matrix system.

Procedure 12.1

Separate plant pigments by paper chromatography

1. Observe the contents of the provided container labeled "Plant Extract." You'll use paper chromatography to separate its pigments.

CAUTION

Keep all solvents away from hotplates and flames at all times. Extinguish all flames and hotplates before you do this experiment.

Question 1
What color is the plant extract, and why is it this color?

2. Obtain a strip of chromatography paper from your lab instructor. Handle the paper by its edges so that oil on your fingers does not contaminate the paper.
3. Use a pencil to mark a faint line across the paper approximately 2 cm from the tip of the paper. Now use a Pasteur pipet or a fine-tipped brush to apply a stripe of plant extract over the pencil mark (fig. 12.3). Blow the stripe dry and repeat this application at least 15 times. For this separation to work well, you must start with an extremely concentrated application of extract on the paper.
4. Place the chromatography strip in a test tube containing 2 mL of chromatography solvent (9 parts petroleum ether : 1 part acetone). Position the chromatography strip so that the tip of the strip (but not the stripe of plant extract) is submerged in the solvent. You can do this by hooking the strip of paper with a pin inserted in the tube's stopper (fig. 12.4).

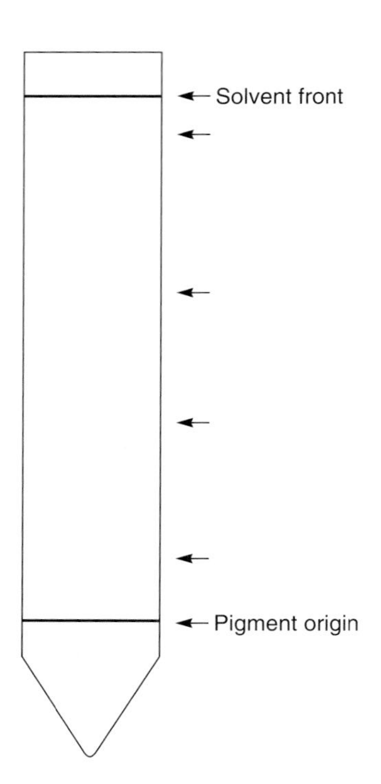

Figure 12.5

Completed chromatogram. On the chromatogram indicate the color of the pigment band to the left of the arrows. To the right of the arrows write the name of the pigment.

5. Place the tube in a test-tube rack and watch as the solvent moves up the paper. Keep the tubes capped and undisturbed during solvent movement.
6. Remove the chromatography strip before the solvent front reaches the top of the strip (i.e., after 2–3 minutes). Mark the position of the solvent front with a pencil and set the strip aside to dry. Observe the bands of color, then draw your results on figure 12.5.
7. Measure the distance from the pigment origin to the solvent front and from the origin to each pigment band. Calculate the R_f number for each pigment; record your data in table 12.1.

Question 2

a. What does a small R_f number tell you about the characteristics of the moving molecules?

Table 12.1

R_F Numbers for Four Plant Pigments

Pigment	R_f
Carotene	
Xanthophyll	
Chlorophyll *a*	
Chlorophyll *b*	

b. Which are more soluble in the chromatography solvent, xanthophylls or chlorophyll *a*? How do you conclude this?

c. Would you expect the R_f number of a pigment to change if we altered the composition of the solvent? Why or why not?

d. If yellow xanthophylls were present in the extract, why did the extract appear green?

e. Is it possible to have an R_f number greater than 1? Why or why not?

ABSORPTION OF LIGHT BY CHLOROPHYLL

A **spectroscope** is an instrument that separates white light into its component colors. These colors range from red to violet and appear as a spectrum when separated (fig. 12.6). Observe this spectrum by looking through the spectroscope provided in the lab. Now insert a chlorophyll sample between the light and spectroscope, and observe the resulting spectrum. Light not visible through the extract has been absorbed.

Question 3

What colors are diminished or absent?

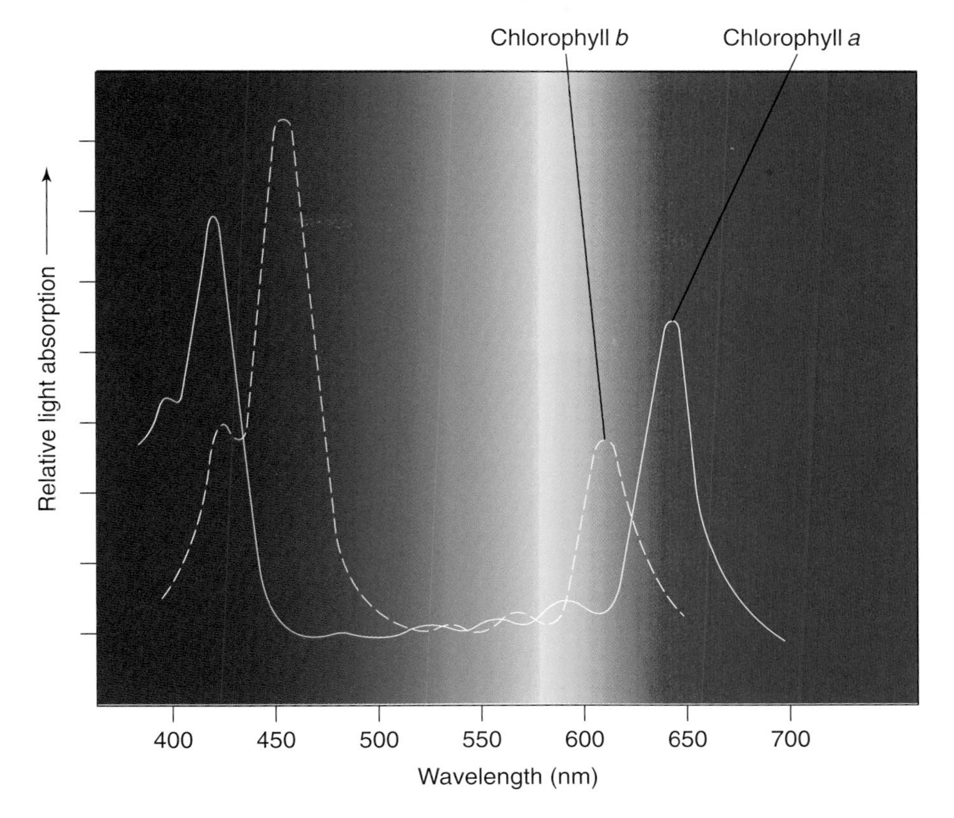

Figure 12.6

Absorption spectra for chlorophylls *a* and *b*. The peaks represent the wavelengths that the respective chlorophylls absorb best. The chlorophylls absorb predominately violet-blue and red light in two narrow bands of the spectrum. The chlorophylls reflect the greenish-yellow light in the middle of the spectrum.

Based on this observation, complete the following absorption spectrum for chlorophyll. For each color, estimate the relative absorbance of that color by placing an X above the color name at the appropriate position along the *y*-axis. Connect the Xs for all colors to complete the absorption spectrum.

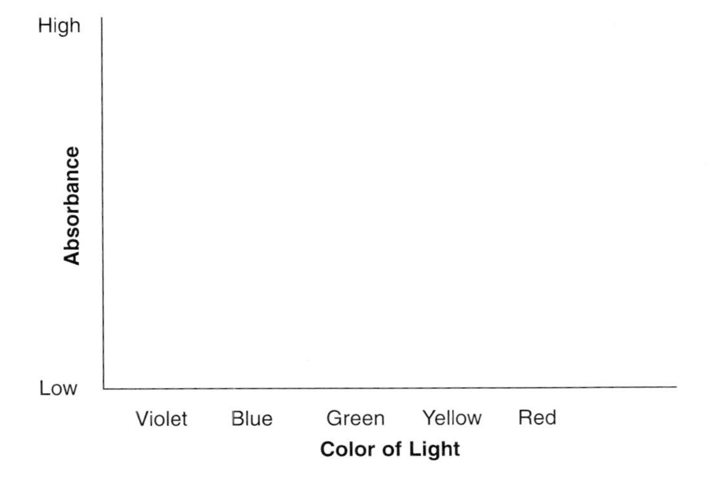

Question 4

What color of light would be least effective for plant photosynthesis? Why?

FLUORESCENCE

Light produces reactions only if it is absorbed by a molecule. When sunlight strikes a plant, the chlorophyll absorbs some of the light and reflects some of the light. The green light is reflected, and is responsible for the plant's green color. The light that is absorbed "excites" the chlorophyll by boosting electrons to a higher-energy orbital. During photosynthesis, the energy of these excited electrons from chlorophyll and chlorophyll's central magnesium atom is passed efficiently to another pigment molecule and photosynthesis proceeds. However, to easily observe these energized electrons, we can disrupt the photosynthetic system by blending the cells during the preparation of the plant extract. The chlorophyll

TABLE 12.2

SOLUTIONS FOR COMPARISON OF PHOTOSYNTHETIC REACTION RATES

Tube	Chloroplasts	0.1 M PO_4 Buffer (pH 6.5)	H_2O	0.2 mM DCPIP
1	0.5 mL	3 mL	1.5 mL	0
2	0.5 mL	3 mL	0.5 mL	1 mL
3	0	3 mL	1.0 mL	1 mL
4	0.5 mL	3 mL	0.5 mL	1 mL

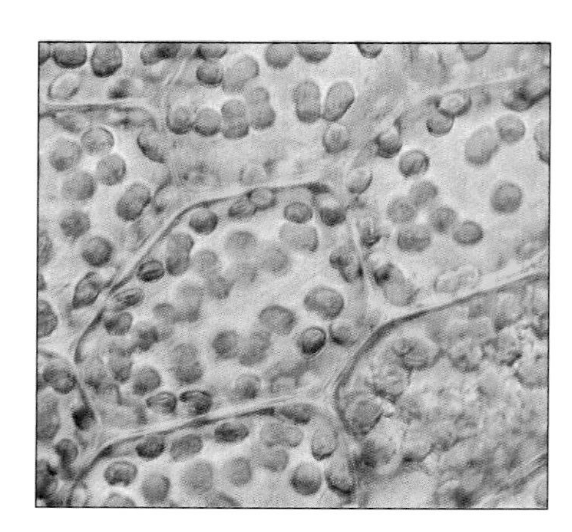

Figure 12.7

These photosynthetic cells of a moss are packed with bright-green chloroplasts.

electrons in the extract are still energized if you shine light on them, but they are left with nowhere to go. They quickly release their energy by falling back to their original orbitals rather than continuing photosynthesis. As they fall back, they emit a photon of red light. This release of light energy is **fluorescence.** The wavelength of light that is reemitted is determined by the structure of the molecule that is reemitting the light.

Procedure 12.2

Observe fluorescence by chlorophyll

Place a glass test tube of chlorophyll extract in front of a bright light. View the extract from the side.

Question 5

What color light does the extract fluoresce?

ELECTRON TRANSPORT IN CHLOROPLASTS

The photochemical reactions of photosynthesis transfer electrons among various compounds within chloroplasts (fig. 12.7). In 1937, Robin Hill demonstrated that isolated chloroplasts could transfer electrons in the absence of CO_2 if provided with an alternate or artificial electron-acceptor. This observation indicated that electron transport does not require CO_2-fixation to occur. That is, electron transfer and CO_2-fixation involve separate sets of reactions.

You can detect electron transfer using a dye called 2,6-dichlorophenol-indolephenol (DCPIP). In its oxidized state, DCPIP is blue. After accepting electrons, DCPIP becomes reduced and colorless. DCPIP can accept electrons released in chloroplasts during photosynthesis. The rate of DCPIP-decoloration depends on its concentration and the rate of electron flow. By measuring decoloration of DCPIP we can indirectly measure the rate of some reactions of photosynthesis. Because the rates of many chemical reactions are pH-dependent, a constant pH of approximately 6.5 is necessary for this experiment. The phosphate buffer used in this experiment maintains a constant pH of the incubation mixture.

Procedure 12.3

Observe electron transport in chloroplasts

1. Prepare test tubes according to table 12.2. Metabolically active chloroplasts will be provided by your instructor.
2. Mix the contents of each tube well and place tubes 1–3 approximately 15 cm in front of a high-intensity light bulb. Wrap tube 4 in aluminum foil and place it with the other three tubes. Do not position tubes behind each other. Keep all tubes directly in the path of the light.
3. Observe the contents of the tubes intermittently; describe the changes in color that you see.
4. If you have time, prepare a replicate of tube 2 in which water is replaced by 1 mL of 0.1 mM simizane or monuron, both of which are herbicides.

Question 6

a. What was the purpose of each of the tubes used in this experiment? Which tubes were controls?

b. What happens when you illuminate the tube containing herbicide?

c. Based on this result, what do you think is the mode of action of these herbicides?

Question 7

a. What happens to the color of the indicator?

b. What is the reason for the color change?

c. Considering the summary equation for photosynthesis, what is the basis for this change in color?

UPTAKE OF CARBON DIOXIDE DURING PHOTOSYNTHESIS

Phenol red (phenol-sulfonphthalein) is a pH-indicator that turns yellow in an acidic solution ($pH < 7$) and becomes red in a neutral to basic solution ($pH > 7$). (For more about pH and pH indicators, see Exercise 4.) In this experiment you will use the pH-indicator phenol red to detect the uptake of CO_2 by a photosynthesizing aquatic plant, *Elodea* (see fig. 3.6). Recall that plants use CO_2 during the light-independent reactions of photosynthesis.

To detect CO_2 uptake you will put a plant into an environment that you have made slightly acidic with your breath. Carbon dioxide in your breath will dissolve in water to form carbonic acid, which lowers the pH of the solution.

$$H_2O + CO_2 \longleftrightarrow H_2CO_3 \longleftrightarrow H^+ + HCO_3^-$$

Water	Carbon Dioxide	Carbonic Acid	Hydrogen Ion	Bicarbonate Ion

As the plant fixes CO_2 the pH rises. When the pH rises above 7, the solution turns red.

Procedure 12.4

Observe the uptake of CO_2 during photosynthesis

1. Fill two test tubes half full with a dilute solution of phenol red provided by your laboratory instructor.
2. Blow slowly into each solution with a straw. Because excess carbonic acid will lengthen this experiment, stop blowing in the tubes as soon as the color changes to yellow.
3. Add pieces of healthy *Elodea* totaling about 10 cm to a tube. Pour off excess solution above the *Elodea*.
4. Place both tubes approximately 0.5 m in front of a 100-watt bulb for 30–60 min.
5. Observe the tubes about every 10 min.

USE OF LIGHT AND CHLOROPHYLL TO PRODUCE STARCH DURING PHOTOSYNTHESIS

The light-dependent reactions of photosynthesis occur on photosynthetic membranes. In photosynthetic bacteria, these membranes are the cell membrane itself (see fig. 3.2). In plants and algae, photosynthetic membranes are called **thylakoids,** which are located within a special organelle called a **chloroplast** (fig. 12.8; also see fig. 3.8). Thylakoids are stacked to form columns called **grana,** which are held in place by **lamellae.** A semiliquid **stroma** bathes the interior of the chloroplast and contains the enzymes that catalyze the light-independent reactions of photosynthesis.

Sugars produced by photosynthesis are often stored as starch. Thus, starch production is another indirect measure of photosynthesis. To produce this starch, photosynthesis requires light as an energy source. In the absence of light, sugars and starch are not produced. Photosynthesis also requires chlorophyll to capture light energy. In the absence of chlorophyll, sugars and starch are not produced.

In the following procedures you will detect the presence of starch by staining it with a solution of iodine and observe the requirement of light and chlorophyll for photosynthesis.

Procedure 12.5

Stain starch with iodine

1. Place separate drops of water, glucose, and starch solutions on a glass slide.
2. Add a drop of iodine to each and describe your results.

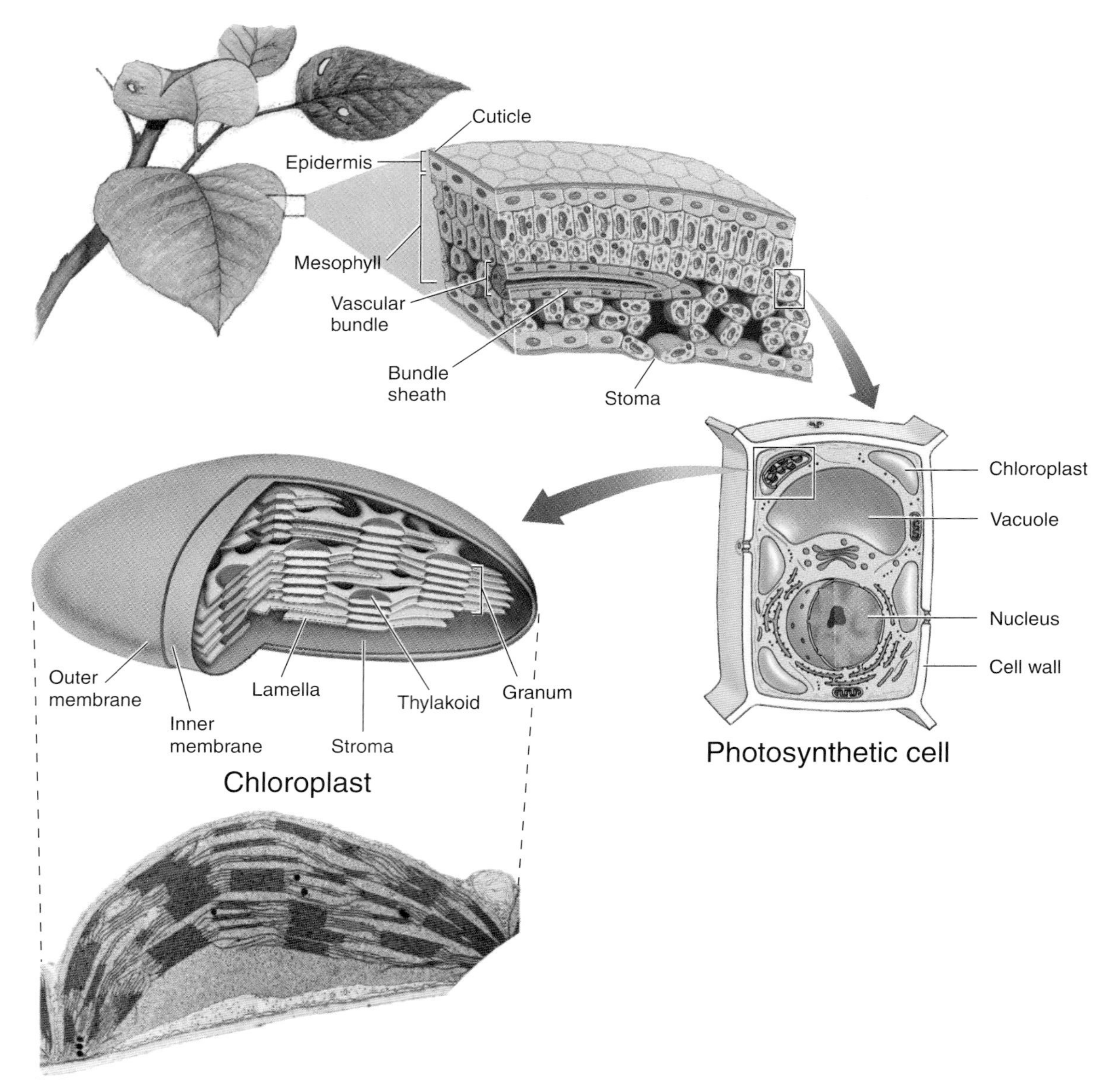

Figure 12.8

The structure of a leaf and chloroplast. Chloroplasts are bounded by a double membrane and contain photosynthetic membranes called thylakoids. Stacked one on top of the other, a column of thylakoids is a granum. Grana are held in place by lamellae, and the interior of the entire chloroplast is bathed by a semiliquid, the stroma. The openings that enable CO_2 to enter the leaf are stomata (singular, stoma).

Procedure 12.6

Observe starch production during photosynthesis

1. Remove a leaf from a *Geranium* plant that has been illuminated for several hours.
2. After immersing the leaf in boiling water for 1 min, bleach the pigments from the leaf by boiling the leaf in methanol for 3–5 min.

CAUTION

Exercise extreme caution when you heat methanol.

3. Place the leaf in a petri dish containing a small amount of water, and then add five to eight drops of iodine.
4. Observe any color change in the leaf.
5. Record in figure 12.9*a* the color of the leaves after each successive treatment.

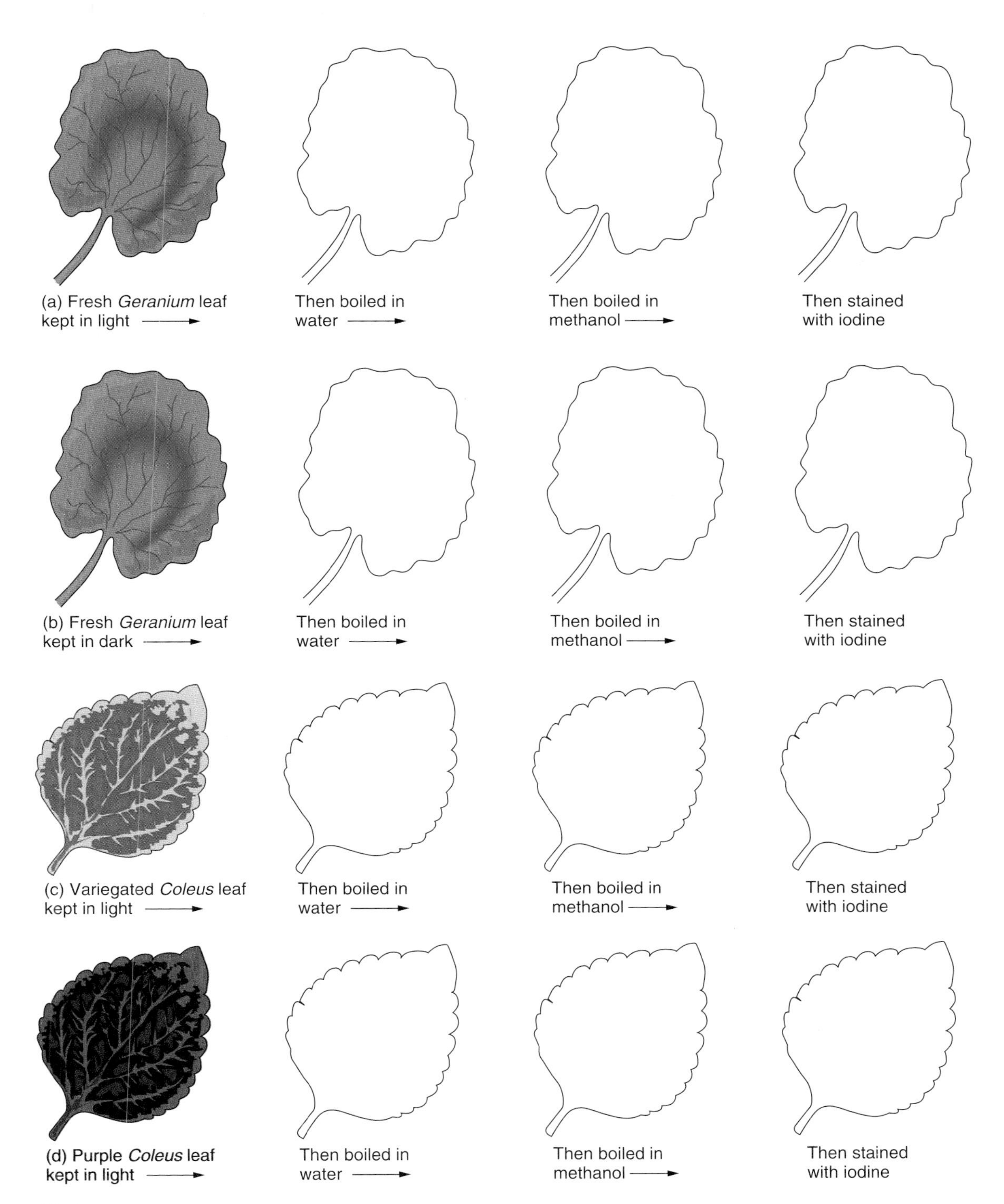

Figure 12.9

The requirement of light and chlorophyll and the production of starch during photosynthesis. Within each diagram, record the color of the leaf following the treatments to indicate (*a*) the production of starch, (*b*) the need for light, and (*c*, *d*) the need for chlorophyll for photosynthesis. Record your results from the appropriate procedure by writing the resulting color of each treated leaf directly onto the leaf outline.

(a)

(b)

Figure 12.10

Coleus plants. (*a*) Leaves of this variegated plant have green, white, purple, and pink areas resulting from combinations of chlorophylls and anthocyanin (red) pigments. (*b*) Leaves of this purple *Coleus* have the same pigment combination throughout the leaf.

Question 8

a. Was starch stored in the leaf? How can you tell?

b. Would you expect leaves to be the primary area for starch storage? Why or why not?

Procedure 12.7

Observe the requirement of light for photosynthesis

1. Obtain a *Geranium* leaf that has been half or completely covered with metal foil or thick paper for three or four days.
2. Repeat the bleaching and staining steps described in procedure 12.6.
3. Describe and explain any color change in the leaf.
4. Record in figure 12.9*b* the color of the leaves after each successive treatment.

Procedure 12.8

Observe the requirement of chlorophyll for photosynthesis

1. Obtain leaves of a variegated *Coleus* plant (fig. 12.10*a*) and a purple-leafed *Coleus* plant (fig. 12.10*b*). Make sketches of their original pigmentation patterns in figure 12.9*c*, *d*. Indicate which areas are green, red, green/red, and white.
2. Extract the pigments and stain for starch according to procedure 12.6. Boiling the leaf in water will remove the water-soluble pigments such as the red cyanins, and boiling the leaf in alcohol will remove chlorophyll. These pigments must be removed for you to see the color changes of the iodine starch test.
3. Record in figure 12.9*c*, *d* the color of the leaves after each successive treatment.

Question 9

a. How does the pattern of starch storage relate to the distribution of chlorophyll?

b. Photosynthesis requires chlorophyll (green), but some of the *Coleus* leaves that you tested were purple. How do you explain your results?

INVESTIGATION

Effects of Herbicides and Other Environmental Variables on Photosynthesis

Many aspects of photosynthesis are sensitive to herbicides and related compounds, as well as naturally occurring chemicals, salts, and nutrients. These chemicals may kill the plant quickly or subtly affect rates of photosynthesis. How would you detect and measure the effects of a herbicide on photosynthetic rates?

a. Decide which compound you will test.

b. Form a hypothesis about the result you expect. Your instructor will advise you about how to write a testable hypothesis. Write your hypothesis here:

c. Decide how you will test your hypothesis. Describe your experimental design here:

d. Do your experiment. What did you conclude? Do your data support your hypothesis?

Questions for Further Thought and Study

1. Why does chlorophyll appear green?

2. In darkness, what happens to starch formed in a leaf?

3. What causes leaves to turn from green to yellow and red in autumn?

4. Of what value to plants is starch?

5. What is the significance of electron transport in the photochemical (i.e., light-dependent) reactions of photosynthesis?

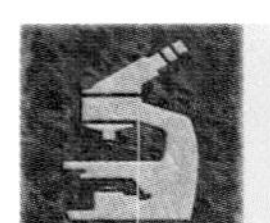

DOING BIOLOGY YOURSELF

Recall that respiration produces CO_2, which combines with water to form carbonic acid that lowers the pH of a surrounding solution. Design an experiment to measure the relative dynamics (mass balance) of photosynthesis versus respiration for *Elodea*.

WRITING TO LEARN BIOLOGY

Use a reference to determine the relative penetration of different wavelengths of light through water. Describe how this could affect the existence and distribution of submerged plants.

13

Mitosis
Replication of Eukaryotic Cells

Objectives

By the end of this exercise you should be able to:

1. Describe events associated with the cell cycle.
2. Describe events associated with mitosis.
3. Distinguish the stages of mitosis on prepared slides of mitotic cells.
4. Stain and examine chromosomes in mitotic cells.
5. Estimate the duration of various stages of mitosis from experimental observations.

Cells grow, have specialized functions, and usually replicate during their life. All of these activities are part of a repeating set of events called the **cell cycle.** A major feature of the cell cycle is cellular replication, and a major feature of cellular replication is mitosis. **Mitosis** is the replication and division of the nucleus of a eukaryotic cell in preparation for cytokinesis. During mitosis, replicated chromosomes within the cell are separated into two identical sets—each set is then surrounded by a nuclear membrane. Each of the two new nuclei has a full set of chromosomes containing a copy of all of the genetic information for the organism. Prokaryotic cells lack nuclei and do not undergo mitosis. Instead, they replicate their chromosome during a process called binary fission (described in Exercise 23).

Mitosis is usually associated with **cytokinesis,** the division of the cell and cytoplasm into halves that each contain a nucleus. In some tissues, cytokinesis is delayed or does not occur at all, and the cells are multinucleate. Mitosis and cytokinesis are important because they provide a mechanism for orderly growth of living organisms.

THE CELL CYCLE

This exercise emphasizes events associated with mitosis, but mitosis is only part of the cell cycle (fig. 13.1). The remainder of the cycle is called the **interphase** and is subdivided further into **cytokinesis (C), gap 1 (G_1), synthesis (S),** and **gap 2 (G_2)** phases.

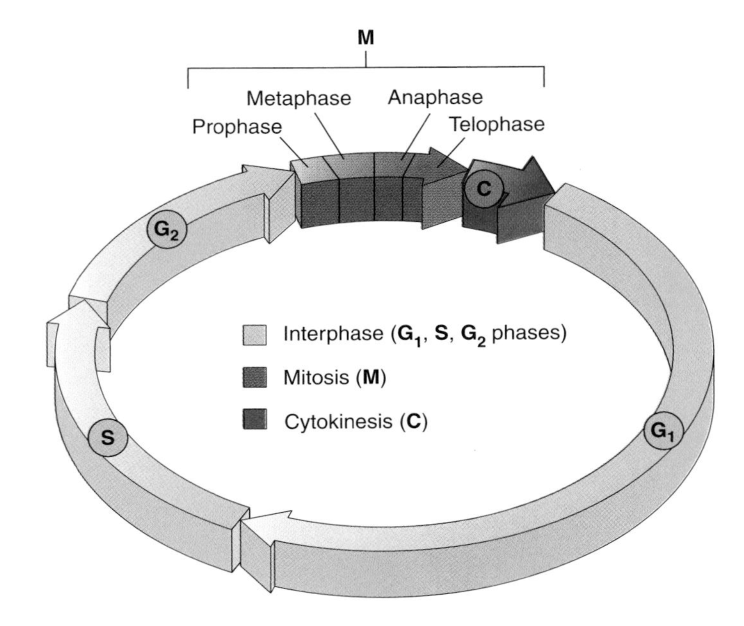

Figure 13.1
The cell cycle. Each wedge represents one hour of the 22-hour cell cycle in human cells growing in culture. G_1 represents the primary growth phase of the cell cycle, S the phase during which a replica of the genome is synthesized, and G_2 the second growth phase.

The cell cycle begins with the formation of a new cell and ends with replication of that cell. The G_1 phase of the cell cycle occurs after mitosis and cytokinesis, and is when the majority of cellular activity for the functions of the cell occurs. Many cell-specific proteins and other molecules are produced for the metabolism of the cell during G_1. During the S phase, the DNA composing the chromosomes is duplicated. At the end of the S phase each chromosome consists of an identical pair of chromosomal DNA strands called **chromatids** that are attached at a **centromere** (fig. 13.2). During the G_2 phase, molecules and structures necessary for mitosis are synthesized.

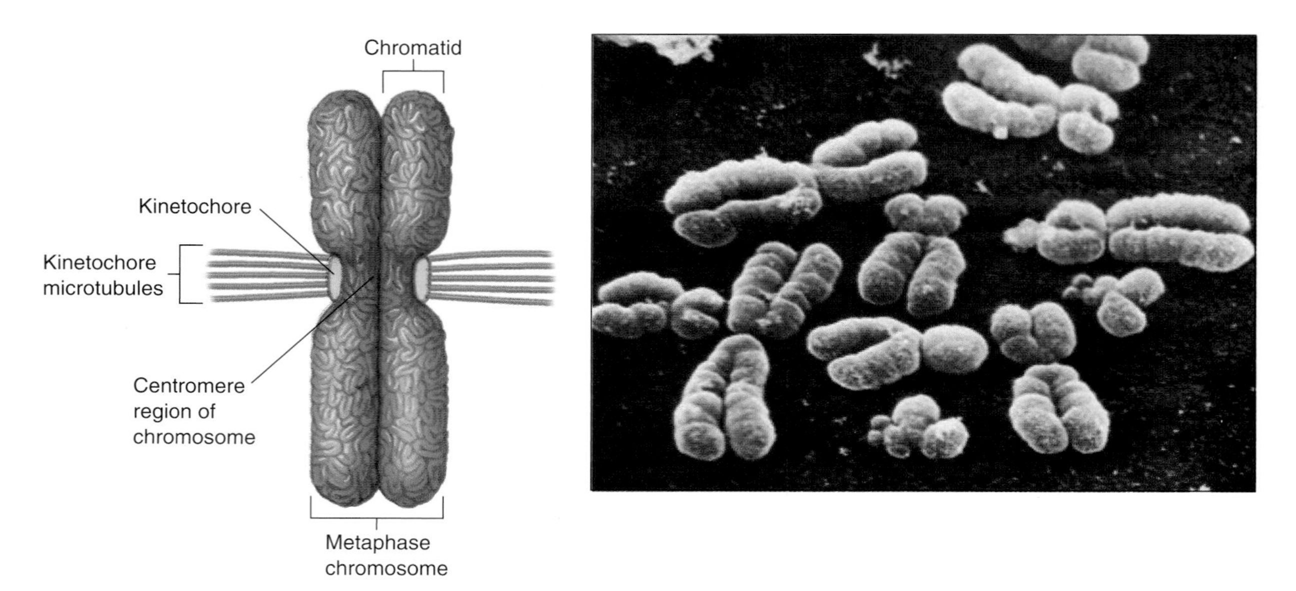

Figure 13.2

Chromosomes. (*a*) In a metaphase chromosome, kinetochore microtubules are anchored to proteins at the centromere. (*b*) The electron micrograph shows how human chromosomes appear during the early stages of nuclear division. Each strand of DNA has already been replicated to form sister chromatids that are identical to each other and held together by a centromere.

Mitosis (M phase) usually lasts for less than 10% of the time of the cell cycle, which usually lasts 10 to 30 h. Actively dividing cells such as those in rapidly growing tissues may spend more than 10% of their time in mitosis, whereas static cells such as bone cells or neurons may rarely enter M phase. Cytokinesis may begin during mitosis but is highly variable in length and timing. Tissues such as striated muscle fibers, and some algal filaments, may undergo mitosis without cytokinesis and produce multinucleate cells.

Question 1

a. Mitosis and cytokinesis are often referred to collectively as "cellular division." Why are they more accurately called cellular replication?

b. Does the cell cycle have a beginning and an end? Explain.

STAGES AND EVENTS OF MITOSIS

Mitosis (1) separates the genetic material that was duplicated during interphase into two identical sets of chromosomes, and (2) reconstitutes a nucleus to house each set. As a result, mitosis produces two identical nuclei from one.

Mitosis is traditionally divided into four stages: **prophase, metaphase, anaphase,** and **telophase** (fig. 13.3). The actual events of mitosis are not discrete but occur in a continuous sequence; separation of mitosis into four stages is merely convenient for our discussion and organization. During these stages, important cellular structures are synthesized and perform the mechanics of mitosis. For example, in animal cells two microtubule organizing centers called **centrioles** replicate. The pairs of centrioles move apart and form an axis of proteinaceous microtubules between them called **polar fibers.** The centrioles continue to move apart until they reach opposite poles of the cell and have a bridge of microtubules called the **mitotic spindle** (or spindle apparatus) extending between them (fig. 13.3). **Kinetochore fibers** attach to each chromosome's **kinetochore,** a complex of proteins that binds to the centromere. At the poles of the cell, the centrioles radiate an array of microtubules outward in addition to the spindle apparatus. These microtubules brace the centrioles against the cell membrane. This arrangement of microtubules radiating from a centriole is called an **aster.** Plant cells lack centrioles and asters, but spindle fibers still form between opposite poles of the cell.

Interestingly, animal cells that are deprived of centrioles will still form a spindle apparatus. Chromosomes will eventually distribute themselves on the spindle apparatus and are moved and separated to opposite poles. The distribution of chromosomes will also occur if the cell is haploid (i.e., has a single set of chromosomes). The vegetative cells of many organisms such as fungi are haploid (have a single set of chromosomes) rather than diploid (have a double set of chromosomes). However, the steps of mitosis are the same as for diploid cells.

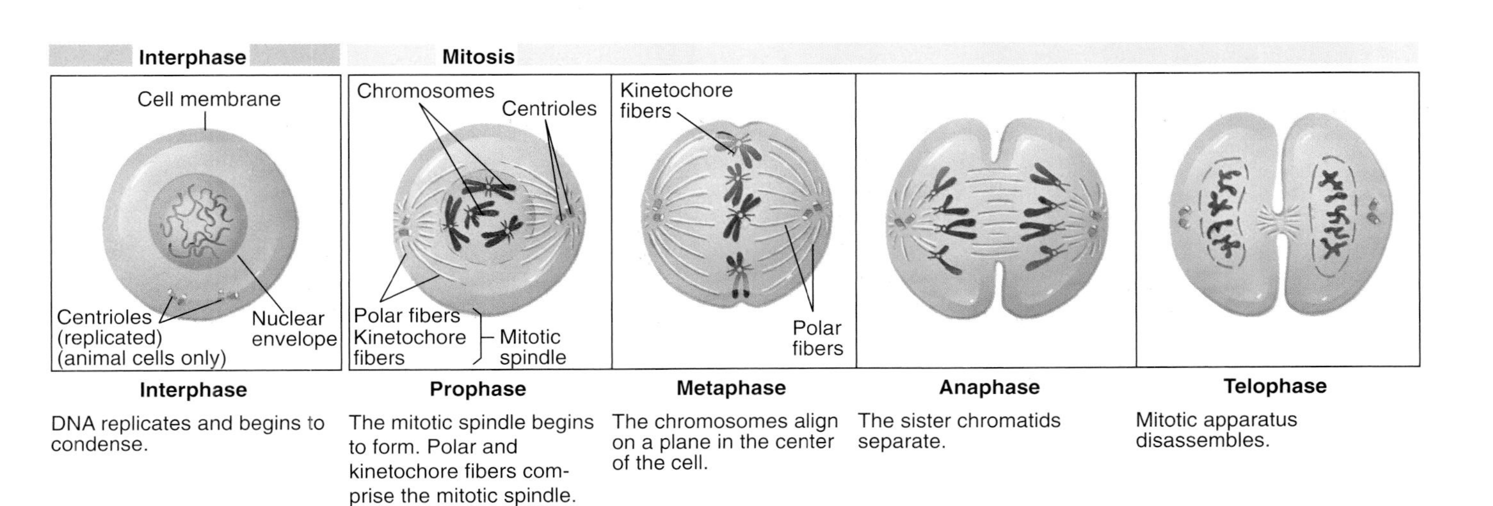

Figure 13.3

Interphase and the stages of mitosis in an animal cell. The cleavage furrow signifying cytokinesis may first appear during anaphase or more typically during telophase.

Procedure 13.1

Describe the specific events of mitosis

Before your lab meeting, review in your textbook the events associated with each stage of mitosis in addition to the preparatory stage, interphase. List these events in table 13.1. Some events and structures occur only in plant cells and some occur only in animal cells. Mark these events in your list with an asterisk. This list can serve as an excellent study guide, so be as complete as possible. One event for each stage is provided in figure 13.3.

Question 2

a. If a nucleus has eight chromosomes during interphase, how many chromosomes does it have during metaphase?

b. How many does it have after mitosis is complete?

Understanding the movements of chromosomes is crucial to understanding mitosis. You can simulate these movements easily with chromosome models made of pipe cleaners or popsicle sticks. This is a simple procedure, but a valuable one. It will be especially helpful when you are comparing the events of mitosis to the events of meiosis, which you will simulate in the next exercise.

Procedure 13.2

Simulate chromosomal replication and movement during mitosis

1. Examine the materials to be used as chromosome models provided by your instructor.
2. Identify the differences in chromosomes represented by various colors, lengths, or shapes of materials. Also identify materials representing centromeres.
3. Place a sheet of notebook paper on your lab table to use in representing the boundaries of the mitotic cell.
4. Assemble the chromosomes needed to represent nuclear material in a cell of a diploid organism with a total of six chromosomes. Place the chromosomes in the cell.
5. Arrange the chromosomes to depict the position and status of chromosomes during interphase G_1. (During G_1 the chromosomes are usually not condensed, as the chromosome models imply, but the models are an adequate representation.)
6. Depict the status of chromosomes after completing interphase S. Use additional "nuclear material" if needed.
7. Move the chromosome models appropriately to depict prophase.
8. Move the chromosome models appropriately to depict metaphase.
9. Move the chromosome models appropriately to depict anaphase.
10. Move the chromosome models appropriately to depict telophase.

TABLE 13.1

EVENTS OF MITOSIS AND INTERPHASE

Interphase	Although interphase is not actually part of nuclear replication, understanding its events is essential to understanding mitosis.
Prophase	
Metaphase	
Anaphase	
Telophase	

11. Draw the results of cytokinesis and the re-formation of nuclear membranes.
12. Chromosomal events occur as a continuous process of movements rather than in distinct steps. Therefore, repeat steps 4–11 as a continuous process and ask your instructor to verify your simulation.

MITOSIS IN ANIMAL CELLS

The most distinctive features of cellular replication in animal cells are the formation of asters with centrioles at their center (discussed earlier) and cytokinesis. Cytokinesis includes formation of a **cleavage furrow** that begins on the periphery of the cell, pinches inward, and eventually divides the cytoplasm into two cells (fig. 13.4). Cells of a whitefish blastula provide good examples of the stages of mitosis and cytokinesis. Whitefish are commonly cultured fish whose eggs and early developmental stages undergo rapid cell divisions (as do all embryonic cells). A blastula is an early embryonic stage of a vertebrate and consists of a sphere of 25–100 cells with a high frequency of different mitotic stages. Exercise 49 (Embryology) details the formation of a blastula during embryonic development.

Procedure 13.3

Observe and describe mitosis in animal cells

1. Obtain a prepared slide of a cross section through the blastula of a whitefish.
2. Examine the cells first on low (10×) then high (40× or 100×) magnification. Note that some of the cells contain condensed and stained chromosomes.
3. Refer to figure 13.3 for a summary of the stages of mitosis. Identify examples of each stage on your prepared slide (fig. 13.5). Verify these stages with your lab partner or teaching assistant.
4. Also identify cells that you believe are between stages. Don't hesitate to designate a cell as representing an early or late part of a stage.

5. Examine the whitefish cells for signs of cytokinesis.
6. Prepared cross sections of cells show only two dimensions, but mitosis is a three-dimensional process. In the space below draw two cells in metaphase: one in which the cross section is parallel to the axis of the spindle apparatus and one in which the cross section is perpendicular to the spindle apparatus.

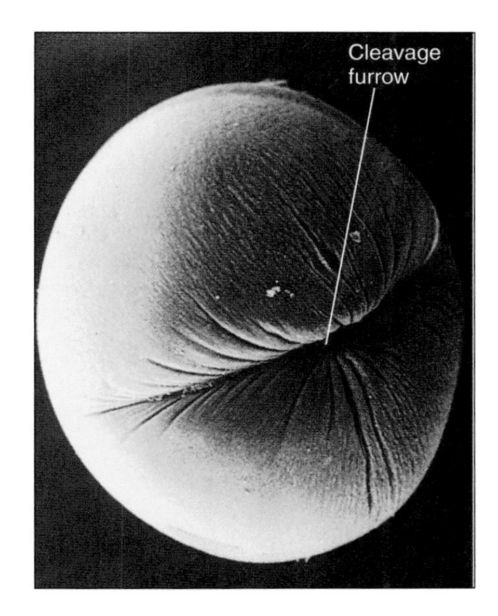

Figure 13.4
Cytokinesis in an animal cell. Cytokinesis, which is the physical division of the cell's cytoplasm, usually occurs after nuclear division is complete. A cleavage furrow is forming around this dividing sea urchin egg.

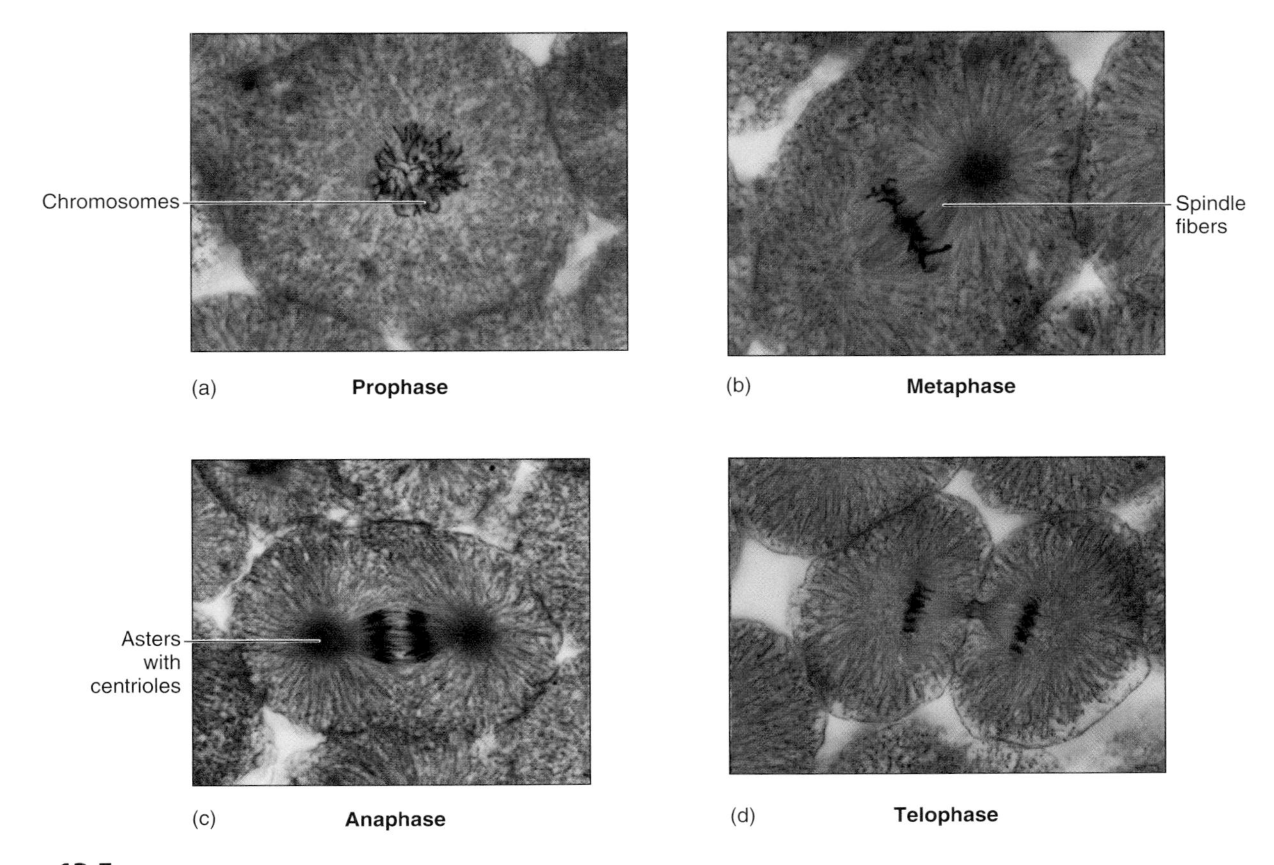

Figure 13.5
Stages of mitosis in the cells of a whitefish embryonic blastula (400×).

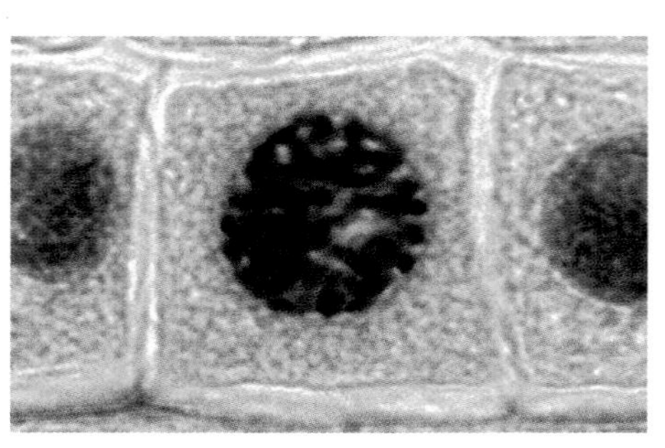
(a) Prophase

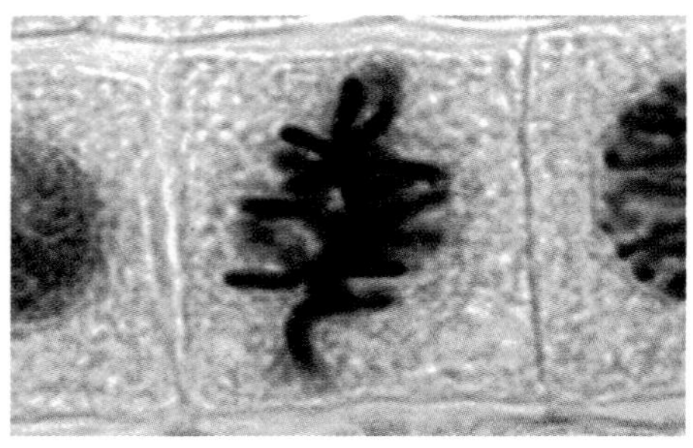
(b) Metaphase

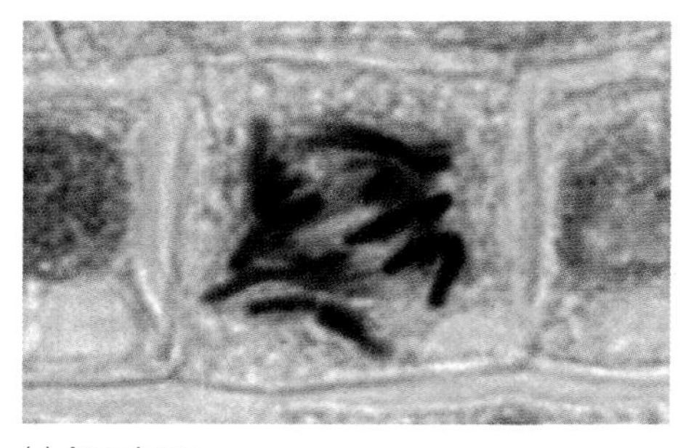
(c) Anaphase

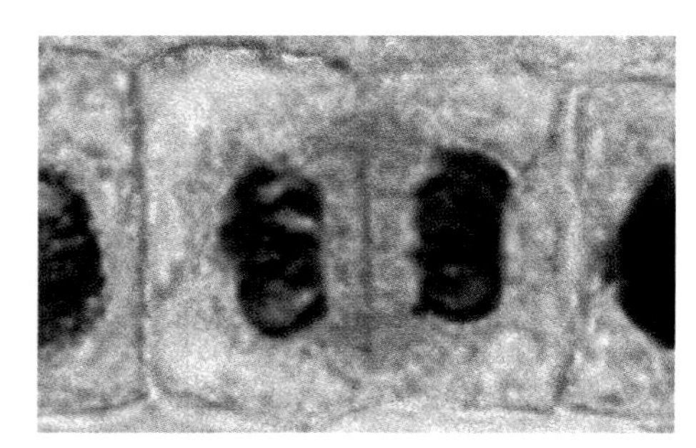
(d) Telophase

Figure 13.6
Stages of mitosis in a plant cell (1000×). The dark structures are chromosomes.

Question 3

a. Why would we choose an embryonic mass of cells for procedure 13.3 in which to study the stages of mitosis?

b. Which stage of mitosis most often is associated with the beginning of cytokinesis?

MITOSIS IN PLANT CELLS

Our model to study cellular replication in plants is the root tip of *Allium* (onion). Root tips of plants contain **meristems,** which are localized areas of rapid cell division due to active growth at the root tips. In plant cells, cytokinesis includes formation of a partition called a **cell plate** that is perpendicular to the axis of the spindle apparatus. The cell plate forms in the middle of the cell and grows out to the periphery. It will separate the two new cells.

Procedure 13.4

Observe and diagram mitosis in plant cells

1. Examine a prepared slide of a longitudinal section through an onion root tip.
2. Search for examples of all stages of mitosis (fig. 13.6).
3. Search for signs of cell plate formation.
4. In figure 13.7, diagram a plant cell with a diploid number of three pairs of chromosomes in each of the stages of mitosis. **Diploid** refers to a nucleus with two of each type of chromosome. Be sure to label the cell wall and cell plate. Two extra cells are provided for you to diagram midphases to demonstrate the continuity of the events of mitosis.

Figure 13.7

Diagram of the stages of mitosis in a plant cell with six chromosomes. The outlines represent the cell walls of each of six cells.

5. Prepared cross sections of cells show only two dimensions, but mitosis is a three-dimensional process. In the space below draw two cells in metaphase: one in which the plane of section is parallel to the axis of the spindle apparatus and one in which the cross section is perpendicular to the spindle apparatus.

Question 4

a. What region of a root has the most mitotic activity?

b. Why is pinching of the cytoplasm inadequate for cytokinesis in plant cells?

c. Locate a plant cell in late telophase. What is the volume of the two new cells relative to a mature cell?

Procedure 13.5

Determine the time elapsed during stages of cell replication

1. The length of the cell cycle in actively dividing root tips of *Allium* is approximately 24 h. With this in mind, predict the time that a cell spends in each stage of the cell cycle. Record your estimates in table 13.2.
2. Reexamine a prepared slide of an onion root tip. In a single field of view at high magnification count the number of cells in each stage listed in table 13.2. Record your data in table 13.2.
3. Repeat step 2 for two more fields of view and record your data in table 13.2.
4. Total the values for the number of cells in each stage for the three replicated fields of view, and record the total in the appropriate column in table 13.2.
5. Combine your results with that of your class. Then record the combined class data in the appropriate column of table 13.2.
6. Use the class data to calculate the time spent by a cell in each stage based on a 24-h cell cycle. Divide the number of cells in each stage by the total number of cells counted. Then multiply this fraction by 24. Record your results in table 13.2.

$$\text{Hours of stage}_i = \frac{24 \times \text{Number of cells in stage}_i}{\text{Total number of cells counted}}$$

TABLE 13.2

DURATION OF EACH STAGE OF MITOSIS IN AN ONION ROOT CELL BASED ON A 24-HOUR CELL CYCLE

Stage of Mitosis	Predicted Duration (hours)	Number of Cells in Each Stage	Total Number of Cells in Each Stage	Class Total Number for Each Stage	Calculated Duration (hours)
Interphase	________	____ ____ ____	________	________	________
Prophase	________	____ ____ ____	________	________	________
Metaphase	________	____ ____ ____	________	________	________
Anaphase	________	____ ____ ____	________	________	________
Telophase	________	____ ____ ____	________	________	________
Totals	________	____ ____ ____	________	________	________

Question 5

a. Why are the combined data from all the class members more meaningful than your results alone?

b. How accurate were your predictions for the length of each stage of mitosis?

c. What sources of error can you list for this technique to determine the time elapsed during each stage of mitosis?

PREPARING AND STAINING CHROMOSOMES

Your instructor has prepared some living onion root tips for you to process further and use to observe the stages of mitosis.

Procedure 13.6

Stain chromosomes

1. Obtain an onion root tip and place it in a small vial with Schiff's reagent for 30 min. Handle Schiff's reagent carefully because it is a colorless liquid that becomes bright red after reaction.

   Keep the vial in the dark and at room temperature until the root tip becomes purple. Your instructor may have already stained some root tips for you.
2. Place the root tip in a drop of 45% acetic acid on a slide and cut away all except the terminal 1 mm of the tip.

CAUTION

Acetic acid is corrosive. Do not spill it.

3. Crush the root tip with a blunt probe and cover the tissue with a coverslip.
4. Smash the tissue by pressing on the coverslip with the eraser of your pencil. Your instructor will demonstrate this procedure.
5. Scan your preparation at low magnification to locate stained chromosomes. Then switch to high magnification and locate formations of chromosomes that indicate each of the stages of mitosis.
6. Add a drop of acetic acid to the edge of the coverslip to avoid desiccation.
7. Locate as many stages of mitosis as you can. Be sure to look at preparations done by other students.

Questions for Further Thought and Study

1. Interphase has sometimes been called a "resting stage." Why is this inaccurate?

2. Most general functions of a cell occur during G_1 of interphase. What events that occur during other phases of the cell cycle might inhibit general metabolism?

3. Read in your textbook about prokaryotic cellular replication; list the fundamental cellular/structural differences between it and eukaryotic cellular replication. What is the basis for these differences?

4. Some specialized cells such as neurons and red blood cells lose their ability to replicate when they mature. Which phase of the cell cycle do you suspect is terminal for these cells? Why?

WRITING TO LEARN BIOLOGY

Refer to your textbook to review the properties of a chemical called colchicine. Describe how colchicine affects dividing cells. What is the mechanism of this effect? How might colchicine be used as a tool in scientific research or medicine?

Meiosis
Reduction Division and Gametogenesis

14

Objectives

By the end of this exercise you should be able to:
1. Describe the events of meiosis.
2. Compare and contrast meiosis and mitosis.
3. List the most significant events of meiosis.
4. Explain the relevance of meiosis to sexual reproduction.
5. Explain the relationship of meiosis and gametogenesis.
6. Describe the events of spermatogenesis and oogenesis.

Sex is one of the most experienced and scrutinized processes of life. Biologists know that the significance of sex, and meiosis in particular, is the recombination of a parent's genes and the packaging of these genes as a gamete. During sexual reproduction a gamete and its genes are combined with another parent's gamete and genes to endow the new offspring with new genetic combinations.

Chromosomes in typical, eukaryotic nuclei occur in pairs; that is, the nuclei are **diploid** (2N). The two chromosomes of a pair are called **homologous chromosomes,** and each homologue of a pair has the same sites, or **loci,** for the same genes, although the homologues may carry different alleles at homologous loci. A nucleus, such as that in a gamete, with only one chromosome of each homologous pair is **haploid** (1N).

Meiosis produces haploid daughter nuclei and is sometimes called "reduction division." Reduction of chromosomes in the nucleus of a gamete to only one of each pair is important because such a haploid nucleus can fuse with another haploid nucleus during sexual reproduction and restore the original, diploid number of chromosomes to the new individual (fig. 14.1).

Question 1

a. Why would shuffling genetic material to produce new combinations of characteristics be advantageous to a species?

b. When would it be deleterious?

Meiosis, like mitosis, is preceded by the replication of each chromosome to form two chromatids attached at a centromere. However, two events that do *not* occur in mitosis include final reduction of the chromosome number by half and production of new genetic combinations. Meiosis reduces the chromosome number by including *two* rounds of chromosome separation called **Meiosis I** and **II.** Thus, the genetic material is replicated once just before meiosis but divided twice during meiosis. This allocates half the original number of chromosomes (one of each original pair) to each daughter cell; that is, the nuclei are haploid.

To produce new genetic combinations each chromosome (composed of two chromatids) initially pairs along its length with its homologue. This pairing of homologous chromosomes is called **synapsis,** and the four chromatids exchange homologous segments of genetic material called **alleles.** Alleles are alternate states of a gene, such as a blue allele or a brown allele composing the gene for eye color. This exchange of genetic material among chromatids is called **crossing-over** and produces new genetic combinations. During crossing-over there is no gain or loss of genetic material. But afterward, each chromatid of the chromosomes contains different segments (alleles) that it exchanged with other chromatids.

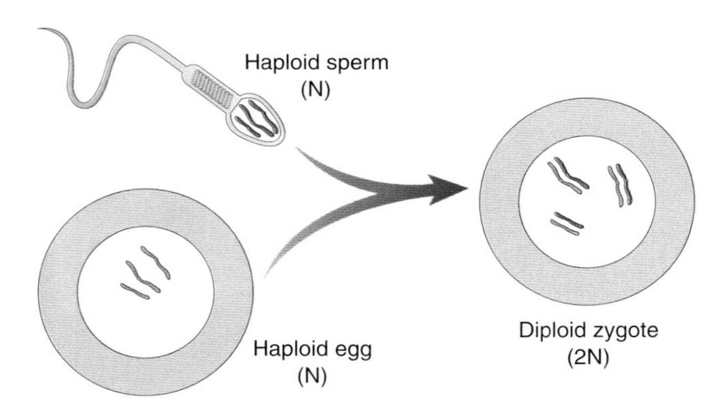

Figure 14.1

In preparation for sexual reproduction, meiosis reduces the number of chromosomes by half to form haploid gametes. Subsequent fusion (fertilization) of the gametes forms a diploid zygote and restores the original diploid chromosome number. In this drawing, (N) stands for haploid and (2N) stands for diploid.

Question 2

a. Synapsis occurs after chromosomal DNA has replicated. How many chromatids are involved in crossing-over of a homologous pair of chromosomes?

b. Suppose synapsis occurred between two homologous chromosomes, and one had alleles for blue eyes and brown hair where the other had alleles for green eyes and blonde hair. How many different combinations of these alleles would be possible?

STAGES AND EVENTS OF MEIOSIS

Although meiosis is a continuous process, we can study it more easily by dividing it into stages just as we did for mitosis. Indeed, meiosis and mitosis are similar, and their corresponding stages of prophase, metaphase, anaphase, and telophase have much in common. However, meiosis takes longer than mitosis because meiosis involves two processes of division instead of one. These two reductions are called meiosis I and meiosis II. Homologous chromosomes are separated at the end of meiosis I, and chromatids composing each chromosome are separated during meiosis II. Each reduction involves the events of prophase, metaphase, anaphase, and telophase (fig. 14.2).

Before your lab meeting, review in your textbook the events associated with each stage of meiosis, including the preparatory stage, interphase. List these events in table 14.1. This list can serve as an excellent study guide, so be as complete as possible. Ask your instructor to check for errors. One event for each stage is provided in figure 14.2.

Premeiotic Interphase

Meiosis I is preceded by an interphase that is similar to the G_1, S, and G_2 of mitotic interphase, including replication of the chromosomes. Each chromosome is replicated.

Compare the outline in table 14.1 with your outline in table 13.1 on mitosis.

Question 3

a. If a nucleus has eight chromosomes when it begins meiosis, how many chromosomes does it have after telophase I? Telophase II?

b. What are the major differences between the events of meiosis and mitosis?

c. What are some minor differences, and why do you consider them minor?

Understanding the movements of chromosomes is crucial to understanding meiosis. You can simulate these movements easily with chromosome models made of pipe cleaners or popsicle sticks. This is a simple but valuable exercise. It is especially instructive if you compare your simulation of meiosis with your simulation of mitosis from the previous exercise. There are important differences between them.

Procedure 14.1

Simulate chromosomal replication and movement during meiosis

1. Examine the materials for chromosome models provided by your instructor.
2. Identify the differences in chromosomes represented by various colors, lengths, or shapes of the materials. Also identify materials representing centromeres.
3. Place a sheet of notebook paper on your lab table. Use it to represent the boundaries of the meiotic cell.

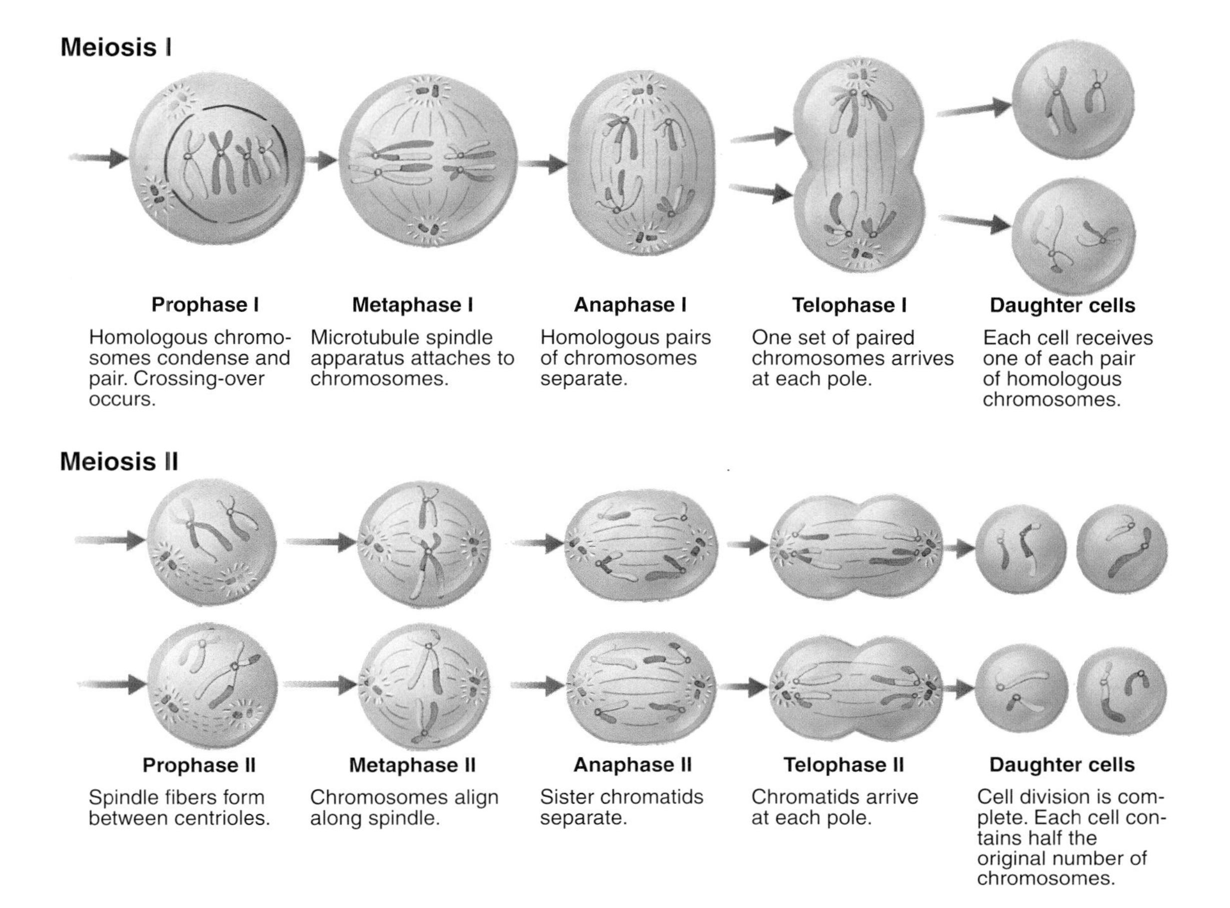

Figure 14.2
Stages of meiosis.

4. Assemble the chromosomes needed to represent the nuclear material in a cell of a diploid organism with a total of six chromosomes (three homologous pairs). Place the chromosomes in the cell.
5. Arrange the chromosomes to depict the position and status of chromosomes during interphase G_1. (During G_1 the chromosomes are usually not condensed as the chromosome models imply; nevertheless, the models are an adequate representation.)
6. Depict the chromosomes after completing interphase S. Use additional "nuclear material" if needed.
7. Depict the chromosomes during prophase I, during metaphase I, during anaphase I, and during telophase I.
8. The interval between meiosis I and meiosis II is called **interkinesis.** Draw the results of cytokinesis, which occurs at this stage in some organisms.
9. Depict the status of chromosomes during prophase II for both daughter nuclei. Repeat this for metaphase II, during anaphase II, and telophase II.
10. Draw the results of cytokinesis and the re-formation of nuclear membranes.
11. Chromosomal events are a continuous process rather than distinct steps. Therefore, repeat steps 4–10 as a continuous process and ask your instructor to verify your simulation.

GAMETOGENESIS

Meiosis occurs in all sexually reproducing eukaryotes and produces haploid nuclei. However, organisms vary in the timing and structures associated with producing functional gametes. **Gametes** are reproductive cells with haploid nuclei resulting from meiosis, and the formation of gametes is called **gametogenesis.** Meiosis is the primary element of gametogenesis in animals, but the cells must mature and usually change their morphology before a newly formed cell from meiosis becomes a functional gamete.

In this exercise you will examine mammalian gametogenesis. Male gametes of mammals are different from female gametes. Gametogenesis is divided into **spermatogenesis,** the formation of sperm cells, and **oogenesis,** the formation of egg cells (fig. 14.3).

Table 14.1

Events of Meiosis
Prophase I
Metaphase I
Anaphase I
Telophase I
Prophase II
Metaphase II
Anaphase II
Telophase II

Mammalian Spermatogenesis

Spermatogenesis occurs in male testes, which are made of tightly coiled tubes called **seminiferous tubules** (fig. 14.4). Examine a prepared slide of a cross section through the seminiferous tubules of a monkey, rat, or grasshopper. Packed against the inner walls of the tubules are diploid cells called **spermatogonia,** which constantly replicate mitotically during the life of males. Some of the daughter cells move inward toward the lumen of the tubule and begin meiosis. These cells are called **primary spermatocytes.** Meiosis I of a primary spermatocyte produces two **secondary spermatocytes,** each with a haploid set of double-stranded chromosomes.

Meiosis II separates the strands of each chromosome and produces two haploid cells called **spermatids.** Spermatids mature and differentiate into **sperm** cells as they move along the length of the tubule. Review these basic stages of spermatogenesis in figure 14.3. Then examine some prepared slides of sperm cells from vertebrates such as guinea pig, rat, and human.

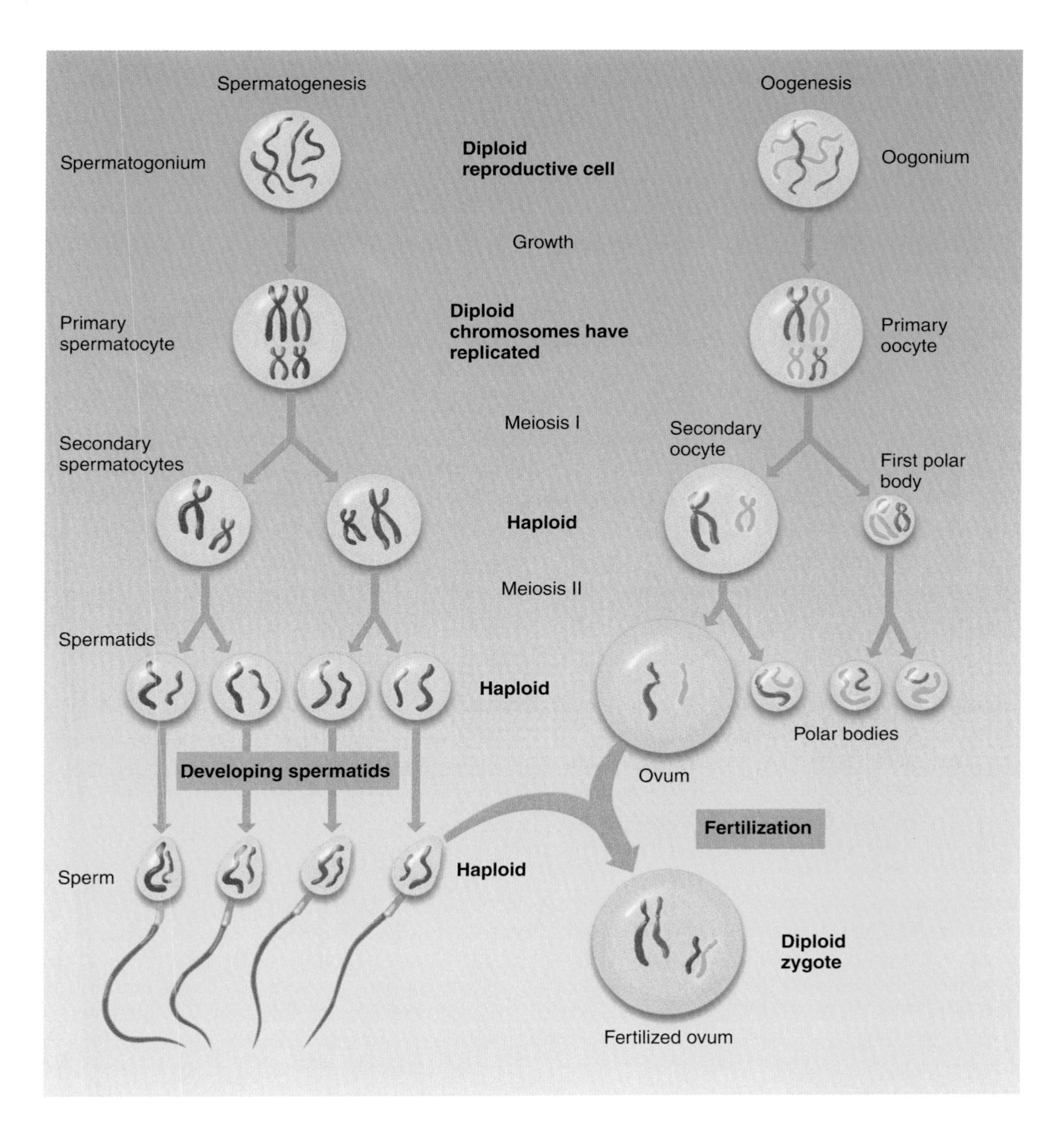

Figure 14.3
Gametogenesis is the formation of sperm and egg; it includes meiosis followed by cellular maturation to produce functional haploid gametes.

Question 4

a. During gametogenesis a sperm cell undergoes considerable structural change. What are the basics of sperm structure and how does it relate to function?

b. What is the advantage of producing sperm in a system of tubes rather than in solid tissue?

c. What is each strand of a double-stranded chromosome called?

Mammalian Oogenesis

Oogenesis occurs in ovaries of females (fig. 14.5). Cells of the ovary that produce female gametes are called **oocytes.** However, oocytes are not produced continually by the ovary, as spermatocytes are produced by the testes. During early fetal development, oogonia are produced in the

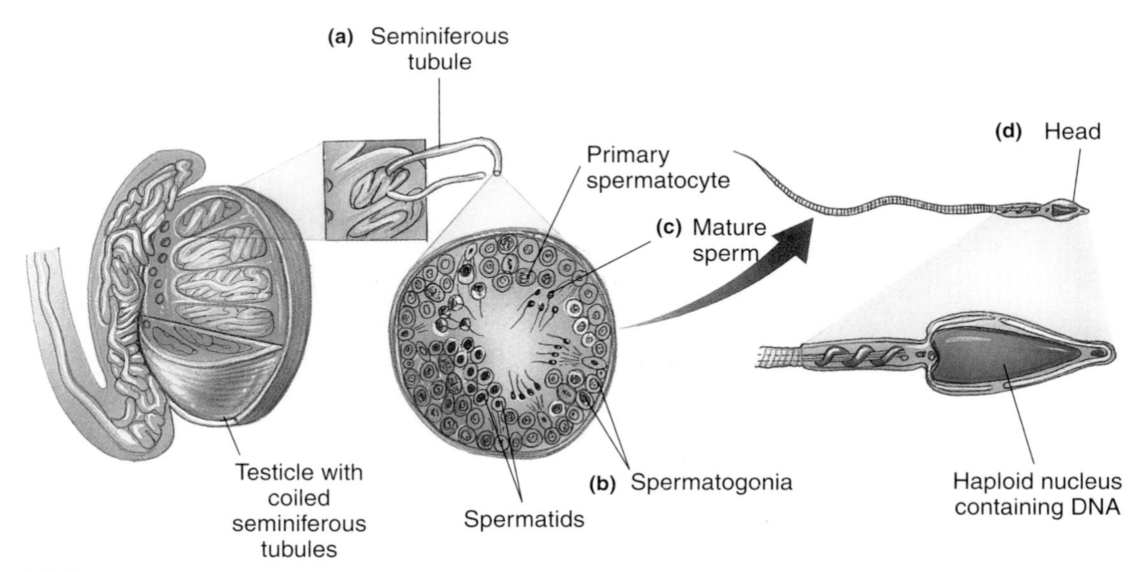

Figure 14.4
The interior of the testis, which is the site of spermatogenesis. Within the seminiferous tubules of the testis (*a*), cells called spermatogonia (*b*) pass through the spermatocyte and spermatid stages to develop into sperm. Each sperm (*c*) possesses a long tail coupled to a head (*d*), which contains a haploid nucleus.

ovaries. These oogonia replicate mitotically to produce as many as two million **primary oocytes.** In humans, ovaries of a newborn female contain all of the primary oocytes that she will ever have (i.e., oogonia produce no more primary oocytes). At birth the primary oocytes in a female have begun meiosis I but are arrested in prophase I. They are surrounded by supportive **follicular cells,** and together they are called **follicles.** At puberty, circulating hormones stimulate growth of one or two of these dormant follicles (and their primary oocytes) each month. The oocyte enlarges and the number of follicular cells increases. Just before **ovulation** (release of the oocyte from the ovary) the oocyte completes meiosis I, which produces a **secondary oocyte** and a **polar body.** This mature follicle is called a **Graafian follicle** and contains a secondary oocyte (fig. 14.6). Each secondary oocyte contains a haploid set of double-stranded chromosomes (two chromatids), but cytoplasmic cleavage is unequal. The secondary oocyte retains most of the cytoplasm and the polar body usually disintegrates.

Examine a prepared slide of a mammalian ovary cross section. In the space below sketch a Graafian follicle and two or three less mature stages.

Question 5
How would retaining extra cytoplasm enhance survival of a developing oocyte?

Meiosis II proceeds but is not completed until after a sperm cell penetrates the egg. Completion of meiosis II produces another polar body and a haploid egg-cell ready for **fertilization** (fusion of nuclei). Review these basic stages of oogenesis in figure 14.3. Then examine a cross section of a cat ovary.

Question 6
a. What are the relative sizes of oocytes in a dormant follicle, a growing follicle, and a Graafian follicle?

b. Are polar bodies visible in your prepared slide of a cat ovary? Why or why not?

After ovulation the remaining follicle cells form the **corpus luteum** on the surface of the ovary. The corpus luteum produces hormones that prepare the uterus for the potential arrival of a fertilized egg.

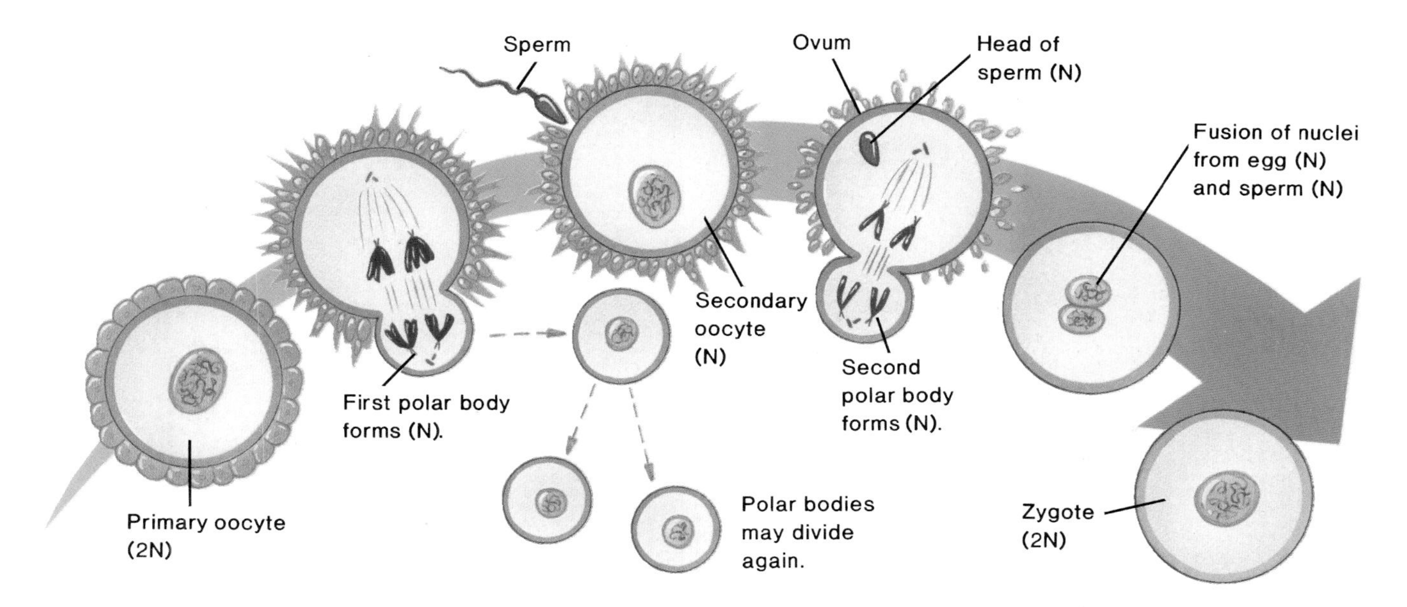

Figure 14.5
Oogenesis. A primary oocyte is diploid (2N). After its first meiotic division, one product is eliminated as a polar body. The other product, the secondary oocyte, is released during ovulation. Sperm penetration stimulates the second meiotic division, and a second polar body and a haploid ovum are produced. Fusion of the haploid (1N) ovum nucleus with a haploid (1N) sperm nucleus produces a diploid (2N) zygote which subsequently forms an embryo.

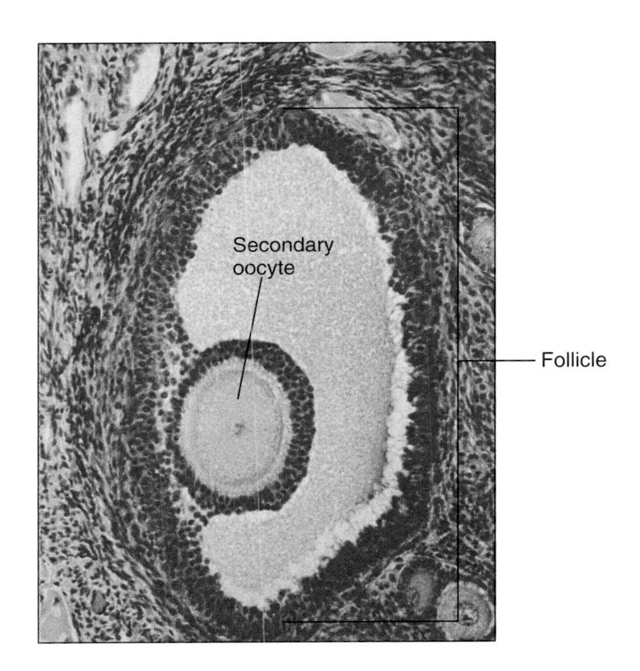

Figure 14.6
A mature secondary oocyte in an ovarian follicle of a cat. This secondary oocyte awaits ovulation.

Plant Gametogenesis

The formation of gametes in plants is somewhat different because their sexual life cycle includes an alternation of generations between haploid and diploid forms. However, meiosis is still the critical process by which plants reduce the number of chromosomes by half to prepare for gamete production.

In flowering plants, meiosis occurs in the anthers and ovary of the flowers. In the anther the spores resulting from meiosis produce a stage of the life cycle (pollen) that will eventually produce male gametes. In the ovary the spores resulting from meiosis produce a stage of the life cycle (ovule) that will eventually produce female gametes. You'll learn more about these events in Exercise 30. In this procedure you will observe prepared slides showing stages of the beginning, middle, and end of meiosis I and II in a representative plant.

Procedure 14.2

Diagram and observe stages of meiosis

1. In figure 14.7, diagram a plant cell with three pairs of chromosomes in each of the stages of meiosis. Be sure to label the cell wall and cell plate. Two extra cells are provided for you to diagram midphases to demonstrate the continuity of the events of mitosis.
2. Examine the following prepared slides of stages of meiosis in a *Lilium* anther (see figs. 30.10, 30.11).
   - a. *Lilium* anther—early prophase I
   - b. *Lilium* anther—late prophase I
   - c. *Lilium* anther—first meiotic division
   - d. *Lilium* anther—second meiotic division
   - e. *Lilium* anther—pollen tetrads. Each of these cells will produce a pollen grain.

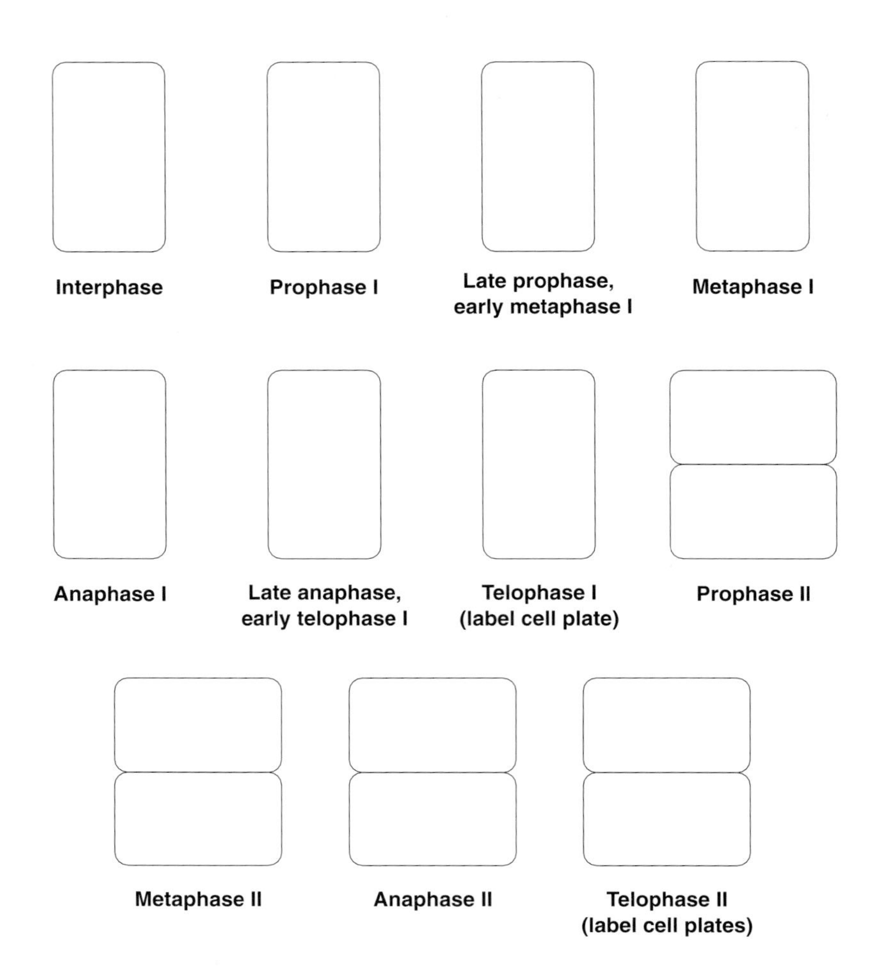

Figure 14.7
Stages of meiosis in plants.

Table 14.2
A Comparison of the Major Features of Mitosis and Meiosis

Mitosis	Meiosis
Performed by all body cells at some stage in their development	
Produces two cells	
Includes one nuclear division	
Resulting cells are diploid	
Resulting cells are genetically identical	
No pairing of homologous chromosomes or crossing-over	

3. Examine the following prepared slides of stages of meiosis in a *Lilium* ovary.

   a. *Lilium* ovary—"mother cell," prophase I

   b. *Lilium* ovary—binucleate stage, end of meiosis I

   c. *Lilium* ovary—four nucleate stage, end of meiosis II

MITOSIS VS. MEIOSIS

Mitosis and meiosis are both forms of cellular replication but they play different roles in the life cycle of animals and plants. Mitosis may occur in either haploid or diploid cells and is necessary for cell production and growth. Meiosis occurs in diploid cells. Its role is to produce cells with a reduced number of chromosomes and shuffle the genetic material so an organism can reproduce sexually. To compare mitosis and meiosis, review table 14.2 and complete the column with the contrasting features of meiosis.

Questions for Further Thought and Study

1. Would evolution occur without the events of meiosis and sexual reproduction? Why or why not?

2. What are the general characteristics of sexual reproduction in humans and other vertebrates that are associated with continuous production of many sperm cells but intermittent, finite production of egg cells?

3. Which process is most accurately referred to as nuclear division: meiosis or mitosis?

4. What special event does interkinesis lack compared to premeiotic interphase?

5. How are mammalian sperm cells produced and incubated at a lower temperature than body temperature?

6. How old is an ovulated oocyte of a 35-year-old woman? What consequences does this have?

WRITING TO LEARN BIOLOGY

Wouldn't it be easier for a cell simply to divide the chromosomes once rather than duplicating them and then dividing them twice during meiosis? Why do you suppose this isn't done?

15

Molecular Biology and Biotechnology

DNA Isolation and Bacterial Transformation

Objectives

By the end of this exercise you should be able to:

1. Isolate DNA from a bacterium.
2. Understand how temperature and pH affect DNA.
3. Insert a gene for resistance to ampicillin into a bacterium.

Biotechnology is the manipulation of organisms to do practical things and to provide useful products. Biotechnology has been around for centuries: humans have selectively bred livestock for meat products, controlled pollination to produce more productive food crops, and used bacteria and fungi to make wine and cheese. But recent progress in the science of molecular biology has revolutionized biotechnology. The revolution stems from new molecular techniques that have made genetic engineering possible.

Genetic engineering is the direct manipulation of genes for practical purposes. Genetic engineers can intervene *directly* in the genetic fate of organisms. We can isolate genes, move them from one organism to the next, and even move genes from one species to the next. The most common goals of this engineering are to harvest the valuable proteins made by engineered genes and to benefit from the new characteristics the genes provide to the target organisms, including humans. Indeed, the impact on society of the current revolution in genetic engineering may soon surpass that of such historical changes as the industrial revolution of the past two centuries.

Genetic engineering got its start in 1973 when Stan Cohen and Herb Boyer transplanted a gene for antibiotic resistance from a frog into a bacterium, and thus "engineered" an antibiotic-resistant organism. In 1980 molecular biologists successfully inserted a human gene for interferon, an antiviral drug, into a bacterium. When the "transformed" bacterium reproduced, it generated billions of progeny, each of which was a miniature drug factory. As a result of this engineering, biologists could cheaply harvest a drug that was previously expensive and generally unavailable. Genetic engineering has since been applied to medicine (gene therapy, drug production) and the production of new foods and environmentally benign pesticides.

At the heart of genetic engineering is the science of **molecular biology** (i.e., the study of molecules critical to life). Molecular biologists recently have concentrated their efforts on manipulating "information molecules" such as DNA and proteins because all outward characteristics of organisms have their basis in proteins from genes made of DNA. In addition to providing techniques for genetic engineering, molecular biology has also impacted fields such as forensics (e.g., linking suspects to crimes, settling paternity disputes) and hiring practices (pinpointing employees at high risk for cancer) just to name a few. To introduce yourself to the core information of molecular biology, you should review DNA structure in your textbook and in Exercise 5, "Biologically Important Molecules."

In this exercise you'll learn two techniques used routinely by molecular biologists: (1) isolation of DNA, and (2) genetic transformation.

ISOLATION OF DNA

Isolation of DNA is a routine and important procedure for molecular biologists. Once isolated, the DNA and the sequence of its subunits can be determined, manipulated, or altered. Bacteria each contain only about 10^{-14} g of DNA, which accounts for approximately 5% of the organism's dry weight. However, molecules of this DNA can be very long; for example, the DNA in an *E. coli*, if strung out, would be approximately 1 mm long. (By analogy, if the bacterium was the size of a grapefruit, then its DNA would be more than 80 km long.) The ability to pack this much DNA into a tiny cell is impressive, especially because DNA is a rather stiff molecule (fig. 15.1).

When DNA or other large molecules are expelled and isolated from cells, the surrounding solution becomes very viscous (i.e., thick, syrupy, and resistant to flow). This is because DNA molecules are long and tend to stick to each other due to cohesion among molecules and hydrogen

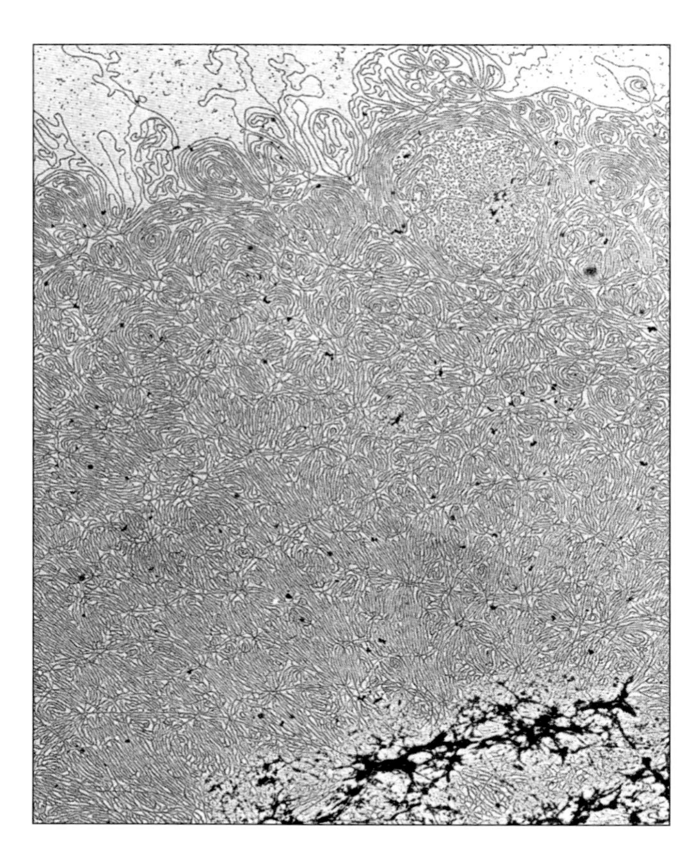

Figure 15.1

A human chromosome contains an enormous amount of DNA. The dark element at the bottom of the photograph is part of the protein matrix of a single chromosome. All of the surrounding material is the DNA of the chromosome.

bonding (recall that hydrogen bonding also holds the two strands of the double helix together; see figure 5.9). Harsh chemicals will nonspecifically disrupt hydrogen bonds, including those between corresponding nitrogenous bases of DNA. Molecular collisions produced by high heat can tear molecules apart. Any breakdown in molecular structure will reduce viscosity.

In this exercise, you will isolate DNA from *Halobacterium salinarum*, a halophilic ("salt-loving") bacterium that grows only in salty environments (4–5 M NaCl). It's especially easy to isolate DNA from this organism because its cell walls disintegrate when placed in low-salt environments (0–2 M NaCl).

Procedure 15.1

Isolate DNA from *Halobacterium salinarum*

1. Several days ago your laboratory instructor prepared petri dishes of culture media and inoculated them with *Halobacterium salinarum*. Obtain one of these cultures from your instructor.
2. Use a flexible plastic ruler to scrape the bacterial growth from the surface of the agar and collect it in a test tube.
3. Add 1 mL of distilled water to the tube. One milliliter equals about 20 drops.
4. Place a small piece of plastic film over the top of the test tube and hold it securely in place with your thumb. Invert the tube several times to mix the contents. This mixing with water lyses the cells, and the liquid will become viscous.
5. Use an eyedropper to slowly and gently pour 1 mL of ice-cold 95% ethanol down the side of the tube. Do not shake or disturb the tube. DNA will precipitate at the interface between the layers of water and ethanol.
6. Insert a glass rod into the liquid. Rotate the rod. DNA will adhere to the rod as it is twirled.

Question 1

What is the texture of the DNA you've isolated?

The DNA that you've isolated is not pure; rather, it is contaminated with small amounts of protein and RNA.

The Influence of Heat and pH on DNA

Heat and pH strongly affect the properties of DNA. For example, DNA typically denatures at alkaline pH and at 80–97°C. Test these effects with the following procedures.

Procedure 15.2

Test the influence of heat on DNA

1. Precipitate DNA in a test tube following steps 1–5 in procedure 15.1 for isolating DNA.
2. Place the tube into a boiling water-bath for 10 min.
3. Place the tube in an ice bath.
4. Insert and twirl a glass rod in the tube.
5. Compare the viscosity of the heat-treated DNA with untreated DNA.

Question 2

a. What effect does heat have on the viscosity of DNA?

b. What do you think is the mechanism for this change in viscosity?

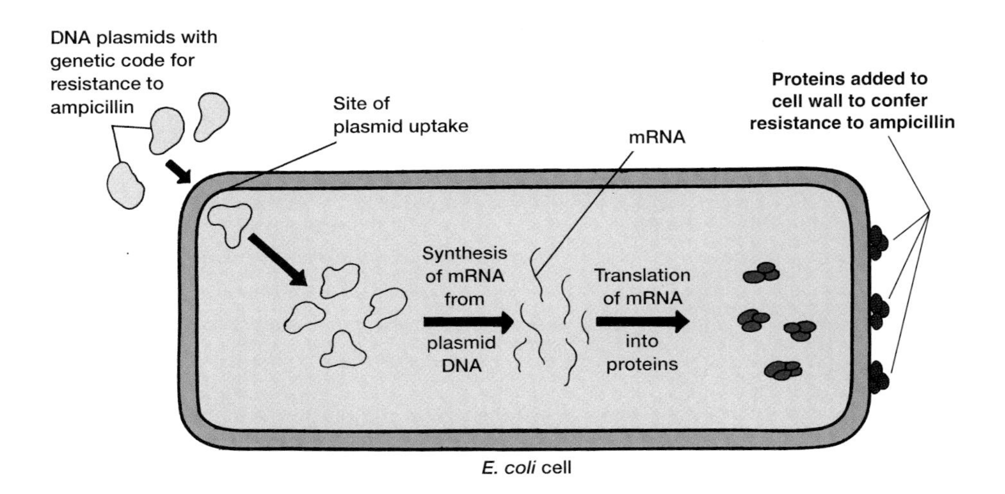

Figure 15.2

Transformation of an *E. coli* cell with plasmid DNA. In this example, the DNA plasmid contains the genetic code for resistance to the antibiotic ampicillin. After uptake of the plasmid, the code is transcribed to messenger RNA, which is translated during protein synthesis. Addition of these proteins to the cell wall will retard attack by ampicillin.

Procedure 15.3

Test the influence of pH on DNA

1. Precipitate DNA in a test tube following steps 1–5 in procedure 15.1 for isolating DNA.
2. Add 2 mL of 1.0 N NaOH to the tube.

CAUTION

NaOH is caustic. Don't spill it on yourself or your clothes.

3. Insert and twirl a glass rod in the tube.
4. Compare the viscosity of the alkali-treated DNA with untreated DNA and with heat-treated DNA.

Question 3

a. What effect does alkaline pH have on the viscosity of DNA?

b. What do you think is the mechanism for this pH-induced change in viscosity?

GENETIC TRANSFORMATION

Much of biotechnology is based on **genetic transformation,** which is the uptake and expression of DNA by a living cell. A successful transformation, summarized in figure 15.2, requires three conditions: (1) a host into which DNA can be inserted, (2) a means of carrying the DNA into the host, and (3) a method for selecting and isolating the organisms that were successfully transformed.

The Host: *Escherichia coli*

The **host organism** you'll use is *Escherichia coli,* one of the most intensively studied organisms in the world. *E. coli* has the following properties that make it ideally suited for transformation:

- It contains only one chromosome made of five million base pairs. This is less than 0.2% of that of the human genome.
- *E. coli* grows rapidly. Transformations in bacteria are rare and occur in only about 0.1% of cells. Therefore, transformations are observed most easily in large, rapidly growing populations. *E. coli* is ideal for transformation studies because, in ideal conditions, it divides every 20 min. As a result, in 10 h a bacterium can produce a billion progeny (30 generations) in only 1 mL of nutrient broth.

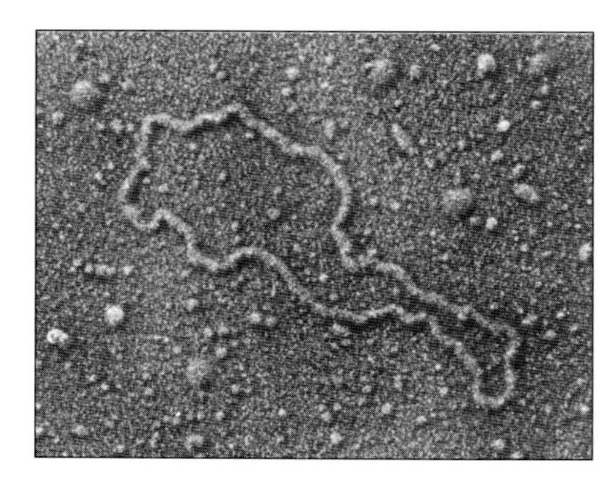

Figure 15.3
A famous plasmid. The circular molecule in this electron micrograph was the first plasmid used successfully to clone a vertebrate gene, pSC101. Its name refers to it being the hundred-and-first plasmid isolated by Stanley Cohen.

Only a small percentage of bacterial cells in a culture can be transformed. Also, small lengths of DNA are taken up more readily than long lengths. However, competence of the bacteria (the chances for successful transformation) increases during the early and middle stages of its growth. Competence also increases when cells suspended in a cold solution of $CaCl_2$ are heat-shocked. Yield is usually about 10^6 transformants per milligram of DNA available for insertion.

A Vector to Move DNA into the Host

A biological **vector** is a DNA molecule that carries DNA sequences into a host. The simplest bacterial vectors are **plasmids,** which are circular pieces of DNA made of 1,000 to 200,000 base pairs (fig. 15.3). Plasmids exist separately from the bacterial chromosome, and they must contain a gene that confers some selective advantage (e.g., resistance to an antibiotic) to remain in the host. We don't completely understand how plasmids enter host cells, but they seem to enter consistently.

Selecting Transformed Organisms

You'll insert into *E. coli* a plasmid (pAMP) containing a gene for resistance to ampicillin, an antibiotic lethal to many bacteria (fig. 15.4). Refer to Exercise 23 for information on bacterial cell wall structure and how ampicillin might kill bacteria. Then you'll select transformed bacteria based on their resistance to ampicillin by spreading the transformed organisms onto nutrient medium containing ampicillin. Organisms that grow on this medium have been transformed. Because *E. coli* grows so fast, you can check for transformed organisms only 12–24 h after completing the experiment.

Procedure 15.4

Transform *E. coli*

A. Preparation

1. Carefully read this procedure and review figure 15.4 before beginning. Wash your hands. Your instructor will demonstrate the steps of the procedure that require sterile technique.
2. Fill a 250 mL beaker with about 50 mL of ethanol and place a glass-rod bacterial spreader in the ethanol to soak.

B. Preincubation to increase competency

3. Obtain a test tube with 1 mL of sterile, yellow nutrient broth, and a tube with 1 mL of clear, colorless 50 mM $CaCl_2$. Label these tubes NB and $CaCl_2$, respectively.
4. Fill a 250 mL beaker half full with crushed ice. Place the tube of $CaCl_2$ in the beaker of ice.
5. Obtain two sterile, plastic transformation tubes and label one of them (+)P, meaning with plasmid, and the other (−)P, meaning without plasmid. Place them in the beaker of ice.
6. Obtain a packaged, sterile, plastic pipet and locate the graduation indicating a volume of 0.25 mL.
7. Open the packaged pipet without touching and contaminating the pipet's open end. Use this pipet to add 0.25 mL of a sterile, ice-cold solution of 50 mM $CaCl_2$ to each of the two transformation tubes. Use sterile technique. This is shown in step 1 of figure 15.4.
8. Several days ago your lab instructor streaked starter plates of nutrient agar with *E. coli*. Scrape a colony (3-mm diameter) from one of these plates with a sterile, plastic inoculating loop. Use sterile technique. This is shown in step 2 of figure 15.4.
9. Place the loop full of bacteria into the transformation tube labeled (−)P. Rinse the bacteria from the loop by gently twirling the loop handle between your fingers.
10. To mix the bacterial suspension, open a sterile, packaged pipet. Insert the pipet into the suspension in the bottom of the tube and gently use a rubber bulb to suck the fluid in and out of the tip three or four times. Be sure the tip is empty before withdrawing the pipet.
11. For tube (+)P, repeat steps 8–10. Use a fresh loop and pipet to inoculate and mix the bacteria. Try to get the same amount of bacteria into each tube. Replace the two tubes in the ice bath and chill the tubes for at least 5 min.

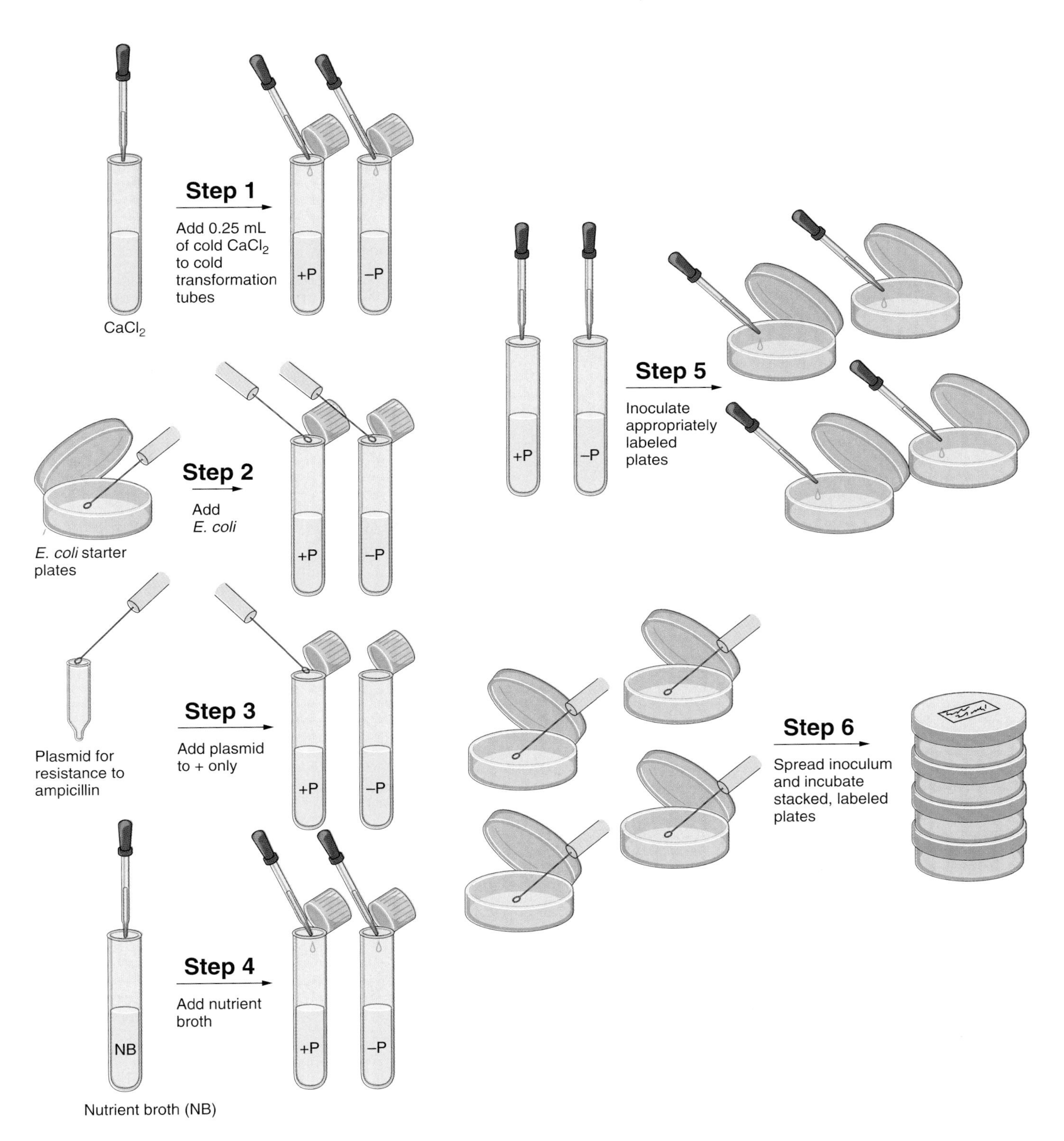

Figure 15.4
Summary of the procedure to transform bacteria by exposing *E. coli* to a plasmid.

C. Incubation

12. Use a sterile loop to obtain one loopful (10 μL) of an ice-cold solution of DNA plasmids from a vial kept by your instructor. Add this loopful of plastids to tube (+)P and gently rotate the loop to rinse the plasmids from the loop. These plasmids contain the gene for resistance to ampicillin. Do **not** add plasmids to tube (−)P. This is shown in step 3 of figure 15.4.

13. Place both tubes in ice for 15 min.

14. While the tubes are cooling, obtain two agar plates labeled (−)AM, meaning nutrient agar without ampicillin, and two plates labeled (+)AM, meaning nutrient agar with ampicillin. Label one of each pair of plates as (−)P and the other two plates as (+)P.

D. Heat shock

15. Heat-shock the transformation tubes (−)P and (+)P by placing them in a 42°C water-bath for 2 min. Heat shock increases the uptake of the plasmid by the bacterial cells.

16. Chill the tubes in ice for 5 min.

17. Remove the tubes from the ice bath; place them in a test-tube rack or empty beaker.

E. Recovery and plating the samples

18. Using a sterile pipet, add 0.25 mL nutrient broth to each tube and tap gently to mix the contents. This is shown in step 4 of figure 15.4.

19. Using sterile technique and a sterile pipet, transfer 0.10 mL of the (−)P cell suspension onto the agar's surface in the middle of the plate labeled (+)AM/(−)P. Transfer another 0.10 mL onto the plate labeled (−)AM/(−)P. Close the plates. This is shown in step 5 of figure 15.4.

20. Using another sterile pipet, transfer 0.10 mL of the (+)P cell suspension onto the middle of the plate labeled (+)AM/(+)P. Transfer another 0.10 mL onto the plate labeled (−)AM/(+)P. Close the plates.

21. Light an alcohol lamp. Dip the bacteria spreader into the ethanol and then into the flame of an alcohol lamp. Let the ethanol burn away; then count to 10 to let the spreader cool. Before spreading the bacteria in the first of your four plates, further cool the spreader by touching it to the agar at the edge of the plate. Then touch the spreader to the cell suspension in the middle of the plate and gently drag it back and forth three times. Rotate the plate 90° and repeat. This is shown in step 6 of figure 15.4. Remember to sterilize the bacteria spreader between each plate. Repeat this procedure to spread the bacteria on the other three plates.

F. Selecting the transformed organisms

22. Put your name and date on each of the four plates, and tape the plates together. Incubate the plates upside down at 37°C.

23. Place all the tubes, loops, etc., in a central location for disposal. Wipe your work area with a weak bleach solution and wash your hands before leaving the laboratory.

24. In the space below indicate your predictions for growth (+) or no growth (−). Explain your reasoning for each prediction in the provided space.

(−)AM/(+)P

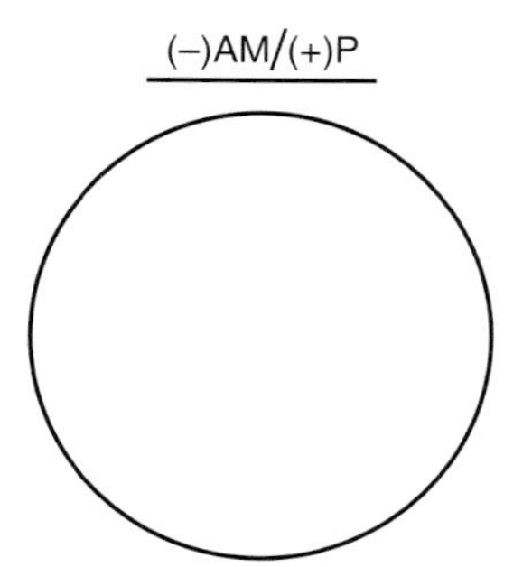

Explain your prediction:

(−)AM/(−)P

Explain your prediction:

(+)AM/(+)P

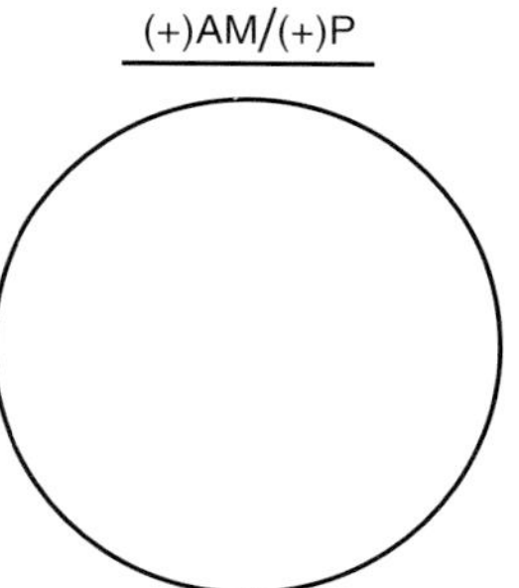

Explain your prediction:

(+)AM/(−)P

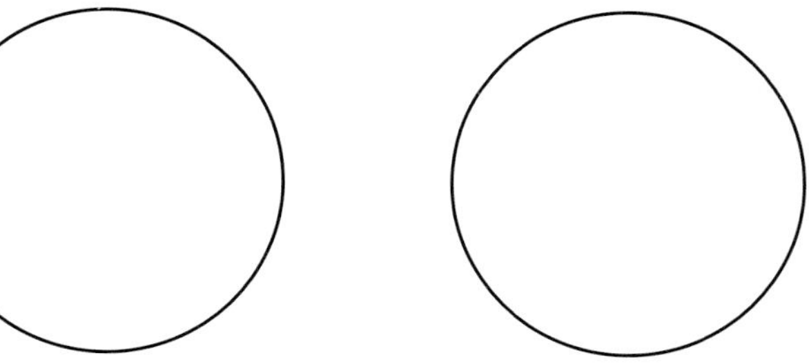

Explain your prediction:

25. In the space below indicate with a (+) which plates had bacterial growth after 24 h. Draw the appearance and coverage of bacterial colonies and explain possible reasons for growth *and* possible reasons for no growth.

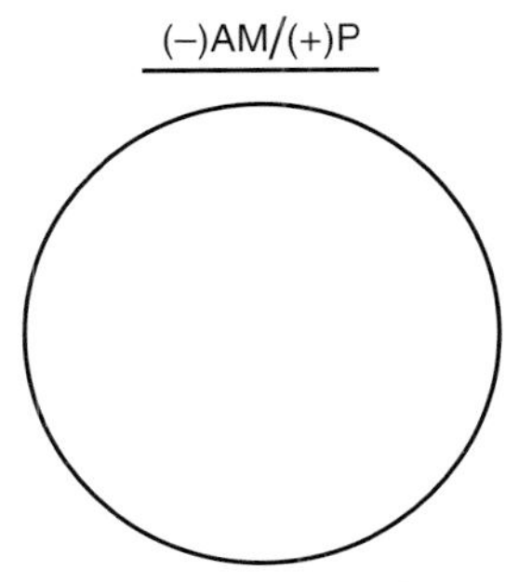

Reason for growth:

Reason for no growth:

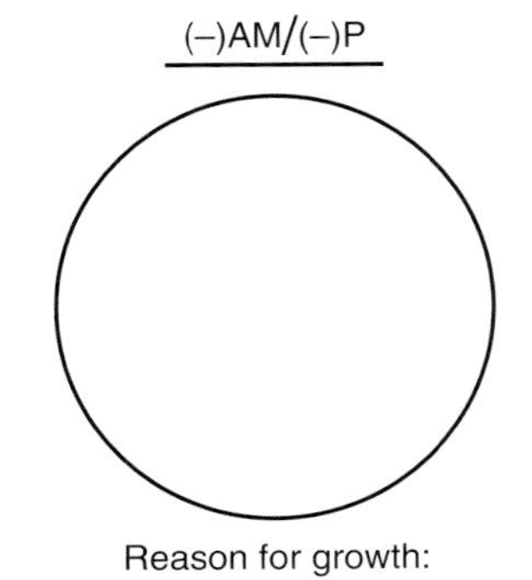

Reason for growth:

Reason for no growth:

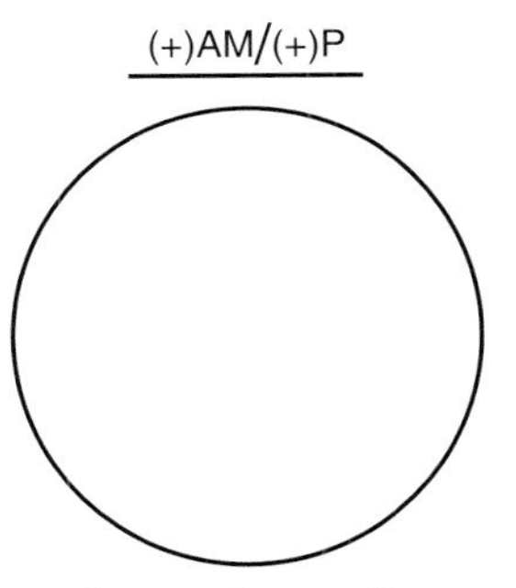

Reason for growth:

Reason for no growth:

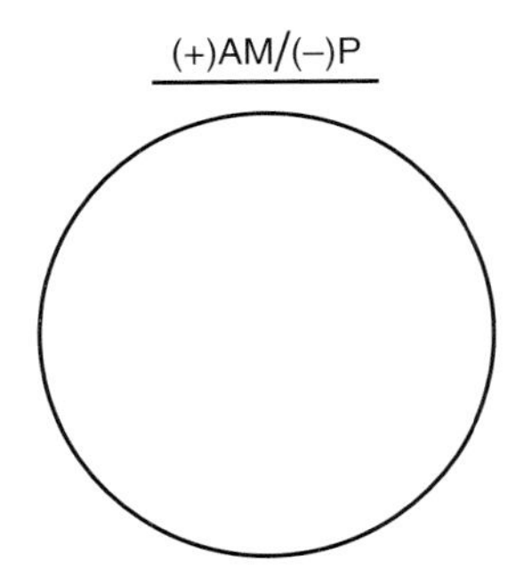

Reason for growth:

Reason for no growth:

Question 4

a. Which treatment produced transformed bacteria?

b. How many transformed colonies grew on each plate?

c. What was the purpose of tube 2 without plasmid?

Questions for Further Thought and Study

1. Consider the ubiquitous occurrence of bacteria in nature, along with the constant fragmentation and release of DNA as cells decompose. How frequently do genetic transformations occur in nature? Explain your answer.

2. How could genetic transformations improve our quality of life? How could they decrease our quality of life?

3. Why is molecular biology often referred to as genetic engineering or biotechnology?

DOING BIOLOGY YOURSELF

Design and conduct an experiment to test the effects of acid pH on the integrity of isolated DNA. How do the results compare to the effects of basic solutions on DNA?

WRITING TO LEARN BIOLOGY

Many people resist the use of genetic engineering to alter organisms. What are their arguments? Do you agree?

16

Genetics
The Principles of Mendel

Objectives

By the end of this exercise you should be able to:

1. Describe simple genetic dominance, incomplete dominance, and lethal inheritance.
2. Describe possible genotypes for some of your personal traits that are inherited as dominant and recessive genes.
3. Explain the importance of Mendel's two laws.
4. Distinguish between an organism's phenotype and genotype.

Published papers are the primary means of communicating scientific discoveries (see Appendix III). One of the most famous of these papers, entitled "Experiments in Plant Hybridization," was written in 1866 by Gregor Mendel, an Austrian monk. Although this paper later became the basis for genetics and inheritance, it went largely unnoticed until it was rediscovered independently by several European scientists in 1900. The experiments and conclusions in Mendel's paper now form the foundation of **Mendelian genetics,** the topic of today's exercise.

Mendel's greatest contribution was to replace the blending theory of inheritance, which stated that all traits blend with each other, with the **particulate theory.** Mendel's particulate theory states that (1) inherited characters are determined by particular factors (now called genes), (2) these factors occur in pairs (i.e., genes occur on maternal and paternal homologous chromosomes), and (3) when gametes form, these genes segregate so that only one of the homologous pair is contained in a particular gamete. Recall from Exercise 14 (Meiosis) that each gamete has an equal chance of possessing either member of a pair of homologous chromosomes. This part of the particulate theory is collectively known as Mendel's First Law, or the **Law of Segregation.** Mendel's Second Law, or the **Law of Independent Assortment,** states that genes on nonhomologous or different chromosomes will be distributed randomly into gametes.

Before you start this exercise, briefly review in your textbook some principles and terms pertinent to today's exercise. A gene is a unit of heredity on a chromosome. A **gene** has alternate states called **alleles,** which are contributed to an organism by its parents. Alleles for a particular gene occur in pairs. Alleles that mask expression of other alleles but are themselves expressed are **dominant;** these alleles are usually designated by a capital letter (for example, *T*). Alleles whose expression is masked by dominant alleles are **recessive,** and they are designated by a lowercase letter (for example, *t*). The **genotype** of an organism includes all the alleles present in the cell, whether they are dominant or recessive. The physical appearance of the trait is the **phenotype.** Thus, if tallness (*T*) is dominant to dwarfness (*t*), a tall plant can have a genotype *TT* or *Tt*. A dwarf plant can only have a genotype *tt*. When the paired alleles are identical (*TT* or *tt*), the genotype is **homozygous. Heterozygous** refers to a pair of alleles that are different (*Tt*). With this minimal review, you're prepared to apply this information to solve some genetics problems.

SIMPLE DOMINANCE

Assume that tallness is dominant to dwarfness. If a homozygous tall plant is crossed (mated) with a homozygous dwarf plant, what will be the phenotype (physical appearance) and genotype of the offspring?

Parents:	*TT* (homozygous tall) × *tt* (homozygous dwarf)
Gametes:	*T* from one parent and *t* from the other parent
Offspring:	genotype = *Tt* phenotype = tall

This first generation of offspring is called the **first filial** or **F_1 generation.**

Each of the F_1 offspring can produce two possible gametes, *T* and *t*. Mendel noted that the gametes from each of the parents combine with each other randomly. Thus, you can simulate the random mating of gametes from the F_1 generation by flipping two coins simultaneously. Assume

TABLE 16.1

RESULTS OF COIN-FLIPPING EXPERIMENT SIMULATING RANDOM MATING OF HETEROZYGOUS (*Tt*) INDIVIDUALS

Response	Number
Heads-heads = *TT* = tall	
Heads-tails = *Tt* = tall	
Tails-tails = *tt* = dwarf	

that heads designates the tall allele (*T*), and tails designates the dwarf allele (*t*). Flipping one coin will determine the type of gamete from one parent and flipping the other will determine the gamete from the other parent. To demonstrate this technique, flip two coins simultaneously 64 times and record the occurrence of each of the three possible combinations in table 16.1.

Question 1

What is the ratio of tall (*TT* or *Tt*) to dwarf (*tt*) offspring?

Keep these results in mind and return to the original problem: What are the genotypes and phenotypes of the offspring of the F_1 generation?

Parents:	*Tt* × *Tt*
Gametes:	(*T* or *t*) × (*T* or *t*)
Offspring:	*TT Tt tT* *tt*
	3 tall 1 dwarf

Thus, the theoretical genotypic ratio for the offspring of the F_1 generation is 1 *TT* : 2 *Tt* : 1 *tt*, and the phenotypic ratio is 3 tall : 1 dwarf.

Question 2

a. How do these ratios compare with your data derived from coin-flipping?

b. Would you have expected a closer similarity if you had flipped the coins 64,000 times instead of 64 times? Why or why not?

Procedure 16.1

Determine genotypic and phenotypic ratios for albinism

Albinos are homozygous recessive for the pair of alleles that produce pigments of skin, hair, and eyes. Suppose a woman with normal colored skin and having an albino mother marries an albino man. Record the genotypic and phenotypic ratios of their children.

Genotype of children's mother _______

Genotype of children's father _______

Possible gametes of mother _______

Possible gametes of father _______

Possible offspring _______

Genotypic ratio of children _______

Phenotypic ratio of children _______

Procedure 16.2

Determine color and height ratios for corn plants

Color of grains (karyopses) and height of *Zea mays* (corn) plants are often determined by a single gene.

1. Examine (a) the ears of corn having red and yellow grains, and (b) the tray of tall and dwarf plants on demonstration.
2. Record your observations below and determine the probable genotypes of the parents of each cross.
   Probable genotypes of parents:

   Color of Corn Grains

   Number of red grains _______

   Number of white grains _______

   Ratio of red : white grains _______

   Probable genotypes of parents _______ × _______

   Height of Plants

   Number of tall plants _______

   Number of dwarf plants _______

   Ratio of tall : dwarf plants _______

   Probable genotypes of parents _______ × _______
3. The preceding crosses involved only one trait and thus are termed **monohybrid crosses.** Let's now examine a cross involving two traits; that is, a **dihybrid cross.** Your instructor will review with you the basis for working genetics problems involving dihybrid crosses.

In corn, red (R) seed color is dominant to white (r) seed color, and smoothness (S) is dominant to wrinkled (s) seed. Observe the cobs of corn derived from a cross between parents having genotypes *RRSS* and *rrss*.

Question 3

a. What is the expected genotype for the F_1 generation?

b. Will all F_1 offspring (seeds) have the same genotype?

Question 4

a. What are the predicted genotypes for the F_2 generation?

b. In what ratio will they occur?

4. To test your prediction in Question 4 count the number of kernels for five rows for each of the following phenotypes:

Red, smooth ________

Red, wrinkled ________

White, smooth ________

White, wrinkled ________

Question 5

a. What are the genotypes of the F_1 generation?

b. How did your data compare with those that you predicted?

INCOMPLETE DOMINANCE

Some traits such as flower color are controlled by incomplete dominance. In this type of inheritance, the heterozygous genotype results in an intermediate characteristic. For example, if a plant with red flowers (*RR*) is crossed with a plant having white flowers (*rr*), all of the offspring in the first filial (F_1) generation will have pink flowers (*Rr*).

Parents:	*RR* (red) × *rr* (white)
Gametes:	*R* × *r*
Offspring:	*Rr* (pink)

Question 6

What are the expected ratios of red, pink, and white flowers in a cross involving two pink-flowered parents?

LETHAL INHERITANCE

Lethal inheritance involves inheriting a gene that kills the offspring. Observe the tray of green and albino seedlings of corn. The albino plants cannot photosynthesize and therefore die as soon as their food reserves are exhausted.

Question 7

a. What is the ratio of green to albino seedlings?

b. Based on this ratio, what might you expect were the genotypes of the parents?

Question 8

Why is it impossible to cross a green and an albino plant?

OTHER SOURCES OF GENETIC DIVERSITY

Genetic diversity can also result from multiple alleles, gene interactions (epistasis), continuous variation, pleiotropy, environmental effects, linkage, and sex linkage. Although time limitations prohibit exercises about these topics, be sure to review them in your textbook.

BLOOD TYPE

Blood type of humans provides an excellent example of **codominance,** another type of Mendelian inheritance. All individuals have one of four blood types: A, B, AB, and O (fig. 16.1). These blood groups are determined by the

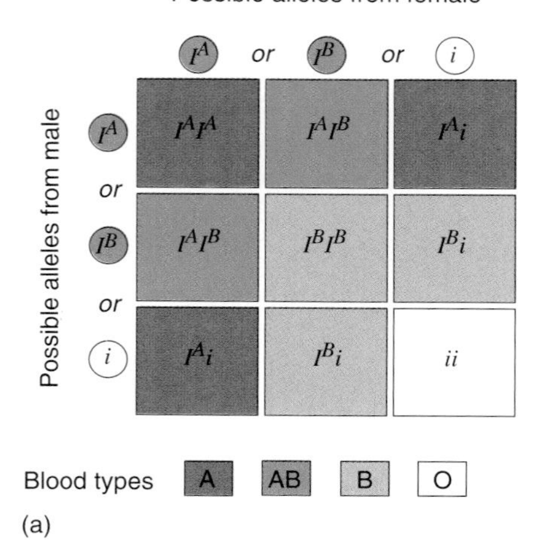

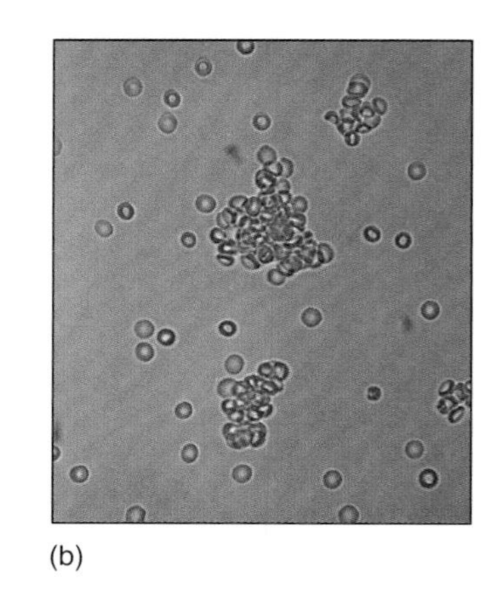

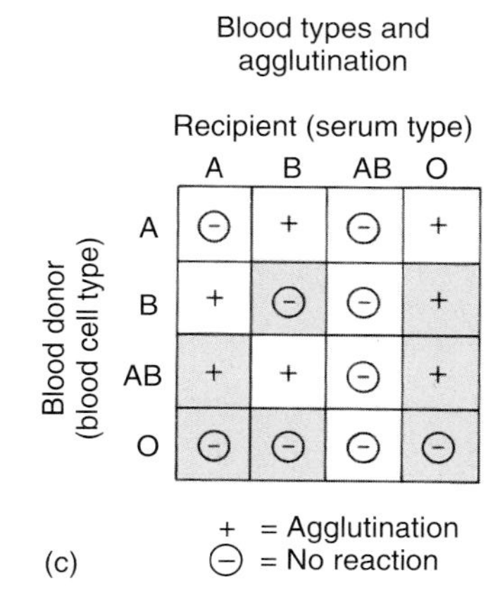

Figure 16.1
ABO blood groups. (*a*) Multiple alleles control the ABO blood groups. Different combinations of the three *I* gene alleles result in four different blood type phenotypes: type A (either I^AI^A homozygotes or I^Ai heterozygotes), type B (either I^BI^B homozygotes or I^Bi heterozygotes), type AB (I^AI^B heterozygotes), and type O (ii homozygotes). (*b*) The blood agglutination reaction. Agglutination occurs when blood cells stick and clump together. (*c*) Agglutination will occur when donor blood cells are incompatible with recipient serum, as designated by a +.

TABLE 16.2

CHARACTERISTICS OF INDIVIDUALS WITH THE FOUR MAJOR BLOOD TYPES

Blood Type	Antigen on Red Blood Cell	Antibody	Genotype
A	A	anti-B	I^AI^A or I^Ai
B	B	anti-A	I^BI^B or I^Bi
AB	A and B	none	I^AI^B
O	O (none)	anti-A and anti-B	ii

presence of **antigens** (proteins) on the surfaces of their red blood cells. If a person has antigen-A on his or her blood cells, then the person has type A blood and possesses blood **antibodies** (proteins) that agglutinate type B blood cells. Similarly, a person having antigen-B on his or her blood cells has type B blood and has antibodies that agglutinate type A blood cells. If a person has antigen-A and antigen-B on his or her blood cells, then the person has type AB blood and lacks A and B antibodies. If a person has no A or B antigens on his or her blood cells, the blood type is O and the person possesses antibodies against both A and B antigens (table 16.2). This system is rather unusual in that individuals have antibodies against the blood antigens that they do not possess.

Blood typing is often important for establishing the possible identity of an individual in forensic work and paternity suits. For example, assume that a woman with type O blood has a child having type O blood. The suspected father has type AB blood.

Could the suspected father with type AB blood be the child's father? The answer is no, because the cross would have the following results:

Parents:	ii (type O) × I^AI^B (type AB)
Gametes:	i and i × I^A and I^B
Offspring:	I^Ai or I^Bi

Half of the offspring from the mother and the suspected father would have type A blood (genotype = I^Ai), and the other half would have type B blood (genotype = I^Bi). Thus, the suspected father could not have fathered a child with blood type AB or with type O blood with this mother.

Suppose there is a mixup of children in the maternity ward of a hospital after the genotypes of the children are determined from the parents' blood types. The following unidentified children have these blood types:

Child 1: type A (genotype I^AI^A or I^Ai)
Child 2: type B (genotype I^BI^B or I^Bi)
Child 3: type AB (genotype I^AI^B)
Child 4: type O (genotype ii)

Question 9
Which child or children could belong to a couple having AB and O blood types?

Blood typing is also important for determining the safety of blood transfusions. Your body automatically produces antibodies for antigens you do not carry (fig. 16.1). For example, people with type A blood have antibodies against B antigens, and people with type B blood produce antibodies against A antigens. If someone having type B blood received blood from someone having type A blood, the recipient's antibodies would react with and agglutinate the red blood cells received from the donor (fig. 16.1*b*). As a result, the recipient would die.

Question 10

a. Can a person with type O blood safely donate blood to a person having type A blood? Why or why not?

b. Which blood type would be a universal donor?

c. Which blood type would be a universal recipient?

Procedure 16.3

Determine blood type for ABO system

1. You will be provided with various samples of synthetic blood. This material is designed to simulate the blood type characteristics of human blood, and it is safe. Also obtain two bottles of antisera.
2. Obtain a clean slide and label the ends A and B. Near one end of the slide place a drop of antiserum A (containing antibodies against antigen-A), and near the other end of the slide place a drop of antiserum B (containing antibodies against antigen-B).
3. Place drops of blood near (but not touching) the two drops of antisera.
4. Mix one of the drops of blood with antiserum A and one with antiserum B. Use a different toothpick to mix each antiserum.
5. Dispose of all used materials properly.
6. Observe any agglutination of blood cells in either of the two antisera.

Agglutination of blood mixed with an antiserum is indicated by a grainy appearance. Agglutination indicates the presence of the respective antigen on red blood cells (fig. 16.1). Determine and record the blood type from your sample based on the presence of antigens.

Question 11

a. What antigens are present on the artificial red blood cells that you tested?

b. What is the blood type of your sample?

You are probably familiar with another characteristic of blood called **Rh factor.** Although more than two alleles determine Rh, we'll use "positive" and "negative" for simplicity and convenience.

Procedure 16.4

Determine Rh

1. Place a drop of anti-Rh serum on a clean slide.
2. Using the procedure just described, mix a drop of blood from the synthetic blood sample provided with the antiserum.
3. Label the slide with your initials and place it on the warming plate in the lab.
4. The blood sample will agglutinate within a few minutes if it is Rh-positive. The absence of agglutination indicates the blood is Rh-negative.
5. Dispose of all materials properly.

Rh Incompatibility

You've probably heard of the incompatibility (agglutination) problems that Rh-negative women may have with their Rh-positive babies (the Rh-positive trait is inherited from the child's father). This problem usually occurs with the second and subsequent children, because women with the Rh blood system must be sensitized to the antigen before antibody production begins. This sensitization usually occurs during birth of the first child.

If you are a woman having Rh-negative blood, you should be concerned but not alarmed. Rh incompatibility is handled routinely by injections of anti-Rh antibodies after delivery of the first child. These antibodies destroy Rh-positive red cells and thus eliminate the Rh-associated risk of subsequent childbirth.

TABLE 16.3

PHENOTYPES AND GENOTYPES OF HUMAN TRAITS

Characteristic	Your Phenotype	Your Genotype	Phenotypes of Class	
			Dominant	Recessive
Widow's peak				
Bent little finger				
Albinism				
Pigmented iris				
Attached earlobes				
Hitchhiker's thumb				
Interlacing fingers				
PTC tasting				
Mid-digital hair				

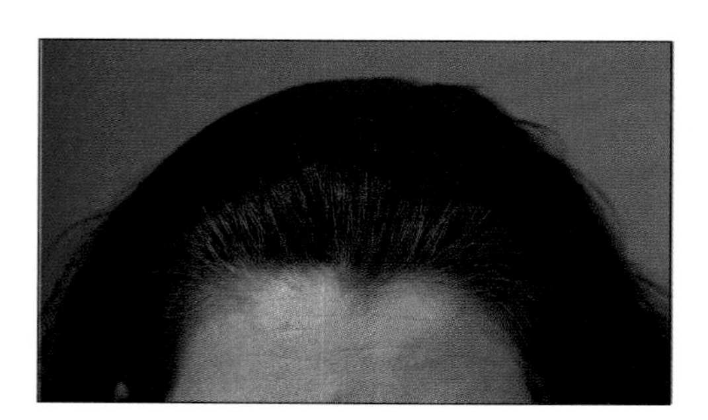

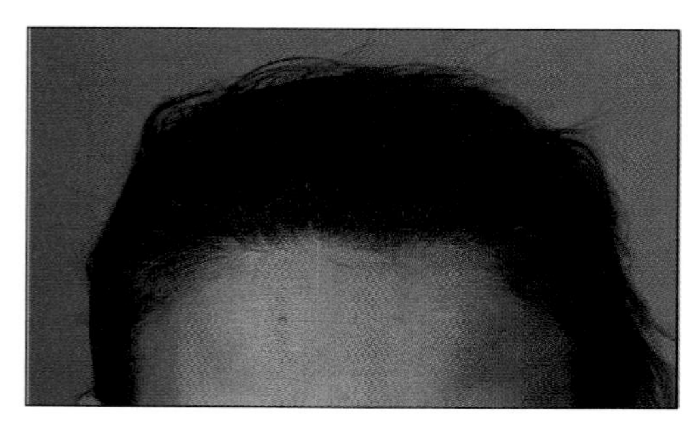

Figure 16.2
Widow's peak hairline (*top*). People lacking a widow's peak have a relatively straight hairline (*bottom*).

OTHER HUMAN TRAITS

The following traits are determined by a single gene. List your phenotype for each trait in table 16.3 and, if possible, list your genotype. If you have the recessive trait for gene G, for example, your genotype is homozygous recessive (*gg*). If you have the dominant trait, your genotype could be GG or Gg, in which case you should enter G in table 16.3. If you have the dominant trait and one of your parents shows the recessive trait, you must be heterozygous (*Gg*) for that trait. Give your results to your instructor so that she or he can provide you with the phenotypic results for your class.

Widow's peak—The *W* allele for widow's peak (i.e., a pointed hairline) is dominant to the *w* allele for a straight hairline (fig. 16.2).

Bent little finger—Lay your hands flat on the table and relax them. If the last joint of your little finger bends toward the fourth finger, you have the dominant allele *B* (fig. 16.3).

Albinism—The *A* allele is dominant and leads to production of melanin, a pigment. Individuals with an *aa* genotype lack pigment in their skin, hair, and iris.

Pigmented iris—If you are homozygous for the recessive allele *p*, you do not produce pigment in the front layer of your iris, and your eyes are either blue or gray (i.e., your eyes are the color of the back layer of the iris). The *P* allele produces pigment in the front layer of the iris (green, hazel, brown, or black), which masks the blue or gray color of the back layer of the iris.

Attached earlobes—The *A* allele for free earlobes is dominant to the recessive *a* allele for attached earlobes (fig. 16.4).

Hitchhiker's thumb—Bend your thumb backward as far as possible. If you can bend the last joint of the thumb back at an angle of 60° or more, you are showing the recessive allele *h* (fig. 16.5).

Interlacing fingers—Casually fold your hands together so that your fingers interlace. The *C* allele for crossing the left thumb over the right thumb when you interlace your fingers is dominant over the *c* allele for crossing your right thumb over your left.

PTC tasting—Obtain a piece of paper impregnated with phenylthiocarbamide (PTC). Taste the paper by chewing on it for a few seconds. If you detect a bitter taste, you have the dominant allele *T*.

Mid-digital hair—The allele *M* for hair on the middle segment of your fingers is dominant to the *m* allele for no middigital hair. If hair is present on the middigit of any finger you have the dominant allele.

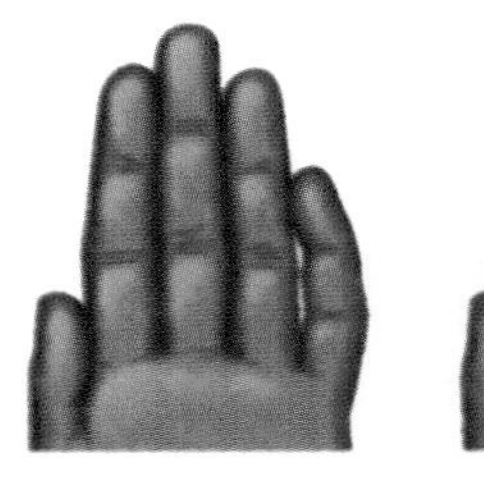
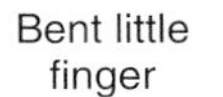

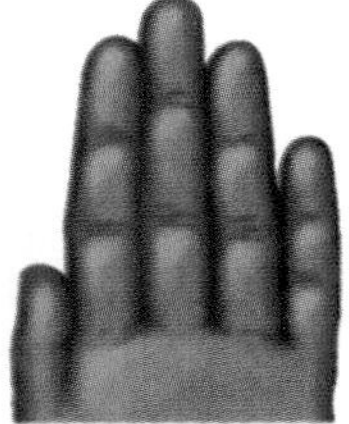

Figure 16.3
Bent little finger.

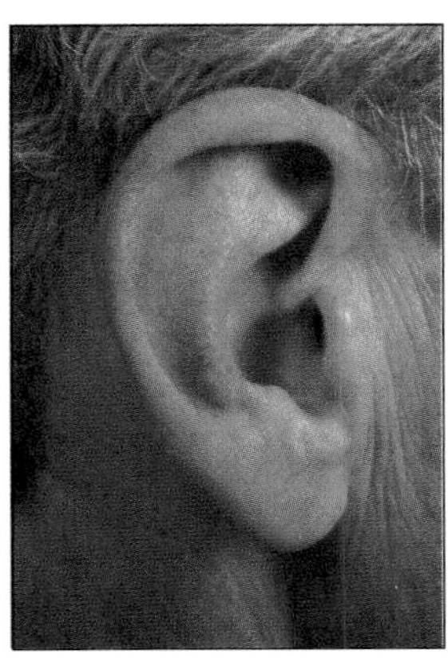
(a)

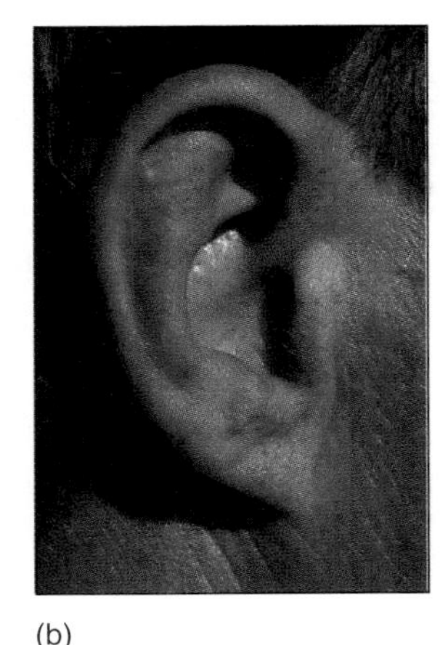
(b)

Figure 16.4
(*a*) Unattached earlobe. (*b*) Attached earlobe.

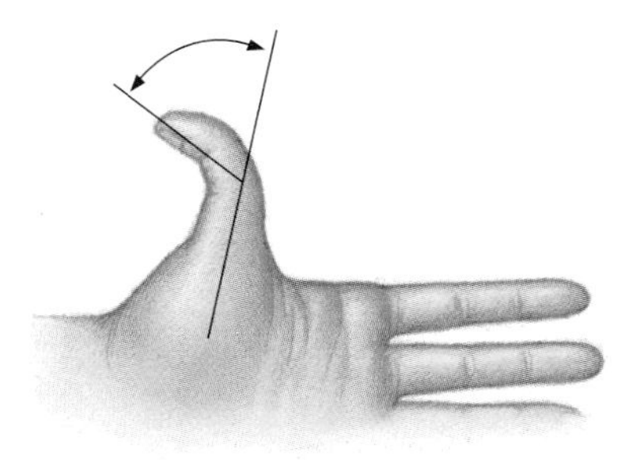
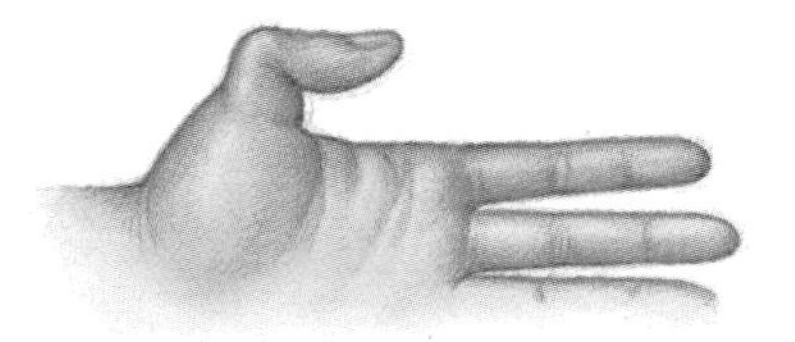

Figure 16.5
Hitchhiker's thumb.

Several diseases are inherited as single-gene traits. These include*

Cystic fibrosis, a disease characterized by chronic bronchial obstruction and growth reduction.

Galactosemia, an inability to metabolize galactose, a sugar in human milk. Inherited as an autosomal recessive trait. Approximately five cases occur per million births. Prenatal diagnosis can be performed on cells obtained through amniocentesis or chorionic villi sampling.

Phenylketonuria (PKU), an inability to metabolize the amino acid phenylalanine. Approximately 100 cases occur per million births. If untreated, this disease produces mental retardation.

Huntington's disease, a mental disorder involving uncontrollable, involuntary muscle movements. The disease occurs relatively late in life, so many affected individuals bear children before they realize that they are carriers. Approximately 100 cases occur per million births.

Juvenile retinoblastoma, a cancer of the retina. The allele is located on chromosome 13.

Question 12
What conclusion about your genotype is evident if one of your siblings, but neither parent, shows the recessive trait?

ANALYZING PEDIGREES

If you understand the simple patterns of inheritance presented in this lab, you can trace a trait in a pedigree (i.e., family tree) to determine if it is inherited in a dominant or recessive pattern of inheritance.

Question 13

a. What features would characterize pedigrees of dominant traits?

b. What features would characterize pedigrees of recessive traits?

Biologists use the following symbols in pedigrees:

Male

Female

Affected male

Affected female

Mating

Parents

Siblings

Procedure 16.5

Analyze a pedigree of inheritance of cystic fibrosis

1. Among Caucasians, about 1 of every 2,500 newborn infants is born with cystic fibrosis. In these individuals, a defective membrane protein results in the production of unusually thick and dry mucus that lines organs such as the tubes in the respiratory system. People having cystic fibrosis often have recurrent and serious infections, and most die in their 20s or 30s.
2. Determine whether the allele for cystic fibrosis is inherited as a dominant or recessive allele.

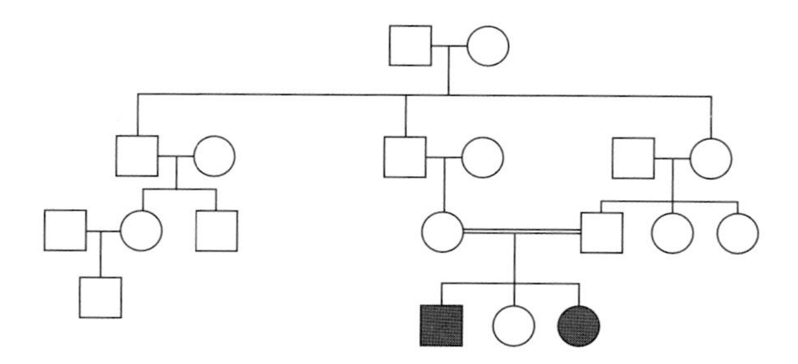

Question 14
What is the inheritance pattern for the cystic fibrosis allele? What is your reasoning for this conclusion?

* Information from B. A. Pierce, *The Family Genetic Sourcebook*. New York: John Wiley and Sons, Inc., 1992.

Procedure 16.6

Analyze a pedigree of inheritance of Huntington's disease

1. Huntington's disease is a severe disorder of the nervous system that usually causes death.
2. Determine whether the allele for Huntington's disease is inherited as a dominant or recessive allele.

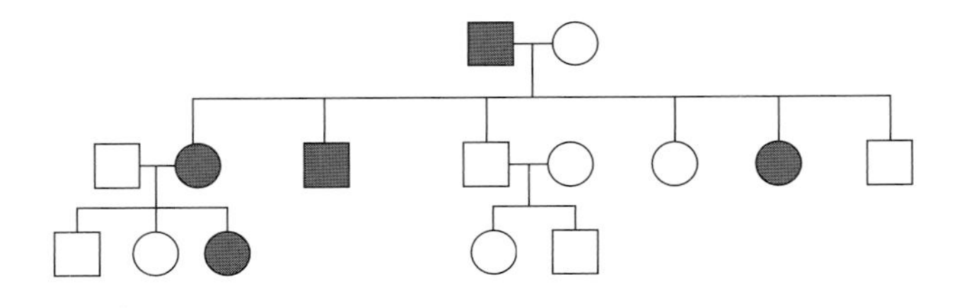

Question 15

What is the inheritance pattern for the Huntington's disease allele? What is your reasoning for this conclusion?

Procedure 16.7

Analyze a pedigree of inheritance of phenylketonuria

1. Phenylketonuria, or PKU, results from an inability to metabolize the amino acid phenylalanine. If untreated, PKU leads to mental retardation.
2. Determine whether the allele for phenylketonuria is inherited as a dominant or recessive allele.

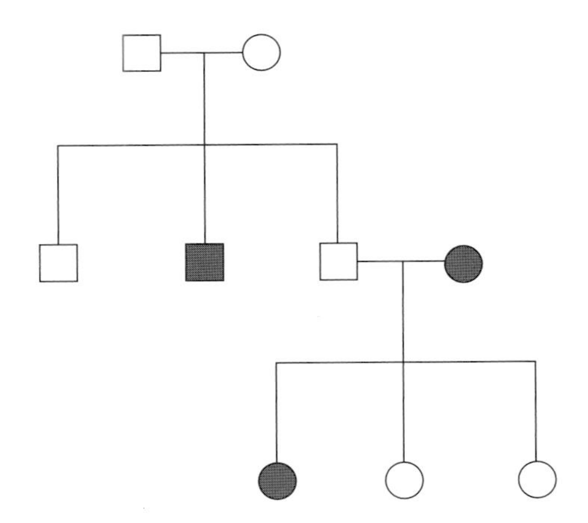

Question 16

What is the inheritance pattern for the phenylketonuria allele? What is your reasoning for this conclusion?

TRANSPOSONS

For much of this century, geneticists thought that genes do not move in cells. However, in 1947 Barbara McClintock proposed that genes could move within and between chromosomes. McClintock based her conclusion on a series of experiments involving genetic crosses in corn. Specifically, McClintock showed that there is a fragment of DNA that can move to and be inserted at the locus for the production of pigments in corn kernels. Because this insertion renders the cell unable to make the purple pigment, the resulting kernel is yellow or white. However, subsequent removal of the DNA fragment results in the cell resuming production of the purple pigment; therefore, the resulting kernel is purple. Thus, Indian corn often has kernels with varying pigmentation, depending on when the DNA fragment was inserted or removed.

A similar phenomenon occurs with the production of other pigments in corn kernels. The translocation to and from the locus for production of these pigments several times during kernel development produces the red-orange swirls characteristic of many kernels of Indian corn.

The fragments of DNA that McClintock studied are now called **transposons.** Transposons are a useful tool for genetic engineering because they provide a way of inserting foreign DNA into a host cell's chromosome. For her work McClintock received the Nobel Prize in 1983.

Procedure 16.8

Observe corn kernels to understand the effects of transposons

1. Work in a group of two to four. Obtain an ear of Indian corn for your group.
2. Look for examples of kernels with (a) purple or white spots, (b) red-orange swirls, and (c) other unusual color patterns. Sketch the pigmentation patterns in 2–3 kernels.
3. Use the information presented in this exercise and in your textbook to determine how transposons could produce such unusual patterns of pigmentation.

Questions for Further Thought and Study

1. What determines how often a phenotype occurs in a population?

2. Are dominant characteristics always more frequent in a population than recessive characteristics? Why or why not?

3. Is it possible to determine the genotype of an individual having a dominant phenotype? How?

4. Why is hybrid seed so expensive to produce?

5. What blood types are not expected for children to have if their parents have AB blood? O blood?

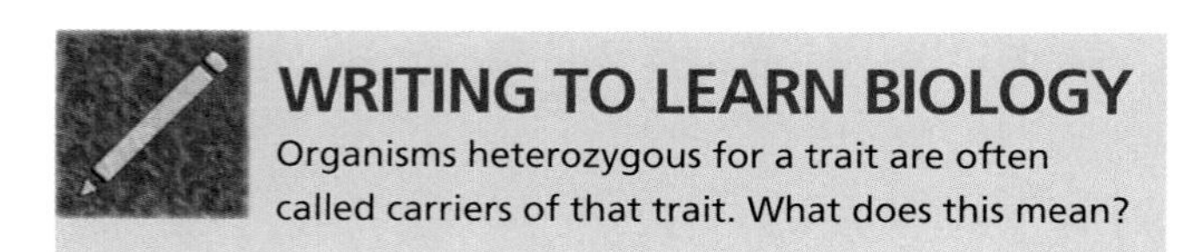

WRITING TO LEARN BIOLOGY

Organisms heterozygous for a trait are often called carriers of that trait. What does this mean?

17

Evolution
Natural Selection and Morphological Change in Green Algae

Objectives

By the end of this exercise you should be able to:
1. Give a working definition of evolution, fitness, selection pressure, and natural selection.
2. Determine the genotypic and phenotypic frequency of a population while properly using the terms allele, dominant, recessive, homozygous, and heterozygous.
3. Explain the Hardy-Weinberg Principle and use it to demonstrate negative selection pressures on a population.
4. Describe the significance of the Volvocine line, particularly in the areas of cellular specialization and colonial complexity.
5. Describe examples of how a mutation affecting the plane of cellular division could result in the evolution of morphologically different body plans.

Figure 17.1
Mutations produce new alleles and genetic combinations that may be adaptive. One branch of this peppermint peach tree has grown from a meristem that had a mutation for red petals. If this mutation is adaptive, it will probably persist in the gene pool of subsequent generations.

The theory of **evolution** broadly describes genetic change in populations. The existence of genetic change (and therefore evolution) is universally accepted by biologists. We know that many mechanisms can change the genetic makeup of populations, but the relative importance of each mechanism is controversial. Events such as **mutations** (changes in the genetic message of a cell, fig. 17.1) and catastrophes (e.g., meteor showers, ice ages) are all responsible to some degree for genetic change. However, Charles Darwin formulated a theory that explains a major force behind genetic change (fig. 17.2).

Darwin postulated that organisms that survive and reproduce successfully have genetic traits aiding survival and reproduction. These traits enhance an organism's **fitness,** which is its tendency to produce more offspring than competing individuals, and therefore contribute more genes to the next generation. Darwin noticed that fit individuals (that is, ones that reproduce the most) produce more offspring because their traits are better adapted for survival and reproduction than are those of their competitors. He further reasoned that if the traits of the more fit individuals are transmitted to the next generation more often, then more of these traits will be found in the next generation. After many generations, the frequency of these traits will increase in the population, and the nature of the population will gradually change. Darwin called this overall process **natural selection** and proposed it as a major force that guides genetic change and the formation of new species. Review in your textbook the theories of evolution and the mechanism of natural selection.

Showing the effects of natural selection in living populations is usually time-consuming and tedious. Therefore, in this exercise you will simulate reproducing populations with nonliving, colored beads representing organisms and

Figure 17.2

Darwin greets his "monkey ancestor." In his time, Darwin was often portrayed unsympathetically, as in this drawing from an 1874 publication.

their gametes. With this artificial population you can quickly follow genetic change over many generations. Before you begin work, review the previous exercise on genetics, especially the terms **gene, allele, dominant alleles, recessive alleles, homozygous,** and **heterozygous.**

You will begin your experiments using a "stock population" of organisms consisting of a container of beads. Each bead represents a haploid (that is, having one set of chromosomes) gamete, and the color of the "gamete" (colored or white) represents an allele it is carrying. Individual organisms from this population are diploid (that is, they each have two sets of chromosomes) and therefore are represented by two beads.

UNDERSTANDING ALLELIC AND GENOTYPIC FREQUENCIES

Frequency is the proportion of individuals in a certain category relative to the total number of individuals considered. The frequency of an allele or genotype is expressed as a decimal proportion of the total alleles or genotypes in a population. For example, if 1/4 of the individuals of a population are genotype *Bb*, the frequency of *Bb* is 0.25. If 3/4 of all alleles in a population are *B*, then the frequency of *B* is 0.75.

In this exercise you will simulate evolutionary changes in allelic and genotypic frequencies in an artificial population. The trait you will work with is fur color. A colored bead is a gamete with a dominant allele (complete dominance) for black fur (*B*), and a white bead is a gamete with a recessive allele for white fur (*b*). An individual is represented by two gametes (beads). Individuals with genotypes *BB* and *Bb* have black fur and those with *bb* have white fur.

Procedure 17.1

Establish a parental population

1. Obtain a "stock population" of organisms consisting of a container of colored and white beads.
2. Obtain an empty container marked "Parental Population."
3. From the stock population select 25 homozygous dominant individuals (*BB*) and place them in the container marked "Parental Population." Each individual is represented by two colored beads.
4. From the stock population select 50 heterozygous individuals (*Bb*) and place them in the container marked "Parental Population." Each individual is represented by a colored and a white bead.
5. From the stock population select 25 homozygous recessive individuals (*bb*) and place them in the container marked "Parental Population." Each individual is represented by two white beads.
6. Calculate the total number of individuals and the total number of alleles in your newly established parental population. Use this information to calculate and record in table 17.1 the correct genotypic frequencies for your parental population.
7. Complete table 17.1 with the number and frequency of each of the two alleles.

Question 1

a. How many of the total beads are colored and how many are white?

b. What color fur do *Bb* individuals have?

c. How many beads represent the population?

THE HARDY-WEINBERG PRINCIPLE

The **Hardy-Weinberg Principle** enables us to calculate and predict allelic and genotypic frequencies. We can compare these predictions with actual changes that we observe in natural populations and learn about factors that influence gene frequencies.

TABLE 17.1

FREQUENCIES OF GENOTYPES AND ALLELES OF THE PARENTAL POPULATION

Genotypes	Frequency	Alleles	Frequency
BB ●●	______	*B* ●	______
Bb ●○	______	*b* ○	______
bb ○○	0.25		

This predictive model includes two simple equations first described for stable populations by G. H. Hardy and W. Weinberg. Hardy-Weinberg equations (1) predict allelic and genotypic frequencies based on data for only one or two frequencies, and (2) provide a set of theoretical frequencies that we can compare to frequencies from natural populations. For example, if we know the frequency of *B* or *BB*, we can calculate the frequency of *b*, *Bb*, and *bb*. Then we can compare these frequencies with those of a natural population that we might be studying. If we find variation from our predictions, we can study the reasons for this genetic change. This comparison is important because biological characteristics of natural populations rarely correspond exactly to theoretical calculations. Furthermore, deviations are important because they often reveal unknown factors influencing the population being studied.

According to the Hardy-Weinberg Principle, the frequency of the dominant allele of a pair is represented by the letter p, and that of the recessive allele by the letter q. Also, the genotypic frequencies of *BB* (homozygous dominant), *Bb* (heterozygous), and *bb* (homozygous recessive) are represented by p^2, $2\,pq$, and q^2, respectively. Examine the frequencies in table 17.1 and verify the Hardy-Weinberg equations:

$$p + q = 1$$
$$p^2 + 2\,pq + q^2 = 1$$

The Hardy-Weinberg Principle and its equations predict that frequencies of alleles and genotypes will remain constant from generation to generation in stable populations. Therefore, these equations can be used to predict genetic frequencies through time. However, the Hardy-Weinberg prediction assumes that

- The population is large enough to overcome random events.
- Choice of mates is random.
- Mutation does not occur.
- Individuals do not migrate into or out of the population.
- There is no selection pressure.

Question 2

a. Consider the Hardy-Weinberg equations. If the frequency of a recessive allele is 0.3, what is the frequency of the dominant allele?

b. If the frequency of the homozygous dominant genotype is 0.49, what is the frequency of the dominant allele?

c. If the frequency of the homozygous dominant genotype is 0.49, what is the frequency of the homozygous recessive genotype?

d. Which Hardy-Weinberg equation relates the frequencies of the alleles at a particular gene locus?

e. Which Hardy-Weinberg equation relates the frequencies of the genotypes for a particular gene locus?

f. Which Hardy-Weinberg equation relates the frequencies of the phenotypes for a gene?

To verify the predictions of the Hardy-Weinberg Principle, use the following procedure to produce a generation of offspring from the parental population you created in the previous procedure.

Procedure 17.2

Verify the Hardy-Weinberg Principle

1. Examine figure 17.3 for an overview of the steps of this procedure.

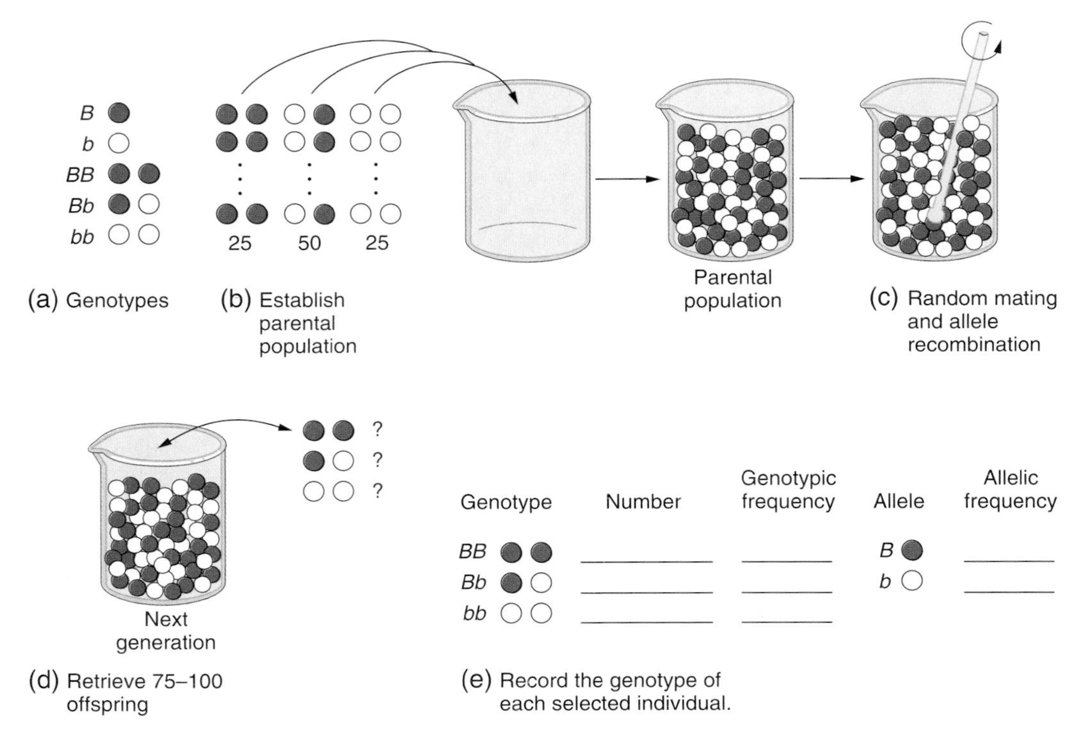

Figure 17.3

Verification of the Hardy-Weinberg Principle. See procedure 17.2 for an explanation of each step.

2. Establish the parental population established in procedure 17.1 (fig. 17.3*a*, 17.3*b*).
3. Simulate the random mating of individuals by mixing the population (fig. 17.3*c*).
4. Reach into the parental container (without looking) and randomly select two gametes. Determine their genotype (fig. 17.3*d*). This pair of gametes with colored or white alleles represents an individual offspring.
5. Record the occurrence of the genotype in figure 17.3*e* as a mark under the heading "Number," or temporarily on a second sheet of paper and return the beads to the container.
6. Repeat steps 4 and 5 (100 times) to simulate the production of 100 offspring.
7. Calculate the frequency of each genotype and allele, and record the frequencies in figure 17.3*e*. Beside each of these new-generation frequencies write (in parentheses) the original frequency of that specific genotype or allele from table 17.1.

Question 3

a. The Hardy-Weinberg Principle predicts that genotypic frequencies of offspring will be the same as those of the parental generation. Were they the same in your simulation?

b. If the frequencies were different, then one of the assumptions of the Hardy-Weinberg Principle was probably violated. Which one?

EFFECT OF A SELECTION PRESSURE

Selection is the differential reproduction of phenotypes (fig. 17.4); that is, some phenotypes (and their associated genotypes) are passed to the next generation more often than others. In positive selection, genotypes representing adaptive traits in an environment increase in frequency because their bearers are more likely to survive and reproduce. In negative selection, genotypes representing nonadaptive

Figure 17.4

Reproduction. These snakes hatching from eggs in a Costa Rican rain forest represent successful reproduction. Some of these newly hatched snakes will mature to reproduce, and others will not. All species reproduce, although not all hatch from eggs. Some organisms reproduce several generations each hour; others only reproduce once every hundred years.

TABLE 17.2

GENOTYPIC FREQUENCIES FOR 100% NEGATIVE SELECTION

Genotype	Generation: First	Second	Third	Fourth	Fifth
BB ●●	______	______	______	______	______
Bb ●○	______	______	______	______	______
bb ○○	______	______	______	______	______
Total	1.0	1.0	1.0	1.0	1.0

traits in an environment decrease in frequency because their bearers are less likely to survive and reproduce.

Selection pressures are factors such as temperature and predation that affect organisms and result in selective reproduction of phenotypes. Some pressures may elicit 100% negative selection against a characteristic and eliminate any successful reproduction of individuals having that characteristic. For example, mice with white fur may be easy prey for a fox if they live on a black lava field. This dark environment is a negative selection pressure against white fur. If survival and reproduction of mice with white fur was eliminated (i.e., if there is 100% negative selection), would the frequency of white mice in the population decrease with subsequent generations? To test this, use the following procedure to randomly mate members of the original parental population to produce 100 offspring (fig. 17.5).

Procedure 17.3

Simulate 100% negative selection pressure

1. Establish the same parental population that you used to test the Hardy-Weinberg prediction.
2. Simulate the production of an offspring from this population by randomly withdrawing two gametes to represent an individual offspring.
3. If the offspring is *BB* or *Bb*, place it in a container for the accumulation of the "Next Generation." Record the occurrence of this genotype on a separate sheet of paper.
4. If the offspring is *bb*, place this individual in a container for those that "Cannot Reproduce." Individuals in this container should not be used to produce subsequent generations. Record the occurrence of this genotype on a sheet of paper.
5. Repeat steps 2–4 until the parental population is depleted, thus completing the first generation.
6. Calculate the frequencies of each of the three genotypes recorded on the separate sheet and record these frequencies for the first generation in table 17.2. Individuals in the next generation will serve as the parental population for each subsequent generation.

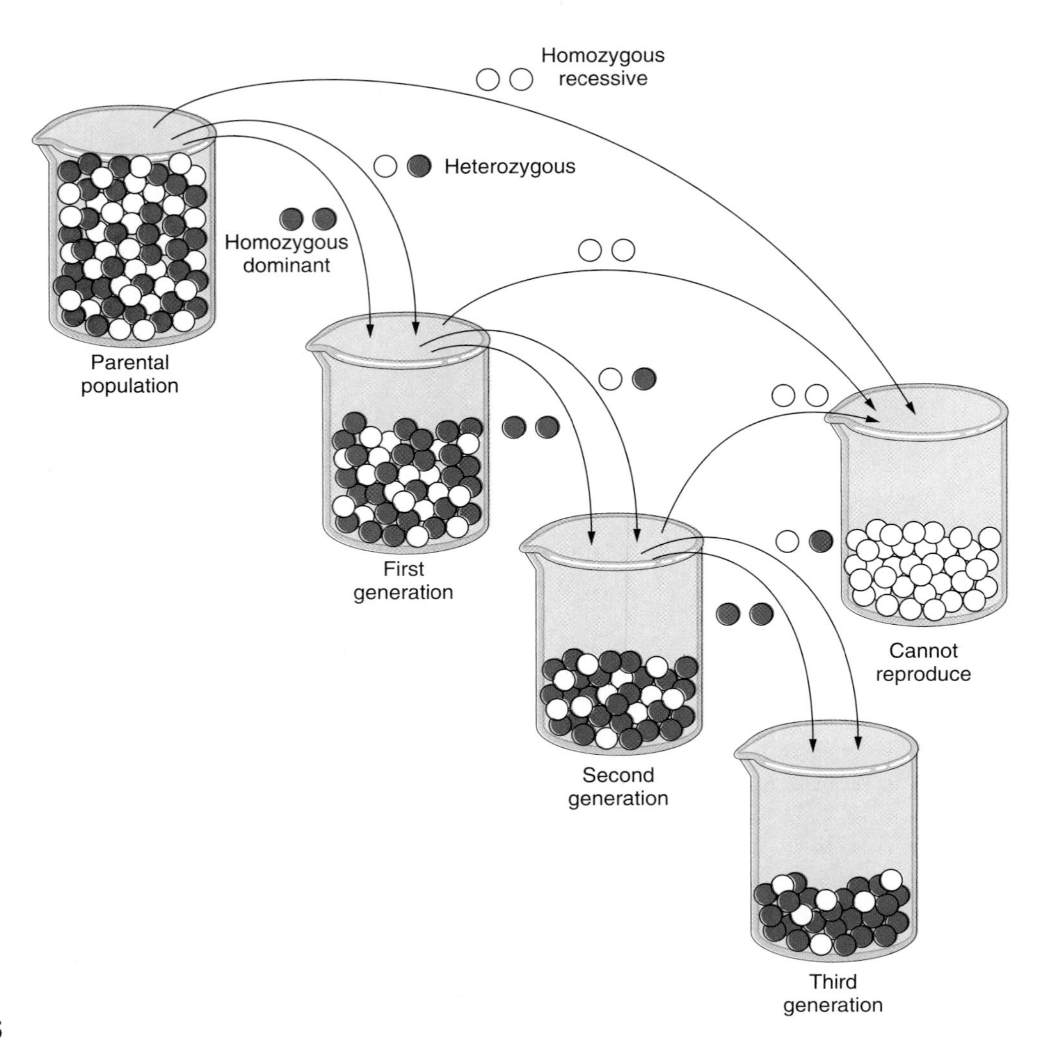

Figure 17.5

Demonstrating the effect of 100% selection pressure on genotypic and phenotypic frequencies across three generations. Selection is against the homozygous recessive genotype. Random mating within the parental population is simulated by mixing the gametes (beads), and the parental population is sampled by removing two alleles (i.e., one individual) and placing them in the next generation. Homozygous recessive individuals are removed (selected against) from the population. The genotypic and phenotypic frequencies are recorded after the production of each generation. The production of each generation depletes the beads in the previous generation in this simulation.

7. Repeat steps 2–5 to produce a second, third, fourth, and fifth generation. After the production of each generation, record your results in table 17.2.
8. Graph your data from table 17.2 using the graph paper at the end of this exercise. *Generation* is the independent variable on the *x*-axis and *Genotype* is the dependent variable on the *y*-axis. Graph three curves, one for each genotype.

Because some members (i.e., the *bb* that you removed) of each generation cannot reproduce, the number of offspring from each successive generation of your population will decrease. However, the frequency of each genotype, not the number of offspring, is the most important value.

Question 4

a. Did the frequency of white individuals decrease with successive generations? Explain your answer.

b. Was the decrease of white individuals from the first to second generation the same as the decrease from the second to the third generation? From the third to the fourth generation? Why or why not?

c. How many generations would be necessary to eliminate the allele for white fur?

TABLE 17.3

GENOTYPIC FREQUENCIES FOR 20% NEGATIVE SELECTION

Genotype	Generation: First	Second	Third	Fourth	Fifth
BB ●●	______	______	______	______	______
Bb ●○	______	______	______	______	______
bb ○○	______	______	______	______	______
Total	1.0	1.0	1.0	1.0	1.0

Note: Total of frequencies for each generation must equal 1.0.

Most naturally occurring selective pressures do not eliminate reproduction by the affected individuals. Instead, their reproductive capacity is reduced by a small proportion. To show this, use procedure 17.4 to eliminate only 20% of the *bb* offspring from the reproducing population.

Procedure 17.4

Simulate 20% negative selection pressure

1. Establish the same parental population that you used to test the Hardy-Weinberg prediction.
2. Simulate the production of an offspring from this population by randomly withdrawing two gametes to represent an individual offspring.
3. If the offspring is *BB* or *Bb*, place it in a container for production of the "Next Generation." Record the occurrence of this genotype on a separate sheet of paper.
4. If the offspring is *bb*, place every fifth individual (20%) in a separate container for those that "Cannot Reproduce." Individuals in this container should not be used to produce subsequent generations. Place the other 80% of the homozygous recessives in the container for production of the "Next Generation." Record the occurrence of this genotype on a sheet of paper.
5. Repeat steps 2–4 until the parental population is depleted, thus completing the first generation.
6. Calculate the frequencies of each of the three genotypes recorded on the separate sheet and record these frequencies for the first generation in table 17.3.
7. Repeat steps 2–5 to produce a second, third, fourth, and fifth generation. Individuals in the "Next Generation" will serve as the parental population for each subsequent generation. After the production of each generation, record your results in table 17.3.
8. Graph your data from table 17.3 using the graph paper at the end of this exercise. *Generation* is the independent variable on the *x*-axis and *Genotype* is the dependent variable on the *y*-axis. Graph three curves, one for each genotype.

Because some members of each generation cannot reproduce, the number of offspring from each generation of your population will decrease. However, the frequency of each genotype, not the number of offspring, is the most important value.

Question 5

a. Did the frequency of white individuals decrease with successive generations?

b. Was the rate of decrease for 20% negative selection similar to the rate for 100% negative selection? If not, how did the rates differ?

AN EXAMPLE OF EVOLUTION: THE VOLVOCINE LINE

The evolution of most species is too slow to witness in the lab, but we can examine modern species to learn about changes that likely occurred over evolutionary time. The **Volvocine line** of algae is a group of modern species that reflects an easily recognized sequence of changes as their common ancestors evolved. In this case the changes were in colony complexity.

Studies of morphology and molecular genetics indicate that an ancient species similar to today's flagellated *Chlamydomonas* was probably the original and most ancient common ancestor to the Volvocine line. The probable sequence of events was that an ancestor of unicellular *Chlamydomonas* evolved a novel colonial morphology that was successful and gave rise to *Gonium* (fig. 17.6). In turn that ancestor evolved greater colonial complexity to give

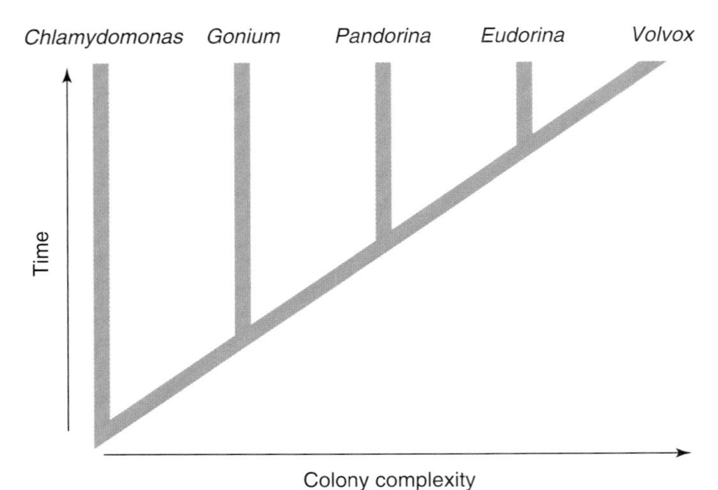

Figure 17.6
A cladogram representing the simplified phylogeny (family tree) of the Volvocine line. Proposed common ancestors are represented by the branching points called nodes.

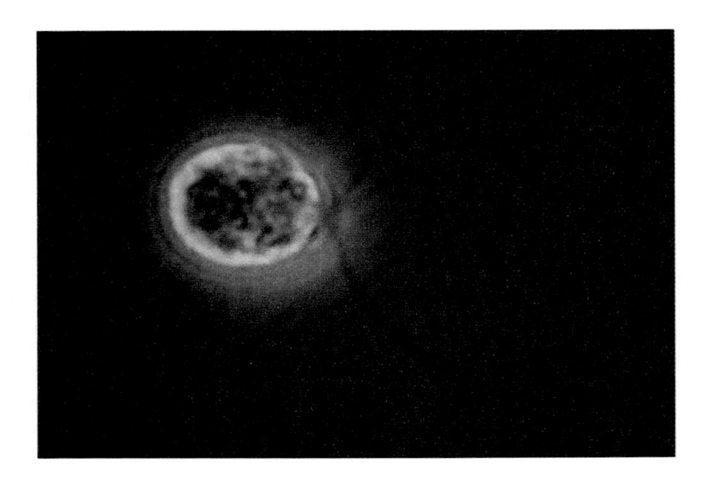

Figure 17.7
Chlamydomonas, a unicellular green alga (1700×). *Chlamydomonas* has two flagella.

rise to today's *Pandorina* and then *Eudorina*. That colonial ancestor later gave rise to *Volvox*, the most complex alga of the Volvocine line. These five genera are modern representatives of a lineage of species that evolved along a path of colonial complexity.

Procedure 17.5

Examine members of the Volvocine line of algae

1. Sequentially examine each of the following organisms with your microscope. When preparing each of the colonial specimens, try both a standard microscope slide and a deep-well or depression slide. Determine which works best for colonies of cells.
2. *Chlamydomonas* is among the most primitive and widespread of the green algae. It is a unicellular biflagellate alga (fig. 17.7). Most species of *Chlamydomonas* are **isogamous,** which means that all of their gametes are identical in size and appearance. All species of the Volvocine line consist of cells similar to *Chlamydomonas*, but the cells are in different configurations.

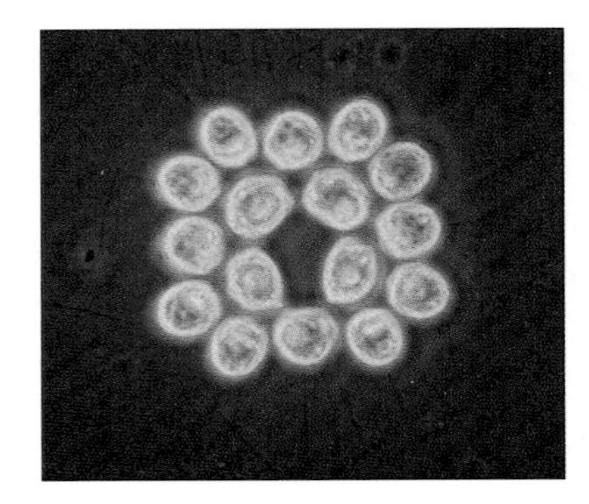

Figure 17.8
Gonium, a colonial green alga composed of 16 cells (200×).

3. *Gonium* is the simplest colonial member of the Volvocine line (fig. 17.8). A *Gonium* colony consists of 4, 8, 16, or 32 *Chlamydomonas*-like cells held together in the shape of a disk by a gelatinous matrix. Each cell in the *Gonium* colony can divide to produce cells that produce new colonies. Like *Chlamydomonas*, *Gonium* is isogamous.

Question 6
Why do colonies of *Gonium* consist of only 4, 8, 16, or 32 cells? That is, why are there no 23-celled colonies?

4. *Pandorina* consists of 16 or 32 *Chlamydomonas*-like cells held together by a gelatinous matrix (fig. 17.9). Examine how *Pandorina* moves. Flagella on *Pandorina* move the ellipsoidal alga through the water like a ball. Eyespots are larger in cells at one end of the colony than in cells at the other end. After attaining its maximum size, each cell of the colony divides to form a new colony. The parent matrix then breaks open like Pandora's box (hence the name *Pandorina*) and releases the newly formed colonies. *Pandorina* is isogamous.

Question 7
What is the significance of a specialization at one end of the colony?

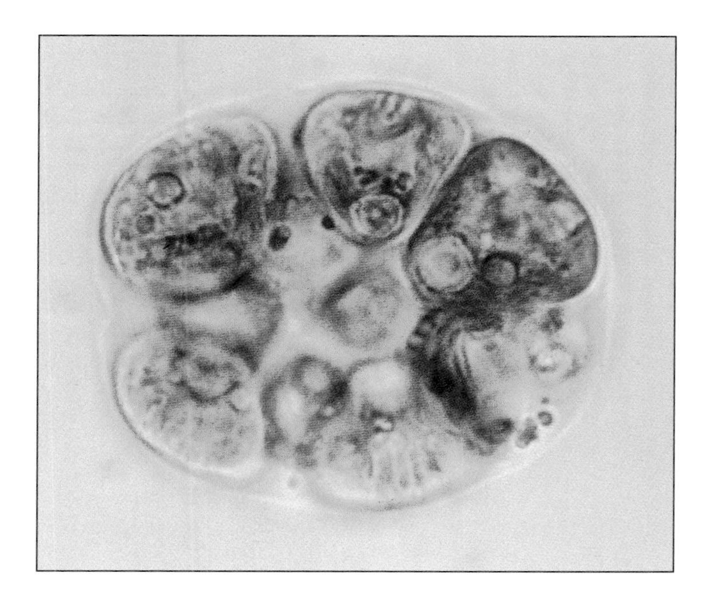

Figure 17.9
Pandorina, a colony of 16 or 32 flagellated green algal cells (150×).

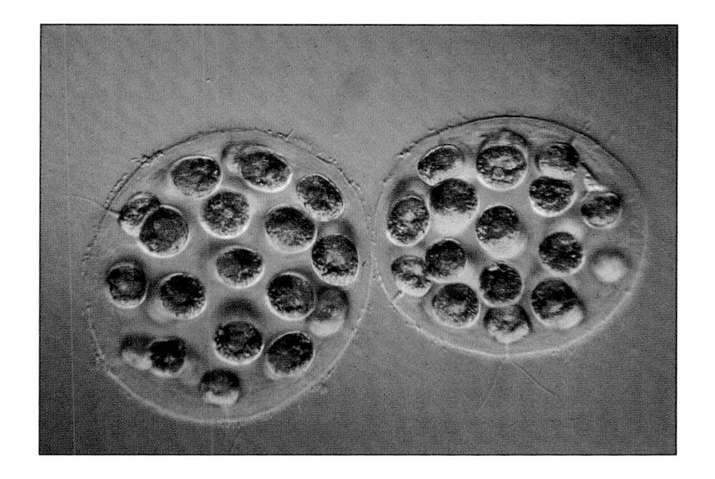

Figure 17.10
Eudorina, a colony of 32 flagellated green algal cells.

5. *Eudorina* is a spherical colony composed of 32, 64, or 128 cells (fig. 17.10). Cells in a colony of *Eudorina* differ in size; smaller cells are located at the anterior part of the colony. These smaller cells cannot reproduce to form new colonies. The anterior surface is determined by the direction of movement.

Question 8
What is the significance of these structural and functional specializations of *Eudorina?*

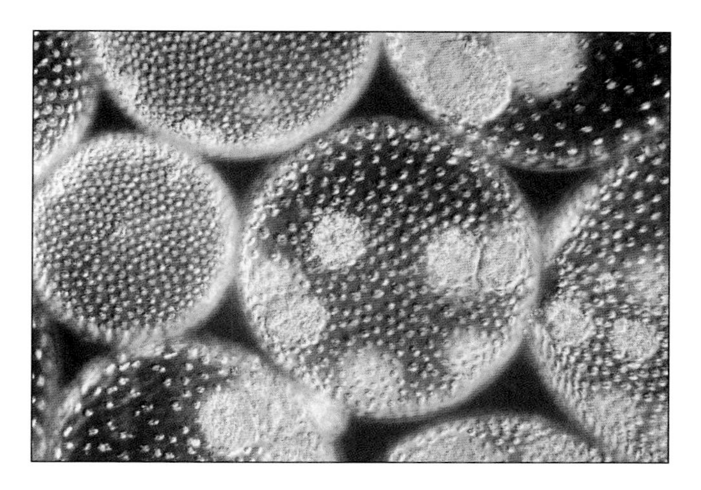

Figure 17.11
Volvox, a common green alga (100×). Colonies of *Volvox* often consist of hundreds of cells. Daughter colonies are visible within the larger parent colony.

6. *Volvox* is the largest and most spectacular organism of the Volvocine line. *Volvox* is a spherical colony made of thousands of vegetative cells and a few reproductive cells (fig. 17.11). *Volvox* is **oogamous** (i.e., the female gametes are large, nonmotile egg cells and sperm are smaller and motile) and polar. Polar organisms have anterior and posterior poles (relative to their direction of movement). Flagella spin the colony on its axis. In some species of *Volvox* and *Gonium*, cytoplasmic strands form a conspicuous network among the cells.

Question 9
a. Does the *Volvox* colony spin clockwise or counterclockwise?

b. What is the significance of the cytoplasmic network in *Volvox?*

To organize your information and observations complete table 17.4.

TABLE 17.4

EVOLUTIONARY SPECIALIZATION OF MEMBERS OF THE VOLVOCINE LINE

Characteristic	*Chlamydomonas*	*Gonium*	*Pandorina*	*Eudorina*	*Volvox*
Number of cells					
Colony size					
Structural and functional specializations of cells					
Reproductive specialization (isogamy vs. oogamy)					

EVOLUTION OF BODY FORM

Division Chlorophyta also provides several examples demonstrating the evolution of body form. Examine the following algae and take special note of their differing body forms:

- *Ulothrix* is a simple, unbranched, filamentous alga characterized by cellular division in one plane.
- *Stigeoclonium* is an elongate, branching, filamentous alga characterized by additional cellular divisions in a second plane, but in only a few cells.
- *Percursaria* is a broadened, flattened, filamentous alga that is one cell thick. It is formed when each cell divides in two planes.
- *Ulva* has a "leafy" plant body that results from cellular division in three planes. Because division in the third plane is restricted to a single division, the flattened blade is only two cells thick.

Question 10

a. How could evolutionary changes in planes of cellular division produce different body forms?

b. What would be the body form of an organism whose cells divided randomly in one plane? In two planes?

Questions for Further Thought and Study

1. How would selection against heterozygous individuals over many generations affect the frequencies of homozygous individuals? Would the results of such selection depend on the initial frequencies of p and q? Could you test this experimentally? How?

2. How are genetic characteristics associated with nonreproductive activities such as feeding affected by natural selection?

3. Why do you think *Volvox* is considered an evolutionary "dead end"?

4. Although Charles Darwin wasn't the first person to suggest that populations evolve, he was the first to describe a credible mechanism for the process. That mechanism is natural selection. What is natural selection and how can it drive evolution?

5. Do you suspect that evolutionary change always leads to greater complexity? Why or why not?

6. Is natural selection the only means of evolution? Explain.

7. What change in a population would you expect to see if a selection pressure was against the trait of the dominant allele?

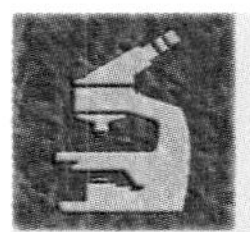

DOING BIOLOGY YOURSELF

Design an experiment to determine the phylogenetic relationships among members of the Volvocine line of algae. What information about their DNA sequences would be useful?

WRITING TO LEARN BIOLOGY

The Hardy-Weinberg equilibrium assumes that pollination and subsequent fertilization must be random. Is that true for most wildflower populations? What characteristics of these plants influence pollination patterns?

17–12

17–14

18

Human Evolution
Skull Examination

Objectives

By the end of this exercise you should be able to:
1. Describe the parts of a modern human skull.
2. Distinguish between skulls of males and females.
3. Distinguish among skulls of apes and modern humans.

Throughout time, no issue has interested humans more than learning about our origins. Where did we come from? What did our ancestors look like? Where did they live? Today, we are beginning to understand those questions, thanks to evidence provided by biologists and anthropologists. Nevertheless, the topic remains controversial and often elicits strong responses from people.

The purpose of this exercise is to present some of the information underlying recent ideas about human evolution. Specifically, this exercise will introduce you to some of the evidence that links modern humans with other primates, both ancient and modern.

THE MODERN HUMAN SKULL

The human skull, including the lower jaw, consists of 22 bones, 8 of which are paired. All of the bones fit together at immovable joints called sutures, which appear as wavy lines. Projections and raised lines are sites of muscle attachment.

Although skulls from males and females share many features, they usually can be distinguished. However, such a diagnosis—even when done by experts—is only about 90% reliable (80% if the lower jaw is missing).

Procedure 18.1

How do skulls of females differ from those of males?

1. Examine the modern human skulls available in the lab to familiarize yourself with their structure and geometry.
2. Compare these skulls with those shown in figure 18.1. The letters beside each feature in the table correspond to those shown in the diagrams.

Procedure 18.2

Study the skulls of apes

Use figure 18.2 and the specimens in the lab to study the following features of ape and human skulls.

Face

Prognathism is the extent to which the face and jaws protrude forward when viewed from the side. Their larger teeth and jaws cause apes to exhibit more prognathism than do humans.

Braincase

The **brow ridge,** which is the mass of bone over the eye sockets, supports the upper facial skeleton against forces produced by chewing. The brow ridge in apes is prominent. In humans, the brow ridge of modern humans is largely internalized because our frontal bone has expanded outward to a more vertical angle.

The **sagittal crest** is a thin ridge of bone atop and down the middle of the braincase. The sagittal crest is associated with having a small braincase and powerful jaws. In apes, the sagittal crest is an attachment site for the large temporalis muscle.

The **foramen magnum** is the large opening in the base of the skull through which the spinal cord passes. The position of the foramen magnum reflects the posture of the body (and, indirectly, the pattern of movement) of **hominoids.** Humans stand erect and walk with the head directly over the vertical spinal column. Conversely, the knuckle-walking

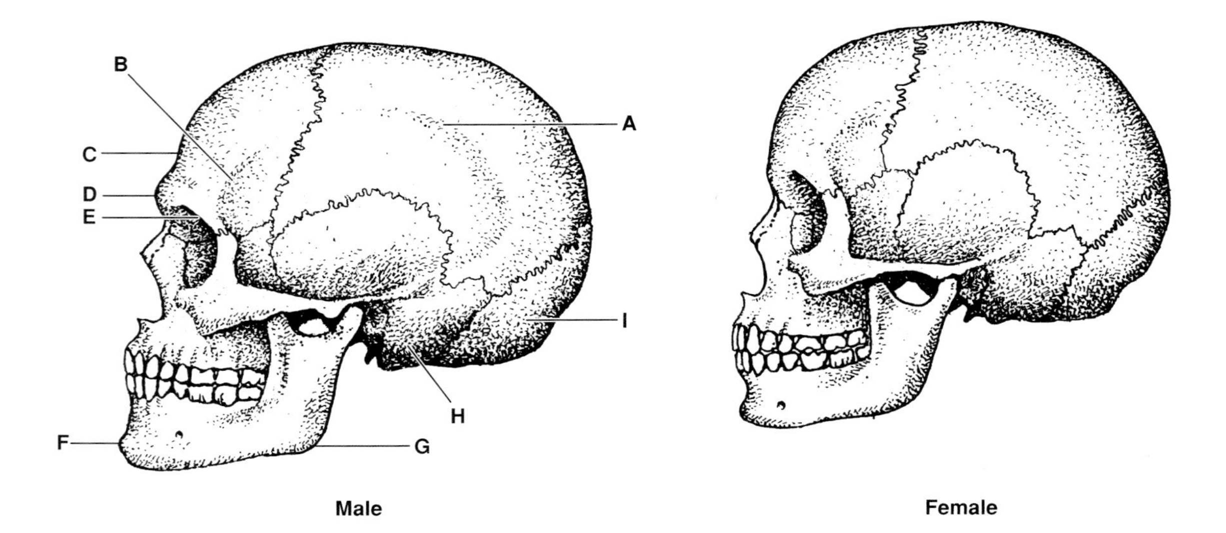

Male Skull	Feature	Female Skull
Large	A. Braincase	Smaller than male
Marked	B. Muscle lines	Slight
Retreating	C. Forehead	Bulging
Developed	D. Brow ridges	Absent
Rounded	E. Upper rim of eye socket	Sharp
Square	F. Chin	Rounded
Nearly a right angle	G. Angle of jaw	Angle more obtuse (over 120°)
Large	H. Mastoid process	Small
Present	I. External occipital protuberance	Absent

Figure 18.1
A comparison of male and female skulls of modern *Homo sapiens*.

apes hold their heads forward, with the foramen magnum toward the rear. Thus, the foramen magnum is located in a more rear position in apes than in humans.

Teeth and Jaws

Adult apes and humans have the same number and types of teeth: 4 canines, 8 premolars, 12 molars, and 8 incisors. Identify these bones on the skulls and diagrams. In apes the canine teeth are longer and more pointed than others. In humans, the canine teeth seldom project above the others.

In humans, the four front teeth (incisors) are smaller, more vertical, and flatter than in apes. In non-human primates, the canine diastema is the gap in the teeth corresponding to the canines of the opposite jaw.

Question 1

a. Between which teeth does the gap occur? Why are these gaps essential in nonhuman primates?

b. Why are they usually absent in humans?

Humans have an outward projection on the lower part of their lower jaw (i.e., chin). Apes lack this feature; instead, they have a smooth, even slant to the front part of their jaw. In humans, teeth are arranged in a relatively continuous curve from the third molar around to the other third molar. The arrangement of teeth in apes is straighter, with a slight curve in front. This is primarily because of the larger size of the incisors and canines.

To summarize the differences between skulls of modern humans and apes, complete table 18.1.

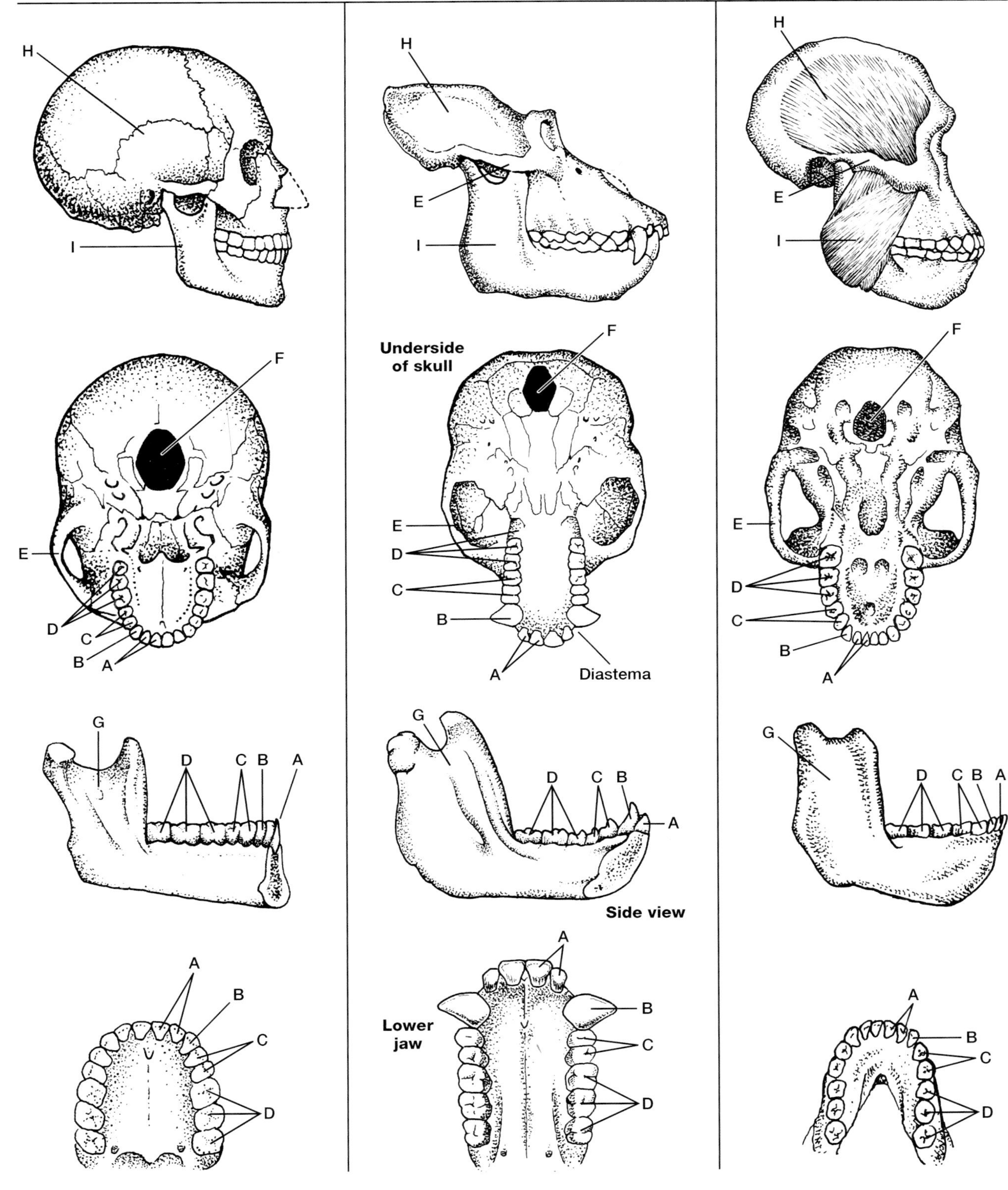

Figure 18.2

Comparison of skull features of modern *Homo sapiens*, gorilla, and *Australopithecus* skull and jaws:

A. Incisors
B. Canines
C. Premolars
D. Molars
E. Zygomatic arch
F. Foramen magnum
G. Vertical ramus for muscle attachment
H. Skull surface for muscle attachment
I. Jaw surface for muscle attachment

TABLE 18.1

PROMINENT FEATURES OF SKULLS OF APES AND HUMANS

Feature	Apes	Humans
Sagittal crest		
Brow ridge		
Foramen magnum		
Prognathism		
Canines		
Canine diastema		
Incisors		
Chin		
Arrangement of teeth		

Procedure 18.3

Study the skulls of fossil primates

Use the information and diagrams in figures 18.3–18.6 to learn about skulls of humans and ancient primates.

Questions for Further Thought and Study

1. Is the skull shown in figure 18.6 that of a man or a woman? Explain your answer.

2. Modern humans and Neanderthal Man are considered to be the same species, *Homo sapiens*. Do you think that they should be the same species? Why or why not? What does this judgment imply about the ability of these two groups of people to interbreed when they lived together in Europe 34,000 years ago?

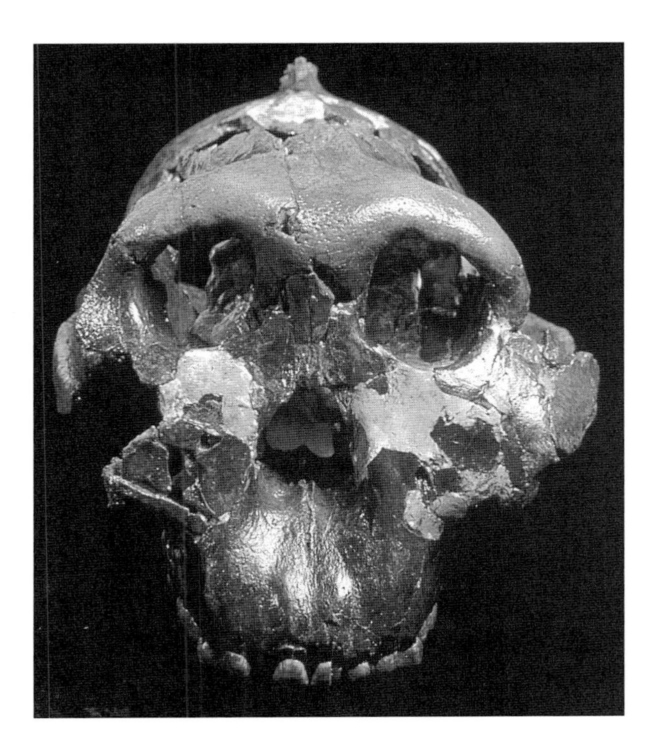

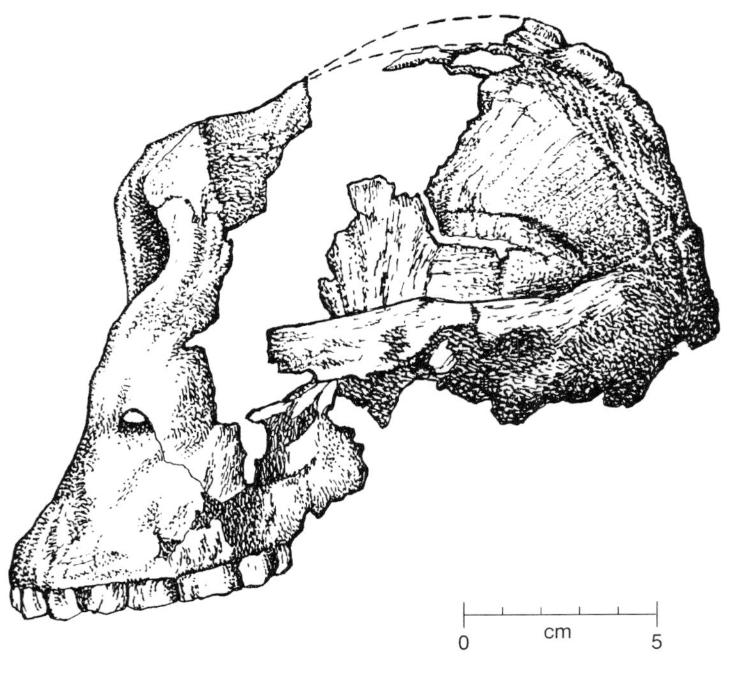

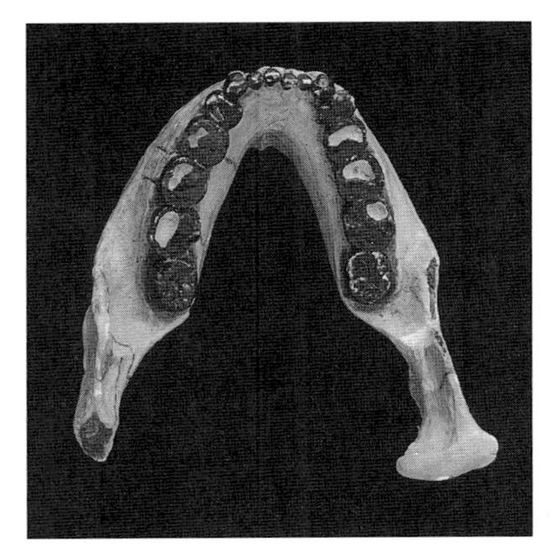

Figure 18.3

Views of the skull and teeth of *Australopithecus boisei*. Age: 1.8 million years. This skull is nearly complete except for the lower jaw. It is commonly known as "Zing," an abbreviation of the original genus name *Zinjanthropus*. Zing was discovered in the Olduvai Gorge in Tanzania by Mary and Louis Leakey in 1959.

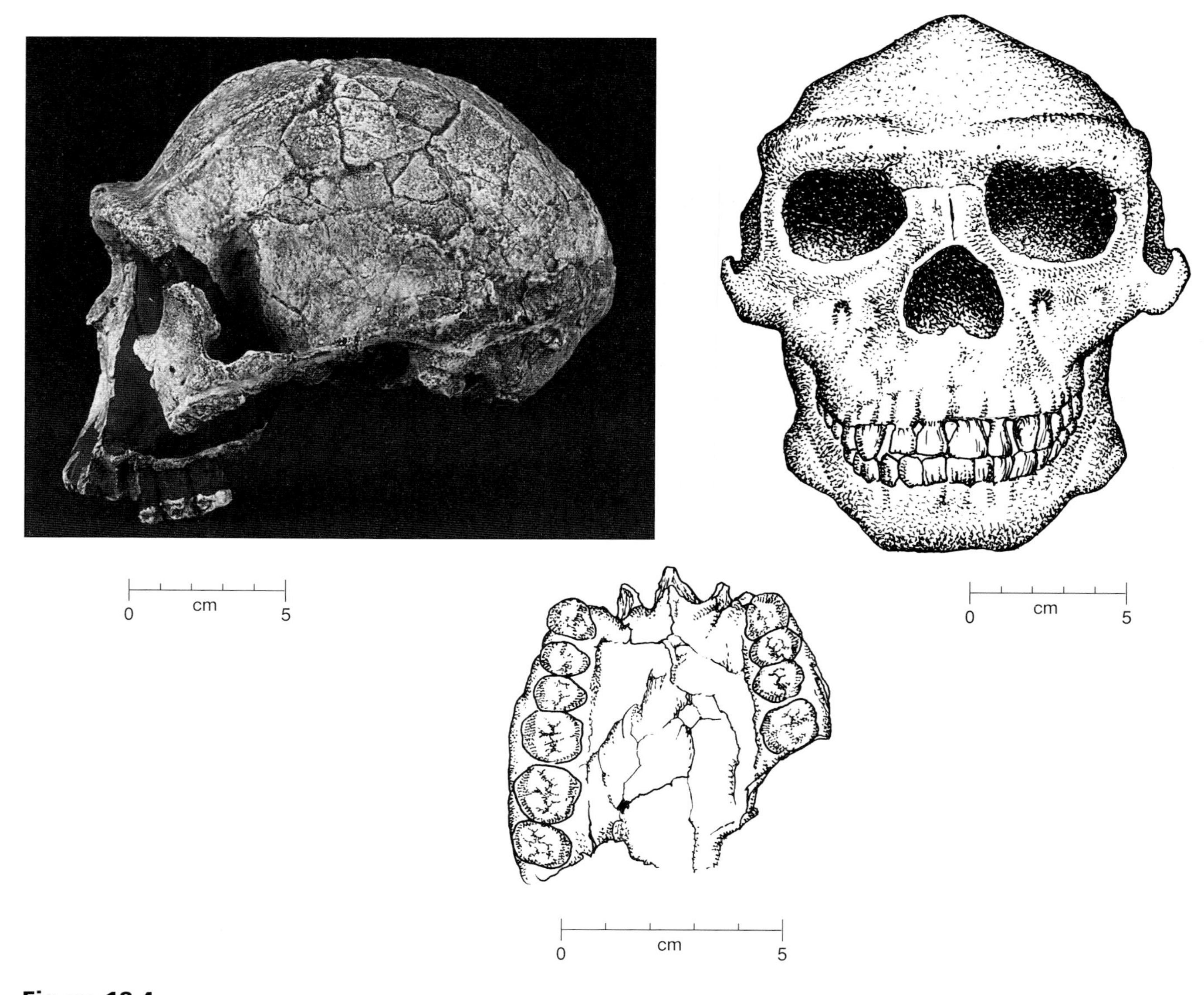

Figure 18.4

Views of the skull and teeth of *Homo erectus*. Age: Less than 1 million years. This diagram shows a reconstruction that includes parts of skulls discovered in 1937 and 1939 in Sangiran.

***Homo sapiens* ("Neanderthal")**

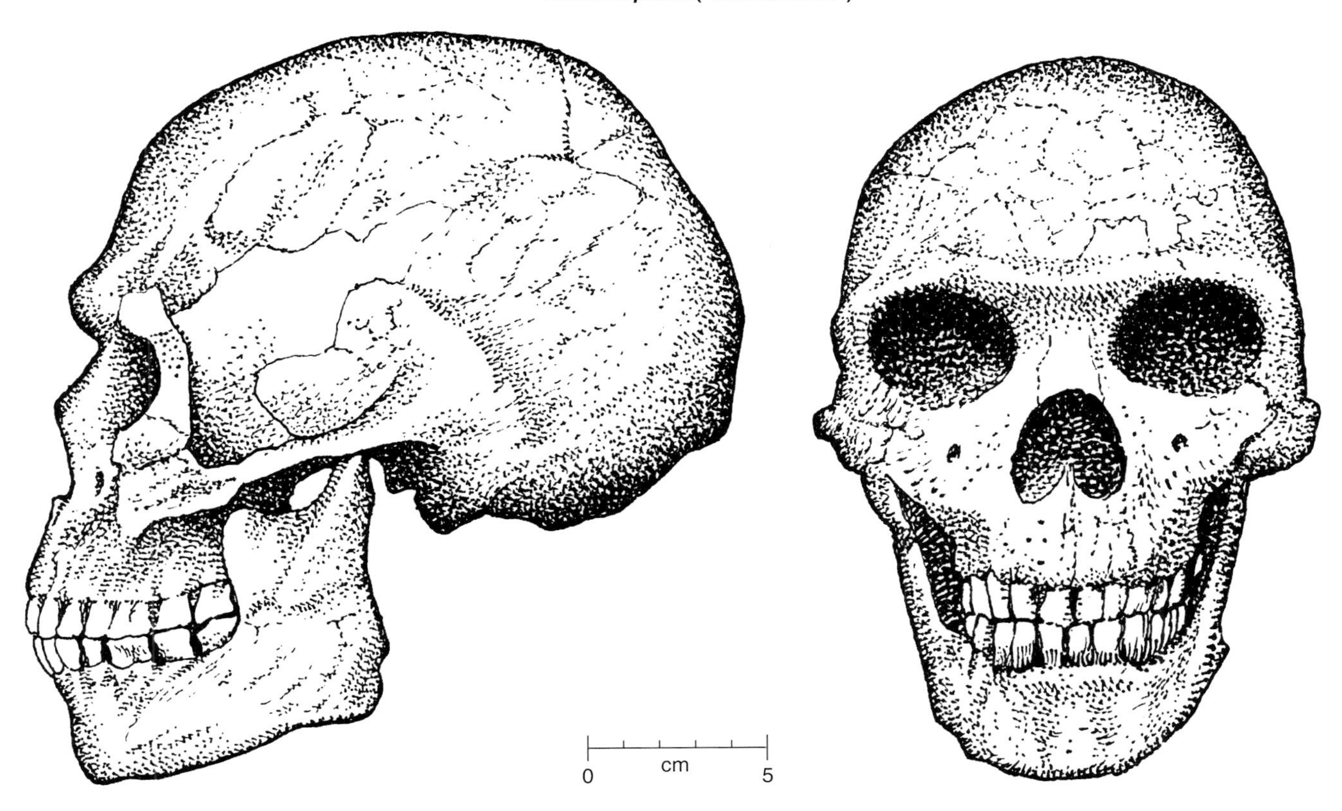

Figure 18.5
Views of the skull and teeth of *Homo sapiens*. Age: About 32,000 years. Also known as Neanderthal Man, this type was recovered in 1932 from Mugharet-es-Skhull, Wadi el-Mughara, Israel. The skull is nearly complete.

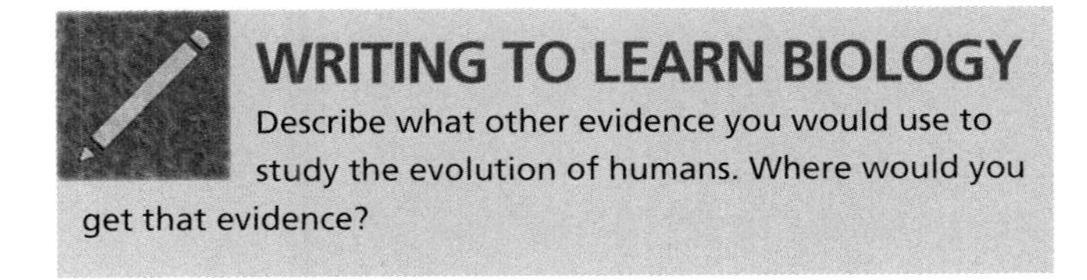

WRITING TO LEARN BIOLOGY

Describe what other evidence you would use to study the evolution of humans. Where would you get that evidence?

***Homo sapiens* (modern)**

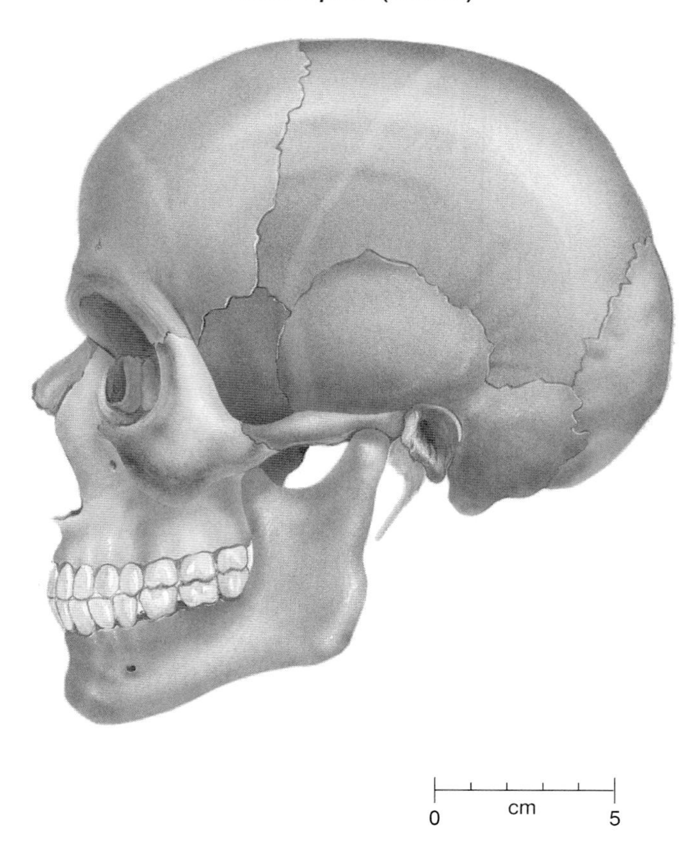

Figure 18.6

Skull of modern *Homo sapiens*. Age: About 10,000 years. This skull is from one of at least 50 skeletons recovered at Oued Agrious, Algeria, during the late 1920s.

19

Ecology
Diversity and Interaction in Plant Communities

Objectives

By the end of this exercise you should be able to:

1. Observe the physical factors, plant dominance, and interactions among organisms in a terrestrial community and characterize the community according to these features.
2. Quantify the distribution and abundance of plants in a community.
3. Detect the experimental effects of competition and allelopathy among plants.
4. Explain four different ways of quantifying the importance of different plant species in a plant community, using transect data.

Ecological communities are extraordinarily complex. The assemblage of plants that you observe at any given time results from interactions among plants and their physical surroundings, among different species of plants, and among plants and animals. All of these interactions are driven by a flow of energy captured by green plants and passed to herbivores, predators, and decomposers.

It is beyond the scope of this exercise to explain all of the processes occurring in a plant community, but we can assemble some basic observations that characterize and distinguish communities.

QUALITATIVE COMMUNITY ASSESSMENT

Procedure 19.1

Observe and assess the ecological characteristics of a terrestrial community

1. Locate and visit a terrestrial community designated by your instructor.
2. Characterize the community according to the criteria and questions that follow. After you've answered the questions, discuss your observations with your instructor and other groups. Be prepared to use your observations as a basis for describing your assessment of energy flow through the community, diversity of the community, and interactions among organisms.

Physical Factors

Observations

1. What levels of light intensity occur throughout the community?

2. Does the community include shade-tolerant as well as shade-intolerant plants?

3. How does light differ among different vertical levels of vegetation? Why could this be important?

4. What is the temperature 2 m above ground?

5. What is the temperature at the soil surface?

6. How does the ground slope? Why could this be important?

7. How would you characterize the soil? Loam? Clay? Sand?

8. What is the nature of the ground cover? Grasses? Bare soil?

9. Is there a layer of leaf litter on the ground?

10. Is the community generally a moist, moderate, or dry environment?

Interpretations

1. How might shade affect the temperature of the community?

2. How might different amounts of light at different vertical levels within the community be important?

3. What parts of the community might be cooler than others? Why could this be important?

4. Why would ground slope be important?

5. Based on your observations of slope and soil type, would you expect the soil to retain moisture?

6. How long has this community been left to develop naturally? That is, what is the age of the community?

Plant Dominance

Observations

1. Which plant species are most abundant (numbers)?

2. Which plant species are most abundant (biomass)?

3. What general categories of plant types (shrubs, trees, etc.) are apparent?

4. What is the vertical distribution of vegetation?

Interpretations

1. Would you describe this community as diverse? Why or why not?

2. What comparison community in your local area would you consider to be more diverse? Less diverse?

3. What observations led you to your conclusion for the previous question?

4. What specific factors make your comparison community more or less diverse? Human impact? Stressful environmental factors? Geology?

Interactions Among Organisms

Observations

1. What evidence do you see of resident vertebrates?

2. What evidence do you see of resident invertebrates?

3. What evidence do you see of plant–animal interactions?

4. What evidence do you see of plant–plant interactions?

5. What adaptations do the plants have to discourage herbivores?

6. Do you see any obvious or subtle evidence of competition by plants for available resources?

Interpretations

1. If you don't see any vertebrates, does that mean they are not around? Explain your answer.

2. Reexamine the observations that you just listed. How would each observation affect the type and growth of plants in the community that you studied?

3. What kinds of competitive interactions are apparent in the community?

4. What kinds of mutually beneficial interactions are apparent in the community?

QUANTITATIVE COMMUNITY ASSESSMENT

Many techniques have been developed to measure the numbers, densities, and distributions of organisms in terrestrial plant communities. One widely used technique is to count organisms within randomly distributed **quadrats** (sometimes called **plots**) of uniform size. Other techniques involve measuring distances between plants or the distance from randomly chosen points to nearest plants. In the **line-intercept method** a **transect,** or line, is established and laid out within the community. Organisms that touch this line are counted and measured. Calculations based on these measurements of quadrats and transects reveal the relative abundances, frequencies, and distributions of the plant species that compose the community.

Procedure 19.2

Assessing a community with the line-intercept method

1. With the help of your instructor, locate a suitable field site with a plant community to be examined.
2. Obtain a measuring tape 10–13 m long, a meterstick, and a notepad. If a measuring tape is unavailable, use a measured piece of string or rope.
3. Assess the general layout of the community to be sampled. With the aid of your instructor decide on a reasonable set of criteria to govern the placement of a transect for each group of students. You may establish a baseline along one side or within the community. Points along this line can be used to randomly establish starting points for the transects used to measure plant distributions.

TABLE 19.1

SUMMARY OF RAW DATA FOR SPECIES OCCURRING ALONG A TRANSECT

Total transect length = ________ Total number of intervals ________

Species i	f_i = Number of Intervals in Which Species i Occurs	n_i = Number of Individuals Encountered	c_i = Length of Transect Intercepted
	F = Total of all Frequencies = ________	N = Total of all Individuals = ________	C = Total length of transect Intersected = ________

Question 1

What concepts or ideas should govern the placement of your transect to obtain a representative sample of the community? Are there any "wrong" places to put a transect? Why or why not?

4. You and your lab partners will work on a single transect. Stretch the measuring tape on the ground to establish a transect.
5. Divide the transect into 1- or 5-m intervals. Each interval represents a separate unit of the transect.
6. At one end of the transect begin counting plants that touch, overlie, or underlie the transect line. For each plant encountered record the type of plant (species) and the length of the line that the plant intercepts. For plants that overhang the line, record the length of the imaginary vertical plane of the line that the plant would intercept. Record this raw data for each interval in your field notepad. Also record any uncovered (bare) lengths within the transects.
7. When all plants from all intervals have been recorded, summarize your data in table 19.1.
8. Sum the values in each of the three data columns of table 19.1 to calculate F, N, and C. Record the calculations at the bottom of each column.
9. Use the data in table 19.1 to calculate the following four parameters for each species within the community. Record your results in table 19.2.

Relative frequency = f_i/F

Relative density = n_i/N

Relative coverage = c_i/C

Importance value of species i = **Relative density + Relative coverage + Relative frequency**

Question 2

What is the meaning of an importance value? Why would we calculate this in addition to density, coverage, and frequency?

TABLE 19.2

RELATIVE VALUES OF EACH SPECIES IN A SELECTED COMMUNITY USING PARAMETERS OF THE LINE-INTERCEPT METHOD

Species	Relative Density	Relative Coverage	Relative Frequency	Importance Value

PLANT INTERACTIONS

Competition

Competition is the interaction among individuals seeking a common resource that is scarce (fig. 19.1). The interaction is usually negative (disadvantageous) for both competitors. The intensity of competition for a resource such as light, food, water, space, and nutrients depends on the amount of resource, the number of individuals competing, and the needs of each individual for that resource. Competition by members of the same species is **intraspecific competition,** whereas competition by members of different species is **interspecific competition.**

Figure 19.1
Competition in a forest. Each plant in this forest competes with all the individuals around it for light, soil nutrients, and moisture.

Procedure 19.3

Examine competition by sunflower seedlings

1. Obtain five pots containing enough potting soil to plant seeds. Measure the area at the surface of the potting soil.
2. Follow your lab instructor's directions to plant 2, 4, 8, 16, and 32 sunflower seeds in each of the five pots, respectively. In each of the pots plant at least 20% more seeds so you can later reduce the number of seedlings to the numbers listed above. Label each pot with the number of seeds, the date, and your name.
3. Water each of the pots gently with a consistent amount of water.
4. Place the pots in the greenhouse or growth area so that each pot has the same environmental conditions of light, temperature, etc. After the seedlings are established remove excess seedlings so the treatments will have the correct number of seedlings listed in step 2.
5. Examine the pots at regular intervals as directed by your instructor. Record measurements of the parameters called for in table 19.3.

TABLE 19.3

EFFECTS OF COMPETITION ON SUNFLOWER SEEDLINGS

	Plants per Pot				
	2	4	8	16	32
General observations: Date ________					
Date ________					
Date ________					
Date ________					
Mean height of individuals: Date ________					
Date ________					
Date ________					
Date ________					
Range of height of individuals: Date ________					
Date ________					
Date ________					
Date ________					
Mean width of ten widest leaves: Date ________					
Date ________					
Date ________					
Date ________					
Mean fresh weight of aboveground biomass (measured at the end of the experiment): Date ________					

Question 3

a. Was competition greater in the more crowded pots?

b. Which parameters best showed the effects of competition?

c. What other characteristics of the competitors might you have measured?

d. Did competition more noticeably affect the number of individuals or the biomass of each individual?

e. In what environments would you expect that competition among sunflowers would be most intense?

f. Would you expect different results if different potting soil was used? Why?

g. Would you expect inter- or intraspecific competition to be the most intense? Why?

h. Would plants and animals compete for the same resources? What might be some differences?

Allelopathy

Allelopathy is the inhibition of a plant's germination or growth by exposure to compounds produced by another plant, and is a form of competition. These compounds may be airborne or leach from various plant parts. Rainfall, runoff, and diffusion typically distribute inhibitory compounds in the immediate area surrounding the producing plant.

Question 4

a. What are the benefits of producing allelopathic compounds?

b. What are the possible disadvantages of producing allelopathic compounds?

INVESTIGATION

Allelopathic Chemicals from Plant Tissues

Not all organs (i.e., roots, stems, leaves) of allelopathic plants produce equal amounts of allelopathic chemicals.

a. Decide which organs you will test.
b. Form a hypothesis about the result you expect. Your instructor will advise you about how to write a testable hypothesis. Write your hypothesis here:

c. Decide how you will test your hypothesis. Describe your experimental design here:

d. Do your experiment. What do you conclude? Do your data support your hypothesis?

Procedure 19.4

Demonstrate allelopathy

1. Determine from your instructor the overall experimental design for the class; that is, how many plants your group will test and how many replicates you will set up for each plant.
2. Obtain tissue (stems and leaves) from the variety of plants provided by your instructor. Some of these plants are suspected to produce allelopathic compounds.
3. For each plant: Mix 10 g of tissue with 100 mL of water in a blender and homogenize. Let the slurry soak for 5–10 min to leach chemicals from the disrupted tissue.
4. Filter or strain the slurry to remove large particulates. Drain the filtrate into a beaker.
5. Obtain a petri dish and line its bottom with a circular piece of filter paper.
6. Label the petri dish appropriately for the plant extract and replicate being tested in that dish.
7. Saturate the filter paper with a measured amount (5–8 mL) of the extract.

TABLE 19.4

EFFECTS OF ALLELOPATHY ON GERMINATION AND GROWTH RATE OF SEEDS

Plant Extract ________ Seed Species ________________ Total No. of Seeds ________

	Time	% Germinated	Length of Radicle	Observations
Rep 1				
Rep 2				
Rep 3				

Plant Extract ________ Seed Species ________________ Total No. of Seeds ________

	Time	% Germinated	Length of Radicle	Observations
Rep 1				
Rep 2				
Rep 3				

Plant Extract ________ Seed Species ________________ Total No. of Seeds ________

	Time	% Germinated	Length of Radicle	Observations
Rep 1				
Rep 2				
Rep 3				

Plant Extract ________ Seed Species ________________ Total No. of Seeds ________

	Time	% Germinated	Length of Radicle	Observations
Rep 1				
Rep 2				
Rep 3				

8. Repeat steps 3–7 for each replicate and each plant being tested.
9. Obtain seeds of radish, lettuce, or oat. Distribute 50 seeds uniformly on the filter paper in each dish.
10. Your instructor may enhance the experimental design by asking you to set up replicate dishes for each extract and to test the effects on different kinds of seeds. Follow the directions given by your instructor.
11. Incubate the covered dishes at room temperature in the laboratory or in the greenhouse. After 24 and 48 h, count the number of seeds that have germinated and calculate the percent germinated.
12. After 72 h (or the length of time specified by your instructor) measure the length of the radicle and make relevant observations about the apparent rate of growth of the emerged embryos. Extend your observations for as many days as specified by your instructor.
13. Record your results in table 19.4.

Question 5

a. Did your observations reveal any differences in allelopathy among tested plants?

b. Which plant species demonstrated the most intense allelopathy?

Questions for Further Thought and Study

1. Diverse plant communities have many species representing a variety of plant types such as grasses, shrubs, succulents, hardwood trees, softwood trees, vines, ferns, and so on. What factors increase a community's diversity? Age of the community? Energy input? Moisture? Nutrients? Disturbance? Human activity? How do they do so?

2. What characteristics of a community make it more resilient than other communities after disturbance?

3. What characteristics would indicate that a community has not been disturbed for a few years?

4. How does competition influence natural selection? Is the presence of competitors a selective force?

20 Community Succession

Objectives

By the end of this exercise you should be able to:

1. Define community succession.
2. Describe how succession occurs.
3. Describe how the environment and resources influence succession.

As time passes, most environments are inhabited by a **succession** or sequence of different communities. A **community** includes all the organisms that live and interact in the same area at the same time. The growth of the bacterial community in milk produces a succession of changes in milk that, in turn, create conditions for other organisms to grow. This phenomenon is called **community succession**. During community succession, each community of organisms (i.e., each stage in succession) changes the environment. Ironically, during succession a community can change the environment and subsequently inhibit its own long-term growth. Then a different community, i.e., a new successional stage, takes its place. Thus, communities change over time.

Milk is good for you; it contains carbohydrates (lactose, or milk sugar), protein (casein, or curd), and lipids (butterfat). Not surprisingly, these nutrients can support the growth of a variety of microbes. Pasteurization retards this growth and involves heating milk to about 170°C to kill pathogenic bacteria. Then the milk is rapidly cooled. However, pasteurization does not kill *all* nonpathogenic bacteria such as *Bacillus*, which produce endospores. Although these nonpathogenic bacteria divide slowly when they are refrigerated, they will ultimately "spoil" the milk, even if the milk is refrigerated. Leaving the milk at room temperature greatly speeds the spoilage.

Community succession occurs at multiple scales and levels, ranging from large-scale ecosystems (e.g., the forests surrounding Mt. St. Helen volcano) to local areas affected by timber harvest, crop harvest, and development.

Question 1
What are some examples of community succession that are occurring on campus or in adjacent areas?

In large-scale, so-called "old-field succession" that occurs in places such as forests and fields, most of the organisms are multicellular, sexually reproducing, relatively long-lived species that colonize the site at various times. During this type of succession, the composition of the community changes and is dominated by autotrophs.

In contrast, small-scale communities such as carcasses or milk cartons are dominated by unicellular, asexually reproducing, relatively short-lived individuals that are largely descended from individuals present at the start. This type of succession is "degradative" because the availability of energy and nutrients is highest at the early stages and declines over time.

SUCCESSIONAL CHANGES IN MILK

In this exercise, you'll study community succession in milk. This process of succession is often more complex than you might suspect. For example,

1. *Pseudomonas* and *Achromobacter* (both Gram-negative rods; see Exercise 23) are common bacteria that digest butterfat and give milk a putrid smell.
2. *Lactobacillus* (a Gram-positive rod) and *Streptococcus* (a Gram-positive coccus) survive pasteurization. These bacteria ferment lactose to lactic acid and acetic acid.
3. Acidity sours the milk and converts the casein to curd.

4. Under acidic conditions, fungi such as yeast grow extremely well. These organisms often metabolize the acids into nonacidic compounds.
5. Finally, *Bacillus* metabolize proteins into ammonia products and raise the milk's pH. The odor of spoiled milk becomes apparent when this occurs.

Succession in milk, under the right conditions and with the proper bacteria, can lead to the formation of cheese. Similarly, bacteria added to milk and the subsequent succession can produce buttermilk, yogurt, and sour creams. However, the most common instance of community succession in milk usually occurs in dairy cases at supermarkets and in our kitchen refrigerators. There, bacteria ferment lactose into acid, thereby spoiling the milk.

Procedure 20.1

Compare community succession in different types of milk

1. Work in small groups as instructed by your lab instructor.
2. Each group will be given one of the following sets of samples:

   Sample 1: Whole milk, kept at room temperature (25°C); 2, 5, and 8 days old.

   Sample 2: Whole milk, kept in a refrigerator (4°C); 2, 5, and 8 days old.

   Sample 3: Whole milk, incubated at 37°C; 2, 5, and 8 days old.

   Sample 4: Whole milk, boiled, then cooled to room temperature (25°C) and sealed; 2, 5, and 8 days old.

   Sample 5: Chocolate milk, kept at room temperature (25°C); 2, 5, and 8 days old.

   Sample 6: Skim milk, kept at room temperature (25°C); 2, 5, and 8 days old.

   Sample 7: Buttermilk, kept at room temperature (25°C); 2, 5, and 8 days old.

   Sample 8: Optional treatment, determined by the instructor.
3. Use pH paper to measure the pH of each sample. Record your data in table 20.1.
4. Note each sample's odor, color, and consistency. Pay particular attention to physical changes in the milk's composition and record any odors, particles, "growths," or new liquids forming in the flask.
5. Use the Gram stain procedure (see Exercise 23) to identify the shapes and staining properties of the bacteria in the milk.
6. Record your observations in table 20.1.
7. Plot your pH data for Samples 1–3 in figure 20.1, Samples 1 and 4 in figure 20.2, and Samples 1, 5, 6, and 7 in figure 20.3.

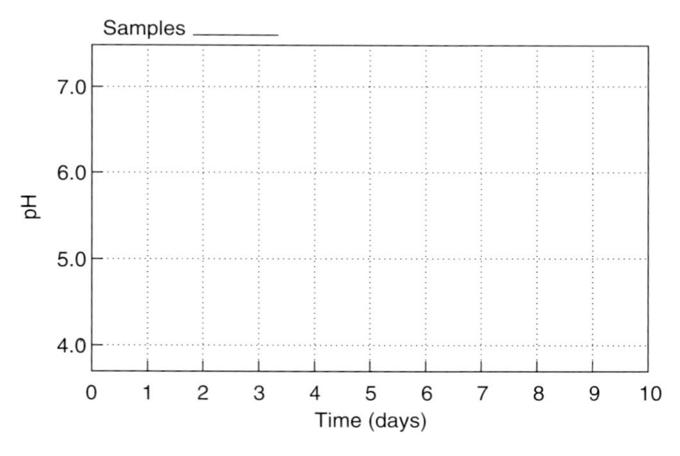

Figure 20.1
The effect of temperature on succession (as measured by changes in pH) in a biological community in whole milk.

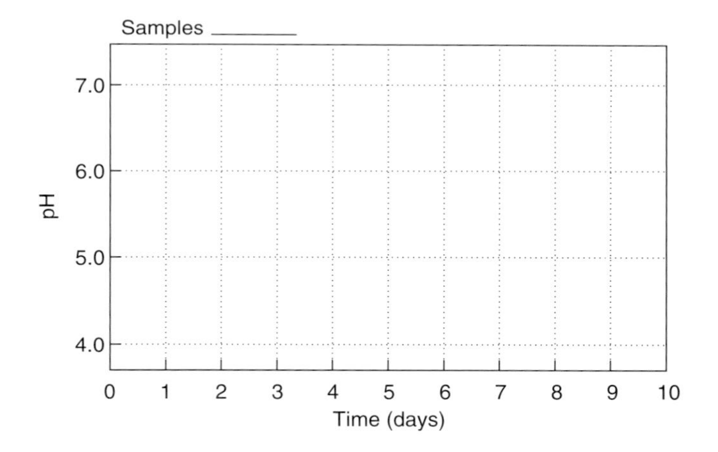

Figure 20.2
The effect of boiling and sealing on succession (as measured by changes in pH) in a biological community in whole milk.

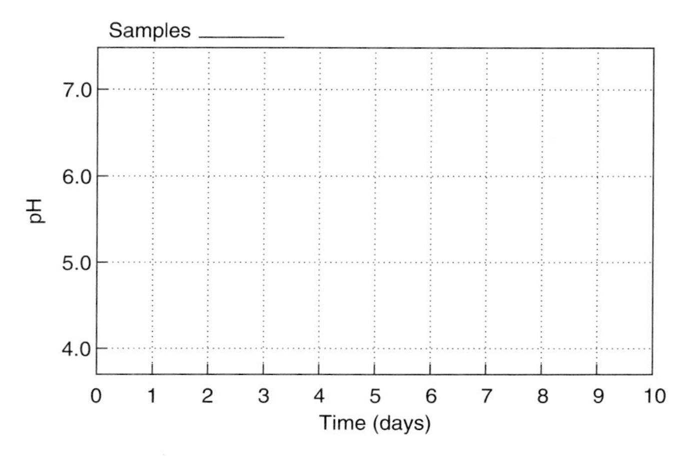

Figure 20.3
Community succession (as measured by changes in pH) in chocolate milk, skim milk, and buttermilk.

TABLE 20.1

COMMUNITY SUCCESSION IN MILK

Sample 1

	pH	Odor	Color	Type of Bacteria Present	Comments
Day 0					
Day 2					
Day 5					
Day 8					

Sample 2

	pH	Odor	Color	Type of Bacteria Present	Comments
Day 0					
Day 2					
Day 5					
Day 8					

Sample 3

	pH	Odor	Color	Type of Bacteria Present	Comments
Day 0					
Day 2					
Day 5					
Day 8					

Sample 4

	pH	Odor	Color	Type of Bacteria Present	Comments
Day 0					
Day 2					
Day 5					
Day 8					

Sample 5

	pH	Odor	Color	Type of Bacteria Present	Comments
Day 0					
Day 2					
Day 5					
Day 8					

Sample 6	pH	Odor	Color	Type of Bacteria Present	Comments
Day 0					
Day 2					
Day 5					
Day 8					

Sample 7	pH	Odor	Color	Type of Bacteria Present	Comments
Day 0					
Day 2					
Day 5					
Day 8					

Sample 8	pH	Odor	Color	Type of Bacteria Present	Comments
Day 0					
Day 2					
Day 5					
Day 8					

Question 2
Which milk changed the slowest? Why?

Question 3
How did the pH of milk change over time in your different samples?

Question 4
Of what value is pasteurization?

Question 5
How did the abundance and type of organisms in the milk change over time? Why is this important?

Question 6
Acidity sours the milk and converts the casein to curd. How long does it take for this change to occur at room temperature? When the milk is refrigerated?

Question 7
Which milk changed the quickest? Why?

Questions for Further Thought and Study

1. Question 1 in this exercise asked you to describe local examples of community succession. How are these examples similar to the "milk community" model that you studied in this lab? How are they different?

2. What is biological succession, and why is it important?

3. How does succession occur? What factors influence the rate of succession?

4. What human activities are based on preventing (or slowing) biological succession?

21

Population Growth
Limitations of the Environment

Objectives

By the end of this exercise you should be able to:

1. Describe how populations grow.
2. Show the effects of resources and environmental conditions on population growth.
3. Graphically analyze how the human population is increasing.

In optimal conditions (e.g., plenty of food, etc.), populations grow in a predictable pattern. Early growth is slow (the so-called "lag" phase of growth), after which it is extremely rapid (fig. 21.1). This rapid, logarithmic (i.e., "log") phase of growth represents the organism's **biotic potential,** which is its maximal reproductive capacity if given unlimited resources. The number of individuals at a given generation during logarithmic growth can be determined from the following formula:

$$N = (a)(2^n)$$

where

N = the number of individuals at a given generation
a = the number of individuals present initially
n = number of generations

This equation shows how fast populations can grow. For example, consider *Escherichia coli,* a common bacterium that can divide every 20 min in ideal conditions. In only one day (1440 min), these bacteria can go through 72 (1440/20) generations. Therefore, if we start our experiment with one bacterium, the number of bacteria present after one day would be:

$= (1)(2^{72})$ **bacteria**
$= 40{,}000{,}000{,}000{,}000{,}000{,}000{,}000$ **bacteria**
$= 4 \times 10^{22}$ **bacteria**

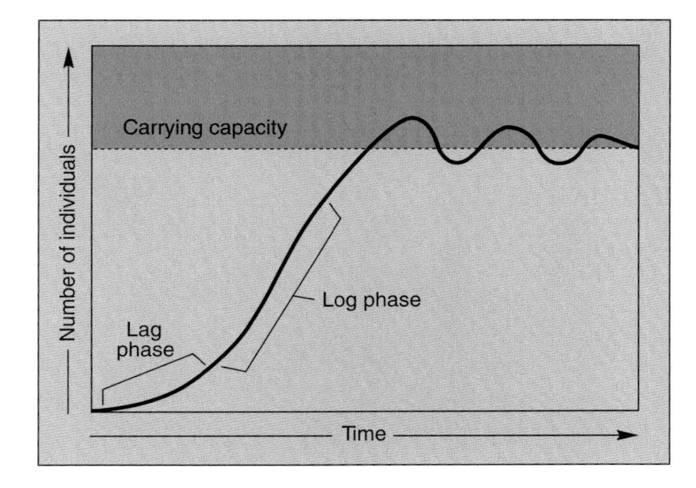

Figure 21.1
The theoretical sigmoid curve of population growth. The dotted line represents the ecosystem's carrying capacity.

This many bacteria would weigh about 3.2 million kg. Bacteria aren't the only organisms having such an incredible biotic potential. For example,

- Oysters each produce about 50 million eggs per year.
- A single pair of Atlantic cod and their descendants reproducing without hindrance would in six years completely fill the Atlantic Ocean.
- The 80 offspring produced every six months by a pair of cockroaches produce 130,000 roaches in only 18 months—enough to overrun any apartment (fig. 21.2).

TABLE 21.1

THEORETICAL AND ACTUAL GROWTH OF *E. COLI* BACTERIA

Generation	Time		Size of Population (10^3 bacteria per mL)	
	Hours	Minutes	Theoretical	Actual
1	0	0	8	8
2	0	20	16	15
3	0	40	32	28
4	1	0	________	48
5	1	20	________	120
6	1	40	________	220
7	2	0	________	221

Figure 21.2

The consequences of exponential growth. All organisms have the potential to produce populations larger than those that actually occur in nature. The German cockroach (*Blatella germanica*), a major household pest, produces 80 young every six months. If every cockroach that hatched survived for three generations, kitchens might look like this culinary nightmare concocted by the Smithsonian Museum of Natural History.

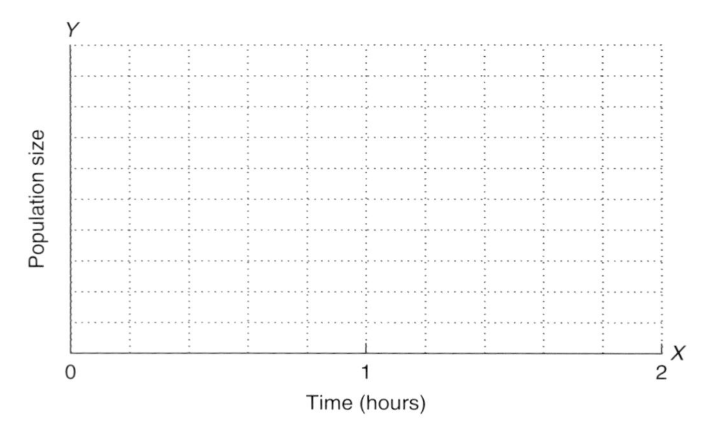

Figure 21.3

Theoretical and actual population growth of *E. coli*.

ENVIRONMENTAL RESISTANCE AND CARRYING CAPACITY

Organisms in the "real world" do not always reproduce at maximum rates, and populations do not grow at ever-increasing rates. Maximum logarithmic growth cannot be sustained because **environmental resistance** includes factors such as disease, accumulation of waste products, and lack of food. Ultimately the size and growth of a population is balanced by the environment: environmental resistance prevents a population from continuing to reproduce at its biotic potential.

To understand the effects of environmental resistance, complete table 21.1. This table provides actual data for a growing but limited population of bacteria. For comparison, you must calculate the size of a theoretical population of *E. coli* if given unlimited resources. Every 20 min the population could potentially double. After doing these calculations, plot the growth of the theoretical and actual populations on figure 21.3.

Question 1

a. How did growth of the actual population compare with that of the theoretical population during early stages of the experiment? At later stages?

TABLE 21.2

GROWTH OF BACTERIA IN A LIMITED-NUTRIENT MEDIUM

Time (hours)	Turbidity Intensity (0–10)	Absorbance Value
0	______	______
4	______	______
8	______	______
12	______	______
24	______	______
48	______	______

b. How long did it take the real populations to double during early stages of the experiment? Middle stages? Later stages?

c. When was growth of the actual population most rapid?

d. At what stage was growth slowest? Why?

The growth of most populations levels off after a logarithmic phase because of **limiting factors** such as disease and lack of food. This stability occurs when the birth rate equals the death rate and is referred to as the **carrying capacity** of the ecosystem (fig. 21.1). In most populations, population size remains near the carrying capacity as long as limiting factors are constant.

In the laboratory, you can measure the growth of real populations such as bacteria that reproduce quickly. As bacteria reproduce in a clear nutrient broth, the broth becomes turbid. You can't accurately count individual bacteria in this broth, but you can measure the increase in turbidity of a growing culture. More turbidity means more bacteria, and turbidity values roughly estimate population size.

Your instructor has previously inoculated some test tubes of culture media with *E. coli*, a common bacterium. At regular time intervals some of the tubes were put into a refrigerator to stop growth. Examine the cultures by using procedure 21.1.

Procedure 21.1

Measure population growth of bacteria

1. Examine cultures of *E. coli* grown for 0, 4, 8, 12, 24, and 48 h.
2. Quantify the relative turbidity of each culture between 0 (clear) and 10 (most turbid).
3. Record your results in table 21.2.
4. If turbidometers are available, measure the turbidity of the solutions according to procedures demonstrated by your instructor. Spectrophotometers may also be used at 600 nm. Record your results in table 21.2.

Procedure 21.2

Measure how resources and environmental conditions affect the size of a population

1. Examine cultures of *E. coli* that have grown for 10 days in the following environments:
   Distilled water, pH 7
   Nutrient broth, pH 3
   Nutrient broth, pH 5
   Nutrient broth, pH 7
   Nutrient broth, pH 9
   Nutrient broth, pH 11
2. Quantify the relative turbidity of each culture between 0 (clear) and 10 (most turbid).
3. Record your results in table 21.3.
4. If turbidometers are available, measure the turbidity of the solutions according to procedures demonstrated by your instructor. Record your results in table 21.3.

TABLE 21.3

GROWTH OF BACTERIA IN A LIMITED-NUTRIENT MEDIUM

Media	Turbidity Intensity (0–10)	Absorbance Value
Distilled water, pH 7	________	________
Nutrient broth, pH 3	________	________
Nutrient broth, pH 5	________	________
Nutrient broth, pH 7	________	________
Nutrient broth, pH 9	________	________
Nutrient broth, pH 11	________	________

Question 2

Compare your data for populations grown in nutrient broth and in distilled water. Does the presence of nutrients ensure rapid growth of bacteria? Why or why not?

Procedure 21.3

Measure population growth of duckweed (*Lemna*)

1. During the first week of this term your instructor placed 10 duckweed (*Lemna*) plants in an illuminated aquarium. Each week since then, he or she has counted the number of plants in the aquarium. Those data are posted by the aquarium.
2. From now until the end of the term, count the number of duckweed plants in the aquarium each week. Plot your data in figure 21.4.

Question 3

a. What do you conclude about population growth of duckweed?

b. What will eventually happen to the size of the population? Why?

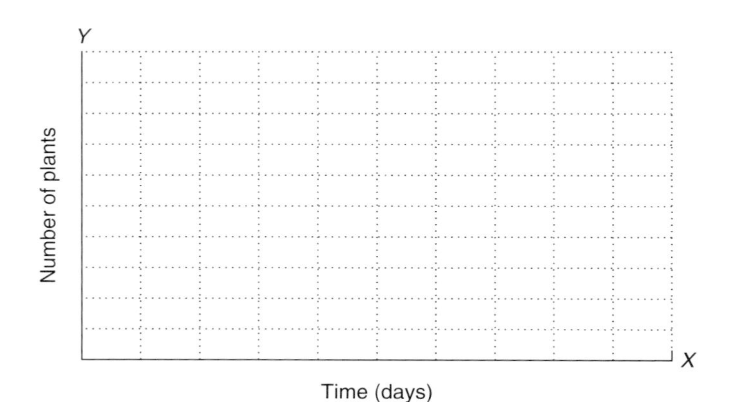

Figure 21.4

Population growth of duckweed (*Lemna*).

Growth of Human Populations

Our population is growing extremely fast. Consider these data:

Year	*Human Population (millions)*
8000 B.C.	5
4000 B.C.	86
A.D. 1	133
1650	545
1750	728
1800	906
1850	1130
1900	1610
1950	2400
1960	2998
1970	3659
1980	4551
1990	5300
2000	6200
2040	13,000 (projected)

Plot these data in figure 21.5.

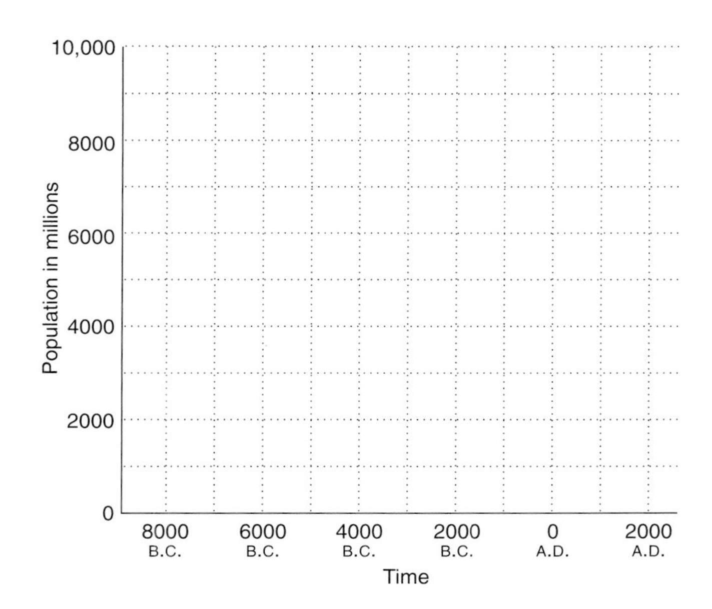

Figure 21.5
Growth of the human population.

Question 4

a. How does the shape of this graph compare with those you made for the bacteria?

b. What do you conclude from this?

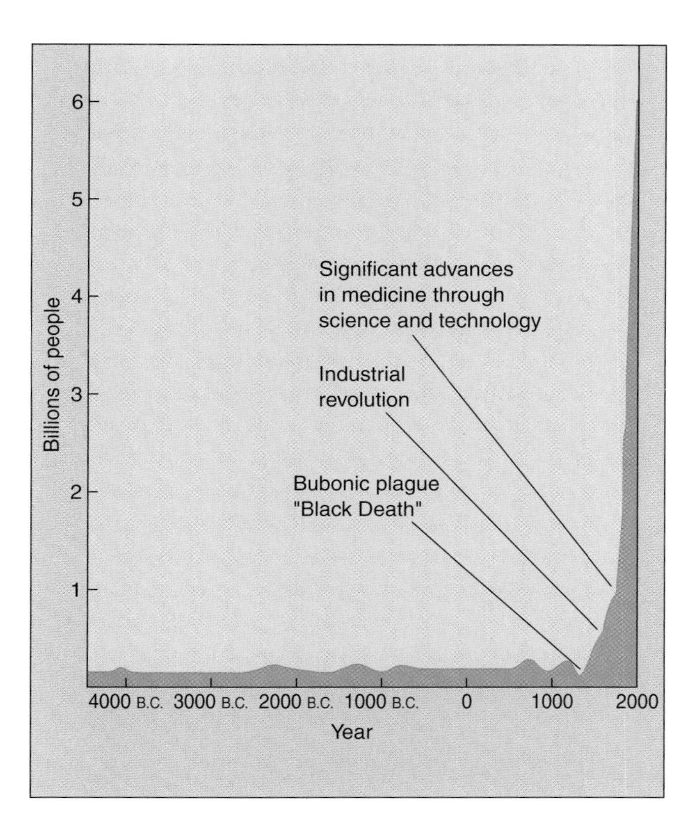

Figure 21.6
History of human population size. Temporary increases in death rate, even severe ones like the Black Death of the 1400s, have little lasting impact. Explosive growth began with the industrial revolution in the 1700s, which produced a significant long-term lowering of the death rate. The current population exceeds 6 billion, and at the current rate will double in 39 years.

The population data just listed have tremendous implications. For example, if our population would have stabilized after World War II, today we could provide all of our energy needs (and have a higher standard of living) without having to burn any coal or import any oil.

Another important feature of a population is its **doubling time.** In 1850, the doubling time for our population was 135 years. Today, the doubling time is about 40 years. This means that during the next 40 years we must double our resources if we are to maintain our current standard of living. *Improving* our standard of living will require that we more than double our resources.

Question 5

a. What is the importance of population growth to you?

b. The doubling time for populations in developed countries is about 120 years but in developing countries it is about 30 years. What is the significance of this?

The human population has increased explosively during the past three centuries (fig. 21.6). Although the birth rate has remained constant (at about 30 per 1000), the death rate has fallen from about 30 per 1000 per year to its current level of about 13 per 1000 per year. The difference between birth and death rates (17 per 1000) means that the human population is growing at a rate of about 1.7% per year. Here's what that means:

- Each hour the world's population grows by 11,000. Each year, the world's population grows by about 90,000,000.

- Each year there are about 90 million more people on Earth. That annual increase in our population equals the combined population of Great Britain, Ireland, Iceland, Belgium, The Netherlands, Sweden, Norway, and Finland.
- In the six seconds it takes to read this sentence, 18 more people will be added to our population. Each of these people eats food, generates wastes, and, in his or her own way, affects our Earth.
- At our present rate of growth (which is less than our biotic potential), in 2000 years our population will weigh as much as the entire earth. And 4000 years later, it would weigh as much as the visible universe.

Question 6

a. What is the significance of this?

b. Can this continue? Why or why not?

c. What will happen when our population exceeds the Earth's carrying capacity?

d. How might the growth of the human population affect the growth of other populations?

INVESTIGATION

The Environmental Effects of Population Growth

The growth and size of a population is strongly influenced by the environment. Test this using bacteria and/or duckweed.

a. Decide what environmental factor (i.e., nutrients, light, acidity, salinity) you will test.

b. Form a hypothesis to determine how this factor affects population growth and size. Write your hypothesis here:

c. Decide how you will test your hypothesis. Describe your experimental design here:

d. Do your experiment. What do you conclude? Do your data support your hypothesis?

e. How does the growth of the human population affect ecosystems?

Questions for Further Thought and Study

1. How can a population be slowed by its own numbers?

2. Some people are now realizing the significance of population growth. Although this exercise treated the problem only in biological terms, the reality of population growth is much more complex because it involves many political, social, and economic problems. What are some of these problems? How do they affect you now? How will they affect you later in life (e.g., when you want to retire)?

3. Should we do anything to slow the "population explosion"? If so, what? If not, why?

4. From a purely ecological standpoint, can the problem of world hunger ever be overcome by improved agriculture alone? What other components must a hunger-control policy include?

5. How are problems such as deforestation, pollution, and world hunger linked with population growth?

6. The late Garrett Hardin, a famous biologist, wrote that "Freedom to breed will bring ruin to us all." Do you agree with him? Explain your answer.

WRITING TO LEARN BIOLOGY

Charles Darwin's ideas about natural selection as a driving force for evolution were strongly influenced by ideas in an essay entitled "Essay on Population" by Thomas Malthus. Go to the library and read about Malthus's ideas. How did they influence Darwin? How are they relevant to this exercise?

21–8

22

Pollution

The Effects of Chemical, Thermal, and Acid Pollution

Objectives

By the end of this exercise you should be able to:

1. Describe how chemical pollution, thermal pollution, and acid rain affect the growth and reproduction of selected organisms.
2. Determine whether a water sample is polluted by algae.

Seldom a day passes in which we don't hear a pollution-related story in the news. Oil spills, leaks of toxic chemicals, and noise pollution are just a few of the pollution-related problems that affect our lives. The effects of pollution can be disastrous, as exemplified by the nuclear explosion at Chernobyl, Russia; oil spills at the coasts of Alaska and Spain; and the dumping of toxic wastes into our rivers, lakes, and oceans. A **pollutant** is any physical or chemical agent that decreases the aesthetic value, economic productivity, or health of the biosphere. There are many kinds of pollutants, such as noise, chemicals, radiation, and heat. Whether or not it makes headlines, pollution of any kind affects fundamental aspects of life, namely our health and how organisms grow and behave. Because populations of organisms interact among themselves and with their environment, pollution always affects more than one organism. For example, consider the following food chain:

algae → zooplankton → small fish → large fish → humans

A pollutant that reduces a population at any step of this food chain will affect all other levels of the food chain, because all steps of the chain are linked. Thus, water polluted with chemicals, such as herbicides that kill algae, will decrease populations of zooplankton that eat the algae; this ultimately affects populations of fish and other organisms that are part of the food chain.

In this exercise, you will study several types of pollution. Specifically, you will

- Simulate the effects of **acid precipitation** by examining seed germination and survival of organisms at acidic pH.
- Study chemical pollution by examining the effects of differing concentrations of nutrients and **pesticides** on growth of algae and shrimp.
- Study **thermal pollution** by examining how organisms survive at unnaturally high temperatures.
- Examine water polluted with an overabundance of algae.
- Use the Allium Test to assay the effects of a variety of pollutants on plant growth.

SIMULATING THE EFFECTS OF ACID RAIN

Acid rain is a worldwide problem caused by atmospheric pollution. Compounds such as nitrates and sulfates released into the atmosphere by automobiles and industries combine with water to form nitric and sulfuric acids that are deposited across the countryside in precipitation. There, the acid decreases the pH of the soil, affects the availability of nutrients, and usually diminishes plant growth. The effects of acid rain are often subtle, but significant; for example, the cumulative pollution of decades of acid rain are thought to reduce forests and crop yields in many areas.

In this exercise, you will study the influence of acid rain on (1) seed germination, and (2) growth and reproduction of brine shrimp. You will simulate acid rain with solutions of sulfuric acid.

CAUTION

Be careful while you handle the cultures and do not get the sulfuric acid solutions on yourself. If you do, wash yourself thoroughly and immediately.

Procedure 22.1

Observe seed germination

1. Examine the three petri dishes labeled "Acid Rain and Seed Germination." All of the dishes were inoculated several days ago with 50 seeds of corn (*Zea mays*). Dish 1 contains seeds soaked in a solution having a pH of 4, and dish 2 contains seeds soaked in a solution having a pH of 2.
2. Compare these seeds with those grown in dish 3 at neutral pH (pH = 7).
3. Use the following formula to calculate the percentage germination for each treatment

$$\% \text{ Germination} = \frac{\text{Number of germinated seeds}}{\text{Total number of seeds}} \times 100$$

4. Record your results in table 22.1.

Question 1

a. How did acidity affect seed germination?

b. What does this tell you about the effect of acid rain on seed germination in natural environments?

c. How would the decreased growth of plants in response to acid rain affect animals in the same environment?

Procedure 22.2

Observe growth of brine shrimp

1. Examine the three cultures labeled "Acid Rain and Brine Shrimp." Several days ago all of the cultures were inoculated with the same number of brine shrimp eggs. Culture 1 contains shrimp growing in a solution having a pH of 4, and culture 2 contains shrimp growing in a solution having a pH of 2. The pH of typical environments for brine shrimp range from 7–8.5.
2. Compare these cultures with that in dish 3 at neutral pH (pH = 7). Use your dissecting microscope to count the number of living shrimp in five randomly selected fields of view of each culture.
3. Record your results in table 22.2.

TABLE 22.1

GERMINATION OF SEEDS IN ENVIRONMENTS OF DIFFERENT pH

Treatment	Total Number of Seeds	Number Germinated	% Germination
pH = 7	______	______	______
pH = 4	______	______	______
pH = 2	______	______	______

TABLE 22.2

GROWTH OF BRINE SHRIMP IN ENVIRONMENTS OF DIFFERENT pH

	Counts of Number of Living Shrimp						
Treatment	1	2	3	4	5	Total	Mean
pH = 7	______	______	______	______	______	______	______
pH = 4	______	______	______	______	______	______	______
pH = 2	______	______	______	______	______	______	______

Question 2

a. How does acidity affect hatching and growth of brine shrimp in your experiment? What does this tell you about the effect of acid rain on survival of brine shrimp?

b. Consider the food chain described in the introduction of this exercise. How would acid rain affect fish and humans in this food chain?

c. Comment on the validity of extrapolating the results of your experiments with seed germination and brine shrimp survival to the natural world.

NUTRIENT ENRICHMENT: EUTROPHICATION

As naturally occurring nutrients are washed from the soil by rain, bodies of water such as lakes and ponds undergo a nutrient enrichment process called **eutrophication** (fig. 22.1). In extreme situations, this enrichment often stimulates so much growth that the lake or pond may eventually fill in with algae and detritus and disappear. Human-generated pollutants such as raw sewage and fertilizers dramatically speed eutrophication. In this exercise, you'll study eutrophication by studying the growth of organisms in nutrient-enriched media.

Several days ago algal cultures were started with similar amounts of algae. Culture 1 contains an excess of nitrogen, a condition similar to that of lakes enriched with raw sewage. Culture 2 contains large amounts of nitrogen and minimal oxygen to simulate anaerobic conditions that occur naturally when bacterial populations in water increase rapidly in response to added sewage. As the bacteria grow and reproduce, they rapidly deplete the oxygen supply in the water, thus producing anaerobic conditions. Culture 3 is an aerobically grown control containing normal amounts of nitrogen.

Procedure 22.3

Examine the effects of simulated eutrophication

1. Examine the two algal cultures labeled "Eutrophication" and the culture labeled "Control."
2. Record your observations in table 22.3.

Figure 22.1
A eutrophic pond. The surface bloom of green algae indicates the abundance of nutrients in the water.

Question 3

a. Which culture has the most algal growth?

b. Which has the least?

c. What does this tell you about the influence of eutrophication on algal growth?

d. How could you reverse the effects of eutrophication?

e. Can eutrophication ever be beneficial? If so, when?

PESTICIDE POLLUTION

Pesticides are common pollutants; however, the increasing demands for food by our expanding population have caused farmers to increase their reliance on pesticides to increase crop yields (fig. 22.2). These pesticides are often effective, but residues washed from the soil by rain contaminate lakes, ponds, and the underground water supply. These pollutants can harm organisms that drink and live in these waters.

Table 22.3

Effects of Nutrient Enrichment on Algal Cultures

Culture	Observations
1 (nutrient-enriched)	
2 (nutrient-enriched, anaerobic)	
3 (control)	

Figure 22.2

Huge amounts of pesticides and herbicides, such as those being applied by this crop duster, are used in agriculture each year. One of the most important tasks facing agriculturists and other biologists is the development of integrated pest management and systems involving genetically engineered organisms that will avoid the need for such chemical excesses.

Procedure 22.4

Examine the effect of a pesticide on the survival of brine shrimp

CAUTION

Do not open the pesticide cultures or spill the liquids on yourself; the pesticides are toxic.

1. Examine the four cultures labeled "Pesticides." Several days ago these cultures were started with the same number of brine shrimp.
2. Culture 1 contains the recommended dosage of the pesticide, culture 2 contains one-tenth the recommended dosage, and culture 3 contains ten times the recommended dosage. Culture 4 is a control that lacks pesticide.
3. Record your observations of each culture in table 22.4.

Question 4

a. Which culture contains the most living shrimp?

b. Which culture contains the least?

c. What does this tell you about how pesticides affect the growth of aquatic organisms?

d. How would this pollution affect terrestrial organisms that depend on the water for drinking water?

Table 22.4

Effects of Pesticide on Survival of Brine Shrimp

Culture	Observations
Pesticide culture 1	
Pesticide culture 2	
Pesticide culture 3	
Control culture 4	

Table 22.5

Effects of Temperature on Survival of Brine Shrimp

Culture	Observations
Culture 1 (room temperature)	
Culture 2 (35°C)	

THERMAL POLLUTION

Excessive heat is a common pollutant. Many factories use water from lakes, reservoirs, or rivers to cool heat-generating equipment, and release the hot water to reservoirs or ponds, where it raises the water temperature. Thermal pollution

- Speeds biochemical reactions, thereby altering the growth of organisms and composition of communities.
- Speeds evaporation of water, thereby concentrating other pollutants in the water.
- Decreases the oxygen supply in the water, because warm water holds less oxygen than does cool water.

Procedure 22.5

Examine the effect of temperature on the survival of brine shrimp

1. Examine the two cultures labeled "Thermal Pollution." Both cultures were started several days ago with the same number of brine shrimp.
2. Culture 1 was grown at room temperature, and culture 2 was grown at 35°C in oxygen-depleted water. Record your observations of each culture in table 22.5.

Question 5

a. Which culture contains the most living shrimp?

b. Which culture contains the least?

c. What does this tell you about the influence of thermal pollution on growth of aquatic organisms?

d. How would thermal pollution of a lake or pond affect nearby terrestrial organisms?

THE ALLIUM TEST

Many pollutants directly affect the growth and development of organisms. Thus, the most informative assay involves measuring growth of living organisms exposed to controlled treatments of suspected pollutants. The **Allium Test** provides this kind of assay. In the Allium Test, bulbs of the common onion, *Allium*, are subjected to solutions of water being investigated for pollutant toxicity. Roots develop quickly, and their number and length after five days estimates the effect of the potential pollutant on a common aspect of plant growth.

TABLE 22.6

GROWTH OF *ALLIUM* ROOTS EXPOSED TO SOLUTIONS CONTAINING DIFFERENT POLLUTANTS

Solutions Assayed	Mean Number of Roots		Mean Length of Roots	
	Treatment	Control	Treatment	Control
____________	______	______	______	______
____________	______	______	______	______
____________	______	______	______	______
____________	______	______	______	______
____________	______	______	______	______
____________	______	______	______	______
____________	______	______	______	______

TABLE 22.7

ROOT GROWTH OF REPLICATE *ALLIUM* BULBS EXPOSED TO A SPECIFIC TREATMENT SOLUTION

Treatment: ____________		Control: ____________	
Number of Roots	Length of Roots	Number of Roots	Length of Roots
______	______	______	______
______	______	______	______
______	______	______	______
______	______	______	______
______	______	______	______
______	______	______	______
______	______	______	______
______	______	______	______
______	______	______	______
______	______	______	______
Mean = ______	Mean = ______	Mean = ______	Mean = ______

CAUTION

Do not open the pesticide cultures or spill the liquids on yourself. Pesticides are toxic.

Procedure 22.6

Use the Allium Test to assay a variety of pollutants

1. Examine the solutions provided by your instructor for your class to assay with the Allium Test. Record the solutions in table 22.6.
2. You may need to work in small groups to assay all of the solutions. Determine which solutions or treatments are your group's responsibility. Some of the treatments may involve different temperatures and some may involve solutions with unknown contents. Record in table 22.7 the treatment and control solutions tested by your group.
3. Obtain onion bulbs. You will need five to ten bulbs for each treatment and five to ten bulbs for controls for each treatment.

4. Trim the outer, loose layers of each bulb. Use a razor blade to trim exposed tissue from the root crown. Your instructor will demonstrate how to do this.
5. Obtain a beaker (or test tube) that is wide enough to support the onion but will allow the root crown to protrude into the solution.
6. Fill 5 to 10 replicate beakers with the solution being tested.
7. Fill 5 to 10 beakers with the appropriate control solutions.

Question 6
What is the appropriate control solution for your treatment?

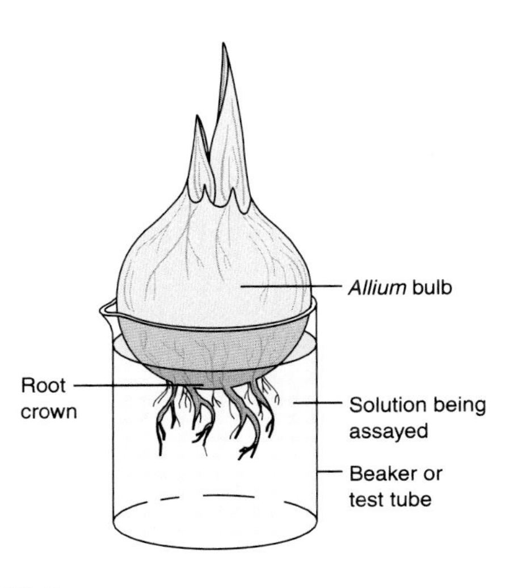

Figure 22.3
Allium Test setup. The number and length of the growing roots estimate the toxicity of the solution being assayed.

8. Put a trimmed onion bulb on your beaker so the root crown is completely submerged (fig. 22.3).
9. Incubate the treatments and controls in the dark at room temperature for three to five days.
10. Check and supplement the solution levels in the beakers periodically if needed.
11. After three to five days, determine the mean root length and number of roots for each bulb. Record the values in table 22.7. These are the results for the treatment solution that your group tested.
12. Record your mean values from table 22.7 in table 22.6.
13. In table 22.6 record the results for the solutions tested by the other groups.
14. Compare the number and length of roots on treated bulbs with those of controls.

Question 7

a. Why did each group need to run controls, rather than one set of controls for everyone?

b. The Allium Test is a common test in environmental science. What are some disadvantages to this test?

c. The Allium Test for thermal pollution may show that higher temperatures accelerate growth. Is this bad? Why or why not?

ALGAL POLLUTION OF WATER SUPPLIES

Growing human populations have created great demands on groundwater supplies. To circumvent these demands, many areas have started using surface waters such as ponds and lakes to meet their water needs. Unlike groundwater, which is usually free of algae, surface water usually contains large amounts of algae that produce tastes and odors (*Anabaena*, *Nitella*, *Pandorina*, *Hydrodictyon*), clog pipes (*Synedra*, *Anacystis*), produce toxins (*Anabaena*), and form mats covering a lake or reservoir (*Oscillatoria*, *Spirogyra*).

Work in small groups to assess algal pollution in the water samples that are available in lab.

Procedure 22.7

Determine algal pollution of water samples

1. Dip a filter base and funnel (fig. 22.4) in 70% ethanol for 1 min. Shake off the excess alcohol and let the apparatus air dry.
2. Use sterilized forceps to place a filter having pores 0.45 μm in diameter on the filter base. Screw the funnel onto the support (fig. 22.4) so that the filter is held securely in place.
3. Add 200 mL of the water sample to the funnel. If the water sample is heavily polluted, decrease the sample to 100, 50, or 25 mL.
4. Use a hand vacuum pump or water aspirator (fig. 22.5) to evacuate the receiving flask. Algae will be trapped on the filter.

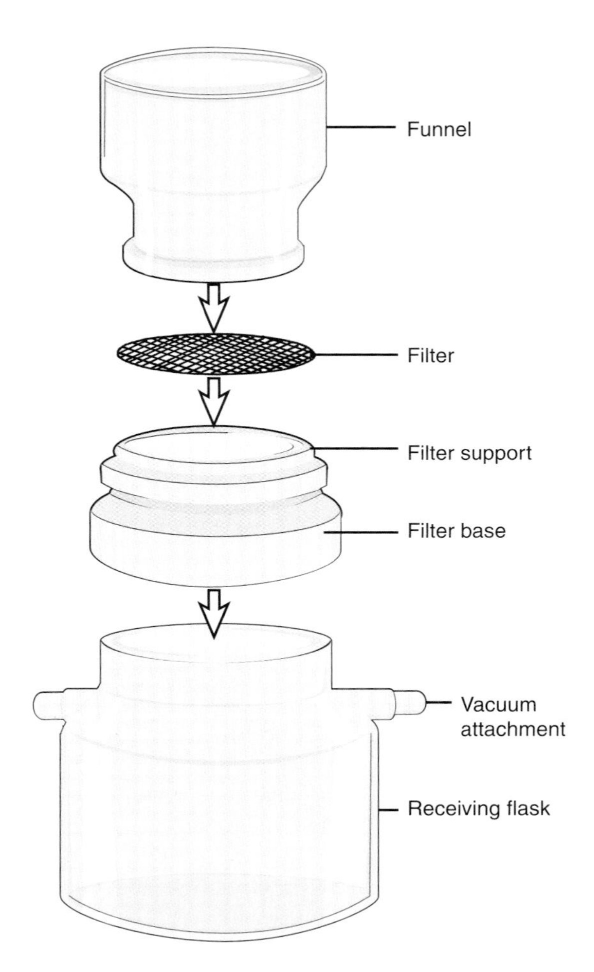

Figure 22.4

Diagram of filtration apparatus used to collect organisms from a water sample.

5. Remove and dry the filter in an oven at 45°C for 8 min.
6. Float the dried filter on approximately 5 mL of immersion oil in a small petri dish. The dry filter will become transparent in the oil. If the filter remains opaque, place the petri dish on a warm surface until the filter clears.
7. Remove the filter from the petri dish by dragging it across the side of the dish.
8. Cut the filter into fourths and place all of these pieces of filter on one clean microscope slide.
9. Examine the filter with reflected light at low total magnification (10–14×).
10. Count the algae in each of ten randomly selected fields of view and record the number in the following table:

No. of Algae in Each Field of View

Field	1	2	3	4	5	6	7	8	9	10	Total
Number of Algae	___	___	___	___	___	___	___	___	___	___	___

11. Use the following formula to calculate the total number of algae on your filter:

$$\text{Total no. of algae on filter} = \frac{1380\ \text{mm}^2}{\text{Area of one field (mm}^2)} \times \frac{\text{Total no. of algae counted}}{\text{No. of fields counted}}$$

Total no. of algae on filter = ________

Calculate the area of the field of view with a stage micrometer or ruler (see Exercise 2). The value 1380 mm^2 is the area of the 47-mm filter that you used to collect the algae.

Area of field of view = ________

12. Use the following formula to calculate the number of algae per milliliter of sample:

$$\text{No. of algae mL}^{-1} = \frac{\text{No. of algae on filter}}{\text{Volume of sample (mL)}}$$

No. of algae mL^{-1} = ________

Algal populations exceeding 1000 organisms mL^{-1} indicate that the water is over-enriched by sewage or other nutrients.

Question 8

a. How many algae per milliliter are in your water sample?

b. Is your sample polluted?

c. In what ways might these algae benefit other organisms in the water?

d. In what ways might they harm other organisms?

e. Suppose you are responsible for "cleaning up" a lake that is polluted with algae. What would you do, and why?

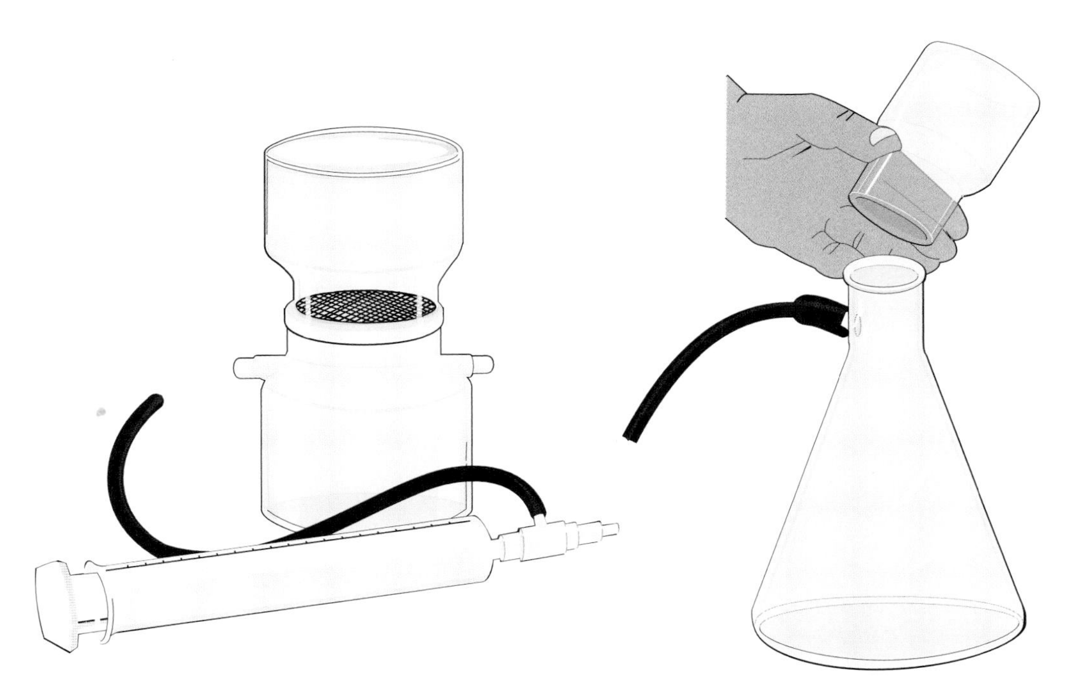

Figure 22.5
Vacuum filtration using (*a*) a hand vacuum-pump, or (*b*) a water aspirator. Applying a vacuum speeds the filtration of samples.

Questions for Further Thought and Study

1. What causes pollution?

2. Is all pollution bad? Why or why not?

3. What are some ways that you could "purify" water polluted with pesticide?

4. What are the consequences of using water from ponds and rivers to cool industrial processes? What are the consequences of stopping the thermal pollution (forcing the industry to stop releasing the heated water)?

5. Polluting lakes with laundry water and sewage often produces an algal "bloom." Why don't populations of predators (i.e., zooplankton and fish) increase to offset this bloom and keep the algal population in check?

6. Diseases such as typhoid fever (caused by *Salmonella typhosa*) and dysentery are often associated with lakes polluted by sewage. Suppose that you live near a large lake in which sewage is dumped, and that as a local health official you are in charge of reducing the incidence of typhoid in your area. How would you do this, assuming that you had the complete cooperation of city and other local officials? How would you do this if you couldn't prevent sewage from being dumped in the lake? What would be the consequences of your actions?

7. Define pollution. Can a material be a pollutant in some situations but not in others?

8. Does pollution always result from the demands of expanding populations? Why or why not?

DOING BIOLOGY YOURSELF

Design your own assay similar in concept to the Allium Test. Use a living plant and test potential pollutants of your choice.

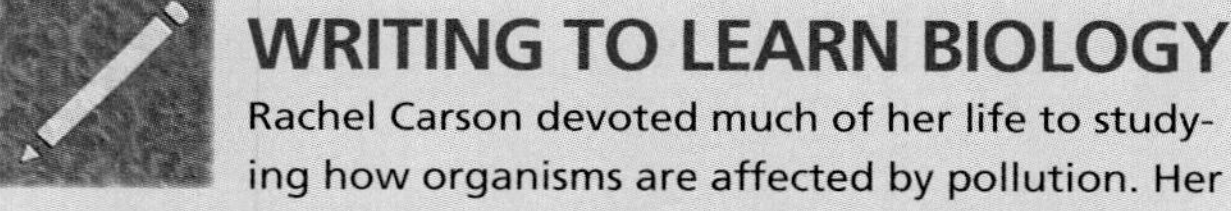

WRITING TO LEARN BIOLOGY

Rachel Carson devoted much of her life to studying how organisms are affected by pollution. Her masterpiece, *Silent Spring,* raised public awareness of pollution and made the word *ecology* a household word. Go to the library and read an article about this remarkable woman. What was her message and why is it still important?

23

Survey of Bacteria
Kingdoms Archaebacteria and Bacteria

Objectives

By the end of this exercise you should be able to:
1. Describe distinguishing features of members of kingdoms Archaebacteria and Bacteria.
2. Describe differences between bacteria and cyanobacteria.
3. Identify representative examples of archaebacteria, bacteria, and cyanobacteria.
4. Perform a Gram stain.

Cellular organisms have evolved along two lines. Species with cells lacking membrane-bound organelles are **prokaryotes** (table 23.1). Those with membrane-bound organelles are **eukaryotes** and include plants, animals, fungi, and protists. About 5000 species of prokaryotes have been described, and many more await identification and description.

Prokaryotes were long thought to be a unified group commonly called bacteria. However, genetic analysis as recently as 1996 of the DNA of prokaryotes has revealed two groups with surprisingly different DNA sequences, both of which are strikingly different from the DNA sequences of eukaryotes. This has led to recognition of three **domains** of organisms.

Domain Archaea includes **kingdom Archaebacteria,** all species of which are prokaryotes. Archaebacteria often inhabit but are not restricted to extreme and stressful environments on earth. **Domain Bacteria** includes **kingdom Bacteria,** all species of which are prokaryotes and are the most abundant organisms on earth (fig. 23.1). **Domain Eukarya** includes **kingdoms Protista, Fungi, Plantae,** and **Animalia.** These kingdoms are all eukaryotes and are described in Exercises 24–30 and 35–39. This classification of living organisms into three domains and six kingdoms is now widely accepted, but much phylogenetic information remains to be revealed. Classification is an exciting and ongoing process.

KINGDOM ARCHAEBACTERIA

Archaebacteria of domain Archaea may be the oldest forms of life on earth, and domains Bacteria and Eukarya probably diverged from Archaebacteria independently. Archaebacteria are diverse prokaryotes that share ribosomal RNA sequences as well as several important biochemical characteristics that are quite distinctive from those of all other kinds of organisms. Archaebacteria are significantly different from the prokaryotes of kingdom Bacteria. Archaebacteria have distinctive membranes, unusual cell walls, and unique metabolic cofactors.

Today's Archaebacteria are probably survivors of ancient lines that have persisted in habitats similar to habitats found throughout the world when bacteria first evolved. These environments are often extremely acidic, hot, or salty. Thus, many Archaebacteria are called **extremophiles.** Many Archaebacteria can live in an anaerobic atmosphere rich in carbon dioxide and hydrogen as well as the more benign environments typical of bacteria and eukaryotes. See Exercise 15 for a procedure describing *Halobacterium salinarum*, a common archaebacterium that inhabits salty environments.

KINGDOM BACTERIA

Bacteria of kingdom Bacteria are distributed more widely than any other group of organisms. Individual bacterial cells are microscopic (1 μm or less in diameter, fig. 23.2); a single gram of soil may contain over a billion bacteria. Bacteria have cell walls, which give them three characteristic shapes (fig. 23.3).

- Bacillus (rod-shaped)
- Coccus (spherical)
- Spirillum (spiral)

Most bacteria are **heterotrophic,** meaning that they derive their energy from organic molecules made by other organisms. Heterotrophic bacteria are **decomposers** because they

TABLE 23.1

PROKARYOTES COMPARED TO EUKARYOTES

Feature	Example
Unicellularity. All prokaryotes are basically single-celled. Even though some bacteria may adhere together or form filaments, their cytoplasm is not directly interconnected, and their activities are not integrated and coordinated as is the case in multicellular eukaryotes.	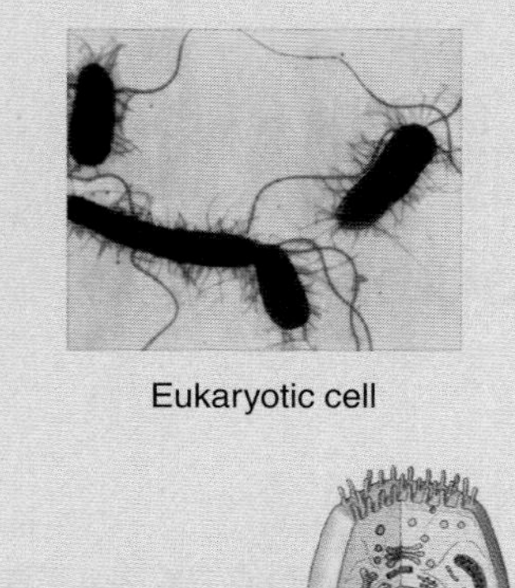 Eukaryotic cell
Cell Size. Most bacterial cells are only about 1 micrometer in diameter, while most eukaryotic cells are over 10 times that size.	Bacterial cell 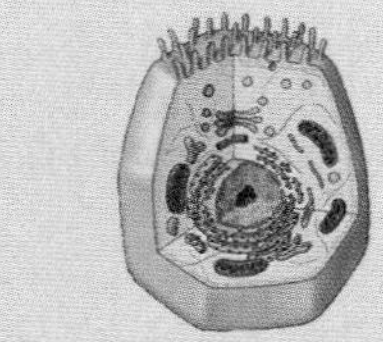 Eukaryotic cell
Chromosomes. Prokaryotic DNA exists as a single circle in the cytoplasm, while in eukaryotes, proteins are complexed with the DNA into multiple chromosomes.	 Bacterial genome 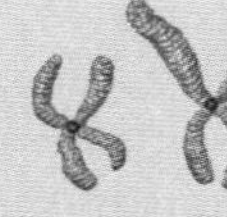 Eukaryotic chromosomes
Cell Division. Prokaryotic cells divide by binary fission. The cells simply pinch in two. In eukaryotes, microtubules pull chromosomes to opposite poles during the cell division process, called mitosis.	 Binary fission in bacteria 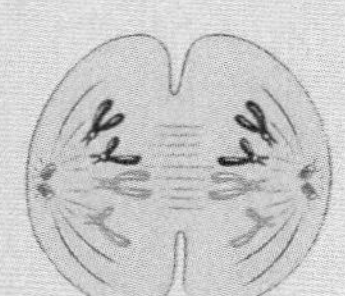 Mitosis in eukaryotes
Internal Compartmentalization. Unlike eukaryotic cells, bacterial cells contain no internal compartments, no internal membrane system, and no cell nucleus.	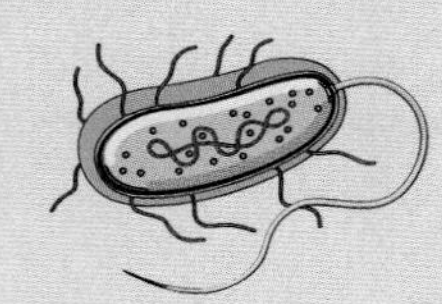 Bacterial cell
Flagella. Prokaryotic flagella are simple, composed of a single fiber of protein that spins like a propeller. Flagella in eukaryotes are complex structures that whip back and forth, rather than rotating.	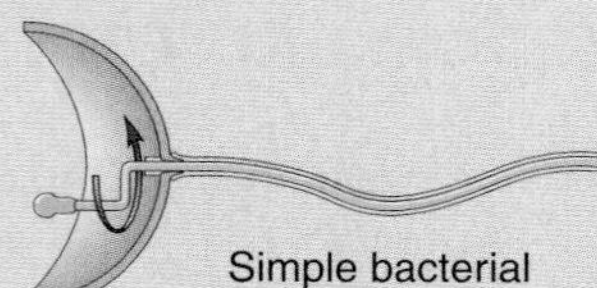 Simple bacterial flagellum
Metabolic Diversity. Prokaryotes possess many metabolic abilities that eukaryotes do not; some prokaryotes can perform several different kinds of anaerobic and aerobic photosynthesis, obtain their energy from oxidizing inorganic compounds, or fix atmospheric nitrogen.	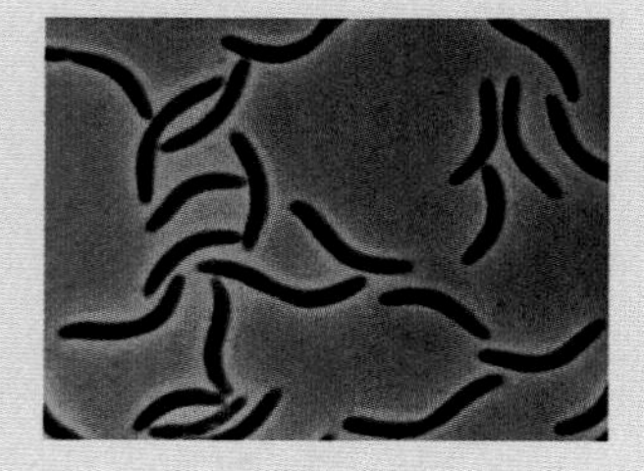 Chemoautotrophs

feed on dead organic matter and release nutrients locked in dead tissue.

Question 1

a. Why is it important that decomposers such as bacteria release nutrients?

b. What term best describes heterotrophic bacteria that feed on living tissue?

Bacteria that derive their energy from photosynthesis or the oxidation of inorganic molecules are **autotrophic.** However, photosynthesis in bacteria is often different from that in eukaryotes, because sulfur rather than oxygen is sometimes produced as a by-product.

A laboratory culture of bacteria usually consists of a tube of liquid nutrients (broth) containing growing bacteria or a tube or plate of solidified agar with bacteria growing on the surface.[1] The jellylike agar is melted, mixed with nutrients, and poured into tubes or plates to solidify. Many species of bacteria can be cultured in nutrient broth or on a layer of nutrient-rich agar. It may surprise you to know that most species of bacteria are *not* culturable in vitro. We just don't know enough about their nutrient and environmental requirements of these bacteria to grow them in the lab.

Bacteria reproduce asexually via **fission,** in which a cell's DNA replicates and the cell pinches in half without the nuclear and chromosomal events associated with mitosis (see Exercise 13) (fig. 23.4). Some bacteria have genetic recombination via **conjugation,** in which all or part of the genetic material of one bacterium is transferred to another bacterium and a new set of genes is assembled.

Some bacteria are pathogenic (table 23.2); that is, they cause diseases such as pneumonia and tuberculosis. However, most bacteria are harmless to humans. Indeed, many beneficial bacteria live in and on your body. Nevertheless, you should handle bacteria with care. The preparation of wet mounts of bacterial cultures requires proper use of a transfer loop and sterilizing flame. Your instructor will demonstrate this aseptic technique.

[1] Agar is a gelatinous polysaccharide used in culture media for microbiology labs. You'll learn more about agar and the red algae it comes from in Exercise 24.

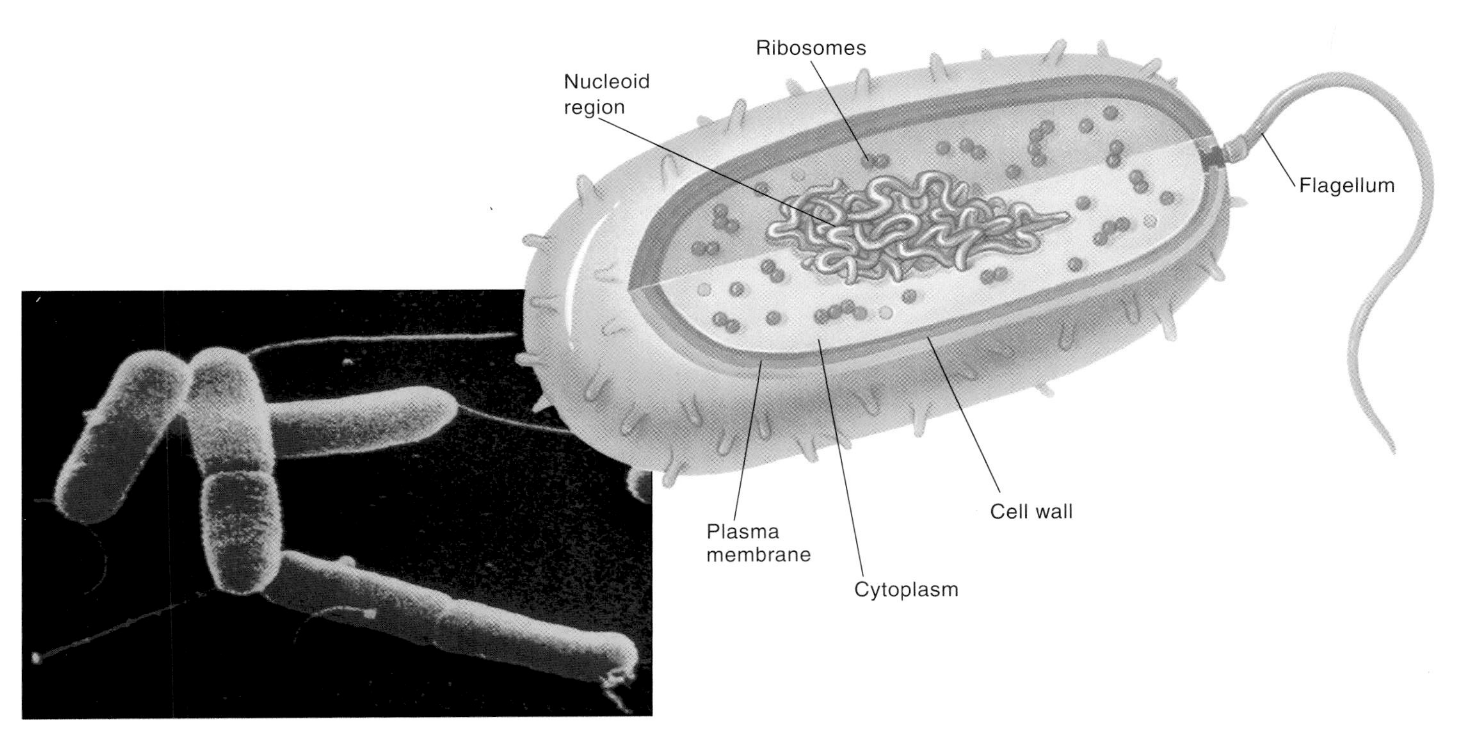

Figure 23.1
The structure of a bacterial cell.

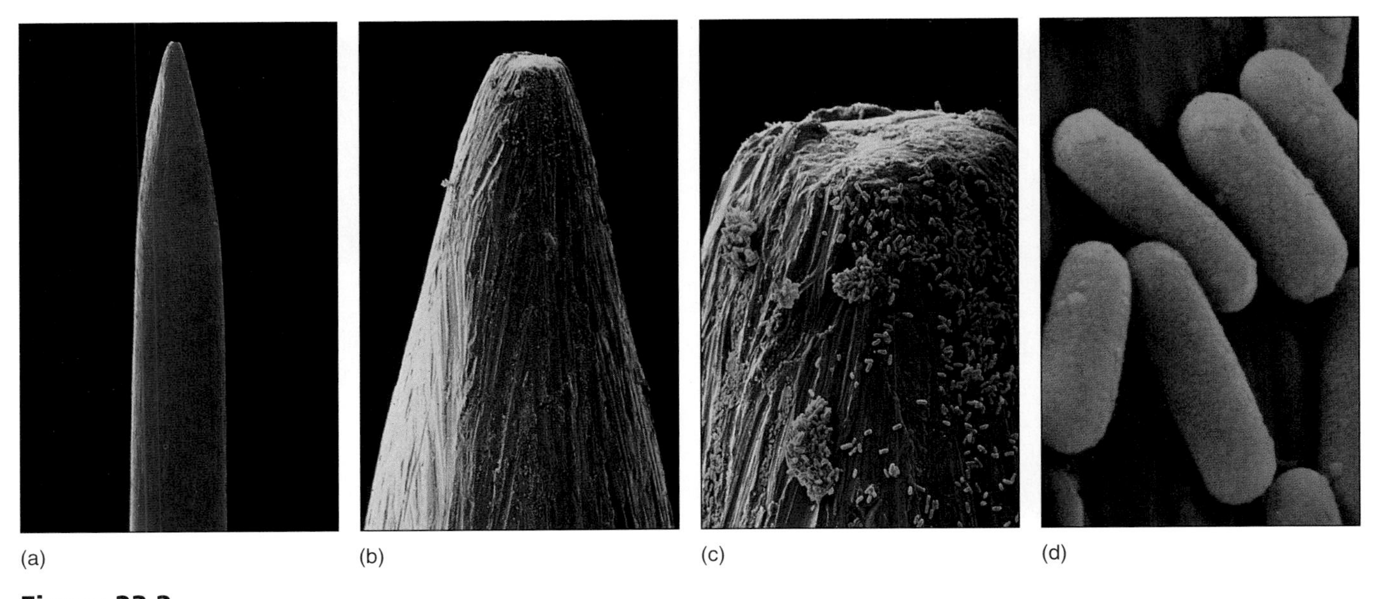

Figure 23.2
Four views of a contaminated pin, which would seem an unlikely site for bacteria to grow. (*a*) The tip of the pin, magnified 7×. When scanning electron micrographs are shown at increasing magnifications (*b*) 35×, (*c*) 178×, and (*d*) 4375×, you see rod-shaped bacteria growing there.

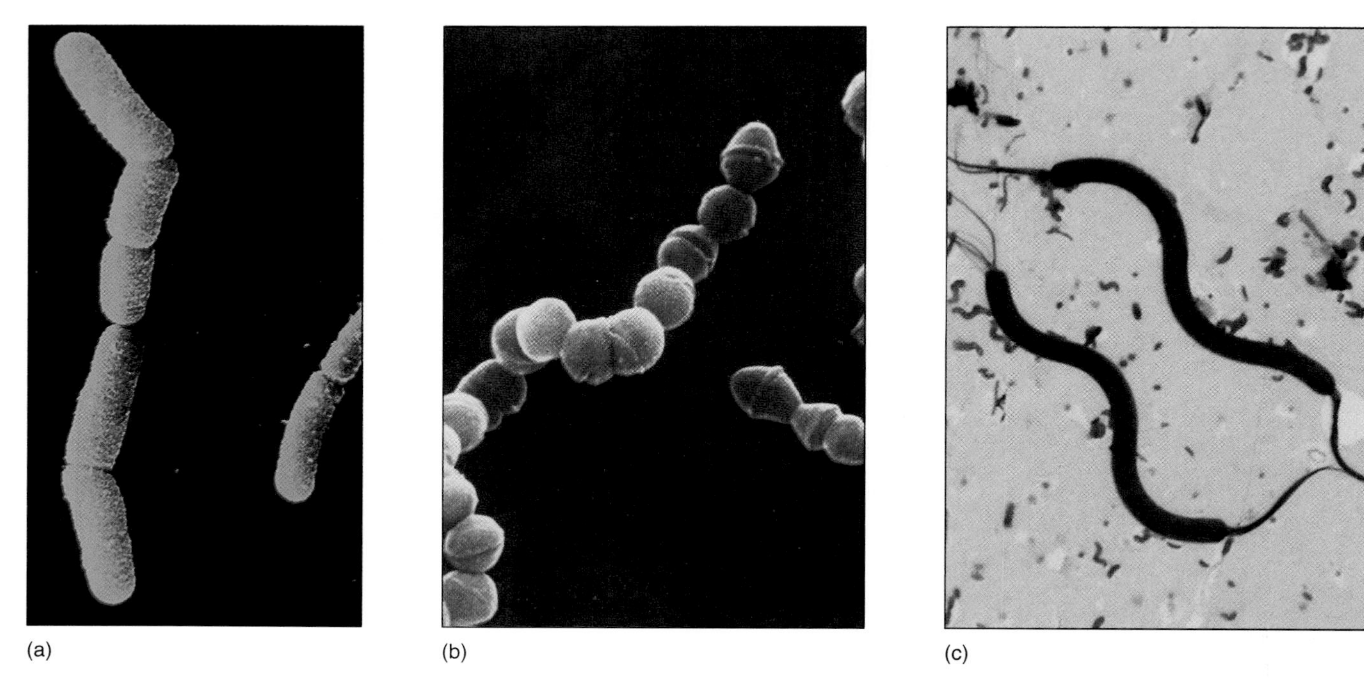

(a) (b) (c)

Figure 23.3
The three basic shapes of bacteria: (*a*) bacillus (*Pseudomonas*); (*b*) coccus (*Streptococcus*); and (*c*) spirillum (*Spirilla*), 400×.

Procedure 23.1

Culture common bacteria

1. Obtain a sterile cotton swab and a closed petri dish containing sterile nutrient agar.
2. Open the packaged swab and drag the tip over a surface such as your teeth, face, or tabletop.
3. Open the petri dish and drag the exposed swab over the surface of the agar in the manner demonstrated by your instructor.
4. Close the lid and tape it shut. Label the dish with a wax pencil and place the dish in an incubator or in a warm area.
5. After 24–48 h examine the agar for bacterial growth. Record your observations.

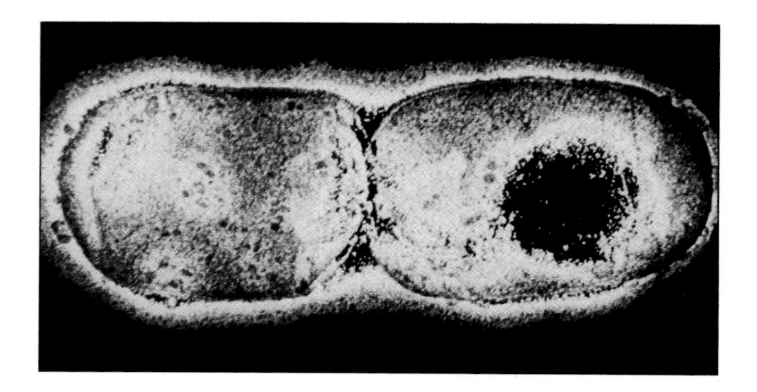

Figure 23.4
Electron micrograph showing fission of *Escherichia coli*, a common bacterium. Fission is a form of asexual reproduction.

Question 2
What is the shape and size of each bacterial colony?

Gram Stain

One of the most important techniques to classify bacteria is the **Gram stain,** which is based on the different structural and chemical compositions of bacterial cell walls. Gram staining is important because it often correlates with the sensitivity of a bacterium to antibiotics. **Gram-positive** bacteria (e.g., *Streptococcus*, *Micrococcus*) have a thick cell wall that retains a purple dye, whereas **Gram-negative** (e.g., *Escherichia coli*, *Serratia*) bacteria have a much thinner cell wall that does not retain the dye.

During the Gram stain technique, crystal violet and iodine are applied to stain all of the bacteria purple. Then alcohol is used to remove the stain from the surface of the Gram-negative cell walls that do not bind the stain. Finally, safranin is used to counterstain the Gram-negative cells with a red color contrasting to purple Gram-positive cells (fig. 23.5). In the following procedure, you will perform a Gram stain on some of your bacteria to see the difference between Gram-negative and Gram-positive organisms.

TABLE 23.2

IMPORTANT HUMAN BACTERIAL DISEASES

Disease	Pathogen	Vector/Reservoir	Epidemiology
Anthrax	*Bacillus anthracis*	Animals, including processed skins	Bacterial infection that can be transmitted through contact or ingested. Rare except in sporadic outbreaks. May be fatal.
Botulism	*Clostridium botulinum*	Improperly prepared food	Contracted through ingestion or contact with wound. Produces acute toxic poison; can be fatal.
Chlamydia	*Chlamydia trachomatis*	Humans, STD	Urogenital infections with possible spread to eyes and respiratory tract. Occurs worldwide; increasingly common over past 20 years.
Cholera	*Vibrio cholerae*	Human feces, plankton	Causes severe diarrhea that can lead to death by dehydration; 50% peak mortality if the disease goes untreated. A major killer in times of crowding and poor sanitation; over 100,000 died in Rwanda in 1994 during a cholera outbreak.
Dental cavities	*Streptococcus*	Humans	A dense collection of this bacteria on the surface of teeth leads to secretion of acids that destroy minerals in tooth enamel—sugar alone will not cause cavities.
Gonorrhea	*Neisseria gonorrhoeae*	Humans only	STD, on the increase worldwide. Usually not fatal.
Hansen's disease (leprosy)	*Mycobacterium leprae*	Humans, feral armadillos	Chronic infection of the skin; worldwide incidence about 10–12 million, especially in Southeast Asia. Spread through contact with infected individuals.
Lyme disease	*Borrelia bergdorferi*	Ticks, deer, small rodents	Spread through bite of infected tick. Lesion followed by malaise, fever, fatigue, pain, stiff neck, and headache.
Peptic ulcers	*Heliobacter pylori*	Humans	Infects the stomach, where it causes ulcers. About 40% of the world's population harbors *H. pylori*. Barry Marshall, who isolated the bacterium in 1982, drank a culture of *H. pylori;* he got an ulcer.
Plague	*Yersinia pestis*	Fleas of wild rodents: rats and squirrels	Killed ¼ of the population of Europe in the 14th century; endemic in wild rodent populations of the western U.S. today.
Pneumonia	*Streptococcus*, *Mycoplasma*, *Chlamydia*	Humans	Acute infection of the lungs, often fatal without treatment
Tuberculosis	*Mycobacterium tuberculosis*	Humans	An acute bacterial infection of the lungs, lymph, and meninges. Its incidence is on the rise, complicated by the development of new strains of the bacteria that are resistant to antibiotics.
Typhus	*Rickettsia typhi*	Lice, rat fleas, humans	Historically a major killer in times of crowding and poor sanitation; transmitted from human to human through the bite of infected lice and fleas. Typhus has a peak untreated mortality rate of 70%.

Procedure 23.2

Observe stained bacteria with oil-immersion magnification

1. Obtain a microscope and a small bottle of immersion oil. Recall from Exercise 2 that the resolving power of a lens depends, among other things, on the amount of light that it gathers. More light improves resolution, and light is scattered when it passes through air. If a drop of immersion oil, a fluid with the same refractive index (ability to bend light) as glass, is placed between the objective lens and the specimen, then the lens can gather more light.
2. Examine the microscope, and verify with your instructor that the microscope is equipped with an oil-immersion objective. This objective can resolve micrometer-sized particles such as bacteria.
3. Bring the low-power objective into observation position.
4. Obtain some prepared slides of stained bacteria from your instructor. These may be commercially prepared slides or slides with bacteria that your instructor has stained to help you practice with your microscope.
5. Place a slide on the stage with the specimen centered over the light path through the hole in the stage.

Deadly Food! Beware!

The bacterium *Clostridium botulinum* can grow in food products and produces a toxin called botulinum, the most toxic substance known. Microbiologists estimate that 1 gram of this toxin can kill 14 million adults! The good news is that *C. botulinum* requires anaerobic conditions for growth, which limits its prevalence. The bad news is that *C. botulinum* is extremely tolerant to stress; it can withstand boiling water (100°C) for short periods, but is killed at 120°C in 5 min. This tolerance makes *C. botulinum* a serious concern when people can vegetables. If home canning is not done properly, this bacterium will grow in the anaerobic conditions of the sealed container and be extremely poisonous. Several adults and infants die every year from botulism in the United States.

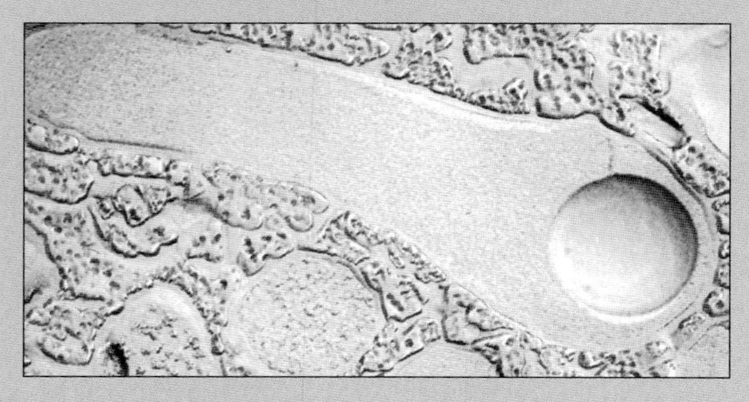

Figure 23.A

Endospores. The round circle at the lower right is an endospore forming within a cell of *Clostridium botulinum*, the bacterium that causes the disease botulism. These resistant endospores enable the bacterium to survive in improperly sterilized canned and bottled foods.

Tolerance to stress is enhanced in *C. botulinum* and many other bacteria by the formation of thick-walled **endospores** that surround their chromosome and a small portion of the surrounding cytoplasm. These highly resistant endospores (fig. 23.A) may later germinate and grow after decades or even centuries of inactivity. The endospores of *Clostridium botulinum* can germinate in poorly prepared canned goods, so never eat food from a swollen (gas-filled) can of food; you risk contracting botulism leading to nerve paralysis, severe vomiting, and death.

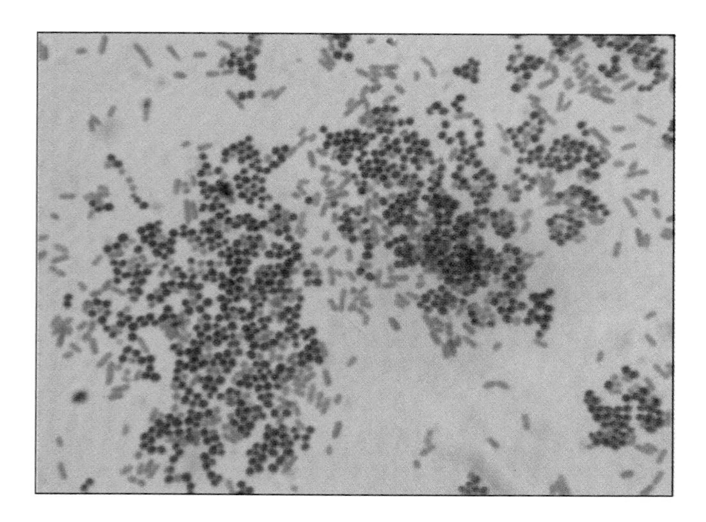

Figure 23.5

The Gram stain. Photomicrograph of bacteria that are Gram positive (purple) and Gram negative (red). The species are *Staphylococcus aureus* (+) and *Escherichia coli* (−) (also see fig. 23.9).

6. While watching from the side, bring the objective as close as possible to the slide without the objective touching the slide. Adjust the diaphragm for medium light intensity.
7. Look through the oculars and slowly adjust the coarse adjustment to increase the working distance. Stop when you see the color of the stained bacteria and are roughly focused on the smear of bacteria.
8. Improve the illumination and sharpen the image as much as possible with the fine-adjustment knob. At this low magnification you will only see small dots, at best.
9. Rotate a higher-power objective into position and refocus.
10. Rotate the nosepiece so that the alignment is halfway between the oil-immersion objective and the next lowest-power objective. There should not be an objective in correct position for observation. This position will allow you to place a drop of oil on the slide.
11. Put one drop of immersion oil on the coverslip directly over the spot of the light path. Do not touch the dropper to the slide or it will contaminate the oil when the dropper is returned to the bottle.
12. Rotate the oil-immersion lens directly into observation position and directly into the drop of oil.

13. While looking from the side, use the fine-adjustment knob to lower the objective until it *gently* touches the coverslip.
14. Look through the oculars and *slowly* rotate the fine-adjustment knob to increase the working distance. This rotation should be counterclockwise. Stop when the stained bacterial color appears. Slowly rotate the fine-adjustment knob back and forth until the bacteria are in focus.
15. Improve your resolution by adjusting the diaphragm.
16. Examine the sizes, shapes, and stains of the bacteria on the slide.
17. Repeat this entire procedure for each of the slides offered by your instructor.
18. When you finish your work, clean the oil from the slides and objectives with the lens paper provided.

Procedure 23.3

Use known bacterial cultures to prepare and observe a Gram stain

1. Obtain a slide, coverslip, transfer loop, alcohol burner, and a culture of living bacteria.
2. Available cultures should include the following bacteria, among others:

   Bacillus megaterium—a large bacterium that is resistant to radiation, desiccation, and heat.

   Rhodospirillum rubrum—a photosynthetic purple bacterium.

   Escherichia coli—found in the human intestine; the most intensively studied of all bacteria.

   Staphylococcus epidermidis—a cocci found among the normal flora of skin.
3. Apply a loop of bacteria to a drop of water on a slide. If your cultures are in liquid broth, then add one drop of culture medium to your slide. Your instructor will demonstrate how to use sterile technique to open, sample, and close the culture of bacteria. Do not add a coverslip to the slide.
4. Heat the slide gently by holding it with a clothespin and passing it over the top of a flame three to four times. Be careful not to break the slide. If hot plates are available, hold the slide to the hot surface for 10 sec. This heat will adhere the bacteria to the slide.
5. Examine figure 23.6 and become familiar with the staining procedure. When the slide has cooled, cover it with crystal violet for 20 sec.

CAUTION

Be careful not to inhale or spill crystal violet or any other biological stain on your skin.

6. Gently rinse the slide with water.
7. Cover the slide with iodine for 1 min.
8. Drop 95% alcohol (decolorizer) on the smear with an eyedropper until no purple shows in the alcohol coming off the slide. Quickly rinse the slide with water to remove the alcohol.
9. Cover the smear with safranin for 20 sec.
10. Gently rinse the slide with water. Air dry the slide, or blot gently if necessary. Add a coverslip.
11. Observe the smear with your microscope using low power and then high power and/or the oil-immersion objective. You will not need a coverslip for heat-fixed slides with oil immersion.
12. Determine if the bacteria are Gram positive or Gram negative.
13. Repeat the Gram stain procedure using other known bacteria that you obtain from culture tubes.
14. Record your observations in table 23.3.

Procedure 23.4

Use a Gram stain to observe living bacteria from your teeth

1. Obtain a slide, coverslip, transfer loop, alcohol burner, and a culture of living bacteria.
2. Use the wide end of a toothpick to scrape your teeth near the gum line.
3. Thoroughly mix what's on the tip of the toothpick in a small drop of water on a microscope slide.
4. Allow this bacterial smear to dry.
5. Repeat steps 4–13 of procedure 23.3.

Question 3

a. Which type of bacteria is most prevalent in the sample from your teeth?

b. Is *Bacillus megaterium* Gram positive or Gram negative? How do you know?

Bacterial Colony Morphology

A bacterial colony is a visible speck or patch of millions of bacterial cells that are typically the progeny of a single cell that reproduced on the agar's surface. A bacterial colony growing on the surface of nutrient agar often has distinctive characteristics depending on the species. Careful observation of the shape, color, size, texture, and margins of a colony is important for any bacterial identification. Figure 23.7 illustrates the most common features of bacterial colony morphology.

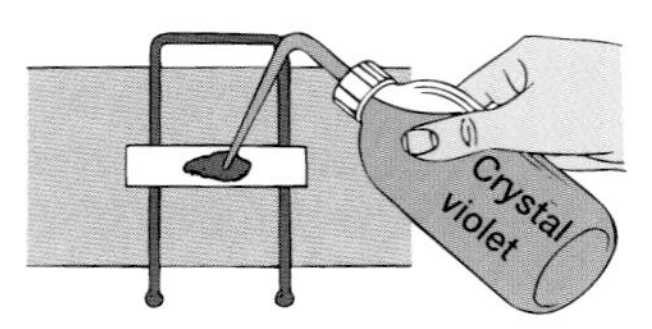

(a) Crystal violet; 20 sec

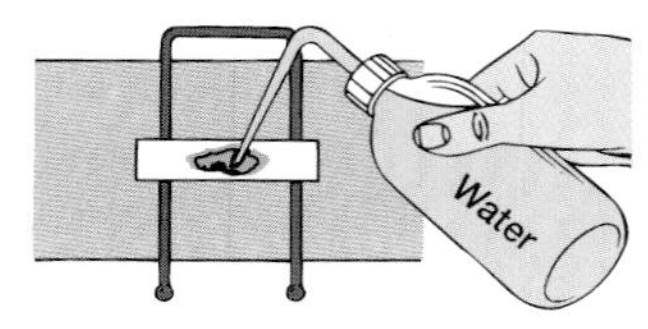

(b) Rinse with water for 2 sec

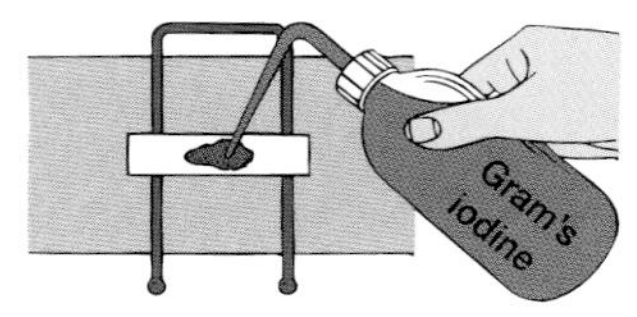

(c) Cover with Gram's iodine for 1 min

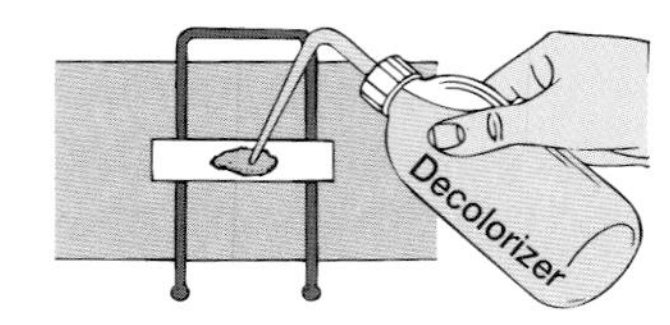

(d) Decolorize for 5–15 sec or until solvent runs colorlessly

(e) Rinse with water for 2 sec

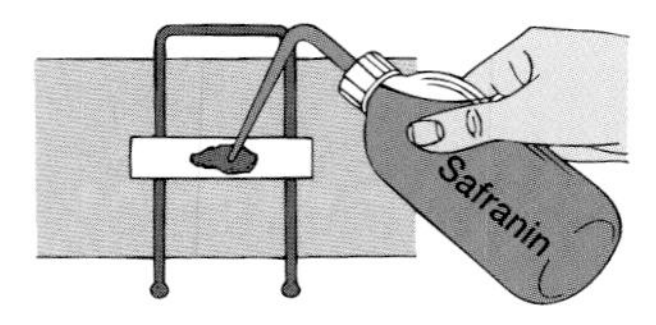

(f) Counterstain with safranin for about 20 sec

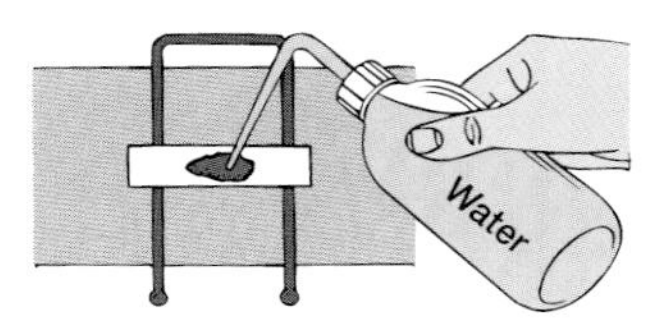

(g) Rinse with water for 2 sec

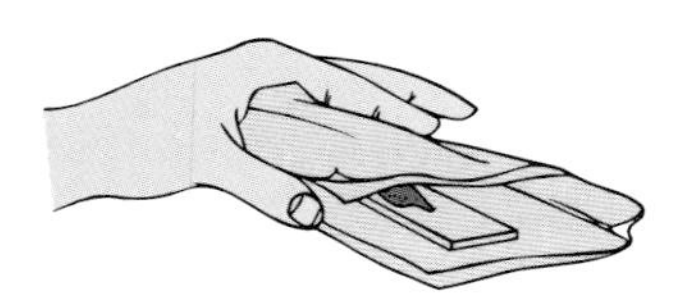
(h) Blot dry with bibulous paper

Figure 23.6
Gram staining procedure.

TABLE 23.3

THE RELATIVE SIZE AND SHAPE OF SOME COMMON BACTERIA

Bacterial Species	Gram Stain (+/−)	Relative Size	Shape

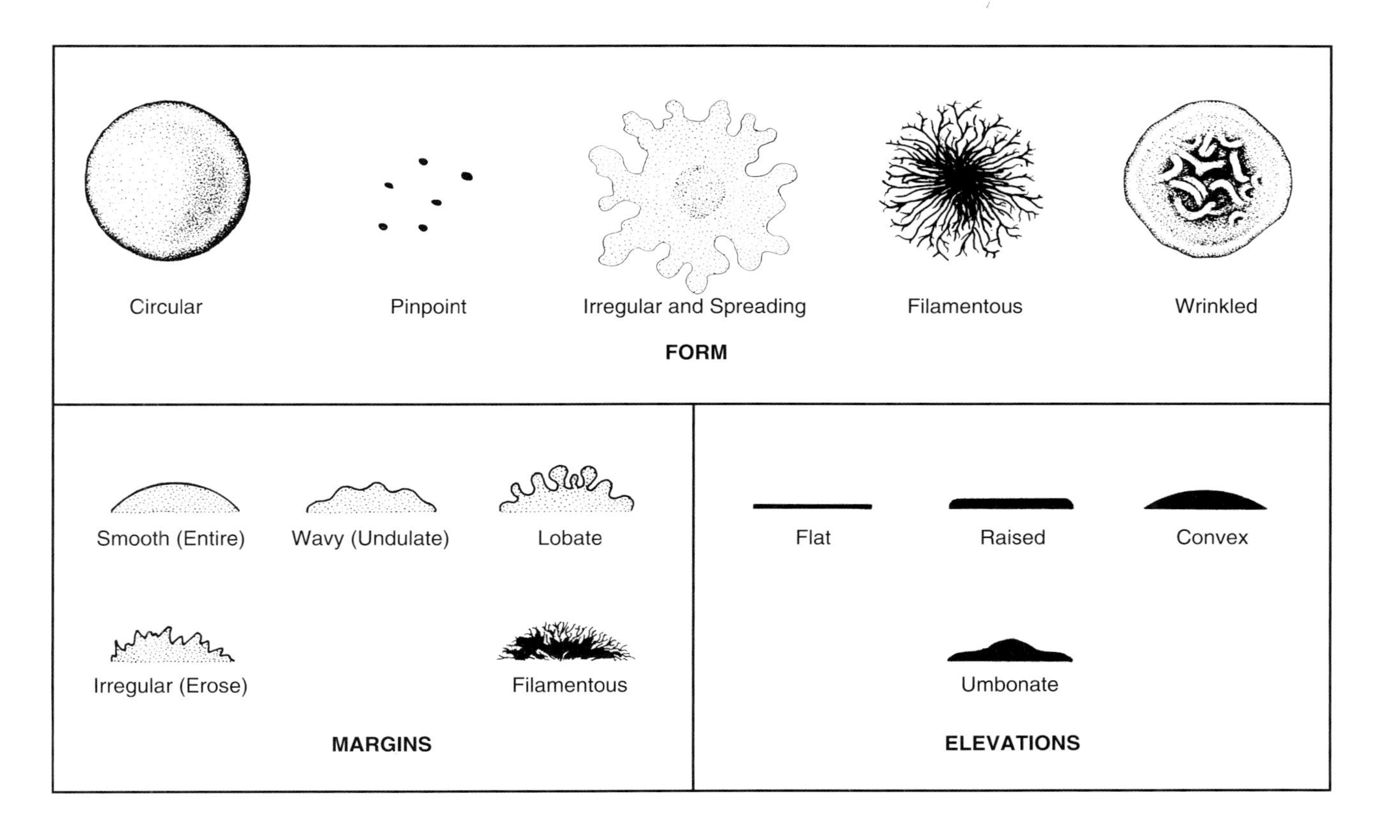

Figure 23.7
Colony characteristics. Colonies are classified by form, margin, and elevation as well as color and diameter.

Procedure 23.5

Evaluate the colony morphology of bacterial species

1. Obtain 48-h growth plates of 4–6 bacterial species provided by your instructor for your evaluation.
2. Familiarize yourself with the colony morphologies shown in figure 23.7.
3. Use a stereomicroscope or hand lens to evaluate a representative colony on each of the plates. Record your observations in table 23.4.
4. When you have completed your analysis, your instructor will provide the species names associated with each of the plates and the accepted colony descriptions for that species. Use this information to evaluate the accuracy of your observations.

Nitrogen Fixation by Bacteria

Certain bacteria and cyanobacteria transform atmospheric nitrogen (N_2) into other nitrogenous compounds that can be used as nutrients by plants. This process is called **nitrogen fixation.** All organisms need nitrogen as a component of their nucleic acids, proteins, and amino acids. However, chemical reactions capable of breaking the strong triple-bond between atoms of atmospheric nitrogen are limited to certain bacteria and cyanobacteria. This process uses an enzyme called nitrogenase along with ATP, energized electrons, and water to convert N_2 to ammonia (NH_3). Ammonia can be absorbed by plants and used to make proteins and other macromolecules.

Rhizobium is a bacterium that can fix nitrogen and can grow intimately with roots of some plants called legumes (e.g., clover, alfalfa, and soybeans). Such associations between *Rhizobium* and host roots form **nodules** on the roots (fig. 23.8). These resident nitrogen-fixers provide ammonia to the plant while the plant provides sugars and other nutrients to the bacteria.

Procedure 23.6

Observe root nodules

1. Observe the root systems on display and note the nodules.
2. Examine a prepared slide of a cross section of a nodule.

Table 23.4

An Evaluation of Bacterial Colony Morphology

Plate ID	Species	Colony Diameter (mm)	Color	Form	Elevation	Margin
A						
B						
C						
D						
E						
F						

Figure 23.8
Root nodules containing nitrogen-fixing bacteria.

Question 4

a. Where are the bacteria? Are they between cells or inside cells?

b. Why is this relationship between a plant and bacterium called a mutualism?

c. How does *Rhizobium* benefit from this association?

d. How does the host plant benefit from the association?

Bacterial Sensitivity to Inhibitors

Growth of some bacterial species is more sensitive to inhibitors such as antibiotics than is growth of other species. For example, an antibiotic may be more effective against *Staphylococcus* than against *Streptococcus*. This is important information to a physician who must select one of many available antibiotics to treat a bacterial infection.

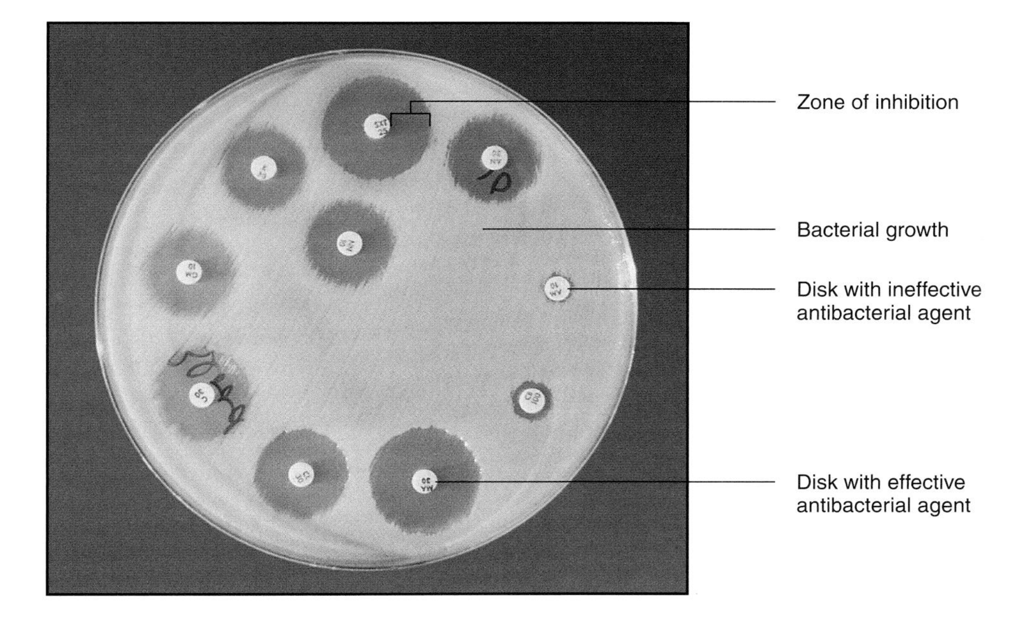

Figure 23.9
A sensitivity plate is used to determine the effectiveness of one or more antibiotics, each of which is on the surface of a paper disk. Any disk containing an effective antibacterial agent inhibits nearby bacterial growth on the agar. If the agent is ineffective, then bacteria will grow up to the disk.

To determine the most effective antibiotic, a medical laboratory may set up a **sensitivity plate.** A sensitivity plate is a petri dish of solid medium that has been uniformly inoculated on its entire surface with a known bacterium or an unknown sample from an infected patient. After inoculation, four to eight small paper disks are placed equidistant from each other on the culture surface. Each disk was soaked in a different antibiotic. After 24 h, an effective antibiotic will produce a visible halo of clear surface around the disks where it inhibited growth of the bacteria (fig. 23.9). If the antibiotic was ineffective, the bacteria will grow to the edge of the paper disk.

You've probably seen television commercials for products such as mouthwash or disinfectant that "kills germs on contact." Many of these products are developed and tested using sensitivity plates. The mouthwash is effective if no bacteria grows around a paper disk that was soaked in the mouthwash.

Examine the sensitivity plates of bacterial cultures on demonstration.

Procedure 23.7

Examine sensitivity plates

1. Obtain from your instructor one of each type of sensitivity plate that has been prepared.
2. Examine each plate, note the bacteria used to inoculate the plate, and note the kinds of disks distributed on the plate.
3. Determine which disks inhibited bacterial growth strongly, weakly, or not at all.
4. Record the bacterial species and your observations in table 23.5.

Question 5
Based on their appearance, which drugs or chemicals inhibit the growth of bacteria?

Cyanobacteria (Blue-Green Algae)

Cyanobacteria are a major group of photosynthetic bacteria that grow in many environments. Most cyanobacteria are free-living, whereas others live symbiotically with plants and other organisms. Cyanobacteria are photosynthetic, and their pigments include **chlorophyll *a*** and the accessory pigments **phycocyanin** (blue) and **phycoerythrin** (red). Because of various proportions of these pigments, only about half of the cyanobacteria are actually blue-green in color; many cyanobacteria range in color from brown to olive green.

TABLE 23.5

INHIBITION OF FOUR BACTERIAL SPECIES BY VARIOUS GROWTH INHIBITORS

Antibiotic/Antiseptic	Plate 1 ______	Plate 2 ______	Plate 3 ______	Plate 4 ______

Cyanobacteria reproduce by fission and are often surrounded by a jellylike **sheath.** Because cyanobacteria are prokaryotes, they are not related to other algae, which are all eukaryotic. Cyanobacteria such as *Oscillatoria* often cause many of the disagreeable tastes, colors, and odors in water.

Procedure 23.8

Examine cyanobacteria

1. Examine living material and prepared slides of *Oscillatoria* with your microscope (fig. 23.10*a*). *Oscillatoria* grows as long chains of cells called **trichomes.**

Question 6

Do all cells of a trichome of *Oscillatoria* appear similar?

2. Examine living material and a prepared slide of *Nostoc* (commonly called witch's butter or starjelly) (fig. 23.10*b*). *Nostoc* forms large, grapelike colonies. Trichomes of *Nostoc* consist of small vegetative cells and larger, thick-walled **heterocysts,** in which nitrogen fixation occurs.
3. Examine a wet mount of living *Gloeocapsa,* characterized by a thick, gelatinous sheath (fig. 23.10*c*).
4. Add a drop of dilute India ink to the slide of *Gloeocapsa* so that the sheath will stand out against the dark background. Often *Gloeocapsa* forms clusters of cells and therefore has a colonial body form. Locate one of these colonies.

Question 7

a. Do adjacent cells share a common sheath?

b. What do you suppose is the function of the sheath?

c. Do clusters of *Gloeocapsa* represent multicellular organisms? Why or why not?

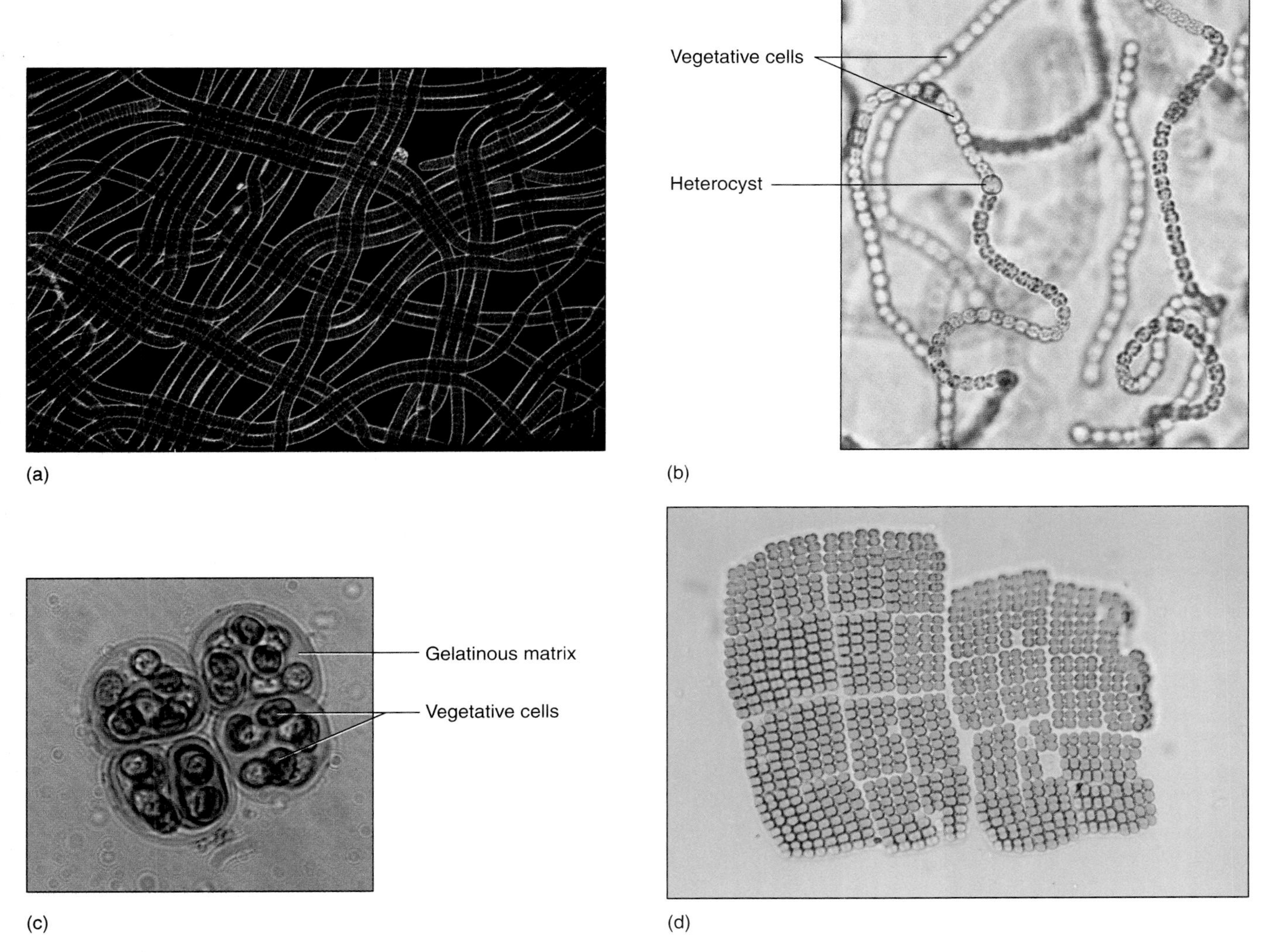

Figure 23.10
Cyanobacteria. (*a*) *Oscillatoria;* (*b*) *Nostoc;* (*c*) *Gloeocapsa;* and (*d*) *Merismopedia.*

5. Examine some living *Merismopedia* (fig. 23.10*d*), which also form colonies. Sketch these cyanobacteria below.

Question 8

a. How is the shape of *Merismopedia* different from other cyanobacteria you studied in this exercise?

b. How would a colony attain this shape?

INVESTIGATION

Search for Antibiotic-Resistant Bacteria

Medical researchers and physicians are becoming increasingly aware that bacteria are extraordinarily diverse. Many disease-causing species have evolved resistance to antibiotics, and many commonly occurring and nonpathogenic bacteria are also resistant. Sensitivity plates are a reliable way to determine the effectiveness of various antibiotics.

a. Determine where in the environment you might find resistant bacteria. Explain below why you chose that environment.

b. Design a simple experiment to select for some resistant species of bacteria growing in that environment.

c. Which of the many antibiotics available would you include in your test for resistance? Why did you choose these?

d. Do your experiment.

e. What did you conclude about the environment you chose to sample? Were your antibiotics effective? Did you isolate any bacteria that resisted all of the antibiotics that you tested?

Questions for Further Thought and Study

1. What is meant by Gram positive?

2. What happens when milk is pasteurized?

3. What causes milk to sour?

4. What ecological roles are performed by cyanobacteria?

5. How do antibiotics kill bacteria? Why do they not affect viruses?

6. How could bacteria become resistant to an antibiotic?

DOING BIOLOGY YOURSELF

Obtain two nutrient agar plates. Touch and drag the tip of your finger on the agar surface of one plate. Wash your hand and repeat the procedure on the other plate. Incubate the plates for 24–48 h. Then compare colony appearances and number. What can you conclude about the presence and diversity of bacteria on your plates? Is the appearance of a colony a good way to distinguish species of bacteria? Why or why not?

WRITING TO LEARN BIOLOGY

There is a great diversity of roles in a typical ecosystem, some of which are shared by a variety of organisms. What ecological roles are performed by cyanobacteria?

Survey of the Kingdom Protista

The Algae

24

Objectives

By the end of this exercise you should be able to:

1. Discuss the distinguishing features of different groups of algae.
2. Appreciate the economic importance of algae.
3. Outline the events of "alternation of generations" in green algae.

Kingdom Protista is the first of four kingdoms of domain Eukarya that you will examine. **Eukaryotes** are organisms composed of cells having membrane-bound nuclei. These organisms are commonly divided into four kingdoms: Fungi, Animals, Plants, and Protista.

Fungi have cell walls and are heterotrophic. **Heterotrophic** organisms feed on organic matter produced by other organisms because they cannot make their required organic compounds from inorganic substances; that is, heterotrophs require organic nutrients produced by other organisms.

Animals are also heterotrophic but lack cell walls and can respond rapidly to external stimuli. Animals are multicellular.

Plants are multicellular, **autotrophic** organisms, meaning that they can synthesize all required organic compounds from inorganic substances using external energy, which is usually sunlight.

Protista is the oldest and most diverse of the four kingdoms of eukaryotes. In a sense, protists include all eukaryotes that lack the distinguishing characteristics of plants, animals, or fungi. Protists include simple eukaryotes such as amoebas, as well as multicellular forms such as the brown alga, kelp. Protists are mostly microscopic, unicellular organisms and probably share common ancestry with fungi, multicellular plants, and animals. Members of kingdom Protista can be studied as three general groups: algae, protozoans, and slime molds. You will study protozoa and slime molds in the next exercise.

INTRODUCTION TO THE ALGAE

One way to determine the importance of something is to remove it and see what happens. If we did that with algae most people would be shocked by the result. Global oxygen production would immediately decline. A major food source, perhaps *the* major food source for the world's ecosystems would be gone. Tens of thousands of irreplaceable algal species would be lost, along with their unique diversity of potentially useful chemicals, many with pharmaceutical value. The absence of algae certainly would lead to rapid extinction of many hundreds of thousands of invertebrate animal species. Ecosystems would collapse. Fortunately we won't lose algae anytime soon. They have tenaciously thrived for 1.5 billion years!

Algae are photosynthetic, eukaryotic organisms typically lacking multicellular sex organs. The major groups of algae are distinguished in part by their energy storage products, cell walls, and color, resulting from the type and abundance of **pigments** (substances that absorb light) in their plastids (table 24.1). Thus, we refer to the major groups of algae as green algae, brown algae, golden-brown algae, and red algae. Red algae owe their color to water-soluble pigments called **phycobilins.** Other algal pigments such as **chlorophylls** and **carotenoids** are insoluble in water but can be extracted with organic solvents such as acetone and alcohol.

Algae are also distinguished by their cellular organization. **Unicellular** algal species occur as single, unattached cells that may or may not be motile. **Filamentous** algal species occur as chains of cells attached end to end. These filaments may be few to many cells long and may be unbranched or branched in various patterns. **Colonial** algae occur as groups of cells attached to each other in a nonfilamentous manner. For example, a colony may include several to many cells adhering to each other as a sphere, flat sheet, or other three-dimensional shape. Multicellular organization is not typical of protists but describes algae of more complex design than simple colonies. Multicellular species have cells of different kinds and functions and show significant interdependence.

TABLE 24.1

THE COMMON PIGMENTS IN ADDITION TO CHLOROPHYLL *a*, STORAGE PRODUCTS, AND CELL WALL COMPONENTS CHARACTERISTIC OF ALGAL PHYLA

Phylum	Predominant Organization	Pigment	Storage Product	Cell Wall
Chlorophyta	Unicellular, filamentous, colonial	Chlorophyll *b*	Starch	Mainly cellulose
Phaeophyta	Filamentous, multicellular	Chlorophyll *c*, fucoxanthin	Laminarin, mannitol, lipids	Cellulose, alginates
Rhodophyta	Multicellular	Chlorophyll *d*, phycobilins	Modified starch	Cellulose, agar, carrageenan
Chrysophyta	Unicellular	Chlorophyll *c*, fucoxanthin	Chrysolaminarin, lipids	Silica, calcium carbonate
Pyrrhophyta	Unicellular	Chlorophyll c	Starch, lipids	Pectin, cellulose plates
Euglenophyta	Unicellular	Chlorophyll *b*	Paramylon, lipids	Protein

PHYLUM CHLOROPHYTA (GREEN ALGAE)

Green algae are the most diverse and familiar algae in freshwater (fig. 24.1). However, a few genera live in salt-water. Although their name (Chlorophyta) means "green chlorophyll plants" (*chloro* + *phyta*), these algae are not classified as plants. Green algae are considered to be ancestral to land plants and share many characteristics with land plants, such as

- Chlorophyll *a*, which occurs in algae and green plants
- Chlorophyll *b*, which occurs in land plants and in the green and euglenoid algae
- Starch as the carbohydrate storage material
- Cell walls made of cellulose

Let's examine a few representatives of green algae, a major phylum.

Unicellular Green Alga: *Chlamydomonas*

Chlamydomonas is a motile, unicellular alga found in soil, lakes, and ditches (fig. 24.1). It probably has the most primitive structure and type of reproduction among green algae. The egg-shaped cells of *Chlamydomonas* contain a large chloroplast and a pyrenoid involved in the production and storage of starch.

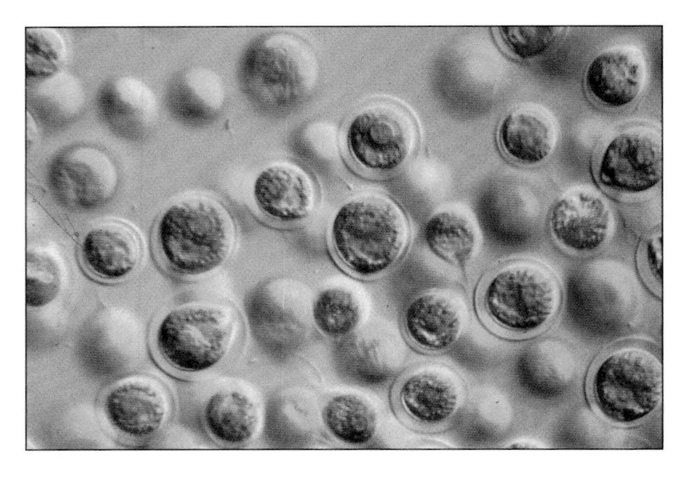

Figure 24.1
Chlamydomonas is a common alga of kingdom Protista that is rich in chlorophylls *a* and *b*. It is a single-celled green alga that is less than 100 μm long.

Procedure 24.1

Observe *Chlamydomonas*

1. Observe a drop of water containing living *Chlamydomonas* and note the movement of the cells.
2. If the movement is too fast, make a new preparation by placing one or two drops of methylcellulose on a slide and adding a drop of water containing *Chlamydomonas*.
3. Mix gently and add a coverslip.
4. Note the **stigma,** which appears as a reddish, light-absorbing spot at the anterior end of the cell.
5. Examine prepared slides of *Chlamydomonas*.

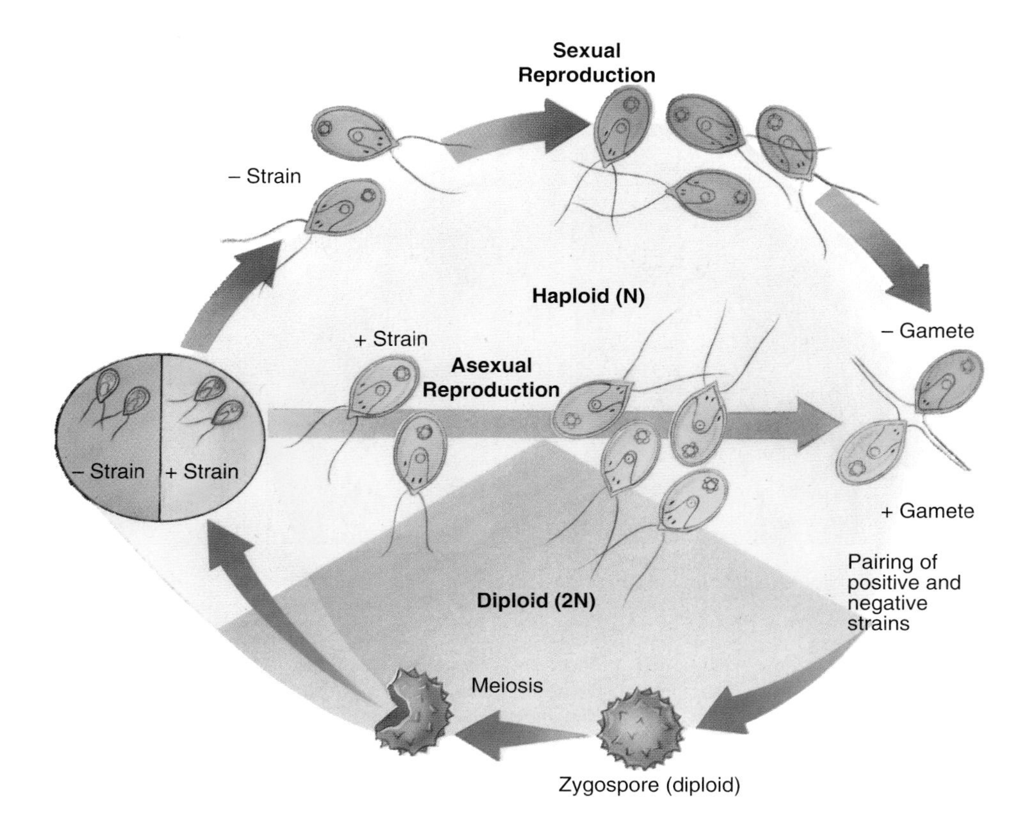

Figure 24.2

Chlamydomonas life cycle. Individual cells of *Chlamydomonas* (a microscopic, biflagellated alga) are haploid and divide asexually, producing copies of themselves. At times, such haploid cells act as gametes, fusing to produce a zygote. The zygote develops a thick, resistant wall, becoming a zygospore (as shown in the lower right-hand side of the diagram). Meiosis then produces four haploid individuals. Only + and – strains can mate with one another, although both may also divide asexually and reproduce themselves.

Question 1

a. Is the movement of *Chlamydomonas* smooth or does it appear jerky?

b. Can you see both flagella? (You may need to reduce the light intensity to see flagella.)

c. How does methylcellulose affect the movement of *Chlamydomonas?*

d. How does the stigma help *Chlamydomonas* survive?

Asexual and Sexual Reproduction in *Chlamydomonas*

Chlamydomonas usually reproduces asexually via mitosis. Sexual reproduction in *Chlamydomonas* is a response to unfavorable environmental conditions. For sexual reproduction, vegetative cells of *Chlamydomonas* undergo mitosis to produce gametes. The gametes fuse to form a diploid **zygote,** which is the resting stage of the life cycle. In most species of *Chlamydomonas*, the gametes of two strains are **isogamous,** meaning they have an identical shape and appearance. For convenience, isogametes of *Chlamydomonas* are referred to as + or –. Gametes unite to form a diploid zygote. **Syngamy** is the pairing and fusion of haploid gametes to form diploid cells. The zygote surrounds itself with a resistant surface and is called a **zygospore.** Under favorable conditions the zygote undergoes meiosis to produce haploid individuals called **spores.** Spores are reproductive cells capable of developing into an adult without fusing with another cell. Study the life cycle of *Chlamydomonas* shown in figure 24.2.

Procedure 24.2

Observe syngamy

1. Place drops of + and – gametes of *Chlamydomonas* provided by your instructor next to each other on a microscope slide, being careful not to mix the two drops. Do not add a coverslip.
2. While you observe the drops through low power of the microscope, mix the two drops of isogametes.

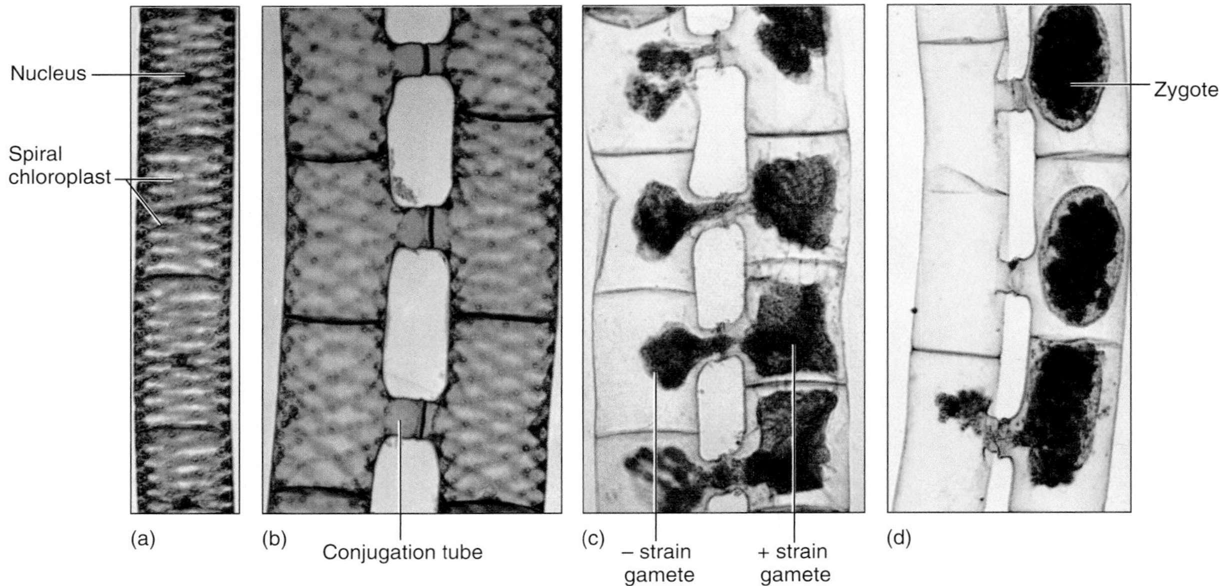

Figure 24.3

Spirogyra (watersilk). (*a*) Ribbonlike chloroplasts are spirally arranged in a vegetative filament. (*b*) Conjugation tubes have formed from papillae that grow out of opposing cells in adjacent filaments. (*c*) The + strain gametes (right) are condensed protoplasts. (*d*) Fusion of gametes in some of the cells has produced zygotes. The zygotes then undergo meiosis to produce haploid cells which can each divide to produce a filament.

3. Switch to high magnification and note the clumping gametes. Try to locate cells that have paired.

Question 2

a. Under what environmental conditions would a zygote not undergo meiosis immediately?

b. Are spores of *Chlamydomonas* haploid or diploid?

c. Which portions of the life cycle of *Chlamydomonas* are haploid?

d. Which are diploid?

Filamentous Green Algae: *Spirogyra* and *Cladophora*

Two of the most common genera of filamentous green algae are *Spirogyra* and *Cladophora*. *Spirogyra* grows in running streams of cool freshwater and secretes mucilage that makes it feel slippery. *Cladophora* is also common is streams and has a much coarser appearance and texture.

Spirogyra reproduces sexually by a process called **conjugation.** During conjugation, filaments of opposite mating types lie side by side and form projections that grow toward each other. These projections touch and the separating wall dissolves, thus forming a **conjugation tube** (fig. 24.3). The cellular contents of the − strain then migrate through the conjugation tube and fuse with that of the nonmotile + strain. These cellular contents function as nonflagellated isogametes. The zygote resulting from the fusion of gametes develops a thick, resistant cell wall and is termed a zygospore. The zygospore is released when the filament disintegrates, at which time the zygospore undergoes meiosis to form haploid cells that become new filaments.

Procedure 24.3

Examine *Spirogyra* and *Cladophora*

1. Obtain and examine a living culture of *Spirogyra*.
2. Prepare a wet mount with a small amount of living *Spirogyra*. Examine it with low and then high magnification. Notice the shape of the chloroplasts and whether filaments are branched.
3. Examine a prepared slide of *Spirogyra*.
4. Sketch below a filament of *Spirogyra* and note the shape of its chloroplasts.

5. Examine a prepared slide of filaments of *Spirogyra* that are undergoing conjugation. Locate the conjugation tubes, gametes, and zygotes.
6. Prepare and examine a wet mount of living *Cladophora*, and then examine a prepared slide.
7. In the space below draw a few cells of *Cladophora* showing their general shape and the filament's branching pattern.

Question 3

a. Are the filaments of *Spirogyra* branched?

b. What is the shape of the chloroplasts of *Spirogyra*?

c. Can you see any conjugation tubes? If you can't, examine the prepared slides that show these structures.

d. How do you think *Spirogyra* reproduces asexually?

Examine some living cultures and prepared slides of *Cladophora* (fig. 24.4). In the space below draw a few cells showing their general shape and the filament's branching pattern.

Question 4

a. How is *Cladophora* morphologically similar to *Spirogyra*? How is it different?

b. What is the shape of its chloroplasts?

Mature *Cladophora* exists in diploid and haploid forms. The diploid stage of the life cycle produces spores and is called the **sporophyte.** The haploid stage of the life cycle produces gametes and is called the **gametophyte.** This phenomenon of alternating haploid and diploid stages of a life cycle is called **alternation of generations.** Alternation of generations is a reproductive cycle in which the haploid gametophyte produces gametes that fuse to form a zygote that germinates to produce a diploid sporophyte. Within the sporophyte, meiosis produces spores that germinate into gametophytes, thus completing the cycle (fig. 24.5). Alternation of generations occurs in many green algae, including *Cladophora*, and in all land plants.

You should become familiar with the concept of alternation of generations because it occurs frequently in the plant kingdom and we will refer to it repeatedly in this and future exercises. Refer to your textbook or instructor for more information.

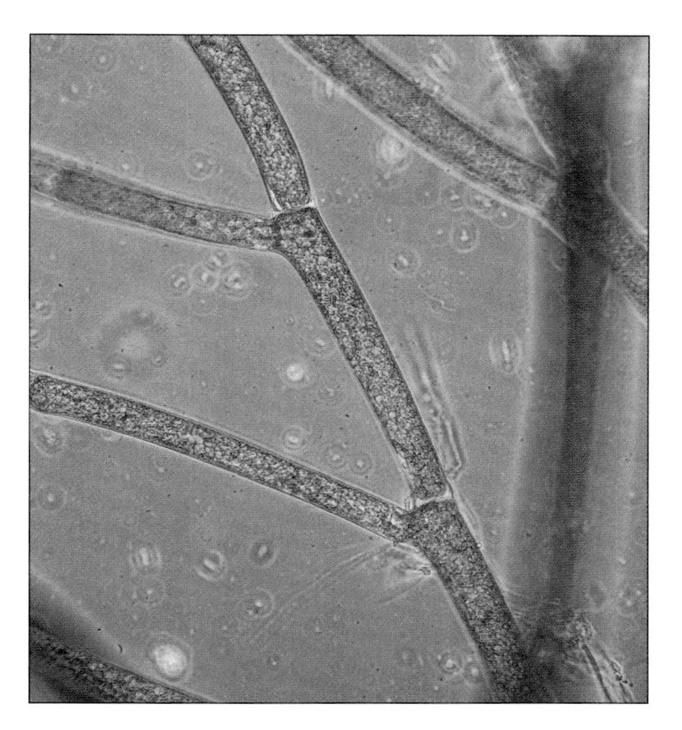

Figure 24.4

The green alga *Cladophora* forms branched filaments consisting of multinucleate cells (100×).

Colonial Green Alga: *Volvox*

Volvox consists of many *Chlamydomonas*-like cells bound in a common spherical matrix (fig. 24.6). Each cell in the sphere has two flagella extending outward from the surface of the colony. Synchronized beating of the flagella spin the colony through the water like a globe on its axis. *Volvox* is one of the most structurally advanced colonial forms of algae, so much so that some biologists consider *Volvox* to be multicellular. Some of the cells of a *Volvox* colony are functionally differentiated; a few specialized cells can produce new colonies, and eggs and sperm are formed by different cells in the colony.

Volvox reproduces by **oogamy.** Motile sperm swim to and fuse with the large nonmotile eggs to form a diploid zygote. The zygote enlarges and develops into a thick-walled zygospore, which is released when the parent colony disintegrates. The zygospore then undergoes meiosis to produce haploid cells that subsequently undergo mitosis and become a new colony. During asexual reproduction, some cells of *Volvox* divide, bulge inward, and produce new colonies called **daughter colonies** that initially are held within the parent colony.

Procedure 24.4

Observe *Volvox*

1. Examine a prepared slide of *Volvox*.
2. Obtain a living culture of *Volvox*. Place a drop of living *Volvox* on a depression slide.
3. Under low magnification, observe the large, hollow, spherical colonies for a few minutes just to appreciate their elegance and beauty.
4. Search for flagella on the surface.
5. Complement your observations of this alga by re-examining prepared slides of *Volvox*.

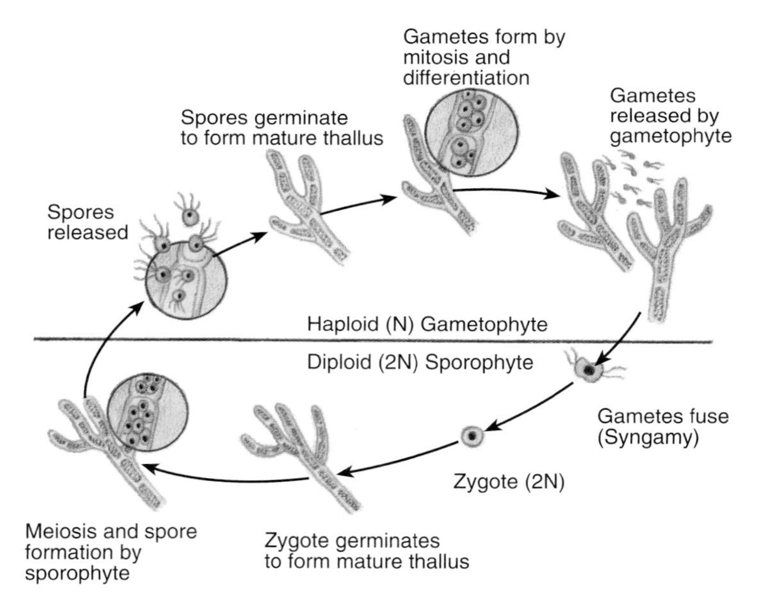

Figure 24.5

Alternation of generations in *Cladophora*. During alternation of generations, the haploid (N) gametophytes alternate with diploid (2N) sporophytes. Gametophytes produce haploid gametes that fuse to form a diploid zygote, which is the first cell of the sporophyte generation. The zygote germinates and undergoes meiosis to form haploid spores, which are the first cells of the gametophyte generation. The gametophyte then uses mitosis to produce gametes, thereby completing the life cycle.

Question 5

a. What is oogamy?

b. What are the tiny spheres inside the larger sphere of *Volvox?*

c. How do you suppose they get out?

d. How do you think the number of cells in a young *Volvox* colony compares to the number in a mature colony?

PHYLUM PHAEOPHYTA (BROWN ALGAE)

Phaeophytes are primarily marine algae that are structurally complex; there are no unicellular or colonial brown algae. Brown algae usually grow in cool water and obtain their name from the presence of a brown pigment called **fucoxanthin.** Brown algae range in size from microscopic forms to kelps over 50 m long. Thalli of some phaeophytes are similar to those of land plants. Review table 24.1 for the characteristics of brown algae.

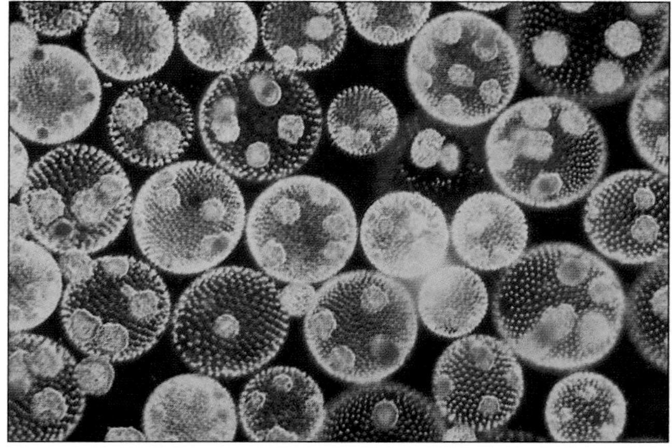

Figure 24.6

Colonies of *Volvox*. Many parent colonies contain asexually produced daughter colonies.

Figure 24.7

Sargassum, a floating brown alga from which the Sargasso Sea got its name. *Sargassum* also lives in other oceans.

Among the larger brown algae is *Macrocystis*, a kelp reaching 100 m in length. The flat blades of this kelp float on the surface of the water, while the base is anchored far below the surface. Another ecologically important brown alga is sargasso weed (*Sargassum*; fig. 24.7), which forms huge floating masses that dominate the vast Sargasso Sea in the Atlantic Ocean northeast of the Caribbean. These mats are microhabitats for a variety of highly adapted and cryptically colored animals.

Fucus

Fucus is another common genus of brown algae (fig. 24.8). *Fucus* (rockweed) typically attaches to rocks in the intertidal zone via a specialized structure called a holdfast. The outer surface of *Fucus* is covered by a gelatinous sheath. Tips of *Fucus* branches, called **conceptacles,** may be swollen and contain reproductive structures called **oogonia** (female) and **antheridia** (male), as shown in figure 24.8. Oogonia are multicellular sex organs that produce eggs. Antheridia are multicellular sex organs that produce sperm. Most protists do not have multicellular reproductive organs.

The life cycle of *Fucus* is similar to the common life cycle of animals. The mature thallus is diploid, and cells within reproductive structures undergo meiosis to produce gametes, thereby skipping the multicellular haploid stage common to many protists, plants, and fungi.

Procedure 24.5

Examine *Fucus*

1. Refer to figure 24.8 as you examine *Fucus* in the lab. Use your dissecting microscope to examine a cross section of the flattened, dichotomously branched thallus of *Fucus*.

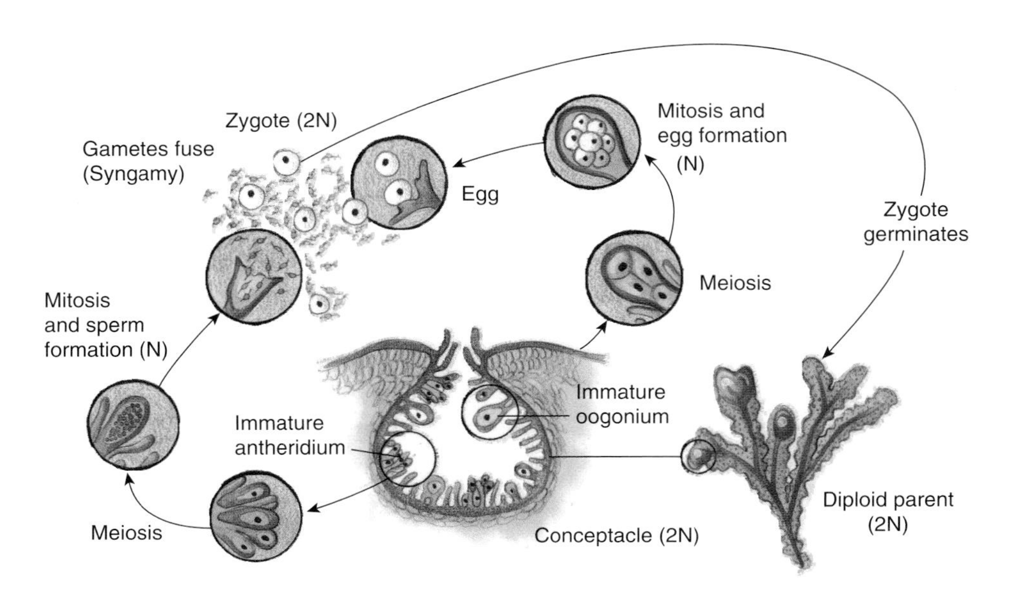

Figure 24.8

Sexual reproduction in a monoecious species of *Fucus*. Some species of *Fucus* have separate male thalli and female thalli containing conceptacles with only antheridia, and only oogonia, respectively.

2. Notice the presence of swollen areas on the thallus of *Fucus*.
3. Work in a group to dissect one of these structures.
4. Examine prepared slides of antheridia and oogonia of *Fucus*.

Question 6

a. How does the structure of *Fucus* differ from the green algae that you examined earlier in this exercise?

b. Are all portions of the thallus photosynthetic? How can you tell?

c. Considering where *Fucus* lives, what do you think is the function of its gelatinous sheath?

d. Are the swollen structures solid masses or are they empty?

Question 7

a. Are the gametes of *Fucus* isogamous or oogamous?

b. How does the structure of tissue surrounding the reproductive structures compare with that of green algae?

Economic Importance of Phaeophyta

Brown algae have considerable economic importance. In the Orient some brown algae are used as food. One of these algae is *Laminaria,* a kelp marketed as "kombu." Brown algae are also important sources of **alginic acid,** which is a hydrophilic substance (i.e., it absorbs large quantities of water). Alginic acid is used as an emulsifier in dripless paint, ice cream, pudding mixes, and cosmetics.

Procedure 24.6

Examine some commercial products of brown algae

1. Taste a small piece of kelp packaged as a food product. How would you describe its taste and texture?
2. Observe the products on display and, in the case of foods, read their contents labels.

PHYLUM RHODOPHYTA (RED ALGAE)

Red algae obtain their color from the presence of red phycobilins in their plastids. Red algae typically live in warm marine waters. The thallus of a red alga can be attached or free-floating, filamentous, or parenchymatous (fleshy) (fig. 24.9*a*).

Procedure 24.7

Examine *Polysiphonia, Porphyra,* and commercial products of red algae

1. Obtain prepared slides of *Polysiphonia* and *Porphyra* and any living cultures that are available.
2. Examine a prepared slide of *Polysiphonia*. Notice the thickness of the filaments compared with that of filamentous green algae.
3. Examine living *Polysiphonia*. This genus is highly branched and filamentous. As with other red algae, their life cycles can be quite complex. Gametophytes of these organisms are dioecious (i.e., they are either male or female).
4. Examine some prepared slides and living *Polyphyra* if available. Compare the structure of *Polysiphonia* with that of *Porphyra*. "Blades" of *Porphyra* consist of two layers of cells separated by colloidal material.
5. Study the display of carrageenan, agar, and other products derived from red algae.
6. Pick up and feel a piece of agar, noting its texture. Agar is a gelatinous polysaccharide from red algae used as a solidifying agent in culture media for microbiology labs (fig. 24.9*b*). A 1% suspension of hot agar remains liquid until it cools to about body temperature.

PHYLUM CHRYSOPHYTA (DIATOMS)

Diatoms are unicellular algae containing chlorophylls *a* and *c* and xanthophyll pigments that give them their golden-brown color. Although diatoms are tiny, their great numbers, rapid rates of reproduction, and photosynthetic capacity make them vitally important as a primary link in the food chain of the oceans.

Diatoms have a hard cell wall made of silicon dioxide (glass) (fig. 24.10). These walls are arranged in overlapping halves, much like the halves of a petri dish. The glass walls of diatoms persist long after the remainder of the cell disintegrates (fig. 24.11). These walls may accumulate in layers of **diatomaceous earth** several hundred meters deep. This depth indicates how many diatoms have existed through the ages.

(a) (b)

Figure 24.9

A red alga and a common extract, agar. (*a*) Irish moss (*Chondrus crispus*) is a red alga that is commercially important as a source of carrageenan. Carrageenan is used as a stabilizer in paints and cosmetics, as well as in foods such as salad dressings and dairy products. (*b*) Microbiologists grow a variety of organisms on media solidified with agar (shown here), which is extracted from seaweeds such as *Gracilaria*. The bacteria shown in figure 23.9 are growing on agar. Agar is also used to make drug capsules, cosmetics, and gelatin desserts.

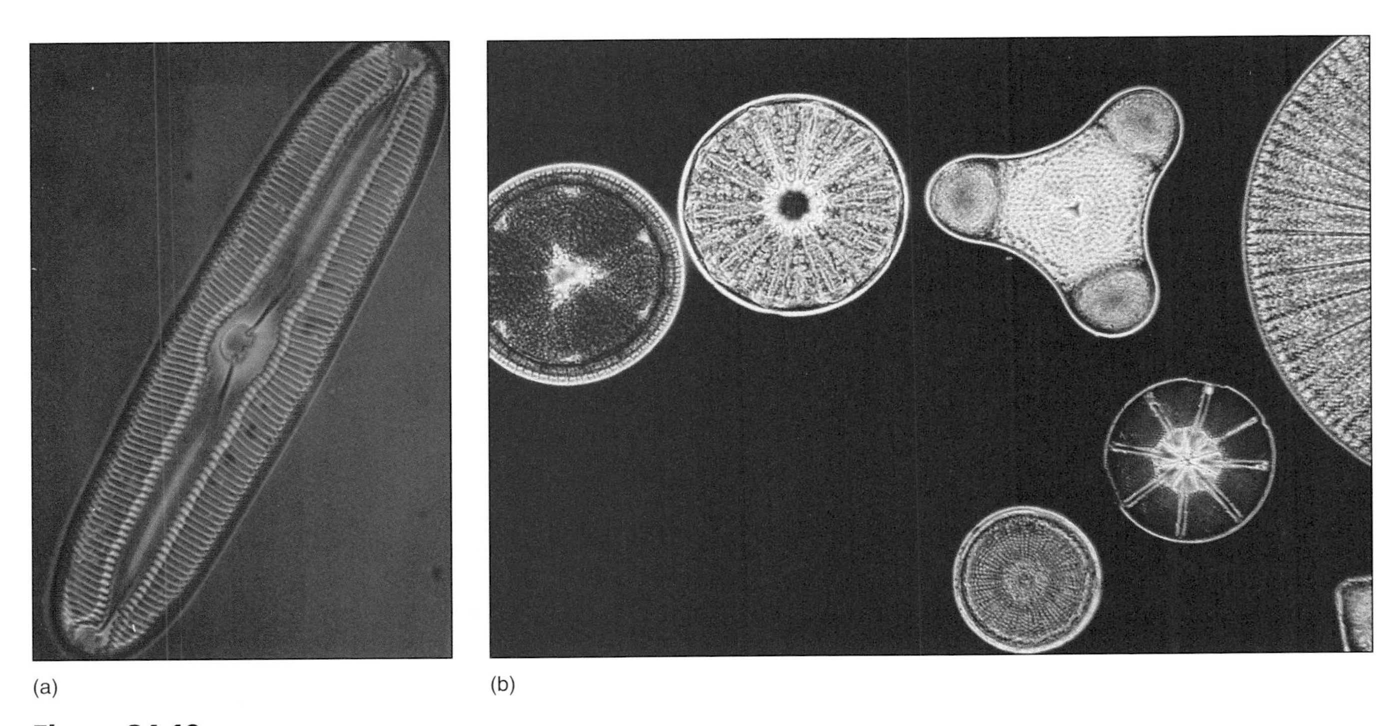

(a) (b)

Figure 24.10

Diatoms (phylum Chrysophyta). (*a*) A pennate (bilaterally symmetrical) diatom. (*b*) Several different kinds of diatoms, including some centrate (round) species.

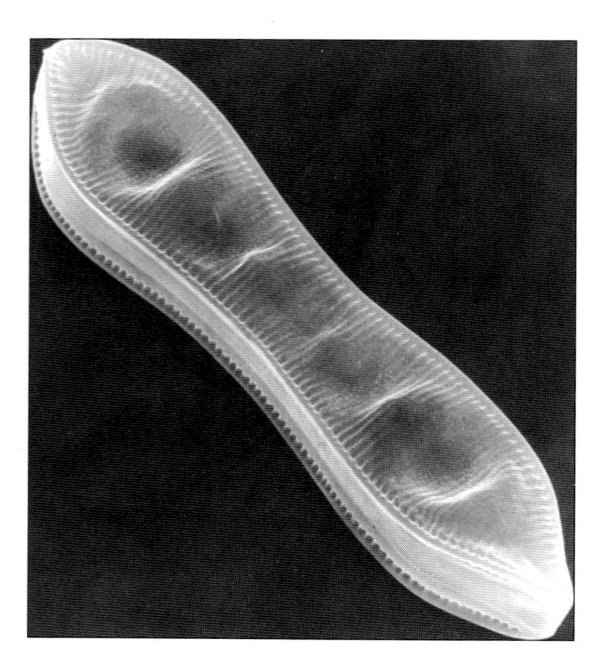

Figure 24.11
A diatom showing its ornate, silicon cell wall.

Procedure 24.8

Examine prepared slides of diatoms, living diatoms, and diatomaceous earth

1. Examine a prepared slide of diatoms. Sketch some of the cells below. Some cells are long and thin, whereas others are disklike.
2. Prepare a wet mount slide from a culture of living diatoms. Compare the shapes of the cells with those on the prepared slide.
3. Mount a small amount of diatomaceous earth in water on a microscope slide.
4. Examine the diatomaceous earth with your microscope.
5. Note the variety of shells, some broken and others intact. A mass of these shells is clean, insoluble, and porous.

Question 8

a. Can you see any pores in the walls of diatoms?

b. Are any of the diatoms moving?

c. If diatoms lack flagella, how do you explain their motility?

d. How would diatomaceous earth compare to sand as a material for swimming pool filters? Which would be better and why?

PHYLUM PYRRHOPHYTA (DINOFLAGELLATES)

Members of this phylum, like those of Chrysophyta, are all unicellular. Dinoflagellates are characterized by the bizarre appearance of their cellulose plates and by the presence of two flagella located in perpendicular grooves (fig. 24.12). A red tide is caused by blooms of a red-pigmented dinoflagellate called *Ptychodiscus bruvis*. Toxin production and oxygen depletion by these blooms of algae can kill massive numbers of fish. Dinoflagellates are important primary producers in oceans (second only to diatoms), and include many autotrophic and heterotrophic forms. Some dinoflagellates are bioluminescent, and others live symbiotically with corals.

Procedure 24.9

Examine dinoflagellates

1. Examine a prepared slide of *Peridinium* or *Ceratium*. Look for longitudinal and transverse flagella and flagellar grooves.
2. Prepare a wet mount slide from a living culture of dinoflagellates. Dinoflagellates are quite small, so be patient while searching for organisms.

Question 9

How do the shapes of dinoflagellates compare with other unicellular algae that you have observed in this exercise?

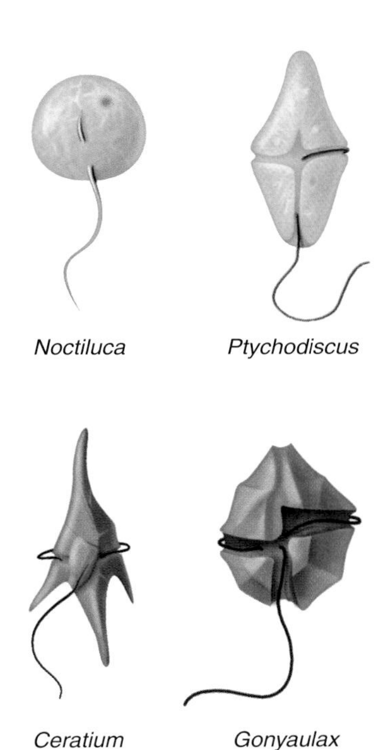

Figure 24.12

Some dinoflagellates: *Noctiluca, Ptychodiscus, Ceratium,* and *Gonyaulax. Noctiluca,* which lacks the heavy cellulose armor characteristic of most dinoflagellates, is one of the bioluminescent organisms that cause the waves to sparkle in warm seas. In the other three genera, the shorter, encircling flagellum is seen in its groove, with the longer one projecting away from the body of the dinoflagellate.

PHYLUM EUGLENOPHYTA (EUGLENOIDS)

This small phylum includes mostly freshwater unicellular algae. Although plastids of euglenoids contain chlorophylls *a* and *b* (like the green algae), euglenoids are distinctive because their cell walls are made largely of protein. The protein makes the cell more flexible. Euglenoids are motile and have two flagella (fig. 24.13).

Procedure 24.10

Observe *Euglena*

1. Observe living and prepared slides of *Euglena* available in the lab referring to figure 24.13.
2. You may want to use a drop of methylcellulose in your preparation to slow the *Euglena*.
3. Note the colored **eyespot** near the base of the flagella.
4. Observe movement and changing shapes.

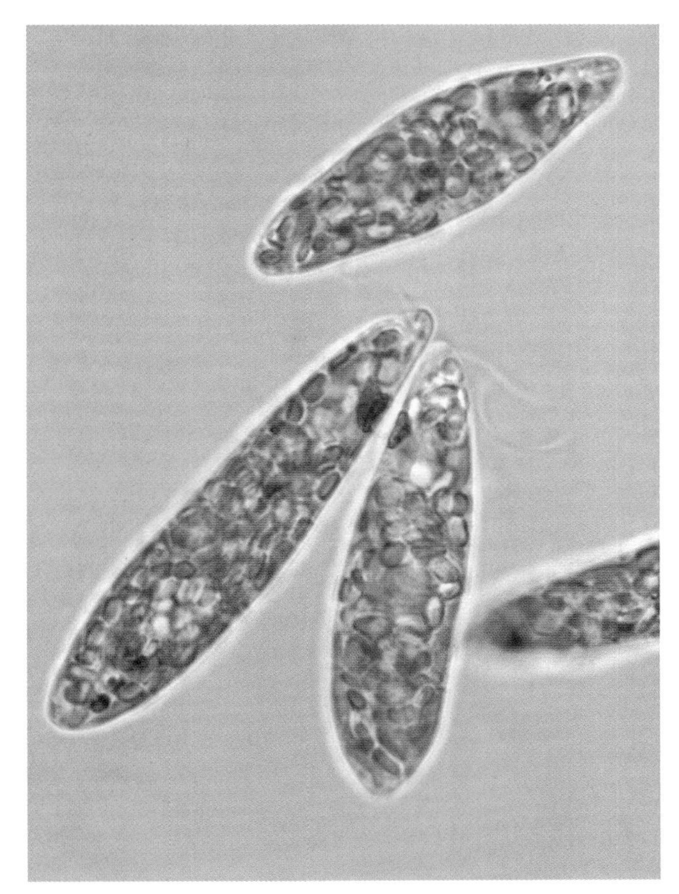

(a)

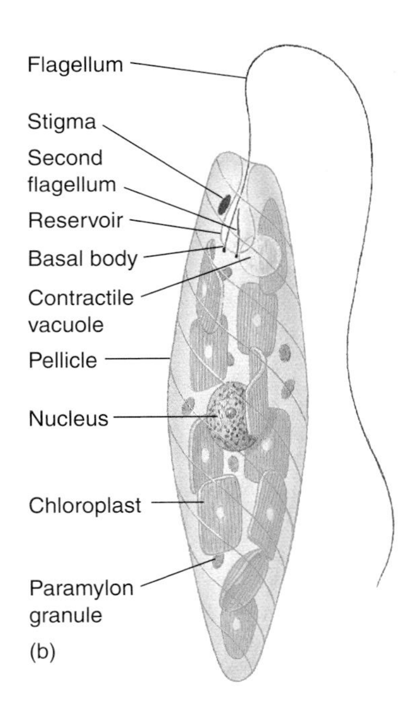

Figure 24.13

Euglenoids. (*a*) Micrograph of individuals of the genus *Euglena* (Euglenophyta). 400× (*b*) Diagram of *Euglena*. Paramylon granules are areas where food reserves are stored.

Chloroplasts of *Euglena* may contain a single pyrenoid, which appears as a clear, circular area within the plastid. *Euglena* is best known for its ability to be autotrophic, heterotrophic, and saprophytic. Its specific mode of nutrition is determined by current environmental conditions. This phenomenon illustrates why it is often impossible to distinguish plant from animal at the cellular level and why kingdom Protista is necessary. Our classification schemes for these and other organisms will improve as we learn more about them.

Question 10
What is the function of the eyespot of *Euglena?*

Questions for Further Thought and Study

1. What are examples of unicellular, filamentous, and colonial green algae?

2. How are green algae different from cyanobacteria?

3. What is meant by "alternation of generations"?

4. What is meant by the term "kelp"?

5. Are the stem, holdfast, and blade of brown algae the same as stems, roots, and leaves of land plants? Why or why not?

6. Brown algae contain chlorophyll, so why do they appear brown and not green?

7. What are the main differences and similarities among the major groups of algae?

8. How do algae affect your life?

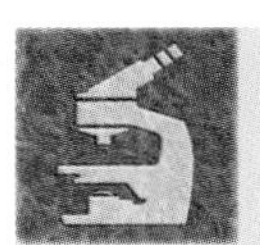

DOING BIOLOGY YOURSELF

Would you expect environmental conditions to influence syngamy? Design two experiments to investigate the effects of two environmental conditions on the frequency of syngamy. How would you measure syngamy? Its frequency?

WRITING TO LEARN BIOLOGY

Describe the plantlike and animal-like characteristics of *Euglena*. Which characteristics conclusively define a plant? Which ones define an animal?

25

Survey of the Kingdom Protista
Protozoa and Slime Molds

Objectives

By the end of this exercise you should be able to:

1. Describe the features characterizing each phylum of kingdom Protista, including slime molds.
2. List examples, habitats, reproductive methods, and unique features of the phyla of kingdom Protista, including slime molds.

Protozoans (*proto* = first, *zoan* = animal) are among the most versatile of all organisms on earth. Protozoa, however, like algae, is a descriptive term rather than a taxonomic group. Protozoans have an animal-like lifestyle, which means they are active consumers and not photosynthetic. Typically, protozoans have food vacuoles to enclose food particles for digestion and contractile vacuoles to expel excess water. Their single cells employ a variety of features for motility and occupy virtually every microhabitat. Review table 25.1 for the characteristics of the major phyla of protozoans.

PHYLUM RHIZOPODA (AMOEBAS)

Amoebas occur throughout the world in marine, freshwater, and terrestrial environments. The unifying characteristic of this phylum is the presence of **pseudopods,** which are movable extensions of cytoplasm used for locomotion and gathering food. Amoebas lack flagella, and most reproduce asexually.

Amoeba

Amoeba is a genus among many organisms commonly called amoebas, and has a structure and physiology typical of most amoeboid genera (fig. 25.1). *Amoeba* are **phagocytic,** meaning they engulf food particles and form a **food vacuole** surrounded by a membrane. They then secrete enzymes into the food vacuole for **intracellular digestion.** A **contractile vacuole** maintains the cell's water balance by accumulating and expelling excess water. Other common amoebas include *Difflugia*, which makes a protective case of sand grains called a **test.** Test is a general term referring to a secreted or partially secreted covering, much like a shell (fig. 25.2). Examine a prepared slide of *Difflugia*. Also examine the amoeba *Entamoeba histolytica*, a parasite that causes dysentery in humans.

Procedure 25.1

Observe *Amoeba* movement and structure

1. Use a dissecting microscope to examine a culture of living *Amoeba*. Locate individuals on the bottom.
2. Prepare a wet mount of living *Amoeba* by using an eyedropper to remove a few drops from the bottom of the culture of organisms.
3. Put the drops in a depression slide if one is available or use a standard slide.
4. Cover the preparation with a coverslip and examine it under low power (10×). Soon the *Amoeba* should move by extending their pseudopods.
5. If nutrient broth is available, add a drop to the preparation and observe the *Amoeba's* response.
6. Examine a prepared slide of stained *Amoeba* and locate the structures shown in figure 25.1.
7. Examine any other live amoebas and prepared slides that are available. Draw the basic structure of these organisms.

Question 1

a. Can you detect moving cytoplasm in the extending pseudopods of *Amoeba?*

b. What do you suppose the living *Amoeba* is moving toward or away from?

TABLE 25.1

KINDS OF PROTISTS

Group	Phylum	Typical Examples	Key Characteristics
HETEROTROPHS WITH NO PERMANENT LOCOMOTOR APPARATUS			
Amoebas	Rhizopoda	*Amoeba*	Move by pseudopodia
Forams	Foraminifera	Forams	Rigid shells; move by protoplasmic streaming
HETEROTROPHS WITH FLAGELLA			
Zoomastigotes	Sarcomastigophora	Trypanosomes	Heterotrophic; unicellular
Ciliates	Ciliophora	*Paramecium*	Heterotrophic unicellular protists with cells of fixed shape possessing two nuclei and many cilia; many cells also contain highly complex and specialized organelles
NONMOTILE SPORE-FORMERS			
Sporozoans	Apicomplexa	*Plasmodium*	Nonmotile; unicellular; the apical end of the spores contains a complex mass of organelles
HETEROTROPHS WITH RESTRICTED MOBILITY			
Water molds	Oomycota	Water molds, rusts, and mildew	Terrestrial and freshwater
Cellular slime molds	Acrasiomycota	*Dictyostelium*	Colonial aggregations of individual cells; most closely related to amoebas
Plasmodial slime molds	Myxomycota	*Fuligo*	Stream along as a multinucleate mass of cytoplasm

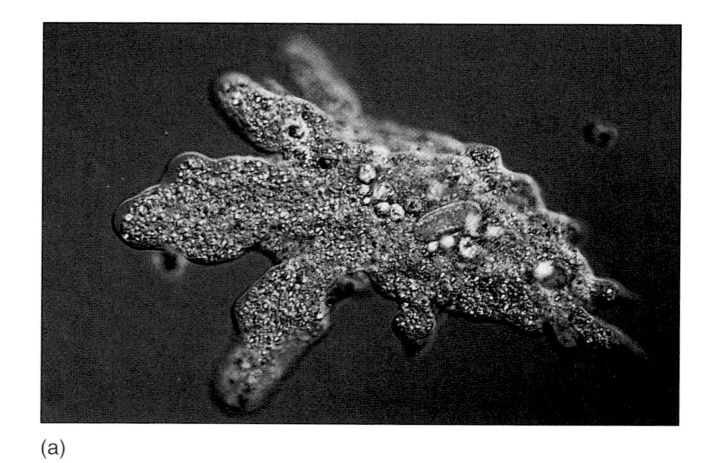

(a)

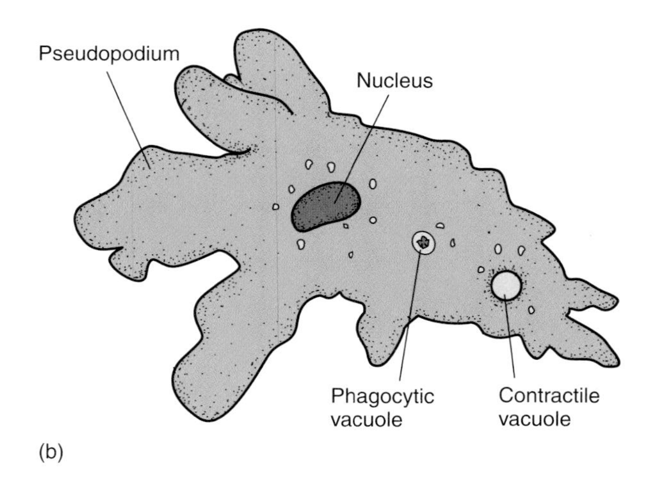

(b)

Figure 25.1

Phylum Rhizopoda. (*a*) Light micrograph of *Amoeba proteus* showing blunt pseudopodia (160×). (*b*) Drawing showing the anatomy of *Amoeba proteus*.

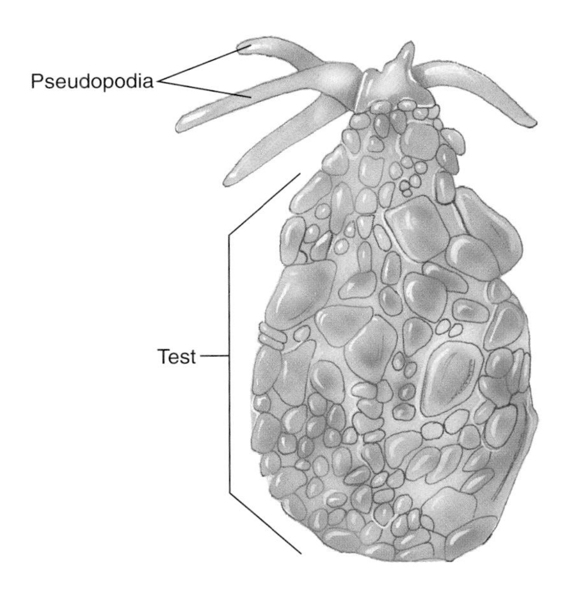

Figure 25.2
Difflugia oblongata, a common freshwater amoeba with a sand-grain case. The case consists of cemented mineral particles collected by the amoeba.

c. How does the *Amoeba* respond to nutrient broth?

d. About how long would it take an *Amoeba* to move across the field of view on low power?

e. Why is a contractile vacuole of a protozoan often more difficult to see than a food vacuole?

f. Why would excess water tend to accumulate in *Amoeba?*

PHYLUM FORAMINIFERA (FORAMS)

These marine organisms are called "shelled amoebas" because they surround themselves with a secreted test and have pseudopods (fig. 25.3). The test is made of calcium carbonate and is perforated with pores allowing thin pseudopods to protrude. The marine fossil record is replete with old and well-preserved tests of forams, and is often used by oil companies to locate oil-bearing strata.

Procedure 25.2
Examine foram tests

1. Obtain a prepared slide of foram tests.

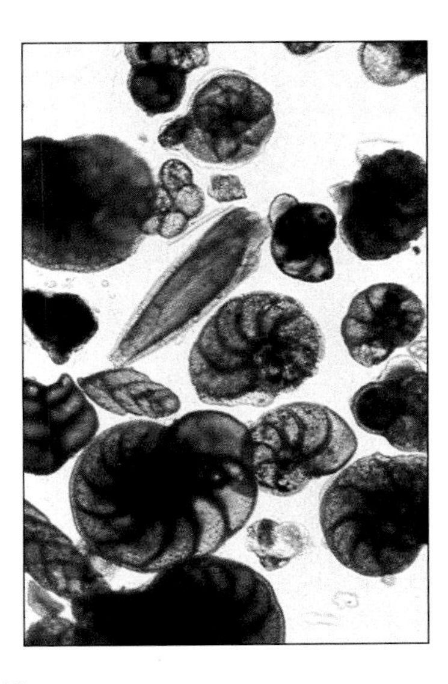

Figure 25.3
The calcareous tests of some representative Foraminifera.

2. Search with low magnification the edges of the cover slips for the foraminiferan tests. They are relatively heavy and shift to the side easily.
3. Draw a few tests below.

Question 2
How could fossilized forams in different geological layers of rock or sediment indicate the probability of finding oil?

PHYLUM SARCOMASTIGOPHORA (ZOOMASTIGOTES)

Zoomastigotes are unicellular, heterotrophic, and have at least one **flagellum.** They are also the most primitive protozoans, and are the likely ancestors of multicellular animal phyla. Parasitic as well as free-living heterotrophic forms are represented in phylum Sarcomastigophora. Heterotrophs get their nutritional energy from organic molecules made by other organisms.

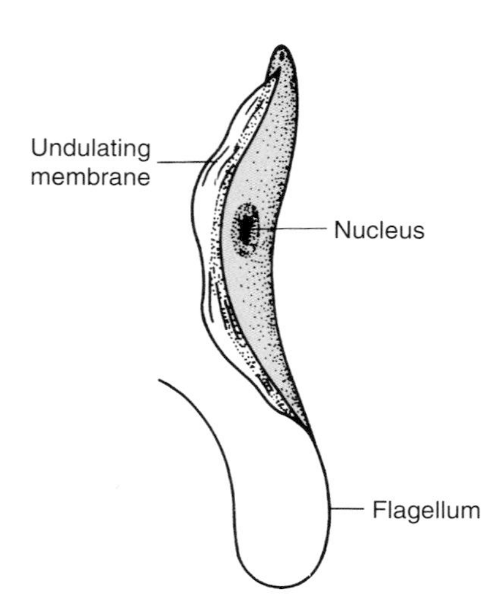

Figure 25.4
Trypanosoma, the protist that causes sleeping sickness.

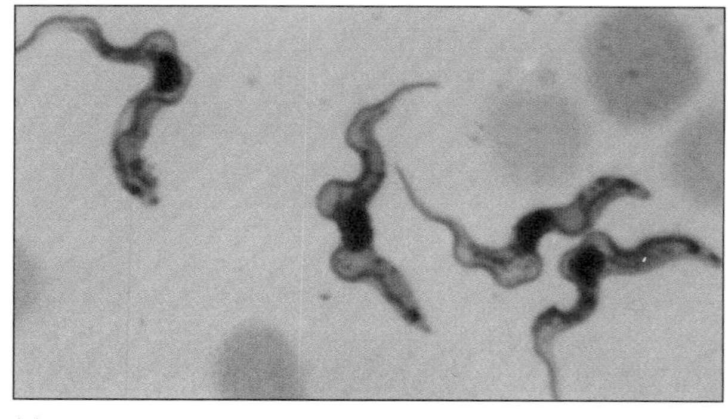

(a)

(b)

Figure 25.5
(*a*) *Trypanosoma* among red blood cells. The nuclei (dark-staining bodies), the anterior (forward-projecting) flagella, and the undulating, changeable shape of the trypanosomes are visible here. (*b*) A tsetse fly, in Tanzania, East Africa, shown here sucking blood from a human arm. Tsetse flies transmit trypanosomes, which are protists that cause sleeping sickness in humans.

Trypanosoma

Trypanosomes are pathogenic and cause African sleeping sickness and Chagas' disease. Charles Darwin probably died from Chagas' disease, for during his later years he suffered from general fatigue, irregular fever, and heart damage. All of these are symptoms of Chagas' disease transmitted by the bite of an assassin bug, which resembles a common stinkbug. Trypanosomes are common in the tropics and spread by infection from biting insects such as mosquitos, sand flies, and tsetse flies.

Procedure 25.3

Examine a prepared slide of *Trypanosoma* and compare its size to that of *Amoeba*

1. Obtain a prepared slide of *Trypanosoma* among blood cells.
2. Locate the organisms intermingled with the blood cells (figs. 25.4, 25.5). The organisms are not inside the blood cells; they are in the surrounding plasma.
3. Try to distinguish the flagellum and undulating membrane of an individual. The **undulating membrane** is a thin, flat surface that can be undulated or waved for locomotion. A rippling wave travels along the membrane and pushes the organism forward.
4. Trypanosomes are quite small. To estimate its size, first note the magnification of the objective you are using.
5. Refer to your notes from Exercise 2 (The Microscope) concerning the diameter of the field of view associated with the current magnification.
6. Estimate the portion of the diameter of the field of view occupied by a *Trypanosoma*. Calculate the size of a trypanosome.
7. Make a similar estimation of the size of *Amoeba* with a prepared slide at the same magnification.

Question 3

a. How large is a trypanosome relative to *Amoeba?*

b. What alga does a trypanosome superficially resemble?

PHYLUM CILIOPHORA (CILIATES)

More than 8000 species of ciliates have been described, all having characteristically large numbers of **cilia.** Review in your textbook the difference between cilia and flagella. Most ciliates also have two types of nuclei: **micronuclei** and **macronuclei** (fig. 25.6). A micronucleus divides by mitosis and contains the genetic information of the cell in normally shaped chromosomes. As many as 80 micronuclei may occur in a single cell. The single macronucleus in a cell contains multiple copies of DNA divided into small pieces. The macronucleus replicates by elongating and constricting. Macronuclei are essential for routine cellular functions.

Paramecium

This free-living, freshwater genus is widely studied and easily observed. *Paramecium*, like most ciliates, undergoes a sexual process called **conjugation** (fig. 25.7). During conjugation, individuals from two different strains align longitudinally and exchange nuclear material (fig. 25.7*b*). This exchange seems to rejuvenate the individuals and is usually followed by frequent mitosis. Asexual reproduction is more common than conjugation and includes mitosis of the micronucleus and transverse fission (fig. 25.7) of the macronucleus and cell body.

Procedure 25.4

Examine conjugating and dividing *Paramecium*

1. Obtain a prepared slide of conjugating *Paramecium* and one of transversely dividing (binary fission) *Paramecium*.
2. Notice the alignment of the conjugating cells. Their nuclei are in close proximity.
3. Notice the plane of division of the transversely dividing cells.

Procedure 25.5

Observe living *Paramecium*

1. Prepare a wet mount from a culture of living organisms.
2. Add a drop of methylcellulose to your wet mount to slow the *Paramecium* and make it easier to examine.
3. Describe aspects of their movement in comparison to *Amoeba*.

4. Identify as many structures as possible in figure 25.6.

Question 4

a. Are cilia visible on living or prepared *Paramecium?*

b. Does *Paramecium* rotate as it moves?

c. How does movement of *Paramecium* compare with that of *Amoeba?* With a flagellated alga?

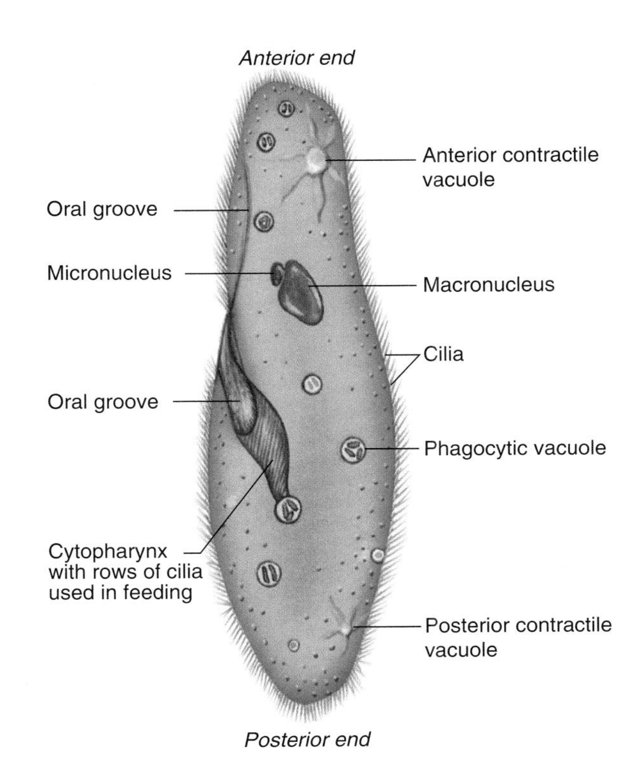

Figure 25.6
Diagram of *Paramecium*, a representative of phylum Ciliophora.

Question 5

a. Why is the division of *Paramecium* cells called "transverse" fission?

b. Why is transverse fission not a sexual process?

c. What are the advantages and disadvantages of conjugation in *Paramecium?*

Vorticella and Other Ciliates

This freshwater ciliate is sessile (i.e., attached to a substrate) and has two notable features: 1) a contractile stalk that attaches the organism to the substrate, and 2) a cell body with a corona of cilia. To feed, *Vorticella* extends its contractile stalk to push the cell body as far as possible from the substrate and from other individuals. Then it rapidly beats its cilia to capture food particles. This is a type of filter feeding (fig. 25.8).

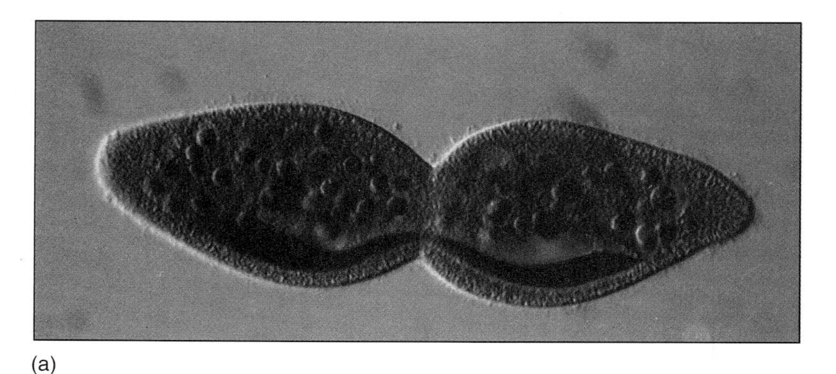

(a)

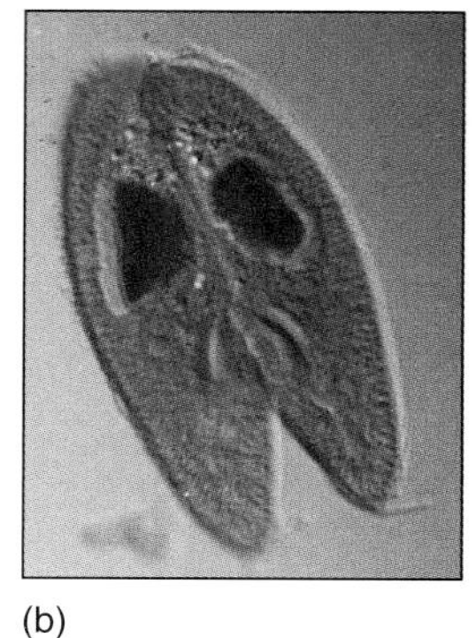

(b)

Figure 25.7

Reproduction among paramecia. (*a*) A mature *Paramecium* divides asexually by transverse fission. (*b*) During conjugation, individuals exchange genetic material. Conjugation is a sexual process.

Procedure 25.6

Examine *Vorticella* and other ciliates

1. Use a dissecting microscope to examine a living colony of *Vorticella,* which often grows on the glass and other hard substrates in stagnant aquaria.
2. Tap the sides of the dish and observe the contraction of each stalk.
3. When the stalks are extended, notice the whirling action of the cilia at the open end of each bell-shaped individual.
4. Examine a prepared and stained slide of *Vorticella* and draw its general shape.
5. Examine other cultures or prepared slides of ciliates available in the lab.
6. Draw the general shape of these organisms and describe the movement of living specimens.

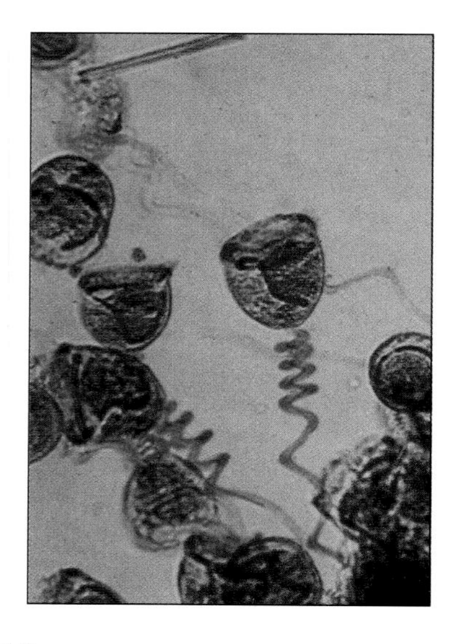

Figure 25.8

Vorticella, phylum Ciliophora, are heterotrophic, feed largely on bacteria, and have a retractable stalk.

Question 6

a. What is the value or function of rapid contraction of the stalk of *Vorticella?*

b. What is the probable function of the moving cilia of *Vorticella?*

PHYLUM APICOMPLEXA (SPOROZOANS)

Sporozoans are all nonmotile, spore-forming parasites of animals. Their "spores" are small, rather featureless stages in their life history that are transmitted from host to host.

Plasmodium

This pathogen is the best-known sporozoan and has been the most common killer of humans in history. *Plasmodium* cause malaria, and mosquitos of the genus *Anopheles* transmit *Plasmodium* from human to human (fig. 25.9). These malarial parasites infect and rupture red blood cells, causing cycles of fever and chills.

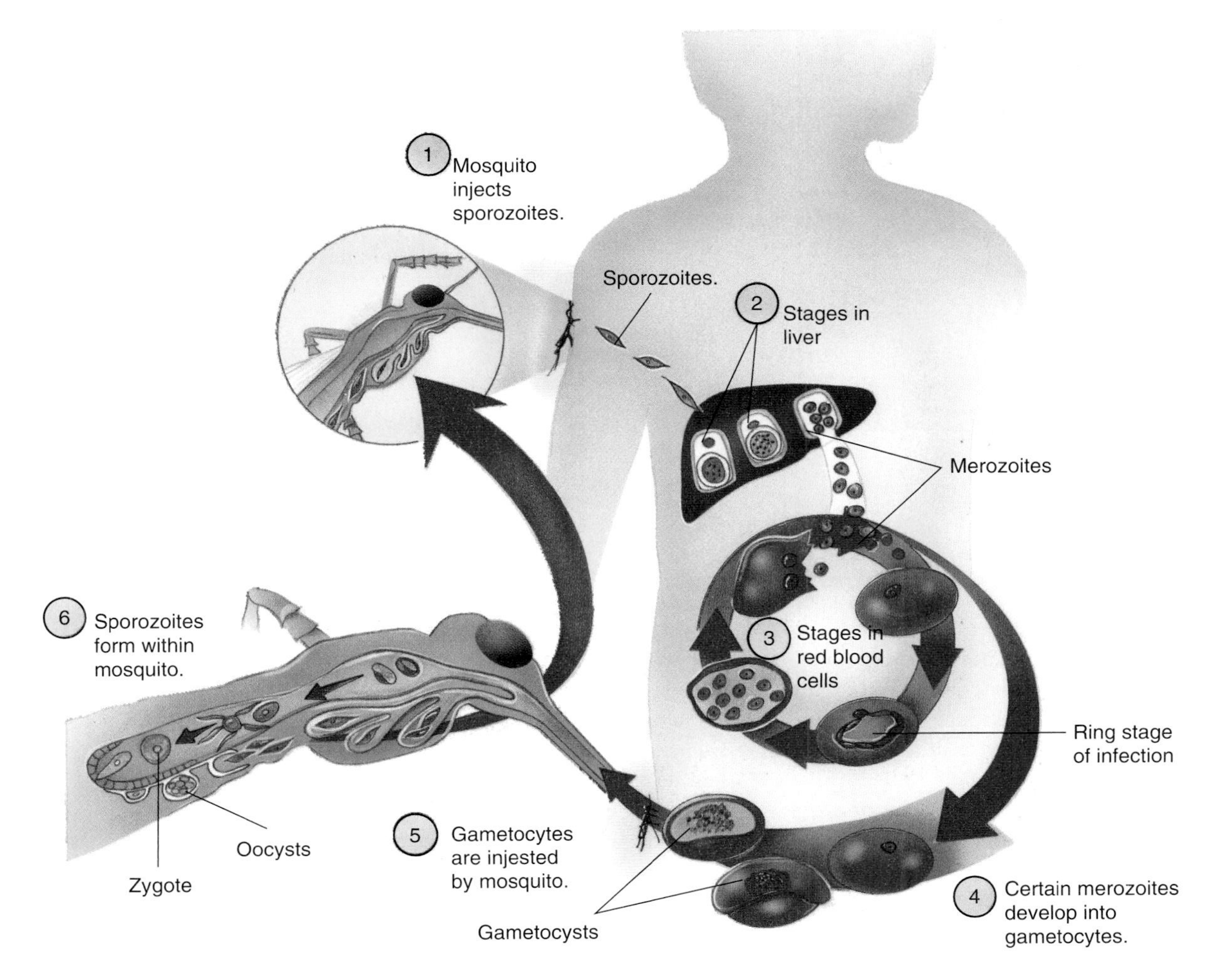

Figure 25.9

Plasmodium is the sporozoan that causes malaria. The life cycle stages include sporozoites, merozoites, and gametocytes. When a mosquito inserts its proboscis into a human blood vessel, it injects about a thousand sporozoites. These sporozoites travel to the liver within a few minutes, where they are no longer exposed to antibodies circulating in the blood. If even one sporozoite reaches the liver, it will multiply rapidly there and cause malaria. The number of malarial parasites increases roughly eightfold every 24 h after they enter the host's body. Although malaria is rare in the United States, worldwide there are about 400 million new cases every year, almost two million of which are fatal. This means that about 2.4 billion people—that's 40% of the Earth's population—are affected by malaria. A compound vaccination against sporozoites, merozoites, and gametocytes would probably be the most effective preventive measure against malaria. Such a vaccine is now being developed.

Procedure 25.7

Examine a blood smear from a victim of malaria

1. Obtain a prepared slide with *Plasmodium*.
2. Locate the infected blood cells. The organisms are inside the infected blood cells, not in the surrounding plasma as are trypanosomes (figs. 25.4, 25.5).
3. Locate and compare infected and uninfected blood cells. The infected cells will exhibit only faint signs of the circular ring stage of the parasite.
4. Review in your textbook the life cycle of *Plasmodium* (fig. 25.10).

SLIME MOLDS

These protists are divided into three phyla: Myxomycota (plasmodial slime molds), Acrasiomycota (cellular slime molds), and Oomycota (water molds). Slime molds have often been classified in kingdom Fungi, but they have amoeboid characteristics such as phagocytic nutrition and unique unicellular forms and assemblages. They also lack the prominent hyphae of fungi. Slime molds also do not contain chitin in their cell walls as do fungi. Hence, we will classify slime molds as protists. In this exercise you will examine *Physarum*, phylum Myxomycota, as a representative slime mold.

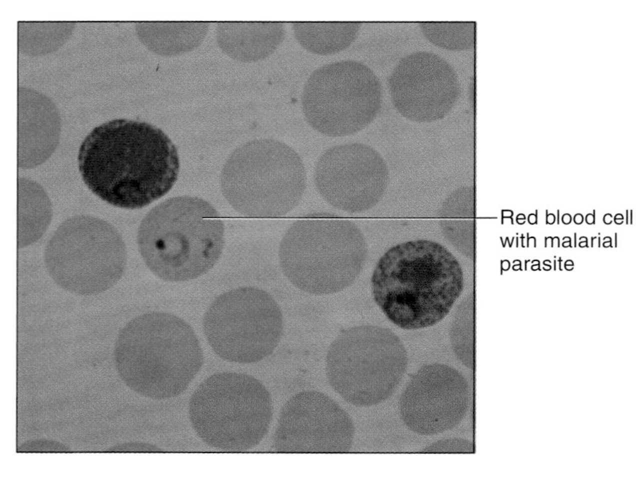

Figure 25.10
Malarial parasite, *Plasmodium*, within a red blood cell (500×).

Phylum Myxomycota (Plasmodial Slime Molds)

Bizarre plasmodial slime molds stream along the damp forest floor in a mass of brightly colored protoplasm called a **plasmodium** in which individual cells are indistinguishable. Plasmodia are coenocytic (multinucleate) because their nuclei are not separated by cell walls, and they resemble a moving mass of slime. The plasmodium of a slime mold should not be confused with the sporozoan genus with the same name.

Often plasmodia of slime molds live beneath detached bark on decomposing tree trunks. Two species occasionally occur on lawns or on mulch beneath shrubs. The moving protoplasm of the plasmodia conspicuously pulsates back and forth as they engulf and digest bacteria, yeasts, and other organic particles.

Procedure 25.8

Examine a culture of *Physarum*

1. Obtain a dissecting microscope and a petri plate containing a culture of *Physarum* growing on oatmeal flakes.
2. Examine the yellowish trails of the colony under the highest magnification of the microscope. Look for cytoplasmic movement.
3. Look for signs of the organism beginning to condense into darkly tipped sporangia. Sporangia are easily visible without a microscope.

Question 7

a. Is cytoplasmic movement of *Physarum* apparent?

b. Is the movement in a particular direction?

c. What is a possible function of cytoplasmic movement in *Physarum?*

The *Physarum* on demonstration is in the vegetative plasmodial stage and feeds on the oatmeal. However, if environmental conditions become less than optimal (e.g., if food or moisture decreases), there are two possibilities. The plasmodium may dry into a hard resistant structure called a **sclerotium** and remain dormant until conditions improve. Or, if light is available, the diploid plasmodium will move to the illuminated area and coalesce. The condensed structure will grow sporangia, and meiosis will produce spores for dispersal (fig. 25.11). Light is associated with an open environment that allows successful reproductive dispersal; dispersal under or within a tree trunk would be difficult. Haploid spores produced by meiosis in the sporangia germinate as amoeboid or flagellated organisms. These haploid stages may later fuse as gametes and grow into a new plasmodium.

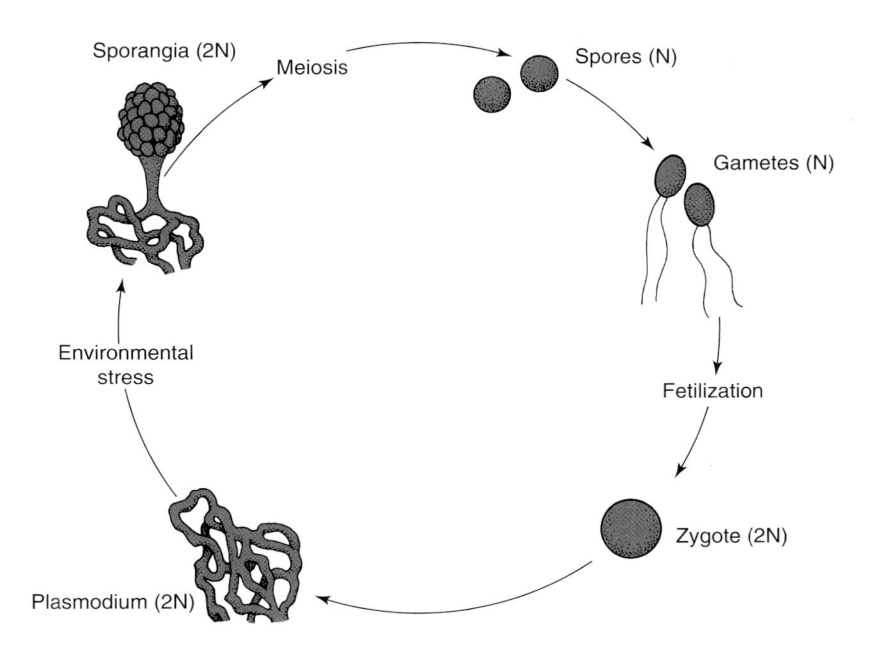

Figure 25.11
Life cycle of *Physarum*, a slime mold.

Questions for Further Thought and Study

1. What requirements might make culturing parasitic zoomastigotes difficult in the lab?

2. Why do some scientists call conjugation "sexual reproduction" and others do not?

3. Is the cell the fundamental unit of life in plasmodial slime molds? Or is the "whole organism," regardless of cellular composition, the fundamental unit? Explain your answer.

4. What functions do cilia, flagella, and pseudopods have in common?

5. What factors may account for the ubiquitous occurrence and great structural diversity of unicellular organisms?

6. In what sense are protists "primitive" and in what sense are they "advanced"?

7. Why are unicellular organisms that reproduce by mitosis considered immortal?

26

Survey of the Kingdom Fungi
Molds, Sac Fungi, Mushrooms, and Lichens

Objectives

By the end of this exercise you should be able to:

1. Describe the characteristic features of the kingdom Fungi.
2. Discuss variation in structures and sequence of events of sexual and asexual reproduction for the major groups of the kingdom Fungi.

Fungi are among the most common and important groups of organisms. Yet they are among the least understood by introductory biology students. Fungi are basically filamentous strands of cells that secrete enzymes and feed on the organic material on which they are growing. That organic material may be humus in the soil where mushrooms grow or a stale loaf of bread where mold thrives. It may be the skin between your toes inhabited by athlete's foot fungus or a decaying animal on the forest floor being decomposed by fungi digesting the animal's dead tissue. Fungi not only cause disease; they are also important decomposers that recycle nutrients from dead organisms.

The basic structure of a fungus is the **hypha** (pl., *hyphae*)—a slender filament of cytoplasm and nuclei enclosed by a cell wall (fig. 26.1). A mass of these hyphae make up an individual organism and is collectively called a **mycelium.** A mycelium can permeate soil, water, or living tissue; fungi certainly seem to grow everywhere. In all cases the hyphae of a fungus secrete enzymes for **extracellular digestion** of the organic substrate. Then the mycelium and its hyphae absorb the digested nutrients. For this reason, fungi are called **absorptive heterotrophs.** Heterotrophs obtain their energy from organic molecules made by other organisms.

Fungi feed on many types of substrates. Most fungi obtain food from dead organic matter and are called **saprophytes.** Other fungi feed on living organisms and are **parasites.** Many of the parasitic fungi have modified hyphae called **haustoria,** which are thin extensions of the hyphae that penetrate living cells and absorb nutrients.

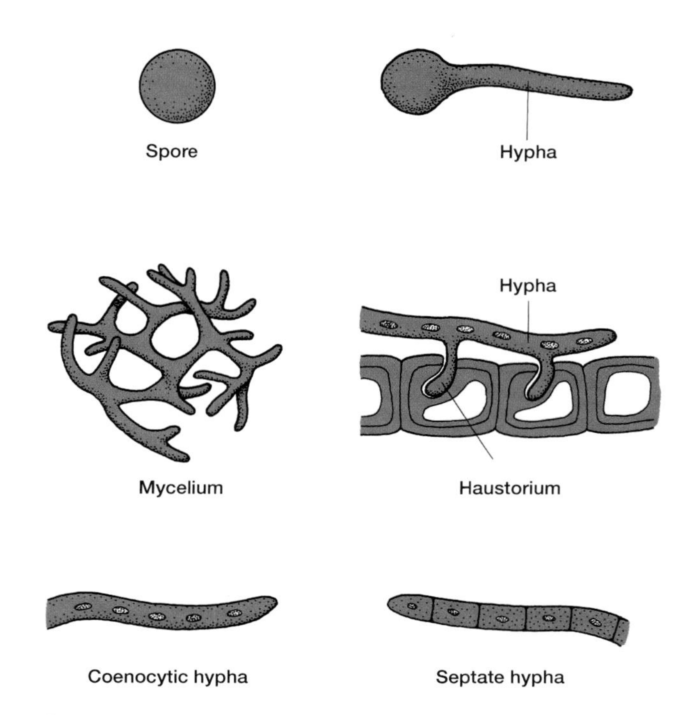

Figure 26.1
Fundamental elements of fungal structure.

Hyphae of some species of fungi have crosswalls called **septa** that separate cytoplasm and nuclei into cells. Hyphae of other species have incomplete or no septa (i.e., are aseptate) and therefore are **coenocytic** (multinucleate). Notably the cell walls of fungi are usually not cellulose but are made of **chitin,** the same polysaccharides that comprises the exoskeleton of insects and crustaceans.

REPRODUCTION IN KINGDOM FUNGI

Reproduction in fungi can be sexual or asexual. All divisions share similar patterns of reproduction and morphology.

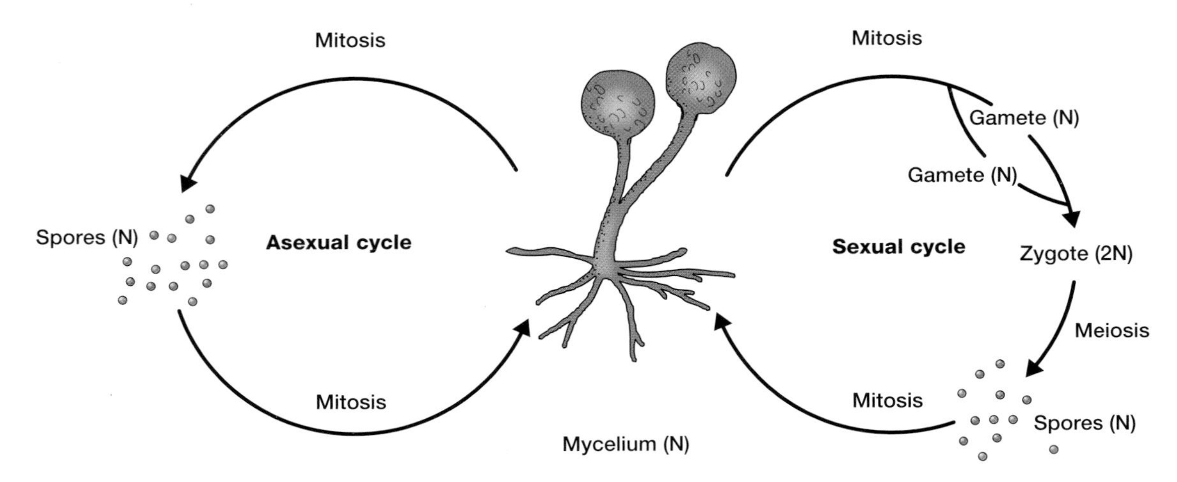

Figure 26.2
Generalized life cycle of a fungus.

Asexual Reproduction

Fungi commonly reproduce asexually by mitotic production of haploid vegetative cells called **spores** in **sporangia, conidiophores,** and other related structures. Spores are microscopic and surrounded by a covering well suited for the rigors of distribution into the environment. *Pilobolus,* an interesting fungus, points its sporangia toward the sun. This orientation of an organism to light is called **phototaxis.** *Pilobolus* ejects its entire sporangium as far as two meters to distribute its spores.

Budding and **fragmentation** are two other methods of asexual reproduction. Budding is mitosis with an uneven distribution of cytoplasm, and is common in yeasts. After budding, the cell with the lesser amount of cytoplasm eventually detaches and matures into a new organism. Fragmentation is the breaking of an organism into one or more pieces, each of which can develop into a new individual.

Sexual Reproduction

The sexual life history of fungi includes the familiar events of vegetative growth, genetic recombination, meiosis, and fertilization. However, the timing of these events is unique in fungi. Fungi reproduce sexually when hyphae of two genetically different individuals of the same species encounter each other. A generalized life cycle of fungi (fig. 26.2) illustrates four important features of fungi:

- Nuclei of a fungal mycelium are haploid during most of the life cycle.
- Gametes are produced by mitosis and differentiation of haploid cells rather than directly from meiosis of diploid cells.
- Meiosis quickly follows formation of the zygote, the only diploid stage.
- Haploid cells produced by meiosis are not gametes; rather, they are spores that grow into a mature haploid organism. Recall that asexual reproduction produces spores by mitosis. In both cases, haploid spores grow into mature mycelia.

None of these features of the sexual cycle are unique to fungi, but together they describe the typical fungal life cycle.

Classification of Fungi

The great diversity of fungi results from modifications of hyphae into varied and specialized reproductive structures often unique to a phylum, genus, or species. The four major phyla of fungi include **Chytridiomycota, Zygomycota, Ascomycota,** and **Basidiomycota.**

As you examine members of these major groups, carefully note variations on the fundamental structure of vegetative mycelia and specialized structures associated with sexual and asexual reproduction. The names of phyla of fungi are derived from sexual reproductive structures rather than asexual structures (table 26.1). However, each phylum has various modifications for both sexual and asexual reproduction.

PHYLUM CHYTRIDIOMYCOTA (CHYTRIDS)

Recent molecular evidence indicates that chytrids may be the most ancient fungi. They are typically aquatic saprobes or parasites on plants, animals, and protists. Although they have flagella, which has been considered a nonfungal characteristic, they also have absorptive nutrition and chitinous cell walls, and share proteins and nucleic acids common to other fungi. Their distinctive reproductive feature is motile spores with flagella.

TABLE 26.1

REPRODUCTIVE FEATURE OF PHYLA OF FUNGI

Phylum	Key Reproductive Feature
Chytridiomycota (chytrids)	Motile spores with flagella
Zygomycota (zygote fungi)	Resistant zygosporangium as sexual stage
Ascomycota (sac fungi)	Sexual spores borne internally in sacs called asci
Basidiomycota (club fungi)	Sexual spores borne externally on club-shaped structures called basidia

Procedure 26.1

Examine a representative of phylum Chytridiomycota

1. Obtain a culture of *Allomyces* or *Chytridium*, two common genera of chytrids.
2. Prepare a wet mount and use your microscope to examine chytrid morphology.
3. Be prepared to compare chytrid morphology with the morphology common to the other major fungal groups.

Question 1

a. Are hyphae apparent?

b. Are the cells motile?

PHYLUM ZYGOMYCOTA (BREAD MOLDS)

Zygomycetes (750 species), which include common bread molds, derive their name from resting sexual structures called **zygosporangia** that characterize the group. Most zygomycetes are saprophytic and their vegetative hyphae lack septa (i.e., they are aseptate).

Procedure 26.2

Examine common bread mold

1. Obtain from your instructor a petri dish containing a piece of moldy bread that was moistened and exposed to air for a few days.
2. Note the velvet texture and various colors of the molds.
3. Use a stereomicroscope to examine the mycelia and notice that they grow as a tangled mass of hyphae.

Question 2

a. How many species of mold are on the bread?

b. Do any of the molds on the bread have hyphae modified as sporangiophores and sporangia?

c. Is pigment distributed uniformly in each mycelium? If not, where is the pigment concentrated in each mold?

d. What is the adaptive significance of spores forming on ends of upright filaments rather than closer to the protective substrate?

A common genus of bread mold is *Rhizopus*. Its hyphae are modified into **rhizoids** (holdfasts), **stolons** (connecting hyphae), and **sporangiophores** (asexual reproductive structures), as shown in figure 26.3. Sporangiophores are upright hyphal filaments supporting asexually reproducing sporangia (fig. 26.4). Within a sporangium, haploid nuclei become spores and are separated by cell walls. These spores are released to the environment when the sporangium matures and breaks open.

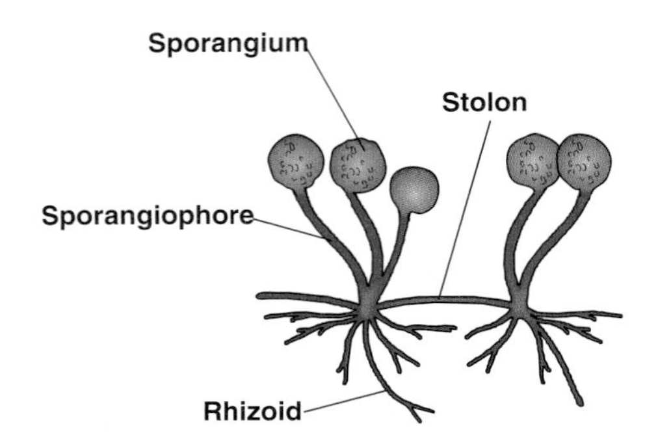

Figure 26.3

Vegetative and asexual reproductive structures of *Rhizopus*, a bread mold.

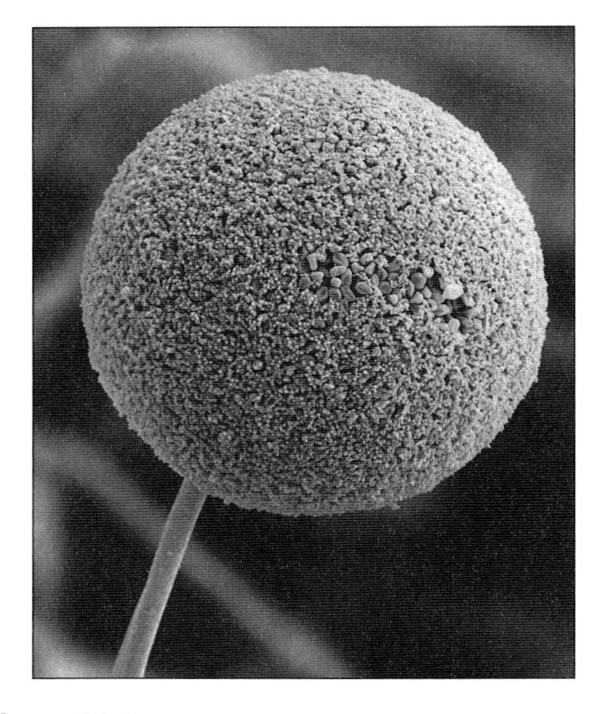

Figure 26.4

A sporangium of *Rhizopus stolonifer* (500×) releases thousands of spores when it matures and breaks open.

Procedure 26.3

Examine *Rhizopus*

1. Obtain a pure culture of *Rhizopus* from your instructor. This culture is growing in a sealed petri plate containing nutrient fortified agar. Do not remove the top of the dish because you may contaminate the culture and unnecessarily release spores into the room.
2. Note that *Rhizopus* is dark; hence, a common name of *Rhizopus* is black bread mold.
3. Your instructor has prepared a demonstration slide of *Rhizopus* stained with lactophenol cotton blue for you to examine. Examine this slide.
4. Your instructor has also prepared some cultures of *Rhizopus* (or *Mucor*, or *Phycomyces*) for observing the structures formed during sexual reproduction. Obtain one of these plates.
5. Because *Rhizopus* is isogamous (meaning that gametes from both strains look alike), the strains are not called male and female, but instead are called + and −. On one side of the plate a + strain of *Rhizopus* was introduced, and on the other side a − strain was introduced. These strains grew toward each other and formed reproductive structures called gametangia where they came into contact. Locate the area of the plate in which the strains have come in contact.
6. Review the description of sexual reproduction in *Rhizopus* listed below. Then locate gametangia and zygosporangia in the living culture.
7. Examine a prepared slide of *Rhizopus* with developed zygosporangia (figs. 26.5 and 26.6). Then observe the zygosporangia where the strains have touched. Be careful not to confuse zygosporangia with dark sporangia.

Sexual Reproduction in *Rhizopus*

1. Sexual reproduction begins when hyphae of each strain touch each other (fig. 26.5).
2. Where the hyphae touch, **gametangia** form and appear as swellings. Within the gametangia many nuclei differentiate to serve as gametes from each strain.
3. The wall between the gametangia breaks down, and the nuclei fuse to form zygotes. Fusion of nuclei (syngamy) occurs in the zygosporangium.
4. A typically massive and elaborate **zygosporangium** differentiates around the zygotes. Except for the zygotes, all other nuclei are haploid.
5. Soon after the zygosporangium forms, the zygotes undergo meiosis. One or more of the resulting haploid cells soon germinates.
6. The hyphae of the germinating cells break out of the zygosporangium, produce a sporangiophore, and form spores asexually.
7. The spores are released and produce a new generation of mycelia.

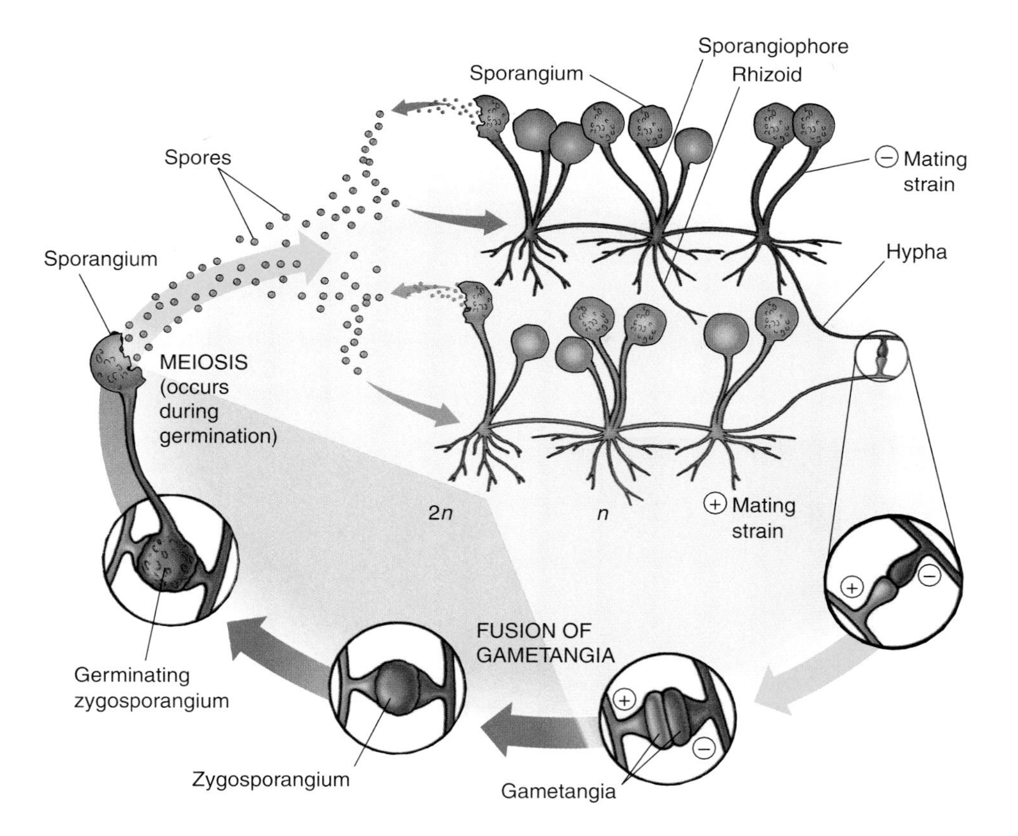

Figure 26.5

Life cycle of *Rhizopus*. Hyphae grow and feed on the surface of the bread or other material and produce clumps of erect, sporangium-bearing stalks. If both + and − strains are present in a colony, they may grow together, and their nuclei may fuse to form, a diploid (2N) zygote. This zygote, which is the only diploid cell of the life cycle, acquires a thick, black coat called a zygosporangium (zygospore). Meiosis occurs during its germination, and vegetative, haploid hyphae grow from the resulting haploid (1N) cells.

Question 3

a. In what structure is the dark pigment of *Rhizopus* concentrated?

b. Is *Rhizopus* reproducing sexually as well as asexually in the same petri dish? How can you tell?

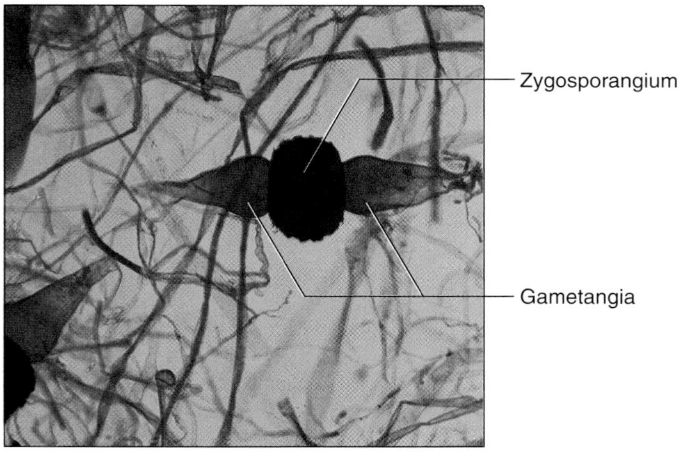

Zygosporangium

Gametangia

Figure 26.6

Sexual conjugation in *Rhizopus*.

PHYLUM ASCOMYCOTA (SAC FUNGI)

Phylum Ascomycota (30,000 species) includes yeasts, some molds, morels, and truffles (fig. 26.7). Its name is derived from a microscopic, sac-shaped reproductive structure called an **ascus.**

Ascomycetes reproduce asexually by forming spores called **conidia.** Modified hyphae called **conidiophores** partition nuclei in longitudinal chains of beadlike conidia (fig. 26.8). Each conidium contains one or more nuclei. Conidia form on the surface of conidiophores in contrast to spores that form within sporangia in *Rhizopus*. When mature, conidia are released in large numbers and germinate to produce new organisms. *Aspergillus* and *Pencillium* are common examples of fungi that form conidia.

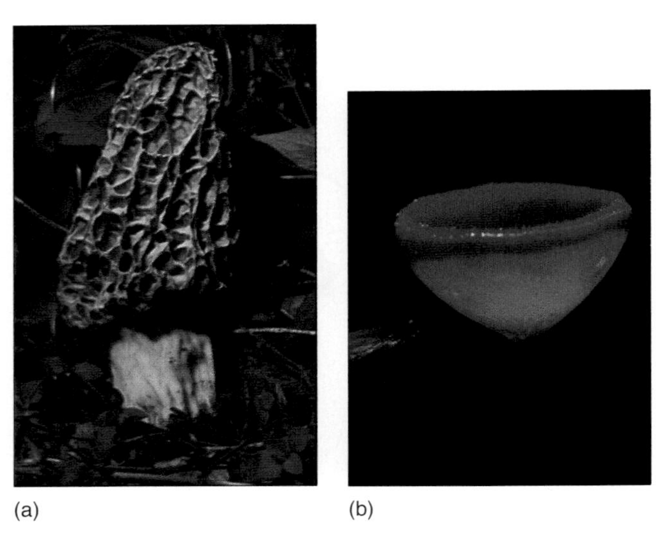

(a) (b)

Figure 26.7

Representatives of Ascomycota. All visible structures of fleshy fungi, such as the ones shown here, arise from an extensive network of filaments (hyphae) that penetrate and interweave with the substrate on which they grow. (*a*) A morel, *Morchella esculenta*, a delicious, edible ascomycete that appears in early spring in the northern temperate woods (especially under oaks). (*b*) A cup fungus in the rain forest of the Amazon Basin.

Many ascomycetes are economically important. For example, species of *Penicillium* are used to produce antibiotics. *Penicillium roquefortii*, which grows only in caves near Roquefort, France, gives a unique flavor to roquefort cheese. *Aspergillus oryzae* is used to brew Japanese saki and to enrich food for livestock.

Procedure 26.4

Examine fungi with conidia

1. Obtain a culture plate of living *Aspergillus*, *Penicillium*, or both. Notice the soft texture of the colonies.
2. Use a dissecting microscope to examine the colonies' hyphae and their reproductive conidia. Note the rounded tufts of these reproductive cells.
3. Conidia are quite small. You will examine them more closely in procedure 26.5.

Procedure 26.5

Examine *Penicillium*

1. Examine a prepared slide of *Penicillium* (fig. 26.9) and a pure, living culture provided by your instructor.
2. Notice the formation of conidia.

Question 4

What is the relative size of *Penicillium* hyphae compared with *Rhizopus* hyphae?

Yeasts are common unicellular ascomycetes and include about 40 genera. Most of their reproduction is asexual by cell fission or budding (i.e., the formation of a smaller

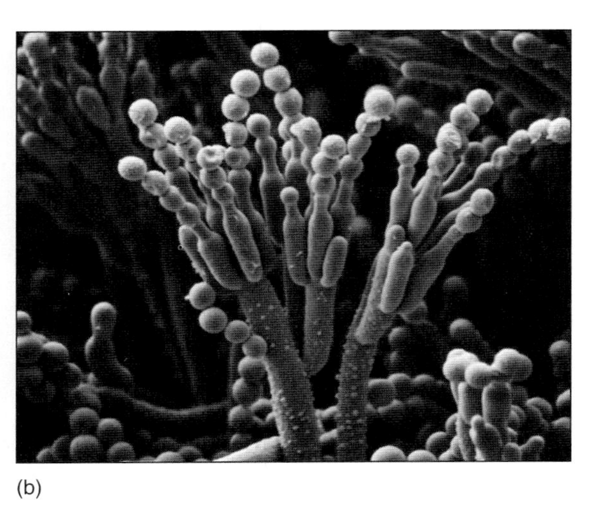

(a) (b)

Figure 26.8

Conidiophores. (*a*) Characteristic conidiophores of an ascomycete, as viewed with a scanning electron microscope. (*b*) A colony of *Penicillium*, another important genus of Fungi Imperfecti, growing on an orange and forming conidia.

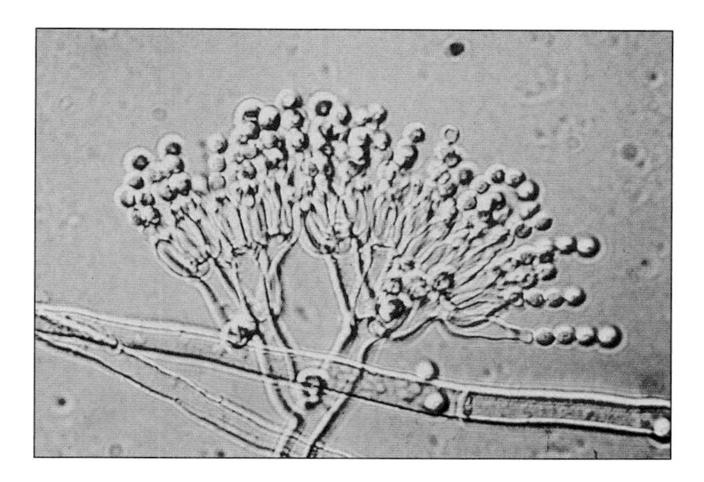

Figure 26.9
Light micrograph of conidiophores of *Penicillium* (790×). A scanning electron micrograph of the conidiophores is shown in figure 26.8*b*.

cell from a larger one). Occasionally, two sexually reproducing yeast cells will fuse to form one cell with two nuclei. This cell functions as an ascus in which syngamy is followed immediately by meiosis. The resulting ascospores function directly as new yeast cells. Yeasts do not form conidia. The yeast used to produce wine and beer is usually a strain of *Saccharomyces cerevisiae*. The fungi growing naturally on grapes used for making wine may impart a unique flavor to a wine more than does the specific variety of grapes.

Procedure 26.6

Examine *Saccharomyces*, a yeast, and *Peziza*, a cup fungus

1. Obtain and examine a stock culture of *Saccharomyces* (fig. 26.10).
2. Prepare a wet mount of the yeast from a culture dispensed by your instructor. Only a small amount of yeast is needed to make a good slide.
3. Review the description of sexual reproduction in cup fungi.
4. Examine a prepared slide of a cross section through the ascocarp of *Peziza* (fig. 26.11). Locate the asci.

Question 5

a. Do you see chains of yeast cells produced by budding?

b. How is the structure of yeast hyphae different from that of molds?

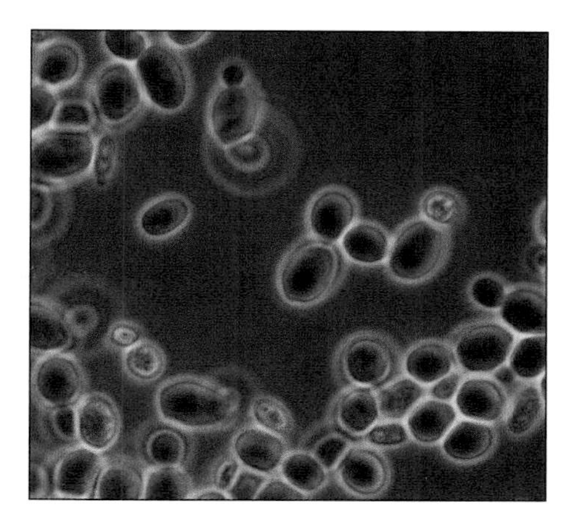

Figure 26.10
Light micrograph of yeast.

Figure 26.11
Asci from the lining of a cup of *Peziza*, a cup fungus. The diploid (2N) zygote in each ascus divides meiotically to produce four haploid (N) nuclei. Each of these nuclei then divides mitotically. These meiotic and mitotic divisions result in a column of eight ascospores in each ascus.

Sexual Reproduction in Ascomycota

1. Sexual reproduction begins with contact of monokaryotic hyphae from two mating strains (fig. 26.12).
2. Where the hyphae touch, large multinucleate swellings appear and eventually fuse. Haploid nuclei of the two strains intermingle in the swelling.
3. A **dikaryotic** mycelium grows from this swelling. Each dikaryotic cell has one nucleus from each parent; the nuclei do not fuse immediately.
4. Tightly bundled dikaryotic hyphae grow and mingle with **monokaryotic** hyphae from each parent to form a cup-shaped **ascocarp.**

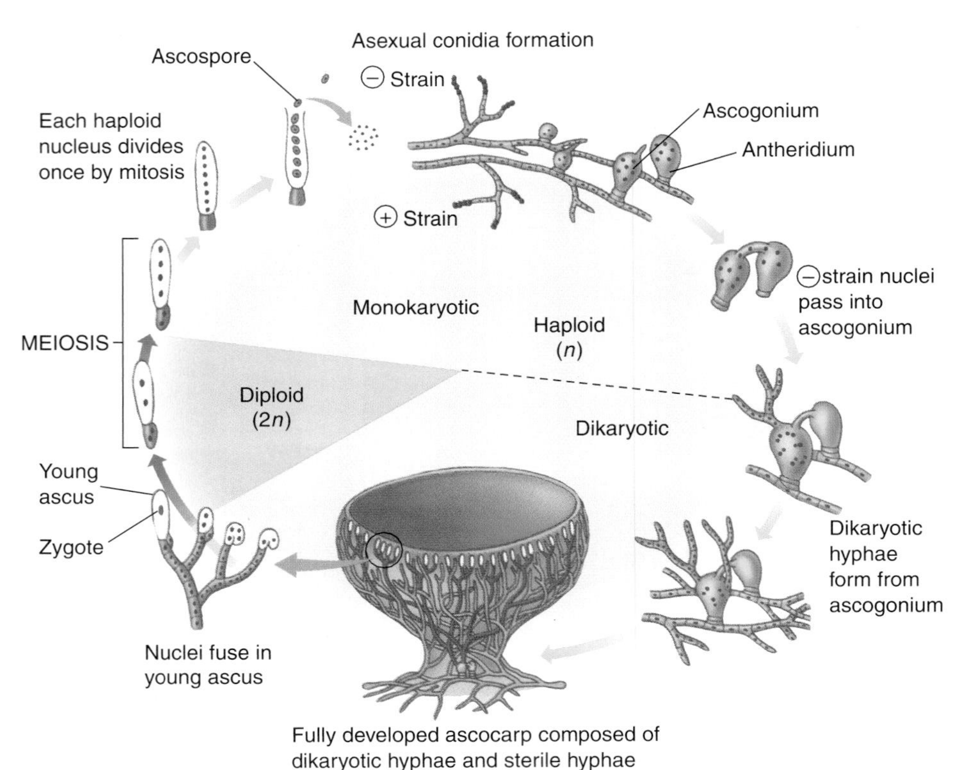

Figure 26.12
The sexual life cycle of an ascomycete.

5. Dikaryotic cells lining the inside of the ascocarp form sac-shaped asci (sing., *ascus*).
6. The nuclei fuse (syngamy) in each ascus to form a zygote.
7. After fusion, meiosis produces four haploid **ascospores.**
8. Subsequent mitosis produces eight ascospores within each mature ascus.
9. The asci on the surface of the ascocarp rupture and release ascospores into the environment.
10. Each ascospore can produce a new mycelium.

Question 6
What is the difference between dikaryotic and diploid cells?

PHYLUM BASIDIOMYCOTA (CLUB FUNGI)

Basidiomycetes (25,000 species) are probably the most familiar fungi (fig. 26.13). They include mushrooms, puffballs, shelf fungi, and economically important plant pathogens such as rusts and smuts. *Agaricus campestris* is a common field mushroom, and its close relative *A. bisporus* is cultivated for 60,000 tons of food per year in the United States. However, just one bite of *Amanita phaloides*, the "destroying angel," may be fatal.

Procedure 26.7

Examine some common mushrooms and their relatives

1. Examine a specimen of an earthstar (*Geaster*). Earthstars are oddly structured basidiomycetes with an array of support structures shaped much like a star (fig. 26.13c).
2. Examine some mushrooms. Mushrooms are familiar examples of the aboveground portions of extensive mycelia permeating the soil. Note the mushroom's **cap** and **pileus.**
3. Find the **gills** on the undersurface of the cap. As you will later see, gills are lined with microscopic, club-shaped cells called basidia where sexual reproduction occurs. Phylum Basidiomycota is sometimes called the "club fungi" and derives its name from these characteristic basidia.

(a) (b) (c)

Figure 26.13

Representative basidiomycetes. (*a*) Fly amanita (*Amanita muscaria*). Many species of *Amanita* are highly poisonous. The cap of this young specimen is not fully opened. (*b*) A common stinkhorn fungus (*Phallus impudicus*). (*c*) Earthstar (*Geaster*).

Sexual Reproduction in Basidiomycota

1. Haploid hyphae from different mating strains permeate the substrate (fig. 26.14).
2. Septa form between the nuclei in the hyphae and form **monokaryotic primary mycelia.** A monokaryotic mycelium has one nucleus in each cell.
3. Cells of the primary mycelia of different mating strains touch, fuse, and produce a **dikaryotic secondary mycelium.**
4. The secondary mycelium grows in the substrate.
5. The dikaryotic hyphae of the secondary mycelium eventually coalesce and protrude above the substrate as a tight bundle of hyphae called a basidiocarp (mushroom).
6. The basidiocarp forms a pileus, cap, and gills.
7. Dikaryotic, club-shaped basidia form on the surface of the gills.
8. The two nuclei in each basidium fuse to form a diploid zygote.
9. The zygote soon undergoes meiosis and produces haploid **basidiospores.**
10. The basidiospores are released from the basidia lining the gills and are dispersed by wind.
11. A basidiospore germinates into a new mycelium.

Procedure 26.8

Examine *Coprinus*, a common mushroom

1. Examine a prepared slide of gills from the cap of *Coprinus*, a common mushroom (fig. 26.15).
2. Note the dark basidiospores in rows along the surface of the gills. Interestingly, the gills form perpendicular to the ground and allow spores to free-fall and disperse. Recent research suggests that gravity influences the orientation of gills.

Question 7

How many spores would you estimate are present on the gills of a single cap of *Coprinus?* Remember that a prepared slide shows only a cross-section.

LICHENS

Lichens (25,000 species) are common, brightly colored organisms in which an ascomycete (rarely other fungi) lives symbiotically with a photosynthetic alga or cyanobacterium. **Symbiosis** means living in a close and sometimes dependent association. About 26 genera of algae occur in different species of lichens. However, the green algae *Trebouxia* and *Trentepohlia* and the cyanobacterium *Nostoc* are in 90% of lichen species.

Lichens reproduce asexually by releasing fragments of tissue or specialized, stress-resistant packets of fungal and algal cells. Each of the two components (fungus and alga) may reproduce sexually by mechanisms characteristic of their phylum, and the new organisms may continue the lichen association.

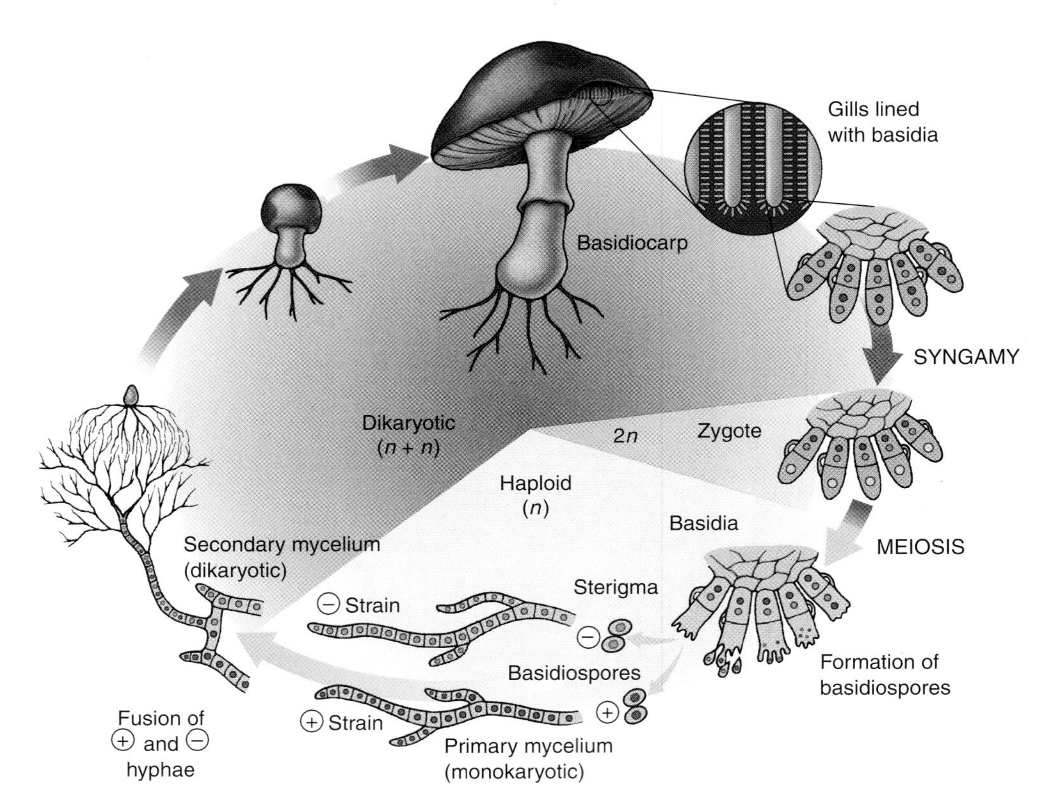

Figure 26.14

The sexual life cycle of a mushroom. Coalescing mycelia from compatible strains produce dikaryotic hyphae, which form the familiar fruiting bodies of mushrooms. On the gills of the mushroom cap the nuclei in dikaryotic cells fuse, undergo meiosis, and produce basidiospores that germinate into a new mycelium. The basidium is the site of syngamy.

The durable construction of fungi, linked with the photosynthetic properties of algae, enables lichens to proliferate into the harshest terrestrial habitats. Lichens have three basic growth forms: **crustose, foliose,** and **fruticose.** The thallus of crustose lichens grows close to the surface of a hard substrate such as rock or bark. The lichen is flat and two-dimensional. Foliose lichens adhere to their substrate, but some of the thallus peels and folds away from the substrate in small sheets. Fruticose lichens are three-dimensional and often grow away from the substrate with erect stalks. The tips of the stalks are often sites of ascus formation by the sexually reproducing ascomycete symbiont.

Lichens are extremely sensitive to air pollution. This is probably because they are adapted to efficiently absorb nutrients and minerals from the air. This makes lichens particularly susceptible to airborne toxins.

Procedure 26.9

Examine lichens

1. Examine a prepared slide of a cross-section of a lichen thallus and note the intimate contact between the symbionts (fig. 26.16).
2. Examine the dried lichens on display and note the three basic growth forms: **crustose, foliose,** and **fruticose** (fig. 26.17).

Question 8

a. What is the value of photosynthetic algae to the growth of a fungus in a lichen?

b. Would you expect lichens to grow best in rural or urban environments? Why?

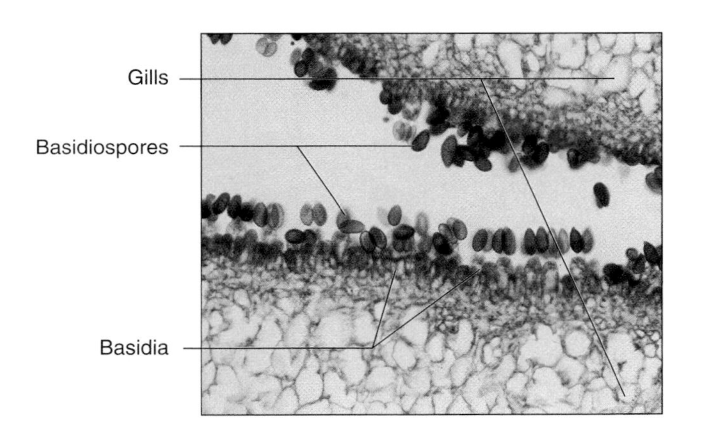

Figure 26.15
Gills of the mushroom *Coprinus*, a basidiomycete.

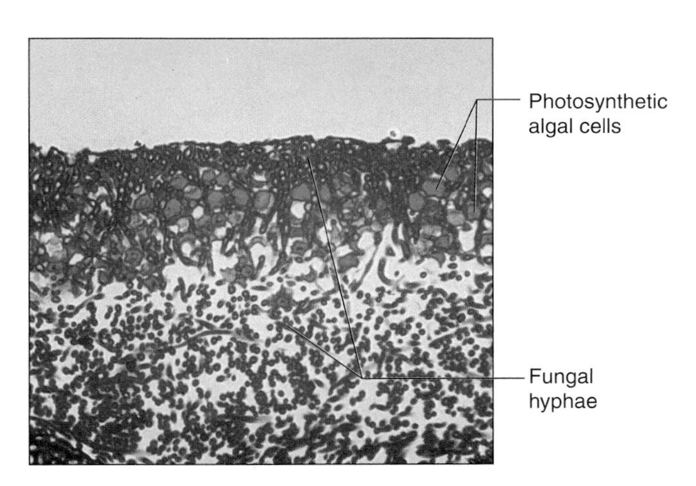

Figure 26.16
Cross-section of a lichen thallus. Lichens consist of algae and fungi growing symbiotically.

Figure 26.17
Growth forms of lichens. (*a*) Crustose lichens. (*b*) Fruticose lichen. (*c*) Foliose lichen.

Questions for Further Thought and Study

1. Mushrooms often sprout from soil in rows or circles that are commonly called fairy rings. How would you explain this formation?

2. What advantages does asexual reproduction have over sexual reproduction?

3. Does dominance of the haploid condition in a fungal life cycle offer an adaptive advantage? Why or why not?

4. Compare and contrast the structure of a fungal mycelium with the structure of a filamentous alga.

5. What is the advantage of maintaining a dikaryotic condition rather than immediate nuclear fusion?

6. For fungi the only distinction between a spore and a gamete is function. Explain.

7. How do fungi affect your life?

DOING BIOLOGY YOURSELF

Wine-making is a multimillion dollar industry, but the biology of wine-making is simple. Make your own special brand of wine by following the instructions in Exercise 11. Good luck!

WRITING TO LEARN BIOLOGY

Compare and contrast the fundamental life cycle of a fungus with that of plants and animals. When does meiosis occur in the sequence of events? Which stages are haploid and which are diploid?

27

Survey of the Plant Kingdom

Liverworts, Mosses, and Hornworts of Phyla Hepaticophyta, Bryophyta, and Anthocerophyta

Objectives

By the end of this exercise you should be able to:

1. Describe the life histories and related reproductive structures of bryophytes.
2. Describe the distinguishing features of mosses, liverworts, and hornworts.

The plant kingdom comprises a remarkably diverse group of multicellular organisms. With few exceptions, plants are autotrophic, contain chlorophyll *a*, and have cell walls containing cellulose. Life cycles of all members of the plant kingdom are variations on alternation of generations. Be sure to review in your textbook this generalized life cycle and be familiar with the major stages presented in figure 27.1.

The major groups of plants that you will examine in this and the next three laboratory exercises are bryophytes (mosses, liverworts, and hornworts), ferns and fern allies, gymnosperms, and angiosperms. These groups of plants are distinguished by morphology, life cycle, and the presence or absence of vascular tissues. Pay special attention to variations in each of these characteristics as you survey the plant kingdom in upcoming weeks.

Bryophytes include liverworts, mosses, and hornworts, and are the most primitive group of terrestrial plants. Bryophytes are green, have rootlike structures called **rhizoids,** and may have stem and leaflike parts. Bryophytes do not generally possess specialized vascular tissues, which transport materials between roots and shoots. This lack of developed vascular tissues in bryophytes typically limits their distribution to moist habitats, because their rhizoids neither penetrate the soil very far nor absorb many nutrients. Also, the lack of vascular tissues necessitates that their photosynthetic and nonphotosynthetic tissues be close together. Because vascular tissues, along with supporting tissues, are often absent, bryophytes are relatively small and inconspicuous. Despite their diminutive size, however, bryophytes occur throughout the world in habitats ranging from the tropics to Antarctica.

There are approximately 24,000 species of bryophytes, more than any other group of plants except the flowering plants. Bryophytes fix CO_2, degrade rocks to soil, stabilize soil, and reduce erosion. Humans have used bryophytes in a number of ways, including as a fuel, in the production of Scotch whiskey, and as packing materials.

The plant body of bryophytes is called a **thallus** (pl., *thalli*). Liverwort thalli are flattened dorsoventrally (from back and front plane, rather than from side to side plane) and are bilaterally symmetrical (i.e., have two equal halves). For comparison, moss thalli are erect and radially symmetrical (circular). Hornwort thalli are similar to those of liverworts.

The life cycle of bryophytes is characterized by a distinct alternation of generations in which the gametophyte is the predominant vegetative phase (fig. 27.2). Bryophytes have multicellular sex organs in which gamete-producing cells are enclosed in a jacket of sterile cells. **Antheridia** are male sex organs that produce swimming, biflagellate sperm. Bryophytes require free water for sexual reproduction because their sperm must swim to eggs. These sperm fertilize eggs produced in **archegonia,** the female sex organs. The fertilized egg is called a **zygote.** This zygote divides and matures in the **archegonium** to produce the **sporophyte,** which remains attached to and nutritionally dependent on the **gametophyte.** The mature sporophyte produces haploid spores (via meiosis), each of which can develop into a gametophyte.

PHYLUM HEPATICOPHYTA: LIVERWORTS

Liverworts are the earliest land plants. Although many liverworts are "leafy," we will restrict our observations to a thallus-type liverwort, *Marchantia*. The gametophytic thallus of this liverwort grows as a large, flat, photosynthetic structure on the surface of the ground (fig. 27.3).

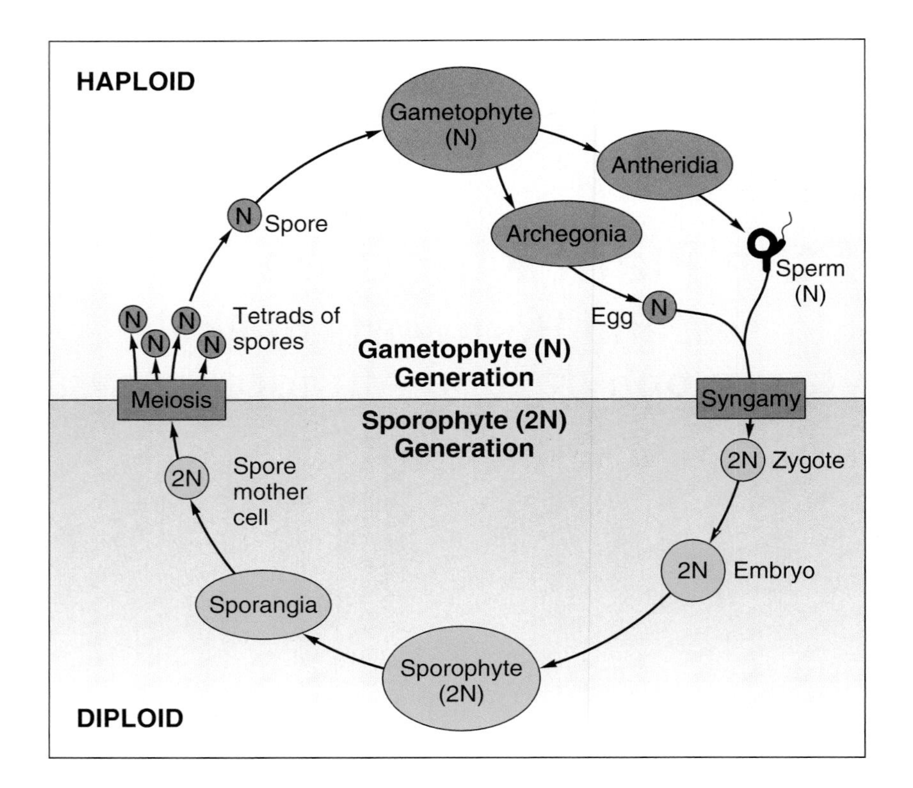

Figure 27.1

In a generalized plant life cycle, gametophytes, which are haploid (N), alternate with sporophytes, which are diploid (2N). Antheridia (male) and archegonia (female) are the sex organs (gametangia) produced by the gametophyte; they produce sperm and eggs, respectively. The sperm and egg fuse during syngamy (fertilization) to produce the first diploid cell of the sporophyte generation, the zygote. Meiosis occurs within the sporangia, which are the spore-producing organs of the sporophyte. The resulting spores are haploid and are the first cells of the gametophyte generation.

Liverwort Gametophyte

Procedure 27.1

Examine the thallus of *Marchantia*

1. Observe some living *Marchantia* and note the Y-shaped (dichotomous) growth. Rhizoids extend downward from the lower (ventral) surface of the thallus.

Question 1
What are the functions of rhizoids?

2. View the upper (dorsal) surface of the thallus with a dissecting microscope and note the pores in the center of the diamond-shaped areas. Obtain a prepared slide of a thallus of *Marchantia* and compare what you see with figure 27.4. Notice that the pores in the dorsal surface of the thallus overlie air chambers containing **chlorenchyma** (chloroplast-containing) cells.

Question 2
What is the function of these pores?

Asexual Reproduction in Liverworts

Liverworts can reproduce asexually via fragmentation. In this process, the older, central portions of the thallus die, leaving the growing tips isolated to form individual plants.

Structures called **gemmae cups** occur on the dorsal surface of some thalli near the midrib (fig. 27.5). Gemmae cups represent another means of asexual reproduction by liverworts. Inside the gemmae cups are lens-shaped outgrowths called **gemmae** (sing., *gemma*), which are splashed out of the cup by falling drops of rain. If a gemma lands in an adequate environment, it can produce a new gametophyte plant. Examine a prepared slide of gemmae cups. Also examine available live or preserved material. In the space below, diagram and label what you see, and compare it to figure 27.5.

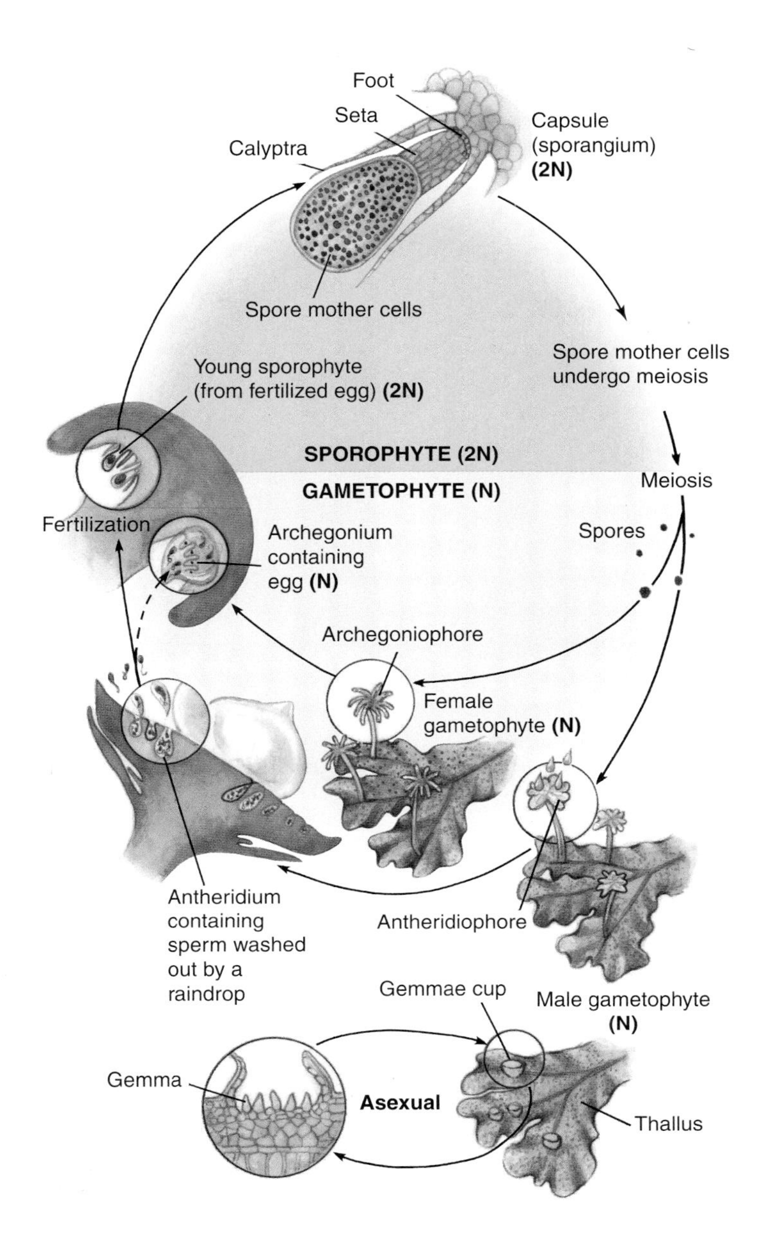

Figure 27.2

Life cycle of *Marchantia*, a liverwort. During sexual reproduction, spores produced in the capsule germinate to form independent male and female gametophytes. The archegoniophore produces archegonia, each of which contains an egg; the antheridiophore produces antheridia, each of which produces many sperm. After fertilization, the sporophyte develops within the archegonium and produces a capsule with spores. *Marchantia* reproduces asexually by fragmentation and gemmae.

Sexual Reproduction in Liverworts

Many species of *Marchantia* are **dioecious,** meaning that they have separate male and female plants. Gametes from each plant are produced in specialized sex organs borne on upright stalks (fig. 27.3). **Archegoniophores** are specialized stalks on female plants that bear archegonia. Each flask-shaped archegonium consists of a **neck** and a **venter,** which contains the **egg** (fig. 27.6*a*). **Antheridiophores** are specialized stalks on male plants that bear antheridia (fig. 27.3). Sperm form in antheridia (fig. 27.6*b*). Flagellated sperm are released and washed from the antheridia during wet conditions and eventually fertilize the egg, which is located in the venter. The zygote remains in the venter and grows into a sporophyte plant.

Procedure 27.2

Examine archegonia and antheridia of liverworts

1. Examine living or prepared liverworts having mature archegoniophores that bear archegonia. Archegonia

(a)

(b)

Figure 27.3

Marchantia. The flat, leafy thallus of this liverwort grows close to the ground. (*a*) A thallus bearing upright male reproductive structures called antheridiophores. (*b*) A thallus bearing upright female reproductive structures called archegoniophores.

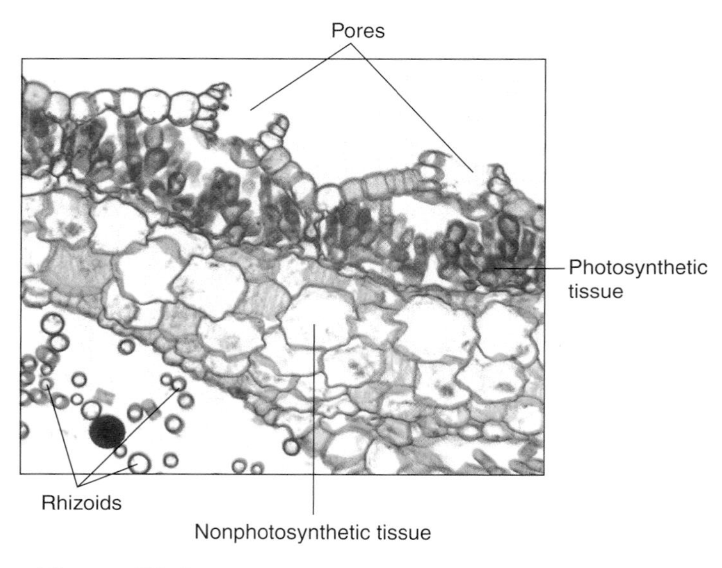

Figure 27.4

Section of *Marchantia* gametophyte showing pores, rhizoids, photosynthetic tissue, and nonphotosynthetic tissue.

at various stages of development are located on the ventral surface.

2. Locate an egg in an archegonium. Notice a pattern of evolution in plants as well as in animals—eggs are larger and fewer in number, while sperm cells are smaller but greater in number.
3. Examine living or preserved liverworts with mature antheridiophores bearing antheridia.
4. Examine a prepared slide of cross sections of an antheridiophore. Antheridia are located just below the upper surface of the disk in a chamber that leads to the surface of the disk through a pore.

Question 3

How do the positions of the archegonium and antheridium relate to their reproductive function?

Liverwort Sporophyte

Procedure 27.3

Examine sporophytes of liverworts

1. Examine a prepared slide of a sporophyte of *Marchantia*. The nonphotosynthetic sporophyte is connected to the gametophyte by a structure called the **foot. Spores** are produced by meiosis in a **capsule** located on a **stalk** that extends downward from the foot.
2. Locate elongate cells called **elaters** among the spores. Elaters help disperse spores by twisting. In humid

(a)

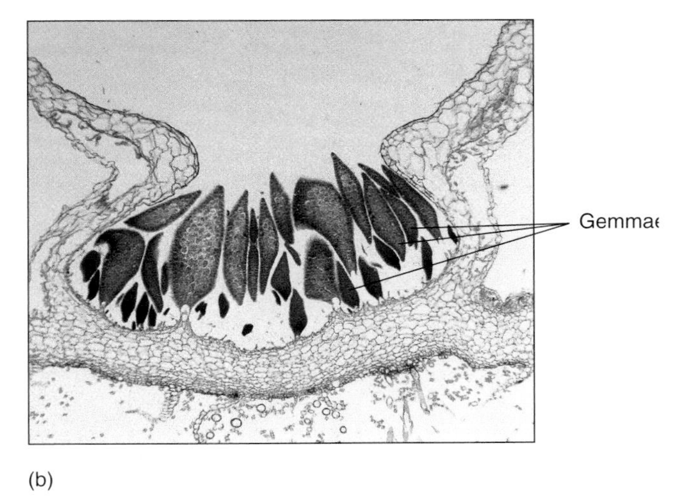

(b)

Figure 27.5
(*a*) Gemmae cups ("splash cups") containing gemmae on the gametophytes of a liverwort. Gemmae are splashed out of the cups by raindrops and can then grow into new gametophytes, each identical to the parent plant that produced it by mitosis. (*b*) Longitudinal section of a gemmae cup (10×).

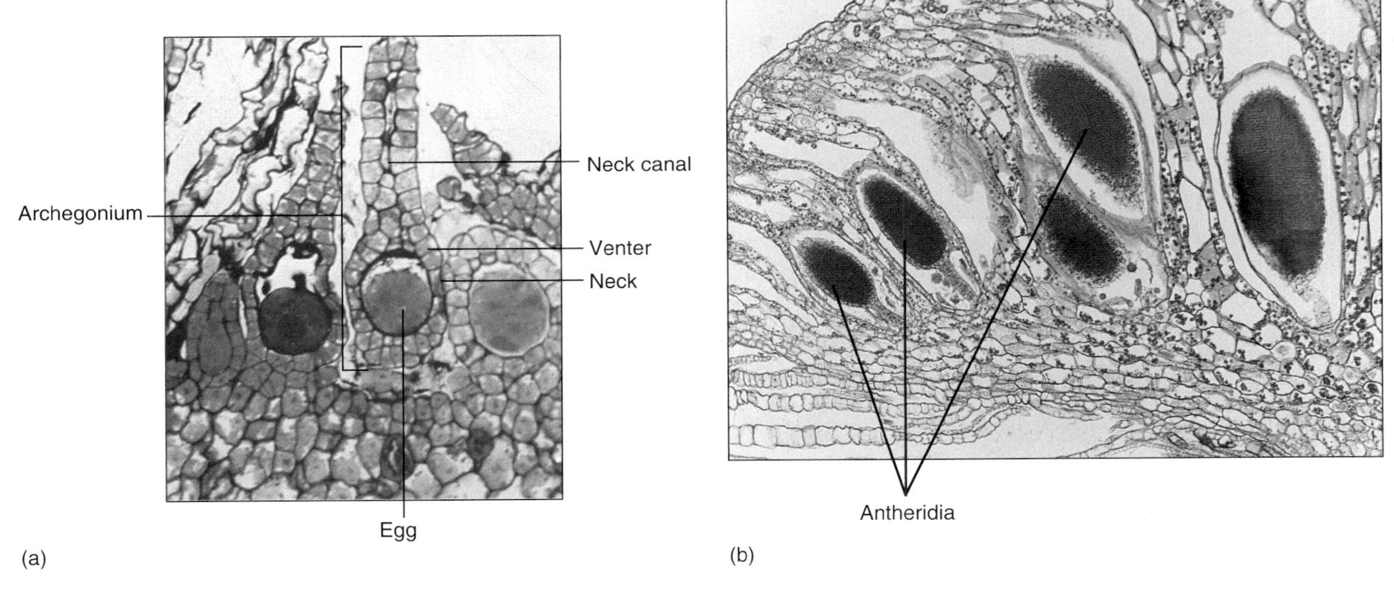

Figure 27.6
(*a*) Section through an archegonium of *Marchantia*. A single egg differentiates within the archegonium. (*b*) Section through an antheridiophore, showing individual antheridia (100×).

conditions the elaters coil, but when it is dry the elaters expand, pushing the spores apart and rupturing the spore case to release the spores.

3. Understand that gamete release, fertilization, spore release, and germination are most efficient in individually specific environmental conditions.

Question 4

a. What is the function of the foot?

b. Are the spores haploid or diploid?

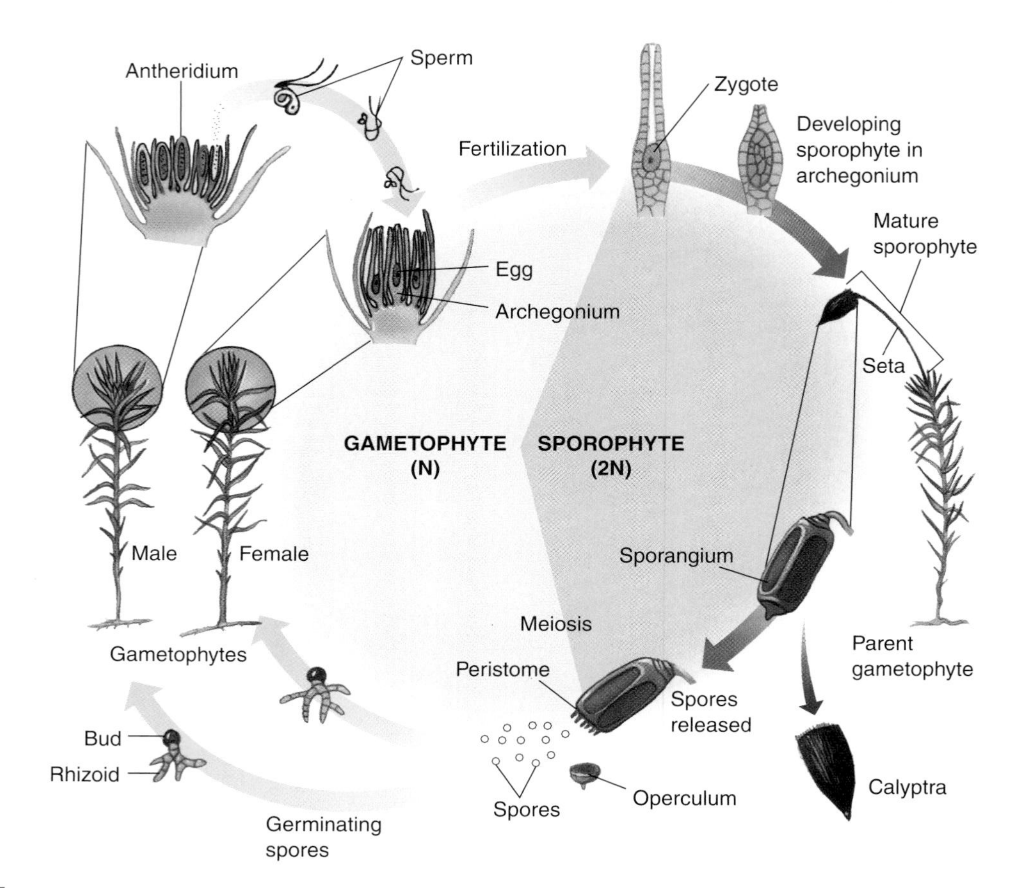

Figure 27.7

Moss life cycle. Haploid (1N) sperm are released from antheridia on the male gametophytes. The sperm then swim through water to the archegonia and down their neck to fertilize the egg. The resulting diploid (2N) zygote develops into a diploid sporophyte. The sporophyte grows out of the archegonium and differentiates into a slender seta with a swollen capsule at its apex. The capsule is covered with a cap, or calyptra, formed from the swollen archegonium. The sporophyte grows on the gametophyte and eventually produces spores after meiosis. The spores are shed from the capsule after a specialized lid—the operculum—drops off. The spores germinate, giving rise to gametophytes. The gametophytes initially grow along the ground. Ultimately, buds produce leafy gametophytes.

c. What is the functional significance of the response of elaters to moisture?

PHYLUM BRYOPHYTA: MOSSES

Mosses are often more visible than liverworts because of their greater numbers, more widespread distribution, and because gametophyte plants of mosses are leafy and usually stand upright. Mosses also withstand desiccation better than do liverworts. Consequently, mosses often grow in a greater diversity of habitats than do liverworts. The moss gametophyte is radially symmetrical and is the most conspicuous phase of the moss life cycle (fig. 27.7).

Moss Gametophyte

Procedure 27.4

Examine mosses

1. Observe the living moss on display called *Polytrichum* (fig. 27.8). The "leafy" green portions of the moss are the gametophytes and are often only one cell thick (except at the midrib).
2. Make a wet mount of a single leaflet and examine it with low magnification.

Question 5

a. How many cells thick is the leaflet?

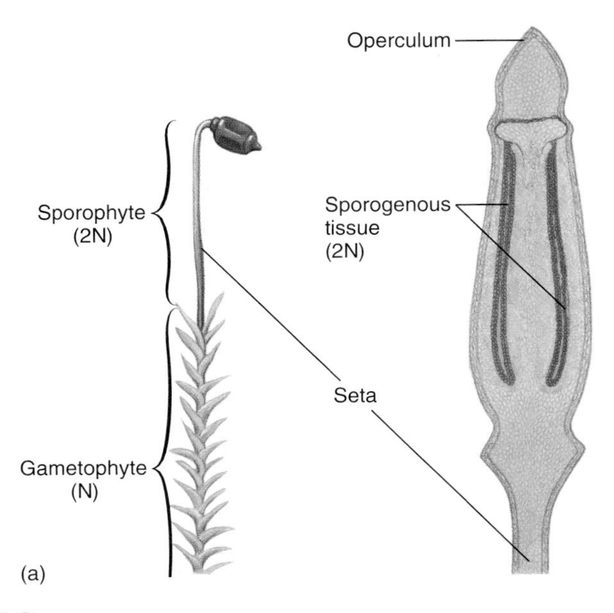

Figure 27.8
The structure of a moss. (*a*) Diagram of the parts of a mature mass sporophyte. (*b*) A hair-cup moss, *Polytrichum*. The leaves belong to the gametophyte. Each of the yellowish-brown stalks, with the capsule at its summit, is a sporophyte. Although moss sporophytes may be green and perform a limited amount of photosynthesis when they are immature, they are soon completely dependent, in a nutritional sense, on the gametophyte.

b. Is there a midrib vein?

c. Are stomata or pores visible on the leaf surface?

d. How does the symmetry of a moss gametophyte compare with that of a liverwort gametophyte?

Moss gametophytes have specialized cells that aid in the absorption and retention of water. Mats of moss act, in effect, like sponges. The following procedure demonstrates the water-absorbing potential of mosses.

Procedure 27.5

Water absorption by moss

1. Weigh 3 g of *Sphagnum*, a peat moss, and 3 g of paper towel.
2. Add the moss and towel to separate beakers each containing 100 mL of water.
3. After several minutes, remove the materials from the beaker.
4. Measure the amount of water left in each beaker by pouring the water into a 100-mL graduated cylinder. Remember that 1 mL of water weighs 1 g.
5. Record your data.

Question 6

a. How many times its own weight did the moss absorb?

b. How does this compare with the paper towel?

c. Why is *Sphagnum* often used to ship items that must be kept moist?

Asexual Reproduction in Mosses

Unlike liverworts, mosses lack structures such as gemmae for asexual reproduction. Mosses reproduce asexually by fragmentation.

Sexual Reproduction in Mosses

Most mosses, like liverworts, are dioecious. Archegonia or antheridia are borne either on tips of the erect gametophyte stalks or as lateral branches on the stalks. The apex of stalks of the female plant (the plant that bears archegonia) appears as a cluster of leaves, with the archegonia buried inside.

Procedure 27.6

Examine archegonia and antheridia of mosses

1. Examine living or preserved mosses having mature archegonia.
2. Examine a prepared slide of moss archegonia (fig. 27.9). Note the canal that leads through the neck and terminates in the venter of the archegonium. When the archegonium matures, cells lining the neck disintegrate and form a canal leading to the egg. Sperm, following a chemical attractant released by the archegonium, swim through this canal to reach the egg.

Question 7
Where is the egg located in the archegonium?

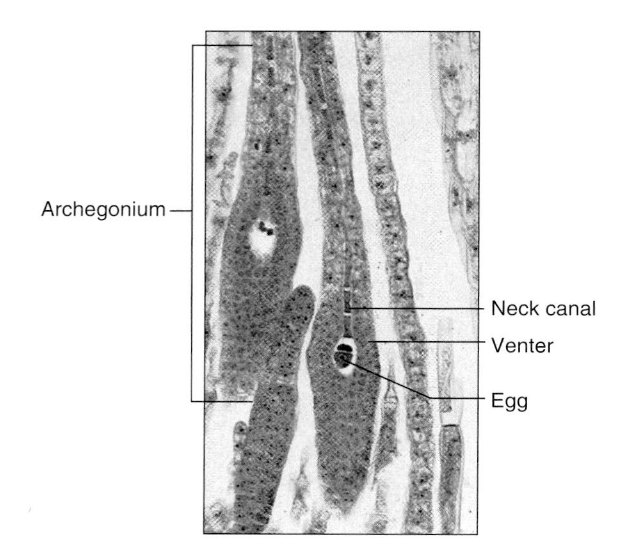

Figure 27.9
A longitudinal section through the tip of a female gametophyte of a moss.

3. Examine living or preserved mosses having mature antheridia. The male plant (i.e., the plant that bears antheridia) has a platelike structure on the tip with the "leaves" expanding outward to form a rosette. This structure is sometimes called a "moss flower" because of its appearance or a "splash cup" because of its function (i.e., the dispersal of sperm by falling raindrops).
4. Examine a prepared slide of moss antheridia, which appear as elongate, saclike structures (fig. 27.10).
5. Locate the outer sterile jacket and the inner mass of cells destined to become sperm.

Question 8
Are sperm haploid or diploid?

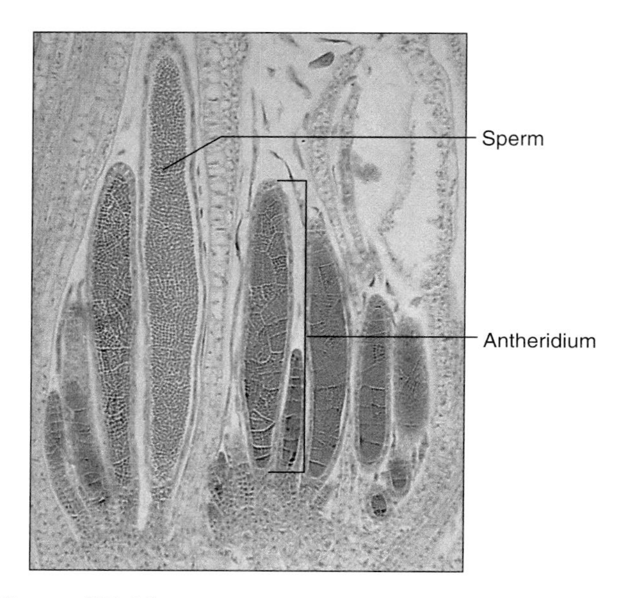

Figure 27.10
A longitudinal section through the tip of a male gametophyte of a moss.

Moss Sporophyte

Moss sporophytes consist of **capsules** located atop stalks, or **seta,** that extend upward the moss gametophyte (fig. 27.8*a*). A sporophyte is attached to the gametophyte by a structure called a **foot.**

Question 9
Is the sporophyte more prominent in mosses or liverworts?

The capsule atop the seta is covered by the **calyptra,** which is the upper portion of archegonium that covers the apex of the capsule. The calyptra falls off when the capsule matures. Inside the capsule are numerous haploid spores formed by meiosis.

Question 10
What is the adaptive significance of the seta of the sporophyte growing well above the mat of the gametophytes?

If enough living moss is available in the lab, remove the calyptra from a sporophyte capsule. On the tip of the capsule is a caplike structure called the **operculum.** Remove the operculum and notice the hairlike teeth lining the opening of the capsule. These teeth help control the release of spores from the capsule. In wet weather these teeth bend inward and prevent release of the spores. In dry weather they bend outward, facilitating distribution of spores by the wind.

Crush the capsule in water on a microscope slide and note the large number of spores that are released.

Question 11
a. What process produces spores?

b. Is the capsule haploid or diploid?

Moss spores germinate and form a photosynthetic **protonema,** which resembles a branching, filamentous alga. Leafy moss plants arise from "buds" located along the protonema.

Question 12
Can you think of any evolutionary implications of the similarity between a moss protonema and a filamentous green alga?

Figure 27.11
Anthoceros, a hornwort.

PHYLUM ANTHOCEROPHYTA: HORNWORTS

The hornworts are the smallest group of bryophytes; there are only about 100 species in six genera. Hornworts have several features that distinguish them from most other bryophytes. The sporophyte is shaped like a long, tapered horn that protrudes from a flattened thallus. Also, archegonia are not discrete organs. Rather, they are embedded in the thallus and are in contact with surrounding vegetative cells.

The most familiar hornwort is *Anthoceros,* a temperate genus (fig. 27.11). Examine some living and preserved *Anthoceros.* Locate the gametophyte thallus and the rhizoids extending from its lower surface. Also locate the hornlike sporophytes extending from the upper surface. Spores are produced in the horn of the sporophyte. If prepared slides are available, locate spores in various stages of development within the sporophyte.

DOING BIOLOGY YOURSELF

Mosses and liverworts are surprisingly diverse. Make a simple collection of living mosses and liverworts from two or three sites. Bring the collection to class and compare yours with those of other students. Use reference books in your lab or library to identify the plants that you've collected.

WRITING TO LEARN BIOLOGY

Because water is required for the swimming sperm to reach the archegonium, would you say that this means that bryophytes are not truly land plants? Why or why not?

Questions for Further Thought and Study

1. List advances in complexity of bryophytes over algae regarding their morphology, habitat, asexual reproduction, and sexual reproduction.

2. What event begins the sporophyte phase of the life cycle? Where does this event occur in mosses and liverworts?

3. What event begins the gametophyte phase of the life cycle? Where does it occur in mosses and liverworts?

4. What features distinguish a moss from a liverwort?

5. Diagram the life cycle of a liverwort, indicating which stages are sporophytic and which are gametophytic.

6. Diagram the life cycle of a moss, indicating which stages are sporophytic and which are gametophytic.

7. How did liverworts obtain their name?

8. What ecological roles do mosses, liverworts, and hornworts play in the environment?

9. Is the sporophyte of mosses ever independent of the gametophyte? Explain.

10. Why do you think that bryophytes are sometimes referred to as the amphibians of the plant kingdom?

28

Survey of the Plant Kingdom

Seedless Vascular Plants of Phyla Pterophyta and Lycophyta

Objectives

By the end of this exercise you should be able to:

1. Discuss similarities and differences between ferns and other plants you have studied in the lab.
2. Describe the life cycles of ferns and their allies.
3. Describe the distinguishing features of true ferns, club mosses, whisk ferns, and horsetails.

Seedless vascular plants include two phyla of nonflowering plants having a vascular system of fluid-conducting xylem and phloem: phylum Pterophyta (true ferns, whisk ferns, and horsetails) and phylum Lycophyta (club mosses) (table 28.1). Their vascular system connects true leaves, roots, and stems. In these plants, the sporophyte is the dominant phase of the life cycle (fig. 28.1).

All ferns and fern allies possess **sporophylls** (*sporo* = spore forming, *-phyll* = leaf). Sporophylls are leaflike structures of the sporophyte generation that bear spores. They may be large **megaphylls** (*mega* = large, *-phyll* = leaf) with several to many veins (as in the megaphylls of true ferns) or they may be smaller **microphylls** (*micro* = small, *-phyll* = leaf) with one vein (as in whisk ferns, scouring rush, and club mosses). **Sporangia,** which form on sporophylls, are where spores are produced by meiosis (review fig. 27.1). Sporangia occur somewhere on all plants. In ferns the sporangia are on the backs of leaves; this is why the leaves are called sporophylls.

FERNS

True ferns (phylum Pterophyta) inhabit almost all kinds of environments and possess characteristics of the more advanced seed plants as well as the less advanced bryophytes. Ferns have an independent sporophyte (fig. 28.1) with well-developed vascular tissue. Unlike bryophytes, ferns have stomata, which are pores that open and close on leaves and, in doing so, regulate gas exchange.

The diversity of ferns is striking; they range from majestic tree ferns to bizarre staghorn ferns. Tree ferns reach heights of up to 16 m. Along with other plants, these ferns once formed forests that were transformed into coal deposits. Today, humans use ferns as decorations, fossil fuels, and in rice cultivation. Review the fern life cycle shown in figure 28.1.

Question 1

a. Which parts of the life cycle are haploid?

b. Which are diploid?

Fern sporophytes grow indefinitely via underground stems called **rhizomes.** Examine the fern rhizomes on display. Also examine the different ferns available in the lab and note the different shapes of the leaflike fronds. Identify the **stalk, blade,** and **pinnae.**

Question 2

a. How many veins are present in each frond?

b. What tissues comprise a vein?

c. What is the function of the stalk? The blade? The pinnae?

TABLE 28.1

THE TWO PHYLA OF SEEDLESS VASCULAR PLANTS

Phylum	Examples	Key Characteristics	Approximate Number of Living Species
Pterophyta	Ferns	Primarily homosporous (a few heterosporous) vascular plants. Sperm motile. External water necessary for fertilization. Leaves are megaphylls that uncoil as they mature. Sporophytes and virtually all gametophytes photosynthetic. About 365 genera.	11,000
	Horsetails	Homosporous vascular plants. Sperm motile. External water necessary for fertilization. Stems ribbed, jointed, either photosynthetic or nonphotosynthetic. Leaves scale-like, in whorls, nonphotosynthetic at maturity. One genus.	15
	Whisk ferns	Homosporous vascular plants. Sperm motile. External water necessary for fertilization. No differentiation between root and shoot. No leaves; one of the two genera has scale-like enations and the other leaf-like appendages.	6
Lycophyta	Club mosses	Homosporous or heterosporous vascular plants. Sperm motile. External water necessary for fertilization. Leaves are microphylls. About 12–13 genera.	1,150

Groups of sporangia called **sori** form on the underside of fern fronds (fig. 28.2). Sporangia may be protected by a shield-shaped **indusium,** which is a specialized outgrowth of a frond. Meiosis in the sporangium produces haploid spores, which are the first stage of the gametophyte.

Procedure 28.1

Examine sori

1. Scrape a sorus into a drop of water and use the low power on your microscope to observe the sorus. You'll notice a row of thick-walled cells along the back of the helmet-shaped sporangium. These cells are termed the **annulus.**

Question 3

a. What is the function of the annulus?

b. Are any spores in the sporangium?

2. Place a few drops of acetone on the sporangium while you observe it with a dissecting microscope.
3. Watch the sporangium for a few minutes, adding acetone as needed.
4. Describe what you see.

Question 4

a. Did the application of acetone cause the spores of the fern to disperse?

b. How is the mechanism for spore dispersal in ferns similar to that of bryophytes?

5. Examine prepared slides of fern sori, referring to figure 28.3. Diagram and label each structure that you see, listing its function.

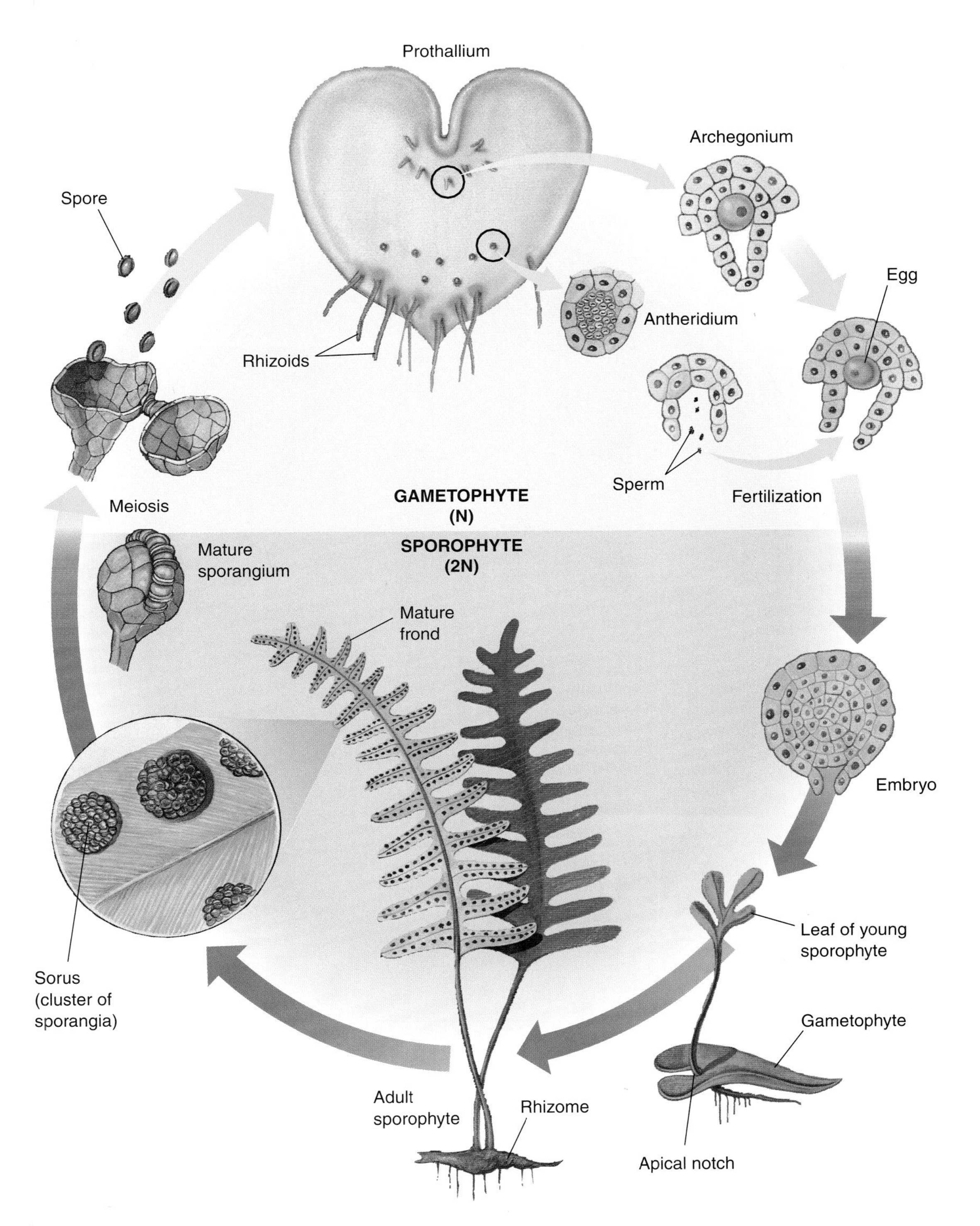

Figure 28.1

Fern life cycle. The haploid gametophytes grow in moist places. Rhizoids are anchoring structures that project from the lower surface of a heart-shaped prothallium. Eggs and sperm develop in archegonia and antheridia, respectively, on gametophytes' lower surface near the apical notch. Sperm released from antheridia swim to archegonia and fertilize the single egg. The zygote—the first cell of the diploid sporophyte generation—starts to grow within the archegonium, and eventually becomes much larger than the gametophyte. Most ferns have horizontal stems, called rhizomes, that grow below the ground. On the sporophyte's leaves (fronds) are clusters of sporangia within which meiosis occurs and spores are formed. When released, the spores germinate and become new gametophytes called prothallia.

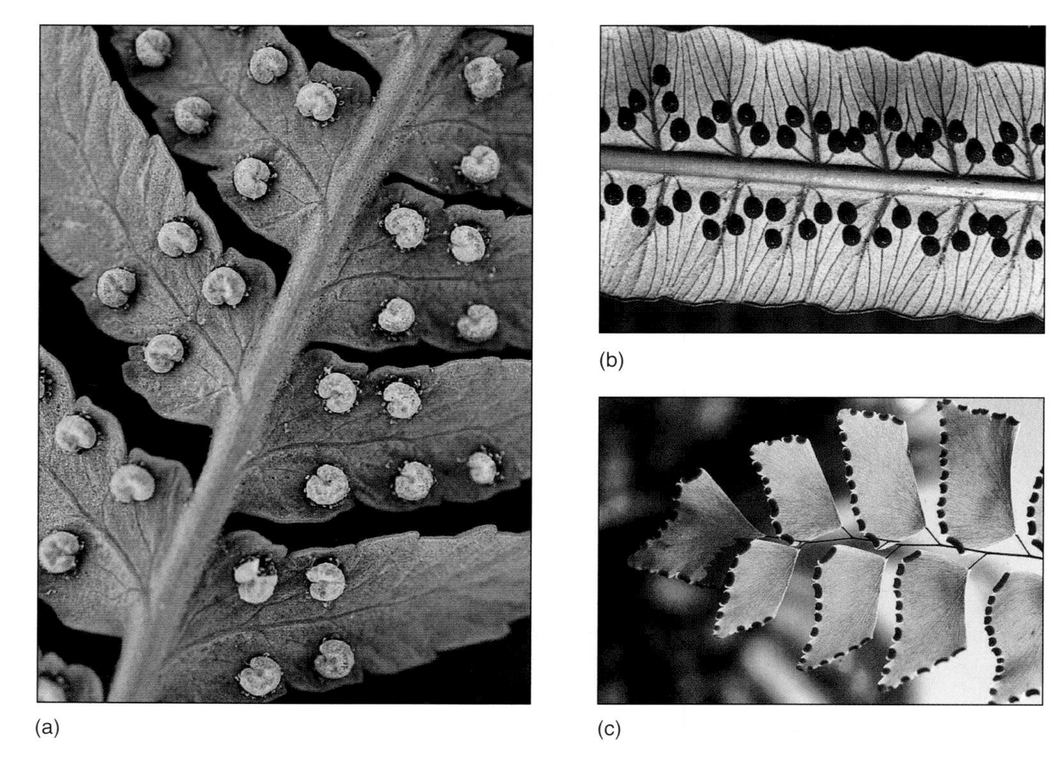

Figure 28.2

Fern sporangia. Most ferns have sporangia aggregated into clusters, called sori, on the undersides of their leaves. (*a*) In some ferns, such as the marginal woodfern (*Dryopteris marginalis*), each sorus is covered by a flap of leaf tissue called an indusium (also see fig. 28.3). (*b*) Other ferns bear uncovered sori, as shown here in *Alsophila sinuata*. (*c*) In still other ferns, as in the giant maidenhair fern (*Adiantum trapeziforme*), sori are enfolded by the edge of the leaf itself.

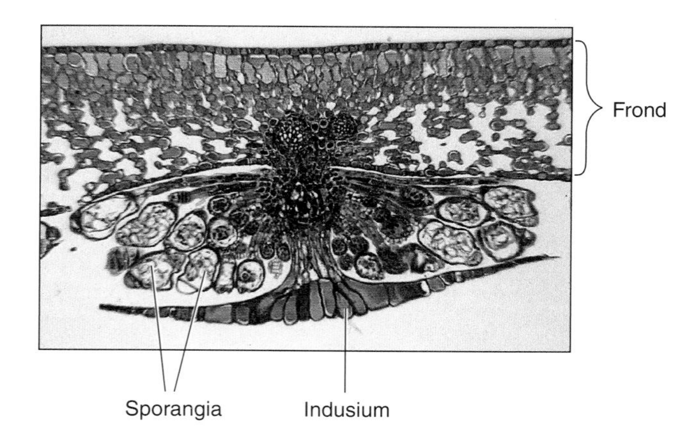

Figure 28.3

Fern sorus.

Fern Reproduction

Fern spores germinate and form a threadlike **protonema.** Subsequent cellular divisions produce an independent, heart-shaped **prothallium** ("valentine plant").

Question 5

a. Is the prothallium haploid or diploid?

b. Is the prothallium sporophyte or gametophyte?

Rhizoids and male and female reproductive structures occur on the underside of the prothallium. However, a prothallium rarely fertilizes itself because the antheridia and archegonia mature at different times. Globe-shaped antheridia form first, followed by archegonia. After producing sperm, antheridia drop off, leaving sperm to swim to the archegonia. Archegonia are vase-shaped and are located near the cleft of the heart-shaped prothallium.

Procedure 28.2

Observe archegonia and antheridia

1. Observe archegonia and antheridia (see fig. 28.1) on prepared slides and on a living prothallium.
2. Observe archegonia and antheridia on living prothallia.

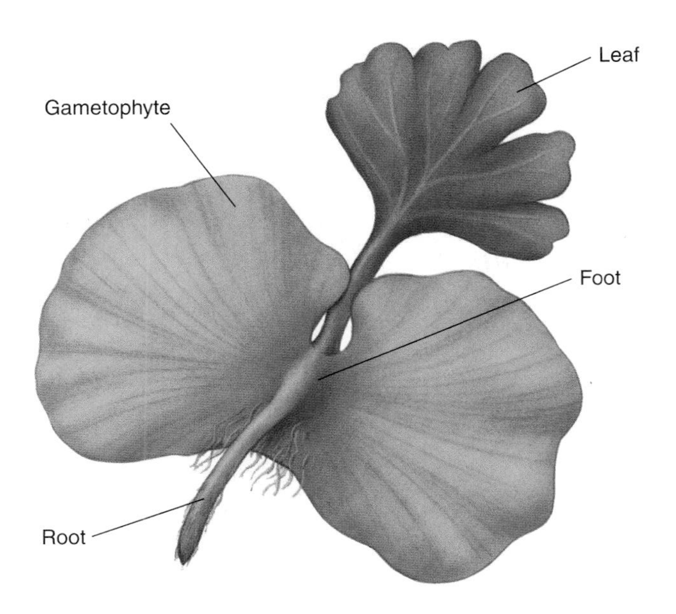

Figure 28.4
Young fern sporophyte growing out of its gametophyte parent. Shortly after this stage, the gametophyte dies and begins to decompose.

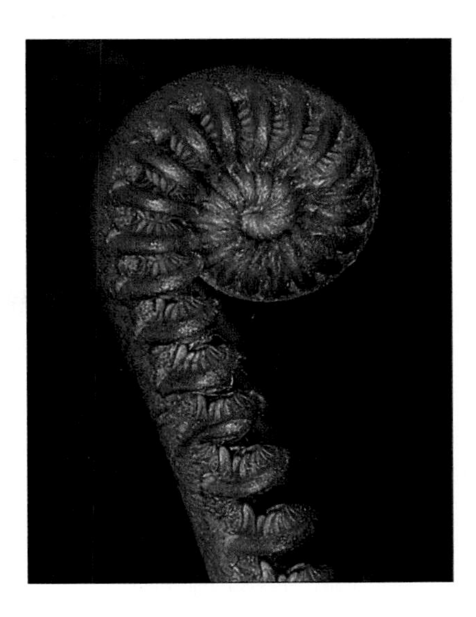

Figure 28.5
A fiddlehead of a tropical tree fern.

Question 6

a. What is the adaptive significance of having these structures on the lower surface of the prothallium rather than on the upper surface?

b. What is the adaptive significance of having sperm and egg produced at different times?

Figure 28.6
Whisk ferns (*Psilotum* sp.) are so called because their branching pattern gives the impression of a whisk broom. The stems bear lobed sporangia (fig. 28.7).

The zygote develops in the archegonium and is nutritionally dependent on the gametophyte for a short time (fig. 28.4). Soon thereafter, the sporophyte becomes leaflike and crushes the prothallium. Fronds of the growing sporophyte break through the soil in a coiled position called a **fiddlehead** (fig. 28.5). The fiddlehead then unrolls to display the frond, which is a single leaf. Fiddleheads are considered a culinary delicacy in some parts of the world.

Most terrestrial ferns are **homosporous;** they produce one kind of spore that develops into a single kind of gametophyte that produces both antheridia and archegonia (see fig. 28.1). Conversely, aquatic ferns such as *Salvinia* and *Azolla* are **heterosporous,** meaning that they produce two kinds of spores: **megaspores** and **microspores.** You will learn more about heterospory in the next exercise. A megaspore forms a gametophyte with only archegonia, and a microspore forms a gametophyte with only antheridia. Examine living *Salvinia* and *Azolla* in the lab.

Question 7

How do *Salvinia* and *Azolla* differ from other ferns you examined earlier?

WHISK FERNS

Whisk ferns (phylum Pterophyta) include only two extant representatives: *Psilotum* (fig. 28.6) and *Tmesipteris*. *Psilotum* has a widespread distribution, whereas *Tmesipteris* is restricted to the South Pacific. *Psilotum* lacks leaves and roots and is homosporous.

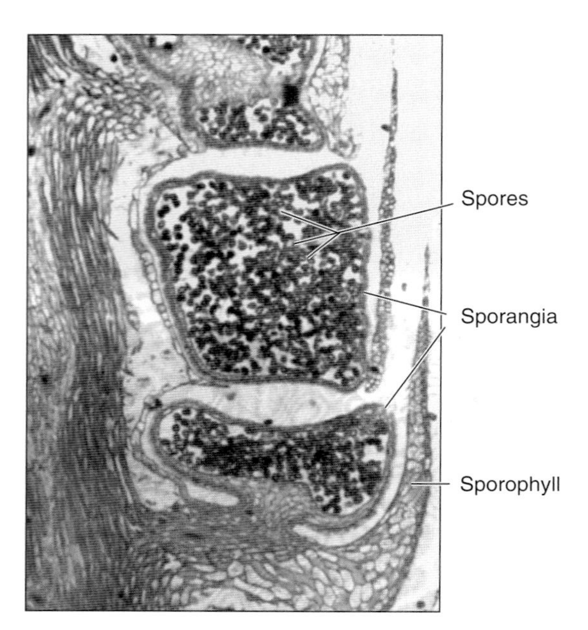

Figure 28.7
Psilotum sporangium.

Figure 28.8
Scouring rush, *Equisetum telmateia,* is the only living genus of the phylum Sphenophyta. This species forms two kinds of erect stems, a green, photosynthetic type and a brownish type terminating in spore-producing cones. The spores produced by meiosis in the cones give rise to a single kind of tiny, green, nutritionally independent gametophyte.

Procedure 28.3
Examine *Psilotum*

1. Examine a prepared slide of a *Psilotum* sporangium (fig. 28.7)
2. Study living *Psilotum* plants in the lab.

Question 8

a. What type of branching characterizes *Psilotum?*

b. Are any roots present?

c. Are any leaves present?

d. Where are the sporangia?

e. Where does photosynthesis occur in *Psilotum?*

HORSETAILS

Equisetum (also called scouring rush) is the only extant genus of horsetails (phylum Pterophyta) (fig. 28.8). *Equisetum* is an example of a plant whose vegetative structure identifies the plant better than does its reproductive structure: *Equisetum* is distinguished by its jointed and ribbed stem. Examine the living *Equisetum* plants.

Question 9

a. Where are the leaves?

b. What part of the plant is photosynthetic?

c. Which part of the life cycle of *Equisetum* is dominant, the sporophyte or gametophyte?

Feel the ribbed stem of an *Equisetum* plant. Its rough texture results from siliceous deposits in its epidermal cells. During frontier times, *Equisetum* was used to clean pots and pans, sand wooden floors, and scour plowshares, thus accounting for its common name of "scouring rush."

(a)

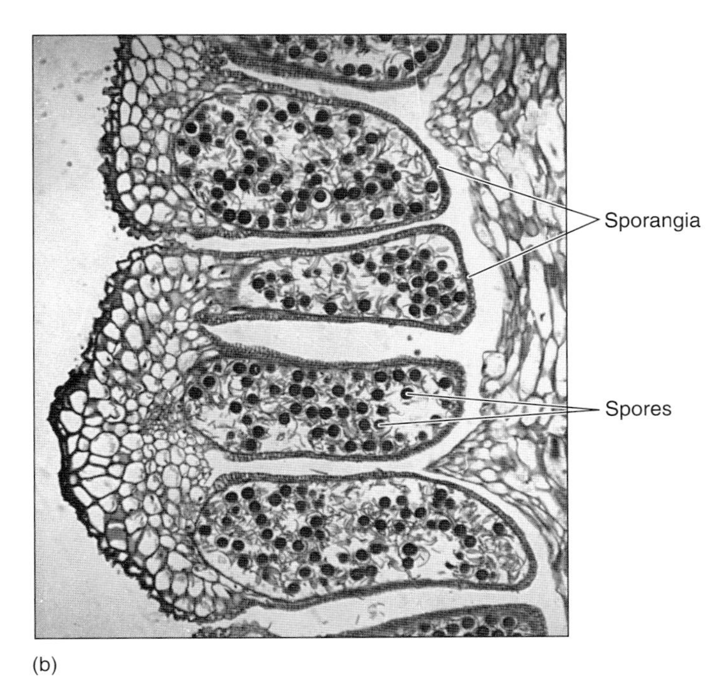

(b)

Figure 28.9
Equisetum. (*a*) Group of four strobili. (*b*) Cross-section of a strobilus showing sporangia and spores.

Strobili of *Equisetum* occur at the tips of reproductive stems. Sporangia form atop umbrella-like structures called **sporangiophores,** which are modified branches. Elaters in sporangia of *Equisetum* help disperse spores. Examine prepared slides of *Equisetum* strobili (fig. 28.9). Diagram, label, and state the function of each major structure composing the strobilus.

Question 10

a. How does the formation of strobili in *Equisetum* compare with that in *Lycopodium* and *Selaginella?*

b. How do elaters aid in the dispersal of spores?

CLUB MOSSES

During the Devonian and Carboniferous periods (300–400 million years ago), club mosses, whisk ferns, and scouring rush were among the dominant plants on earth. Indeed, most of our coal deposits consist largely of these plants. However, the "giant" species of these phyla are now extinct, and modern representatives are relatively small compared to their giant ancestors.

Club mosses (phylum Lycophyta) possess true roots, stems, and leaves. Most asexual reproduction by club mosses occurs via rhizomes. If one is available, locate the rhizome on a *Lycopodium*, a club moss. Study the aboveground portion of the *Lycopodium* plant (fig. 28.10). *Lycopodium* is evergreen, as are some ferns, most gymnosperms, and some angiosperms.

Figure 28.10
Strobili, or cones, are aggregations of closely packed sporangium-bearing branches or leaves. Shown here are strobili of *Lycopodium obscurum*, a club moss.

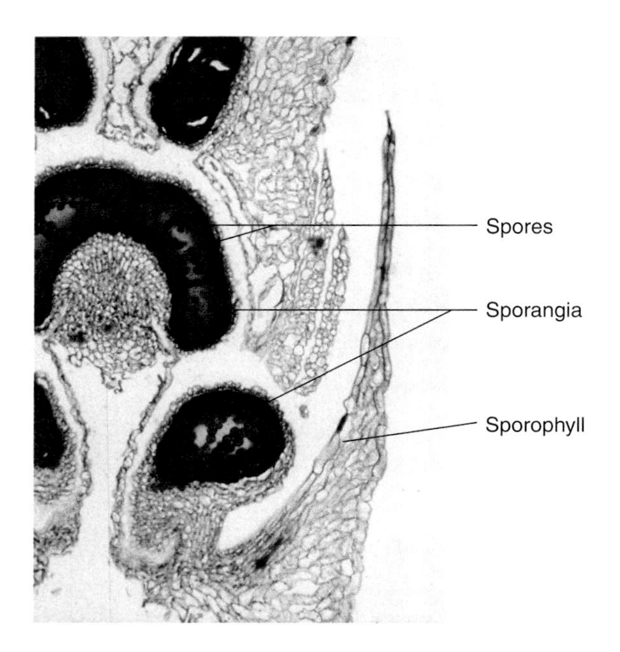

Figure 28.11
Sporophyll and sporangium of *Lycopodium*, a club moss.

Question 11

a. How could a rhizome be involved in asexual reproduction?

b. How is a rhizome different from a rhizoid?

c. Does the rhizome have leaves?

d. What is the shape and size of the leaves?

e. What is the significance of the form of the leaves?

f. Is a midvein visible?

g. What does the term "evergreen" mean?

h. Is being evergreen a good characteristic for classifying plants? Why or why not?

Sporangia of *Lycopodium* occur on small modified leaves called **sporophylls,** which are clustered in **strobili** (cones) that form at the tips of branches. Species with these cones probably shared common ancestry with the familiar cone-bearing gymnosperms such as pine (*Pinus*) (see Exercise 29).

Procedure 28.4

Examine club mosses

1. Examine strobili on a living *Lycopodium* plant. Also examine prepared slides of strobili of *Lycopodium* (fig. 28.11).
2. Diagram, label, and state the function of each major feature of the strobilus.
3. Examine prepared slides of spores of *Lycopodium* as well as those available on plants growing in the lab.

Question 12

a. How many sporangia occur on each sporophyll?

b. Can you see why spores of *Lycopodium* are sometimes called "vegetable sulfur"?

c. Why are the spores a good dry lubricant?

d. Which is the dominant part of the *Lycopodium* life cycle, the sporophyte or gametophyte?

4. Examine living *Selaginella* plants (fig. 28.12). Many species of *Selaginella* produce two kinds of strobili, which are typically red and yellow. If strobili are present, examine spores derived from these cones.
5. Examine hydrated and dehydrated resurrection plants (*Selaginella lepidophylla*).

Question 13

a. Are spores of *Selaginella* similar in size?

b. What is this condition called?

c. What is the functional significance of the difference in the appearance of dehydrated and rehydrated *Selaginella?*

d. Can you see why these plants are sometimes referred to as "resurrection plants"?

Figure 28.12
Selaginella, a club moss.

Procedure 28.5
Examine *Isoetes*

1. Examine living *Isoetes* (quillwort) plants.
2. Compare and contrast the following features of *Isoetes* with *Lycopodium* and *Selaginella:*
   - Shape of aerial portion of plant
   - Branching patterns
   - Shape, size, and arrangement of leaves

At the branching points along the stem of *Isoetes*, you'll see an unusual runnerlike organ. These proplike axes are called **rhizophores** and have structural features that are intermediate between stems and roots.

To review the structures and characteristics of seedless vascular plants, complete table 28.2 on the next page.

Questions for Further Thought and Study

1. Compare ferns with bryophytes. Which represent the best "engineering job"? Explain.

2. What are the advantages of vascular tissues in land plants?

3. What are the distinguishing features of club mosses, whisk ferns, and scouring rush? How are these plants different from ferns?

TABLE 28.2

SUMMARY OF STRUCTURES COMMON TO SEEDLESS VASCULAR PLANTS

Plant Structure	Sporophyte or Gametophyte	Function
Prothallium		
Pinna		
Spore		
Frond		
Annulus		
Sporangium		
Antheridium		
Archegonium		
Microspore		
Megaspore		
Microphyll		
Megaphyll		

WRITING TO LEARN BIOLOGY

What problems did plants face as they moved onto land? What adaptations of mosses, liverworts, ferns, and other seedless plants are relevant to the transition?

29

Survey of the Plant Kingdom
Gymnosperms of Phyla Cycadophyta, Ginkgophyta, Coniferophyta, and Gnetophyta

Objectives

By the end of this exercise you should be able to:
1. Describe the distinguishing features of the gymnosperms.
2. Understand the life cycle of pine, a representative gymnosperm.
3. List some adaptations of pine to cold, dry environments.
4. Identify the parts and understand the function of a cone.
5. Identify the parts and understand the function of a seed.

Gymnosperms are plants with exposed seeds borne on scalelike structures called **cones** (strobili). Like ferns, gymnosperms have a well-developed alternation of generations. Unlike most ferns, however, gymnosperms are **heterosporous;** that is, they produce two kinds of spores (fig. 29.1). **Microspores** occur in male cones and form male gametophytes. **Megaspores** occur in female cones and form female gametophytes. Gametophytes of gymnosperms are microscopic and completely dependent on the large, free-living sporophyte. Gymnosperms include four phyla: Cycadophyta, Ginkgophyta, Coniferophyta, and Gnetophyta (table 29.1). The last of these, the Gnetophyta, is discussed only briefly here because the phylum consists of a few rare genera.

PHYLUM CYCADOPHYTA

The Cycadophyta (cycads) once flourished, but today the phylum consists of only about ten genera and 200 species. Cycads resemble palms because they have unbranched trunks and large, closely packed leaves that are evergreen and tough (fig. 29.2). Sperm of cycads are flagellated. Examine a branch of a cycad such as *Zamia* bearing developing seeds. Note that the seeds are fleshy and exposed to the environment.

Question 1
What is the evolutionary significance of flagellated sperm in cycads?

PHYLUM GINKGOPHYTA

The Ginkgophyta consists of one species, *Ginkgo biloba* (Maidenhair plant), a large dioecious tree that does not bear cones. *Ginkgo* are hardy plants in urban environments and tolerate insects, fungi, and pollutants. Males are usually planted because females produce fleshy, smelly, and messy fruit that superficially resembles cherries. Leaves of *Ginkgo* have a unique shape (fig. 29.3). *Ginkgo* has not been found in the wild and would probably be extinct but for its cultivation in ancient Chinese and Japanese gardens.

Question 2
What does dioecious mean?

PHYLUM CONIFEROPHYTA

The Coniferophyta are a large group of cone-bearing plants that includes the 5000-year-old bristlecone pines, the earth's oldest living individual organisms. The cones they bear are reproductive structures of the sporophyte generation that consist of several scalelike sporophylls arranged about a central axis (fig. 29.4). Sporophylls, also present in ferns and their allies, are modified leaves specialized for reproduction. Sporophylls bear spores. In conifers, sporophylls of male cones are called **microsporophylls.** On the surface of each microsporophyll is a layer of cells called a **microsporangium** that produces spores. Sporophylls of female cones are

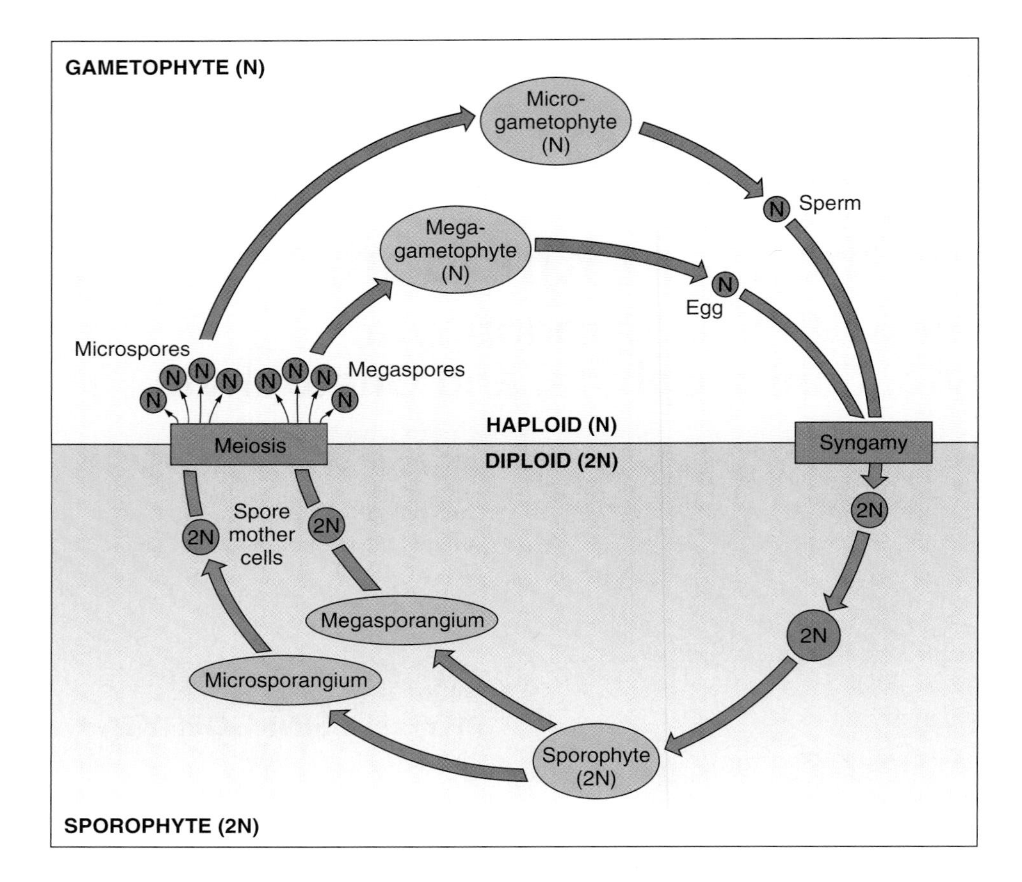

Figure 29.1
Diagram of the life cycle of a heterosporous vascular plant.

megasporophylls; each bears two spore-producing **megasporangia** on its upper surface. Microsporangia and megasporangia are patches of cells near the central axis of the sporophylls composing the cones on the respective sporophylls. Male cones are small and similar in all conifers (fig. 29.5). However, female cones are variable; they may be small (1–3 cm in *Podocarpus*) and fleshy (a *Juniperus* "berry") or large and woody (6–40 cm in *Pinus*).

In this exercise you'll study pine (*Pinus*), a representative and widely distributed conifer. Pine has considerable economic value because it is used to produce lumber, wood pulp, pine tar, resin, and turpentine. Other examples of conifers include spruce, cedar, and fir.

The Pine Sporophyte

The life cycle of *Pinus* is typical of conifers (fig. 29.6). You are already familiar with the sporophyte of pine—it is the tree.

Procedure 29.1

Examine pine twigs and leaves

1. Examine pine twigs having leaves (needles) and a terminal bud. Notice that the leaves are borne on dwarf branches only a few millimeters long. The length and number of leaves distinguish many of the species of *Pinus*.
2. Examine a prepared slide of a cross section of a pine leaf; locate the structures labeled in figure 29.7.

Question 3

a. Where on the branch was the terminal bud last year?

b. How can you tell?

TABLE 29.1

THE FIVE PHYLA OF EXTANT GYMNOSPERMS

Phylum	Examples	Key Characteristics	Approximate Number of Living Species
Coniferophyta	Conifers (including pines, spruces, firs, yews, redwoods, and others)	Heterosporous seed plants. Sperm not motile; conducted to egg by a pollen tube. Leaves mostly needlelike or scalelike. Trees, shrubs. About 50 genera.	601
Cycadophyta	Cycads	Heterosporous vascular seed plants. Sperm flagellated and motile but confined within a pollen tube that grows to the vicinity of the egg. Palmlike plants with pinnate leaves. Secondary growth slow compared with that of the conifers. Ten genera.	206
Gnetophyta	Gnetophytes	Heterosporous vascular seed plants. Sperm not motile; conducted to egg by a pollen tube. The only gymnosperms with vessels. Trees, shrubs, vines. Three very diverse genera (*Ephedra*, *Gnetum*, *Welwitschia*).	71
Ginkgophyta	*Ginkgo*	Heterosporous vascular seed plants. Sperm flagellated and motile but conducted to the vicinity of the egg by a pollen tube. Deciduous tree with fan-shaped leaves that have evenly forking veins. Seeds resemble a small plum with fleshy, ill-scented outer covering.	1

(a)

(b)

Figure 29.2

Cycads. (*a*) A male cycad with a strobilus. (*b*) A female cycad with strobili. *Zamia pumila* is the only species of cycad native to the United States. The starchy roots and stems (which are mostly underground) were used by Native Americans for food.

Figure 29.3

Maidenhair tree, *Ginkgo biloba,* the only living representative of the phylum Ginkgophyta, a group of plants that was abundant 200 million years ago. Among living seed plants, only the cycads and *Ginkgo* have swimming sperm. This photograph shows *Ginkgo* leaves and fleshy seeds.

c. How are needles or leaves arranged?

d. How many leaves are in a bundle?

e. How are pine leaves different from those of deciduous plants?

f. Why are pines called evergreens?

g. What other plants have you studied in the lab that are evergreen?

h. What is the function of each of the structures labeled in figure 29.7?

i. How do the structural features of pine leaves adapt the tree for life in cold, dry environments?

(a)

(b)

Figure 29.4

Pine cones. (*a*) First-year ovulate (seed) cones open for pollination. (*b*) Second-year pine cones at time of fertilization.

Figure 29.5

Pollen-bearing cones of *Pinus*.

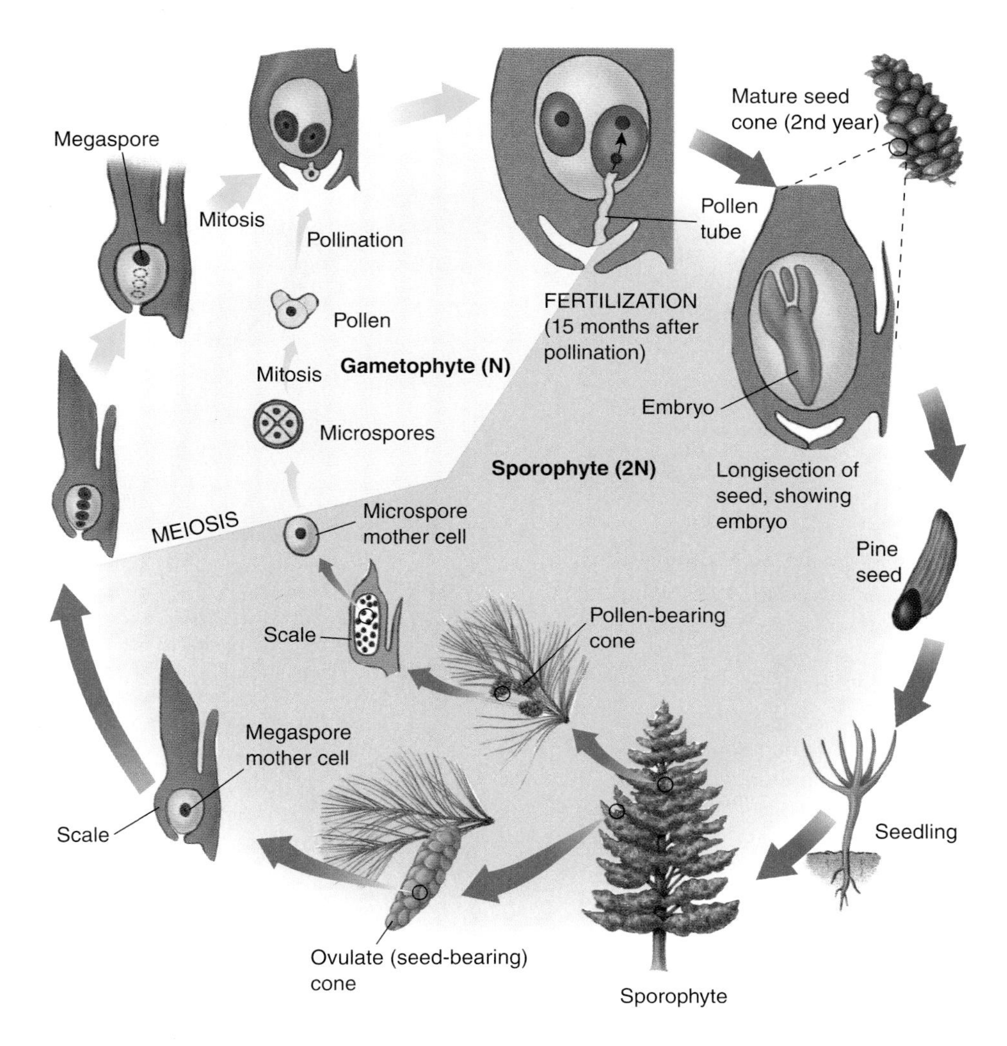

Figure 29.6

Pine life cycle. In seed plants, the gametophyte generation is greatly reduced. A germinating pollen grain is the mature microgametophyte of a pine. Pine microsporangia are borne in pairs on the scales of the delicate pollen-bearing cones. Megagametophytes, in contrast, develop within the ovule. The familiar seed-bearing cones of pines are much heavier than the pollen-bearing cones. Two ovules, and ultimately two seeds, are borne on the upper surface of each scale of a cone. In the spring, when the seed-bearing cones are small and young, their scales are slightly separated. Drops of sticky fluid, to which the airborne pollen grains adhere, form between these scales. These pollen grains germinate, and slender pollen tubes grow toward the egg. When a pollen tube grows to the vicinity of the megagametophyte, sperm are released, fertilizing the egg and producing a zygote there. The development of the zygote into an embryo occurs within the ovule, which matures into a seed. Eventually, the seed falls from the cone and germinates, the embryo resuming growth and becoming a new pine tree.

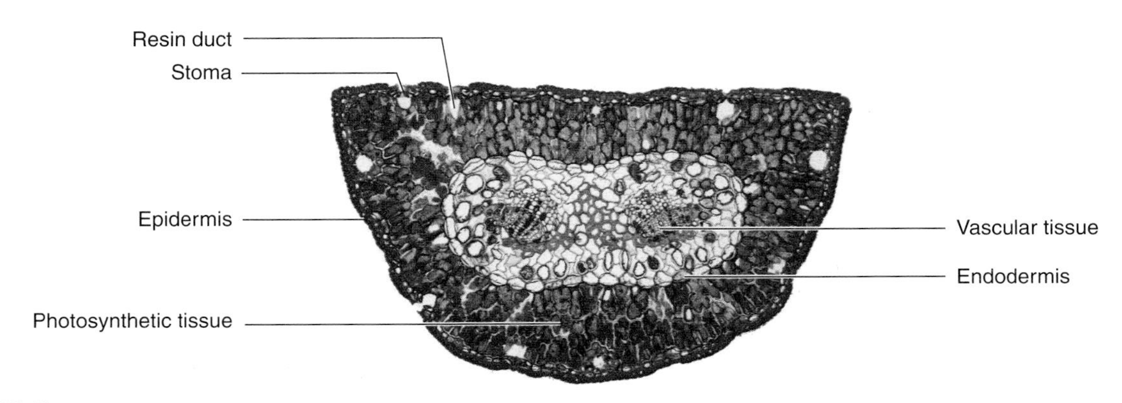

Figure 29.7

Cross-section of a pine leaf (needle).

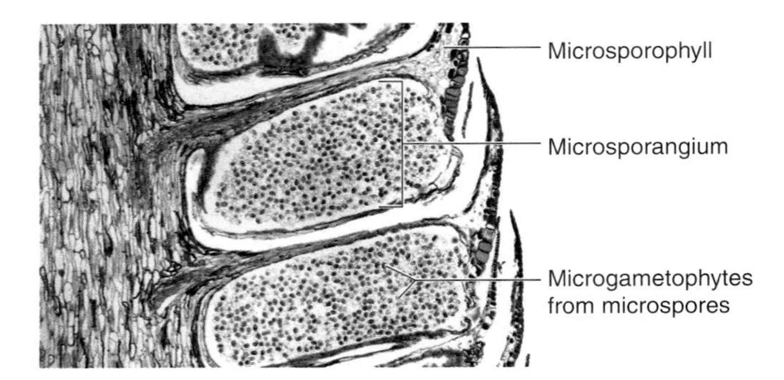

Figure 29.8
A single microsporophyll (with microsporangium) from a male cone. Within the microsporangium, microspore mother cells form microspores via meiosis.

Examine pine branches with staminate (male) and ovulate (female) cones. Examine a prepared slide of a young staminate cone and note the pine pollen in various stages of development. Each scale (microsporophyll) of the male cone bears a microsporangium, which in turn produces diploid **microspore mother cells** (fig. 29.8). Microspore mother cells undergo meiosis to produce **microspores** that develop into microgametophytes called pollen grains (fig. 29.9). Each **pollen grain** consists of four nuclei and a pair of bladderlike wings. Note that the gametophytes of gymnosperms are reduced to only a few cells.

Procedure 29.2

Examine pine pollen

1. Prepare a wet mount of some pine pollen; notice their characteristic shape.
2. Remove a scale from a mature staminate cone and tease open the microsporangium.
3. Prepare a wet mount of the microsporangium and its contents, and examine the contents with your microscope.
4. Draw the shape of a pine pollen grain below.

Question 4

a. Are all of the cones the same size?

b. How might the arrangement and location of staminate cones on the tree help ensure cross-pollination?

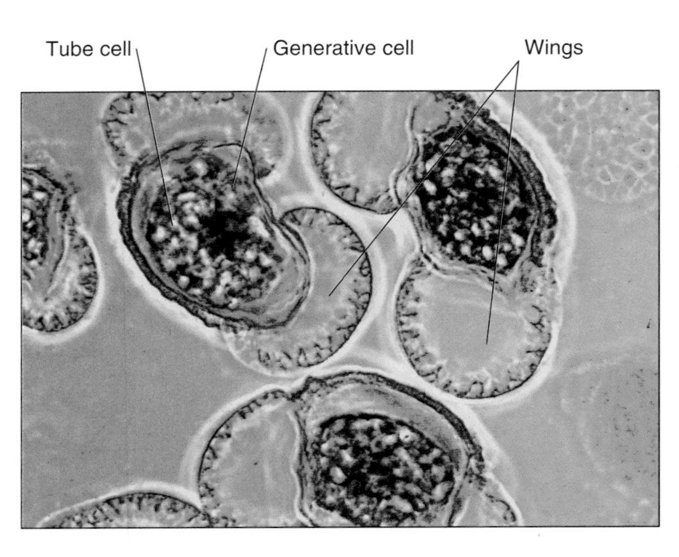

Figure 29.9
Pollen grains of *Pinus*, each with enclosed male gametophyte. Each gametophyte includes a small generative cell and a larger tube cell. When the pollen grain germinates, the pollen tube will emerge between the two bladder-shaped wings.

c. What is the probable function of the wings of a pine pollen grain?

Procedure 29.3

Examine pine cones

1. Examine young living or preserved ovulate cones. These cones will develop and enlarge considerably before they are mature.
2. Examine a prepared slide of a young ovulate cone ready for pollination. Each ovuliferous scale of the female cone bears two megasporangia, each of which produces a diploid **megaspore mother cell.** Each megaspore mother cell undergoes meiosis to produce a **megaspore** that develops into a **megagametophyte.** The tissue of the megasporangium immediately surrounding the megagametophyte is the **nucellus,** which is surrounded by **integuments.** A megagametophyte and its surrounding tissues constitute an ovule and contain at least one archegonium with an egg cell.
3. Examine a prepared slide of an ovulate cone that has been sectioned through an ovule (fig. 29.10). An ovule develops into a seed.
4. Examine a mature ovulate cone and notice its spirally arranged **ovuliferous scales.** These scales are analogous to microsporophylls of staminate cones, but ovuliferous scales are modified branches rather

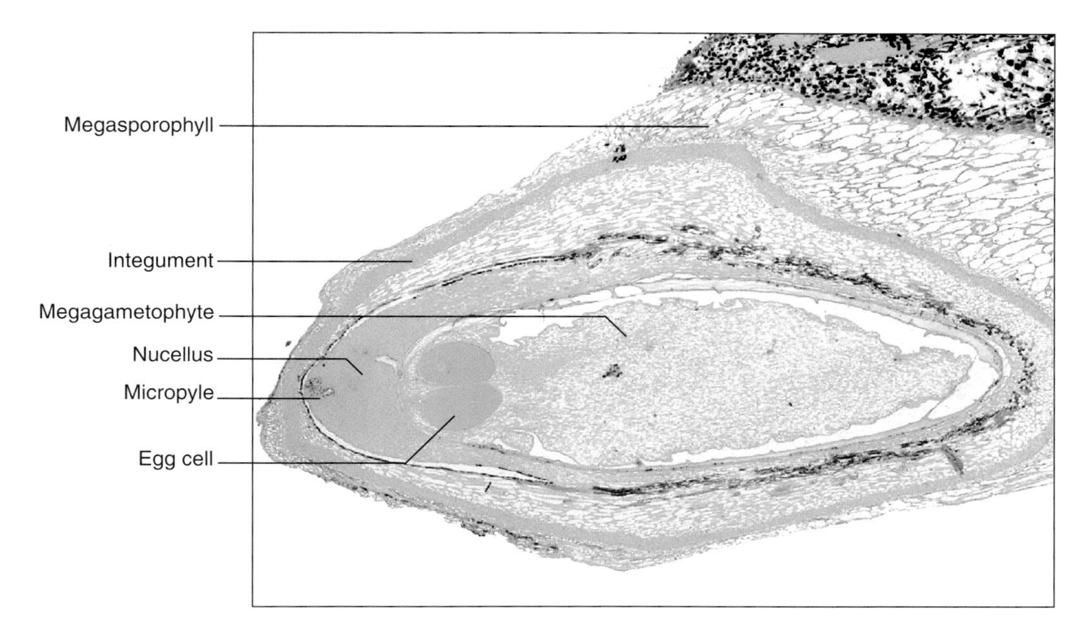

Figure 29.10
Pine ovule section before fertilization. The egg will be fertilized later to form a new sporophyte inside the seed.

than modified leaves. At the base of each scale you'll find two naked seeds. Notice that the seeds are exposed to the environment and are supported (but not covered) by an ovuliferous scale.

Question 5

a. On which surface of the scale are the seeds located?

b. How large is a staminate cone compared to a newly pollinated ovulate cone? A mature ovulate cone?

Pollination and Seed Formation in Pine

Pollination is the transfer of pollen to a receptive surface. Pollen grains released in late spring are carried by wind to ovulate cones, in which pollen sifts through the ovuliferous scales and sticks in a drop of resin at the micropylar end of an ovule. The pollen grain then germinates and grows a tube into the archegonium, where it releases its two non-motile sperm nuclei. One of these sperms disintegrates; the other fuses with the egg to form a zygote. The zygote, while in the ovule, develops into the embryo of a new sporophyte. The ovule is now a seed and consists of an embryo, a seed coat (integuments of the megasporangia), and a food supply (tissue of the megagametophyte).

Procedure 29.4

Examine a pine seed

1. Examine a prepared slide of a pine seed. Locate the embryo, seed coat, and food supply. Seeds are released when the cone dries and the scales separate. This usually occurs 13–15 months after pollination.
2. Examine some mature pine seeds, noting the winglike extensions of the seed coat.

Seed

The evolution of seeds is one of the most significant events in the history of the plant kingdom. Indeed, the evolution of seeds is one of the factors responsible for the dominance of seed plants in today's environment, because a seed permits a small but multicellular sporophyte to remain dormant until conditions are favorable for continued growth. While dormant, the young sporophyte is protected by a seed coat and surrounded by a food supply.

Refer to the pine life cycle shown in figure 29.6 to answer Question 6. If plant material and time are available, examine other gymnosperms in the lab.

Question 6

a. Where are spores located?

b. What is the male gametophyte?

c. What is the female gametophyte?

d. What is an ovule?

e. What is an integument?

f. What is the function of the winglike extensions of a pine seed?

g. How are other gymnosperms similar to pine?

h. How are they different?

PHYLUM GNETOPHYTA

The gnetophytes (71 species in three genera) include some of the most distinctive (if not bizarre) of all seed plants (fig. 29.11). They have many similarities with angiosperms, such as flowerlike compound strobili, vessels in the secondary xylem, loss of archegonia, and double fertilization. The slow-growing *Welwitschia* plants are extraordinary in appearance, and live only in the deserts of southwestern Africa. Most of their moisture is derived from fog that rolls in from the ocean at night. *Welwitschia* stems are short and broad (1 m). Mature, 100-year-old plants have only two leaves that are wide and straplike.

(a)

(b)

(c)

Figure 29.11

Gnetophytes. (*a*) Leaves and immature seeds of *Gnetum*. *Gnetum* grows as a shrub or woody vine in tropical or subtropical forests. (*b*) *Welwitschia* plants in the Namib Desert of southwest Africa. (*c*) *Ephedra* is the only living genus of Gnetophyta found in the United States, and is a common diet supplement.

Questions for Further Thought and Study

1. What is the difference between pollination and fertilization?

2. What does the term gymnosperm mean?

3. How is alternation of generations different in ferns and pines?

4. How are the environmental agents for uniting sperm and egg different in pines and bryophytes?

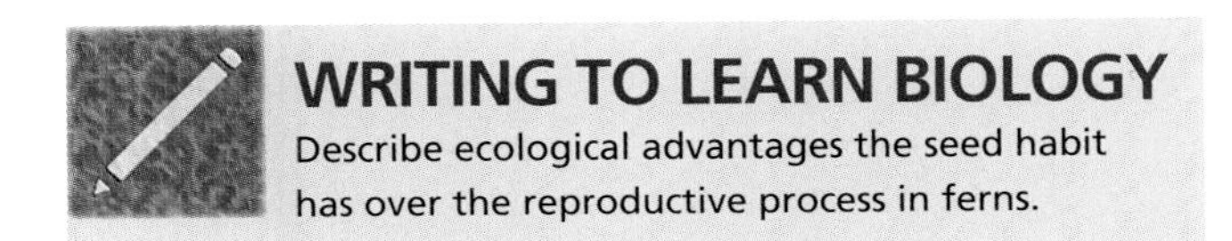

WRITING TO LEARN BIOLOGY

Describe ecological advantages the seed habit has over the reproductive process in ferns.

30

Survey of the Plant Kingdom
Angiosperms

Objectives

By the end of this exercise you should be able to:

1. Relate the life cycle of angiosperms to the phyla discussed previously.
2. Discuss the events associated with development of microspores, megaspores, seeds, and fruit.
3. List and discuss the reasons why angiosperms are considered to be the most advanced land plants.

Flowering plants (phylum Anthophyta—also referred to as angiosperms) are the most abundant, diverse, and widespread of all land plants. They owe their success to several factors, including their structural diversity, efficient vascular systems, mutualisms (especially with fungi and insects), and short generation times (fig. 30.1).

PHYLUM ANTHOPHYTA

Angiosperms range in size from 1 mm (*Wolffia*) to over 100 m tall (*Eucalyptus*). As in gymnosperms, the sporophyte of angiosperms is large and independent of the microscopic gametophyte. This is opposite of the events in bryophytes. The gametophyte of angiosperms depends totally on the sporophyte. Angiosperms are important to humans because our world economy is overwhelmingly based on them. Indeed, we eat and use **vegetative structures** (roots, stems, leaves) as well as **reproductive structures** (flowers, fruits, seeds) of angiosperms.

Figure 30.1
This bee is pollinating a desert poppy in eastern Arizona. The evolution of flowers and the success of flowering plants correlates with the development of mutualisms with insects and other pollinators.

Botanists commonly divide angiosperms into monocots and dicots (fig. 30.2). The "typical" features of these groups of plants are described in table 30.1.

Having studied gymnosperms, you probably concluded that vegetative adaptations of angiosperms and gymnosperms are similar. This is true. However, in today's exercise you will study two unique adaptations of angiosperm reproduction: flowers and fruit.

Figure 30.2
In monocots such as this lily (*Lilium longiflorum*), flower parts usually occur in multiples of three. A lily has six stamens, three petals, three petal-like sepals, and a three-chambered ovary.

TABLE 30.1

CHARACTERISTICS OF THE TWO CLASSES OF ANGIOSPERMS

Monocotyledons
1. One cotyledon per embryo
2. Flower parts in sets of three
3. Parallel venation in leaves
4. Multiple rings of vascular bundles in stem
5. Lack a true vascular cambium (lateral meristem)
Dicotyledons
1. Two cotyledons per embryo
2. Flower parts in sets of four or five
3. Reticulate (i.e., netted) venation in leaves
4. One ring of vascular bundles or cylinder of vascular tissue in stem
5. Have a true vascular cambium (lateral meristem)

STRUCTURE AND FUNCTION OF FLOWERS

Luckily for florists, angiosperms have a seemingly infinite variety of flowers, ranging from the microscopic flowers of *Lemna* (duckweed) to the rare, monstrous blossoms of *Rafflesia,* which measure up to 1 m across. In today's exercise, we'll consider only the "typical" flower depicted in figure 30.3.

Examine the flower model(s) available in the lab and the living flowers on display and identify the following parts:

- **Peduncle**—flower stalk.
- **Receptacle**—the part of the flower stalk that bears the floral organs; located at the base of the flower; usually not large or noticeable.
- **Sepals**—the lowermost or outermost whorls of structures, which are usually leaflike and protect the developing flower; the sepals collectively constitute the **calyx.**
- **Petals**—whorls of structures located inside and usually above the sepals; may be large and pigmented (in insect-pollinated flowers) or absent (in wind-pollinated plants); the petals collectively constitute the corolla.

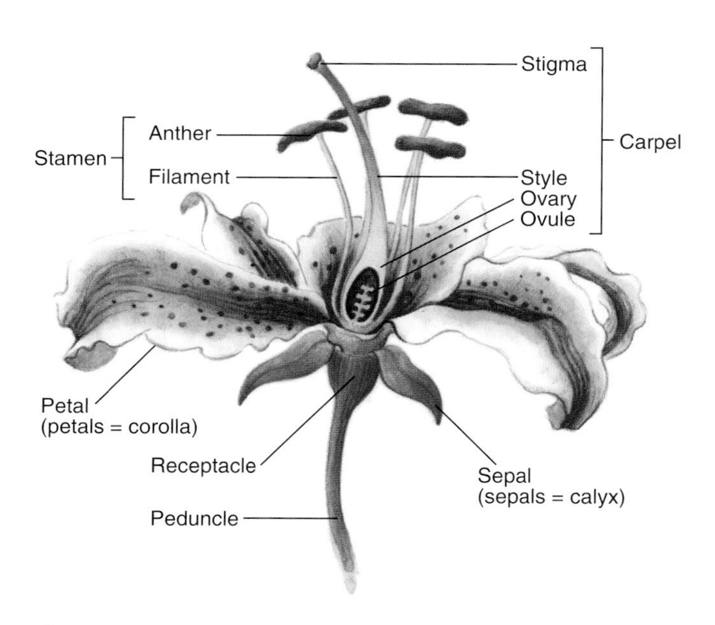

Figure 30.3
The parts of a flower. This is a generalized flower with four primary parts: sepals, petals, stamens, and carpels. The carpels and stamens are the fertile parts of a flower. The roles of carpels and stamens in the life cycle of angiosperms are shown in figure 30.4.

- **Perianth**—the combined calyx and corolla.
- **Androecium**—the male portion of the plant which rises above and inside the petals; consists of stamens, each of which consists of a **filament** atop which is located an **anther;** inside the anthers are **pollen grains,** which produce the male gametes.
- **Gynoecium**—the female portion of the plant which rises above and inside the androecium; consists of one or more **carpels,** each made up of an **ovary, style,** and **stigma;** the ovary contains **ovules** that contain the female gametes; during pollination, pollen grains are transferred to the stigma, where they germinate and grow a tube through the style to the ovary. The term **pistil** is sometimes used to refer to an individual carpel or a group of fused carpels.

The sepals and petals are usually the most conspicuous parts of a flower, and a variety of flower types are described by the characteristics of the perianth. In **regular** (actinomorphic) flowers such as tulips, the members of the different whorls of the flower consist of similarly shaped parts that radiate from the center of the flower and are equidistant from each other. The flowers are **radially symmetrical.** In other flowers such as snapdragons, one or more parts of at least one whorl are different from other parts of the same whorl. These flowers are generally bilaterally symmetrical and are said to be **irregular** (zygomorphic).

Procedure 30.1

Examine flowers and their parts

1. Obtain a flower provided in the laboratory.
2. Remove the parts of the flower, beginning at the lower, outside rosettes of sepals and petals.
3. Locate the petals and sepals and determine their arrangement and point of attachment.

Question 1

How would you describe this flower?

4. Remove the petals and sepals.
5. Locate and remove a stamen and place it on a slide. Examine the stamen with low magnification.
6. Add a drop of water to the preparation and open the anther (if necessary) to disperse pollen. Cover with a coverslip.
7. Use high magnification to locate the generative and tube nucleus of a pollen grain. The **generative nucleus** is usually small, spindle-shaped, and off center (you cannot see it easily). The **tube nucleus** is larger and centered. You'll learn more about pollen nuclei later in this exercise.
8. To enhance visibility of the nuclei, stain them with acetocarmine. Add one or two drops to the preparation and gently heat it. Do not overheat. Reexamine the preparation.
9. Locate the gynoecium of the flower, and make longitudinal and transverse sections. The gynoecium consists of one or more fused carpels, each with an interior cavity called a locule containing ovules.

Question 2

a. How many carpels (locules) are apparent?

b. How many ovules are developing in each locule?

10. Examine the other types of flowers available in the lab. Repeat procedure 30.1 to guide your examination. Preserved specimens of *Lilium* (lily) and *Ranunculus* (buttercup) may also be provided for you to study and dissect.

Alternation of Generations in Flowering Plants

The life cycle of flowering plants involves the alternation of a multicellular haploid stage with a multicellular diploid stage as is typical for all plants (see fig. 29.1, fig. 30.4). The diploid **sporophyte** produces haploid spores by meiosis. Each haploid spore grows into the **gametophyte,** which produces gametes by mitosis and cellular differentiation. In angiosperms the sporophyte is the large, mature organism with flowers that you easily recognize. The gametophyte within the flower is reduced to a pollen grain (that produces a sperm nucleus) or an embryo sac (that produces an egg) within an ovule. Be sure to review your textbook's description of alternation of generations.

Production of spores in the sporophyte by meiosis is part of a larger process called **sporogenesis.** Flowering plants produce two kinds of spores: **microspores** and **megaspores.** Production of gametes by the gametophytes is **gametogenesis.**

Microsporogenesis, Production of Pollen, and Microgametogenesis

Microsporogenesis is the production of microspores within **microsporangia** of a flower's anthers via meiosis of **microspore mother cells** (microsporocytes) (fig. 30.5). These microspores grow and mature into microgametophytes, also known as pollen grains (fig. 30.6). The haploid nuclei in a

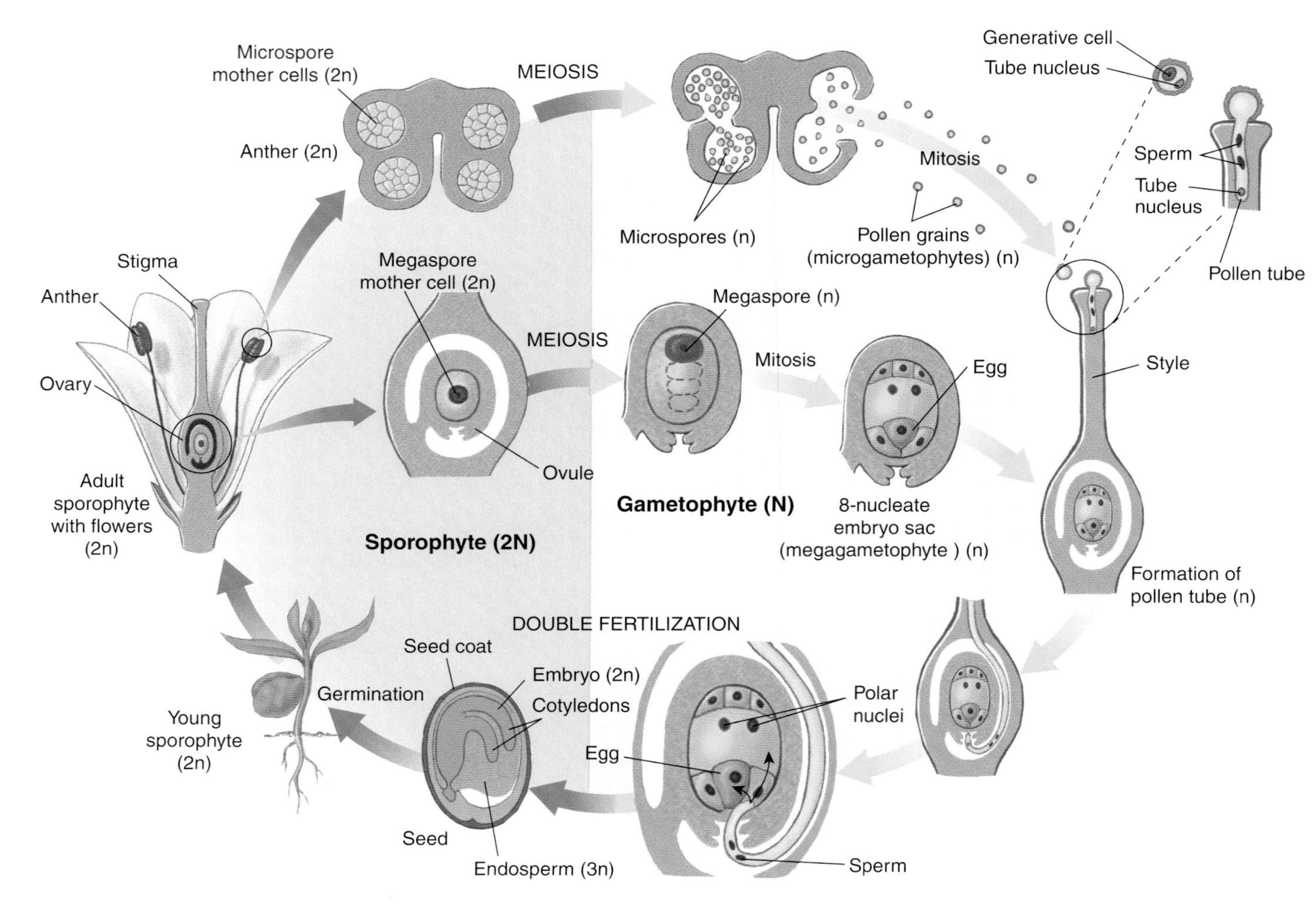

Figure 30.4

Angiosperm life cycle. Eggs form within the embryo sac inside the ovules, which, in turn, are enclosed in the carpels. The pollen grains, meanwhile, form within the sporangia of the anthers and are shed. Fertilization is a double process. A sperm and an egg come together, producing a zygote; at the same time, another sperm fuses with the polar nuclei to produce the endosperm. The endosperm is the tissue, unique to angiosperms, that nourishes the embryo and young plant.

mature pollen grain include a **tube nucleus** (or vegetative nucleus) and a **generative nucleus.** The pollen grain will germinate when it lands on a flower stigma, and the tube nucleus will control the growth of the pollen tube. The generative nucleus will replicate to produce two **sperm nuclei.**

Pollen cause allergies in many people. However, studies of pollen are important to science beyond the treatment of allergies. For example, geologists examine pollen brought up in sediment cores during oil drilling. Dark brown to black pollen indicate temperatures too high for oil deposits and indicate that a well will likely produce natural gas. Orange pollen indicate the less intense heat associated with high-quality oil production. In addition, examination of fossil pollen tells us about the diversity of ancient flora and helps us locate ancient seas and their shorelines where pollen is known to accumulate.

Procedure 30.2

Examine stages of microsporogenesis and microgametogenesis in *Lilium*

1. Examine fresh or preserved specimens of dehiscent and predehiscent anthers (fig. 30.7).

Question 3

a. What differences can you detect between the structures of dehiscent and predehiscent anthers?

b. Which stage is the most mature?

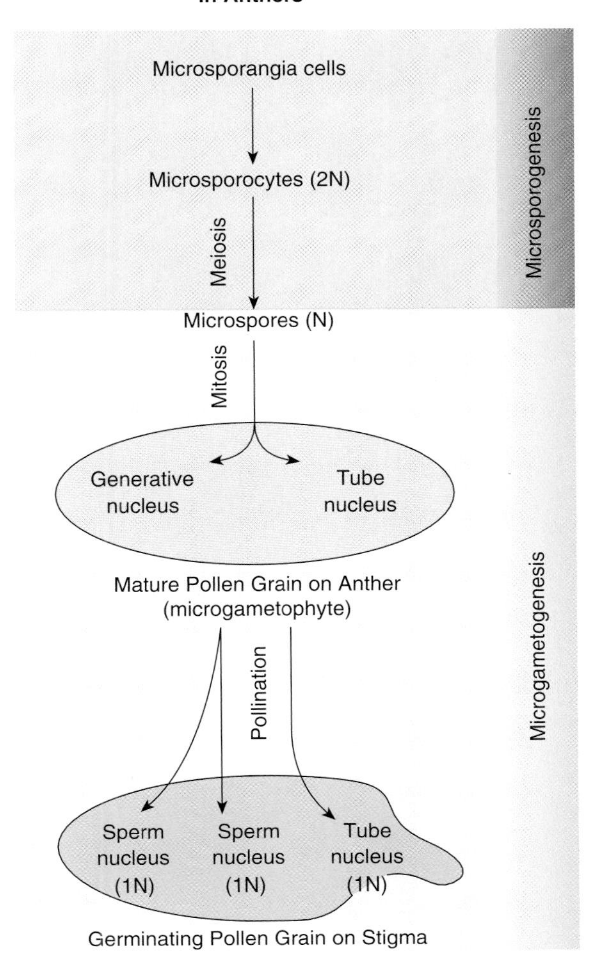

Figure 30.5

Microsporogenesis and microgametogenesis in the anthers of flowers. Mature pollen grains (see fig. 30.6) are microgametophytes. Micrographs of microsporogenesis and microgametogenesis are shown in figure 30.8.

2. Examine a prepared cross-section of a young anther. Note the immature sporangia tissue that will form microsporocytes.
3. Examine a prepared cross-section of a lily anther showing microsporocytes in early prophase I (fig. 30.8*a*).
4. Examine a prepared slide showing stages of meiosis II (fig. 30.8*b*).
5. Examine a prepared slide showing pollen tetrads of microspores produced by meiosis (fig. 30.8*c*).
6. Examine a prepared slide showing mature pollen with two or more nuclei (fig. 30.8*d*).
7. Examine living or prepared pollen from various plants if available. Note any differences among pollen grains and differences between pollen of monocots and dicots.

Figure 30.6

Scanning electron micrograph of pollen grains. Pollen grains of ragweed are spherical and elaborately sculptured. Their outer wall contains proteins that regulate germination of the pollen grain when it lands on a stigma of a ragweed flower. These same proteins also cause allergies in many people.

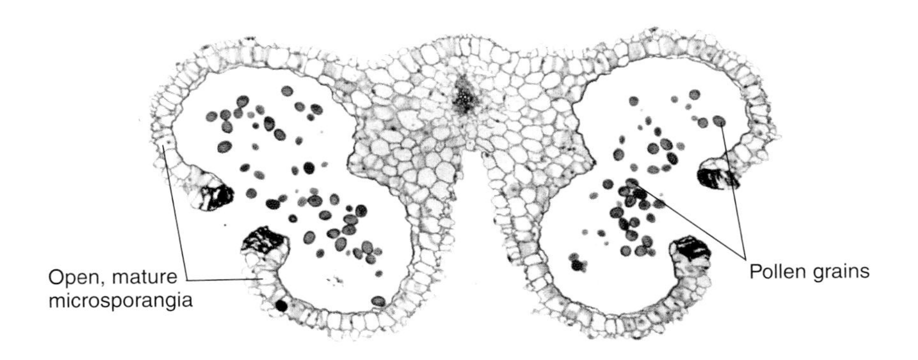

Figure 30.7

Lilium anther (40×). Mature pollen grains are the product of microsporogenesis and microgametogenesis within the microsporangia of flower anthers. A higher-magnification view of developing pollen grains is shown in figure 30.8.

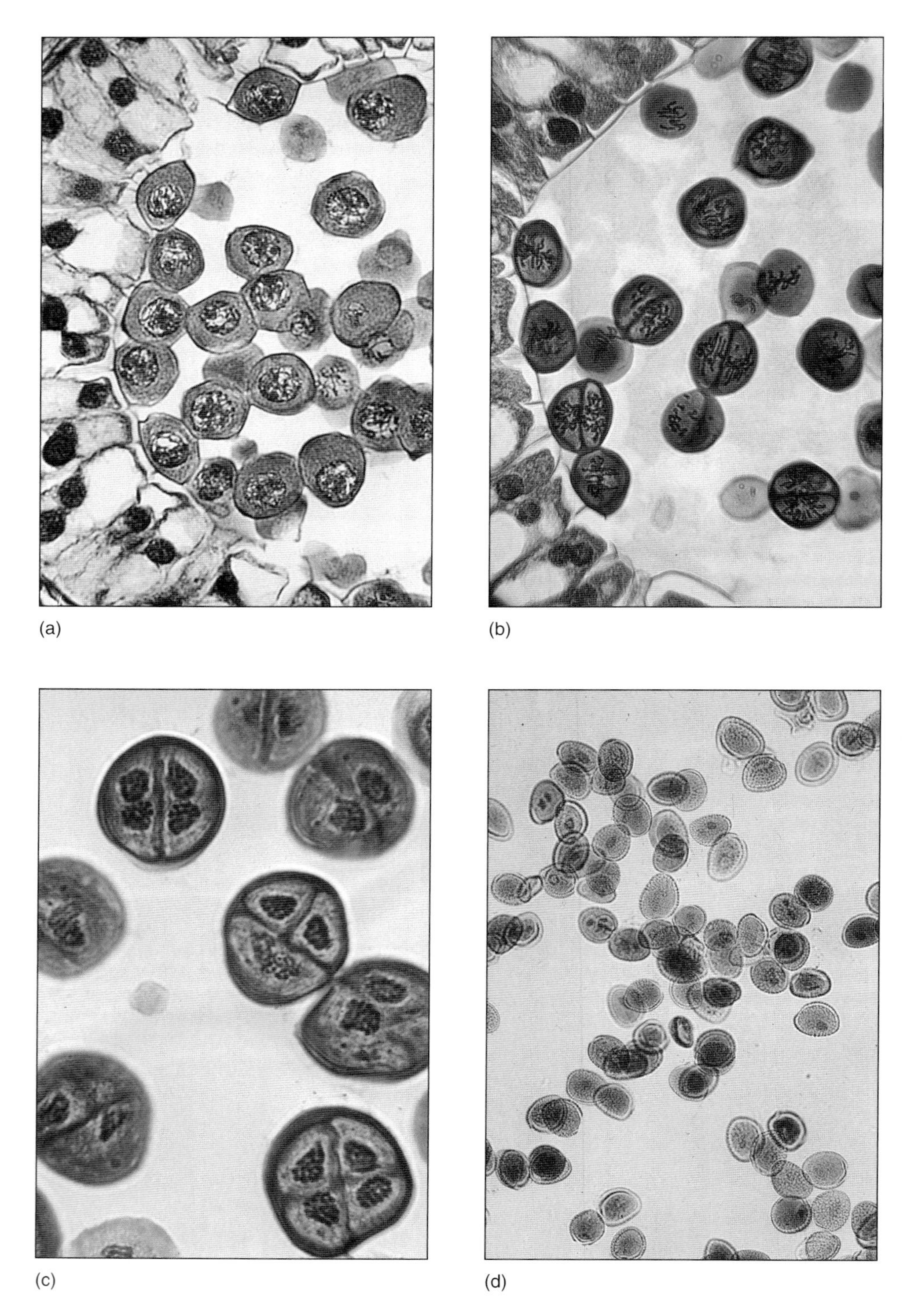

Figure 30.8

Stages of microsporogenesis and microgametogenesis in *Lilium*. (*a*) Cells in early prophase I. (*b*) Cells in second meiotic division. (*c*) Tetrads of microspores from meiosis. (*d*) Mature pollen. For an overview of these stages, see figure 30.5.

Procedure 30.3

Observe germination of pollen grains

NOTE

Begin this experiment at the beginning of the lab period so that the pollen grains will germinate before the lab is over.

1. At the beginning of the lab period, place some pollen in a drop of 1% sucrose within a ring of petroleum jelly on a microscope slide. Alternatively, a preparation of pollen germinating on 5% sugar and agar may be available for observation.
2. Apply a coverslip and examine the pollen intermittently with your microscope throughout the lab period. Incubate the slide in a warm place.
3. Ring the coverslip with petroleum jelly to prevent the preparation from drying.

Question 4

a. How many pollen grains germinated?

b. Can you see vegetative and generative nuclei in the pollen tubes?

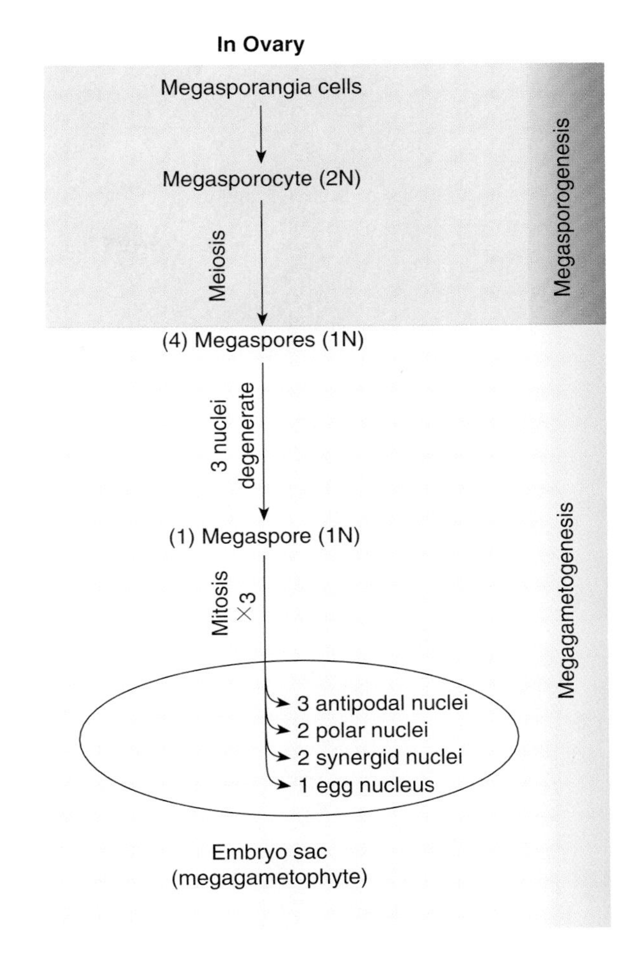

Figure 30.9
Megasporogenesis and megagametogenesis in the ovaries of flowers. The embryo sac is the mature megagametophyte. The locations of these processes are shown in figures 30.10 and 30.11.

Megasporogenesis, Production of Ovules, and Megagametogenesis

Megasporogenesis is the production of megaspores (fig. 30.9); it occurs in the sporangia of the flower ovary by meiosis of **megaspore mother cells** (megasporocytes). These megaspores undergo **megagametogenesis;** that is, they develop into **megagametophytes** that produce egg gametes. The megagametophyte and its surrounding tissues are called an **ovule.** Ovules usually have two coverings called **integuments.** The entire haploid structure is called the **embryo sac** and consists of only six to ten nuclei, one of which is an egg. A seven- or eight-celled embryo sac is most common (fig. 30.10).

Procedure 30.4

Examine stages of megasporogenesis and megagametogenesis in *Lilium*

1. Examine a cross section of a *Lilium* ovary and locate the six megasporangia. Within each of these megasporangia a megasporocyte will form and develop. The stages for development of a megasporocyte are shown in figures 30.9, 30.10, and 30.11.
2. Examine a prepared slide of a cross section of a *Lilium* ovary showing a diploid megasporocyte within the sporangium before meiosis (fig. 30.11*a*).
3. Examine a prepared slide showing the four-nucleate embryo sac after meiosis (fig. 30.11*b*). Meiosis produced four haploid megaspores in the embryo sac. In *Lilium* these nuclei will undergo mitosis, but this is not typical. In most other angiosperms, three of the four nuclei degenerate, and the single remaining nucleus passes through two mitotic divisions before the next stage.
4. Examine a prepared slide showing the eight-nucleate embryo sac (fig. 30.11*c*). Because of the large size of the embryo sac, it is seldom possible to observe all eight nuclei in the same section.

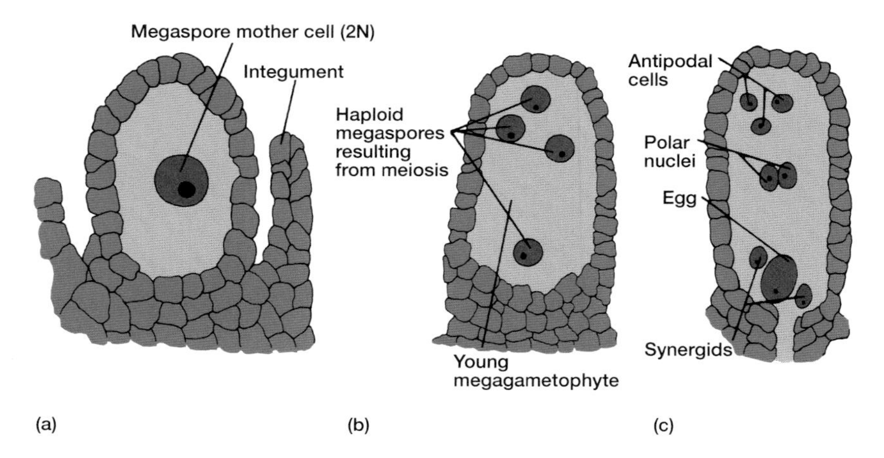

Figure 30.10

Megasporogenesis and megagametogenesis in *Lilium*. (*a*) The megaspore mother cell (megasporocyte) is diploid and undergoes meiosis. (*b*) Four haploid megaspores result from meiosis. (*c*) The mature megagametophyte contains eight nuclei, one of which is the egg.

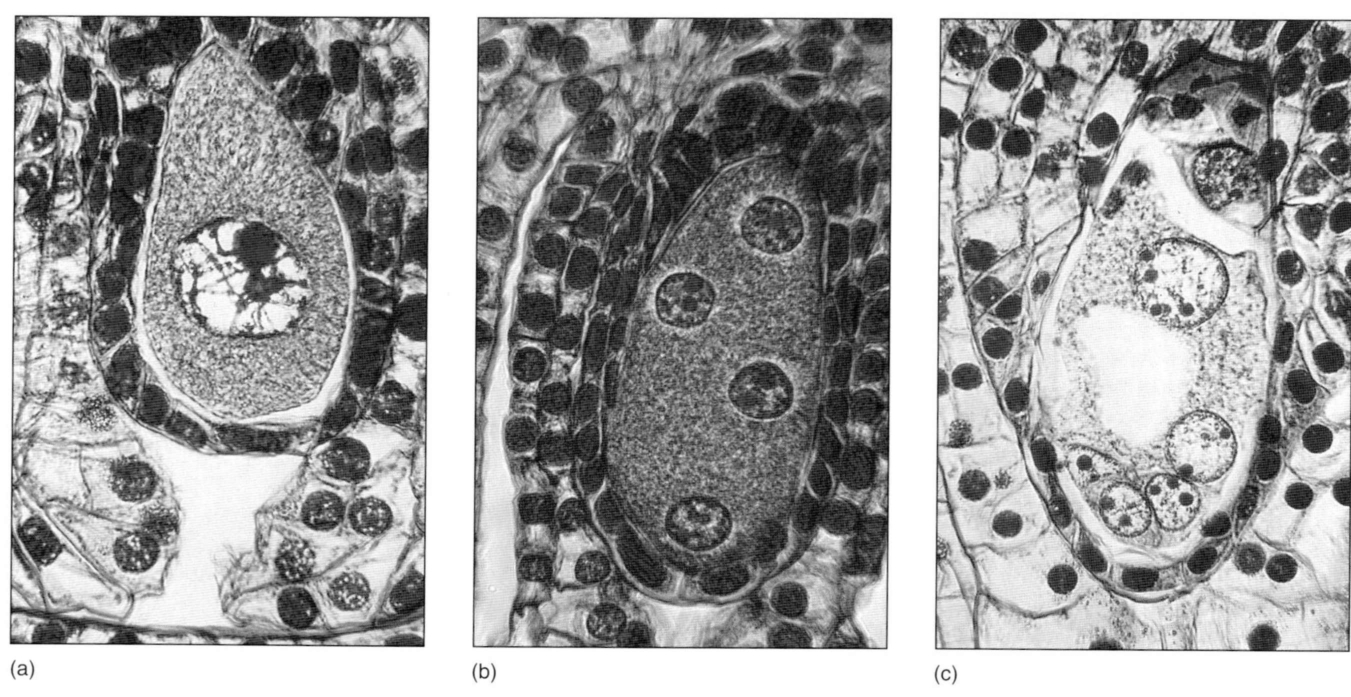

Figure 30.11

Stages of megasporogenesis and megagametogenesis in *Lilium*. (*a*) Diploid megasporocyte in lily ovary. (*b*) Embryo sac with four nuclei produced by meiosis. (*c*) Embryo sac (mature megagametophyte) with eight nuclei. For an overview of these stages, see figure 30.9.

5. At one end of the megagametophyte toward the micropyle (the small opening where the pollen tube enters the ovule), locate the eight nuclei including the egg and two **synergid nuclei** associated with fertilization. At the opposite end locate three **antipodal cells** that usually do not participate in reproduction. In the center are two **polar nuclei** that migrated from each pole of the megagametophyte.

POLLINATION AND FERTILIZATION

Biologists have long been interested in plant pollination, for obvious economic reasons. Angiosperm reproduction depends on pollination. However, you may have a more personal interest in pollen if you suffer from hay fever (fig. 30.6).

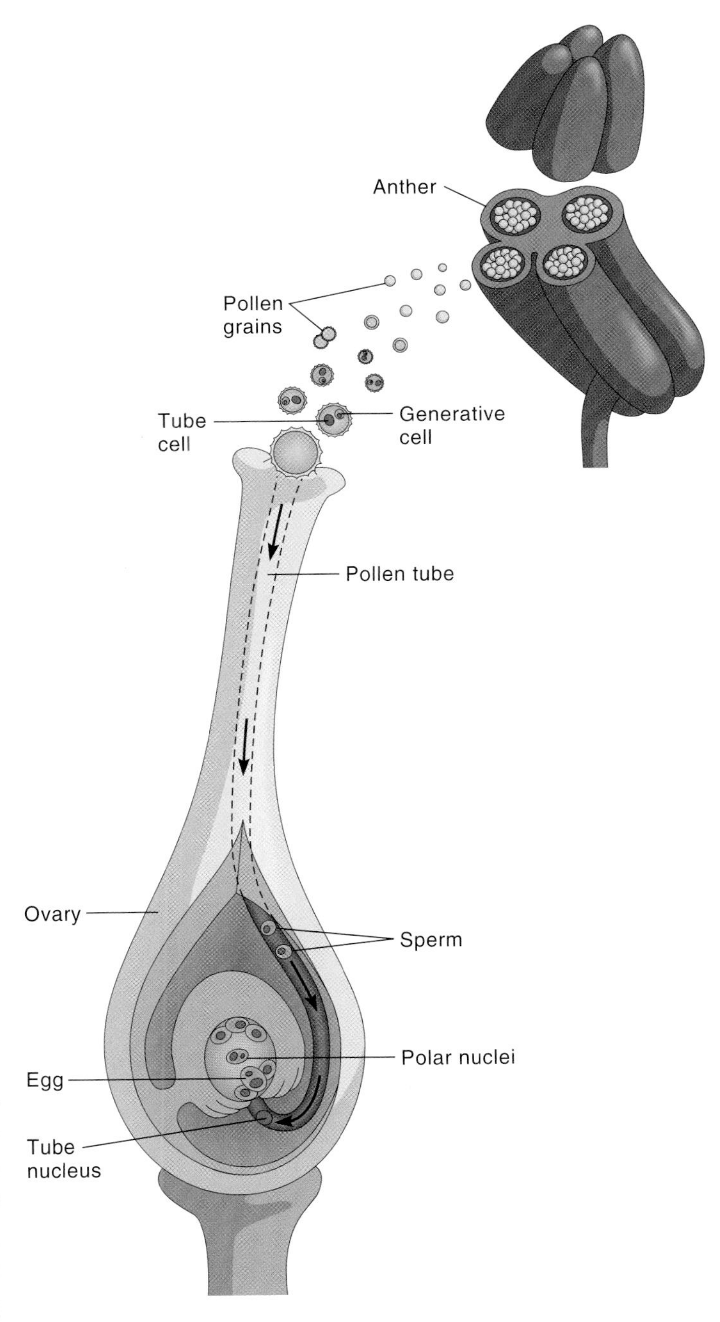

Figure 30.12

Pollination and fertilization. Pollen is transferred from the anther to the stigma of the carpel. Two sperm resulting from division of the generative cell move from the pollen grain through the pollen tube to an opening in the ovule. One sperm fertilizes the egg, and the other fertilizes the polar nuclei.

After the development of a microgametophyte (pollen grain) with sperm and a megagametophyte (ovule) with egg, sexual reproduction in angiosperms occurs as follows:

1. Pollination occurs when pollen is transported to the surface of the flower's stigma (fig. 30.12).
2. The pollen grain germinates, and a pollen tube grows through the stigma and style to the ovary. Its growth is governed by the style and the tube nucleus of the pollen grain.
3. One sperm nucleus fuses with the egg to form the diploid (2N) zygote, and the other sperm fuses with the two polar nuclei to form a triploid (3N) nucleus. This process is called **double fertilization** and is characteristic of angiosperms. The triploid nucleus will give rise to the endosperm.
4. The zygote develops into the embryo. The triple-fusion of the sperm nucleus and two polar nuclei forms the triploid endosperm that provides food for the embryo.
5. The integuments of the ovule form the seed coat, and the fruit develops from the ovary and other parts of the flower.

SEED AND EMBRYO DEVELOPMENT

A **seed** is a mature ovule that includes a seed coat, a food supply, and an embryo. Embryology of the mature zygote occurs within the seed and before germination. This embryology and its controlling factors are complex, but stages of development are easily observed.

The stages of embryo development in the seed of *Capsella* (a dicot) are shown in figure 30.13. The developing embryo grows, absorbs endosperm, and stores those nutrients in "seed leaves" called cotyledons. Development includes the following stages:

- **Proembryo stage.** During development, the **zygote** divides to form a mass of cells called the **embryo.** Initially the embryo consists of a **basal cell, suspensor,** and a two-celled proembryo. The suspensor is the column of cells that pushes the embryo into the **endosperm.** Note that the endosperm is extensive but is being digested.
- **Globular stage.** Cell division of the proembryo soon leads to the globular stage that is radially symmetrical and has little internal cellular organization.
- **Heart-shaped stage.** Differential division of the globular stage produces bilateral symmetry and two **cotyledons** forming the heart-shaped embryo. The enlarging cotyledons store digested food from the endosperm. Tissue differentiation begins, and root and shoot meristems soon appear.
- **Torpedo stage.** The cotyledons and root axis soon elongate to produce an elongate torpedo-stage embryo. Procambial tissue appears and will later develop into vascular tissue.

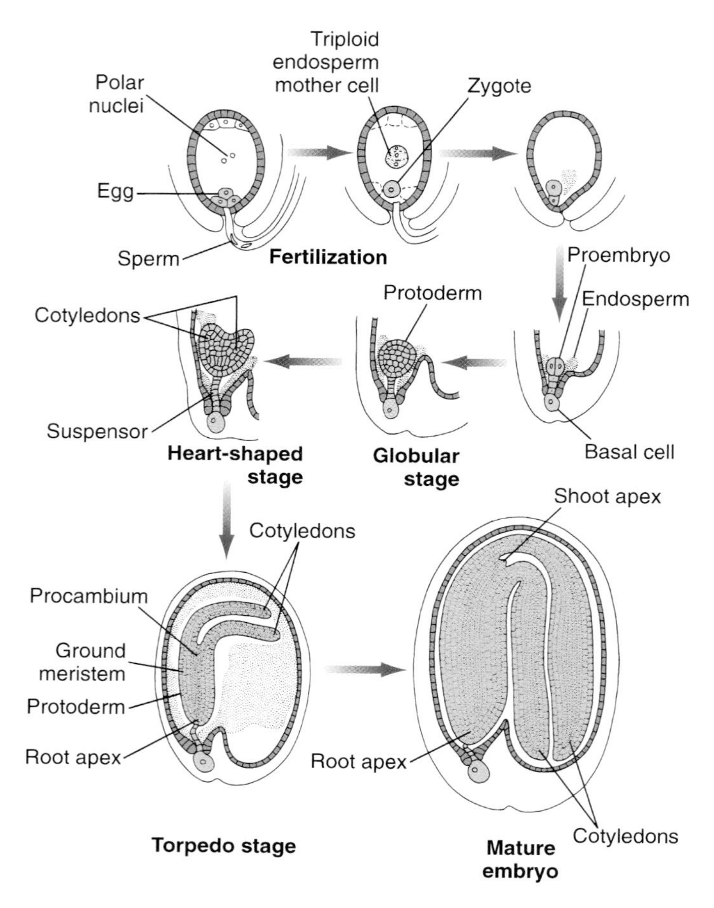

Figure 30.13

Stages of development in a *Capsella* embryo. This development transforms a zygote into a mature embryo.

- **Mature embryo.** The mature embryo has large, bent cotyledons on each side of the **stem apical meristem.** The **radicle,** later to form the root, is differentiated toward the suspensor. The radicle has a **root apical meristem** and **root cap.** The **hypocotyl** is the region between the apical meristem and the radicle. The endosperm is depleted, and food is stored in the cotyledons. The **epicotyl** is the region between attachment of the cotyledons and the stem apical meristem; it has not elongated in the mature embryo.

Procedure 30.5

Examine development of a *Capsella* embryo

1. Obtain and examine prepared slides showing the various stages of embryo development.
2. The cross section most likely passes through an entire fruit of *Capsella* and shows a number of sectioned, developing seeds each in a slightly different stage. Some seeds were sectioned off center and the stage may not be obvious. Nevertheless, each slide should include an example of at least one stage.
3. Locate among all the sectioned seeds examples of the globular, heart, torpedo, and mature embryos. These stages are continuous, so each slide may have multiple stages and intermediates between stages.
4. You may need to examine several slides. Find examples of as many stages as you can.
5. Compare your observations with figure 30.13.

Question 5

a. Why is the endosperm being digested?

b. Is *Capsella* a monocot or a dicot? How can you tell?

SEED STRUCTURE

Procedure 30.6

Observe parts of a bean seed

1. Obtain some beans that have been soaked in water for 24 h.
2. Peel off the seed coat and separate the two cotyledons. Between the cotyledons you'll see the young root and shoot.
3. Examine the opened seed and compare its structure with that shown in figure 30.14. Look for these features:
   - **Micropyle**—a small opening on the surface of the seed through which the pollen tube grew.
   - **Hilum**—an adjacent, eliptical area at which the ovule was attached to the ovary.
   - **Cotyledon**—food for the embryo.
   - **Embryo with young root and shoot**—develops into the new sporophyte.
4. Add a drop of iodine to the cut surface and observe the staining pattern. Indicate this pattern on figure 30.14.
5. Repeat steps 2–4 with soaked peas.

Question 6

a. How are seeds of peas and beans similar? How are they different?

b. Based on the appearance of seeds that you have examined, how do you think each is dispersed (by insects, wind, etc.)?

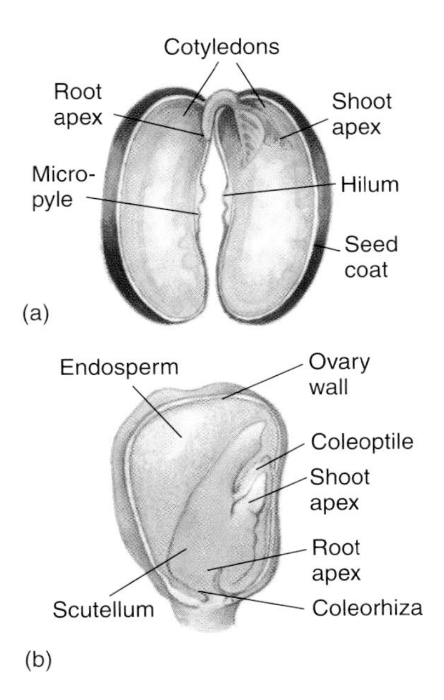

Figure 30.14
Seed structure of a garden bean (dicot) and corn (monocot). (*a*) The two cotyledons in each seed of garden bean (*Phaseolus vulgaris*) absorb the endosperm before germination. (*b*) Corn (*Zea mays*) has seeds in kernels (grains); the single cotyledon is an endosperm-absorbing structure called a scutellum.

Procedure 30.7

Examine a corn grain with embryo

1. Examine a prepared slide of a corn grain (fig. 30.15). Identify the following features:
   - **Endosperm**
   - **Cotyledon** (scutellum)
   - **Coleoptile** (sheath enclosing shoot apical meristem and leaf primordia of grass embryos)
   - **Root**
   - **Root cap**
   - **Coleorhiza** (sheath enclosing embryonic root of grass embryo)
   - **Shoot apical meristem**
2. Use a razor blade to longitudinally split a water-soaked corn grain.
3. Add a drop of iodine to the cut surface and observe the staining pattern. Indicate this pattern on figure 30.15.

Question 7

a. What does this staining pattern tell you about the content of the endosperm and embryo?

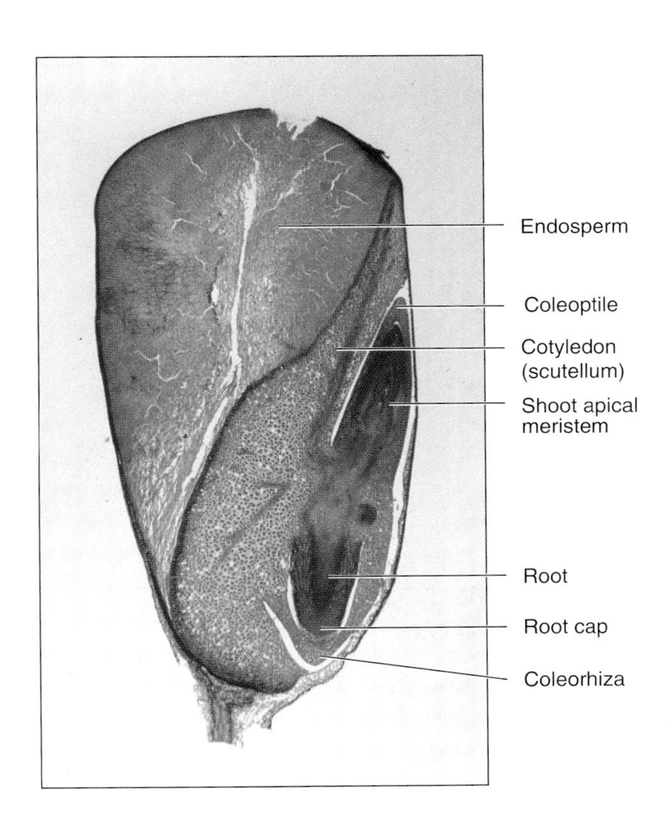

Figure 30.15
The structure of a corn grain, longitudinal section. The embryo is nourished by the starchy endosperm.

b. Do mature seeds of monocots or dicots store most of their food in cotyledons? How can you tell?

FRUIT

A **fruit** is a mature, ripened ovary plus any associated tissues. Therefore, a fruit contains seeds. Are you surprised that tomatoes and okra are fruit? A mature fruit is often larger than the ovary at the time of pollination and fertilization, which indicates that a great deal of development occurs while the seeds are maturing. Most fruits are either dry or fleshy. Dry fruits crack or split at maturity and release their seeds. Sometimes the dry wall surrounds the seed until it germinates. The fruit wall is usually tough and hard and is sometimes referred to as stony. The seeds of fleshy fruits remain in the tissue until germination.

A typical fruit has an outer wall called a **pericarp** composed of an **exocarp, mesocarp,** and **endocarp.** Within the pericarp are seeds, various partitions, and **placental tissues.** The fruit often includes the receptacle of the flower.

Dichotomous Key to Major Types of Fruit

I. Fleshy fruits
  A. Simple fruits (i.e., from a single ovary)
    1. Flesh mostly of ovary tissue
      a) Endocarp hard and stony; ovary superior and single-seeded (cherry, olive, coconut): **drupe**
      b) Endocarp fleshy or slimy; ovary usually many-seeded (tomato, grape, green pepper): **berry**
    2. Flesh mostly of receptacle tissue (apple, pear, quince): **pome**
  B. Complex fruits (i.e., from more than one ovary)
    1. Fruit from many carpels on a single flower (strawberry, raspberry, blackberry): **aggregate fruit**
    2. Fruit from carpels of many flowers fused together (pineapple, mulberry): **multiple fruit**
II. Dry fruits
  A. Fruits that split open at maturity (usually more than one seed)
    1. Split occurs along two seams in the ovary. Seeds borne on one of the halves of the split ovary (pea and bean pods, peanuts): **legume**
    2. Seeds released through pores or multiple seams (poppies, irises, lilies): **capsule**
  B. Fruits that do not split open at maturity (usually one seed)
    1. Pericarp hard and thick, with a cup at its base (acorn, chestnut, hickory): **nut**
    2. Pericarp thin and winged (maple, ash, elm): **samara**
    3. Pericarp thin and not winged
      (sunflower, buttercup): **achene**
      (cereal grains): **caryopsis**

Procedure 30.8

Observe diversity of fruits

1. Use the following descriptions to study and classify fruits available in the lab. This list is not complete but describes a few common types of fruit.

Procedure 30.9

Observe the structure of a bean, sunflower, corn grain, apple, and tomato

1. Examine the pod of a string bean. Notice that this single carpel has two seams that can open to release seeds. Remove the seeds; then locate and remove the seed coat. Split the seed and locate the embryo.

Question 8

a. Does the pod appear to be a single carpel with one cavity containing seeds?

b. Is the micropyle near the attachment of the seed to the pod?

2. Crack the outer coat of a sunflower fruit and remove the seed. Locate the embryo and determine whether sunflower is a monocot or dicot.

Question 9

a. Before today would you have referred to the uncracked sunflower achene as a fruit? Why or why not?

b. Is sunflower a monocot or dicot?

3. Examine an ear of corn from which the husks have been removed without disturbing the "silk." The strands of silk are styles of the gynoecium. Examine a corn grain that has been soaked in water for several hours. Section the grain longitudinally. Identify the endosperm and embryo. Refer to figure 30.15 for more information on the structure of the grain.

Question 10

a. If a corn grain is actually a fruit, where is the pericarp?

b. Does a corn grain have a cotyledon as well as an endosperm?

4. Examine an apple, which is an example of a pome. Section the apple longitudinally and transversely. Compare its structure with figure 30.16. Most of the outer flesh of the apple is derived from tissue other than the ovary. Locate the outer limit of the pericarp and the limit of the endocarp.

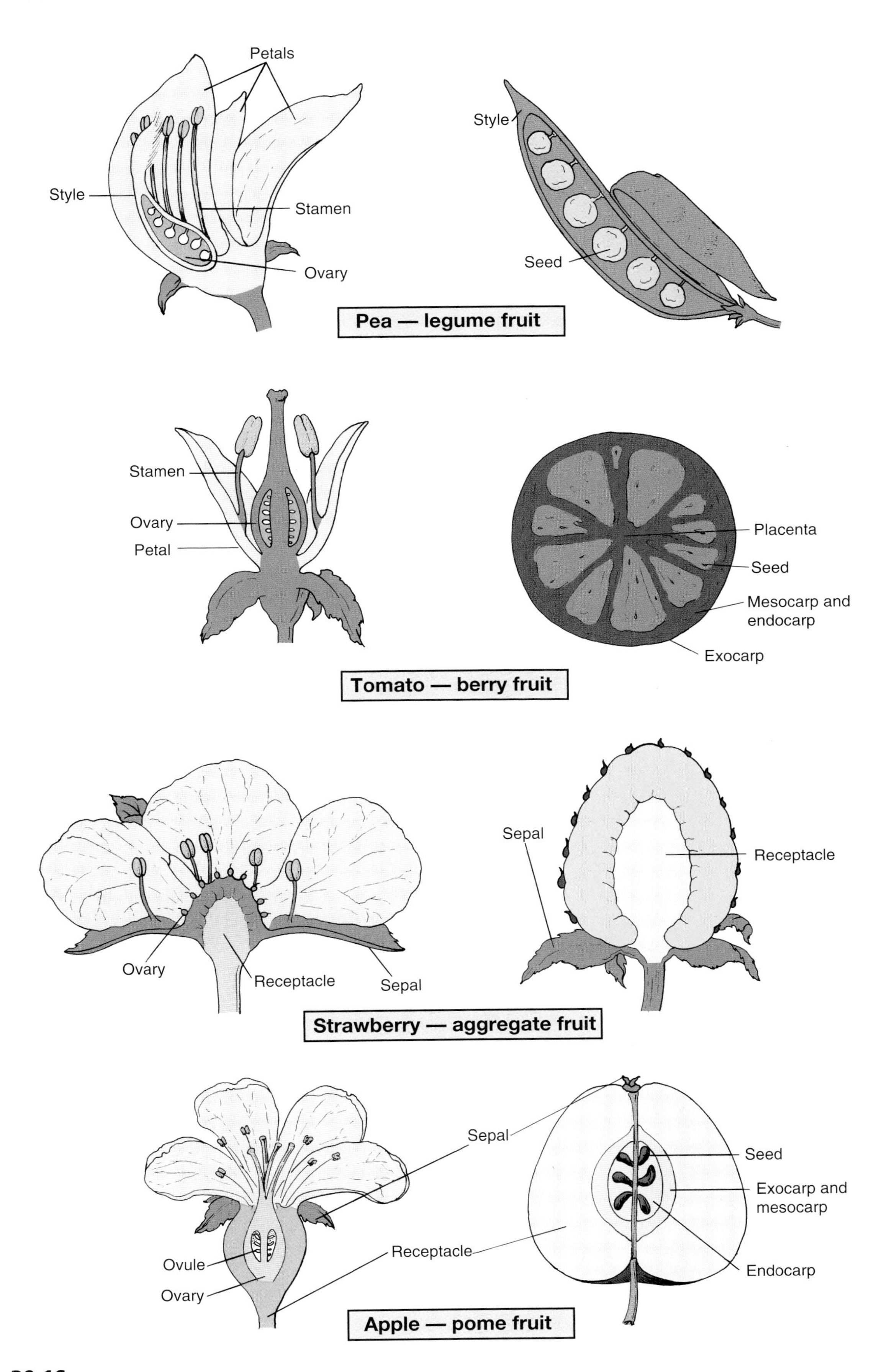

Figure 30.16

Diagrams of fruit and their originating flowers.

Question 11

a. How many carpels are fused to form an apple?

b. How might the fleshy pericarp aid in seed dispersal?

5. Examine a tomato and section it transversely. Compare its structure with that shown in figure 30.16. The jellylike material is the placenta giving rise to the ovules.

Question 12

How many carpels are fused to form a tomato?

Questions for Further Thought and Study

1. What is meant by "double fertilization"?

2. What features of seeds and fruits have enabled angiosperms to become so widespread?

3. What are the similarities and differences between cones and flowers?

4. What is the difference between a fruit and a vegetable?

5. Diagram the life cycle of an angiosperm. Which parts are haploid? Which are diploid?

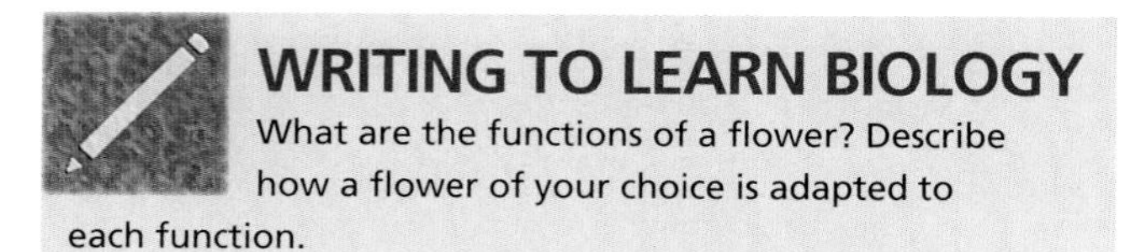

WRITING TO LEARN BIOLOGY

What are the functions of a flower? Describe how a flower of your choice is adapted to each function.

30–14

31

Plant Anatomy
Vegetative Structure of Vascular Plants

Objectives

By the end of this exercise you should be able to:
1. Describe the functions of leaves, stems, and roots.
2. Distinguish between primary and secondary growth.
3. Describe the functional significance of the internal and external structure of leaves, stems, and roots.
4. Explain what causes growth rings in wood.

The structure of plants varies greatly among species; compare, for example, an oak tree with a cactus. However, these structural differences are typically quantitative rather than qualitative; that is, the differences among leaves, roots, and stems result not from unique tissues but rather from different arrangements and proportions of the same tissues. These differences among plants represent different ways of achieving the same "goals": survival and reproduction. This exercise will concentrate on the structure of roots, stems, and leaves of vascular plants.

ROOTS

During seed germination, a **radicle** or young **primary root** emerges from the seed and grows down. The primary root soon produces numerous **secondary roots** and forms a root system that absorbs water and minerals, anchors the plant, and stores food. Root systems have different morphologies. For example, a **taproot system** has a large main root and smaller secondary roots branching from it (e.g., carrot). In a **fibrous root system,** the primary and secondary roots are similar in size (e.g., roots of many grasses) (fig. 31.1). Examine the displays of taproot and fibrous root systems available in the lab.

Primary growth of roots and all primary tissues is formed by **apical meristems.** A meristem is a localized area of cellular division. Apical meristems occur at the tips of roots and stems. Primary growth (i.e., growth in length) produces herbaceous (nonwoody) tissue. **Secondary growth** refers to growth in girth resulting from nonapical meristems, some of which are discussed later in this exercise.

Question 1

a. How do taproot systems and fibrous root systems help plants survive and reproduce?

b. Would one type of root system provide more adaptive advantages in a particular environment such as a rain forest?

The Root Apex

Examine the root tips of two-day-old seedlings of radish (*Raphanus*) and corn (*Zea*) with your dissecting microscope. Refer to figure 31.2 and identify the **root cap, root apical meristem, zone of elongation,** and **zone of maturation.**

The cone of loosely arranged cells at the root cap perceives gravity and protects the root apical meristem. The root cap protects the root by secreting mucilage and sloughing cells as the root grows through the soil (fig. 31.3). The root apical meristem is behind the root cap and produces all of the new cells for primary growth. These cells elongate in the zone of elongation, which is 1–4 mm behind the root tip. This elongation produces primary growth.

(a)

(b)

Figure 31.1

Two common types of root systems of vascular plants. (*a*) The taproot system of dandelion (*Taraxacum*) consists of a prominent taproot and smaller lateral roots. (*b*) The fibrous root system of a grass consists of many similarly sized roots. Fibrous root systems form extensive networks in the soil and successfully minimize soil erosion.

Question 2

a. In the space below, sketch the root tips that you examined. In which area of a root tip are cells largest? In which area are they smallest?

b. Aside from their size, do all cells in the root tip appear similar? Why is this significant?

Reexamine the two-day-old radish seedling with your dissecting microscope. Note the **root hairs** in the zone of maturation. Root hairs are outgrowths of epidermal cells and are short-lived. Root hairs increase the surface area of the root.

Question 3

a. Why do you think root hairs occur only in the zone of maturation?

b. What is the function of root hairs?

Primary Tissues of the Root

The root apical meristem produces cells that differentiate into primary tissues of the root. The outer layer of cells is the **epidermis.** Just inside the epidermis is the **cortex,** whose cells contain numerous **amyloplasts,** which are starch-containing plastids. The inner layer of the cortex is the **endodermis,** which regulates water flow to the vascular tissue in the center of the root. Immediately inside the endodermis is the **pericycle,** which can become meristematic and produce **secondary roots.**

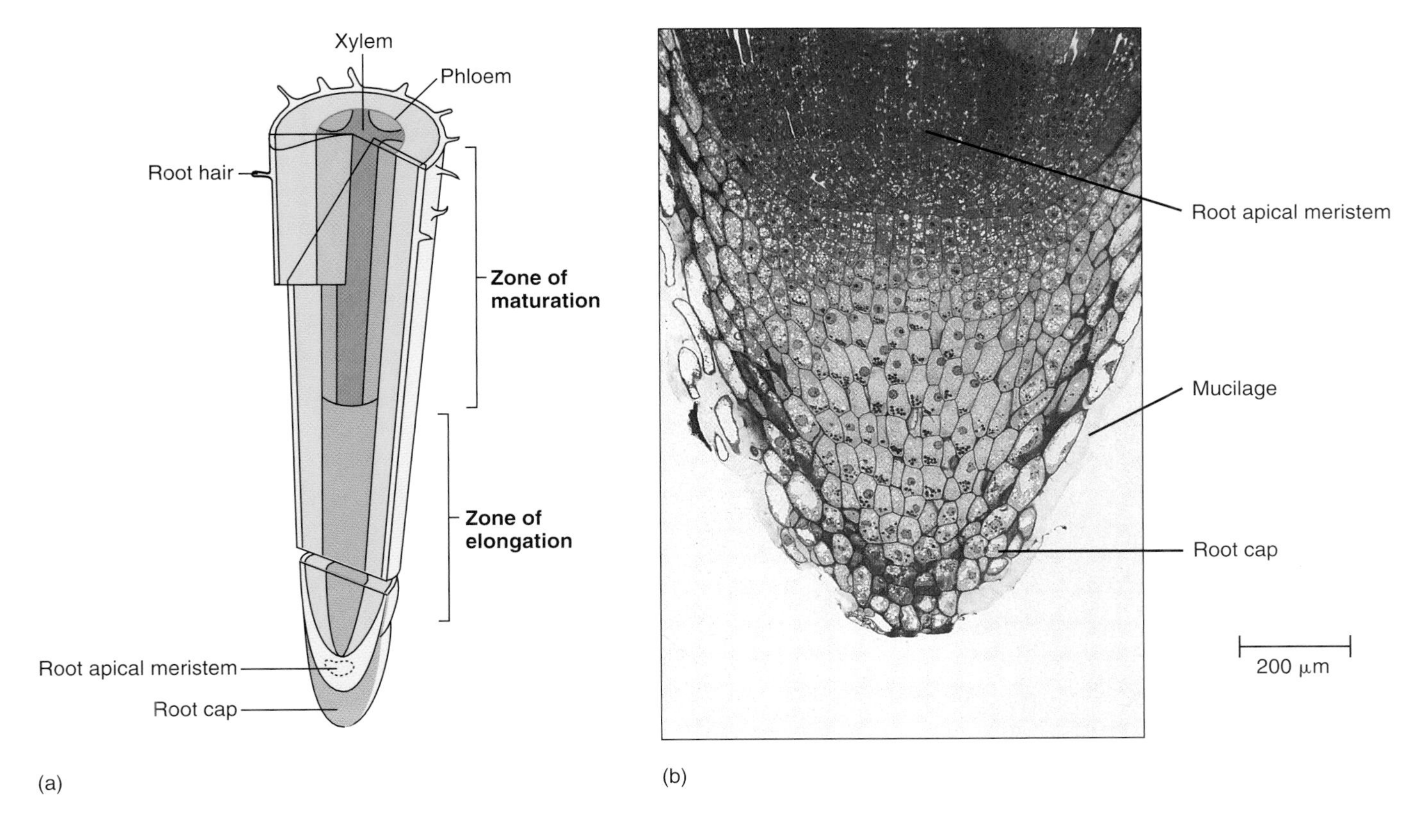

Figure 31.2

(*a*) Root tip showing root cap, root apical meristem, zone of elongation, and zone of maturation. (*b*) The root apical meristem is covered by a thimble-shaped root cap that protects the meristem as the root grows through the soil. Mucilage produced by the root cap lubricates the root as it grows through the soil (see fig. 31.3).

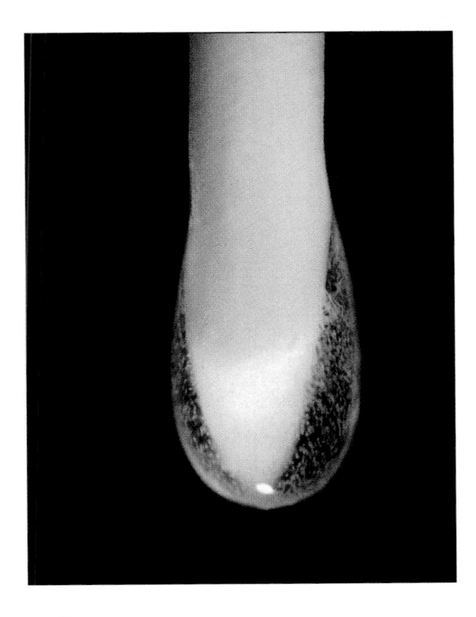

Figure 31.3

Tips of roots secrete large amounts of mucilage, a lubricant that helps the root force its way through the soil. The mucilage is secreted primarily by the root cap. Movement of the root through soil is also aided by the sloughing of root-cap cells. These sloughed cells are visible in the drop of mucilage on the tip of this root.

Procedure 31.1

Examine primary tissues of the root

1. Examine a prepared slide of a cross-section of a buttercup (*Ranunculus*) root (figs. 31.4 and 31.5). Sketch what you see. Label and state the function of each tissue that is present.

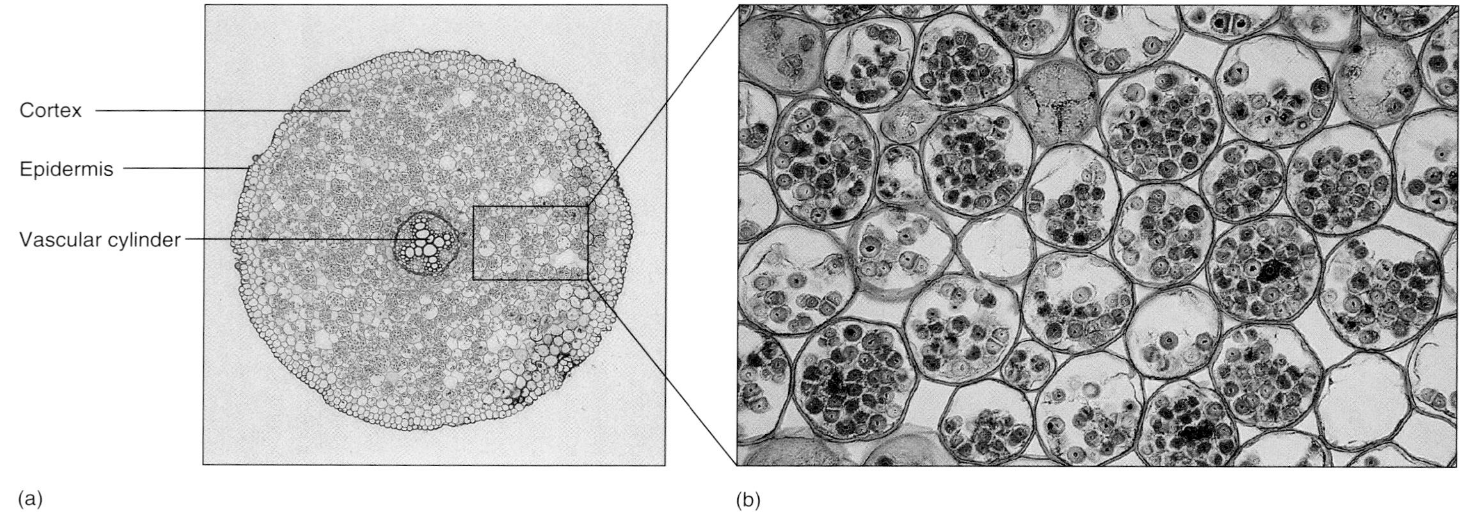

Figure 31.4

Transverse sections of a root of a buttercup (*Ranunculus*). (*a*) Overall view of mature root. The vascular cylinder includes tissues specialized for long-distance transport of water and solutes, whereas the epidermis forms a protective outer layer of the root (16×). (*b*) Detail of cortex (250×). Each parenchyma cell in the cortex contains many amyloplasts, which store starch.

2. Examine a prepared slide labeled "lateral root origin." Locate the epidermis, cortex, pericycle, and newly formed secondary root. Sketch the lateral root and label its parts.

3. If time permits, also examine a prepared slide of a cross-section of a corn (*Zea*) root.

Question 4

a. Based on the presence of amyloplasts, what do you suppose is the primary function of the cortex?

b. Do secondary roots arise inside the primary root or on its surface?

c. How does the structure of a monocot root differ from that of a dicot?

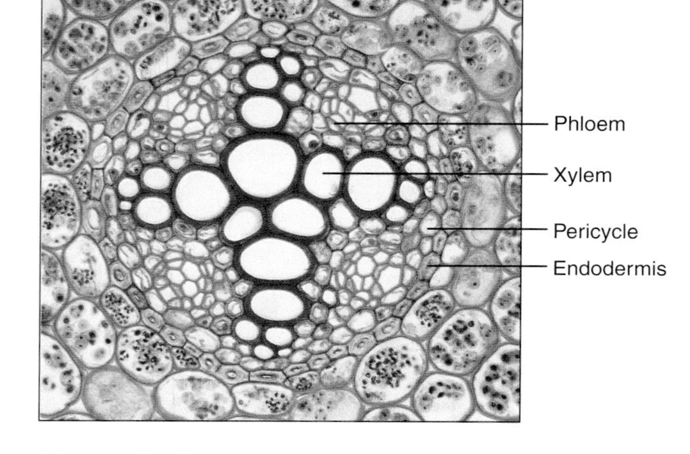

Figure 31.5

A cross-section through the center of a root of a buttercup (*Ranunculus*), a dicot (125×). The phloem and xylem are vascular tissues of the vascular cylinder shown in figure 31.4*a*.

In the center of the buttercup root is the **vascular** (fluid-conducting) **cylinder** composed of **xylem** and **phloem** (fig. 31.5). Xylem transports water and minerals; phloem transports most organic compounds in the plant. Water-conducting cells in the xylem of angiosperms are called **tracheids** and **vessels,** and are dead and hollow at maturity. Tracheids are long, spindle-shaped cells with thin areas called **pits** where the cell walls of adjacent cells overlap (fig. 31.6). Water moves through these pits from one cell to the next. Vessels are stacks of cylindrical cells with thin or completely open end-walls. Water moves through vessels in straight, open tubes. These tubes are usually stained red in slide preparations of buttercup roots.

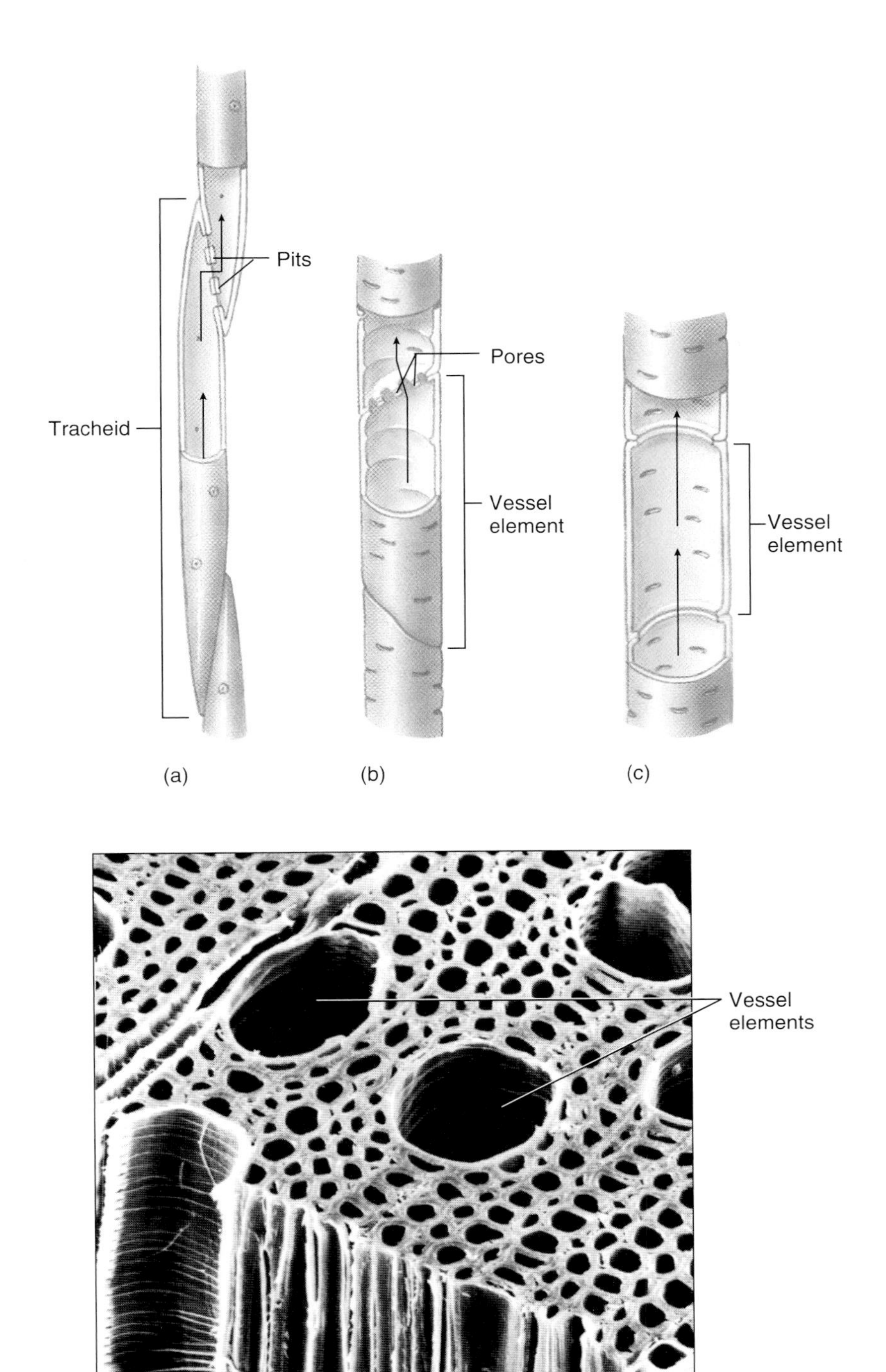

Figure 31.6

Comparison of vessel elements and tracheids. (*a*) In tracheids, water passes from cell to cell through pits. (*b*, *c*) In vessel elements, water moves through pores, which may be simple or interrupted by bars. (*d*) The large openings shown in this scanning electron micrograph of the wood of a red maple (*Acer rubrum*) are vessel elements (350×).

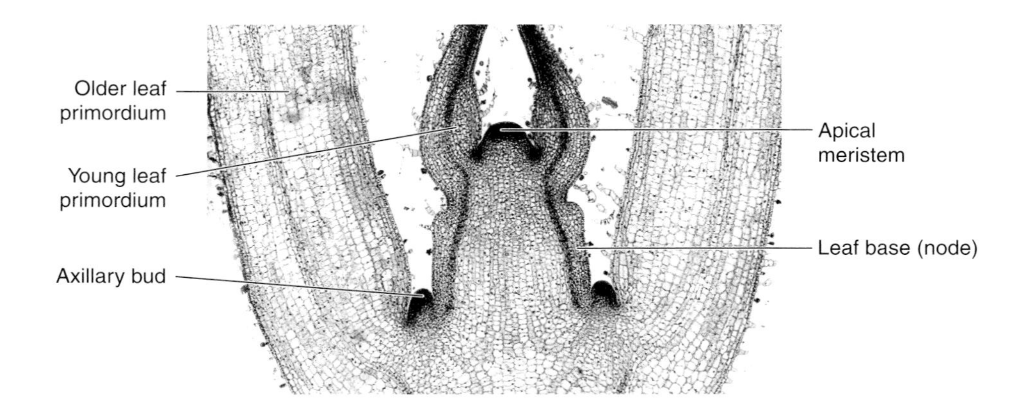

Figure 31.7

Coleus shoot tip. The apical meristem is the site of rapid cell division and primary growth. Young leaves are produced at the tip of the shoot. Shoot tips, unlike root tips, are not covered by a cap.

Conducting cells in phloem are called **sieve cells** and **sieve tube members,** and are alive at maturity. Phloem cells are small, thin-walled, and arranged in bundles that alternate with the poles of xylem. Sieve tube members are usually stained green.

Procedure 31.2

Examine carrot root

1. Prepare two thin cross-sections of a carrot root.
2. Stain one slice with iodine (a stain for starch) and examine it with your microscope.
3. Stain the other section of carrot root with phloroglucinol. Phloroglucinol stains **lignin,** a molecule that strengthens xylary cell walls.

CAUTION

Phloroglucinol contains hydrochloric acid. Do not spill any on yourself or your belongings!

Question 5

a. Where is starch located in a carrot root?

b. What can you conclude from this observation?

STEMS

Stems are often conspicuous organs of plants and function for support and transport of water and solutes. Some stems (e.g., cacti) also photosynthesize and store food.

The Shoot Apex

Examine a living *Coleus* plant and note the arrangement of leaves on the stem. Keeping this observation in mind, examine a prepared slide of a longitudinal section of a shoot tip of *Coleus* (fig. 31.7). Note that the dome-shaped **shoot apical meristem** is not covered by a cap as was the root. The shoot apical meristem produces young leaves (**leaf primordia**) that attach to the stem at a **node.** An **axillary bud** between the young leaf and the stem forms a branch or flower.

Question 6

a. How does the absence of a cap at a shoot apex differ from the apical meristem of a root?

b. How would you explain this difference?

c. Are all cells in the shoot apex the same size? Why is this significant?

External Features of Mature Woody Stem

Examine the features of a dormant twig (fig. 31.8). A **terminal bud** containing the apical meristem is at the stem tip and is surrounded by **bud scales. Leaf scars** from shed leaves occur at regularly spaced **nodes** along the length of the stem. The portions of stem between the nodes are called **internodes. Vascular bundle scars** may be visible within the leaf scars. Axillary buds protrude from the stem just distal to each leaf scar.

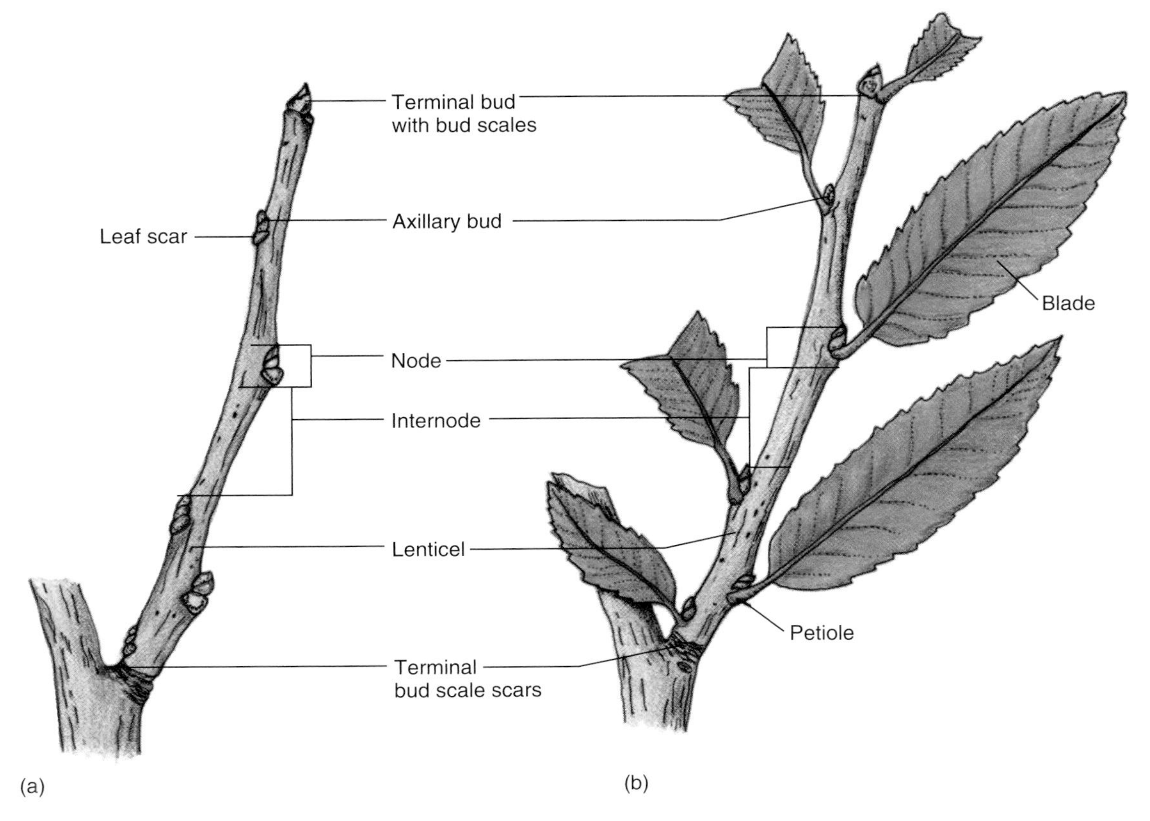

Figure 31.8
A woody twig. (*a*) The twig in its dormant winter condition. (*b*) The same twig as it appeared the summer before.

Search for clusters of **bud scale scars.** The distance between clusters or from a cluster to the terminal bud indicates the length of yearly growth.

Primary Tissues of Stems

Examine a prepared slide of a cross-section of a sunflower (*Helianthus*) stem (fig. 31.9). An **epidermis** covers the stem. The epidermis is coated with a waxy, waterproof substance called **cutin.** Below the epidermis is the **cortex,** which stores food. The **pith** in the center of the stem also stores food. In sunflower stems, the cortex is not uniform. Rather, the three to four cell layers of the cortex just below the epidermis are smaller, rectangular cells with unevenly thickened cell walls. These are **collenchyma** cells; they support elongating regions of the plant.

In stems, xylem and phloem are arranged in bundles (fig. 31.10). Locate and sketch one of these bundles on the slide you are examining.

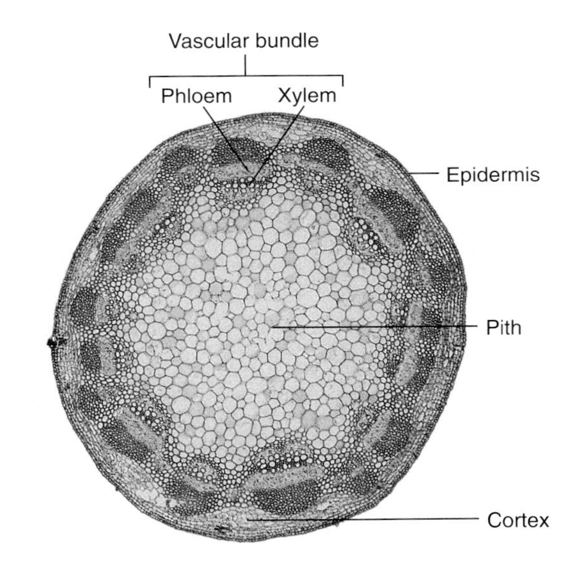

Figure 31.9
Cross-sectional view of a stem of sunflower (*Helianthus annus*), a dicot (10×). Note the ring of vascular bundles surrounding the pith. A high-magnification view of a vascular bundle from this stem is shown in figure 31.10.

Question 7

a. What is the significance of a coating of cutin on the epidermis?

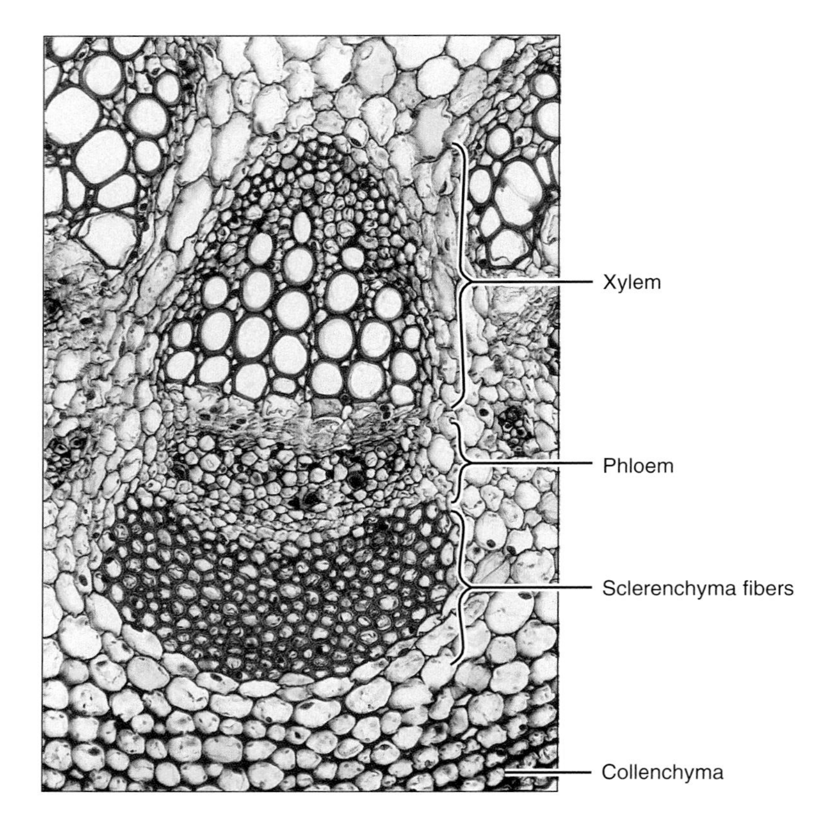

Figure 31.10
Cross-section of a vascular bundle of a sunflower stem (485×). The sclerenchyma fibers support the stem and protect the vascular tissue. Note that the walls of fibers are much thicker than those of adjacent cells.

b. How does the arrangement of xylem and phloem in stems differ from that in roots?

The darkly stained, thick-walled cells just outside the phloem in figure 31.10 are **sclerenchyma fibers,** which function in support. Sclerenchyma fibers from some plants are used to make linen, rope, and burlap.

The ring of vascular bundles in sunflower stems is typical of **dicots,** which are flowering plants with two **cotyledons** (seed leaves). Examine a prepared slide of a cross section of a corn stem (fig. 31.11). Corn is a **monocot** (a flowering plant with only one cotyledon). In the space below sketch a cross section of a sunflower stem and a corn stem. Note the distribution of the vascular bundles.

Question 8
How does the arrangement of vascular bundles differ in stems of monocots as compared to dicots?

Secondary Growth of Stems

Between the xylem and phloem and each vascular bundle in dicot stems is a meristematic tissue called **vascular cambium.** The vascular cambium is a secondary meristem that produces secondary growth (i.e., growth in girth). The vascular cambium is cylindrical and produces secondary xylem to its inside and secondary phloem to its outside.

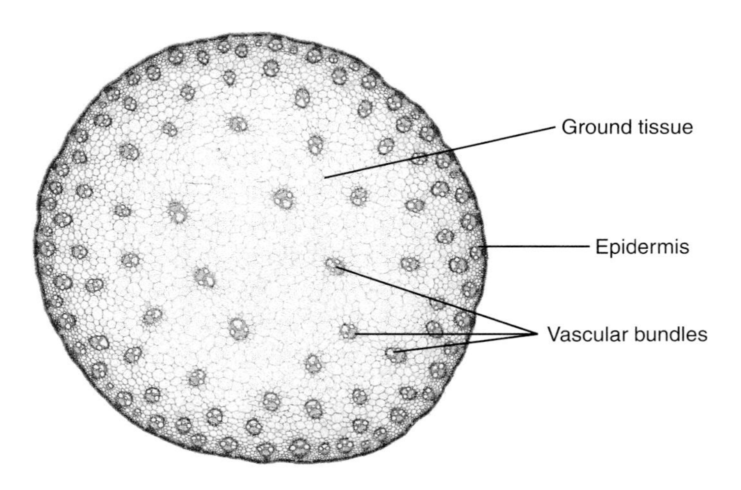

Figure 31.11

Cross-section of a stem of corn (*Zea mays*), a monocot (5×). Unlike in dicots such as sunflower (fig. 31.9), bundles of vascular tissue in monocots occur throughout the ground tissue. The stem is surrounded by an epidermis.

Question 9

How is secondary growth different from primary growth?

Procedure 31.3

Examine secondary growth in woody stems

1. Examine cross-sections of 1-, 2-, and 3-year-old basswood (*Tilia*) stems. The vascular cambium is a narrow band of cells between the xylem and phloem (fig. 31.12).
2. Compare the structures of the three stems. Note that the secondary xylem of older stems consists of concentric annual rings made of alternating layers of large and small cells. The large cells are formed in the spring, and the smaller cells are produced in summer.
3. In the space below draw 1-, 2-, and 3-year-old stem cross-sections.

Question 10

a. How do you account for this seasonal production of different-sized cells?

b. What is the common name for secondary xylem?

c. What is "grain" in wood?

d. Aside from conducting water and minerals, what is another important function of secondary xylem?

4. Examine a prepared slide of secondary xylem of pine (*Pinus*) (fig. 31.13). In cone-bearing plants such as pine, the conducting cells of xylem are all tracheids. The absence of vessel elements gives wood of these plants a relatively uniform appearance.
5. Now examine a prepared slide of secondary xylem of oak (*Quercus*), a flowering plant (fig. 31.14). Sketch what you see.

Question 11

a. What are the large cells in oak wood?

b. What is their function?

c. Which type of wood do you think transports more water per unit area, pine or oak? Why?

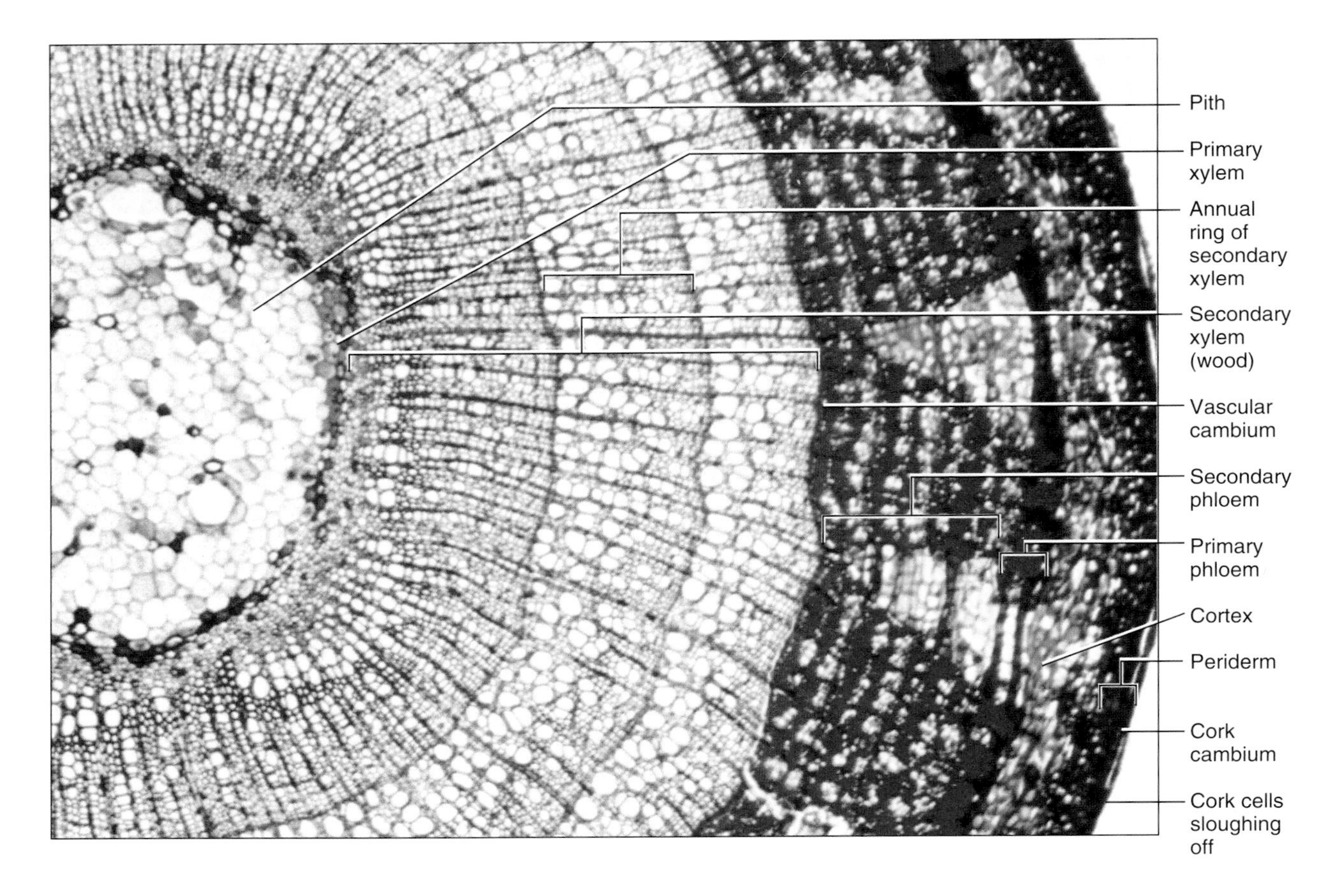

Figure 31.12

A cross-section of a portion of a young linden (*Tilia*) stem showing secondary growth (400×). The vascular cambium produces secondary xylem (wood) to the inside and secondary phloem to the outside. Note the annual rings in the secondary xylem. A close-up of a growth ring from pine is shown in figure 31.13.

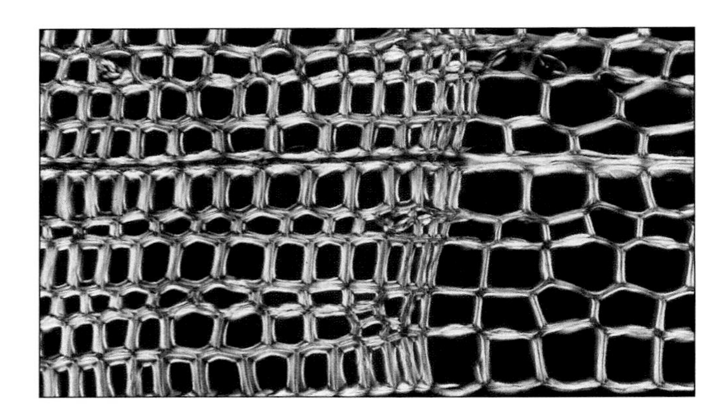

Figure 31.13

The wood of gymnosperms (such as this pine) consists almost exclusively of tracheids. These water-conducting cells are relatively small and help to support the plant. Although water moves slower through tracheids than through vessel elements, tracheids are less likely to be disabled by air bubbles that form in response to freezing and wind-induced bending of branches. The larger cells on the right side of this photo form during the wet days of spring and are called spring wood; the smaller cells at the left form during the drier days of summer and are called summer wood. The change in density between spring and summer wood produces a growth ring, which appears as "grain" in wood.

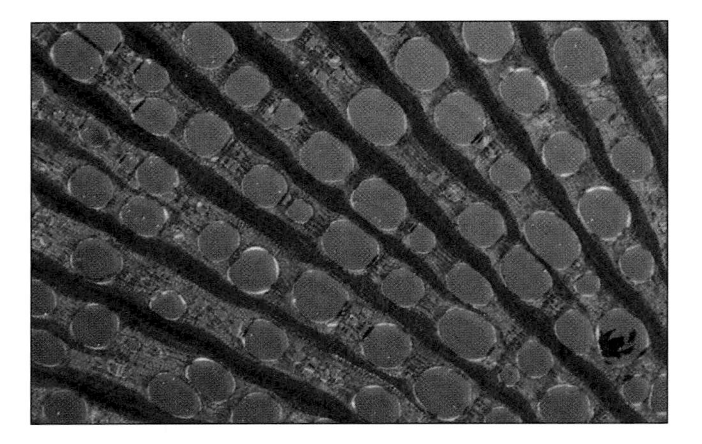

Figure 31.14

Unlike the xylem of gymnosperms, which contains only tracheids, the xylem of angiosperms also contains vessel elements. These vessel elements are much wider than tracheids and appear in this micrograph as large circles. Vessels are an adaptation for increased rates of water flow in angiosperms (see fig. 31.6*d*).

Bark

Bark includes all tissues outside of the vascular cambium, including the secondary phloem (fig. 31.12). Secondary phloem consists of pyramidal masses of thick- and thin-walled cells. The thin-walled cells are the conducting cells.

The increase in stem circumference resulting from activity of the vascular cambium eventually ruptures the epidermis. The ruptured epidermis is replaced by a tissue called the **periderm** that, like the epidermis, functions to minimize water loss. Periderm consists of cork cells produced by another secondary meristem called the **cork cambium.**

Locate the cork cambium in cross-sections of 1-, 2-, and 3-year-old basswood stems. The cork cambium is a band of thin-walled cells located beneath the epidermis. The cork cambium produces cork cells to the outside and cork parenchyma to the inside. Cork cells stain red because of the presence of suberin, a water-impermeable lipid.

What Good Is Bark?

You've doubtless seen chips of bark used as garden mulch to spread around plants. Because bark cells (secondary phloem and periderm) were once alive, they contain many nutrients that are released into the soil as the bark decays. However, sawdust (wood) has few nutrients in it and may result in nitrogen deficiency due to rapid decomposition. Consequently, sawdust is a poor mulch for plants.

Tree Girdling

You've probably heard of "tree girdling," which refers to stripping a ring of bark from a branch. In the wild, porcupines also girdle trees. Girdling removes secondary phloem from a tree, thereby leaving no pathway for photosynthate to move from the leaves to the roots. As a result, the shoot accumulates sugars and grows rapidly. The following spring it is impossible to send sugars from the roots to the girdled branch to renew growth. Thus, the branch or tree dies the year after it is girdled.

Question 12

Is the amount of cork similar in 1-, 2-, and 3-year-old stems? If not, how does it differ? Why is this important?

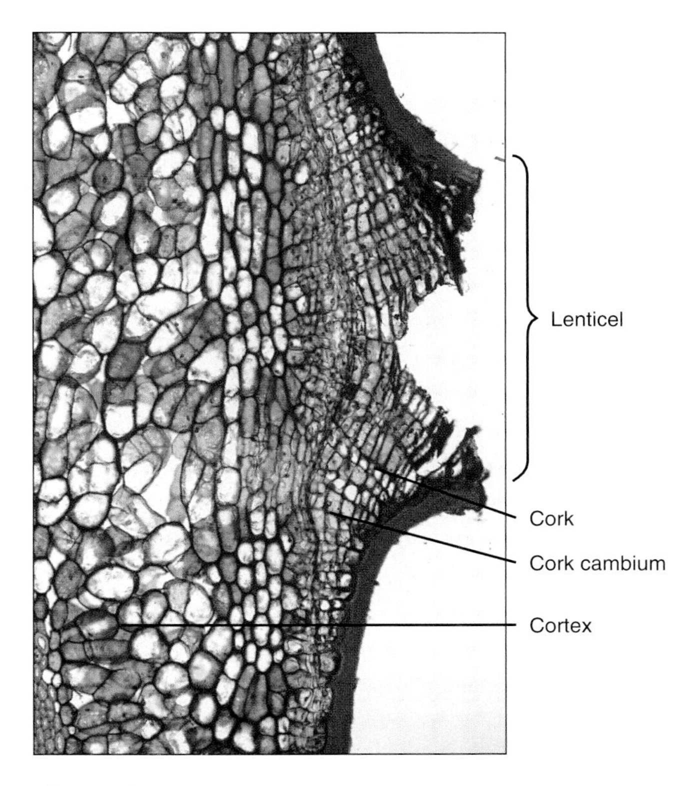

Figure 31.15

A lenticel. Cross-section of part of a young stem of elderberry (*Sambucus*). Gas exchange across the cork layer of the stem occurs through lenticels.

As the stem diameter continues to increase, the original periderm ruptures and new periderms form in the underlying tissues. Tissues outside the new periderm die and form encrusting layers of bark.

Gas exchange through peridermal tissues occurs through structures called **lenticels** (fig. 31.15). Examine a prepared slide of a lenticel and locate lenticels on a mature woody stem.

Question 13

How does a lenticel differ from the remainder of the periderm?

LEAVES

With few exceptions, most photosynthesis occurs in leaves, although some may occur in green stems. Leaves typically consist of a **blade** and a **petiole.** The petiole attaches the leaf blade to the stem. **Simple leaves** have one blade connected to the petiole, whereas **compound leaves** have several **leaflets** sharing one petiole (fig. 31.16). **Palmate** leaflets of a compound leaf arise from a central area, as your fingers arise from your palm. **Pinnate** leaflets arise in rows along a

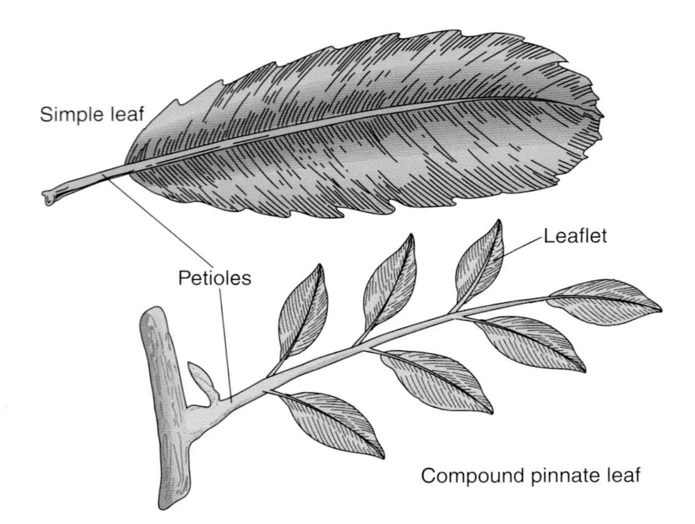

Figure 31.16
Simple and compound leaves.

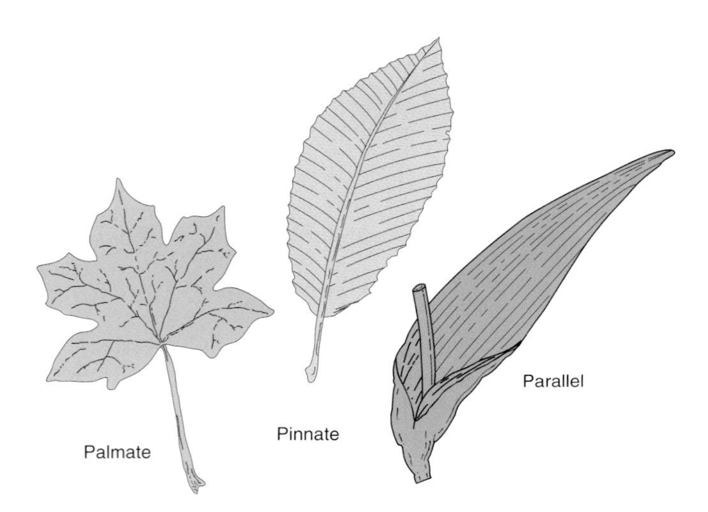

Figure 31.17
Palmate, pinnate, and parallel venation of leaves.

central midline. Examine the simple and compound leaves on display, and sketch a representative of each type of leaf.

Leaves are also classified according to their **venation** (i.e., arrangement of veins) (fig. 31.17). **Parallel veins** extend the entire length of the leaf with little or no cross-linking. **Pinnately veined** leaves have one major vein (i.e., a midrib) from which other veins branch. **Palmately veined** leaves have several veins each having branches. Veins of vascular tissue in leaves are continuous with vascular bundles in stems.

Examine the leaves on display in the lab and determine their venation. Sketch a few of these leaves to show their venation. List the names of common leaves on demonstration and indicate whether each is simple, pinnately compound, or palmately compound. Also indicate whether venation in each leaf is parallel, pinnate, or palmate.

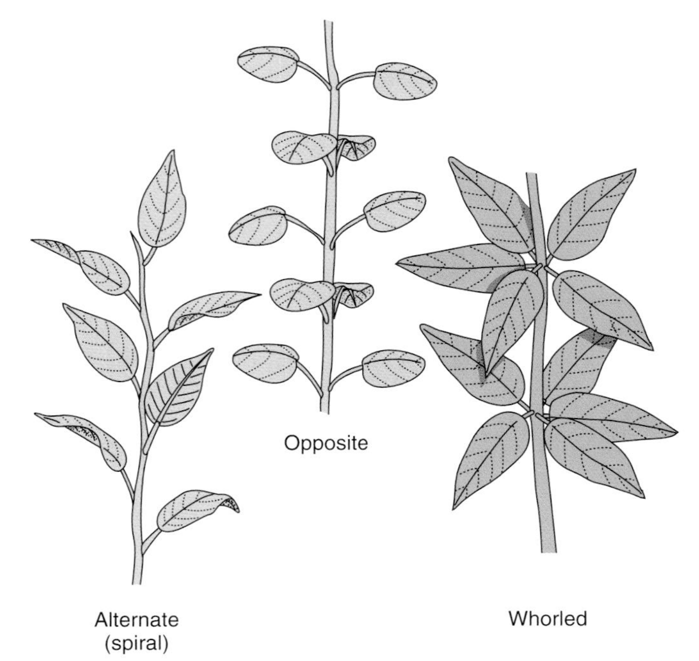

Figure 31.18
Patterns of leaf arrangement.

The arrangement of leaves on a stem is called **phyllotaxis** and characterizes individual plant species (fig. 31.18). **Opposite phyllotaxis** refers to two leaves per node located on opposite sides of the stem. **Alternate phyllotaxis** refers to one leaf per node, with leaves appearing first on one side of the stem and then on another. **Whorled phyllotaxis** refers to more than two leaves per node. Examine the plants on display in the lab to determine their phyllotaxis.

Internal Anatomy of a Leaf

Examine a cross-section of a leaf of *Ligustrum* (privet) (figs. 31.19 and 12.8). Note that the leaf is only 10–15 cells thick—pretty thin for a solar collector! The epidermis contains pores called **stomata,** each surrounded by two guard cells (you will study stomata again in the next exercise). Just below the upper epidermis are closely packed cells called **palisade mesophyll** cells; these cells contain about 50 chloroplasts per cell. Below the palisade layers are **spongy mesophyll** cells with numerous intercellular spaces. Examine and sketch a cross-section of a corn leaf.

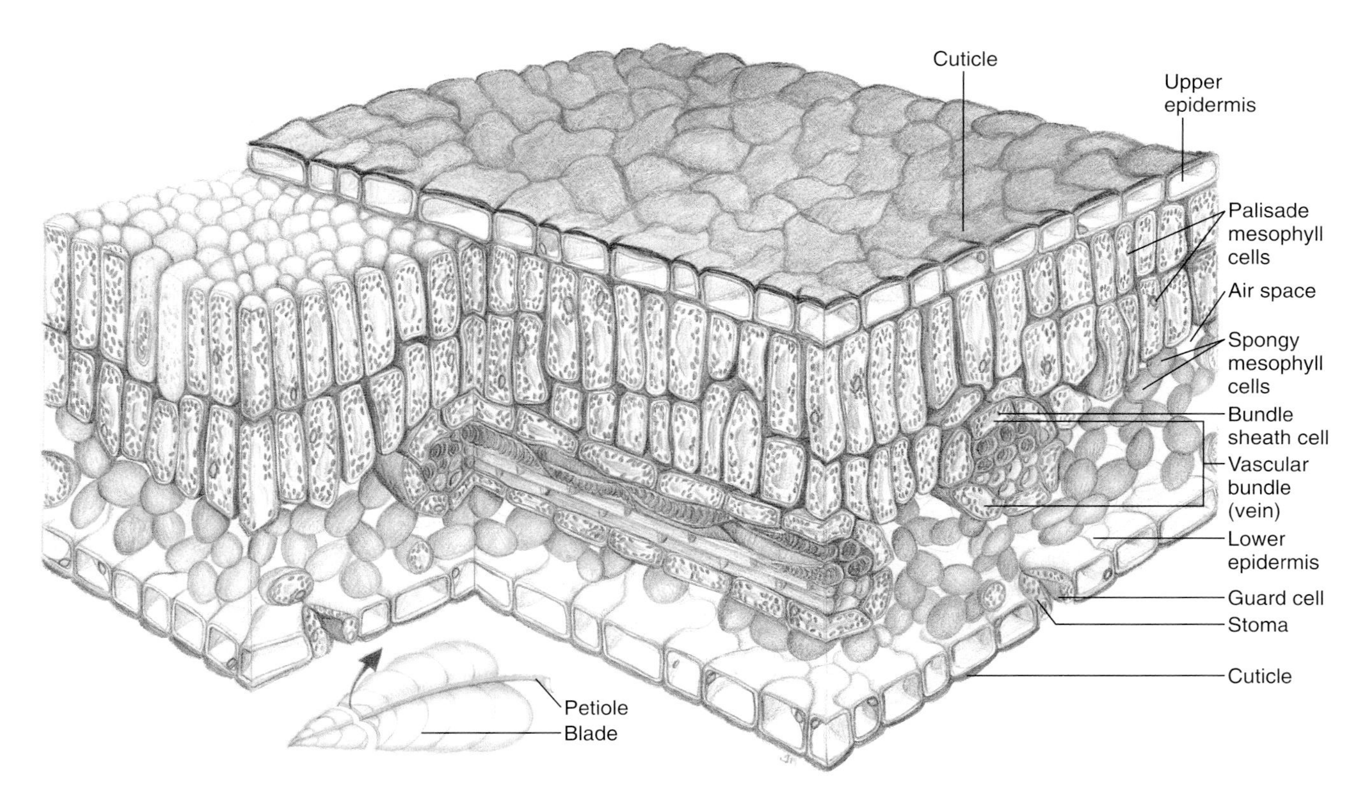

Figure 31.19

Ligustrum leaf, cross-section. Most photosynthesis occurs in the densely packed palisade mesophyll cells, which are just beneath the upper epidermis of the leaf. Gas exchange occurs through stomata, which are usually most abundant on the lower side of the leaf. Water loss is minimized by the waxy cuticle that covers the leaf.

Question 14

a. What is the function of stomata?

b. Do epidermal cells of leaves have a cuticle? Why is this important?

c. What is the significance of chloroplasts being concentrated near the upper surface of the leaf?

d. What are the functions of air spaces near the lower surface of the leaf?

e. How is the internal anatomy of a corn leaf different from that of *Ligustrum*?

f. How is it similar?

g. Based on the arrangement of vascular tissue, how could you distinguish the upper versus lower surfaces of a leaf?

h. If time permits, also examine prepared slides of leaves of corn (*Zea mays*, a monocot) and pine (*Pinus*, a gymnosperm). What differences are there in the structures of these leaves? How do these structural differences correlate with functional differences?

Questions for Further Thought and Study

1. What is the function of xylem? Phloem? Vascular cambium? Epidermis?

2. What are the functions of stomata and lenticels? In what ways do these structures differ?

3. How is the internal anatomy of a stem different from that of a root?

4. How is primary growth different from secondary growth?

5. How is a leaf structurally adapted for its function?

6. Why does a stem typically contain more sclerenchyma and collenchyma than does a leaf?

DOING BIOLOGY YOURSELF

Choose a couple of defined environments nearby, such as a vacant field or riverside. Survey the variety of leaf morphologies of the dominant plants. What can you conclude about the predominance of monocots versus dicots in each environment?

DOING BIOLOGY YOURSELF

Roots grow downward. Design and conduct an experiment using *Zea mays* (corn) seedlings to demonstrate whether this is a negative response to light or a positive response to gravity.

WRITING TO LEARN BIOLOGY

Why is leaf abscission especially important for temperate plants?

WRITING TO LEARN BIOLOGY

Would you expect to find annual rings in wood of a tropical dicot tree? Why or why not?

32

Plant Physiology
Transpiration

Objectives

By the end of this exercise you should be able to:
1. Describe the structure and function of stomata.
2. Describe how environmental conditions such as wind and light influence stomatal opening.
3. Measure the transpiration rate of a plant.

To survive and reproduce, land plants must cope with many environmental problems. Among the biggest of these problems are obtaining water and avoiding desiccation (water loss). "Solutions" to these and other challenges require expenditure of energy and an efficient structure and physiology. Structural and functional solutions to problems such as maintaining water balance are called adaptations, and the study of these adaptations is among the most interesting aspects of biology (fig. 32.1). In this exercise you will study the "water physiology" of plants—how plants are adapted to control their water content and evaporative loss. Before starting today's exercise, review in your textbook the structure and function of xylem and stomata, and review how water moves through plants (fig. 32.2).

The loss of water from plants is called **transpiration.** Most transpiration occurs through stomata of leaves. However, transpiration also occurs in flowers. Demonstrate this with the following procedure.

Figure 32.1
The droplets of water at the edges of these leaves of a strawberry plant are formed by guttation. Guttation results from root pressure (i.e., a positive pressure in the xylem) common in small plants growing in moist soil on cool, damp mornings. Although most water movement in plants results from evaporation of water from leaves, guttation may be an important way of moving water in short, herbaceous plants.

Procedure 32.1
Visualize transpiration in flowers

1. At the beginning of the laboratory period obtain a small beaker containing methylene blue from your lab instructor.

CAUTION
Be careful not to spill methylene blue—it will stain your skin and clothes!

2. Working in small groups, submerge a stem of periwinkle (*Catharanthus*) having a white flower so that only about 0.5 cm of the stem extends below the flower into the dye. Cut the submerged stem.
3. Float the flower in the dye solution so that the cut portion of the stem is submerged in the dye.
4. Set the beaker containing the flower and dye in an illuminated area, and examine the flower after 2–3 h.

Question 1
a. In which part of the flower is the dye located?

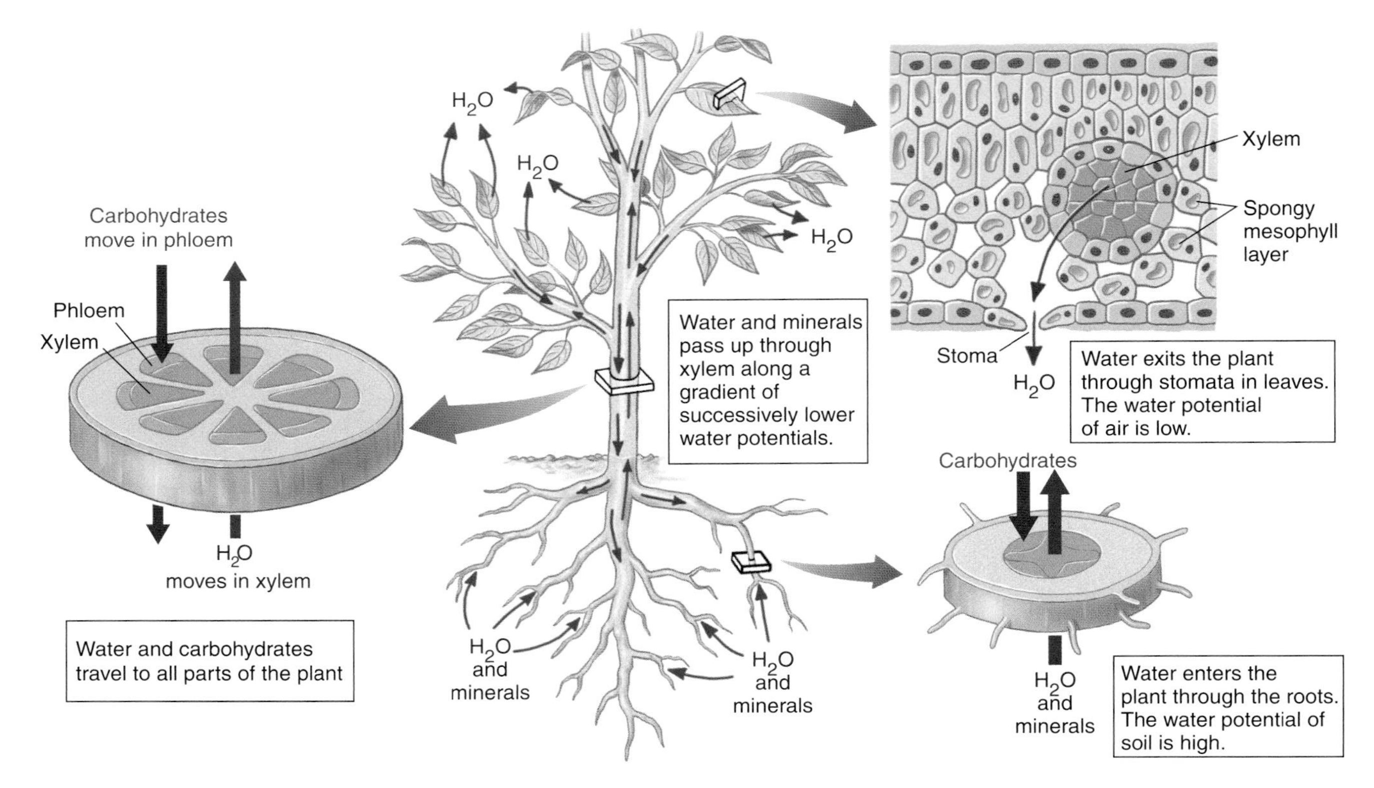

Figure 32.2

Flow of materials in a plant. Water with dissolved minerals enters the plant through root hairs and moves upward through the vascular system. Most of this water is eventually lost through the leaves during transpiration. Dissolved minerals are used by plant cells for metabolic reactions. Carbohydrates made during photosynthesis move throughout the plant in the phloem. These carbohydrates are either stored or used as an energy source for metabolism and growth.

b. What do you conclude from this observation?

PLANT–WATER RELATIONS

Leaves of most plants have epidermal pores called **stomata** (sing., *stoma*), which are formed by the separation of a pair of guard cells (fig. 32.3). Together, the guard cells and the pore between them compose a **stoma** (plural, *stomata*). In a previous exercise, you learned that CO_2 is essential for photosynthesis. Carbon dioxide in the air reaches the photosynthetic cells of the leaf via stomata surrounded by guard cells. If enough water is available, guard cells absorb water and become turgid. When guard cells are turgid they expand and bend to form a stomatal pore through which CO_2 enters and water vapor exits a leaf. When water is scarce, guard cells lose their turgidity and shrink. As a result, guard cells touch each other and the stomatal pore closes (fig. 32.4). Thus, the water status of a plant determines whether stomata are open or closed and indirectly determines whether gases move in and out of the leaf.

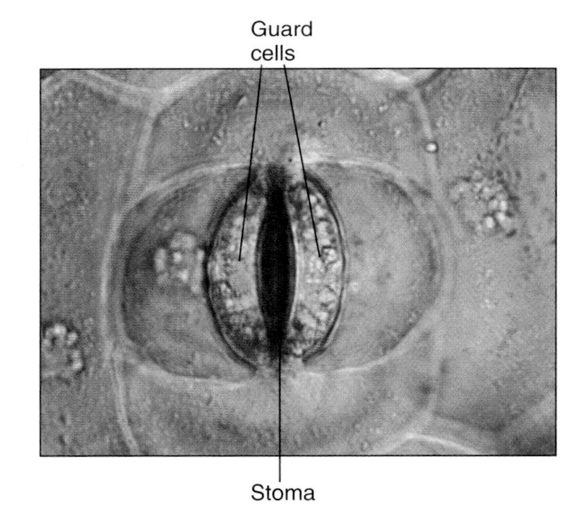

Figure 32.3

A stoma. Unlike the other epidermal cells, the guard cells flanking this stoma contain chloroplasts. Water passes out through the stomata, and carbon dioxide enters by the same portals. The mechanism of stomatal opening and closing is described in figure 32.4.

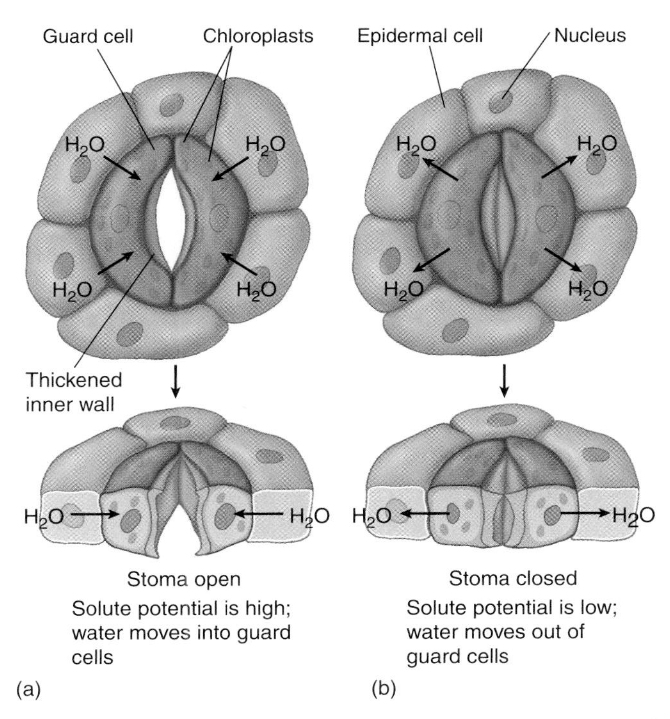

Figure 32.4

The opening and closing of guard cells. (*a*) Two guard cells form each stoma on the surface of a leaf. When the concentration of solutes is high in the guard cells, water diffuses into the cells. This influx of water causes the cells to swell and bow apart, thereby forming a pore and opening the stoma. Water evaporates through these stomatal pores during transpiration. (*b*) When the solute pressure is low in the guard cells, the guard cells become limp and the stomatal pore closes. A variety of environmental factors cause stomata to open and close.

STOMATAL STRUCTURE

Guard cells in dicots are bean-shaped and attached to each other at their ends. Microfibrils form transverse bands around the cell so that when water enters the cells they cannot expand around the middle. Instead, they lengthen and bow apart, thereby opening the stomatal pore. Use procedure 32.2 to examine living guard cells and how they respond to changing environmental conditions.

Procedure 32.2

Stomatal structure

1. Remove part of a leaf from a well-watered *Zebrina* plant. Place a small (~3mm × 3mm) piece of the leaf on a microscope slide. Use your microscope's low-power objective to examine the lower side of the leaf. Locate the stomata; sketch their shape.
2. Obtain a leaf from a well-watered *Kalanchöe* plant that has been illuminated for 4–5 h.
3. Bend the leaf until it snaps; then pull the surface portion of the leaf away from the rest of the leaf in a slow downward motion (your instructor will demonstrate this technique). A thin sheet of transparent epidermis will tear away from the leaf.
4. Quickly place the piece of epidermis in a few drops of distilled water on a microscope slide (do not let the tissue desiccate) and place a coverslip on the tissue. Be careful not to trap any air bubbles beneath the tissue or coverslip.
5. Examine the tissue with your microscope and diagram in the space provided below the epidermis and its stomata.
6. Make a wet mount of epidermis from both sides of the leaf.
7. Place a few drops of 10% NaCl at one edge of the coverslip covering the *Kalanchöe*.
8. Wick this solution under the coverslip by touching the solution at the opposite edge of the coverslip with a piece of paper towel.
9. Reexamine the stomata.
10. Observe the demonstration of an epidermal peel of *Zea mays* (corn), a monocot. Diagram cells of the epidermis of corn.

Question 2

a. Approximately how many stomata are in a mm^2 of the lower surface area of *Kalanchöe?*

b. Are the densities of stomata similar on the upper and lower surfaces of *Kalanchöe?*

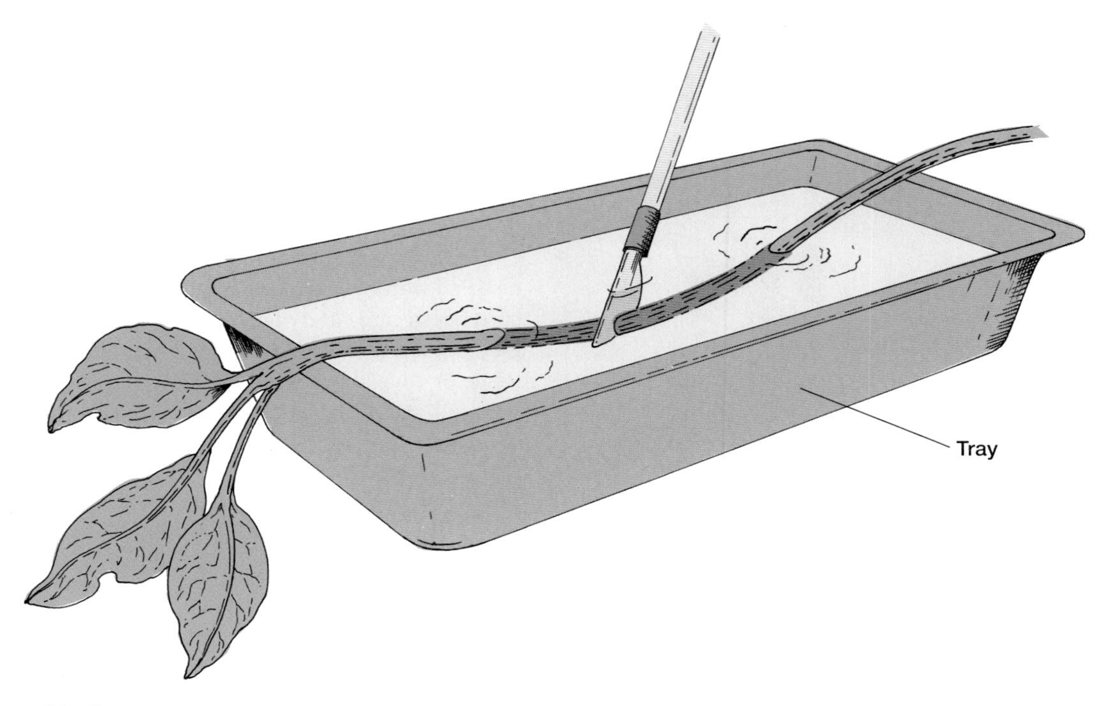

Figure 32.5
Method of cutting submerged stems. Cutting the stem underwater avoids introducing air into the vascular tissue.

c. The guard cells of *Kalanchöe* are turgid and form a stomatal pore. How?

d. Are stomatal densities similar in *Zea mays* and *Kalanchöe?*

e. Are stomata of the two plants arranged in a pattern, or do they occur randomly?

f. How is the structure of individual stomata different in *Zea mays* than in *Kalanchöe?*

Question 3

a. Did the NaCl solution cause stomata to close? Why or why not?

b. What is the advantage of closed stomata when water is in short supply?

c. What are the disadvantages?

TRANSPIRATION

Transpiration is the loss of water from plants; it is strongly influenced by environmental conditions such as CO_2, light, wind, and humidity. Use procedure 32.3 to quantify the effects of light and wind on transpiration.

Procedure 32.3

Quantify transpiration

1. Obtain a sunflower (*Helianthus*) plant from your lab instructor. If sunflower plants are not available, use the pine branches that are available in the lab.
2. Submerge the stem in water and obliquely cut the stem with a sharp scalpel (fig. 32.5).
3. While it is still submerged in water, attach the severed stem to a water-filled tube connected to a 1.0-mL pipet as shown in figure 32.6. Be sure that no

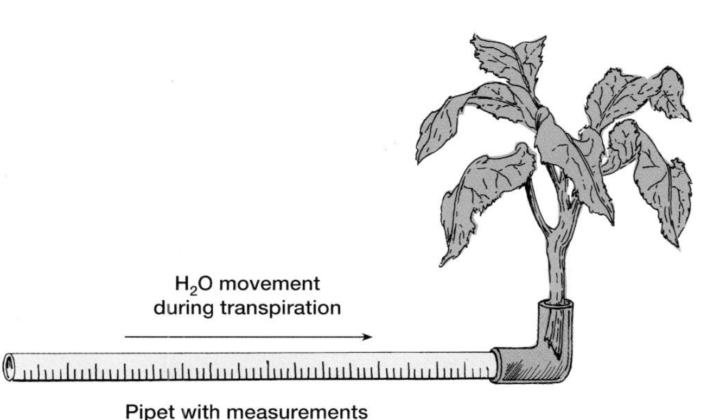

Figure 32.6
Experimental setup for measuring transpiration. As water is transpired from the leaves, the water moves through the pipet.

TABLE 32.1

EXPERIMENTAL CONDITIONS FOR MEASURING TRANSPIRATION

Condition	Transpiration Rate ($mL\ H_2O\ h^{-1}$)
Control (standard illumination)	________
Darkness or reduced light	________
Light; mild breeze (created by a small fan)	________
Dark; mild breeze	________
Light; leaves coated with petroleum jelly	________
Light; all leaves removed	________

air bubbles are in the system and that no water is on the leaves.

4. Place the stem in an illuminated area. Transpiration will move water through the pipet.
5. Allow about 10 min for the rate of water movement in the pipet to stabilize. Note the position of the meniscus in the pipet.
6. Allow transpiration to occur for 15 min.
7. Determine and record the volume of water (mL) transpired by observing the distance that the meniscus moved.
8. Multiply this value by 4 to convert to mL per hour. Record this rate of transpiration in table 32.1 as the control value.
9. Continue the experiment, and determine the transpiration rate under the conditions in the order listed in table 32.1.
10. If data from other groups are available, determine the mean transpiration rates for each condition and record them next to your group's values.

INVESTIGATION

Does Transpiration Vary Among Species of Plants?

a. What features of a plant species might influence its transpiration rate?

b. What effect on transpiration would you predict for each feature?

c. What mechanisms would underlie each feature's effect on transpiration?

d. Design and execute an experiment to test the effects of one or more of the features you have listed.

e. Were your predictions correct? What have you concluded?

Question 4
State your conclusion about the influence of light, dark, breeze, clogged stomata, and removed leaves on water movement in plants.

Light:

Dark:

Breeze:

Clogged stomata:

Removed leaves:

Questions for Further Thought and Study

1. What structural features of plants minimize water loss?

2. Why was it important in today's exercise to sever the stems while they were submerged in liquid?

3. What adaptations help plants conserve water?

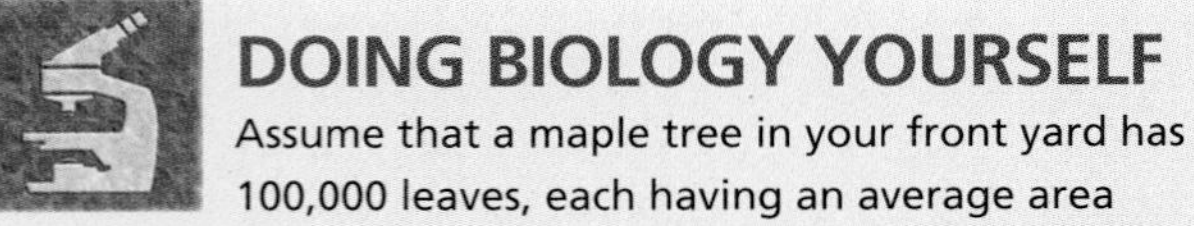

DOING BIOLOGY YOURSELF

Assume that a maple tree in your front yard has 100,000 leaves, each having an average area of 35 cm^2. If the transpiration rate of each leaf is 0.05 mL H_2O h^{-1} cm^{-2} leaf area, how many liters of water move through the plant in a day?

WRITING TO LEARN BIOLOGY

Describe any adaptations that would improve water economy in plants. In which environments might you expect to find the adaptations you described?

33

Plant Physiology
Tropisms, Nutrition, and Growth Regulators

Objectives

By the end of this exercise you should be able to:

1. Define the terms phototropism and gravitropism.
2. Describe symptoms of mineral deficiency in plants.
3. Explain how the quality and quantity of light affect seed germination.
4. Explain the modes of action of auxin and gibberellic acid.

Plants respond to a variety of environmental stimuli such as light and gravity, and require several nutrients such as calcium and potassium. Plant growth and development are controlled by internal chemical signals called **growth regulators.** Auxin and gibberellic acid are examples of growth regulators in plants.

In this exercise you will study a variety of common physiological responses of plants. Throughout this exercise, remember that each of these responses is part of an overall design for the survival and reproduction of plants in varied environments.

PLANT TROPISMS

A **tropism** is a movement in response to an external stimulus. The direction of movement is determined by the direction of the most intense stimulus. Two tropisms you will study in today's exercise are **phototropism,** which is directed growth in response to light (fig. 33.1), and **gravitropism,** which is directed growth in response to gravity (fig. 33.2).

Figure 33.1
Sunflowers (*Helianthus annus*) get their common name because the flowers of some varieties track the sun across the sky, much like radiotelescopes track satellites. This "solar tracking" by sunflowers and other plants (e.g., cotton, alfalfa, beans) helps the plants regulate the amount of light that they absorb. Desert plants such as the "compass plant" use phototropism to orient their leaves parallel to the sun's rays and minimize the amount of light that they absorb; this prevents overheating and desiccation. Other plants orient their leaves perpendicular to the sun's rays to maximize the amount of light absorbed for photosynthesis.

Figure 33.2

Negative gravitropism by a stem of *Coleus*. This plant was placed on its side 24 hours before this picture was taken. The stems have curved away from the pull of gravity.

TABLE 33.1

OBSERVATIONS OF PHOTOTROPISM

Time (h)	Mean Curvature (degrees)
0.5	________
1.0	________
1.5	________
2.0	________

Phototropism

Procedure 33.1

Observe phototropism

1. Obtain from your instructor some 10- to 14-day-old radish (*Raphanus*) seedlings. These seeds have been grown in diffuse, overhead light.
2. At the beginning of the lab period, place your seedlings approximately 25 cm from a 100-watt light so that light strikes the shoots at a right angle.
3. Use a protractor to measure curvature of the seedlings every 30 min for 2 h.
4. Record your results in table 33.1.

Question 1

a. In which direction did the seedlings curve?

b. Is this curvature positive or negative phototropism?

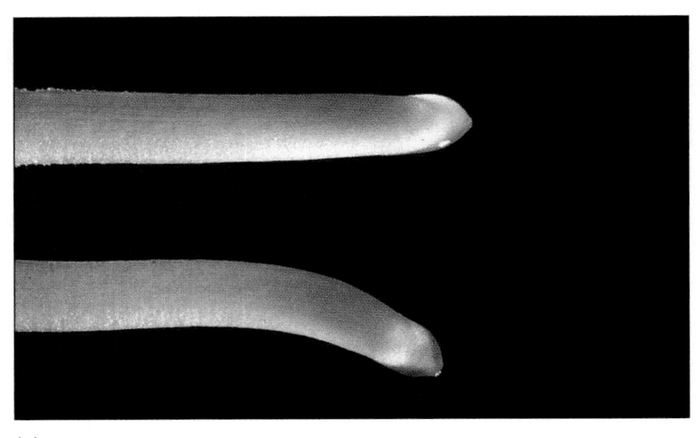

(a)

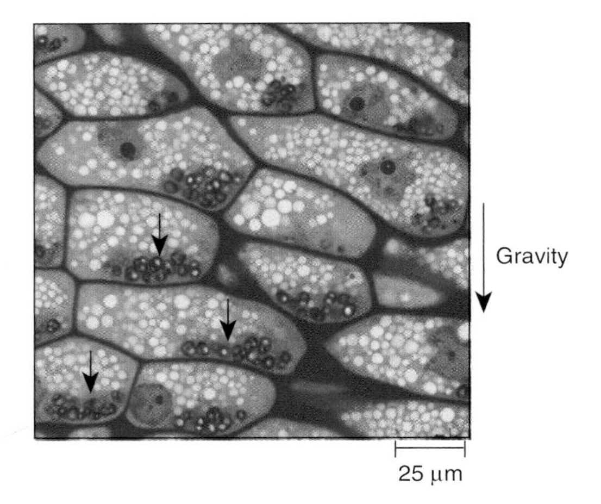

(b)

Figure 33.3

Perception of gravity in root caps. Cells in the center of a root cap contain numerous starch-laden amyloplasts located in the lower part of the cells. Amyloplasts (arrows) sediment to the bottom of the cell when roots are oriented horizontally. This sedimentation of amyloplasts has long been thought to be the basis for how roots perceive gravity (upper photo) (see fig. 33.4). To see a higher-magnification view of the root tip and root cap, see figures 31.2*b* and 31.3.

c. What is the adaptive significance of phototropism?

d. Would you expect roots to react similarly to light? Why or why not?

Gravitropism

Plants may perceive gravity by the movement of starch-laden amyloplasts within cells (fig. 33.3). Growing corn roots show their response to gravity by curving down (fig. 33.4).

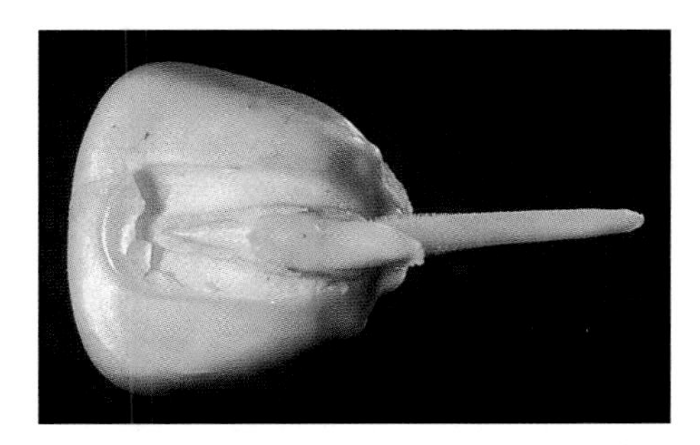

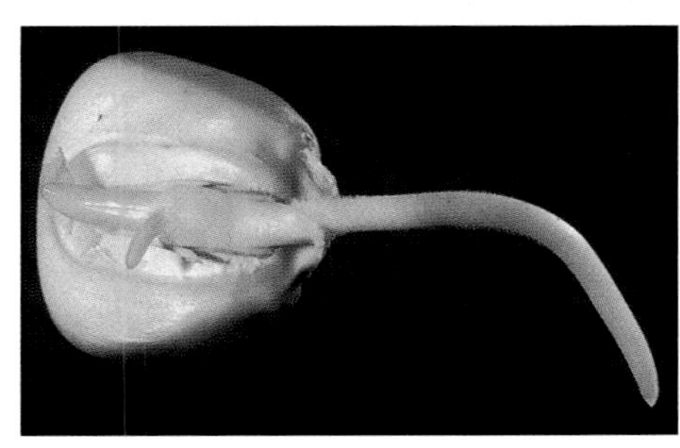

Figure 33.4
Gravitropism by horizontally oriented roots of corn. Downward curvature begins within 30 min and is completed within a few hours. Curvature results from faster elongation of the upper side of the root than of the lower side.

Yesterday, *Zea mays* (corn) seedlings having roots approximately 1 cm long were placed in a glass beaker. Seedlings labeled "H" had their root oriented horizontally, whereas those labeled "D" had their root pointing down. The terminal 3–4 mm was removed from roots of seedlings also labeled with an asterisk (*).

Procedure 33.2
Observe root gravitropism

1. Obtain the containers of corn seedlings oriented horizontally and vertically.
2. Examine the experimental setup and the direction of root growth.
3. Record your observations in table 33.2.

Question 2

a. Which roots grew down?

b. Which one(s) didn't?

c. What do you conclude from these observations?

TABLE 33.2

OBSERVATIONS OF GRAVITROPISM IN ROOTS

Treatment or Orientation	Direction of Growth
Intact roots oriented vertically (V)	________
Intact roots oriented horizontally (H)	________
Detipped roots oriented vertically (V*)	________
Detipped roots oriented horizontally (H*)	________

d. Where in roots does the differential growth occur that produces gravitropism?

e. Are roots positively or negatively gravitropic?

f. What is the adaptive significance of gravitropism?

Procedure 33.3
Examine stem gravitropism

1. Obtain containers of tomato plants (*Lycopersicon*) or sunflower (*Helianthus*) plants.
2. Turn two or three of the potted plants horizontally.
3. Measure the distance from the stem tip to the table's surface. Record your measurements in table 33.3.
4. Every 30 min remeasure the distance between the stem tips and table, and record your results.
5. Some of the plants available in the lab were turned horizontally and inverted yesterday; others were left upright.
6. Examine stem curvature after 24 h and record your observation in table 33.3.
7. Make a simple sketch of the seedling stem curvatures after 24 h.

Question 3

a. How are the stems oriented after 24 h?

b. What is the adaptive advantage of a gravitropic response?

TABLE 33.3

OBSERVATIONS OF GRAVITROPISM IN STEMS

Treatment–Time	Distance of Stem Tip from Table Surface
Horizontal—Time 0 min	________
Horizontal—Time 30 min	________
Horizontal—Time 60 min	________
Horizontal—Time 90 min	________
Horizontal—Time 120 min	________
Inverted—Time 24 hours	Observation:

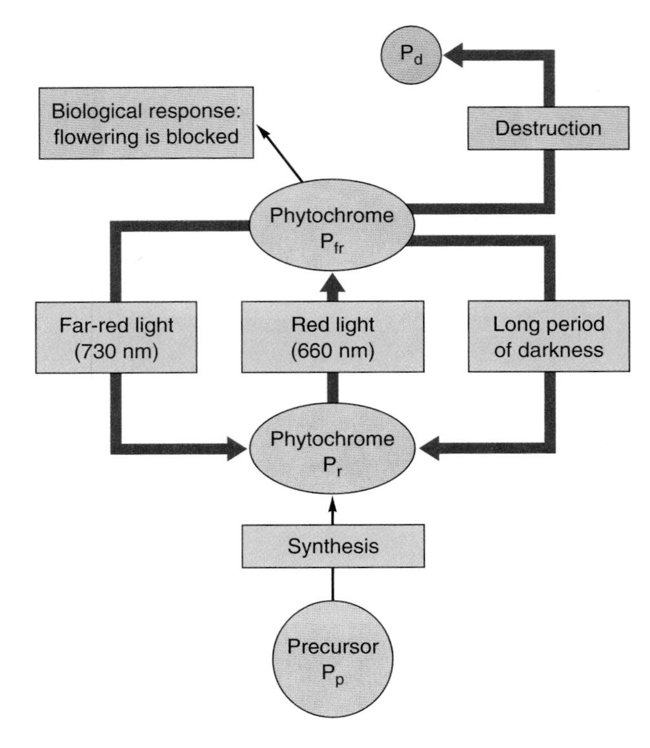

Figure 33.5

How phytochrome works. Phytochrome is synthesized in the P_r form from amino acids, designated P_p for phytochrome precursor. When exposed to red light, P_r changes to P_{fr}, which is the active form that elicits a response in plants. P_{fr} is converted to P_r when exposed to far-red light, and it also converts to P_r or is destroyed in darkness. The destruction product is designated P_d.

SEED GERMINATION

Several environmental factors affect seed germination. For example, seeds of many plants germinate only in response to certain types of light. The pigment that absorbs light affecting seed germination (and several other developmental responses) is **phytochrome;** it alternates between two forms, depending on the light it has absorbed. Phytochrome is activated by red light (660 nm) and inactivated by far-red light (730 nm) and/or darkness (fig. 33.5). The activation and deactivation reactions are reversible, and the ultimate physiological effect induced by the phytochrome depends on which wavelengths were absorbed last. More than 50 different developmental processes are affected by phytochrome absorption of red and far-red light.

Grand Rapids lettuce seeds (*Lactuca sativa*) are excellent models for examining the effects of light on germination because they are sensitive to light and germinate quickly. The seeds are dormant when they are first shed, and germination is poor even with adequate oxygen and heat. Dormancy is broken by light absorption by phytochrome. The phytochrome usually exists in the red-absorbing form. Activation (absorption of red light) causes increased water absorption by the radicle cells, and the radicle grows and elongates. Germination begins.

Procedure 33.4

Observe germination of lettuce seeds

1. Obtain some Grand Rapids lettuce seeds from your laboratory instructor.
2. Working in small groups, place 50 seeds in each of six petri dishes containing water-soaked filter paper.
3. Cover the petri dishes with lids and expose the seeds to the following treatments:

   Treatment 1: Continuous darkness (wrap the dish in metal foil)

   Treatment 2: Continuous light

   Treatment 3: Red light for 10 min, then darkness

   Treatment 4: Far-red light for 10 min, then darkness

   Treatment 5: Red light for 10 min, followed by far-red light for 10 min, then darkness

   Treatment 6: Red light for 10 min, far-red light for 10 min, red light for 10 min, then darkness
4. During your next laboratory period determine the percentage of germination resulting from each treatment.
5. Record your results in table 33.4.

Question 4

What do you conclude about the influence of light on germination of lettuce seeds?

Table 33.4	
Germination of Lettuce Seeds Under Six Different Light Treatments	
Treatment	**Percentage Germination**
1. Dark	________
2. Constant room light	________
3. Red light, then dark	________
4. Far-red light, then dark	________
5. Red light, followed by far-red light, then dark	________
6. Red light, far-red light, red light, then dark	________

Table 33.5	
Germination of Onion Seeds in Light and Dark	
Treatment	**Percentage Germination**
Room light	________
Dark	________

Germinating seeds of different plants often respond differently to light. Germination of onion seeds, for example, is affected by light versus dark. Determine their sensitivity with procedure 33.5.

Procedure 33.5

Observe germination of onion seeds

1. Divide 50 onion seeds, placing 25 in each of two petri dishes containing water-soaked filter paper.
2. Put lids on the petri dishes and label the lids "light" and "dark." Place one petri dish in room light and the other dish in darkness.
3. Examine the seeds during your next lab period and record the percentage germination in table 33.5.

Question 5

a. Does light promote or inhibit germination of onion seeds?

b. Is this response different from that of lettuce?

c. How could seed germination be influenced by depth of planting? How might phytochrome be involved in such a response?

PLANT GROWTH IS AFFECTED BY LIGHT

Many seeds germinate in the dark and push their stems above the soil to receive light. If the seedlings do not receive light they grow abnormally; the leaves grow longer and the plant appears pale and spindly. Such a plant is **etiolated.** Chlorophyll does not develop until the plant is exposed to light. Activation of phytochrome helps promote normal growth. Use procedure 33.6 to observe etiolation.

Procedure 33.6

Examine etiolation

1. Obtain two groups of bean (*Phaseolus*) seedlings. They were planted 10–14 days ago, with one group being grown in the dark and one group in light.
2. Complete the first two lines of table 33.6 with your observations.
3. After completing and recording your observations, reverse the two treatments and examine the plants during your next lab period.
4. Complete the last two lines of table 33.6 with your final observations.

Question 6

a. How did the seedlings manage to grow at all while in the dark?

b. Did etiolation include stem elongation?

c. What is the advantage of rapid stem elongation of a seed germinating in the dark?

PLANT NUTRITION

Growth of green plants requires suitable temperature and adequate amounts of carbon dioxide, oxygen, water, light, and a group of essential elements. Carbon, hydrogen, and oxygen compose 98% of the fresh weight of plants. The remaining 2% is composed of 13 other elements classified either as macronutrients or micronutrients:

TABLE 33.6

PLANT GROWTH UNDER DIFFERENT LIGHT TREATMENTS

Treatment	Leaf Size	Stem Diameter	Height	Color of Shoots	Stem Strength
Light					
Dark					
Light then dark					
Dark then light					

TABLE 33.7

HOAGLAND'S SOLUTION, A COMMON NUTRIENT MEDIUM FOR HEALTHY PLANT GROWTH

Macronutrients	Grams/Liter	Micronutrients	Grams/Liter
$Ca(NO_3)_2 \cdot 4H_2O$	1.18	H_3BO_3	0.60
KNO_3	0.51	$MnCl_2 \cdot 4H_2O$	0.40
$MgSO_4 \cdot 7H_2O$	0.49	$ZnSO_4$	0.05
KH_2PO_4	0.14	$CuSO_4 \cdot 5H_2O$	0.05
		$H_2Mo_4 \cdot 4H_2O$	0.02
		Ferric tartrate	0.50

- **Macronutrients** are nutrients needed in relatively large amounts (100–2000 ppm[1]). Nitrogen (N), phosphorus (P), calcium (Ca), potassium (K), magnesium (Mg), and sulfur (S) are macronutrients.
- **Micronutrients** are nutrients needed in relatively small amounts (<100 ppm). Iron (Fe), chlorine (Cl), copper (Cu), manganese (Mn), zinc (Zn), molybdenum (Mo), and boron (B) are examples of micronutrients.

Although plants need more macronutrients than micronutrients, all are essential for plant growth. The absence of any essential micronutrient or macronutrient will ultimately kill a plant. Before your lab period, seeds were germinated in distilled water and washed sand. Then the young seedlings were transferred to a variety of growth media. Control plants were watered with a solution containing all essential nutrients. Hoagland's Solution is a common mixture of nutrients used to provide for healthy growth (table 33.7). Your control plants were watered with a solution similar to Hoagland's. Other groups of seedlings were watered with solutions that were deficient in an essential element. Follow procedure 33.7 to observe the effects of nutrient deficiency.

Procedure 33.7

Examine plants with symptoms of nutrient deficiency

1. Examine the plants on demonstration that have been grown for their first month in nutrient solutions lacking Ca, N, P, Mg, K, S, and Fe.
2. Record the symptoms for deficiency of each nutrient in table 33.8. Do not refer to table 33.9 until you have completed your observations.
3. After examining all the seedlings, refer to table 33.9 and determine which deficiencies are most obvious and easily diagnosed.

Question 7

How would you "cure" a plant suffering from these deficiency symptoms?

[1] *One part per million (ppm) equals one second in 277 hours (11.5 days), and a minute in two years.*

TABLE 33.8

SYMPTOMS OF NUTRIENT DEFICIENCY

Nutrient Solution	Symptom(s)
Complete (containing all required minerals)	
Complete solution minus Ca	
Complete solution minus N	
Complete solution minus P	
Complete solution minus Mg	
Complete solution minus K	
Complete solution minus S	
Complete solution minus Fe	

FUNCTIONS OF THE MAJOR PLANT HORMONES

Hormone	Major Functions	Where Produced or Found in Plant
Auxin (IAA)	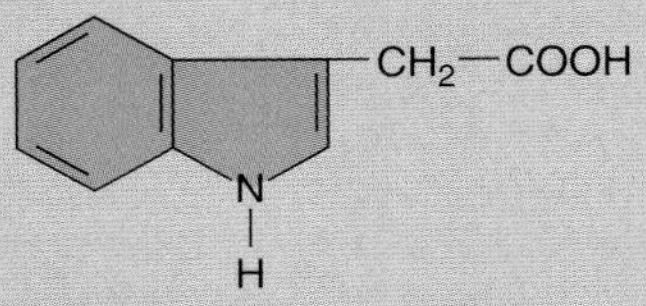 Promotion of stem elongation and growth; formation of adventitious roots; inhibition of leaf abscission; promotion of cell division (with cytokinins); inducement of ethylene production; promotion of lateral bud dormancy	Apical meristems; other immature parts of plants
Cytokinins	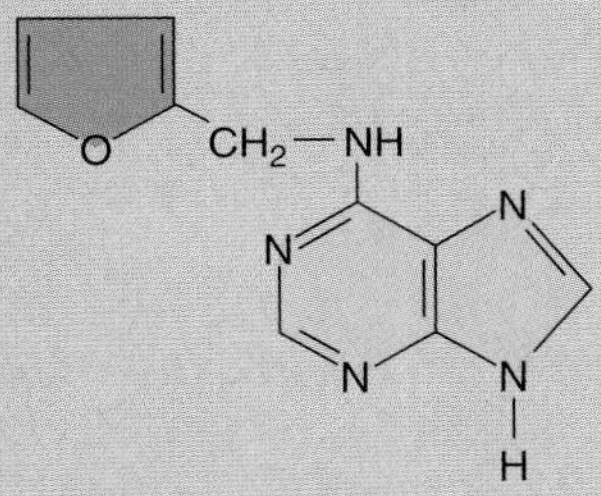 Stimulation of cell division, but only in the presence of auxin; promotion of chloroplast development; delay of leaf aging; promotion of bud formation	Root apical meristems; immature fruits
Gibberellins	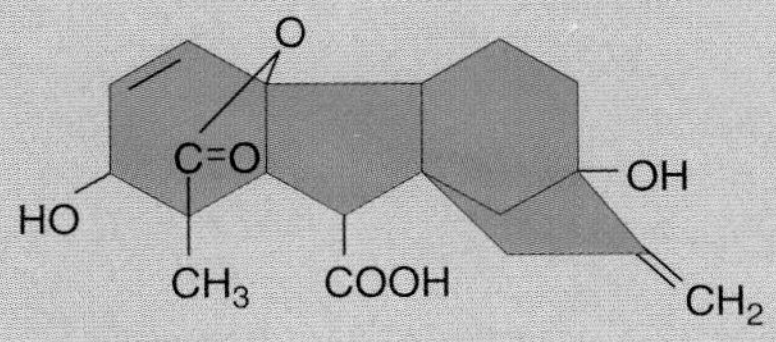 Promotion of stem elongation; stimulation of enzyme production in germinating seeds	Roots and shoot tips; young leaves; seeds
Ethylene	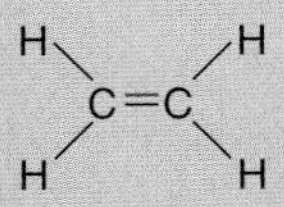 Control of leaf, flower, and fruit abscission; promotion of fruit ripening	Roots, shoot apical meristems; leaf nodes; aging flowers; ripening fruits
Abscisic acid	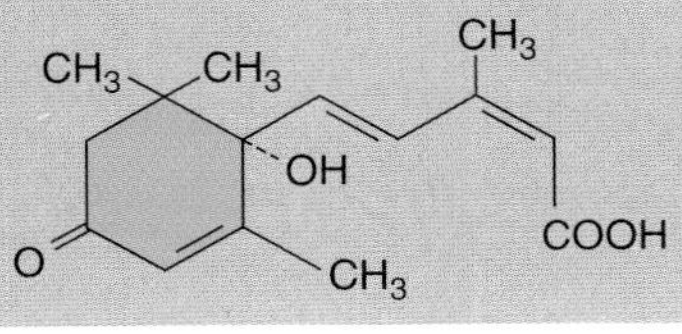 Inhibition of bud growth; control of stomatal closure; some control of seed dormancy; inhibition of effects of other hormones	Leaves, fruits, root caps, seeds

Figure 33.6

Functions of the major plant hormones.

TABLE 33.9

FUNCTIONS AND DEFICIENCY SYMPTOMS OF SOME MAJOR ELEMENTS

Element	Major Functions	Deficiency Symptoms
Nitrogen (N)	Major component of amides, amino acids, and proteins. Present in membranes, organelles, and the cell wall. Balance of carbohydrates and nitrogenous substances necessary.	Leaves often more erect. Stem and leaves stunted with excess root development. Foliage, especially older leaves, chlorotic. Unable to flower.
Phosphorus (P)	Constituent of phospholipids, nucleic acids, and nucleoproteins. Important in respiration and energy transfer.	Small plants, narrow leaves, root system larger but fewer laterals. Accumulation of sugars in older leaves promotes the synthesis of purple anthocyanin pigments. Stiff but weak stems and leaves. Older leaves yellowed. Other leaves dark green.
Potassium (K)	Not known to be structurally part of organic compounds. Role is likely catalytic and regulatory. Needed to activate several enzyme systems.	Internodes short; stems weak. Localized chlorotic or molting of older leaves, particularly at the tips and margins. Later stages may have necrotic mottling. Leaf margins frequently curled under.
Magnesium (Mg)	Constituent of chlorophyll. Important cofactor for enzymes in respiration and in phosphate metabolism.	Older leaves become chlorotic between the veins at the tips and margins. Usually not characterized by necrotic spots. Root system frequently overdeveloped. Leaf margins may cup upward.
Calcium (Ca)	Component in pectin compounds of middle lamella. Present in organic acids bound to proteins. Plays a role in nitrogen metabolism and membrane integrity.	Deficiency may cause ion uptake imbalance, particularly with magnesium. Young leaves are affected first. Tips and margins of leaves become light green and later necrotic. Tips of leaves become limp. Terminal bud often dies.
Sulfur (S)	Component of proteins. Component of iron-sulfur proteins of the electron transport system.	Younger leaves light green. Veins lighter than intervein area.
Iron (Fe)	Electrons transported in cytochromes.	Effects localized on new leaves. Leaves chlorotic. Veins remain green.

Commercial Fertilizers

Commercial fertilizer contains three primary nutrients: nitrogen, phosphorus, and potassium. It may also contain other macronutrients and/or micronutrients. Examine the bag of fertilizer in the lab—its contents are described by three numbers on the bag's label. For example, a bag labeled "10–12–8" is 10% nitrogen (as ammonium or nitrate), 12% phosphorus (as phosphoric acid), and 8% potassium (as mineral potash).

PLANT GROWTH REGULATORS

Plant growth and development are controlled by internal chemical signals called **growth regulators.** In this exercise you will study some common physiological responses of plants to growth regulators such as auxin and gibberellic acid (fig. 33.6). Remember that each of the plant responses is part of an overall design for survival and reproduction in varied environments.

Indoleacetic Acid (IAA)

Indoleacetic acid (IAA), also known as **auxin,** is the best-known growth regulator in plants. It is produced by shoot tips and dramatically affects cell growth (fig. 33.7). Auxin also inhibits development of axillary buds. That is, IAA promotes apical dominance.

Procedure 33.8

Examine the effects of auxin

1. Examine the *Zea* or *Coleus* plants subjected to the treatments described in table 33.10.
2. Record your observations in table 33.10.

Question 8

a. What is the effect of replacing the shoot tip with IAA?

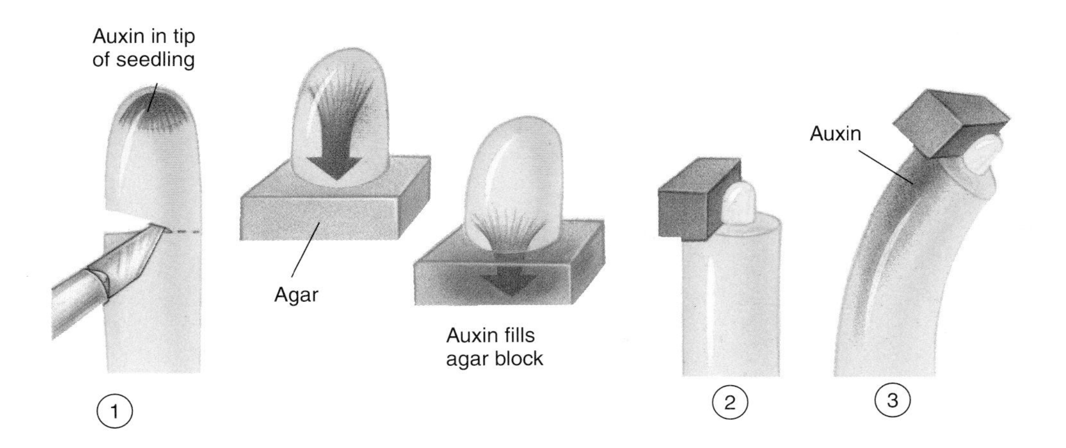

Figure 33.7
Went's demonstration of how auxin affects plant growth. Frits Went, a Dutch plant physiologist, discovered how auxin controls plant growth. Went removed the tips of grass seedlings grown in the light and put them on agar. Auxin flowed from the tips of the seedlings into the agar blocks (1). Went then placed these blocks of agar on one side of the ends of grass seedlings that had been grown in the dark and from which the tips had been removed (2). The seedlings bent away from the side on which the auxin-filled agar block was placed (3). Went concluded that auxin promoted cell elongation and that it accumulated on the side of a grass seedling bent away from the light.

Table 33.10
Effects of Auxin on *Coleus* Plants

Treatment	Observations
Control (no treatment)	
Shoot tip removed	
Shoot tip removed and the cut surface coated with 1% IAA in lanolin	
Shoot tip removed and the cut surface coated with lanolin alone	

b. Why was it necessary to apply only lanolin to one plant?

2,4-Dichlorophenoxyacetic Acid

2,4-dichlorophenoxyacetic acid (2,4-D) is a synthetic auxin used frequently in agriculture as a herbicide. 2,4-D kills weeds in grass lawns by selectively eliminating broad-leaved dicots.

Procedure 33.9
Examine the effects of 2,4-D

1. Examine plants that were treated two weeks ago with 2,4-D. The pots originally contained a mixture of monocots (e.g., corn, oats) and dicots (e.g., peas, beans).
2. Record your observations in table 33.11.

TABLE 33.11

EFFECTS OF 2,4-D ON MONOCOTS AND DICOTS

Type of Plant Treated with 2,4-D	Observations
Monocots	
Dicots	

Question 9
Which plants were affected most by the herbicide?

Figure 33.8
The effect of gibberellic acid on cell elongation in cabbage. Cabbage (*Brassica oleracea*), a biennial that is native to the seacoast of Europe, will "bolt" when the heads are treated with gibberellin. Although more than 80 gibberellins have been isolated from natural sources, apparently only one is active in shoot elongation.

Gibberellic Acid (GA)

Dwarfism often results from a plant's inability to synthesize active forms of **gibberellic acid (GA),** another plant growth regulator. Gibberellic acid promotes stem elongation and mobilizes enzymes during seed germination (fig. 33.8).

Procedure 33.10
Examine effects of gibberellic acid

1. Examine four trays of bean or corn plants that were treated in the following ways:
   Tray 1: Normal plants—untreated
   Tray 2: Dwarf plants—untreated
   Tray 3: Normal plants—treated with two or three drops of GA on alternate days for two weeks
   Tray 4: Dwarf plants—treated with two or three drops of GA on alternate days for two weeks
2. Record your observations in table 33.12.

Question 10
a. What is the effect of GA on dwarfism?

b. Did applying GA to normal plants have any effect? If so, what?

TABLE 33.12

EFFECTS OF GIBBERELLIC ACID ON CORN PLANTS

Plant and Treatment	Observations
Normal, untreated	
Dwarf, untreated	
Normal, treated with GA	
Dwarf, treated with GA	

Questions for Further Thought and Study

1. How are phototropism and gravitropism similar? How are they different?

2. What is the adaptive significance of seed germination being influenced by environmental conditions such as light and water?

3. What comprises commercial "plant food"?

4. What are some commercial uses for plant growth-regulating substances?

5. Some seeds must pass through the digestive tract of an animal before they can germinate. What is the adaptive significance of this feature?

DOING BIOLOGY YOURSELF

Design an experiment to determine whether seed germination by tropical species of plants is more sensitive to light than that by temperate species.

WRITING TO LEARN BIOLOGY

Describe the characteristics of plants you would choose to culture were you to travel in space. Base your choices on nutrient requirements, phototropic sensitivity, and gravitropic sensitivity.

33–12

34

Bioassay
Measuring Physiologically Active Substances

Objectives

By the end of this exercise you should be able to:

1. Explain the concept of a bioassay.
2. Construct a dose-response curve.
3. Use a bioassay to estimate the concentration of a physiologically active substance.

Bioassay is a method used by biologists to determine the concentration (or the relative concentration) of a physiologically active substance. This technique typically measures a substance's biological activity rather than its concentration—the substance's concentration is later determined by comparing the magnitude of this biological activity to the activities elicited by a series of known concentrations of the substance. The unknown substance can be any kind of chemical, such as a nutrient, herbicide, pesticide, vitamin, or pollutant.

Sometimes a bioassay is used to estimate the concentration of a substance; other times a bioassay is used to measure the effect of an unknown substance or mixture *relative* to that of a particular concentration of a known substance. For example, if a leaf extract inhibits seed germination, then the "full-strength" extract could be described in terms of its having an effect *equivalent to* that of a specific concentration of a known germination-inhibitor. Finally, a bioassay could also be useful in cases where an effect cannot be quantified, but only described qualitatively. For example, a bioassay could be used to determine whether soil is contaminated with certain substances by planting certain types of seeds and checking the seedlings for certain symptoms.

In the following procedures you will bioassay for the level of activity of three plant growth regulators: auxin, 2,4-D, and gibberellic acid. Each of these growth regulators has important uses. For example, synthetic auxins such as 2,4-D are used extensively as herbicides because they are inexpensive, relatively nontoxic to humans, and selectively kill (by causing excessive growth) many broadleaf weeds. These compounds are also used to produce roots on cuttings, inhibit the preharvest dropping of fruit, produce seedless cucumbers, and inhibit the sprouting of lateral buds ("eyes") on Irish potatoes. Similarly, gibberellic acid is used by Hawaiian sugarcane growers to improve yields, by California growers of "Thompson seedless" grapes to produce larger grapes, and by brewers to stimulate the germination of barley seeds for the production of alcoholic beverages. These germinated barley seedlings are called malt; their germination involves the conversion of starch to sugars, which are later converted to ethanol during anaerobic fermentation.

In the lab you will assay (i.e., measure) the effect of auxin on the curvature of split stems, the effect of 2,4-D on root elongation, and the effect of gibberellic acid on enlargement of radish cotyledons. All of these bioassays will be done at a constant pH. This will stabilize the reactivities and potency of the chemicals.

BIOASSAY FOR AUXIN

You will measure how much a stem curves in response to known concentrations of auxin (see fig. 33.6). Using these raw data you will construct a standard curve (graph) showing the relationship between known concentrations of auxin and degree of curvature of a stem. You will then bioassay a solution of auxin of unknown concentration by measuring its effect on stem curvature. Using the standard curve you can graphically determine the unknown concentration of auxin based on the amount of curvature induced by the unknown solution.

Procedure 34.1

Bioassay auxin

1. Working in a group of three to five students, obtain a tray of dark-grown pea seedlings that are 8 to 10 days old.
2. Locate the third internode. Cut from the seedlings 35 uniform sections (one per seedling) 3 cm long from just below this internode.
3. Use a razor blade to split 3/4 of the length of each stem segment.

TABLE 34.1

CONCENTRATIONS OF AUXIN USED TO BIOASSAY A SOLUTION OF UNKNOWN CONCENTRATION

Auxin Concentration	Log_{10} Auxin Conc.	Curvature (degrees)					Mean
Buffer alone	—	____	____	____	____	____	____
0.0001 mM auxin	−4	____	____	____	____	____	____
0.001 mM auxin	−3	____	____	____	____	____	____
0.01 mM auxin	−2	____	____	____	____	____	____
0.1 mM auxin	−1	____	____	____	____	____	____
10 mM auxin	1	____	____	____	____	____	____
Unknown	____	____	____	____	____	____	____

4. Soak the split stems in distilled water for 20–30 min to remove some of the endogenous, or naturally occurring, auxin.
5. Label each of seven petri plates to receive auxin at one of the concentrations listed in table 34.1. Add to each petri dish 20 mL of the appropriate solution listed in table 34.1. Each solution has been adjusted to pH 5.9 with 10 mM phosphate buffer.
6. Place five split stems in each of the seven petri dishes.
7. Incubate the soaking sections at room temperature in the dark.
8. Examine the stems after 24 h.
9. Use a protractor to measure the curvature for each stem for each treatment and record your data in table 34.1. Calculate and record the mean curvature for each treatment.
10. Plot the mean curvature (degrees) versus the log_{10} of the known concentrations of auxin on one of the sheets of graph paper at the end of this exercise. The resulting graph is known as a dose-response curve.
11. Determine the unknown concentration of auxin by first locating the mean curvature value of the unknown treatment on the y-axis of your standard curve (refer to fig. 7.4). Draw a line from this point parallel to the x-axis until the line intersects the standard curve. Draw a line from this intersection straight down until it intersects the x-axis. This point on the x-axis marks the concentration of the unknown solution.
12. Record the concentration of the unknown solution in table 34.1.

Question 1

a. What are your conclusions?

b. According to your data, what is the approximate concentration of auxin in the unknown solution?

c. How does auxin affect curvature of stems?

d. What is the relationship between auxin concentration and stem curvature?

BIOASSAY FOR 2,4-D

You will measure the effect of 2,4-D on elongation of cucumber roots. You'll then bioassay a solution of 2,4-D of unknown concentration by comparing its effect on root elongation with that induced by known concentrations of 2,4-D.

Procedure 34.2

Bioassay 2,4-D

1. Working in a group of three to five students, label each of seven petri plates to receive 2,4-D at one of the concentrations listed in table 34.2.
2. Place several pieces of sterile filter paper into each petri dish.
3. Add to each petri dish 10 mL of the appropriate solution listed in table 34.2. Each solution has been adjusted to pH 5.9 with 10 mM phosphate buffer.
4. Place 20 cucumber seeds in each petri dish and store each dish in the dark.

TABLE 34.2

CONCENTRATIONS OF 2,4-D USED TO BIOASSAY A SOLUTION OF UNKNOWN CONCENTRATION

Conc. of 2,4-D	Log_{10} 2,4-D Conc.	Mean Root Length (mm)
Buffer alone	—	______
0.001 mg/L	−3	______
0.01 mg/L	−2	______
0.1 mg/L	−1	______
1 mg/L	0	______
10 mg/L	1	______
Unknown	______	______

TABLE 34.3

CONCENTRATIONS OF GIBBERELLIC ACID USED TO BIOASSAY A SOLUTION OF UNKNOWN CONCENTRATION

Conc. of Gibberellic Acid	Log_{10} Conc. Gibberellic Acid	Mean Weight (mg)
Buffer alone	—	______
0.001 mg/L	−3	______
0.01 mg/L	−2	______
0.1 mg/L	−1	______
1 mg/L	0	______
10 mg/L	1	______
Unknown	______	______

5. After five days, measure the length of the primary root of each seedling. Calculate and record the mean length for roots in each treatment in table 34.2.
6. Plot the mean root length versus the $\log_{10}$ of the known concentrations of 2,4-D on one of the sheets of graph paper at the end of this exercise.
7. Determine the unknown concentration of 2,4-D in a manner similar to that described in step 11 of procedure 34.1 for auxin bioassay. Record the concentration of the unknown solution in table 34.2.

Question 2

a. What are your conclusions?

b. According to your data, what is the approximate concentration of 2,4-D in the unknown solution?

BIOASSAY FOR GIBBERELLIC ACID

You will measure the effect of gibberellic acid on enlargement of radish cotyledons (see figs. 33.6 and 33.8). You'll then bioassay a solution of gibberellic acid of unknown concentration by comparing its effect on cotyledon enlargement with that induced by known concentrations of gibberellic acid.

Procedure 34.3

Bioassay gibberellic acid

1. Working in a group of three to five students, obtain a flat of dark-grown radish seedlings that are approximately 1.5 days old.
2. Label each of seven petri plates to receive gibberellic acid at one of the concentrations listed in table 34.3.
3. Add to each petri dish 5 mL of the appropriate solution listed in table 34.3. Each solution has been adjusted to pH 5.9 with 10 mM phosphate buffer.
4. Excise the smaller cotyledon and all of the hypocotyl (the portion of seedling between the cotyledon and root) from 105 seedlings.
5. Place pads of sterile filter paper in each of seven petri dishes. Moisten the filter paper.
6. Place 15 excised cotyledons on the moistened filter paper in each petri dish. Use cotyledons having similar sizes in all treatments.
7. Place the dishes in a humid environment (e.g., inside a plastic bag) under a constant fluorescent light.
8. After three days, weigh the cotyledons to the nearest milligram and calculate the mean weight of a cotyledon in each of the treatments. Record these mean weights in table 34.3.
9. Plot the mean weight of the cotyledons versus the $\log_{10}$ of the known concentrations of gibberellic acid on one of the sheets of graph paper at the end of this exercise.
10. Determine the unknown concentration of gibberellic acid in a manner similar to that described in step 11 of procedure 34.1 for auxin bioassay. Record the concentration of the unknown solution in table 34.3.

Question 3

a. What are your conclusions?

b. According to your data, what is the approximate concentration of gibberellic acid in the unknown solution?

Questions for Further Thought and Study

1. How could you use a bioassay to study a suspected pollutant in soil? In water? In air?

2. Bioassays are often the first step involved in studying the effects of a potential drug. Why?

3. What are the strengths of a bioassay? What are the limitations?

DOING BIOLOGY YOURSELF

Establishing a dose-response curve for an unknown or unfamiliar substance, or for an extract that may contain more than one substance, is often a goal of a bioassay. Design an experiment to bioassay an extract of leaf litter for the inhibition or promotion of seed germination.

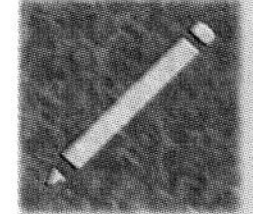

WRITING TO LEARN BIOLOGY

Briefly describe the value of bioassays as opposed to direct chemical measurement. What assumptions must be made to validate that a bioassay for a chemical is accurate?

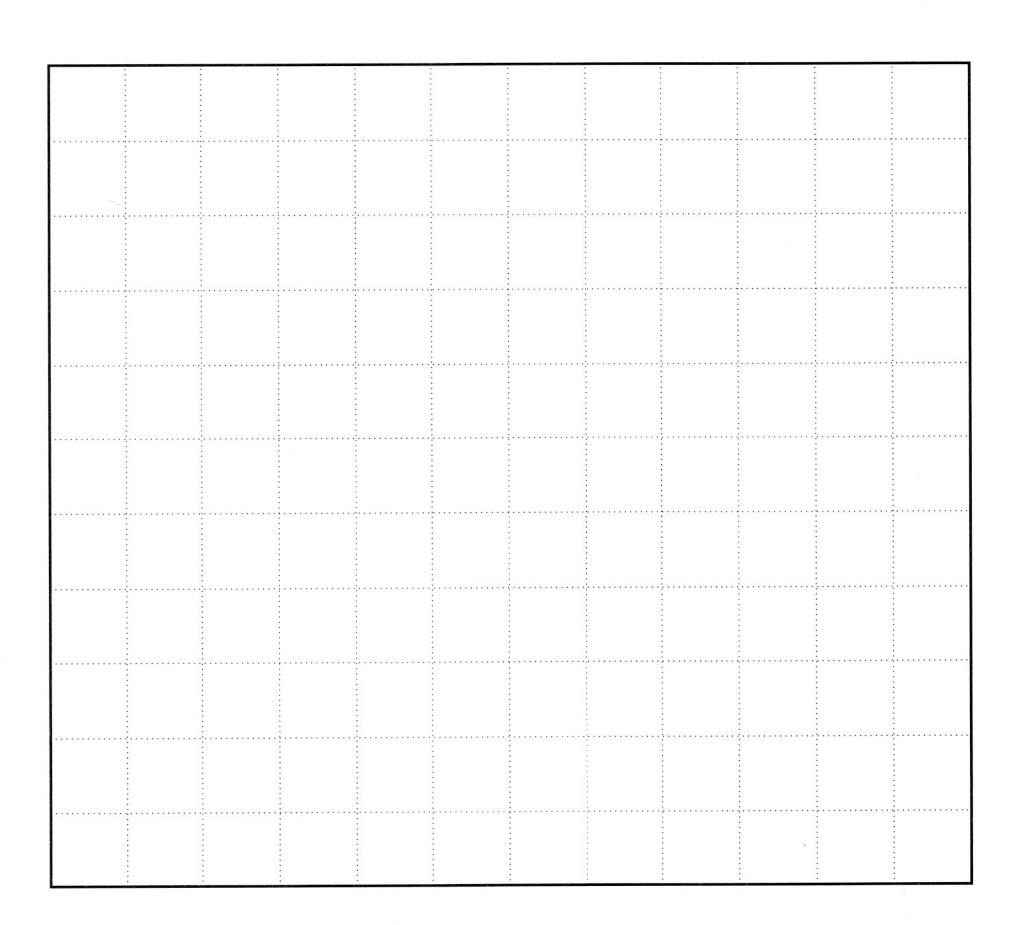

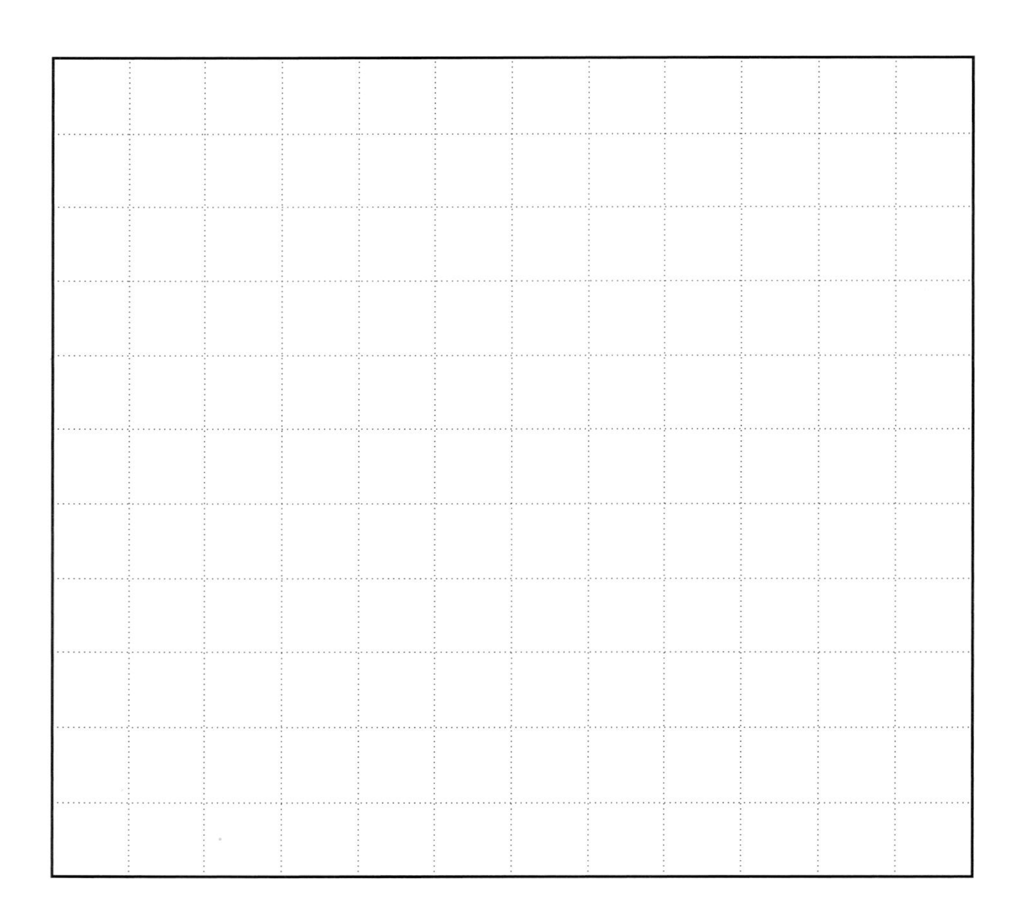

34–6

35

Survey of the Animal Kingdom
Phyla Porifera and Cnidaria

Objectives

By the end of this exercise you should be able to:
1. Describe how structures specific to poriferans and cnidarians help them survive in their environment and promote their evolutionary persistence.
2. List the fundamental characteristics of members of phylum Porifera and phylum Cnidaria.
3. Recognize members of the three major classes of cnidarians.
4. Describe the body forms of cnidarians and describe reproduction of those species alternating between polyps and medusae.
5. Compare the feeding methods of sponges and jellyfish.

To most people, ancient animals such as sponges and jellyfish appear simple and unsophisticated. Their bodies and body cavities are simple and they lack the complex behavior and sensory capabilities of most higher animals. But don't let that simplicity fool you. What sponges and jellyfish lack in complexity is more than compensated for by their extraordinarily elegant design. After all, their overall body plans (table 35.1) have persisted in changing environments for many millions of years; their "simple" morphology accomplishes the primary functions of food getting, reproduction, and adaptive response to their environment as does the morphology of other more familiar animals. Like all animals, sponges of phylum Porifera and cnidarians of phylum Cnidaria are eukaryotic, **multicellular,** and **ingestive-feeding heterotrophs.** Heterotrophs derive their energy from organic molecules made by other organisms.

PHYLUM PORIFERA

Sponges are the simplest of the major animal phyla and comprise 10,000 species (fig. 35.1). Most sponges live in the ocean, but a few encrust rocks and sticks in freshwater. Sponges lack tissues and organs, and are typically **asymmetrical** assemblages of cells. Bodies of asymmetrical organisms have no symmetry or pattern such as left and right halves or anterior and posterior regions. Sponge cells are so loosely assembled that if a sponge is forced through a fine mesh, the disassociated cells will survive. Even more remarkably, the disassociated cells of some species can reassemble as a functioning organism.

Grantia

Examine a preserved specimen of *Grantia,* one of the simplest sponges (fig. 35.2). At first glance, sponges such as *Grantia* appear plantlike because they are **sessile** (i.e., attached to the substrate). Some sponges even appear green because symbiotic algae live in their bodies. However, sponges are **filter-feeding** heterotrophs and have no photosynthetic pigments. Notice that *Grantia* is a tubular, open-ended chamber surrounded by a thin, porous, folded wall of cells. Phylum Porifera gets its name from the many pores in the chamber walls.

Question 1
Do any features of *Grantia* distinguish this organism as an animal? If so, which ones?

Structure of Sponges

Sponge walls filter seawater and remove food particles. In the simplest sponges this wall is lined on the outside by an **epithelial layer** of flat cells (fig. 35.3). Inside the sponge is the **spongocoel,** a central cavity lined by flagellated cells called **choanocytes** (sometimes called collar cells). Their moving flagella draw water through pores within **porocytes** into the spongocoel and across the collars of the choanocytes to trap food particles (fig. 35.3). Choanocytes produce a constant flow of water into a sponge. Eventually the filtered water exits through a large hole in the end of the sponge called the **osculum.**

TABLE 35.1

PHYLA PORIFERA AND CNIDARIA

Phylum	Typical Examples	Key Characteristics	Approximate Number of Named Species
Porifera (sponges)	Barrel sponges, boring sponges, basket sponges, vase sponges	Asymmetrical bodies without distinct tissues or organs; sac-like body consists of two layers breached by many pores; internal cavity lines with food-filtering cells called choanocytes; most marine (150 species live in freshwater)	5,150
Cnidaria (cnidarians)	Jellyfish, hydra, corals, sea anemones	Soft, gelatinous, radially symmetrical bodies whose digestive cavity has a single opening; possess tentacles armed with stinging cells called cnidocytes that shoot sharp harpoons called nematocysts; almost entirely marine	10,000

Question 2

Consider objective 1 listed at the beginning of this exercise. Are choanocytes significant to a fundamental process for sponges? What is the process and how are choanocytes significant?

Grantia has a slightly more complicated wall that is folded. Examine a prepared slide of a cross section through *Grantia* (fig. 35.4). Folds of the wall form **incurrent canals** opening to the outside and **flagellated canals** opening to the central spongocoel. The flagellated canals are lined with choanocytes that move their flagella to draw water into the incurrent canals, through porocytes, through the canals, and on to the spongocoel. A specialized collar of microvilli surrounding the flagellum of a choanocyte traps food particles that are engulfed by the cell body. Digestion is **intracellular,** meaning that it occurs inside cells. Some sponges are more complicated than *Grantia* and have highly folded walls or a complicated series of small chambers lined with choanocytes. A spongocoel may be difficult to distinguish in these complicated sponges (see fig. 35.1).

The wall of a sponge also contains **amoebocytes,** crystalline skeletal structures called **spicules** (fig. 35.5), and a gelatinous matrix called **mesenchyme.** As their name implies, amoebocytes are creeping, mobile cells with a variety of functions including differentiation into other cell types. They also secrete the skeleton of calcareous spicules (containing calcium), siliceous spicules (containing silicon), or proteinaceous **spongin** fibers.

Procedure 35.1

Examine spicules

1. Examine a prepared slide of cleaned and isolated spicules.
2. Also locate spicules in a cross section of *Grantia.*
3. If enough preserved sponge is available, make a wet mount of spicules in a depression slide by gently crushing a small piece of sponge on a slide with a coverslip.

(a)

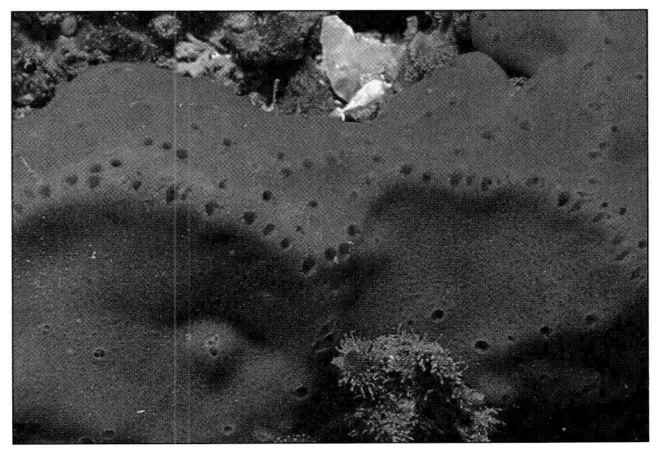

(b)

Figure 35.1

Sponges (phylum Porifera) such as (*a*) the yellow tube sponge, *Verongia*, and (*b*) *Axiomella* have a variety of colors. A seemingly inactive sponge may filter 1000 times its own volume of water per day through walls filled with small canals and chambers. There are up to 18,000 chambers per square millimeter of sponge. Flagellated cells line the chambers to circulate water and filter extremely small particles of food. Eighty percent of the organic matter captured by sponges is too small to be seen with a microscope. Use of this small-size fraction of dissolved matter has made sponges successful for hundreds of millions of years.

4. If bleach is available, add one or two drops to dissolve the protein and expose the spicules.
5. Examine the preparation under low, then high, magnification.

Examine the fused lattice of spicules of *Euplectella* (fig. 35.6). Spicules of different sponges have many shapes and may be fused in an ornate lattice. The beautiful lattice of the sponge *Euplectella* is unusual because it houses several species of shrimp. Interestingly, when a male and female shrimp enter the spongocoel they may grow too large to escape. A dried specimen of *Euplectella* with its permanent residents was formerly used in Japan as a wedding present symbolizing the permanent bond of marriage.

Figure 35.2

Grantia is a common tubular sponge with a folded body wall filled with pores for filtering water.

Spongia

Examine some large, dried sponges on display. Note that *Spongia* has a more complex arrangement of chambers than does *Grantia*.

Examine a prepared slide of spongin fibers. Fibers of spongin compose the skeleton of common bath sponges such as *Spongia*. Spongin fibers are flexible and are not crystalline like spicules. In past years, natural sponges were gathered easily in the Caribbean and Mediterranean, dried, and sold. Today, natural sponges are rare and have been replaced by synthetic products.

Question 3

a. What is the advantage of a folded or convoluted wall in sponges?

b. What function other than support might spicules serve?

c. How many prongs do spicules of *Grantia* have?

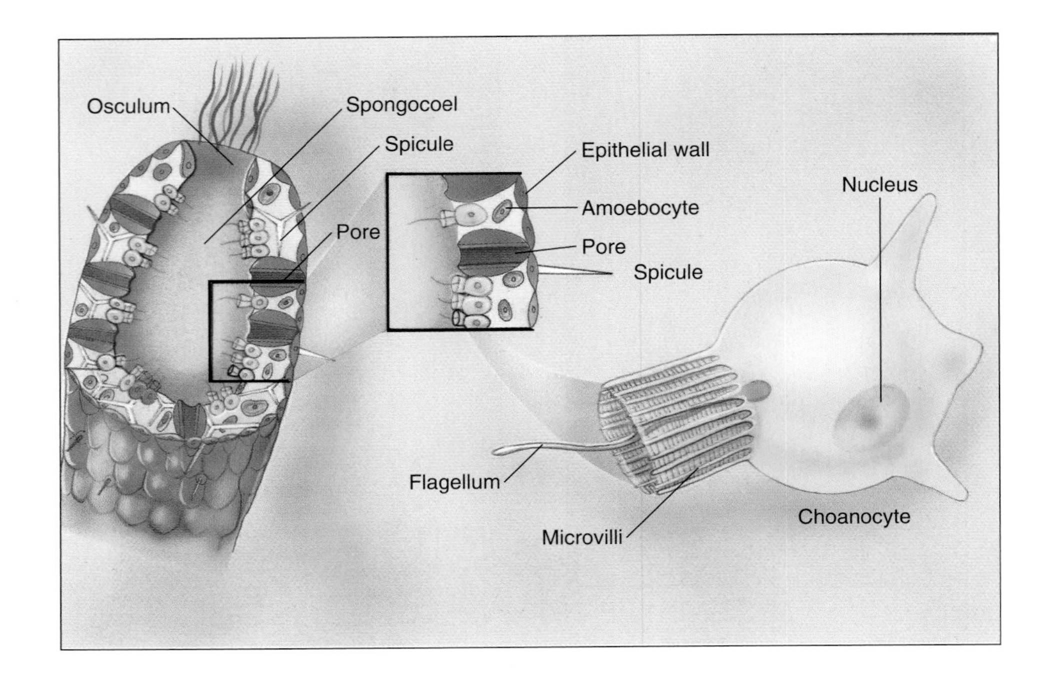

Figure 35.3
Morphology of a simple sponge. An epithelial layer forms the outer wall, reinforced by spicules. Pores are formed by porocytes that extend through the unfolded body wall of this simple sponge. Choanocytes are cells with a flagellum surrounded by a collar of microvilli that traps food particles. Food is moved from the microvilli toward the base of the cell, where the food is incorporated into a food vacuole. The food vacuole is passed to amoeboid cells, where digestion occurs.

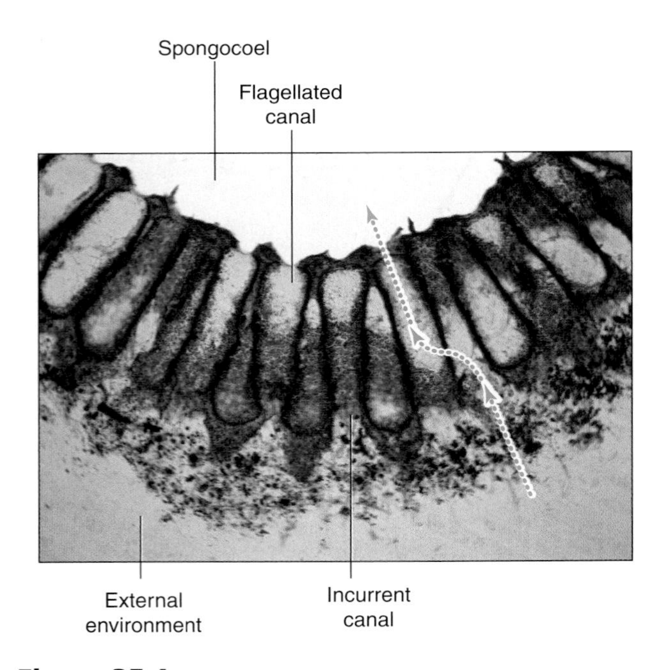

Figure 35.4
Cross section through the folded body wall of *Grantia*. Arrows show the path of water flow. The beating flagella of choanocytes move the water containing food particles from the external environment into an incurrent canal, through the folded wall lined with choanocytes, into the flagellated canal, into the spongocoel, and out the osculum.

Figure 35.5
This photomicrograph shows a variety of sponge spicules (150×).

d. What characteristics of *Spongia* make them useful as a household sponge?

e. Consider objective 1 listed at the beginning of this exercise. Are spicules significant to a fundamental process for sponges? In what way?

Figure 35.6

The fused glass (silicon) spicules of *Euplectella* form a beautiful, intricate lattice. The living tissue has been removed from this dried specimen.

Sponge Reproduction

Sponges reproduce asexually and sexually. Asexual reproduction includes budding and the release of stress-resistant aggregates of amoebocytes called **gemmules.** In favorable conditions, amoebocytes in a gemmule can grow into a mature organism. During sexual reproduction, choanocytes and amoebocytes differentiate into gametes. Eggs remain in the mesenchyme, but sperm are released into the water and are captured by choanocytes or amoebocytes of other sponges. The captured sperm are transported to eggs and fertilization occurs. After a brief development, the embryo is expelled from the sponge. Most species of sponges are hermaphroditic (i.e., have male and female reproductive organs) but produce eggs and sperm at different times.

PHYLUM CNIDARIA (COELENTERATA)

Phylum Cnidaria (also called coelenterates) includes class Hydrozoa (hydras), class Scyphozoa (jellyfish), and class Anthozoa (anemones and corals). Cnidarians are almost all marine carnivores. Their bodies are **radially symmetrical** and more complex than sponges. Radial symmetry describes a body form with repetitive body areas arranged in a circle around a central point such as the pieces of a pie. This body plan has sensory organs exposed to the environment in all directions around the perimeter. This is evolutionarily adaptive for a slow or nonmotile organism. The body wall has two cellular layers, an **ectodermis** on the outside and an **endodermis** (sometimes called the gastrodermis) lining the gastrovascular cavity. A gelatinous **mesoglea** separates the two true body layers. Cells of cnidarians are organized into true tissues (nervous, muscular, and reproductive) but not organs.

Cnidarians have two basic body plans: **polyps** and **medusae** (fig. 35.7). Polyps are cylindrical animals with a mouth surrounded by tentacles atop the cylinder (i.e., the end facing away from the substrate). Polyps are usually attached to the substrate and may be solitary or colonial. In contrast to polyps, medusae are usually free-floating and umbrella-shaped. Their mouths point downward and are surrounded by hanging tentacles. The classes of cnidarians are distinguished primarily by the relative dominance of the polyp stage or the medusa stage in the life cycle. Many cnidarians occur only as polyps, others only as medusae, and still others alternate between these two forms. This alternation is a form of **polymorphism,** which means "many forms."

The life cycle of many cnidarians is characterized by alternation between polyp and medusa (i.e., polymorphism; fig. 35.7). During the life cycle, medusae produce and release eggs and sperm into water for fertilization, although some species retain their eggs. After fertilization the zygote develops into a swimming mass of ciliated cells called a **planula larva.** A planula eventually attaches to the substrate and develops into a polyp. The polyp may reproduce asexually by budding other polyps or may continue the sexual cycle by budding immature medusae called **ephyra.** Ephyra develop into mature medusae.

All cnidarians are carnivores that capture their prey (small fishes and crustaceans) with tentacles that ring their mouth. These tentacles are armed with stinging cells called **cnidocytes** containing small, barbed harpoonlike structures called **nematocysts.** Captured prey are pushed through the mouth into the **gastrovascular cavity** (GVC), where **extracellular digestion** occurs followed by phagocytosis of small food particles and some intracellular digestion (fig. 35.8).

Question 4

a. Consider objective 1 listed at the beginning of this exercise. Are cnidocytes significant to fundamental processes for cnidarians? In what ways?

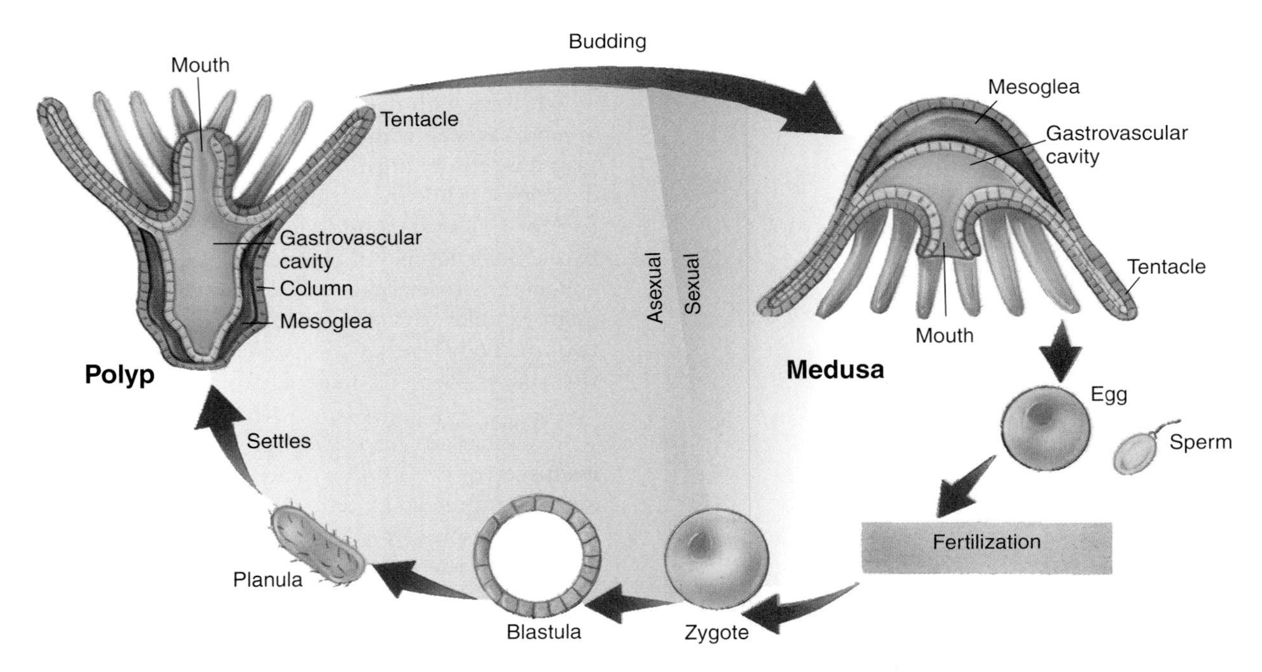

Figure 35.7

A generalized life cycle for Cnidaria. Cnidarians alternate between medusa and polyp forms. Male and female medusae use meiosis to produce gametes that, in water, are fertilized. These gametes are the only haploid (1N) stage; all other stages are diploid (2N). After a short period of free swimming, the planula larva settles to the substrate and forms a polyp. When the polyp buds (an asexual process), additional polyps and medusa buds form. Medusae separate from the polyp and swim away. The polyp or medusa stage has been lost or reduced in many cnidarians, such as anemones and corals.

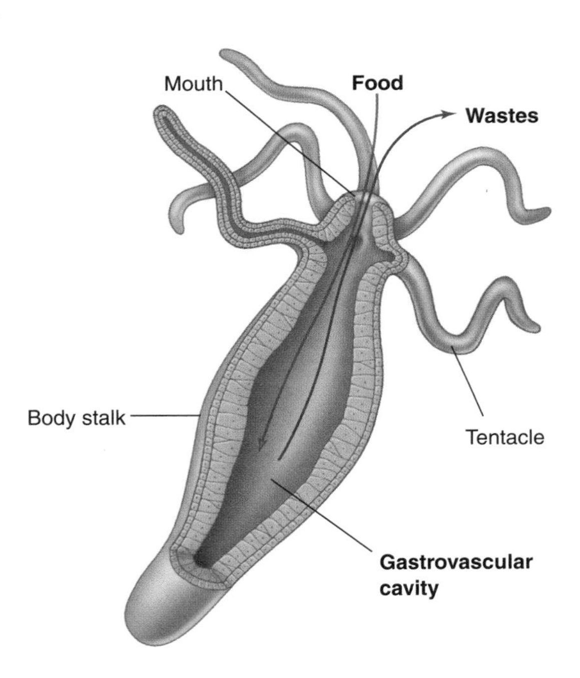

Figure 35.8

The gastrovascular cavity of *Hydra*, a coelenterate. Since there is only one opening, the mouth is also the anus, and no specialization is possible in the different regions that participate in extracellular digestion.

b. Consider objective 1 listed at the beginning of this exercise. How could polymorphism contribute to the evolutionary success of cnidarians in their environment?

Class Hydrozoa

The polyp stage dominates the hydrozoan life cycle, although both polyps and medusae occur in most species. The outer layer of cells, the ectoderm, and the inner layer, the endoderm, surround the gastrovascular cavity; these layers are separated by the gelatinous, acellular mesoglea. Amoeboid cells circulate in the mesoglea. Ectodermal cells include cnidocytes and muscular contractile cells. Endodermal and glandular cells secrete enzymes into the gastrovascular cavity for extracellular digestion. The gastrodermis lacks cnidocytes.

Hydra

Hydra are small, common hydrozoans that live in shallow, freshwater ponds. They are usually less than 1 cm tall and prey on smaller invertebrates among the filaments and leaves of freshwater algae and plants. *Hydra* have no medusa stage.

Procedure 35.2

Observe *Hydra*

1. Obtain a living *Hydra*; observe it in a small petri dish. Polyps of *Hydra* are solitary and occasionally hang from the water's surface with their **basal disks** adhering to the surface of the water. More often, they attach to a hard substrate with their basal disk. However, *Hydra* can detach themselves and move, not by swimming but by somersaulting along their substrate.
2. Allow a few minutes for the animal to relax in the petri dish; then tap the edge and observe the animal's response.
3. If small, living crustaceans such as *Daphnia* or *Artemia* (brine shrimp) are available, place some of these organisms near the tentacles of a *Hydra*. When a prey item touches a tentacle it sticks tightly.
4. Observe the cellular structure of *Hydra* by studying a prepared slide of a cross section of the organism (fig. 35.9).

Question 5

a. How do *Hydra* respond to a tap on their substrate?

b. What tissues must exist for this response?

c. *Hydra* are predators. How actively does *Hydra* stalk their food?

d. What specialized cells of tentacles aid in capturing prey?

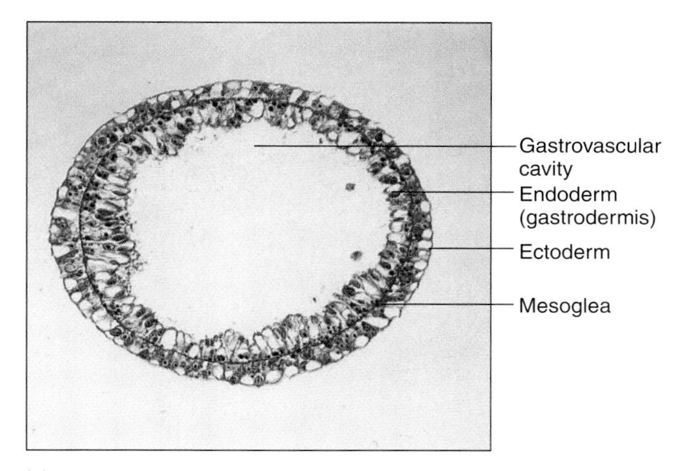

(a)

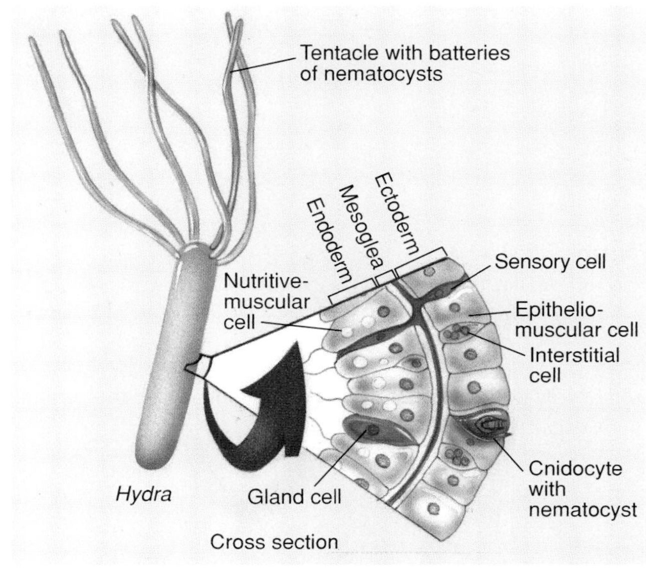

(b)

Figure 35.9

Micrograph of a *Hydra*, a hydrozoan. (*a*) A cross section of *Hydra*. (*b*) Structure of *Hydra*.

Obelia

Examine a prepared slide of another hydrozoan, *Obelia*. Examine a prepared slide of the small medusae of *Obelia* in addition to the polyp colony. *Obelia* typifies most hydrozoans because it has colonial polyps and free-swimming medusae (fig. 35.10). These colonial polyps appear plantlike and branch from a tube. Polyps of *Obelia* are polymorphic because some are specialized feeding polyps called **gastrozooids;** others are reproductive polyps called **gonozoids.**

Question 6

a. What structures determine whether a polyp of *Obelia* is a gastrozooid (feeding polyp) rather than a gonozoid?

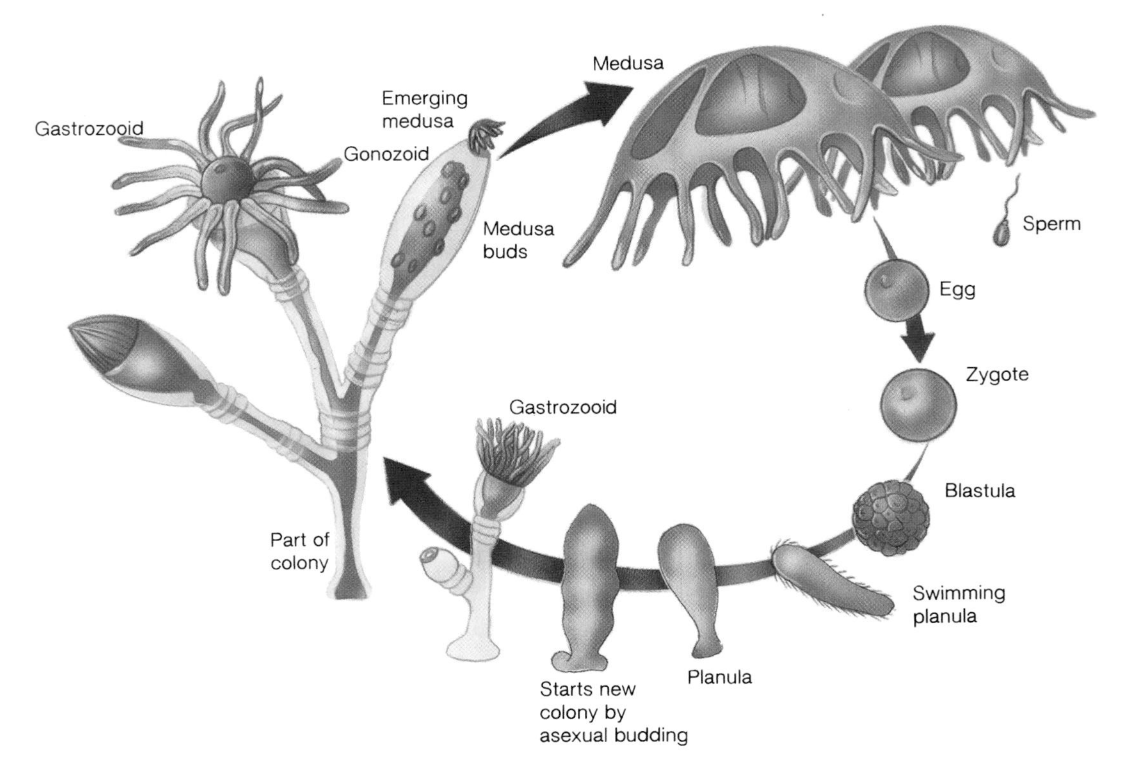

Figure 35.10
Structure and life cycle of *Obelia*, a colonial hydrozoan. Meiosis occurs within the medusae to produce gametes, the only haploid (1N) stage of the life cycle. Production of diploid (2N) medusae by the diploid gonozoids is asexual.

b. How do gonozoids obtain their food in this colonial organism?

c. Gonozoids continue the reproductive cycle by budding medusae. About how many maturing medusae are visible in a typical gonozoid?

Figure 35.11
Portuguese man-of-war, *Physalia utriculus*. The Portuguese man-of-war is a colonial hydrozoan that has adopted the way of life characteristic of jellyfish. This highly integrated colonial organism can ensnare fish by using its painful stings and tentacles, which are sometimes over 15 m long.

Physalia

Examine a preserved specimen of *Physalia,* a common hydrozoan better known as the Portuguese man-of-war (fig. 35.11). *Physalia* is a floating colony of polymorphic polyps. Some of the polyps form a gas-filled sac that floats and suspends the long tentacles of nutritive polyps. Touching the nematocysts on the dangling tentacles is lethal for small fish and painful (but rarely lethal) for swimmers.

Gonionemus

Examine a preserved specimen of *Gonionemus* with your dissecting microscope (fig. 35.12). Examine any other preserved hydrozoans on display. *Gonionemus* is a hydrozoan with large medusae. Medusae are more gelatinous than polyps because the mesoglea is more extensive. Locate the **velum** on the inner periphery of the medusae and the mouth at the end of the **manubrium.** The **gastrovascular cavity** radiates from the center as **radial canals** connected by a **circular canal** around the perimeter. The **gonads** (tissue that produces gametes) attach to the radial canals and

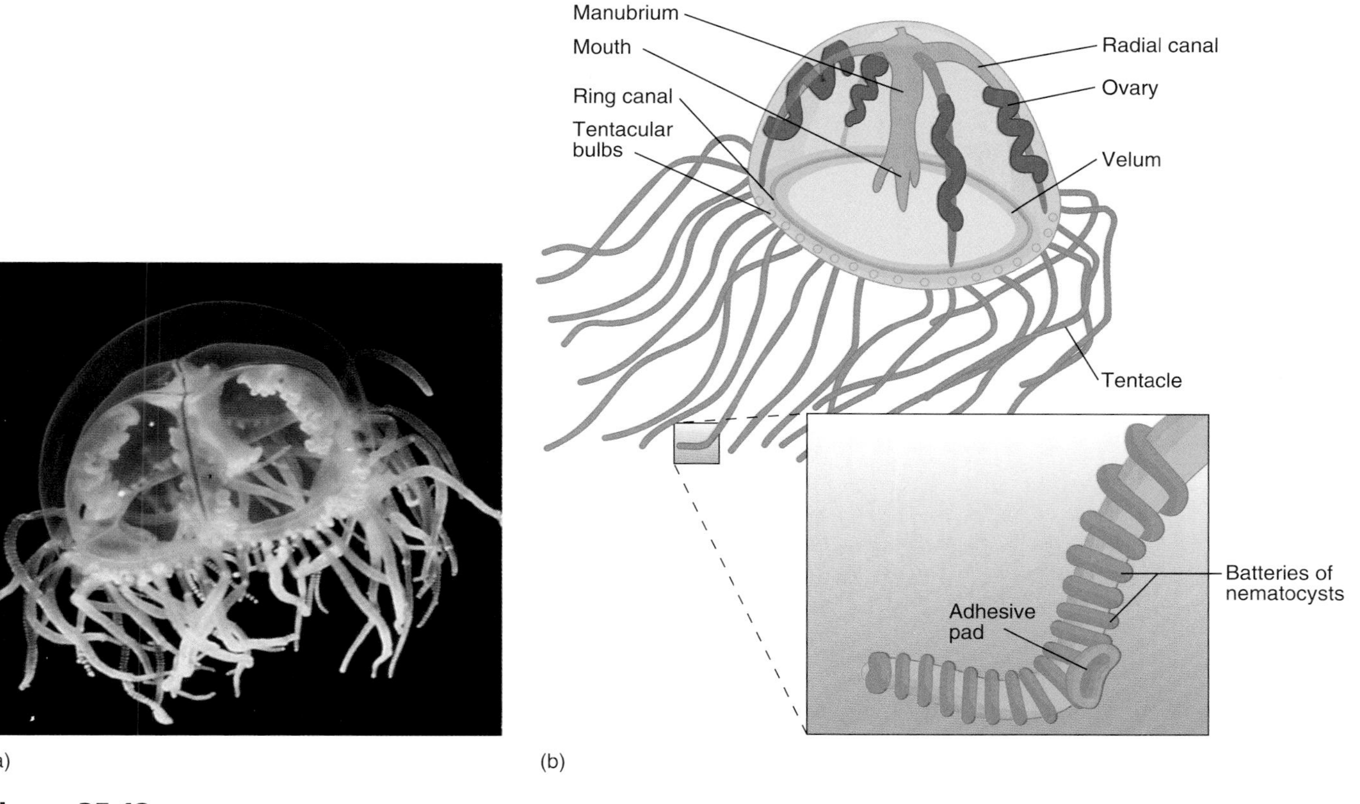

Figure 35.12
Gonionemus, a hydrozoan. (*a*) The medusa. (*b*) Diagram showing the structure of the medusa.

appear similar in males and females. The **tentacles** have a rough surface.

Question 7
Which cells give tentacles of *Gonionemus* their rough surface?

Class Scyphozoa

Observe preserved medusae of *Aurelia* and *Cassiopeia*. Scyphozoans are commonly called jellyfish because the gelatinous medusa dominates their life cycle. The polyp is reduced to a small larval stage. The mesoglea has amoeboid cells. The GVC is divided into four radiating pouches, and the gastrodermis has cnidocytes. This group also includes one of the largest invertebrates in the world, *Cyanea capillota*. *Cyanea* lives in the North Sea and can exceed 2 m in diameter. *Aurelia* (2–6 cm) is a more typical jellyfish for you to examine (fig. 35.13).

Examine prepared slides of (1) planula larvae produced by sexual reproduction of medusae, (2) the polyp stage called a **scyphistoma**, and (3) ephyra (immature medusae) budded from the polyp. Review the life cycle of *Aurelia* (fig. 35.14).

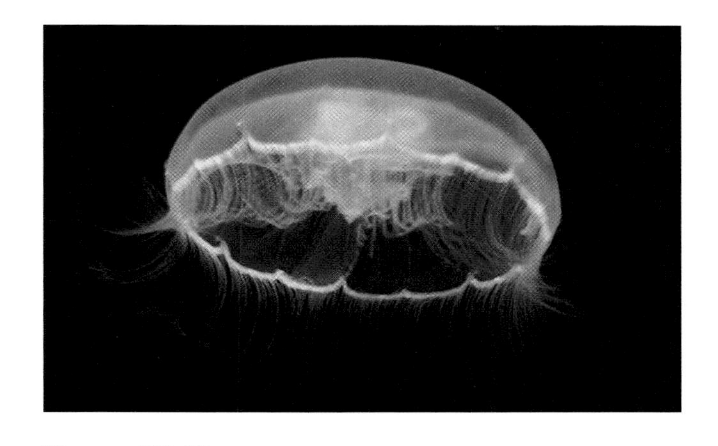

Figure 35.13
Aurelia, the moon jellyfish. The gelatinous medusa is the dominant body-form in the life cycle of these cnidarians.

Question 8
How do medusae of *Aurelia* and *Gonionemus* differ in size, arrangement of tentacles, and shape of manubrium?

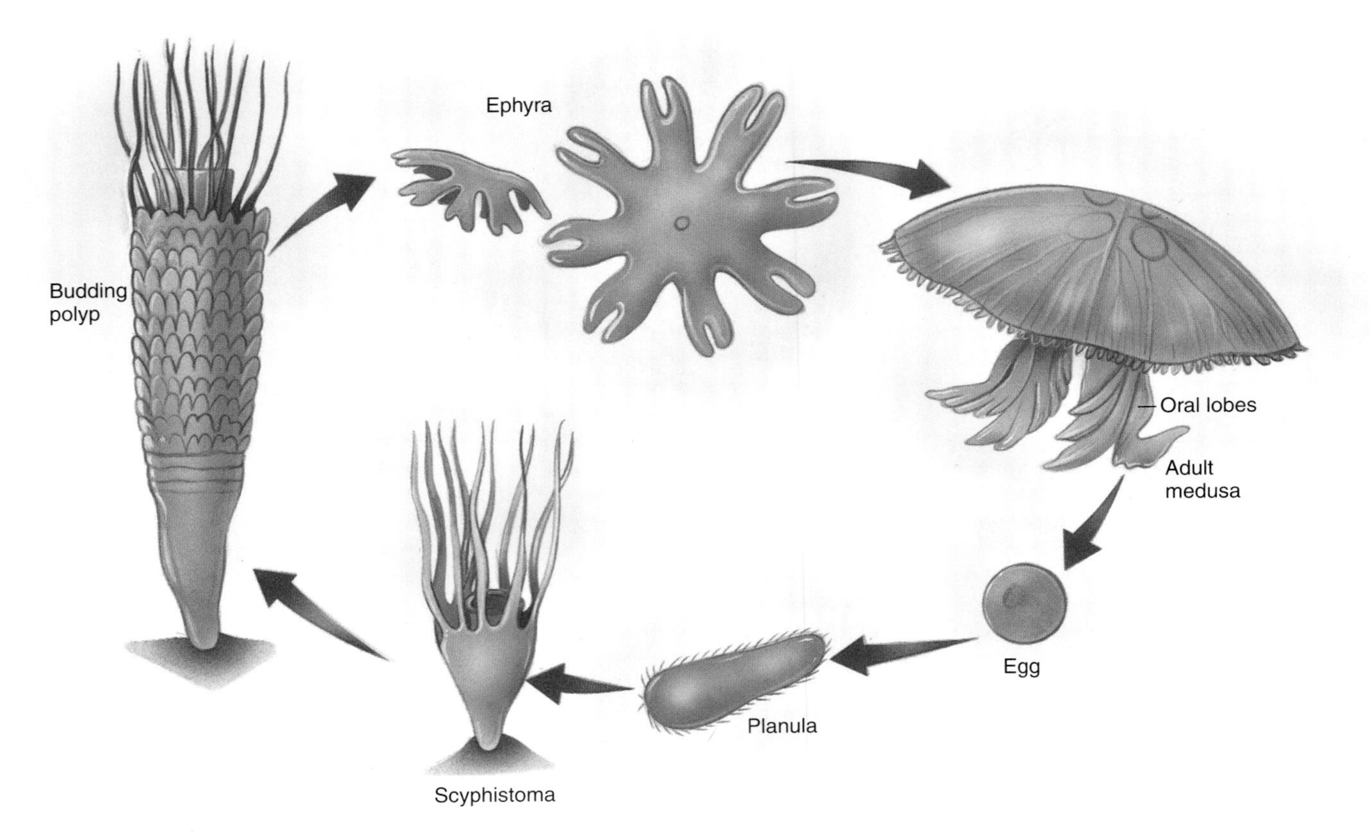

Figure 35.14

Aurelia life history. For the dioecious *Aurelia*, like all scyphozoans, the medusa stage dominates the life cycle. Male and female medusae produce gametes for fertilization in the water. These gametes are the only haploid (1N) stage of the life cycle; all other stages are diploid (2N). The zygote develops into a planula larva that settles to the substrate and forms a scyphistoma (polyp) that produces ephyrae (immature medusae) by budding. Medusae separate from the polyp and swim away.

Class Anthozoa

Anthozoans (anemones and corals) form the largest class of cnidarians with more than 6000 species (fig. 35.15). Anthozoan polyps are solitary or colonial, and there is no medusa. The mouth leads to a tubular pharynx and to a GVC with septate compartments. Gonads are gastrodermal.

Metridium

Obtain a specimen of the common anemone, *Metridium*, and find its mouth and tentacles. Make a cross section through the body of *Metridium* and expose the gastrovascular cavity (this dissection may be on demonstration). Locate the structures shown in figure 35.16. Anemones are sessile and attach themselves to a substrate with their flat and sticky basal disk. However, this attachment is not permanent, and anemones can slowly slide on a film of mucus. When pieces of the basal disk tear away from a moving anemone the pieces form a new individual. This type of asexual reproduction is **fragmentation.**

Notice that the gastrovascular cavity of an anthozoan polyp is partitioned by thin septa. These septa distinguish anthozoan from scyphozoan and hydrozoan polyps.

Figure 35.15

Sea anemone, a common anthozoan.

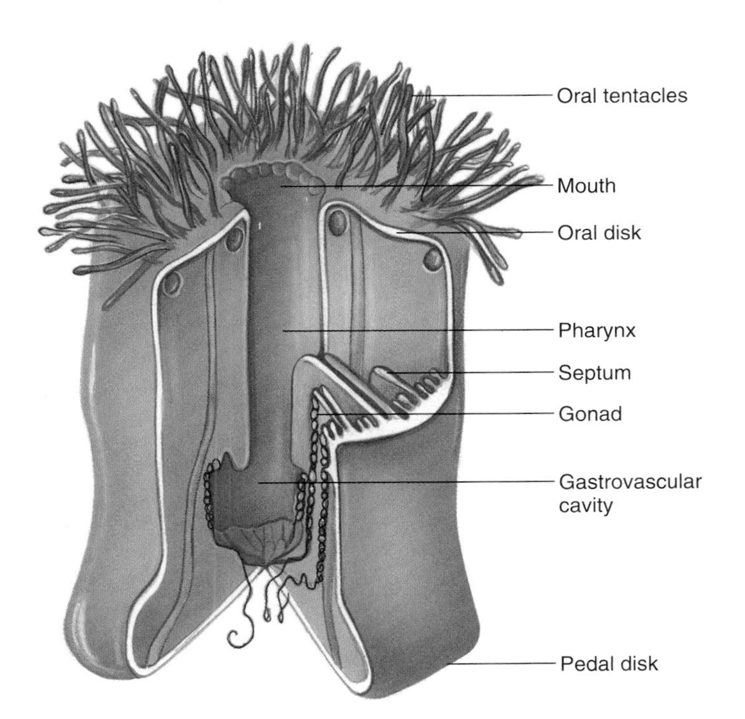

Figure 35.16
Class Anthozoa. The structure of the anemone, *Metridium*. Anthozoans have no medusa stage.

Figure 35.17
Calcium carbonate skeleton of a stony coral.

Corals

Examine a piece of dry, calcareous coral and look for small depressions (fig. 35.17). Coral polyps are structurally similar to anemones, but corals are usually colonial and much smaller. Most corals secrete a hard skeleton of calcium carbonate with many small cups surrounding the polyps. The small, fragile polyps are probably absent from the specimen you are examining, but the depressions in which they lived are numerous.

Examine a piece of *Tubipora* (fig. 35.18) and any other anthozoans on display. The tropical organ pipe coral, *Tubipora*, is organized differently. Long parallel polyps are encased in calcareous tubes connected at intervals by transverse plates. The calcareous tubes are impregnated with iron salts that give the colony an attractive color.

Figure 35.18
Tubipora, organ pipe coral.

Question 9

a. Locate the radial ridges within each depression on a piece of coral. What structures within the polyp did they support?

b. What is the advantage of a partitioned gastrovascular cavity?

c. Consider objective 1 listed at the beginning of this exercise. How does fragmentation contribute to the evolutionary success of anthozoans in their environment?

Questions for Further Thought and Study

1. Which group within kingdom Protista probably gave rise to sponges? On what evidence do you base your answer?

2. Why are spicules used as a primary characteristic in the taxonomy of sponges?

3. Why are sponges considered to be an evolutionary dead end?

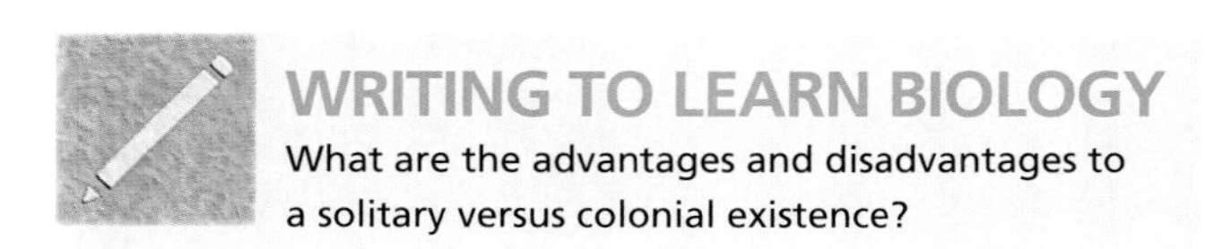

36

Survey of the Animal Kingdom
Phyla Platyhelminthes and Nematoda

Objectives

By the end of this exercise you should be able to:

1. Describe how structures specific to platyhelminths and nematodes help them survive in their environment and promote their evolutionary persistence.
2. Describe the general morphology of flatworms in phylum Platyhelminthes and roundworms in phylum Nematoda.
3. List characteristics which phyla Platyhelminthes and Nematoda have in common with phyla Porifera and Cnidaria.
4. List characteristics of flatworms and roundworms that are more advanced than those of more primitive phyla.
5. List examples of roundworms and examples of each major class of flatworms.

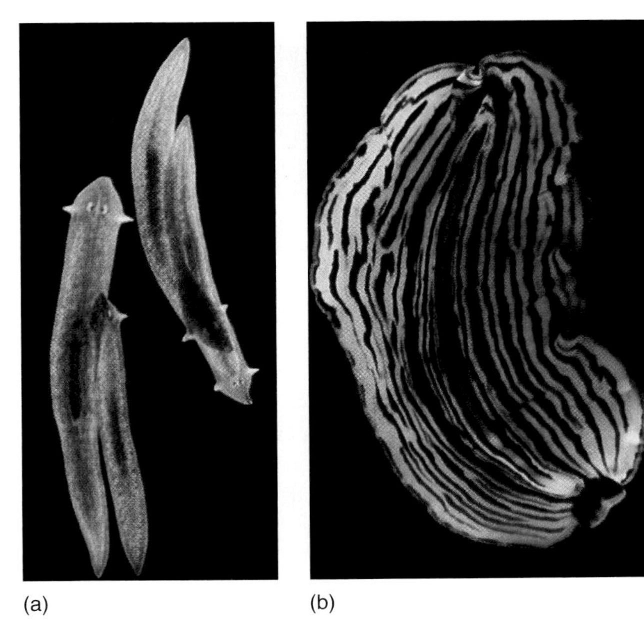

(a) (b)

Figure 36.1
Flatworms (phylum Platyhelminthes). (*a*) A common freshwater flatworm. (*b*) A free-living marine flatworm.

Flatworms of phylum Platyhelminthes and roundworms of phylum Nematoda live in marine, freshwater, terrestrial, and parasitic environments. Their morphology is more complex than that of sponges and jellyfish (table 36.1). For example, flatworms and roundworms have a cellular **mesoderm** in addition to ectoderm and endoderm. Flatworms are **acoelomate,** meaning that their mesoderm is a solid mass of tissue with no internal cavity surrounded by mesoderm. **Coelomate** animals are discussed in the next three exercises and have major organs suspended in a coelomic cavity completely surrounded by mesoderm. Because flatworms have three germ layers, flatworms and roundworms are described as **triploblastic.** They also have **organs** made of interdependent tissues and are the simplest animals having **bilateral symmetry** with distinct anterior and posterior ends.

PHYLUM PLATYHELMINTHES

Flatworms are dorsoventrally compressed, and free-living species have primitive sense organs (fig. 36.1). They also have a gastrovascular cavity with one opening that is both mouth and anus. Their nervous system is more advanced than that of cnidarians (see Exercise 35) and consists of a ladderlike arrangement of nerve cords extending the length of the body.

Class Turbellaria

Turbellarians (3000 species) are free-living flatworms inhabiting freshwater, saltwater, and moist terrestrial environments. Turbellarians scavenge and prey on small animals. They are **hermaphroditic,** meaning that individuals have both male and female sex organs.

TABLE 36.1

PHYLA PLATYHELMINTHES AND NEMATODA

Phylum	Typical Examples	Key Characteristics	Approximate Number of Named Species
Platyhelminthes (flatworms)	*Planaria*, tapeworms, liver flukes	Solid, unsegmented, bilaterally symmetrical worms; no body cavity; digestive cavity, if present, has only one opening	20,000
Nematoda (roundworms)	*Ascaris*, pinworms, hookworms, *Filaria*	Pseudocoelomate, unsegmented, bilaterally symmetrical worms; tubular digestive tract passing from mouth to anus; tiny; without cilia; live in great numbers in soil and aquatic sediments; some are important animal parasites	12,000+

Dugesia

Dugesia, often called planaria, is a common freshwater turbellarian with typical characteristics of flatworms (fig. 36.2). The head has lateral lobes and sensory organs called **eyespots.** *Dugesia* feeds by sucking food through its mouth and into a tubular **pharynx** leading to the gastrovascular cavity. The muscular pharynx is usually retracted in the body but can be everted through an opening in the midventral epidermis (middle of the lower surface). Most digestion in the gastrovascular cavity is extracellular, but phagocytic cells line the cavity and complete digestion of small particles intracellularly.

Planaria are simple organisms having no body cavity other than the digestive cavity. Planarias are acoelomate, a term that will be explained more fully in Exercise 37, figure 37.1. The mesodermal tissue includes the loose mass of cells between the ectoderm on the surface and the endoderm lining the gastrovascular cavity and pharynx.

Procedure 36.1

Observe planaria

1. Obtain a living *Dugesia* and examine its morphology with a stereoscopic (dissecting) microscope.
2. Place the animal in the center of a petri dish and follow its movements for a few minutes.
3. If a spotlight is available, determine how *Dugesia* may respond to strong light.
4. Offer *Dugesia* a small piece of liver or boiled egg; watch it eat.
5. After a few minutes, gently roll the animal from the food and find its protruded pharynx.
6. Examine a whole mount of a stained planaria and a prepared slide of a cross section through a planaria.
7. Locate the structures shown in figure 36.2.
8. Examine cross sections taken through the region of the pharynx and away from the pharynx. Draw and label these cross sections, and have your instructor check them for accuracy.

Question 1

a. What features of *Dugesia* distinguish its head from its tail?

b. What is the difference between the eyes of most animals you are familiar with and the eyespots of *Dugesia?*

c. How does the head of *Dugesia* move differently from the tail?

d. Does *Dugesia* move randomly or in an apparent direction?

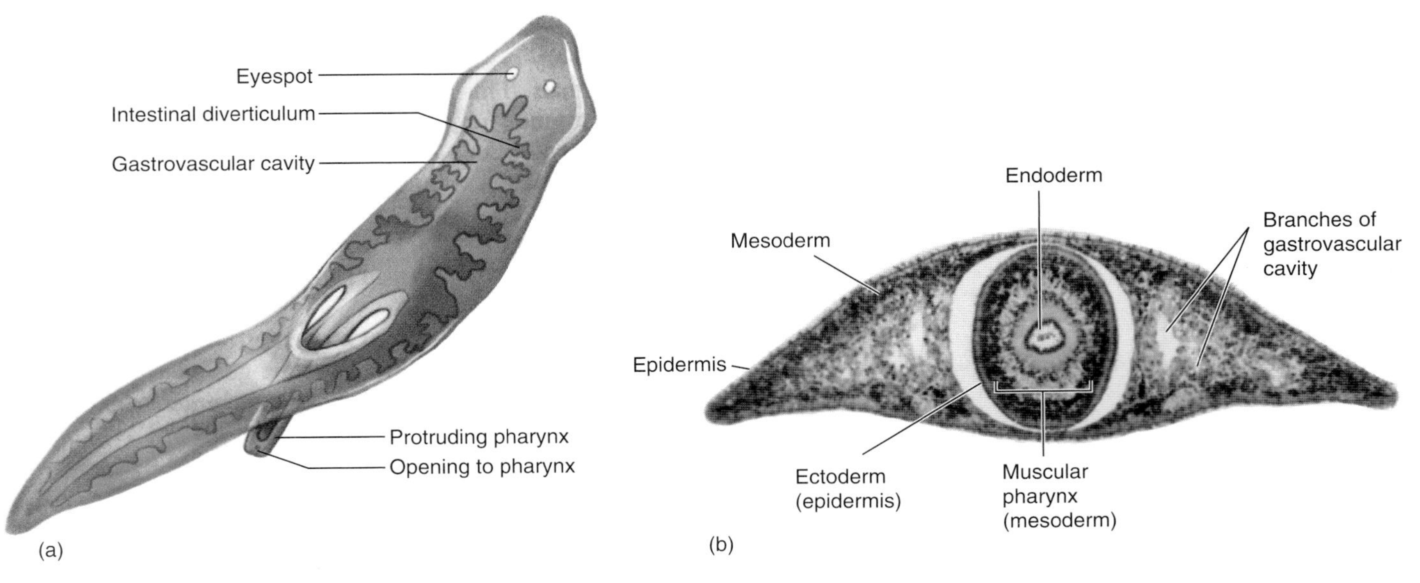

Figure 36.2
Anatomy of a free-living flatworm. (*a*) External structure. (*b*) Cross section of a planarian taken through the pharynx region.

e. How is *Dugesia* adapted for directional movement?

f. How does a flatworm respond when touched with a probe?

g. Where is the feeding tube located? Why is this unusual for bilaterally symmetrical organisms?

h. Is the gastrovascular cavity of *Dugesia* a simple sac? How is it divided and what advantage do these divisions offer?

i. Consider objective 1 listed at the beginning of this exercise. How could being monoecious contribute to evolutionary success of flatworms in their environment?

Class Trematoda

Trematodes, commonly called flukes, are parasites, and their oval bodies are usually a few millimeters long. Flukes infect vertebrates and include both **endoparasites** (parasites inside their host) (fig. 36.3) and **ectoparasites** (parasites on the surface of their host). Trematodes lack an epidermis and are covered by an acellular but metabolically active **epicuticle.** This epicuticle is made of protein and lipids secreted by mesodermal cells, and resists digestive enzymes. The epicuticle helps in respiration and absorbing nutrients. The ventral surface of a fluke usually has two adhesive organs (suckers). The anterior sucker surrounds the mouth.

Opisthorchis

Use a stereoscopic (dissecting) microscope to examine a prepared slide of *Opisthorchis* and locate the structures shown in figure 36.4. *Opisthorchis* (*Clonorchis*) *sinensis*, the Chinese liver fluke, often parasitizes humans in Japan and China. This monoecious, adult fluke attaches to the bile duct and releases eggs that move through the digestive system of the host and exit with the feces. Larvae of flukes typically develop in snails and fish. Humans are infected when they eat raw or poorly cooked fish.

Question 2

a. How does the shape of the digestive sac of *Opisthorchis* compare with that of *Dugesia?*

b. How does the position of the mouth of *Dugesia* and flukes compare?

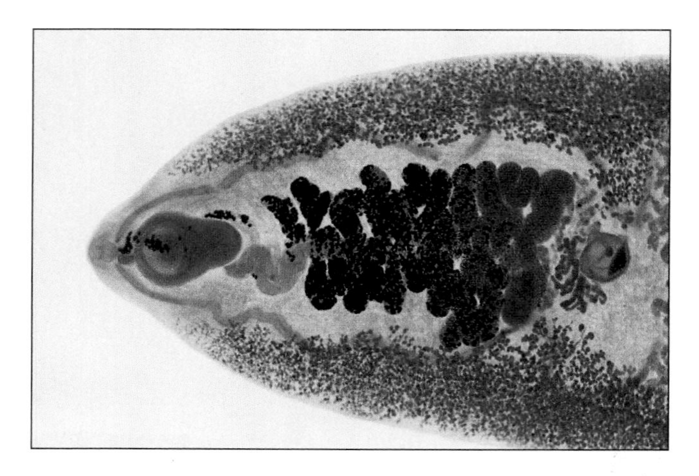

Figure 36.3

Fasciolopsis buski is a parasitic fluke that infects the small intestine of humans and pigs. Humans acquire the fluke by eating water chestnuts and other aquatic plants with larvae attached. The larvae develop in intermediate hosts such as snails and fish. A major adaptation of flukes and other parasites is the prolific production of eggs—as many as 3000 eggs per day for a lifetime of many years. The dark, coiled uterus shown here contains hundreds of eggs, but most of these eggs will never be engulfed by an intermediate host and hatch. To complete a complicated life-cycle through multiple hosts, a fluke must produce immense numbers of eggs to ensure that one or two are successful. To appreciate the structure of the entire organism, see the related genus *Fasciola* in figure 36.5.

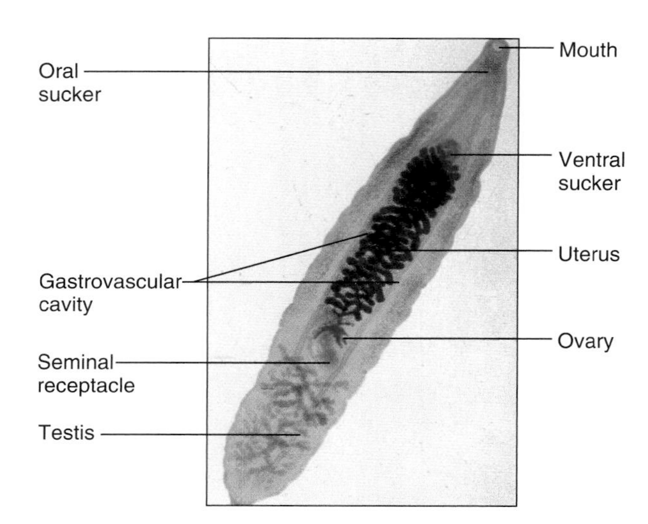

Figure 36.4

Internal structure of *Opisthorchis*, the Chinese liver fluke.

Fasciola

Examine figure 36.3 and read its caption carefully. Use a stereoscopic (dissecting) microscope to examine a prepared slide or plastic mount of *Fasciola* and locate the features shown in figure 36.5. *Fasciola hepatica*, the sheep liver fluke, infects sheep, other vertebrates, and (rarely) humans. It is much larger than *Opisthorchis* but similar in structure. *Fasciola* sucks food (blood, mucus, and cells) through a muscular pharynx located behind the mouth.

Question 3

Consider objective 1 listed at the beginning of this exercise. How could the production of large numbers of eggs contribute to the evolutionary success of flatworms in their environment?

Schistosoma

Examine a demonstration slide of *Schistosoma* and draw its body shape. *Schistosoma*, a blood fluke, inhabits the intestinal veins and other organs of many vertebrates (including humans) and causes the disease schistosomiasis (fig. 36.6). *Schistosoma* is socioeconomically important because it infects more than 200 million people in countries having tropical and temperate climates. Symptoms of infection include an enlarged liver, spleen, and bladder, as well as nutritional deficiency. Some of these symptoms, such as bleeding ulcers in the intestine and other organs, are enhanced by large deposits of eggs (up to 3500 per day from one female!).

Snails are **intermediate hosts** for *Schistosoma*. An intermediate host is an organism harboring immature stages of a parasite, whereas a **definitive host** contains sexually mature, egg-laying stages of the life cycle. Irrigation ditches with snail populations enhance the spread of schistosomiasis in underdeveloped agricultural areas. Immature larvae of blood flukes released from the snails burrow through skin and infect people wading in these ditches.

Unlike most trematodes, *Schistosoma* is dioecious. The male has a ventral groove along the length of his body into which the slender female cradles for copulation. Often the female remains in this groove for the remainder of her lifetime and extends slightly to lay eggs. Your prepared slide may include schistosomes in this position for copulation.

Question 4

How does the shape of *Schistosoma* differ from that of other flukes you have studied?

Class Cestoda

Cestodes, commonly called tapeworms, are the most specialized platyhelminths. They are endoparasites of the gut of vertebrates and are covered by a cuticle similar to that of trematodes (fig. 36.7). However, tapeworms lack a mouth or digestive tract and have a unique body plan (fig. 36.8). Their cuticle efficiently absorbs nutrients from their host.

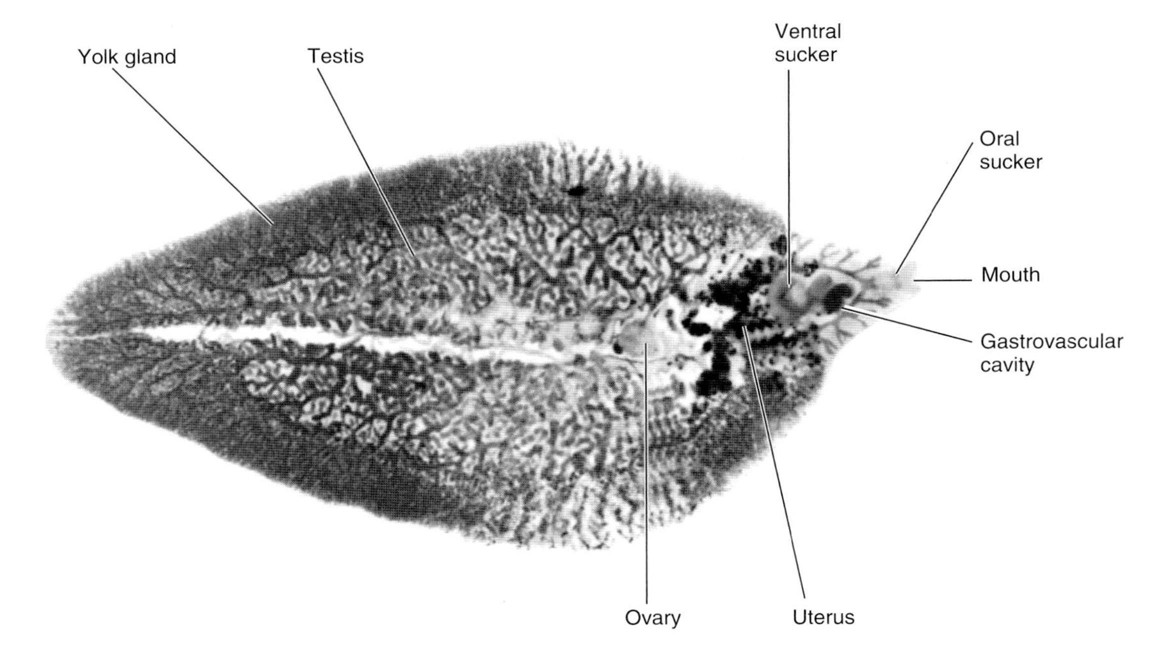

Figure 36.5
Internal structure of *Fasciola,* the sheep liver fluke.

The anterior end, or **scolex,** adheres to the host's intestinal wall with hooks or suckers. Behind the scolex is the **neck** followed by a series of segments called **proglottids.** The scolex and neck are small, but the chain of proglottids may be 10–15 m long.

A tapeworm grows as the scolex and neck continually produce a chain of proglottids as self-contained packets of male and female reproductive organs. Self-fertilization occasionally occurs in a proglottid, but cross-fertilization by copulating proglottids of adjacent worms is more common. **Gravid** (egg-carrying) proglottids eventually break from the end of the worm and pass from the host with its feces.

Examine a prepared slide or whole specimen of a pork tapeworm, *Taenia solium*. Then examine a prepared slide of a scolex. Draw the basic shape of the scolex and note any hooks or suckers. Also examine and sketch from prepared slides young, mature, and gravid proglottids.

An adult *Taenia solium* may be up to 10 m long. Humans infect themselves with *Taenia* by eating uncooked meat from pigs, which are often intermediate hosts.

Compare a mature proglottid with that shown in figure 36.9. Although each proglottid has a complete reproductive system, the excretory ducts and longitudinal nerves are continuous between proglottids.

Examine a demonstration specimen of *Dibothriocephalus latus* if it is available. *Dibothriocephalus latus* is the largest tapeworm to infect humans and can be up to 20 m long. Their intermediate hosts are small crustaceans and fish.

Question 5

a. How does a scolex compare in size to a proglottid near the scolex? Near the posterior end?

b. Tapeworms have no digestive system or mouth. How, then, do they obtain food?

c. Examine figure 36.8 and its caption carefully. Which proglottids, mature or gravid, occur closest to the scolex?

d. What is the difference between a mature and gravid proglottid?

e. Consider objective 1 listed at the beginning of this exercise. Is having a scolex significant to a fundamental process for tapeworms? For which process, and how is it significant?

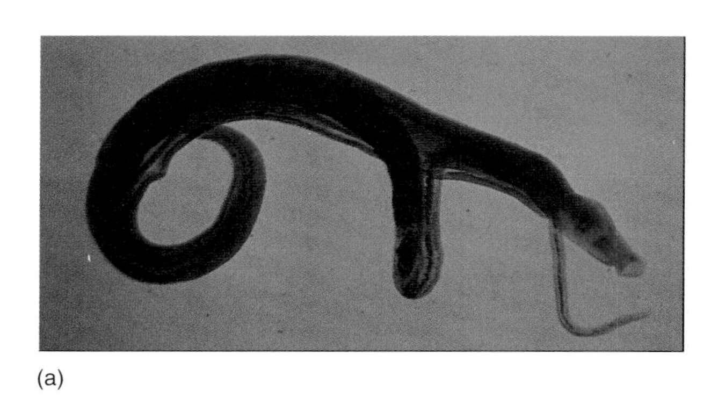

(a)

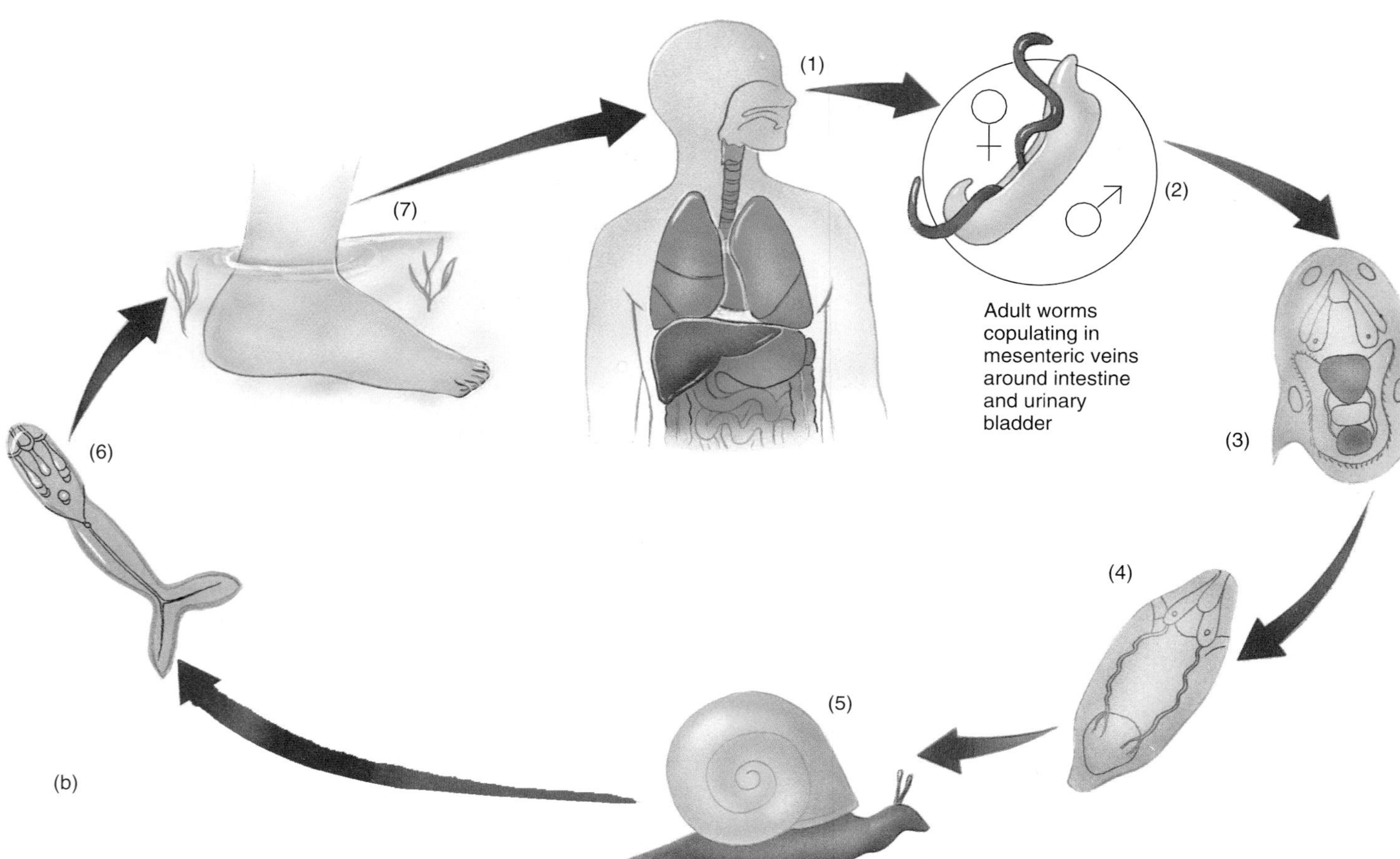

Figure 36.6

Schistosoma, the blood fluke that causes schistosomiasis. *Schistosoma* infects more than 200 million people. (*a*) Adult male and female *Schistosoma mansoni* in copulation. (*b*) Life cycle of *Schistosoma mansoni* begins in a human (1). Adult schistosomes are in permanent copulation (2) and produce thousands of eggs. Eggs leave the host with feces or urine and hatch as larvae (3)(4). A larval stage penetrates a snail (5) and develops the final stage (6). This motile larval stage is released and penetrates the skin of a human (7), where it invades the intestinal veins and develops into an adult.

PHYLUM NEMATODA

Nematodes (fig. 36.10), commonly called roundworms, live in all aquatic, terrestrial, and parasitic environments and often occur in great numbers—a single decomposing apple may contain 100,000 nematodes of different species. If we removed everything but nematodes from the environment, we would still see a ghostly outline of our entire biosphere. Estimates of the global number of species yet to be described often exceed one million. Almost all feeding types (parasitic, predatory, etc.) are represented. Reproductive morphologies include dimorphic species, hermaphroditic species, and even species with males as well as hermaphrodites. Nematode diversity is extraordinary.

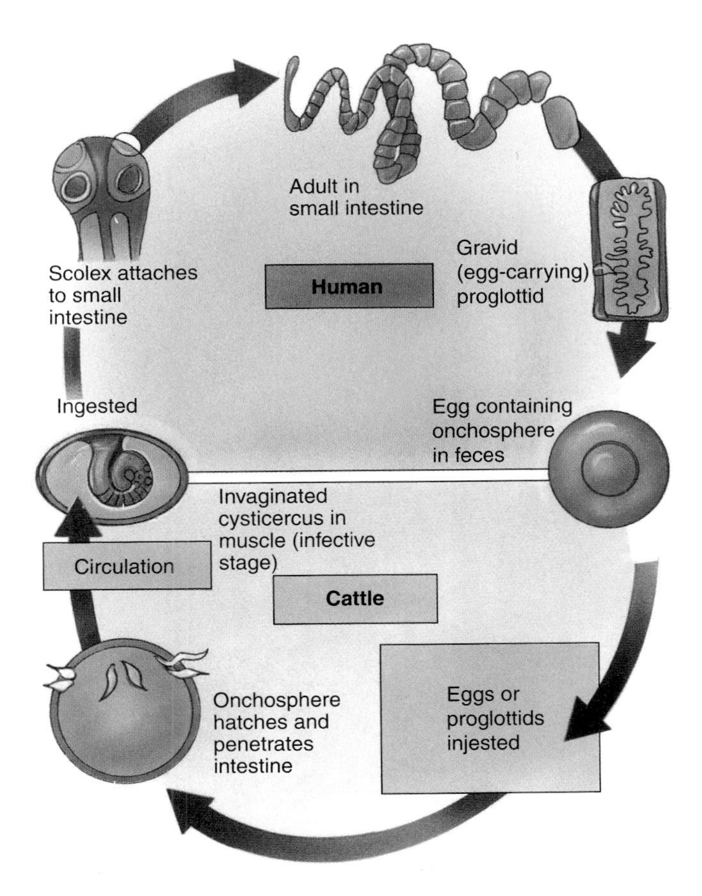

Figure 36.7

Life cycle of a beef tapeworm. Tapeworms, which are members of the class Cestoda, are the most specialized flatworms and require human and cattle hosts. The onchosphere and cysticercus are specialized stages of development.

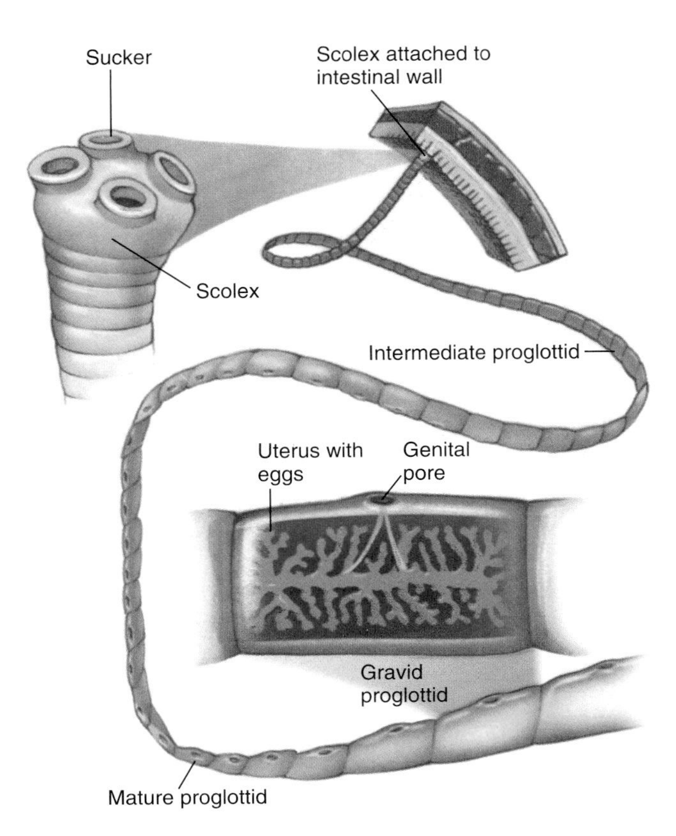

Figure 36.8

Body plan of a tapeworm (class Cestoda). A mature proglottid has mature reproductive organs; a gravid proglottid contains many eggs.

Many nematodes cause diseases in humans, other animals, and plants. One of the most serious of these diseases is elephantiasis, the grotesque swelling of an arm or leg resulting from nematodes (commonly *Filaria*) clogging the lymphatic system that drains the host's appendage (fig. 36.11). Fluid accumulates, and the appendage swells. Another parasitic nematode is the eye worm, *Loa loa*, which lives under the skin of humans and occasionally crawls across the surface of an eye (fig. 36.12).

Nematodes are slender and long with a rather featureless exterior. They lack flagella and cilia, and are covered with a flexible and chemically complex **cuticle.** The cuticle resists digestive enzymes and is permeable only to water, dissolved gases, and some ions. Roundworms have two morphological advances absent in flatworms. First, in addition to their digestive cavity, roundworms have a body cavity called a **pseudocoel** consisting of a fluid-filled space between the body wall and digestive tract (see fig. 37.1). Internal organs are suspended in this cavity. Second, nematodes have a **complete digestive tract** with a mouth and anus.

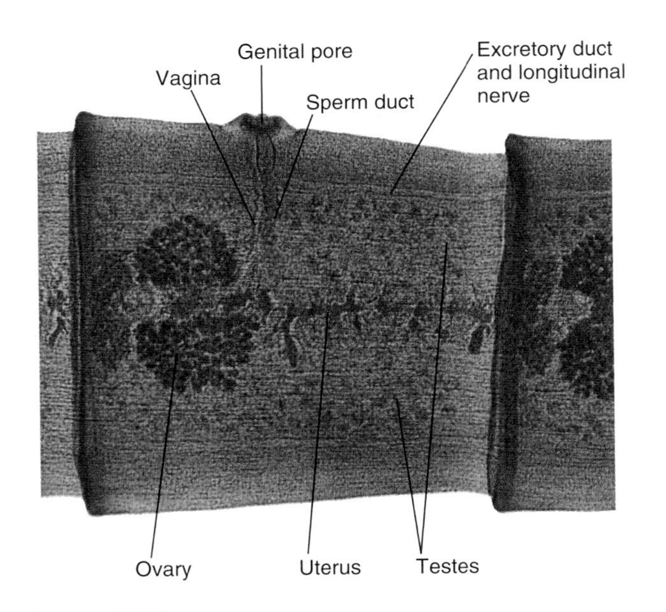

Figure 36.9

Mature proglottid of *Taenia pisiformis*, dog tapeworm. Portions of two other proglottids are also visible.

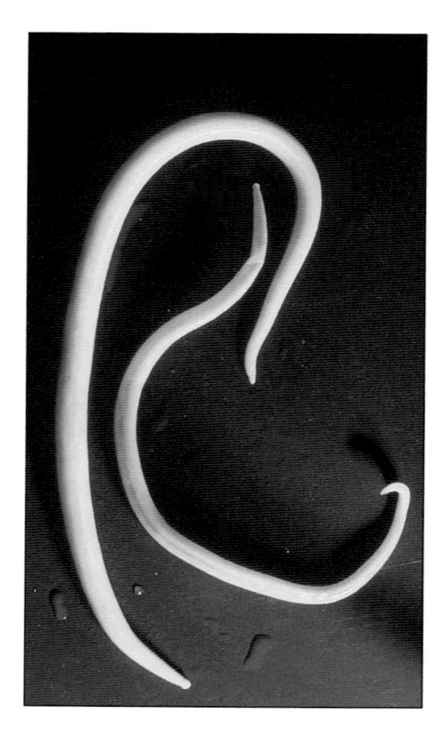

Figure 36.10
Ascaris lumbricoides, a common roundworm, inhabits the intestines of pigs and humans. Male ascarid worms are smaller than females and have a curved posterior end.

Question 6

a. How does the number of body cavities of nematodes compare with that of flatworms?

b. What are the advantages of a digestive tract having a separate entrance and exit?

c. Female *Ascaris* are more numerous than males. Why might this be adaptive?

Rhabditus and *Turbatrix*

Procedure 36.2

Examine living nematodes

1. Examine a culture of living *Rhabditus* or the vinegar eel, *Turbatrix*.

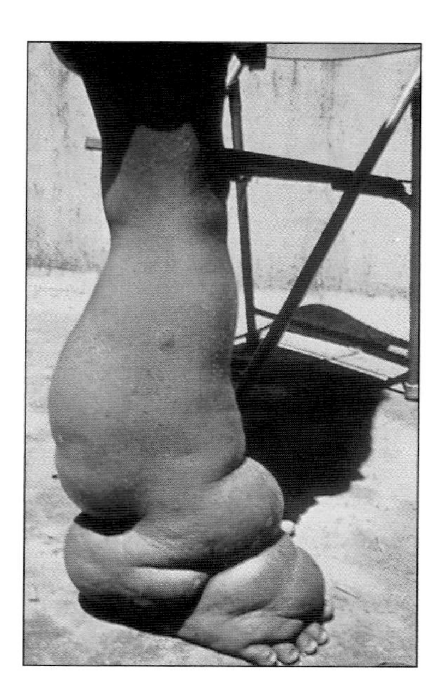

Figure 36.11
Elephantiasis of a leg caused by adult filarial worms that live in lymph passages and block the flow of lymph. Tiny juveniles, called microfilariae, are picked up in a blood meal of a mosquito where they develop to the infective stage and are transmitted to a new host.

2. Using a toothpick, put a small piece of the culture medium on a slide with a drop of water and coverslip, and focus with low magnification of your microscope.
3. Watch the nematodes move.

The flexible cuticle and hydrostatic pressure of fluid in their pseudocoel aid locomotion by resisting antagonistic muscle contraction. Nematodes have only longitudinal muscles and lack circular or diagonal muscles. This combination of features produces a characteristic motion.

Question 7

a. How would you describe the motion of a nematode?

b. How is this movement related to the movement of their muscle layers?

Ascaris

Ascaris is a large nematode that infects the intestinal tract of humans and other vertebrates (fig. 36.10). Males are smaller than females and have a hooked posterior end. The opening in the posterior end is the anus.

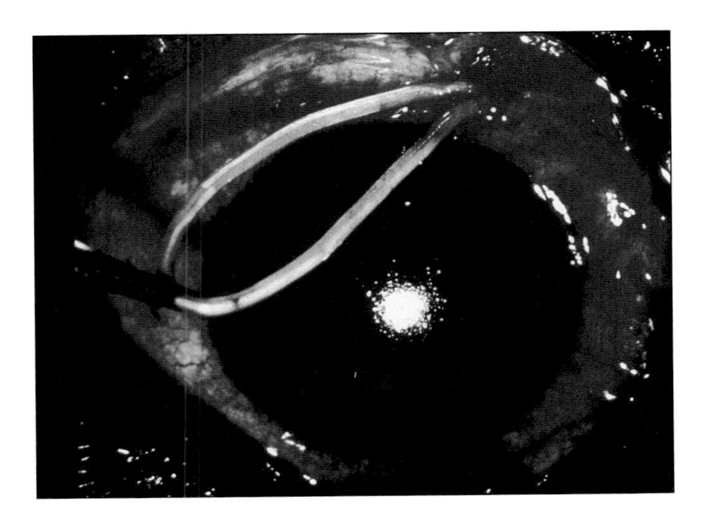

Figure 36.12
The nematode *Loa loa* appears in the eye of its host. This one is being surgically removed from someone's eye.

Procedure 36.3
Dissect *Ascaris*

1. Obtain a preserved *Ascaris lumbricoides* and examine its external features using a dissecting microscope. Compare the external features of males and females.
2. At the anterior end, locate the mouth surrounded by three lobes of tissue.
3. Prick the cuticle with a dissecting probe (teasing needle) and determine its consistency.
4. Pin the ends of a female specimen near the edge of a dissecting pan to permit viewing with a dissecting microscope. Slit the body wall longitudinally with a dissecting needle or sharp-pointed scissors.
5. Pin the body wall open and locate the internal organs shown in figure 36.13. The excretory pore may be small and difficult to find. A dissection of a male may be on demonstration.
6. Examine a prepared slide of a cross section of both a male and female *Ascaris*.
7. Locate the features shown in figure 36.14 and determine from where along the length of your dissected specimen this section was taken.
8. Examine cross sections from different areas of the body if appropriate slides are available. Gametes mature as they move along the length of the tubular reproductive organs.
9. Dispose of waste in labeled containers.

Question 8

a. The cuticle of *Ascaris* is flaky and tough. Consider objective 1 listed at the beginning of this exercise. What might be an adaptive advantage of a thick and tough cuticle?

b. Where do the internal organs of *Ascaris* attach to the body wall?

c. How does the diameter of the female reproductive tract change?

d. Are any sensory organs evident in *Ascaris?* Why would this be adaptive?

Trichinella

Trichinella spiralis causes the disease trichinosis. Adult females of *Trichinella* live in the intestine of their host and release larvae. These larvae migrate through the body to striated muscles, especially in the diaphragm and tongue, where they form calcified cysts that can be painful. The larvae remain encysted until the muscle tissue is eaten by another host, where the larvae mature. Humans infect themselves by eating poorly cooked pork containing encysted larvae. However, the occurrence of trichinosis is decreasing. In the United States fewer than 50 cases of human trichinosis are reported annually to the Centers for Disease Control and Prevention, compared with 500 cases a year in the 1940s.

Procedure 36.4
Examine *Trichinella, Necator,* and *Enterobius*

1. Examine a slide of muscle tissue containing encysted larvae of *Trichinella*.
2. If slides are available, examine specimens of *Necator* (hookworm) and *Enterobius* (pinworm).
3. Note variation in the morphology of the anterior and posterior ends of the organisms.
4. Sketch and note the body sizes of *Trichinella, Necator,* and *Enterobius*.

Hookworm causes anemia in infected livestock, which causes great economic losses. Pinworms, although less dangerous, irritate many children because the pinworms infect the intestine and inflame the anus (fig. 36.15). Children often scratch themselves, put their fingers in their mouths, and thereby reinfect themselves. Pinworms infect about 30% of children and 16% of adults in the United States.

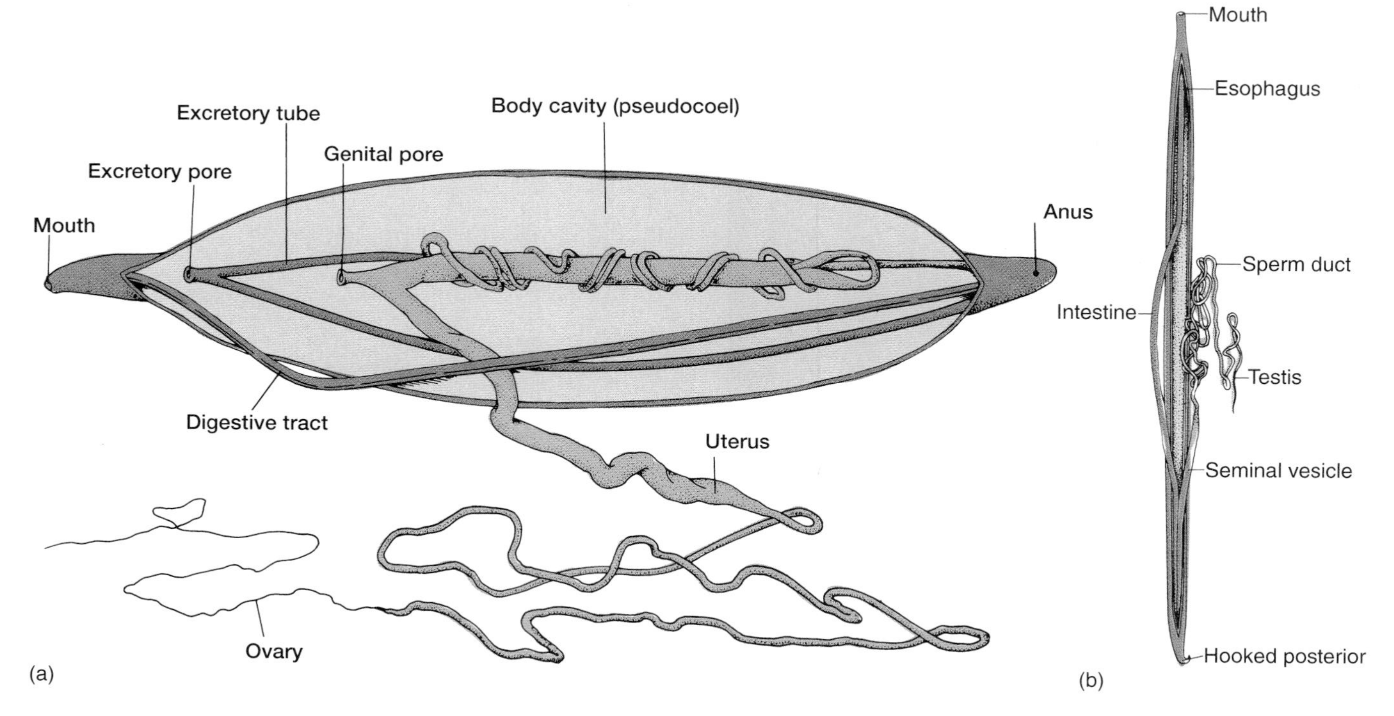

Figure 36.13

Nematodes are commonly called roundworms. Internal anatomy of (*a*) a female and (*b*) a male nematode.

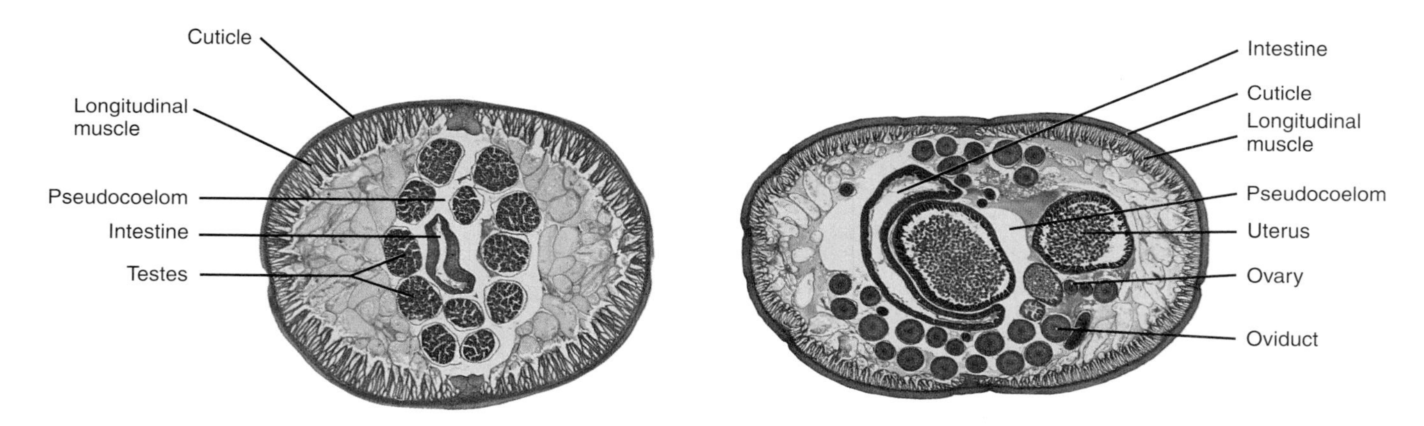

Figure 36.14

Cross sections of male and female *Ascaris*, a large nematode that infects the intestinal tract of a variety of vertebrates.

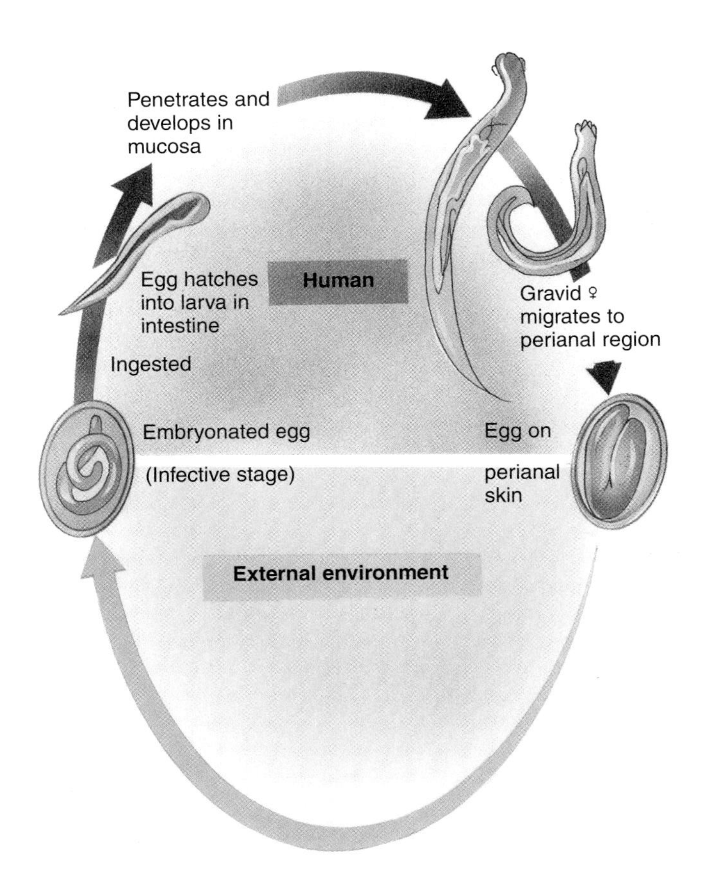

Figure 36.15
The life cycle of the pinworm, *Enterobius vermicularis*. The adult crawls from the intestine and lays eggs on the skin just outside the anus. The host, usually a human child, is irritated by the pinworms and scratches. Eggs are then unknowingly eaten and the larvae hatch in the intestine to complete the life cycle.

Questions for Further Thought and Study

1. Flatworms are the first organisms we have discussed with an anterior-posterior orientation. How does this affect their movement compared to the movement of more primitive organisms?

2. What are the disadvantages of a flatworm's digestive system having only one opening?

3. The complete digestive tract of nematodes and other phyla allows functional specialization. What specializations are common in the digestive tract of higher organisms such as humans?

4. How are some flatworms adapted as parasites?

5. How are some nematodes adapted as parasites?

WRITING TO LEARN BIOLOGY

Are nervous systems—especially sensory systems—more or less complex in endoparasites, ectoparasites, or free-living specimens? Explain.

37

Survey of the Animal Kingdom
Phyla Mollusca and Annelida

Objectives

By the end of this exercise you should be able to:

1. Describe how structures specific to mollusks and annelids help them survive in their environment and promote their evolutionary persistence.
2. Describe the general morphology of organisms of phylum Mollusca and phylum Annelida.
3. List the characteristics that phyla Mollusca and Annelida have in common with phyla Platyhelminthes and Nematoda.
4. Discuss those characteristics of mollusks and annelids that were newly derived from those of their ancestral phyla.
5. List examples of each of the major classes of mollusks and annelids.
6. Understand the differences between acoelomate, pseudocoelomate, and coelomate and know which phyla are associated with each.

Snails, clams, octopuses, and squids of phylum Mollusca and segmented worms of phylum Annelida are **coelomate** organisms (table 37.1, fig. 37.1). That is, they have a coelomic body cavity surrounded by mesoderm and containing complex systems of organs and compartments. Coelomates are further divided into **protostomes** and **deuterostomes.** Phyla Mollusca, Annelida, and Arthropoda (see Exercise 38) are protostomes and have well-developed systems of nerves, circulation, excretion, reproduction, and digestion. Deuterostome phyla include Echinodermata, Hemichordata, and Chordata (see Exercise 39). A detailed comparison of protostomes and deuterostomes is presented in Exercise 39.

PHYLUM MOLLUSCA

Mollusks are soft-bodied animals with a specialized layer of epidermal cells called a **mantle** that secretes a **shell.** Mollusks produce many kinds of external shells—some mollusks have only a remnant internal shell, whereas others have no shell at all (fig. 37.2). This phylum's diversity of 110,000 species is surpassed only by arthropods and probably nematodes. Although mollusks are coelomate, their coelom is often reduced to a small chamber surrounding the heart. The circulatory system is open (except in cephalopods), meaning that blood pools in sinuses and bathes the organs directly. **Open circulatory systems** have a few large vessels and a heart, but no smaller vessels and capillaries.

The basic body plan of a mollusk shows little segmentation, but consistently includes (1) a **visceral mass** of organ systems (digestion, excretion, and reproduction) and sensory structures; (2) a ventral, muscular, and often highly modified **foot** used for locomotion; (3) a calcium-based shell, though occasionally absent; and (4) a mantle that secretes the shell and may aid in respiration and locomotion in some species (fig. 37.2). Some mollusks also have a differentiated **head.**

Class Polyplacophora

Obtain a preserved chiton and examine its external features. Polyplacophorans (*poly* = many, *placo* = plate, *phora* = move), commonly called chitons, are exclusively marine and have a primitive molluskan structure. The dorsal shell is divided into eight plates embedded in the mantle. The ventral foot is a broad oval muscle used to propel chitons slowly over the surface of rocks. The **radula,** a horny toothed organ in the mouth, scrapes food (algae) from rocks.

Class Gastropoda

Most gastropods (*gastro* = stomach, *poda* = foot) (snails) have a single shell that is often coiled and elaborate. Some gastropods, such as marine nudibranchs and the common garden slug, do not produce a shell (fig. 37.3). Most species are marine, but freshwater snails and land snails are common. Gastropods also feed with a rasping band of teeth called a radula (fig. 37.4).

Table 37.1

Phyla Mollusca and Annelida

Phylum	Typical Examples	Key Characteristics	Approximate Number of Named Species
Mollusca (mollusks)	Snails, oysters, octopuses, nudibranchs	Soft-bodied coelomates whose bodies are divided into three parts: head-foot, visceral mass, and mantle; many have shells; almost all possess a unique rasping tongue, called a radula; 35,000 species are terrestrial	110,000
Annelida (segmented worms)	Earthworms, polychaetes, beach tube worms, leeches	Coelomate, serially segmented, bilaterally symmetrical worms; complete digestive tract; most have bristles called setae on each segment that anchor them during crawling	12,000

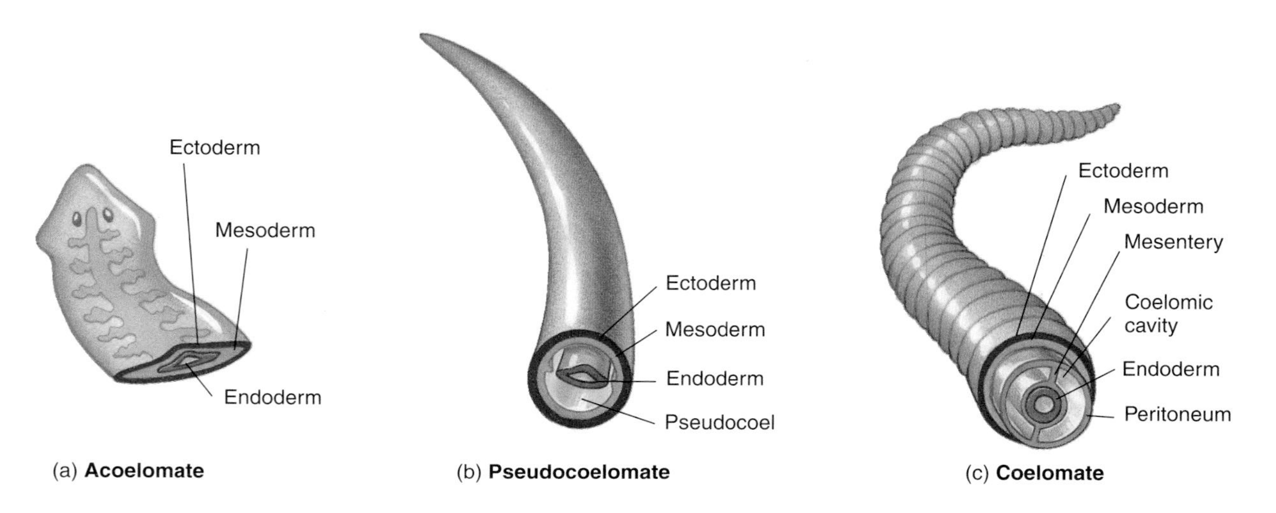

Figure 37.1

Three body plans for bilaterally symmetrical animals. (*a*) Acoelomates (including flatworms) have no body cavity. (*b*) Pseudocoelomates (including nematodes) develop a body cavity between the mesoderm and endoderm. (*c*) Coelomates (including annelids, mollusks, and more advanced phyla) have a body cavity bounded by mesoderm.

Examine preserved gastropods and aquarium snails and compare the relative spiraling of the shells of different species. This spiral growth is common in mollusks and results from unequal growth of the two halves of the larva and mantle.

Class Bivalvia

Clams, oysters, scallops, and mussels are bivalves (*bi* = two, *valve* = door or shell) having a dorsally hinged shell in two parts (figs. 37.2, 37.5). The mantles of the left and right valves join posteriorly to form a ventral **incurrent siphon** and a dorsal **excurrent siphon** that direct water through the clam. When a clam burrows in sediment the siphons extend to the water. The space between the mantle and visceral mass is the **mantle cavity.** Water flows into the mantle cavity through the incurrent siphon, over the gills, and then dorsally into a space between the visceral mass where gills attach to the mantle. Water exits this chamber posteriorly through the excurrent siphon. As water flows over gills, suspended food particles are filtered by cilia and swept along grooves on the gills to the ventral edge. Cilia at the ventral edge move food to the labial palps that surround and direct food to the **mouth.**

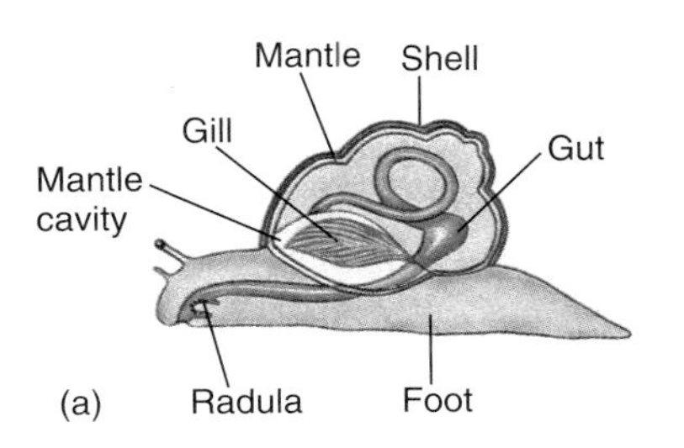

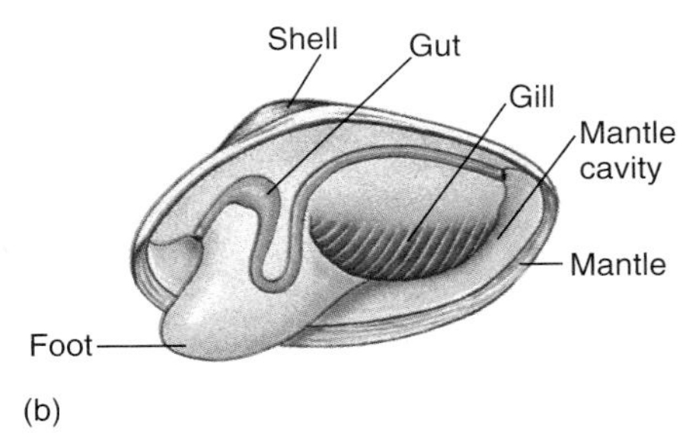

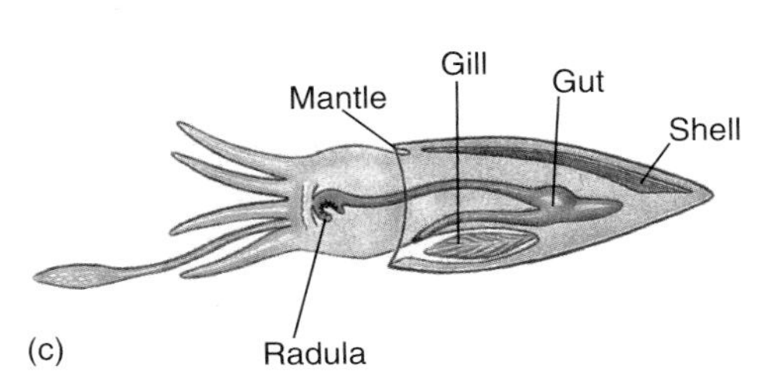

Figure 37.2
The body plans of the three major classes of mollusks: (*a*) Gastropoda, (*b*) Bivalvia, and (*c*) Cephalopoda. A mollusk's shell has an outer layer overlying layers of densely packed crystals of calcium carbonate. A visceral mass includes organs of digestion, excretion, and reproduction and extends as a muscular foot adapted for locomotion, attachment, or food capture (in squids and octopuses). The radula is a scraping and feeding organ characteristic of mollusks (except bivalves, which obtain their food by filter-feeding). Folds of tissue called the mantle arise from the dorsal body wall, line the shell, and enclose a cavity adjacent to the visceral mass. Within the mantle cavity are gills or lungs.

Procedure 37.1

Examine bivalve anatomy

1. Obtain a preserved freshwater bivalve, *Anodonta*, and locate the anterior, posterior, **hinge,** and **umbo** of the shell (fig. 37.6).
2. Open the shell by carefully cutting the anterior and posterior **adductor muscles.**
3. As you pry the shell open, loosen the mantle from the surface of the left valve.
4. Examine the mantle. Remove half of the mantle to expose the visceral mass.
5. Locate the **gills, foot,** and **labial palps.**
6. Cut into the foot to find the stomach and digestive tract.
7. Trace the flow of water and food through your dissected specimen.
8. When you finish your observations, dispose of the preserved tissue and shells in appropriate containers.

Question 1

a. What is the texture of the mantle of *Anodonta*?

b. Are siphons in *Anodonta* as obvious as they are in figure 37.6?

c. Consider objective 1 listed at the beginning of this exercise. In what ways would having a shell contribute to the survival and reproductive success of mollusks in their environment?

Examine a cluster of oyster shells and note how they are stuck together. Mature oysters attach permanently to their substrates. Most bivalves produce pearls, but the finest pearls are made by *Pinctada* in the warm waters of the Pacific. Natural pearls form in oysters when an irritant such as a grain of sand lodges between the mantle and the shell. The mantle responds by surrounding the irritant with layers of the same crystalline material used for the shell. Cultured pearls are produced by oysters about three years after aquaculturists introduce an irritant to their mantle.

Figure 37.3

Colorful nudibranchs, such as *Flaballina iodinae*, are gastropod mollusks without a shell. They have a rather mysterious defensive strategy—nudibranchs use the weapons of their prey. They eat jellyfish, hydrozoans, and corals, all of which have stinging structures called nematocysts. When a nudibranch attacks and eats jellyfish, it can swallow and digest these nematocysts without discharging them. The stingers pass through the digestive tract and are stored in feathery projections on the dorsal surface of the nudibranch. A predator taking a mouthful of nudibranch gets a nasty taste.

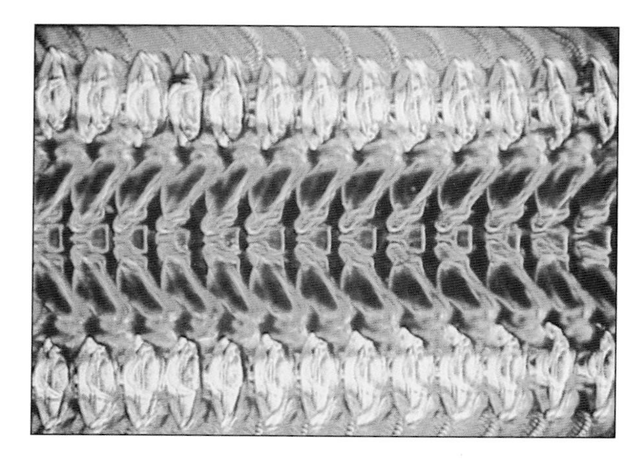

Figure 37.4

These rows of teeth are part of a snail's radula. A radula is a unique rasping tongue of mollusks with rows of chitinous teeth covering its surface. The membranous radula is stretched over a rigid cartilaginous rod that is pressed against a surface to be scraped. Snails use their radula to scrape algae from rocks or tear pieces from plant leaves.

Question 2

a. How does the foot of a bivalve differ from that of snails or chitons?

Figure 37.5

A giant clam (class Bivalvia, *Tridacna* spp.) of the Indopacific may weigh up to 900 lb (400 kg). Its mantle is often intensely colored by blue, green, and brown pigments, and it harbors rich and colorful colonies of symbiotic algae in the blood sinuses of the tissue. Photosynthesis by this algae provides extra food for the clam. Some clams have lenslike structures that focus light deep within their tissues and promote photosynthesis by the algae.

b. How could predators attack an animal closed "tight as a clam"?

c. Is immobility a problem for filter-feeders such as oysters? Why or why not?

Class Cephalopoda

Examine a preserved specimen of the common squid, *Loligo*. Note the number and arrangement of the tentacles and hardened jaws. Most cephalopods (*cephalo* = head, *poda* = foot) (e.g., squid, octopuses, nautilus, and cuttlefish) do not resemble their molluskan relatives because cephalopod shells may be absent or reduced to an internal remnant (fig. 37.2). For example, a cuttlebone is actually the internal shell of a mollusk called a cuttlefish. You may have offered your pet parakeet a cuttlebone to sharpen its beak. The foot of a cephalopod is modified into tentacles. Squid and other cephalopods are predatory, and their external features are appropriately adapted. Unusual among mollusks is the relatively closed circulatory system of cephalopods.

Examine the eyes of *Loligo* closely. Also examine other displayed cephalopods such as the chambered nautilus; its external shell is beautiful and unusual for a cephalopod

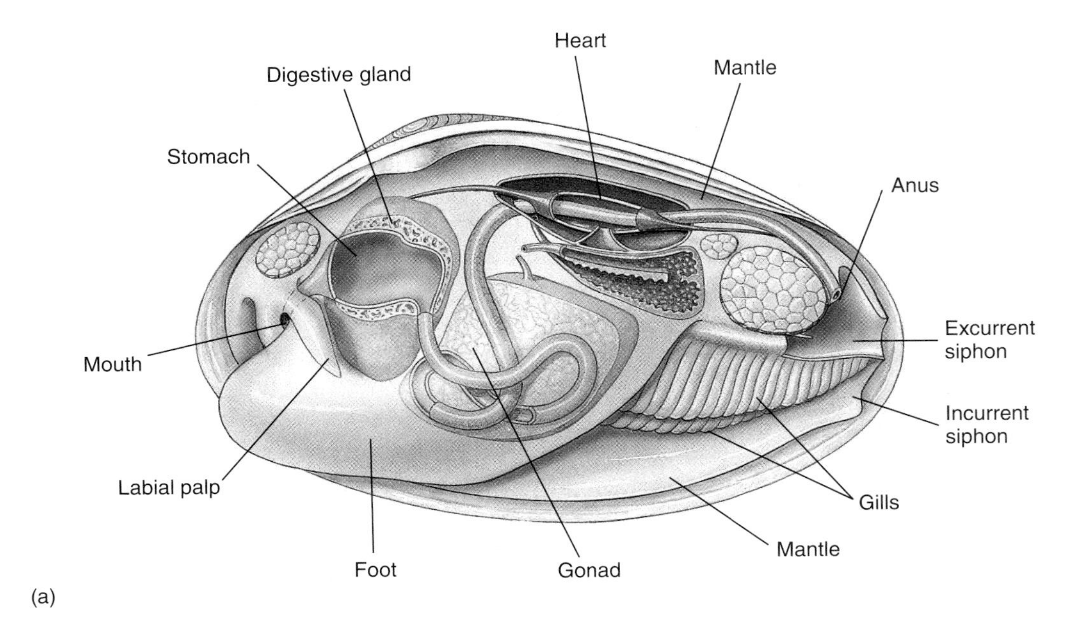

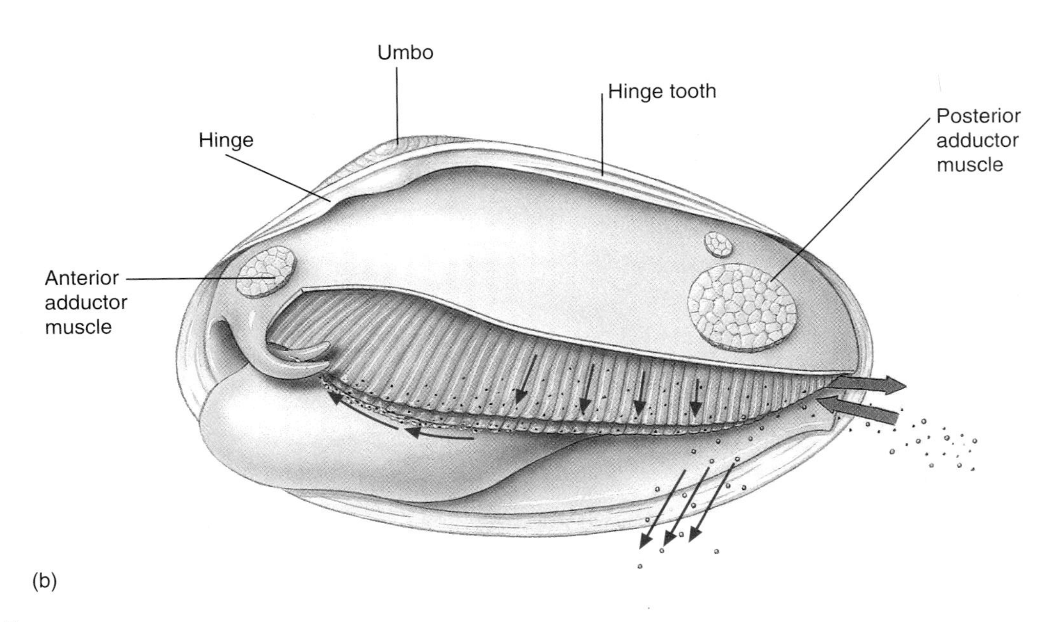

Figure 37.6
(*a*) Clam (*Anodonta*) anatomy. The left valve and mantle are removed. (*b*) During filter-feeding, water enters the mantle cavity posteriorly and is drawn forward by ciliary action to the gills and palps. As water enters the tiny openings of the gills, food particles are filtered out and caught in strings of mucus that are carried by cilia to the palps and directed to the mouth. Arrows show the path of food particles being moved toward the mouth by cilia on the gills. Sand and debris drop into the mantle cavity and are removed by cilia.

(fig. 37.7). Review mollusks by completing table 37.2 with brief descriptions of basic molluskan features. The eyes are probably the most surprising feature of a squid. The largest eyes in kingdom Animalia belong to the giant squid; they resemble mammalian eyes and use a lens to form clear images.

Question 3

a. What features of squid and octopuses are adaptations for predation?

(a)

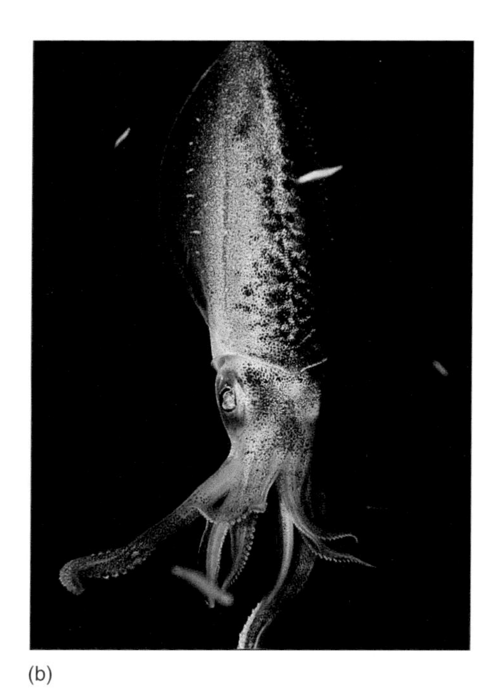

(b)

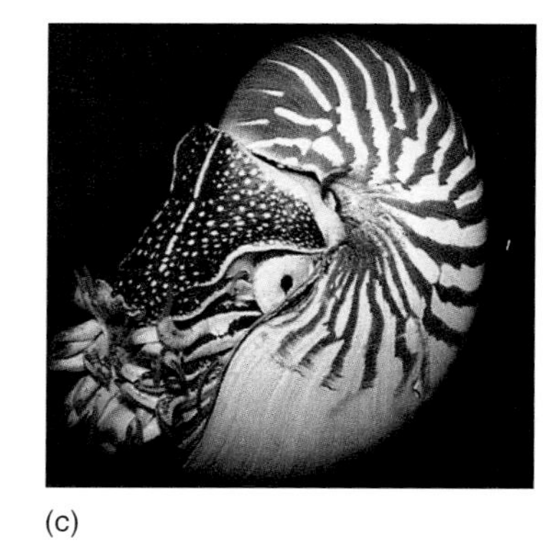

(c)

Figure 37.7

Cephalopod diversity. (*a*) An octopus. Octopuses generally move slowly along the bottom of the sea. (*b*) A squid. Squids are active predators, competing effectively with fish for prey. (*c*) Pearly nautilus, *Nautilus pompilius*.

TABLE 37.2

A COMPARISON OF MAJOR CHARACTERISTICS OF FOUR CLASSES OF MOLLUSKS

	Polyplacophora	Gastropoda	Bivalvia	Cephalopoda
Shell				
Mantle				
Foot				
Locomotion				
Feeding				
Sensory structures				

b. Do all of the tentacles of a squid have suckers?

c. What are some functions of suckers?

d. Find the mouth at the base of the tentacles. What is the shape and consistency of the jaws? You may need to make an incision to expose the mouth and jaws.

e. Why are sensory organs more prominent in cephalopods than in other classes of mollusks?

(a)

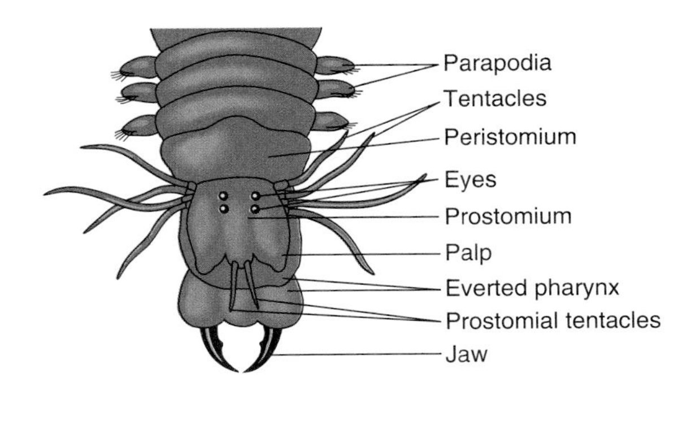

(b)

Figure 37.8
Nereis virens, a common polychaete (phylum Annelida). (*a*) External structure. (*b*) Anatomy of the anterior end.

f. Consider objective 1 listed at the beginning of this exercise. Are image-forming eyes significant to fundamental processes for cephalopods? In what ways?

PHYLUM ANNELIDA

Annelids include earthworms, leeches, and many less familiar marine and freshwater species. The most distinctive characteristic of this phylum is **segmentation.** The body of an annelid is divided into repetitive segments (sometimes called metameres) arranged on a longitudinal axis and divided by **septa.** Each segment contains parts of the circulatory, digestive, nervous, and excretory systems. The circulatory system of annelids is closed, meaning that blood is always retained in vessels. Annelids also have **setae,** which are small, bristle-like appendages often occurring in pairs on lateral and ventral surfaces. The degree of setal development is distinctive for each of the three classes of annelids.

Class Polychaeta

Most polychaetes, such as the clam worm *Nereis*, are marine worms living in sediment (fig. 37.8). *Nereis* is distinctly segmented and each segment bears a pair of fleshy appendages called **parapodia.** These appendages have a large surface area, are highly vascularized with blood vessels, and help the polychaete move and respire. Protruding from the fleshy parapodia are many setae from which the class derives its name (*poly* = many, *chaeta* = setae). In some species the brittle, tubular setae are filled with poison and used for defense; in others the setae help filter food from the water.

Procedure 37.2
Examine polychaetes

1. Examine a preserved *Nereis* and a prepared slide of a parapodium. Sketch the parapodium. Note the tufts of bristles and lobes of tissue.
2. Examine the well-developed head of *Nereis* and locate the features shown in figure 37.8*b*. The mouth and jaws of polychaetes are retractile, so you may have to pull out the jaws with a pair of forceps.
3. Examine other displayed polychaetes such as *Aphrodita*, the sea mouse, and *Chaetopterus*, the parchment worm. These worms have many setae and highly modified parapodia. The parchment worm gets its common name from the paperlike tube it builds.

Question 4

a. The common name of the sea mouse refers to what external feature that is characteristic of polychaetes?

b. List several functions of parapodia and setae.

c. What is the probable function of the tentacles shown in figure 37.8*b*?

d. Consider objective 1 listed at the beginning of this exercise. Are parapodia significant to fundamental processes for polychaetes? In what ways?

Class Oligochaeta

A common oligochaete is *Lumbricus terrestris*, the familiar earthworm.

Earthworm Locomotion

Movement of oligochaetes is not undulatory—rather, movement involves extension, anchoring, and contraction. These motions occur by alternating contractions of circular and longitudinal muscles against hydrostatic (water) pressure within each segment.

Procedure 37.3

Examine locomotion in earthworms

1. Watch a living earthworm move on the hard surface of a pan, then compare and contrast this motion with that of a snake and a vinegar eel.
2. Place the worm on some loose soil and describe its burrowing motion.

Question 5

a. Does the earthworm move randomly?

b. What do you suppose the worm is seeking or avoiding?

c. What muscles allow the worm to change its length and thickness?

d. How does an earthworm's motion differ from that of a snake and nematode?

External Anatomy of Earthworms

Oligochaetes (*oligo* = few or small, *chaeta* = setae), besides the common earthworm, include many freshwater species. They lack parapodia and have few setae. The anus is on the terminal segment, but the mouth is preceded by a fleshy lobe called the **prostomium** (*pro* = before, *stoma* = mouth). Posterior to the mouth is the first body segment, the **peristomium** (*peri* = around, *stoma* = mouth).

The most obvious external feature of an earthworm is the **clitellum,** a series of swollen segments at the anterior third of the body. Copulating worms attach at their clitella and exchange sperm (fig. 37.9). Earthworms are hermaphroditic, and each of the copulating worms produces egg and sperm cells. Sperm mature in **seminal vesicles** and exit the worm through **male gonopores** on segment 15 (fig. 37.10). Sperm then pass to the adjacent worm and move along its body surface to openings of the **seminal receptacles** (each side of segment 10), where they are stored temporarily. After copulation, the worms separate. A few days later the clitellum secretes a mucous band that slides anteriorly and picks up eggs from the **female gonopores** (segment 14) and stored sperm from seminal receptacles. After the eggs are fertilized in the mucous band, the worm releases it as a **cocoon.**

Procedure 37.4

Examine the external features of an earthworm

1. Examine the external features of a preserved *Lumbricus* and notice its segmentation.
2. Although the body lacks parapodia, if you touch the smooth epidermis and cuticle you can feel short setae on the ventral surface.
3. Use a dissecting microscope to determine the number and arrangement of these setae.
4. Locate the mouth and anus of the earthworm.
5. Use a dissecting microscope to locate the male gonopores and seminal receptacle openings on your specimen (fig. 37.10).
6. Locate the female gonopores on your specimen.

Question 6

a. How many setae are on each segment of the earthworm?

b. Are they paired?

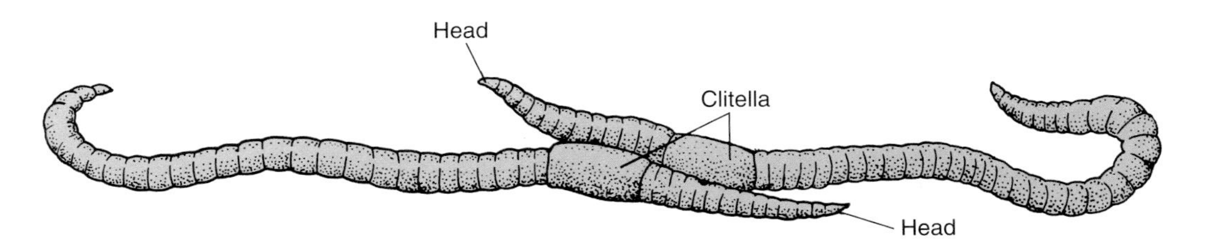

Figure 37.9
Position for copulation and transfer of sperm in earthworms (class Oligochaeta).

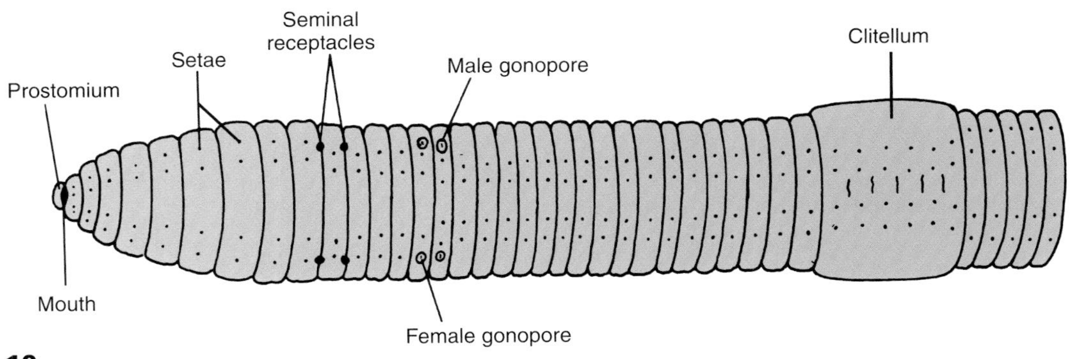

Figure 37.10
External features of an earthworm's ventral surface.

c. What external features indicate the dorsal and ventral surfaces?

d. Consider objective 1 listed at the beginning of this exercise. How could production of a cocoon contribute to the evolutionary success of earthworms in their environment?

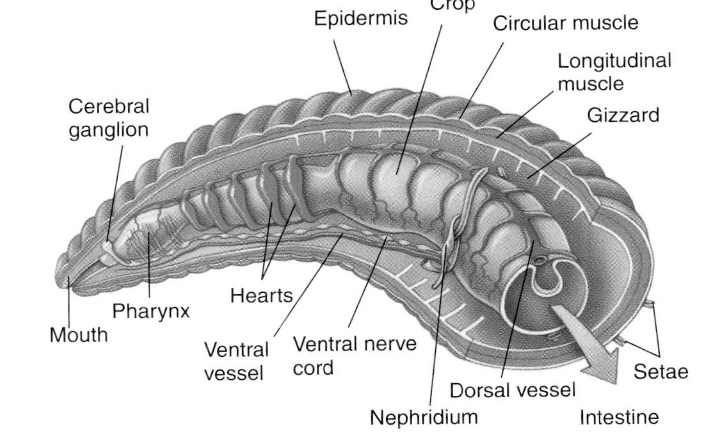

Figure 37.11
Internal anatomy of an earthworm. The long tubular body consists of many similar segments.

Internal Anatomy of Earthworms

Segmentation is best appreciated by looking at internal anatomy (fig. 37.11). Notice that some of the repeating segments are not identical. Segments of the digestive tract are fused and specialized to form a muscular **pharynx** (for suction and ingestion of food), **esophagus** (for transport of food), **crop** (for food storage and some digestion), **gizzard** (for maceration or crushing of food), and **intestine** (for absorption of nutrients). Recall that nematodes also have a linear digestive tract, but it lacks the specialization of that of annelids. Reproductive organs cluster around the anterior segments. Locating small structures such as the **ovaries** and seminal receptacles will require careful work. A rudimentary **brain** is just anterior and dorsal to the pharynx and is continuous with the **ventral nerve cord.**

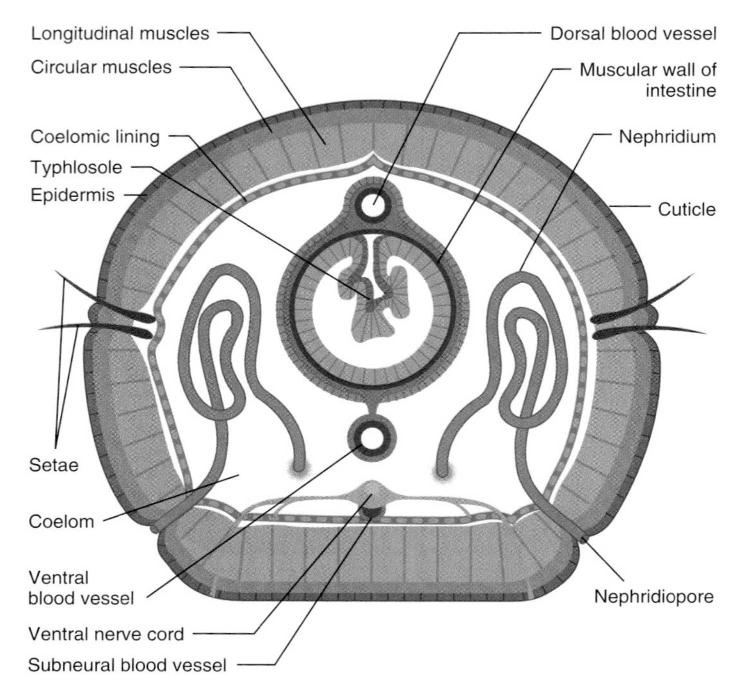

Figure 37.12
Cross section and internal anatomy of an earthworm.

Some structures such as lateral branches of the ventral nerve cord and paired **nephridia** occur in each segment. Nephridia, which are small, white, convoluted tubes, are found on the inner surface of each segment. Nephridia function like kidneys—they collect and release excretory wastes. Ciliated, funnel-shaped **nephrostomes** on the ends of the nephridia gather waste products that are released through external pores called **nephridiopores.**

Procedure 37.5

Examine the internal features of an earthworm

1. Pin the terminal segments of a preserved *Lumbricus*, dorsal side up, near the edge of a dissecting pan.
2. Open the body with a shallow, longitudinal incision. Use sharp-pointed scissors and be careful not to cut too deeply. Hold the skin up with forceps to prevent damage to internal organs.
3. Pin the skin back on both sides of the incision and expose the organs lying in the coelomic cavity. You may have to cut each septum so the body wall lies flat when pinned.
4. When you have completed the dissection, pour a small amount of water into the pan to cover the exposed worm.
5. Locate the structures shown in figure 37.11. You may need to use a dissecting microscope for close examination.
6. Locate the hearts, dorsal blood vessel, and ventral blood vessel.
7. Locate the organs of the digestive system. Slit open the digestive tract and observe its contents.
8. Locate a nephridium in each segment of your dissected specimen.
9. When you have completed your observations, dispose of your specimen in an appropriate container.
10. Examine a prepared slide of a cross section of an earthworm; locate the structures in figure 37.12.

Question 7

a. How many segments of an earthworm have a heart?

b. Does the ventral nerve cord traverse the entire length of the body?

c. Is the inside of the digestive tract the same from the pharynx to the end of the intestine? Explain.

d. How could absorption in the intestine be increased without increasing the intestine's length?

e. How do the layers of musculature in an earthworm differ from those of a nematode?

f. List two or three features of an earthworm cross section that distinguish the dorsal and ventral surfaces.

The digestive tract has an internal fold of tissue called the **typhlosole** arising from the dorsal wall. This creates a U-shaped intestinal lumen and doubles the surface area for absorption.

Class Hirudinea

Hirudineans include leeches, which are primarily freshwater ectoparasites (fig. 37.13). Some leeches eat detritus and small invertebrates such as worms, snails, and insect larvae.

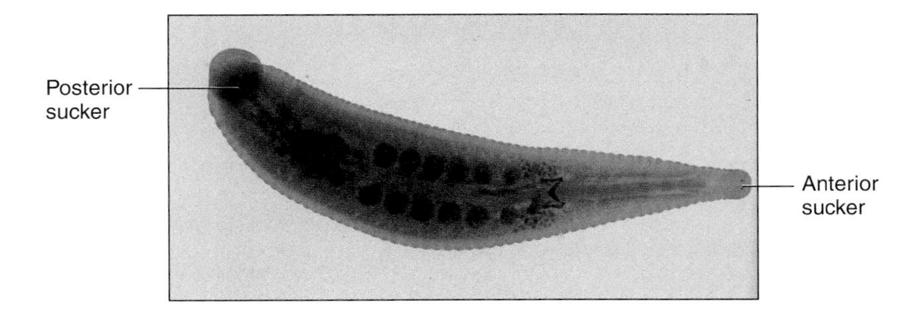

Figure 37.13
External view of a leech. Leeches are primarily freshwater ectoparasites. The mouth is within the anterior sucker.

Leeches are not segmented as distinctly as are other annelids. Leeches lack setae, are dorsoventrally flattened, and have anterior and posterior suckers that hold prey. Bloodsucking leeches eat infrequently, but species such as *Hirudo medicinal* can quickly consume five to ten times their body weight in blood. Many years ago physicians used these leeches for bloodletting from sick patients thought to have "too much blood." Today, laboratory-grown leeches are occasionally used to extract fluid that has accumulated around injuries and the incisions of microsurgery to enhance healing. Leeches extract fluid more efficiently and with less damage than does hypodermic suction.

Leeches reproduce sexually and individuals are hermaphroditic. Two leeches intertwine, and species with copulatory organs inject a packet of sperm called a **spermatophore** into the female gonopore. Other species have no copulatory organ; these leeches copulate by injecting a spermatophore directly through the epidermis of their partner (the spermatophore may have tissue-dissolving enzymes to aid penetration). The introduced spermatophore releases sperm into the coelomic cavity, and the cells move to the ovaries to fertilize the eggs. In two days to many months after copulation, the leech secretes a nutrient-rich cocoon to protect the eggs, and eggs are extruded into the cocoon. The cocoon is brooded by the leech or attached to submerged objects or vegetation.

Procedure 37.6

Observe leeches

1. Observe living leeches; compare their movement with that of earthworms.
2. Examine the external anatomy of a preserved leech and locate the two suckers.
3. Open the body with a pair of scissors and look for signs of segmentation. A dissected leech may be on demonstration.

Question 8

a. What is the difference in general body shape of leeches compared to oligochaetes or polychaetes?

b. What function other than feeding do suckers serve?

c. Are setae visible on a leech?

d. Is internal segmentation of a leech as distinct as that of an oligochaete?

e. Consider objective 1 listed at the beginning of this exercise. How could production of a packetlike spermatophore contribute to the evolutionary success of leeches in their environment?

Question 9

Now that you have examined the unifying characteristics of mollusks and annelids, list three or four characteristics that they share with flatworms and nematodes as their close ancestors.

Questions for Further Thought and Study

1. Mollusks exhibit a variety of feeding methods. List at least four and discuss adaptations and examples for each type.

2. A snail shell is quite different from the familiar bony skeleton of a mammal. In what ways does a shell function as a skeleton?

3. Some land snails have formed a lung from a major layer of tissue. What is that layer?

4. You have examined at least three phyla commonly referred to as "worms." How would you define this term?

5. Of what economic importance are earthworms?

6. Worms have no lungs or gills. How do they breathe?

7. Do you suppose bloodletting by leeches was a good technique to cure psychosomatic illnesses? Why or why not?

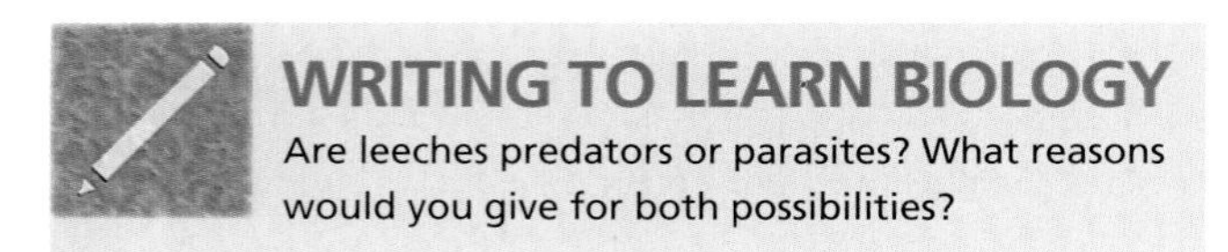

38

Survey of the Animal Kingdom
Phylum Arthropoda

Objectives

By the end of this exercise you should be able to:

1. Describe how structures specific to arthropods help them survive in their environment and promote their evolutionary persistence.
2. Describe the general morphology of organisms of phylum Arthropoda.
3. List characteristics that arthropods share with the phyla discussed previously.
4. Discuss those characteristics of arthropods that were newly derived from those of their ancestral phyla.
5. List examples of the major classes of arthropods.
6. Describe modifications of the exoskeleton and paired appendages of arthropods.

Figure 38.1
Scorpions are the oldest known terrestrial arthropods. Typical of all arthropods, scorpion bodies are segmented, have jointed appendages, and are covered with a hard, chitinous exoskeleton. Shown here is the world's largest scorpion (18 cm), the Emperor scorpion (*Pandinus emperator*) from west Africa. Some extinct species were five times larger than the Emperor. The sting and neurotoxic venom of most species is equivalent to a hornet sting. The most notorious stingers are *Androctonus* of North Africa and species of *Centruroides* in Mexico, Arizona, and New Mexico. Children and weak adults may die from the neurotoxin of a scorpion in 6–7 hours by paralysis of respiratory muscles or cardiac failure. Scorpions locate their prey by detecting vibrations, often through sand rather than air. Desert species can locate and dig out a burrowing cockroach in just a few seconds. Desert species can also withstand temperatures of 46°C (115°F) and tolerate water loss equivalent to 40% of their body weight.

Phylum Arthropoda is the largest phylum of animals and includes spiders, ticks, mites, scorpions, centipedes, millipedes, shrimp, crabs, and insects (table 38.1, fig. 38.1). Estimates of the total species range from 2–10 million. Arthropods live in all major habitats; their tremendous success is due mainly to a rigid external skeleton and jointed appendages (*arthro* = jointed, *poda* = foot or appendage). Appendages are extensions of the main body and are used for locomotion, feeding, reproduction, defense, and sensing the environment. The structure of appendages is often used to classify arthropods. Bodies of arthropods are segmented, although some segments may fuse during development. All organ systems of arthropods are highly developed. Arthropods are coelomate and their circulatory system is open.

The external skeleton, or **exoskeleton,** of arthropods is made of **chitin,** which is a long chain of nitrogen-containing sugar molecules. These fibrous molecules are strong and resist chemicals. Chitin may be as soft as the body of a butterfly or, if impregnated with calcium carbonate, as hard as the shell of a lobster. This exoskeleton provides protection, a moisture barrier, and a place for muscle attachment. Although this tough covering limits growth, arthropods periodically shed their exoskeleton and quickly enlarge before the new exoskeleton hardens.

Each body segment of ancestral arthropods had a single pair of appendages. Recall that this condition of paired appendages also occurs in annelids. In modern arthropods there are extensive variation and elaboration of these

TABLE 38.1

PHYLUM ARTHROPODA

Phylum	Typical Examples	Key Characteristics	Approximate Number of Named Species
Arthropoda (arthropods)	Beetles, other insects, crabs, spiders	Most successful of all animal phyla; chitinous exoskeleton covering segmented bodies with paired, jointed appendages; many insect groups have wings	1,000,000

TABLE 38.2

THE MAJOR CHARACTERISTICS OF ARTHROPODS AND THEIR ADAPTIVE ADVANTAGE

Characteristic	Adaptive Advantage

appendages, which are covered by the rigid exoskeleton and usually have flexible joints. These jointed appendages provide arthropods with such strength and flexibility that the variation in structure and function of appendages is enormous. It is so great that a survey of the phylum and its diversity is primarily a study of variation in appendages.

The prominent characteristics of arthropods conferred an adaptive advantage to this group during their evolution and contributed to their remarkable success. This success is reflected by the high diversity, great numbers, and wide distribution of arthropods. Their adaptive characteristics probably allowed them to replace other species that weren't equipped to deal with their environment as well, and also allowed them to exploit new environments, microhabitats, and niches. Review the introductory information in this exercise (as well as in your textbook) and complete table 38.2 for six or more arthropod characteristics.

SUBPHYLUM CHELICERATA

Chelicerates such as spiders and scorpions are arthropods with the appendages of their most anterior segment modified into feeding structures called **chelicerae.** The second pair of appendages are **pedipalps** and are modified for capturing prey, sensing the environment, or copulating. Body

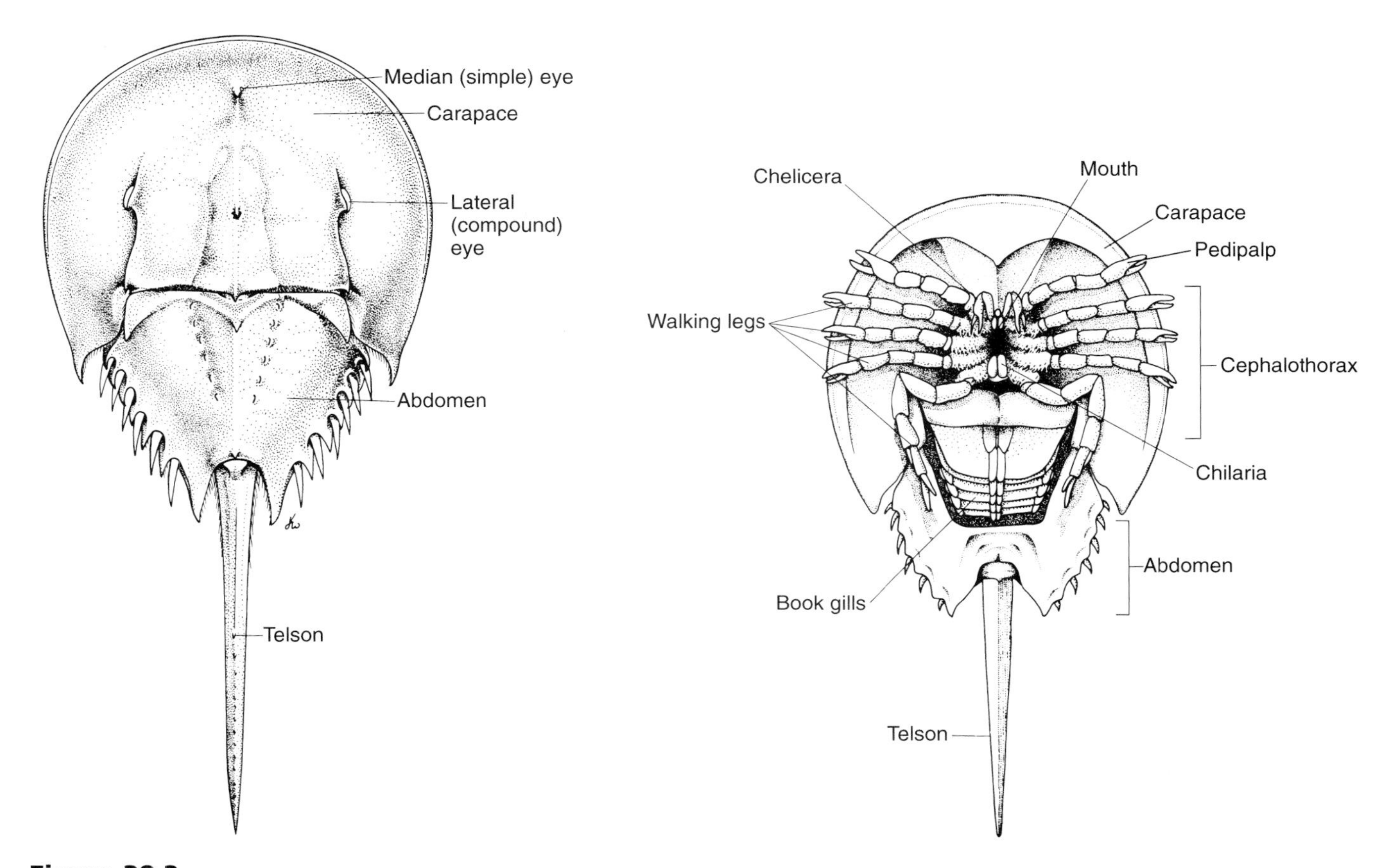

Figure 38.2
Dorsal and ventral views of a horseshoe crab, *Limulus* (class Merostomata). Horseshoe crabs are ancient chelicerates abundant on the Atlantic Coast and Gulf of Mexico.

segments of chelicerates are fused into two body regions: a **cephalothorax** consisting of a fused head and thoracic segments, and an **abdomen** as the most posterior body region. Chelicerates lack antennae.

Class Merostomata (Horseshoe Crabs)

Horseshoe crabs, commonly *Limulus*, are ancient marine chelicerates existing since the Cambrian period (500 million years ago). They are abundant on the Atlantic Coast and Gulf of Mexico.

Examine a preserved *Limulus* and find the structures illustrated in figure 38.2. Notice that a horseshoe-shaped **carapace** covers the cephalothorax and that a flexible joint seen easily from a dorsal view separates the cephalothorax from the abdomen. The ventral surface includes five pairs of appendages modified as walking legs. The most anterior pair of legs are modified pedipalps. Anterior to the pedipalps are the chelicerae. Posterior to the walking legs are a pair of degenerated legs called **chilaria.** The appendages of the abdominal segments are modified as **book gills** and are where gas exchange occurs.

Question 1

a. Why do we call the external chitinous covering of an arthropod a skeleton even though it is not made of bone?

b. What external features of a horseshoe crab are used for burrowing and scavenging in sand?

c. Why are appendages of abdominal segments called "book" gills?

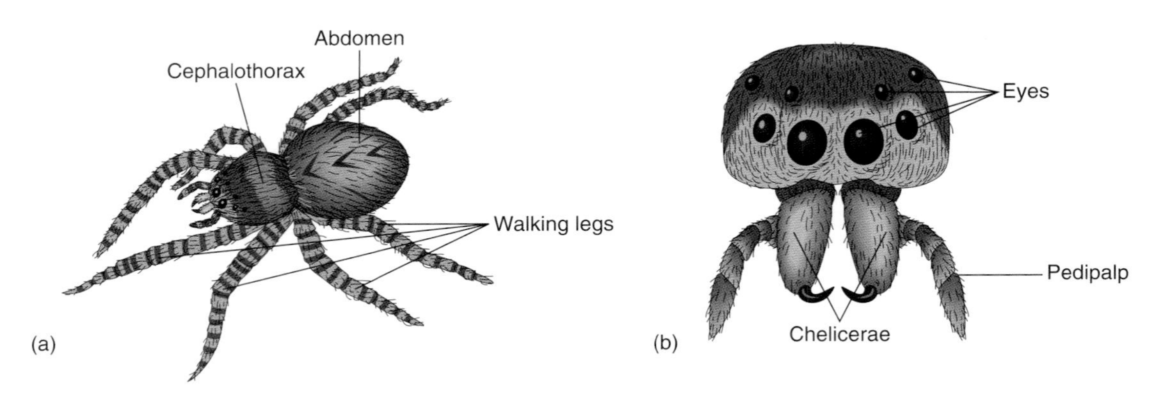

Figure 38.3
Anatomy of a jumping spider. (*a*) External view. (*b*) Anterior view of head.

Class Arachnida (Scorpions, Ticks, Mites, Daddy Longlegs, and Spiders)

Arachnids (57,000 species) are the most diverse class of chelicerates and are mostly terrestrial. The cephalothorax has chelicerae modified as fangs to pierce prey, has pedipalps to manipulate food and sense the environment, and has four pairs of walking legs.

Examine a preserved scorpion, noting the many distinct segments of its thorax and abdomen. Compare the segmentation of scorpions with that of other preserved chelicerates that are available such as mites, ticks, and daddy longlegs. Scorpions were the first terrestrial arthropods and have been around since the Silurian period (425 million years ago). Their lack of fused segments is unusual for chelicerates and indicates the scorpion's ancient origins. These secretive carnivores are 1–8 cm long. Stings of *Centruroides*, a common scorpion in Mexico, Arizona, and New Mexico, have killed many humans, mostly children (fig. 38.1). Its venom is neurotoxic and causes convulsions, paralysis of respiratory muscles, and heart failure. However, most scorpion stings are not fatal.

Examine the external features of a preserved spider and compare them with the features shown in figure 38.3. Spiders are the most familiar arachnids and comprise 30,000 species. They are terrestrial and prey mostly on insects and other small invertebrates (figs. 38.4, 38.5). Spiders are often confused with insects, but spiders have two body regions and eight legs. Insects have three body regions and six legs. The spinnerets of spiders are independently moving nozzles that release silk from internal silk glands. Web silk is made primarily of polypeptides of the amino acids glycine, alanine, and serine. When this fluid is emitted it hardens, not from exposure to air but from polypeptide cross-linkages that form during release. The fluid is not forced out under pressure; rather, it is drawn out by the hind legs or by the weight of the body falling through the air. Spiders can produce silk as strong as nylon, highly elastic, and dry or sticky depending on the construction of the web.

Question 2

a. What external features make scorpions appear so menacing?

b. Which are larger, a scorpion's chelicerae or pedipalps?

c. What is the shape of a spider's chelicerae?

d. Do you see any evidence that a spider's body is segmented?

e. How many eyes do most spiders have? Are they paired and similar in size?

f. Many spiders are hairy. How might this feature be adaptive?

SUBPHYLUM CRUSTACEA

Class Crustacea (Crayfish, Crabs, Shrimps)

Crustaceans (35,000 species) live in marine and freshwater. Only a few species are terrestrial. Crustaceans differ from the

SEND SERVICE INQUIRIES TO:
ENVOYEZ VOS DEMANDES DE RENSEIGNEMENTS À:

MCGRAW-HILL RYERSON
P.O. BOX 635 STN. MAIN
WHITBY, ON L1N 9Z9

PHONE:
TÉLÉPHONE:

MCGRAW-HILL RYERSON

McGraw Hill Education

The McGraw-Hill Companies

SEND RETURNS TO:
ENVOYEZ VOS RETOURS À:

MCGRAW-HILL RYERSON
300 WATER ST
WHITBY ONTARIO L1N 9B6

PHONE:
TÉLÉPHONE:

ACCT.: / N° DU CLIENT: 479788

SHIP TO / EXPÉDIEZ À:
Olaveson
University Of Toronto Scarboro
Biology
1265 Military Trail
Scarborough ON M1C 1A4

ACCT.: / N° DU CLIENT: 479788

BILL TO / FACTUREZ À:
Olaveson
University Of Toronto Scarboro
Biology
1265 Military Trail
Scarborough ON M1C 1A4

PURCHASE ORDER NO.: / N° DE COMMANDE: (PO)

MH Order No.: / MH N° de commande: 3431489 DATE: 05/13/05 (SO)

Control / Invoice No.: / N° DE CONTRÔLE: / N° DE FACTURE: 14466577 (INV)

SHIP METHOD: / MÉTHODE D'ENVOI: UPS STANDARD

ORDER TYPE: / TYPE DE COMMANDE: COMPLIMENTARY COPY

NO. OF CARTONS: / NOMBRE DE BOÎTES: 1

PARCEL ID: / N° D'EXPÉDITION: **13400438**

ISBN	DESCRIPTION TITRE	QUANTITY QUANTITÉ	UNIT PRICE PRIX UNITAIRE	DISC % ESC. %	EXTENDED PRICE PRIX NET
0-07-255287-5	BIOLOGY LABORATORY MANUAL	1			
		SUBTOTAL /	TOTAL PARTIEL --		

** Shipping charges and tax, if applicable, are not reflected on this document. **
** Please see your invoice for final pricing information. **

** Frais de transport et taxes, si applicables, ne sont pas inclus dans ce document **
** S.V.P. Veuillez vous réferer à votre facture pour le montant exact. **

PRE **THIS IS NOT AN INVOICE / CECI N'EST PAS UNE FACTURE**

MESSAGES / INSTRUCTIONS SPECIALES:

COMPLIMENTS OF / CE VOLUME VOUS EST OFFERT PAR VOTRE REPRESENTANT(E)
CATHERINE GATLIN

Figure 38.4

Tarantulas, sometimes called bird spiders, may have a 25-cm leg span. They live on the ground, dwell in trees, and sometimes burrow. The hairy body of this brown tarantula is extremely sensitive to vibration. Many hairs contain chemoreceptors as well as mechanoreceptors, and hairs on the tips of the feet are often iridescent. Captive tarantulas can live 20 years or more.

Figure 38.5

Black widow spiders (*Latrodectus mactans*) are shiny black with a red hourglass on the abdomen's ventral surface. As with other spiders, their chelicerae have poison glands opening at the tip of the fangs. Toxins and enzymes are injected into prey, liquefying the prey's tissue. The nutritious broth is then eaten. Black widows live in most parts of the world, and their venom causes nausea, muscular spasms, respiratory paralysis, and pain in the abdomen and legs. Human death from their bite is rare. Black widows, like other spiders, can withstand months of starvation by reducing their metabolic rate up to 40%. A male is only half the size of the female and is killed soon after mating.

other subphyla because they have fundamentally **biramous** or double-branched appendages (fig. 38.6). Crustaceans have two pairs of antennae and usually have **compound eyes** with multiple lenses. Crustaceans along with insects are commonly called mandibulates because they have opposing mandibles derived from an anterior pair of appendages.

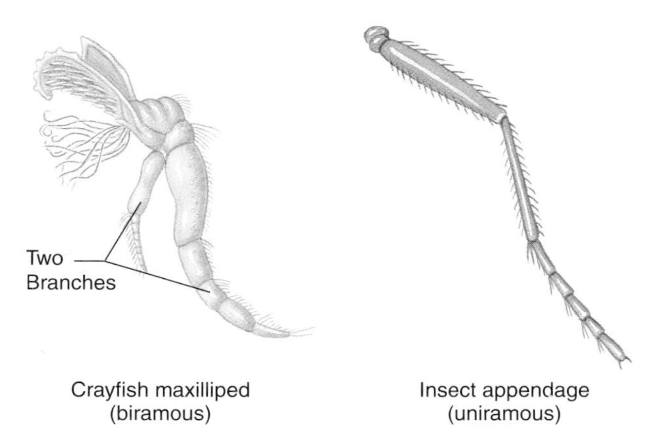

Figure 38.6

Branched and single appendages. A biramous leg in a crustacean (crayfish) and a uniramous leg in an insect.

Crustacean Anatomy

The body of a crustacean such as a crayfish usually has two regions: a cephalothorax covered by a carapace, and an abdomen (fig. 38.7). The five anterior pairs of crustacean appendages are modified into **first antennae, second antennae, mandibles, maxillae,** and **maxillipeds.**

Obtain and examine a preserved crayfish. Appendages 3 (**mandible**) through 9 (**cheliped**) are used for feeding. Notice that these appendages have different shapes. The shape of some appendages such as the cheliped indicate an obvious function.

Question 3
How might mouthparts of various shapes be adaptive for crayfish?

The four pairs of walking legs attach to the thorax at the ventral edge of the carapace. Abdominal appendages are much smaller than walking legs and are called **swimmerets** or **pleopods.** If you study a male crayfish you'll notice that each of the first pair of swimmerets has an odd spatulate shape, modified to transfer packets of sperm to a female during reproduction. The most posterior pair of appendages are broad, flat **uropods.** They surround the terminal abdominal segment called the **telson.** All crustacean appendages are fundamentally biramous (i.e., double-branched).

By now you've probably guessed that studying arthropods consists mostly of examining segments and their associated paired appendages. Appendages of two different species may have similar functions but different embryological origin. Such structures are **analogous.** For example, chelicerae of a spider are analogous to mandibles of a crayfish because they have similar functions even though they are not derived from the same body segment. In contrast,

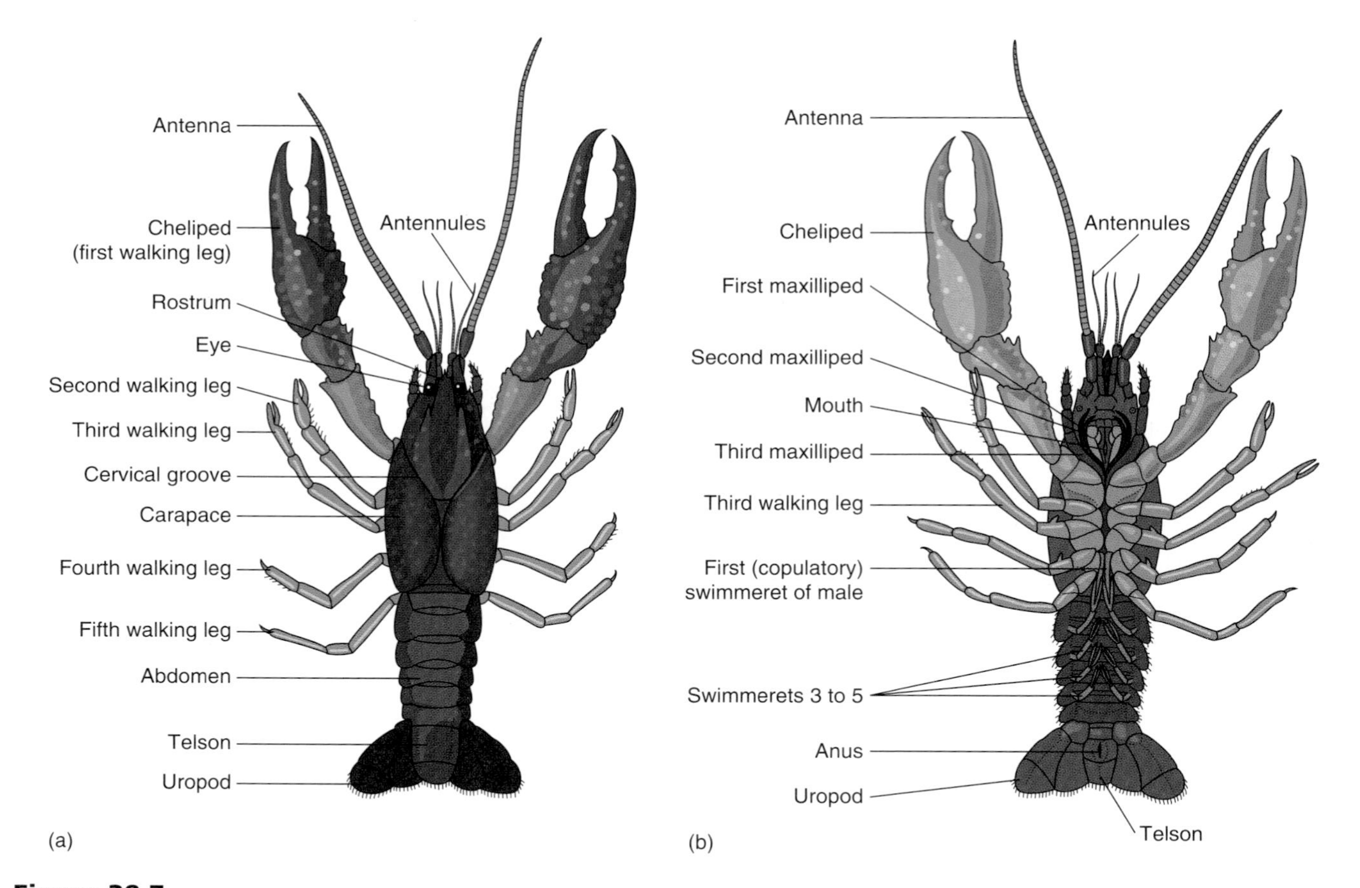

Figure 38.7

External structure of the crayfish (class Crustacea). (*a*) Dorsal view. (*b*) Ventral view.

homologous structures of two different species have similar developmental origin but may or may not serve the same function. Chelicerae of a spider are homologous to the antennae of a crayfish.

Procedure 38.1

Study the external anatomy of a crayfish

1. Obtain a preserved *Cambarus* (the common crayfish) and locate the external features shown in figure 38.7.
2. To appreciate the variation in appendages in *Cambarus*, use forceps to remove one of the first antennae, a cheliped, a walking leg, an anterior swimmeret, a posterior swimmeret, and a uropod.
3. Arrange them in order in a dissecting pan or on a piece of paper and relate each one's structure to its function.
4. Examine the appendages you've removed and note which ones are biramous.
5. Observe a live crayfish (if one is available) and determine its two major methods of locomotion.
6. Examine other available crustaceans such as a crab. The cephalothorax of a crab is obvious, but the abdomen is highly modified.

Question 4

a. Which body region of a crayfish is most obviously segmented?

b. What structures are located just under the carapace and attached to each leg of a crayfish? What is the adaptive advantage of these structures being attached to legs?

c. How many legs are **chelate** (pincerlike with opposing claws)?

d. What is the function of the uropods and telson, and what feature indicates this function?

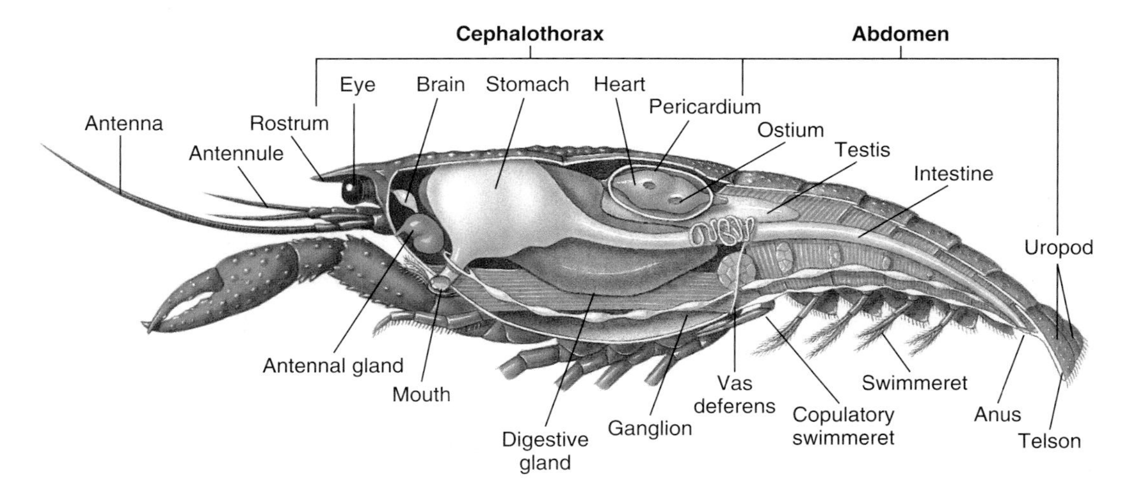

Figure 38.8
Internal structure of a male crayfish.

e. Are the anterior swimmerets different from the posterior pair? Is your crayfish a male or female?

f. Which set of legs (swimmerets or walking legs) appears best adapted to carry an incubating egg mass delicately and in a protected place on the body?

g. In what two ways do crayfish move?

h. How does the shape of a crab's abdomen differ from that of a crayfish?

The internal anatomy of a crayfish features a diamond-shaped **heart** surrounded by a thin **pericardial sac.** The heart lies on the dorsal midline just anterior to the abdominal segments (fig. 38.8). In the open circulatory system of a crayfish, blood flows from the heart through large arteries to the gills, and to sinuses surrounding and bathing the internal organs. Blood returns to the pericardial sac and enters the heart through small openings called **ostia.**

The **gonads** (testes or ovaries) are lateral and just anterior to the heart. Testes are usually white, and ovaries are orange. Sperm from testes exit the body through a pore at the base of the fifth pair of walking legs, and eggs are released at the base of the third walking legs. During reproduction, males and females copulate and sperm cells are passed to the genital openings of the female. Fertilization is internal, and fertilized eggs are extruded and retained for maturation on swimmerets of the female.

The **cardiac stomach** and **pyloric stomach** are continuous, membranous structures along the dorsal midline of the cephalothorax. They are surrounded by muscle and reinforced with ridges of tissue. The cardiac stomach receives food from the **esophagus** and **mouth.** Food moves from the cardiac stomach to the pyloric stomach, through the intestine, and out through the anus at the base of the telson. A large **digestive gland** that secretes enzymes and stores food lies just posterior and laterally to the stomachs.

Beneath the internal organs lies the anterior part of the **ventral nerve cord.** A pair of nerves from the ventral nerve cord pass around the esophagus and come together anteriorly as a **brain** between and beneath the eyestalks. Posterior to the esophagus the nerve cord extends the length of the body as a series of swellings or **ganglia,** each of which controls organs in the immediate segment.

Procedure 38.2
Study the internal anatomy of a crayfish

1. Obtain a preserved crayfish and cut away the dorsal half of the cephalothorax with a longitudinal incision on each side from the midlateral, posterior edge of the carapace toward the eye on either side of the rostrum.
2. Cut across the base of the rostrum and carefully remove the dorsal portion of the carapace.
3. Locate three pairs of ostia on the heart. Remove the heart.

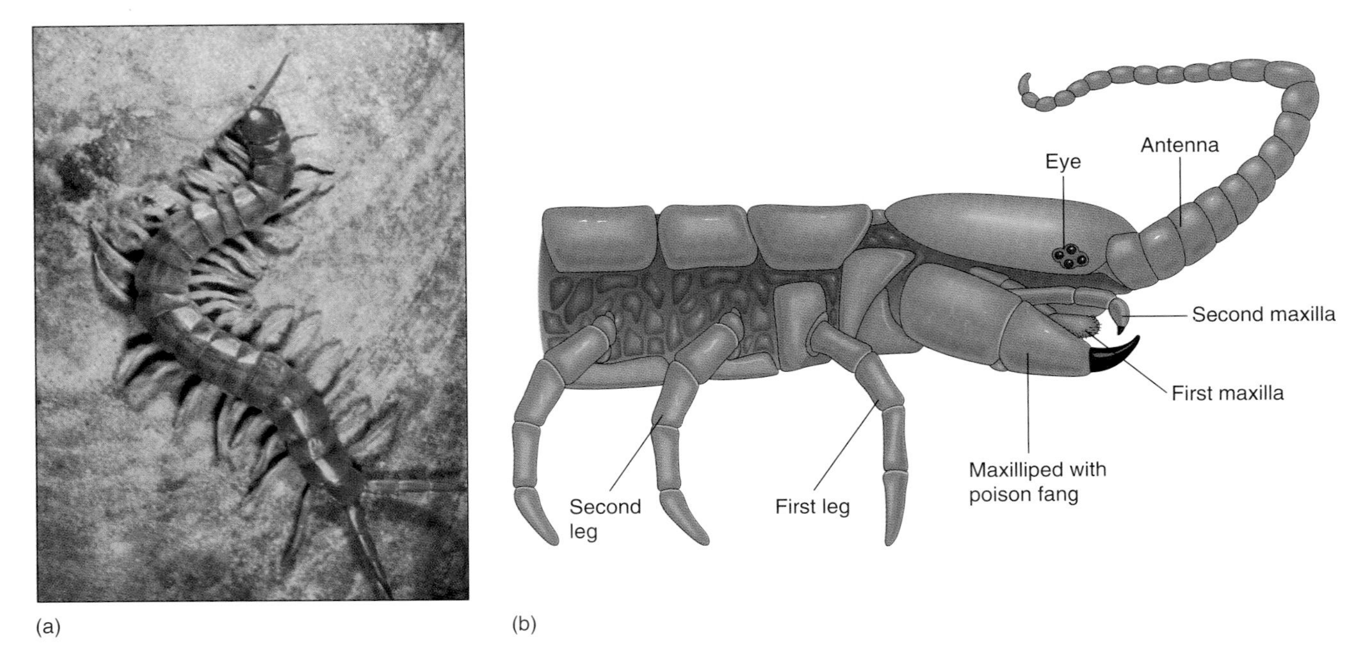

Figure 38.9
Centipede, *Scolopendra* (class Chilopoda). (*a*) Most segments have one pair of appendages each. The first segment bears a pair of poison claws (maxillipeds), which in some species can inflict serious wounds. Centipedes are carnivorous. (*b*) The head of a centipede.

4. Locate the stomachs and digestive gland.
5. Carefully remove the internal organs (stomachs, digestive gland, gonads). This exposes the end of the torn esophagus and the anterior part of the ventral nerve cord.
6. Remove a considerable amount of musculature on the floor of the abdomen (tail) to see the nerve cord more clearly.
7. Locate the **green glands.** They are excretory organs opening at the base of each antennae.
8. When you have finished examining your specimen, dispose of the material as directed by your instructor.

SUBPHYLUM UNIRAMIA

This subphylum of mandibulates includes centipedes, millipedes, and insects. They have **uniramous,** or single-branched, appendages.

Class Chilopoda (Centipedes)

Centipedes (3000 species) live in soil under logs and stones where they prey on small arthropods. Centipedes move rapidly and are dorsoventrally flattened. Some of the larger centipedes such as *Scolopendra* can inflict a painful bite but are not lethal to humans (fig. 38.9).

Examine a preserved centipede. Note that each body segment bears a pair of legs. The large fangs (sometimes called poison claws) on the head are not mandibles, but are maxillipeds, which are appendages modified for feeding. The mandibles are smaller and lie between the maxillipeds.

Question 5

a. Do centipedes have 100 legs as their name suggests?

b. What structures of chelicerates are analogous to the antennae of chilopods (centipedes) and other mandibulates?

Class Diplopoda (Millipedes)

Examine a preserved millipede. Millipedes (8000 species) live in the same environment as centipedes but feed mainly on decaying plant material (fig. 38.10). Millipedes move slowly and are round in cross section. A disturbed millipede will frequently roll up to protect its soft underside. The number of legs on each segment distinguishes millipedes from centipedes.

Figure 38.10

Millipede, *Sigmoria*, in North Carolina. Apparent body segments are actually fused pairs of body segments. Each apparent segment has two pairs of legs.

Question 6

a. How many pairs of legs does a millipede have?

b. Each apparent segment is actually two fused segments. How many legs are on each apparent segment?

Class Insecta (Flies, Grasshoppers, Butterflies, Beetles, and Others)

Insects are by far the largest group of organisms on earth, and probably include 1 to 10 million species. At least 70% of all known species of animals are insects, and they dominate virtually all terrestrial habitats. Three separate body regions and six thoracic legs are the major diagnostic features of insects. Insects possess all the major characteristics of arthropods, but their ability to fly is probably the key to their great success. Although other groups of organisms can fly, insects were the first fliers. There are many advantages to flying, such as avoiding predators. Can you name some others?

Most insects have two pairs of wings. However, insect wings are not modified appendages. Rather, they are evaginations (outgrowths) of the thoracic exoskeleton.

The success of insects on land is further enhanced by an efficient respiratory system of tubes called **tracheae** that conduct air throughout the body. Insects also have highly modified mouthparts. For example, mouthparts of grasshoppers grind coarse plant tissue, whereas mosquito mouthparts puncture and suck fluid from animals.

External Anatomy of a Grasshopper

The grasshopper, *Romalea*, has characteristics typical of insects (fig. 38.11). Insects have three body regions: head, thorax, and abdomen, one pair of antennae, and six legs. The thorax is divided into the **pro-, meso-,** and **metathorax,** each having a pair of legs. Mouthparts are covered by the **labrum,** an extension of the head. Beneath the labrum are the **mandibles,** followed by a pair of **maxillae** with segmented extensions called **palps,** and then the **labium** with palps (fig. 38.12).

Romalea has 10 abdominal segments, each with a **spiracle** or breathing pore opening to the respiratory system of tracheal tubes (fig. 38.13). The terminal abdominal segment bears the reproductive genitalia. The terminal segment of males is blunt, whereas that of females is modified to lay eggs and is called an **ovipositor.**

Procedure 38.3

Examine the external features of a grasshopper

1. Obtain a preserved specimen and locate the external features shown in figure 38.11.
2. Examine the head and find simple eyes called ocelli as well as compound eyes (fig. 38.12).
3. Remove the labrum and locate the remaining mouthparts.
4. Use your probe to locate the stout mandibles. Press the mandibles with your probe to test their rigidity.

Question 7

a. How many ocelli does a grasshopper have?

b. What is the probable function of the maxillary and labial palps?

c. Wing morphology varies among species and orders of insects. Spread the wings of the grasshopper that you are examining. Do all of the wings have the same shape and consistency?

d. Do you suspect that the fore- and hindwings have different functions? Why or why not?

e. Consider objective 1 stated at the beginning of this exercise. Are wings significant to fundamental processes for arthropods? In what ways?

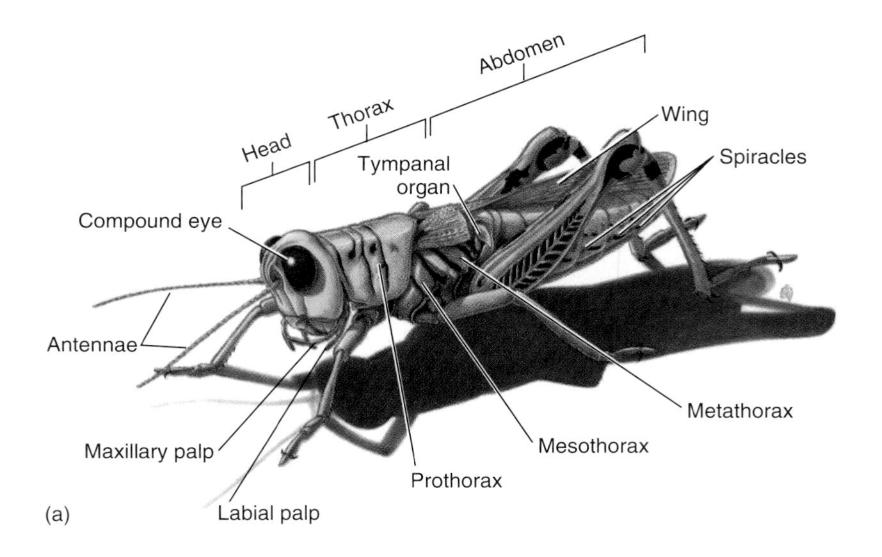

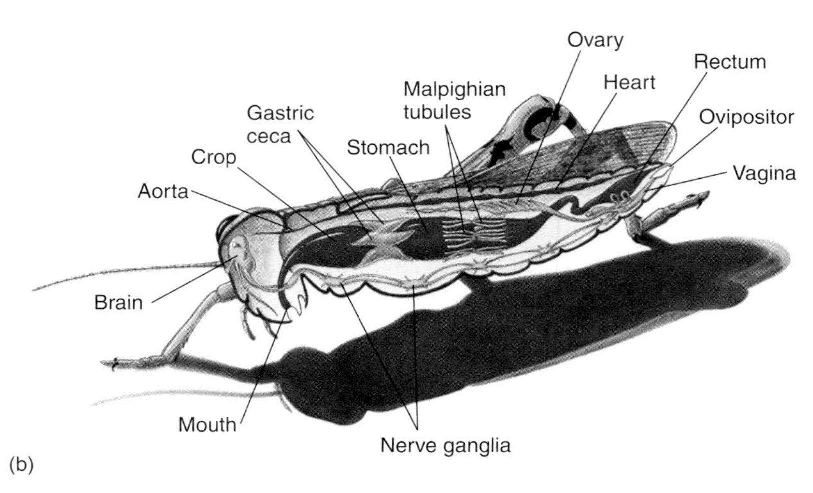

Figure 38.11

Grasshopper, *Romalea*, a member of class Insecta. (*a*) External anatomy. (*b*) Internal anatomy.

Procedure 38.4

Examine the internal anatomy of a grasshopper

1. Use scissors to clip the wings away from the grasshopper's body.
2. Make two lateral incisions (one on each side) from the posterior end. Cut the length of the abdomen and thorax along an incision line just dorsal to the spiracles.
3. Connect the two lateral incisions by cutting across the dorsal surface just behind the head. Then remove the dorsal strip of exoskeleton.
4. Pin the insect to the dissecting pan. Remove the legs if necessary.
5. Locate the heart and aorta just below the strip of exoskeleton that you removed (fig. 38.11*b*).
6. Notice the body cavity between the body wall and digestive tract. It is the hemocoel and is filled with hemolymph, a colorless blood.
7. The internal organs may be covered with a yellow fat body. Remove it and the heart to expose the digestive tract.
8. Search for tubules of the respiratory tracheal system. These tubules conduct air from the spiracles to the tissues.
9. Locate reproductive organs (ovaries and testes) that lie on either side of the digestive tract.
10. Examine the digestive tract and locate the major structures labeled in figure 38.11*b*.

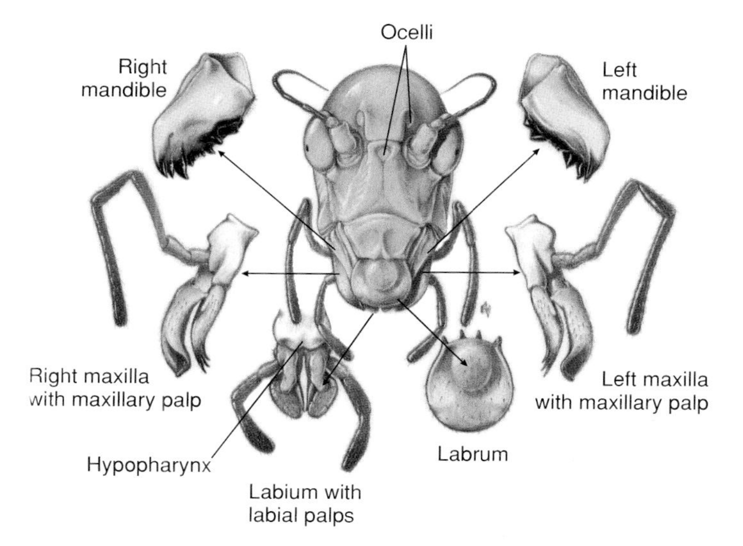

Figure 38.12
Head and mouthparts of a grasshopper.

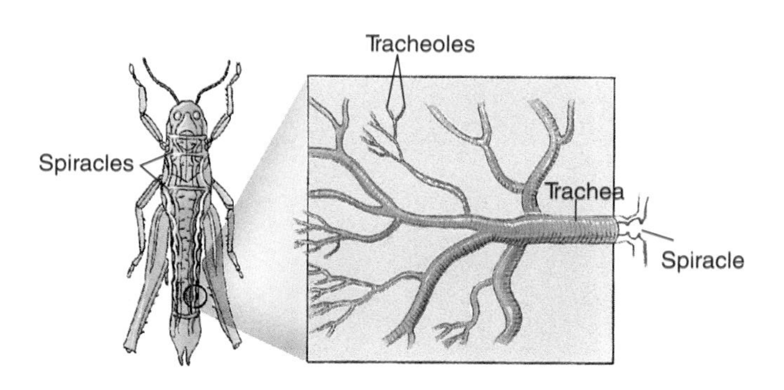

Figure 38.13
The tracheal system of a grasshopper. Tracheae and tracheoles are connected to the exterior by specialized openings called spiracles and carry oxygen to all parts of a terrestrial insect's body.

11. Remove the digestive tract to examine the ventral nerve cord and its series of swollen nerve ganglia.

Question 8

a. How many abdominal segments does a grasshopper have? Is there a nerve ganglion for each segment?

b. Gastric ceca produce digestive enzymes and secrete them into the stomach. How many ceca are there?

c. The spiracles can be opened or closed. Why might this be adaptive?

d. The brain is in the dorsal part of the head. Where does the ventral nerve cord encircle the digestive tract and join with the brain?

Examine all the insects available in the laboratory and note any differences from the basic anatomy of a grasshopper. If you find differences, ask your instructor to explain the variation among these insects. If you look carefully, you should keep your instructor busy for a long time discussing the most diverse animals in the world.

To review phylum Arthropoda, outline in table 38.3 the differences among the classes you have examined.

Question 9

Now that you have examined the unifying characteristics of arthropods, list three or four characteristics that they share with annelids, their most likely ancestors.

DICHOTOMOUS KEY

A common tool for identifying organisms is a **dichotomous key,** such as the one presented in Exercise 30. Dichotomous keys list and describe pairs of opposing traits, each of which leads to another pair of traits until a level of classification of the specimen being identified is reached. By using a key you'll learn the characteristics that distinguish each of the groups identified by the key.

The diversity of insects is immense. They are classified into 26 orders distinguished mainly by the structure of wings, mouthparts, and antennae. You will be provided with five to eight preserved insects. Use this key to identify the order of each specimen.

Procedure 38.5

Use a dichotomous key to identify the order of an insect

The key will eventually lead you to the order of the specimen.

1. Select a specimen and read the first pair of characteristics.
2. Choose the one that best describes your specimen.
3. Proceed according to the number at the end of your choice to the next pair of characteristics.

TABLE 38.3

A COMPARISON OF MAJOR CHARACTERISTICS OF CLASSES OF PHYLUM ARTHROPODA

	Class					
	Merostomata	Arachnida	Crustacea	Chilopoda	Diplopoda	Insecta
Names of body regions						
Number of legs						
Arrangement of legs						
Segmentation						
Number of antennae						
Names of major sensory organs						
Names of major mouthparts						

DICHOTOMOUS KEY TO SOME MAJOR ORDERS OF INSECTS

1. Insects with 2 wings . . . .(flies) **Diptera**
   Insects with 4 wings, a pair of forewings and a pair of hindwings 2
2. Fore- and hindwings are not alike in texture and color. One pair may be hard and dense while the other may be light and transparent 3
   Fore- and hindwings similar, usually clear, thin, and transparent 5
3. Forewings thick and leatherlike at base, tips much thinner and may be transparent; mouthparts pointed and beaklike to puncture prey and suck body fluids . . . .**(bugs) Hemiptera**
   Forewings same texture throughout, biting mouthparts with opposing mandibles 4
4. Forewings leathery and with veins . . . . . .**(grasshoppers, crickets) Orthoptera**
   Forewings hard, without veins . . . .**(beetles) Coleoptera**
5. Wings of same length, antennae usually shorter than head 6
   Wings not of same length, antennae long or enlarged toward end 7
6. Large insects (usually > 3 cm), wings long, transparent and with many strong veins; abdomen long and slender . . . .**(dragonflies) Odonata**
   Smaller insects, wing venation faint, wings extending posterior to the abdomen . . . .**(termites) Isoptera**
7. Wings covered with fine, opaque scales; tubular, coiled, sucking mouthparts . . . .**(butterflies, moths) Lepidoptera**
   Wings thin, transparent, and not covered with scales; mandibles well developed . . . .**(ants, bees, wasps) Hymenoptera**

Questions for Further Thought and Study

1. Arthropods usually have a distinct head. How would you define a "head"? What are the advantages and disadvantages of such a body region?

2. Does an insect's exoskeleton limit growth? Why or why not?

3. Diagram the arrangement of muscles necessary to bend a joint with an exoskeleton versus a joint supported by an endoskeleton.

4. Arthropod body segments are sometimes distinct, sometimes indistinct, and sometimes fused as groups to form body regions. Which groups of arthropods appear the most distinctly segmented? Which appear the least segmented?

5. What effect would 2.5 million spiders per acre have on the insect community?

6. Do you suspect that each eye of a spider provides the same sensory input to the brain? Why or why not?

7. What activities and body functions of arthropods require the most specialized appendages?

8. Do beetles have wings? If so, where are they?

9. What other group of organisms have you studied thus far has chitin as part of their outer covering?

10. What class of arthropods dominates the sea?

WRITING TO LEARN BIOLOGY

Do you think that arthropods constitute a single phylum, or should they be divided into multiple phyla? Describe what divisions you would make and your reasons for them.

38–14

39

Survey of the Animal Kingdom
Phyla Echinodermata, Hemichordata, and Chordata

Objectives

By the end of this exercise you should be able to:

1. Describe how structures specific to echinoderms, hemichordates, and chordates help them survive in their environment and promote their evolutionary persistence.
2. Describe the morphology of organisms of phyla Echinodermata, Hemichordata, and Chordata.
3. List characteristics that echinoderms and chordates share with phyla discussed previously.
4. Discuss characteristics of echinoderms and chordates that are unique or advanced compared to more primitive phyla.
5. Describe the water vascular system of echinoderms.
6. Discuss embryological characteristics that distinguish deuterostomes from protostomes.
7. Understand which phyla are protostomes and which are deuterostomes.

Members of all three phyla remaining in our survey of animals—that is, Echinodermata, Hemichordata, and Chordata—are **deuterostomes.** Deuterostomes are a major departure from the phylogenetic line of **protostomes** such as annelids, mollusks, and arthropods because embryological development of deuterostomes has contrasting features (table 39.1), most notably the blastopore of deuterostomes gives rise to an anus rather than the mouth. The blastopore is the opening to the first cavity formed in a developing embryo, and will be discussed more in Exercise 49.

Read table 39.1 carefully. Refer to your textbook for a more detailed comparison of protostomes and deuterostomes.

PHYLUM ECHINODERMATA

Echinoderms (6000 species) are marine bottom-dwellers and include sea stars, brittle stars, sea urchins, sand dollars, sea cucumbers, and crinoids (table 39.2, fig. 39.1). These organisms are called echinoderms (*echino* = spiny, *derm* = skin) because their internal skeleton of calcareous plates, called **ossicles,** usually has spines protruding through a thin layer of skin. The five classes of echinoderms are distinguished primarily by the arrangement of their ossicles.

Adult echinoderms are radially symmetrical, and their bodies typically consist of a ring of five repetitive parts (that is, they are pentaradial). In contrast, larvae of

TABLE 39.1

A Comparison of Four Major Features of Developing Embryos or Protostomes and Deuterostomes

Feature	Protostomes	Deuterostomes
1. Fate of the first opening (blastopore) to the digestive cavity	Becomes the mouth	Becomes the anus
2. Pattern of early cell division	Spiral	Radial
3. Fate of cells in the early embryo	Determinate, fate is fixed during early development	Indeterminate, fate is not fixed until late development
4. Mesoderm formation	From endodermal cells near the blastopore	From endodermal cells opposite the blastopore
Major phyla	Annelida, Mollusca, Arthropoda	Echinodermata, Hemichordata, Chordata

TABLE 39.2

PHYLA ECHINODERMATA AND CHORDATA

Phylum	Typical Examples	Key Characteristics	Approximate Number of Named Species
Echinodermata (echinoderms)	Sea stars, sea urchins, sand dollars, sea cucumbers	Deuterostomes with radially symmetrical adult bodies; endoskeleton of calcium plates; five-part body plan and unique water vascular system with tube feet; able to regenerate lost body parts; marine	6,000
Chordata (chordates)	Mammals, fish, reptiles, birds, amphibians	Segmented coelomates with a notochord; possess a dorsal nerve cord, pharyngeal slits, and a tail at some stage of life; in vertebrates, the notochord is replaced during development by the spinal column; 20,000 species are terrestrial	42,500

echinoderms are bilaterally symmetrical. This indicates that radial symmetry is secondarily derived and not directly related to the symmetry of more ancient phyla such as Cnidaria.

Echinoderms have a unique **water vascular system** consisting of a series of coelomic water-filled canals with hollow projections called **tube feet.** Muscle contractions and hydrostatic pressure in the water vascular system extend and move the tube feet and other parts of the system, and thereby move the animal (fig. 39.2).

Class Asteroidea (Sea Stars)

Ossicles of class Asteroidea, including the common sea star *Asterias*, are arranged loosely under the skin, and spines are small and blunt. Arms of sea stars are continuous with the central disk. The mouth is at the center of the lower, **oral** surface, and the anus is on the upper, **aboral** surface. Surrounding the blunt spines of sea stars are **dermal gills** for respiration by diffusion, and pincerlike **pedicellariae** used to remove debris from the surface (fig. 39.3). Also on the aboral surface is the **madreporite,** a sieve connecting the water vascular system with the environment.

Sea stars often prey on oysters and clams by using their arms and tube feet to grip the shell and persistently apply pressure to pry it open. Sea stars then evert their stomach inside the clam to digest and engulf the tissue.

Procedure 39.1

Examine the external anatomy of a sea star

1. Examine a preserved *Asterias* and locate the external and internal features shown in figure 39.4.
2. Examine the oral surface of the sea star, and locate the central mouth. Locate the tube feet protruding from the **ambulacral groove.**
3. Use a dissecting microscope to examine the spines on the aboral surface (fig. 39.3).
4. Touch the madreporite with a probe, and note its consistency.
5. Examine figure 39.2 and trace the path of water from the madreporite to the tube feet.
6. If living sea stars are available, observe their locomotion.

Question 1

a. How many tube feet would you estimate are on the oral surface of a sea star?

b. Are sea star spines movable?

c. What is the consistency of the madreporite?

Figure 39.1

Diversity in echinoderms (phylum Echinodermata). (*a*) Sea star, *Oreaster occidentalis* (class Asteroidea), in the Gulf of California. (*b*) Sea cucumber, *Strichopus* (class Holothuroidea), in the Philippines. (*c*) Feather star, *Comanthina* (class Crinoidea), uses its highly branched arms in filter-feeding. Although this probably reflects the original use of echinoderm appendages, most modern echinoderms use their arms for locomotion, capturing prey, and scavenging the substrate for food. (*d*) Brittle star, *Ophiothrix* (class Ophiuroidea). (*e*) Sea urchin, class Echinoidea. Urchins defend themselves with long and sometimes barbed spines that are often filled with toxins. Spines of this pencil urchin (*Heterocentrotus* sp.) are particularly robust and blunt.

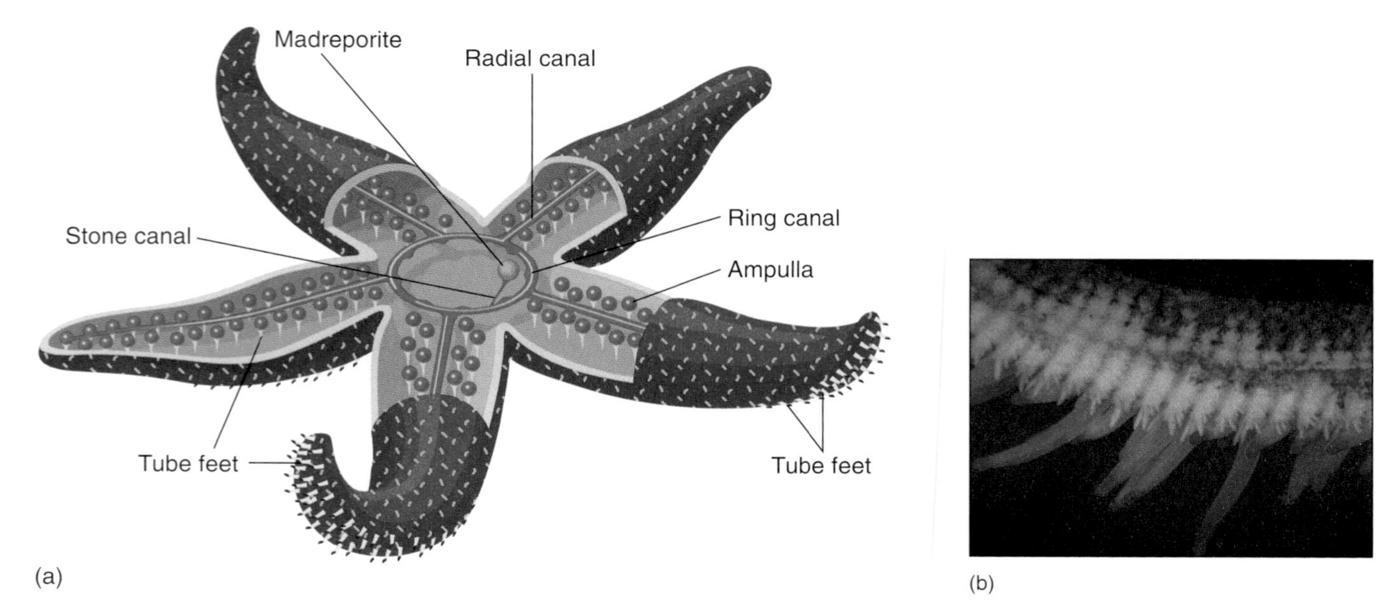

Figure 39.2
Echinoderms. (*a*) The echinoderm body plan of a sea star, emphasizing the water vascular system. (*b*) The extended tube feet of a sea star, *Ludia magnifica*. Tube feet are used for locomotion and to grip and pull apart prey such as clams.

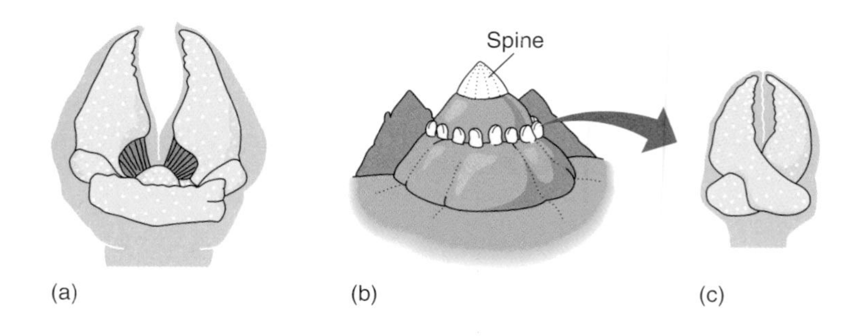

Figure 39.3
Pedicellariae of sea stars. (*a*) Forceps-type pedicellariae of *Asterias*. (*b*) Spine surrounded by (*c*) scissors-type pedicellariae of *Asterias*.

d. How fast does a living sea star move?

e. Do tube feet of a living sea star move in unison?

Procedure 39.2

Examine the internal anatomy of a sea star

1. Obtain a preserved sea star, and cut off at least 2 cm of one arm.
2. Examine the severed end and compare the structures with those shown in figure 39.5.
3. Cut along both sides of the arm up to the central disk. Then cut across the aboral (upper) surface to join the two lateral incisions.
4. Remove the upper body wall and expose the coelomic cavity of the arm. Locate the internal organs shown in figure 39.6.
5. Remove one of the pyloric ceca (digestive gland) so you can examine the gonad.
6. Cut around the perimeter of the central disk. Then cut from the perimeter up to and around the madreporite. This incision will allow you to remove

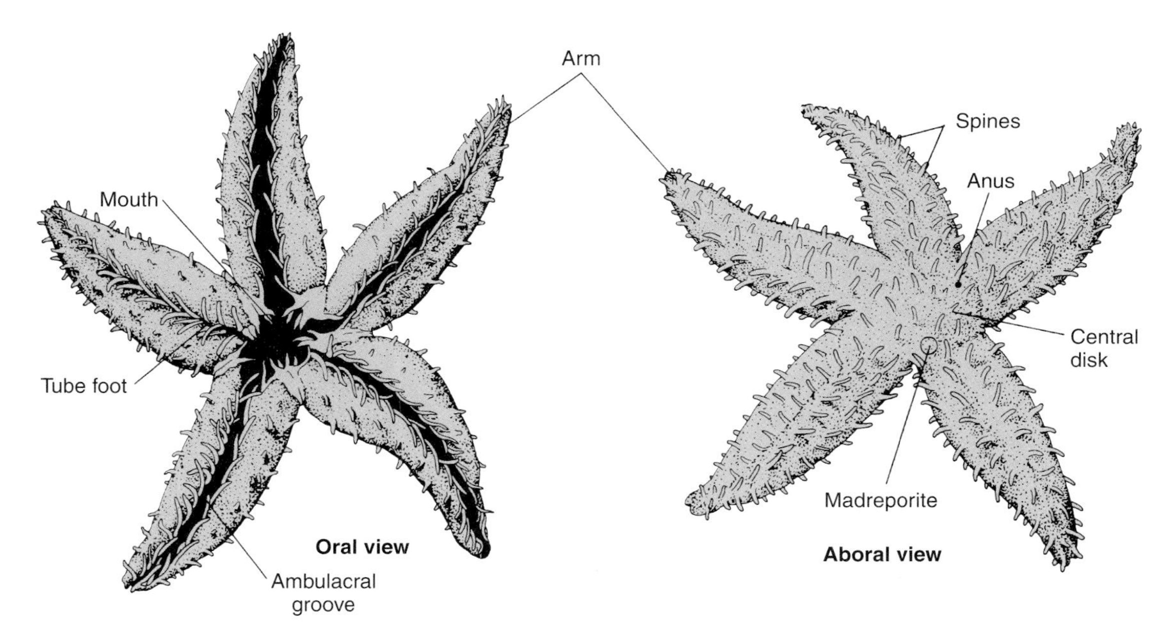

Figure 39.4
External features of sea stars.

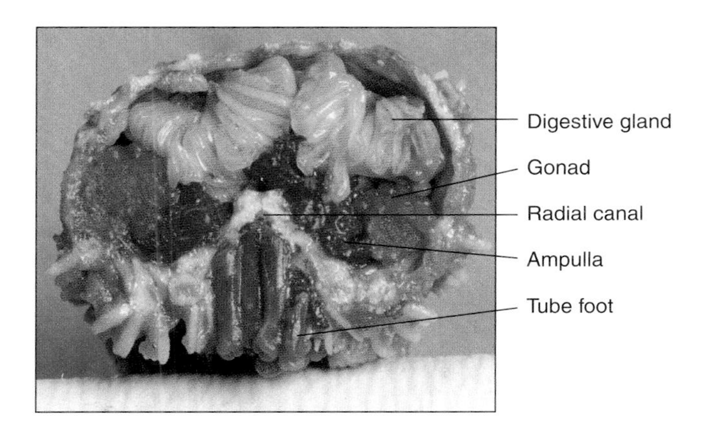

Figure 39.5
Cross section of sea star arm showing internal anatomy.

the upper body wall without tearing away the madreporite.

7. Remove the upper body wall. As you lift it, try to see the delicate connection between the anus on the surface of the thick-walled pyloric stomach below.
8. Locate the structures shown in figure 39.6.

Question 2

a. How many tube feet would you estimate are in one arm?

b. What part of the water vascular system extends into each arm?

c. What other phyla that you have examined rely on "hydraulics" as part of their locomotion system?

d. Does the cardiac stomach wall appear highly folded and extensible? How does that relate to the feeding method of most sea stars?

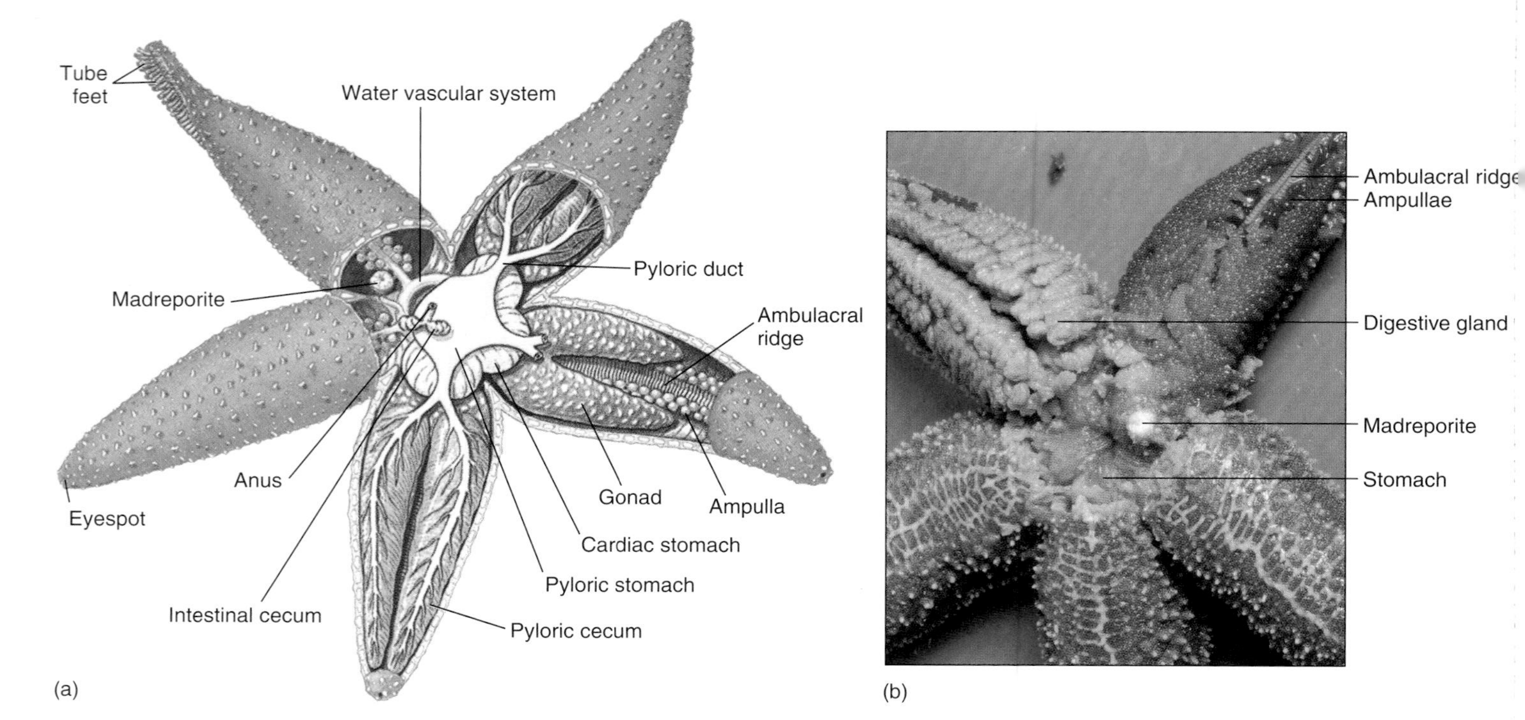

Figure 39.6
Sea star internal anatomy. (*a*) Schematic with aboral surface removed. (*b*) Partially dissected sea star.

Class Ophiuroidea (Brittle Stars)

Examine a preserved brittle star such as *Ophioderma*. Brittle stars have slender, sometimes branched arms that are clearly demarcated from the central disk (fig. 39.1*d*). As the name "brittle" star implies, their arms detach easily, allowing escape from predators.

The ambulacral grooves are closed in brittle stars, and the reduced tube feet are not used for locomotion. The thin flexible arms of brittle stars allow them to crawl rapidly like an octopus rather than creep slowly like sea stars. Brittle stars eat suspended food particles captured with their tube feet and passed to their mouth.

Question 3

a. Between brittle stars and sea stars, which have the most apparent ossicles? Do they overlap?

b. Are tube feet visible in *Ophioderma*?

Class Crinoidea (Sea Lilies and Feather Stars)

Examine a preserved crinoid. Crinoids are the most ancient echinoderms; only a few genera live today (figs. 39.1*c*, 39.7). They differ from other living echinoderms because their oral surface (mouth and anus) usually faces up. Their ossicles are well developed and give the animal a coarse, jointed appearance. Highly branched and feathery arms surround the mouth and anus. Most ancient crinoids were attached to the substrate by a stalk and appeared to be plants. However, most modern species are not stalked or permanently attached. Crinoids filter-feed by capturing food particles on the mucus of their tube feet.

Question 4

How does the position of the mouth and anus of a crinoid relate to a primitive sessile existence?

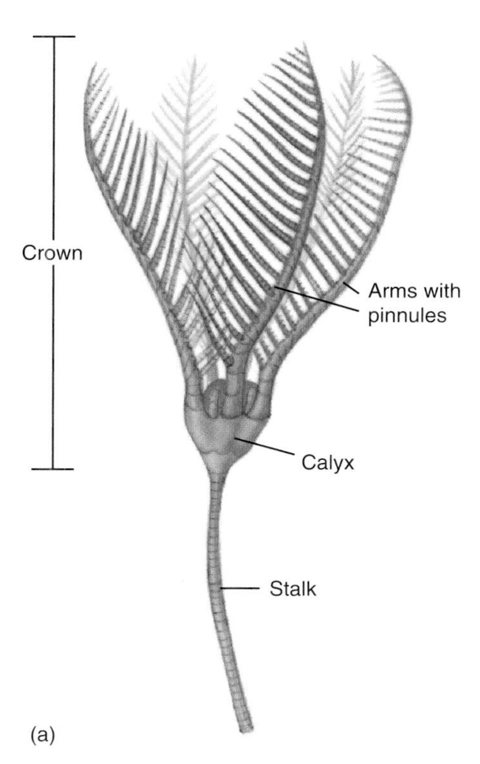

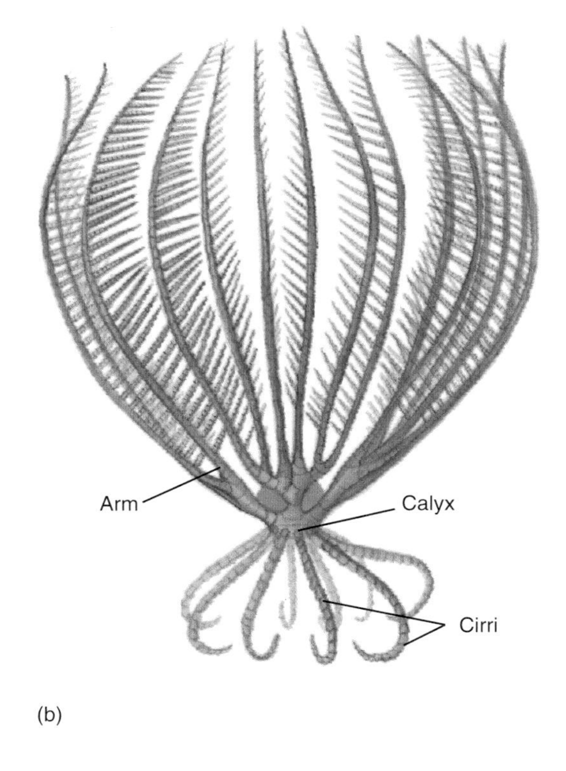

Figure 39.7
Class Crinoidea. (*a*) A sea lily (*Ptilocrinus*). (*b*) A feather star (*Neometra*).

Class Echinoidea (Sea Urchins and Sand Dollars)

Examine an urchin such as *Arbacia* and locate the features shown in figure 39.8. Also examine a dissected Aristotle's lantern if one is available. Also examine a sand dollar and compare its test of fused ossicles to an urchin's test. Urchins lack distinct arms, and their ossicles are fused into a solid shell called a **test** (fig. 39.1*e*). Holes in the test allow long tube feet to protrude. Spines of sea urchins are jointed, movable, and longer than those of other classes of echinoderms. These spines and long tube feet control locomotion of urchins. The mouth contains five ossified plates, or teeth, used to scavenge and scrape surfaces of rocks and gather algae for food. This small internal structure of five teeth is called **Aristotle's lantern.**

Question 5

a. Is an urchin's test pentaradially symmetrical?

b. Urchins and sand dollars lack arms. How do they move?

Class Holothuroidea (Sea Cucumbers)

Examine a sea cucumber *Cucumaria* and determine the orientation of its radial symmetry. Find the mouth at one end; it is surrounded by modified tube feet called **tentacles.** Sea cucumbers look different from other echinoderms because they have soft bodies with reduced ossicles and few if any spines (figs. 39.1*b*, 39.9). Radial symmetry is less evident in sea cucumbers and their body axis is oriented horizontally. This orientation gives sea cucumbers a semblance of cephalization. The tentacles secrete a mucus that captures small floating organisms, which they eat. Interestingly, some sea cucumbers respond to stress by rupturing anteriorly and rapidly expelling their pharynx, digestive tract, and other organs. This process is called evisceration; because of it, the animal must regenerate the lost parts of the organs. Some gourmets consider sea cucumbers a delicacy.

Examine other preserved echinoderms and review in your textbook the major characteristics of each class. Then complete table 39.3.

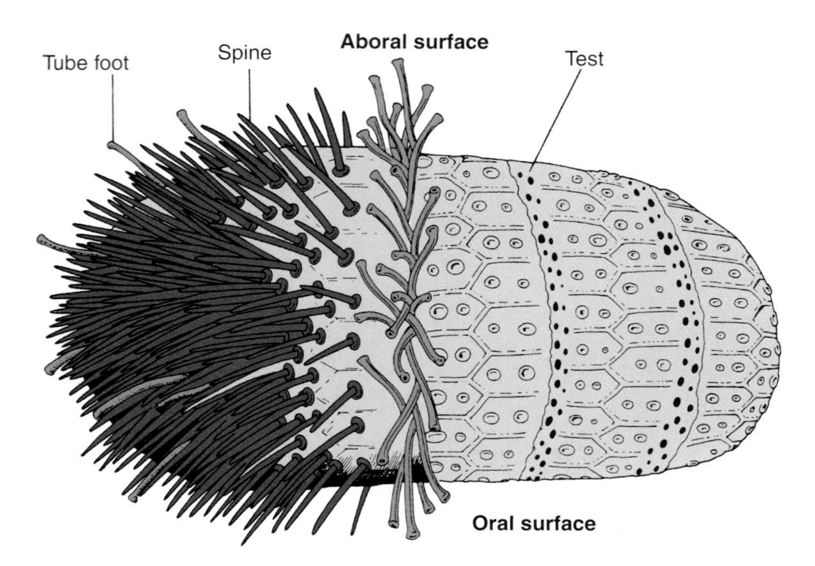

Figure 39.8
External anatomy of a sea urchin. Spines and tube feet are removed on the right half of the diagram to show the test.

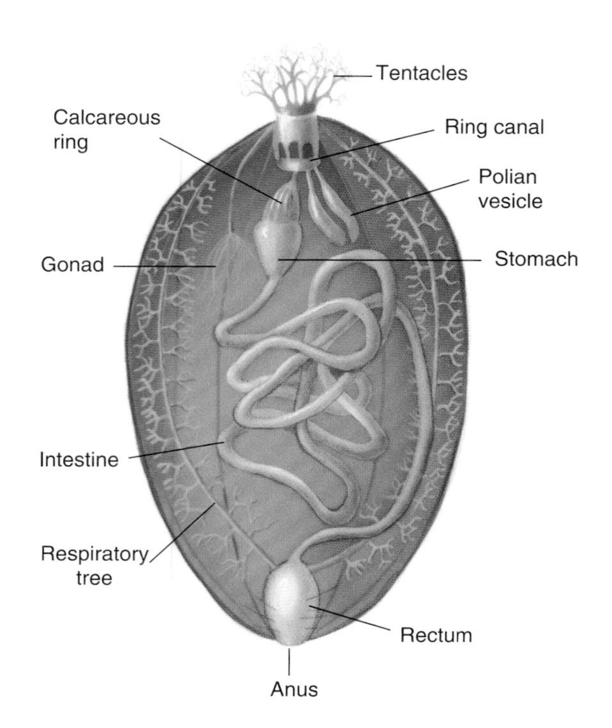

Figure 39.9
Internal structure of a sea cucumber, *Thyone*. The mouth leads to a stomach that is supported by a calcareous ring. The calcareous ring is also the attachment site for longitudinal retractor muscles of the body. Contractions of these muscles pull the tentacles into the anterior end of the body. The stomach leads to a looped intestine.

Question 6

a. Are tube feet visible on the sea cucumber?

b. Hydras, octopuses, and sea cucumbers have tentacles. Do tentacles have a single universal function, or varied functions? What functions are common?

PHYLUM HEMICHORDATA

Examine a preserved acorn worm, such as *Balanoglossus* or *Saccoglossus*, and compare it with figure 39.10. Hemichordates such as acorn worms are inconspicuous animals and include only 90 species. Members of this phylum share two important features with phylum Chordata: (1) a **dorsal nerve cord,** part of which is hollow, and (2) **pharyngeal slits,** which are openings in the throat that filter water that has entered through the mouth. Hemichordates do not, however, have a notochord as once thought or even possess half a notochord as their name implies.

Acorn worms are soft-bodied, marine animals that burrow in sand or mud. Their bodies are fleshy and contractile, and consist of a **proboscis,** a **collar,** and a **trunk.**

Question 7

a. Are pharyngeal gill slits readily apparent?

b. Does the shape and flexibility of the acorn worm body appear to be adapted for burrowing? How so?

PHYLUM CHORDATA

Chordates include 42,500 species of fish, amphibians, reptiles, birds, and mammals. They all are characterized by (1) a dorsal hollow nerve cord, (2) a **notochord,** a cartilaginous rod that forms on the dorsal side of the gut in the embryo, and (3) pharyngeal slits. An internal, bony skeleton is also common and provides sites for muscle attachment for efficient movement.

Table 39.3

A Comparison of the Major Characteristics of the Classes of Echinoderms

	Class				
	Asteroidea	Ophiuroidea	Crinoidea	Echinoidea	Holothuroidea
Shape of arms					
Development of tube feet					
Development of ossicles					
Feeding method					
Spine structure					

Subphylum Urochordata (Tunicates or Sea Squirts)

Examine a preserved, adult tunicate. Urochordates, sometimes called tunicates, are sessile or planktonic marine organisms whose larvae possess the general chordate form—that is, they are elongated with a notochord and dorsal nerve cord. In contrast, the structure of an adult is highly modified to include a sievelike basket perforated with pharyngeal gill slits and surrounded by a cellulose sac called a **tunic.** Water enters through an incurrent siphon, is filtered by the pharyngeal basket, and exits through an excurrent siphon (fig. 39.11). Water is actively filtered; some tunicates only a few centimeters long can filter 170 liters of water per day. Food collected by mucus on the pharyngeal basket is moved by cilia to the stomach and intestine. The intestine empties into the body cavity near the excurrent siphon.

Examine a prepared slide of larval tunicates. A larval tunicate has bilateral symmetry, a dorsal nerve cord, and a notochord, but loses these features when it settles for adult life (fig. 39.11c).

Question 8

What other group of organisms has cellulose in their supporting structures? Does this shared feature surprise you?

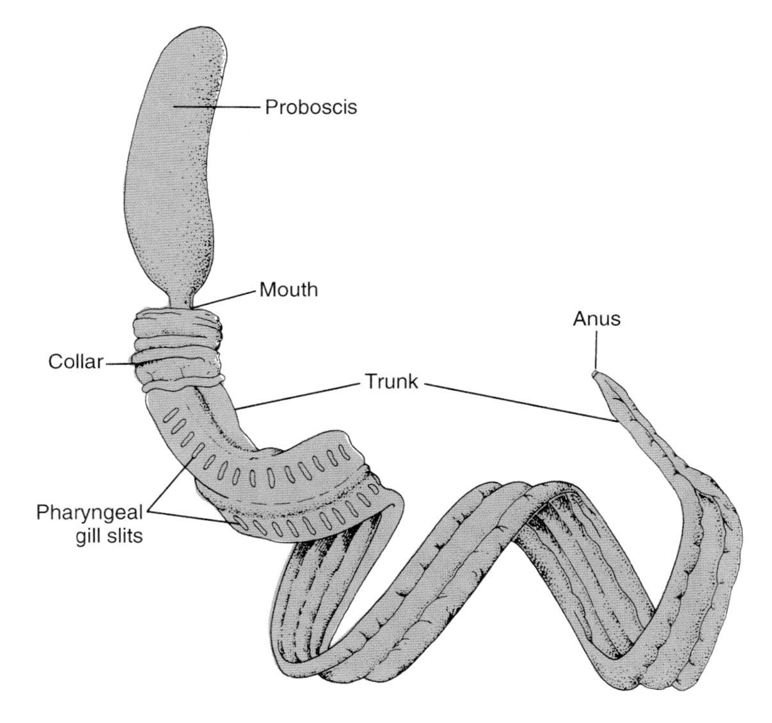

Figure 39.10

External lateral view of an acorn worm (phylum Hemichordata). Acorn worms are marine animals that burrow in sand or mud.

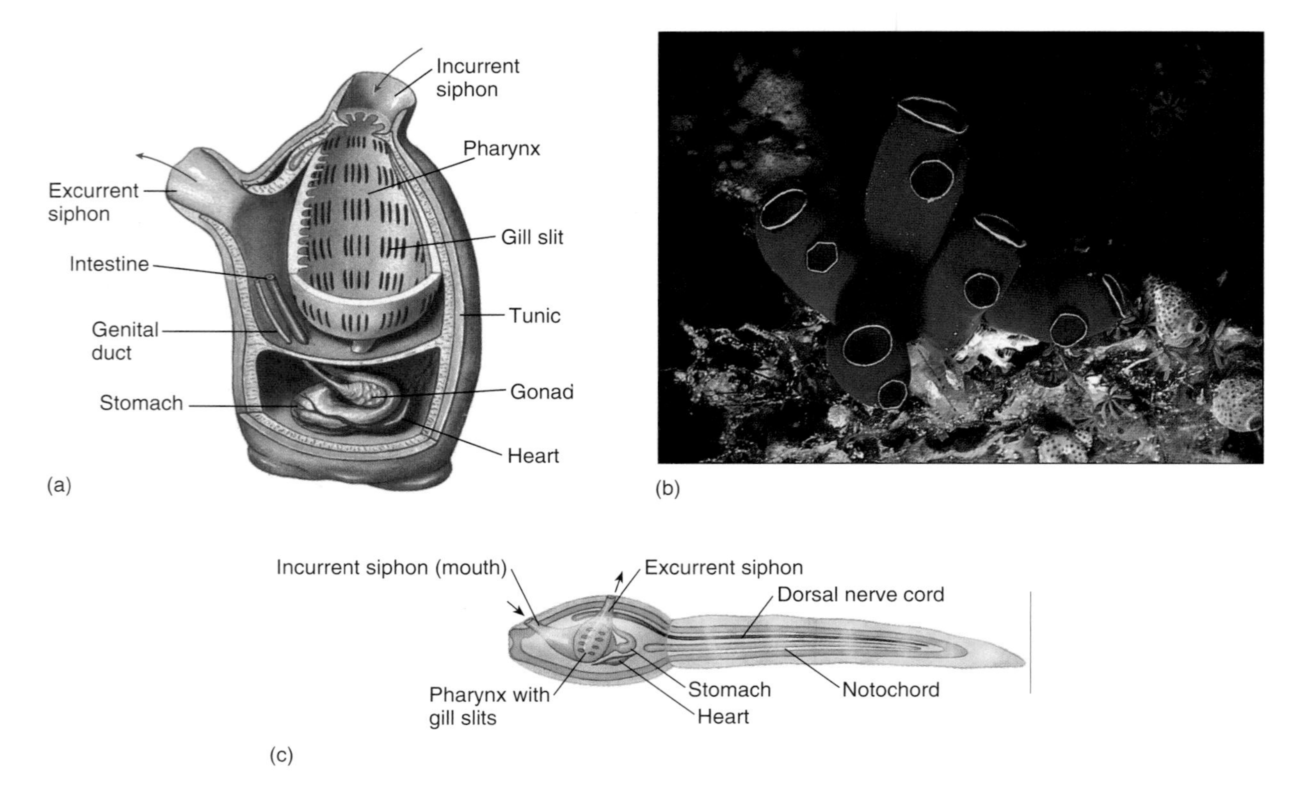

Figure 39.11

Tunicates. (*a*) The structure of an adult tunicate. (*b*) A beautiful blue and gold tunicate. (*c*) Larval structure. The structure of a larval tunicate is much like that of the postulated common ancestor of the chordates.

Subphylum Cephalochordata (Lancelets)

Examine either a preserved lancelet or a slide of a whole mount and compare the specimen with that shown in figure 39.12. Also examine a cross section through the pharynx and try to visualize the paths of food and water (fig. 39.13). Lancelets are small, fishlike, marine chordates that burrow in sand or mud. They are commonly called amphioxus, but the most common genus is *Branchiostoma.*

The dorsal nerve cord and notochord extend the length of the animal. The buccal cavity surrounds the mouth followed by a long pharynx with many gill slits (openings) separated by gill arches of reinforced tissue. As seawater enters the mouth and exits through the slits it must pass over the surfaces of the arches that form the sides of the slits. As this occurs, food particles are caught on the arches, and are eventually swept to the intestine. Notice that the anus is not terminal. Lancelets and vertebrates have a **post anal tail,** which is another diagnostic trait of chordates. After water passes by the arches it moves into a surrounding chamber called an **atrium** and then leaves the body through the **atriopore.**

Subphylum Vertebrata (Fish, Birds, Amphibians, Reptiles, and Mammals)

Vertebrates have a vertebral column that replaces the notochord in adults and surrounds the dorsal nerve cord (fig. 39.14). Vertebrates also have a distinct head. There are seven classes of living vertebrates, three of them fishes and four of them terrestrial tetrapods.

Class Agnatha (Lampreys and Hagfishes)

Examine a preserved lamprey, *Petromyzon* (fig. 39.15). Also examine a prepared slide of an **ammocoete,** the larva of a lamprey. Living agnathans descended from representatives of the earliest stages in the evolution of vertebrates. They lack jaws typical of other vertebrates but have a cartilaginous endoskeleton and a notochord. Seven pharyngeal gill slits are evident near the head. The gill arches separating the gill slits are reinforced with cartilage. The mouth is at the center of the round **buccal funnel** and is armed with horny teeth and a rasping tongue. Most lampreys are parasites. They attach their buccal funnel to the side of a fish, rasp a hole in the body with their tongue, and feed on the body fluids of the fish.

(a)

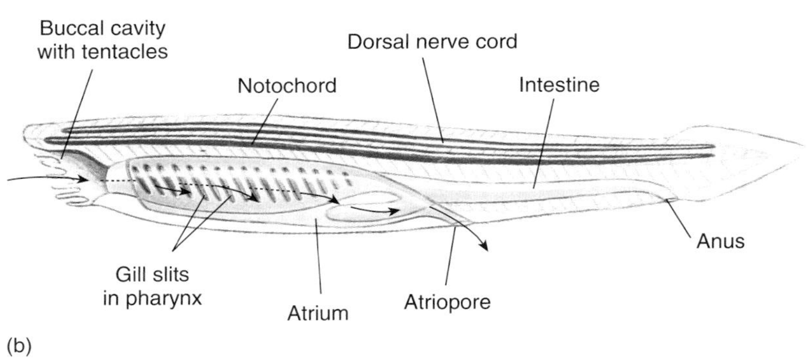

(b)

Figure 39.12
(*a*) Two lancelets, *Branchiostoma lanceolatum* (phylum Chordata, subphylum Cephalochordata), partly buried in shell gravel, with their anterior ends protruding. The muscle segments are visible; the square, pale yellow objects along the side of the body are gonads, indicating that these are male lancelets. (*b*) Internal structure of amphioxus. This bottom-dwelling cephalochordate has the three distinctive features of chordates: notochord, dorsal nerve cord, and pharyngeal gill slits. The vertebrate ancestor probably had a similar body plan.

Question 9
Which closely related subphylum of chordates does an ammocoete resemble?

Class Chondrichthyes (Sharks, Skates, and Rays)

Sharks and their relatives are abundant in oceans as predators and scavengers. Their **endoskeleton** is cartilaginous and the anterior gill arches are modified into jaws. Like agnathans, their cartilaginous skeleton is not necessarily primitive but is probably derived secondarily from an ancestral bony skeleton.

Examine a preserved specimen of *Squalus*, the dogfish shark (fig. 39.16). Its external anatomy illustrates some advanced features appropriate for a predator. Fin structure includes paired pelvic fins (on the ventral surface near the anus) and pectoral fins (behind the gill slits) for stabilization and maneuvering. Jaws are large and powerful, and receptors in the nostrils and epidermis are sensitive to smells and electrical currents. A **lateral line** runs along each side of the body and contains sensory cells to detect slight vibrations.

Question 10
a. Which fins of sharks provide power and speed?

b. Why is the number of pharyngeal gill slits in sharks fewer than that in lampreys?

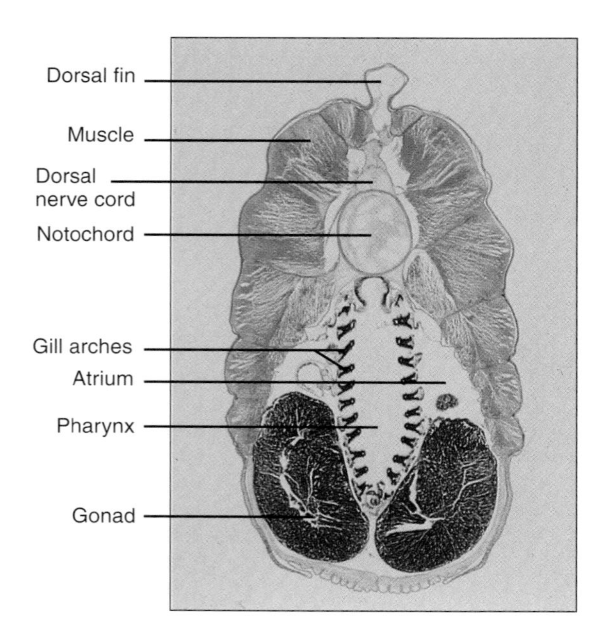

Figure 39.13
Cross section through the pharynx of amphioxus.

c. Consider objective 1 listed at the beginning of this exercise. Is a lateral line system significant to fundamental processes for sharks and bony fish? How so?

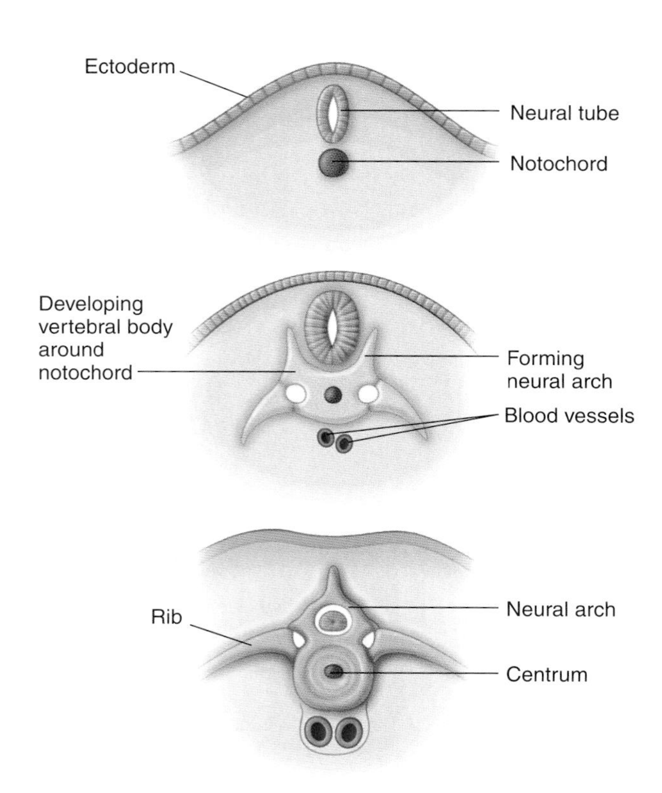

Figure 39.14
Embryonic development of a vertebra. During the course of evolution, or of development, the flexible notochord is surrounded and eventually replaced by a cartilaginous or bony covering, the centrum. The neural tube is protected by an arch above the centrum. The vertebral column is a strong, flexible rod that the muscles pull against when the animal swims or moves.

Class Osteichthyes (Bony Fish)

Examine the external anatomy of a preserved fish (fig. 39.17). Although sharks are built for speed, the maneuverability of bony fishes is much greater. If living fish are available, observe their swimming and "breathing." Bony fish are the most diverse class of vertebrates (20,000 species). Advanced features of bony fish include a bony endoskeleton, modified gill arches, and internal air bladders for balance and buoyancy. Gills are protected by a movable gill cover called an **operculum.** Along each side and branching over the head of most fishes is a **lateral-line system** consisting of sensory pits in the skin. These pits detect water currents and predators or prey that may be moving near the fish.

Question 11

a. How do the number and shape of fins of a bony fish differ from those of a shark?

b. Fins of a bony fish are flexible and diverse in shape. Describe the location of a fin present in bony fish but not in sharks.

c. How does the symmetry of the tail of a fish compare with that of a shark?

d. Does most of the power for movement by a fish come from the tail or from other fins?

(a)

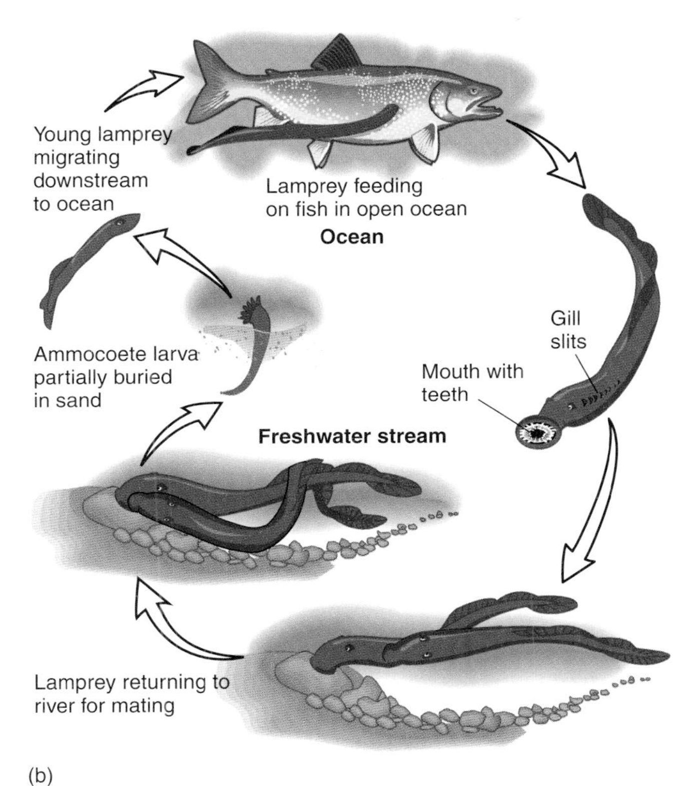

(b)

Figure 39.15

Lamprey (class Agnatha, *Petromyzon marinus*). (*a*) Note the sucking mouth and teeth used to feed on other fish. (*b*) External structure and life history of a sea lamprey. Sea lampreys feed in the open sea; toward the end of their lives lampreys migrate into freshwater streams, where they mate. Females deposit eggs in nests on the stream bottom, and the young larvae hatch three weeks later.

e. Can fish move water over their gills without moving through the water? What role does the operculum play in this movement?

f. How does the buoyancy of an air bladder affect the motion of a fish compared to that of a shark?

Class Amphibia (Frogs, Toads, and Salamanders)

Examine preserved amphibians on display. Amphibians were the first land vertebrates, arising from fish with stout, fleshy fins. Most amphibian adults are terrestrial, but they lay eggs in water (fig. 39.18). The eggs are fertilized externally and each hatches into an aquatic larval stage called a **tadpole.** Tadpoles undergo a dramatic metamorphosis of body shape as they become adults.

Development of legs and lungs in amphibians was a major evolutionary event. However, primitive lungs had already developed in some fish. In addition to lungs, the soft moist skin of some amphibians is highly vascularized and accounts for as much oxygen diffusion as the lungs.

Question 12
How are the legs of a frog different from the fins of a fish to enable movement on land?

Class Reptilia (Turtles, Snakes, and Lizards)

Examine preserved reptiles and note their morphological diversity (fig. 39.19). Reptiles, unlike their ancestors, are independent of aquatic environments and have developed structures for internal fertilization (fig. 39.20). Most reptiles also lay watertight eggs that contain a food source (the yolk) and a series of four membranes—the chorion, the amnion, the yolk sac, and the allantois (fig. 39.21). Each membrane plays a role in making the egg an independent life-support system. The outermost membrane of the egg, the **chorion,** allows oxygen to enter the porous shell but retains water within the egg. The **amnion** encases the developing embryo within a fluid-filled cavity. The **yolk sac** provides food from the yolk for the embryo via blood vessels connecting to the embryo's gut. The **allantois** surrounds a cavity into which waste products from the embryo are excreted.

Reptiles have a dry skin covered with scales that retard water loss. This dry skin does not aid respiration, but the lungs are well developed. Reptiles, fish, and amphibians are *poikilothermic*, meaning that their body temperature depends on the environment.

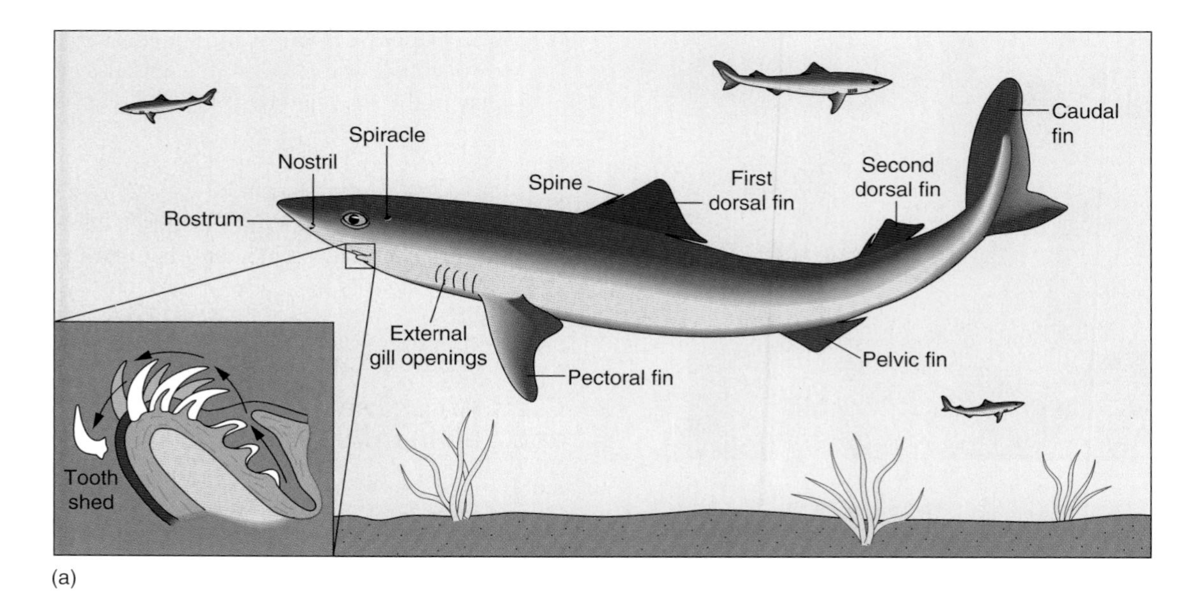

Figure 39.16

Sharks (class Chondrichthyes). (*a*) Dogfish shark, *Squalus acanthias*. Section of lower jaw (inset) shows new teeth developing inside the jaw. These teeth move forward to replace lost teeth. The rate of replacement varies in different species. (*b*) Head of sand tiger shark, *Carcharias* sp., showing a series of successional teeth.

Question 13

a. What is the adaptive significance of internal fertilization and a watertight egg?

b. How do the legs of different reptiles vary in number, size, and function?

c. Would you expect the legs of a terrestrial tetrapod to be more robust than those of an aquatic organism? Why or why not? Is this true for the reptiles and amphibians that you examined?

d. Consider objective 1 listed at the beginning of this exercise. How could poikilothermy contribute to the evolutionary success of reptiles in their environment?

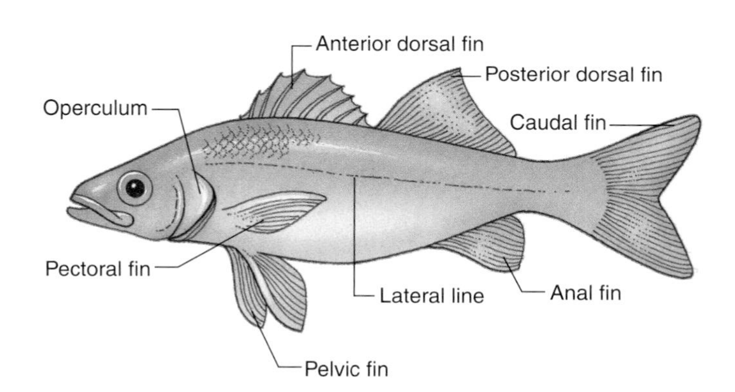

Figure 39.17

External anatomy of a bony fish, class Osteichthyes.

Figure 39.18

A frog (class Amphibia). This poison arrow tree frog (*Dendrobates* sp.) exhibits strong coloration. These colors advertise its powerfully toxic secretions to predators, which quickly learn the frog's noxious taste. Natives of South America use the frog's toxins as a weapon. They kill the frog by piercing it with a sharp stick and holding it over a fire. The heat causes the cutaneous glands to secrete drops of venom, which are scraped into a container and allowed to ferment. Arrows dipped into the poison can paralyze birds and small monkeys. One poison arrow tree frog contains enough toxin to kill about 20,000 mice.

Figure 39.19

Pit vipers are venomous reptiles; they have a pair of heat-detecting pit organs on each side of the head. Pit organs are visible between the eye and the nostril of this golden eyelash viper (*Bothrops schlegeli*). These vipers can locate and strike a motionless warm animal in total darkness by sensing heat from its body. Pit organs are highly sensitive to infrared wavelengths and are especially sensitive to sudden changes of temperature. Pit organs can detect temperature differences of 0.2°C or less, allowing effective hunting of small animals at night.

Figure 39.20

Internal fertilization. The male injects sperm-containing semen into the female's body during copulation. Reptiles such as these turtles were the first terrestrial vertebrates to develop this form of reproduction, which is particularly suited to terrestrial existence.

Class Aves (Birds)

Examine a prepared slide and whole mount of a feather (fig. 39.22). Notice the interlocking structures. Also examine specimens of birds (fig. 39.23). Birds are the only animals with feathers, and they share the ability to fly with only a few groups. Eyes of birds are always prominent, and vision is one of their most highly developed senses. Birds are *homeothermic*, meaning that they maintain a constant body temperature. Other adaptations to flight include a high body temperature for high metabolism, a lightweight skeleton, an efficient respiratory system, and heavy musculature at the breast to move the wings.

Question 14

a. What other groups of animals fly?

b. What are wings of flying animals other than birds made of?

c. Why might birds use keen vision more than reptiles or amphibians do?

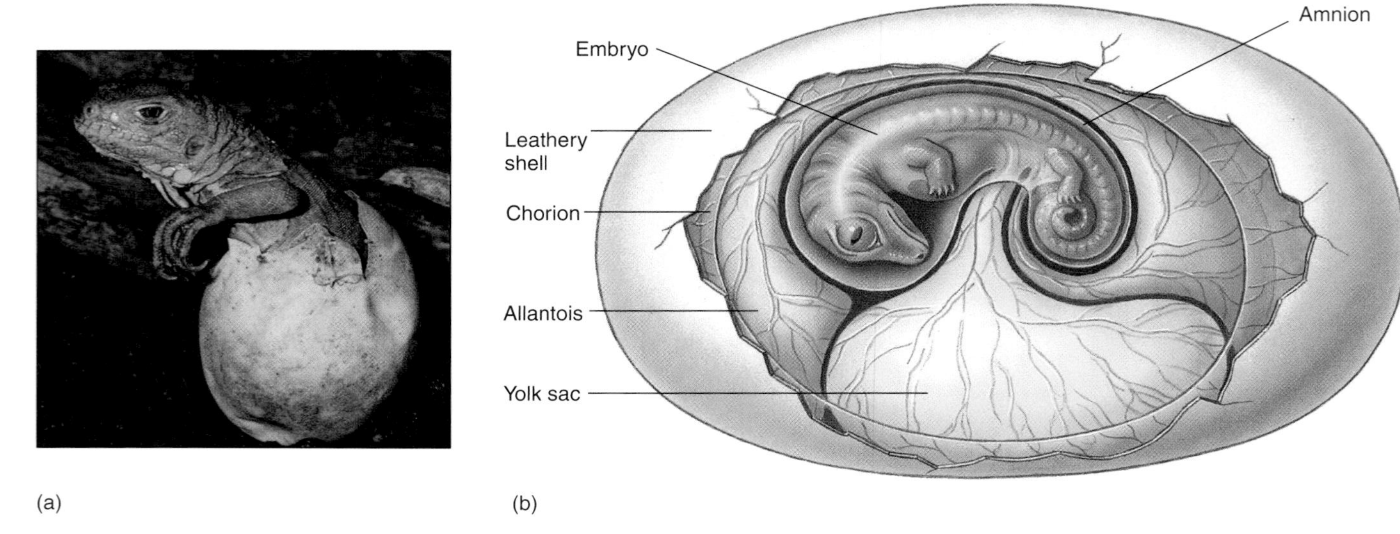

Figure 39.21

The watertight amniotic egg enables reptiles to live in a wide variety of habitats. *(a)* This green iguana from South America is hatching from a typical reptilian egg. Reptiles are the most primitive vertebrates to produce terrestrial eggs. The shell of reptile eggs can be leathery or rigid. *(b)* The major functional membranes of a reptilian egg.

d. Consider what you have learned about enzymes in Exercise 10. What might be the adaptive advantage of homeothermy?

Class Mammalia

Examine some preserved mammals. Also examine a prepared slide of mammary gland and nipple cross sections. Mammals are covered with insulating body fat and hair, and maintain a constant body temperature as birds do (fig. 39.24). Mammals are active and have a well-developed circulatory system with a four-chambered heart. The circulatory system distributes oxygen, nutrients, and heat. Mammals nourish their young with milk produced by the mother's **mammary glands** (fig. 39.25).

Although you are already familiar with the external anatomy of *Homo sapiens*, you will study the anatomy of a rat, another representative mammal, in later exercises. As you examine preserved and living mammals, search for common features such as hair distribution, body orientation, and structures for locomotion.

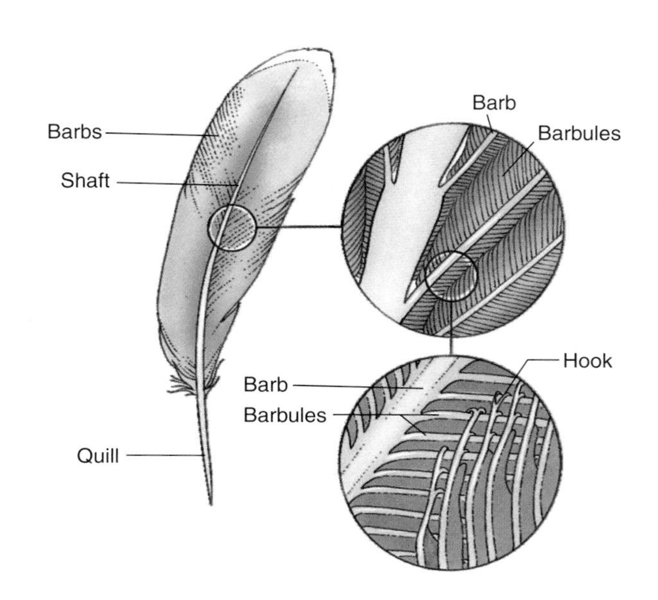

Figure 39.22

Anatomy of a contour feather, showing enlargements of barbs and barbules. The efficiency of flight is closely linked to the structure of feathers.

(a)

(b)

Figure 39.23

Birds (class Aves). (*a*) The emperor penguin, *Aptenodytes forsteri*, is the largest penguin. An adult stands 112 cm tall, weighs up to 30 kg, and is marvelously adapted to life in icy Antarctic water. This is the only penguin that spends its entire life on ice without touching land. Its insulating feathers and paddlelike wings are powerful adaptations for fishing and avoiding leopard seals, their major predator. To gather fish and squid, an emperor penguin may dive an incredible 267 m (875 feet) and remain submerged for 18 minutes. Its metabolic control and adaptations for such deep dives remain a mystery. (*b*) The California condor (*Gymnogyps californianus*) is the largest land bird in North America. Young condors acquire full adult plumage after six years and may live 50 years. They are efficiently adapted to soaring effortlessly in search of carrion. Their bald heads are adapted for reaching deep within the carcass and tearing pieces of meat. Unfortunately, they are in danger of becoming extinct. The remaining three or four wild individuals were captured in 1987. Offspring have been raised in captivity and were reintroduced into their dwindling habitat in 1991. Even efficient survival adaptations of the condor have not prevented a dramatic population decline. Condors are extremely sensitive to human disturbances, and we have steadily encroached on their habitat.

Figure 39.24

This tarsier (*Tarsius syrichia*) is a primitive primate (class Mammalia) found in the Philippines. It is the size of a rat, lives in a tree, and eats insects. The position of its eyes in the front of its head allows full stereoscopic vision. It has nails instead of claws, which indicates a common ancestry with higher primates. Its large eyes are efficient adaptations for nocturnal activity. The retinas lack cones for detecting color but are extra-rich in rods for black/white sensitivity.

Question 15

a. What factors govern the distribution of hair on species such as the human or rat?

b. How do the mammals that you are examining vary in body orientation (resting stance and position during movement)?

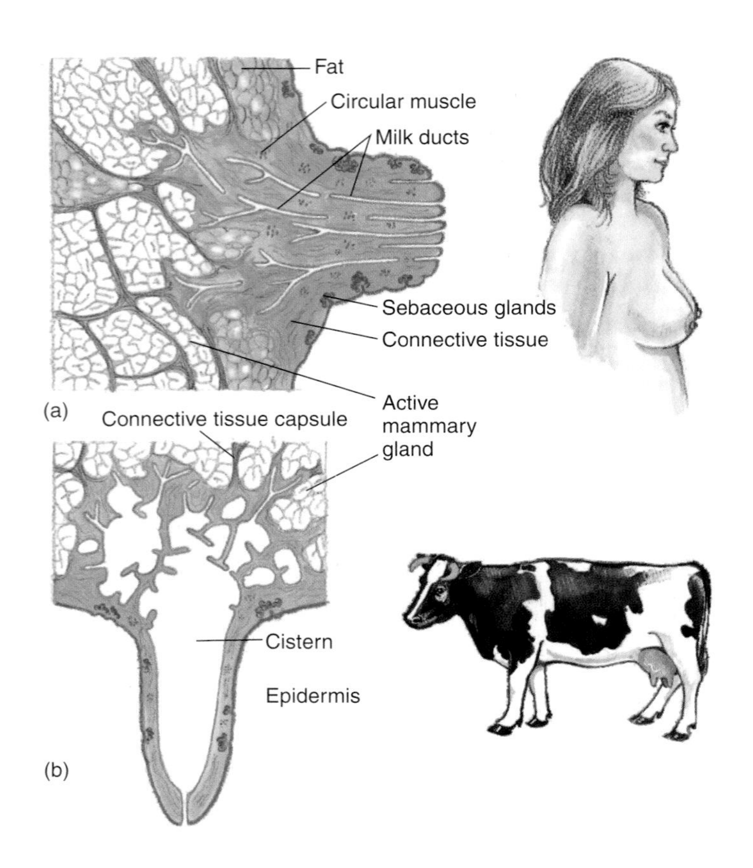

Figure 39.25

Mammary glands are specialized to secrete milk following the birth of young. (*a*) Many ducts lead from the glands to a nipple. Parts of the duct system are enlarged to store milk. Suckling by an infant initiates a hormonal response that causes the mammary glands to release milk. (*b*) Some mammals (e.g., cattle) have teats that are formed by the extension of a collar of skin around the opening of mammary ducts. Milk collects in a large cistern prior to its release.

Questions for Further Thought and Study

1. Does it surprise you that echinoderms are more closely related to our own phylum (Chordata) than are other phyla? Why would you have thought otherwise?

2. Why are embryological features important for distinguishing the major groups of phyla?

3. Echinoderms lack cephalization. What characteristics of this group deemphasize the need for a head?

4. What problems were associated with colonizing land during the evolution of vertebrates?

5. Why do you suppose four rather than five or six appendages is the rule for vertebrates?

6. A cuticle occurs on the surface of organisms of many phyla and appears to be an advantageous feature. Why have higher organisms not retained this structure?

7. What is the difference in the developmental derivation of mandibles among insects, jaws of vertebrates, and the beak of an octopus?

8. Although other groups of vertebrates are more numerous and have existed longer than mammals, mammals are often called the most advanced form of life. Why?

9. Locomotion in mammals is varied. Do you believe that their powers of locomotion are superior to those of birds? Why or why not?

10. Compare the origin and function of reptile scales, bird feathers, and mammal hair. How are they similar? How do they differ?

WRITING TO LEARN BIOLOGY

What external anatomical features of amphibians are associated with their dual life on land and in water?

40

Vertebrate Animal Tissues

Epithelial, Connective, Muscular, and Nervous Tissues

Objectives

By the end of this exercise you should be able to:

1. Understand the general classification scheme for vertebrate tissues.
2. List examples, functions, and distinguishing features of each type of tissue.
3. Associate structure with function for each type of tissue that you examine.

Cells with similar structure and function constitute a **tissue,** such as muscle tissue or nervous tissue. Tissues and their functions integrate to form **organs.** Organs are structures such as the stomach, lungs, and liver composed of several different tissues grouped together and having an integrated function. Organs work together as **systems,** such as the respiratory system or digestive system (fig. 40.1). Thus, we can define an organism at various levels of biological organization. At the cellular level, vertebrates contain between 50 and several hundred different kinds of cells, depending on how finely you differentiate cell types. These diverse cells are traditionally grouped into four tissue types, based on structure and function: **epithelial, connective, muscular,** and **nervous tissues** (fig. 40.2).

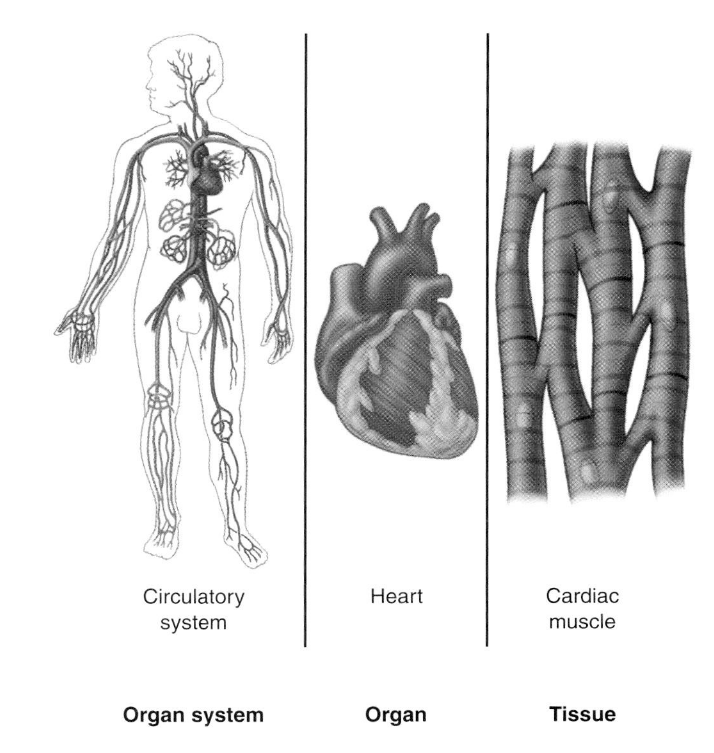

Figure 40.1

Levels of organization within the body. Similar cell types operate together and form tissues. Tissues functioning together form organs. Several organs working together to carry out a function for the body are called an organ system. The circulatory system is an example of an organ system.

EPITHELIAL TISSUE

Epithelial cells protect the body. They cover the exterior of an organism, line the gut, and line the coelomic cavity. Specifically, epithelial cells (1) protect underlying tissues from dehydration and mechanical damage, (2) provide a selectively permeable barrier that facilitates or impedes passage of materials, (3) provide sensory surfaces, and (4) secrete fluids. Three classes of epithelial tissue include **simple epithelium, stratified epithelium,** and **glandular epithelium.**

You will examine many microscope slides throughout this exercise. Notice that each request for you to view a slide has been highlighted in the text with cyan to help you be complete and organized. As you view each of the slides complete table 40.1 at the end of this exercise.

Simple Epithelium

Simple epithelial tissues are a single cell layer thick and are classified according to the shapes of their cells (fig. 40.3).

- **Squamous** epithelial cells are irregular and flattened. Thus, one cell layer of simple squamous epithelium is

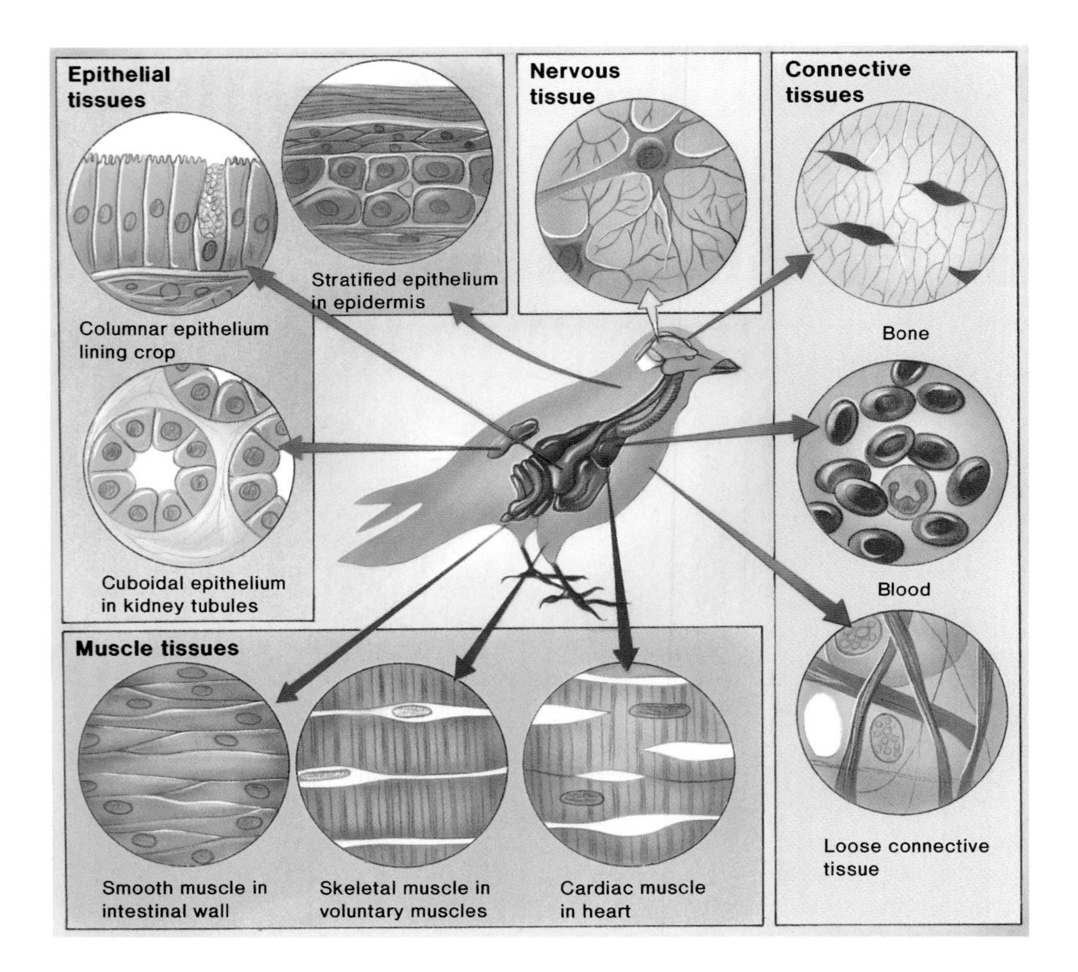

Figure 40.2

Vertebrate tissue types. Epithelial tissues are indicated by *blue* arrows, connective tissues by *green* arrows, muscle tissues by *red* arrows, and nervous tissue by a *yellow* arrow.

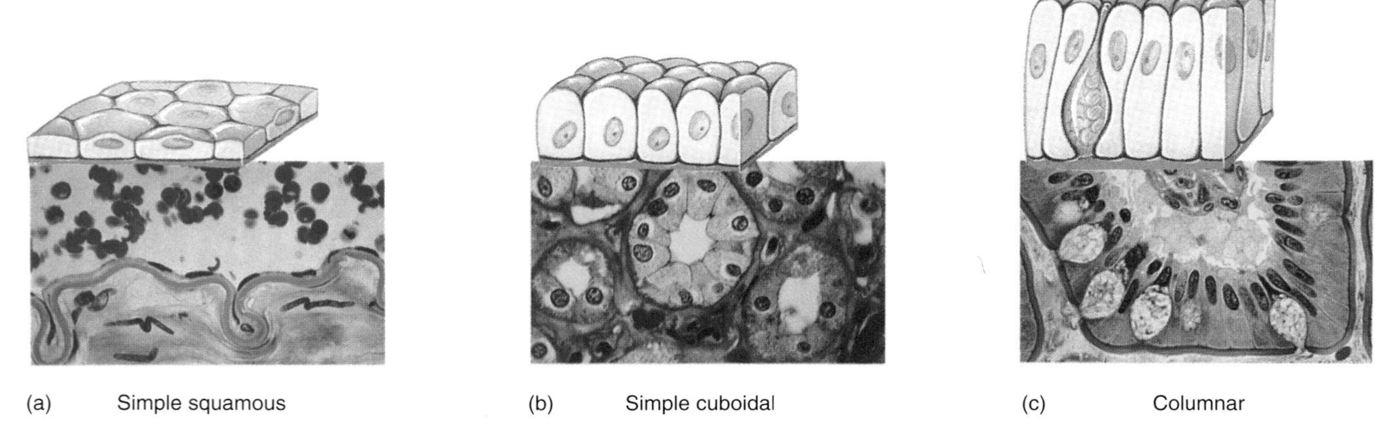

Figure 40.3

Types of epithelial tissue, based on shape. (*a*) Squamous epithelium lines the artery shown here. The nuclei are flat. The round cells above the epithelium are blood cells in the interior of the artery (also see fig. 40.4). (*b*) Cuboidal epithelium forms the walls of these kidney tubules, seen in cross section (also see fig. 40.4). (*c*) Columnar epithelium forms the outer cell layer of this human intestine. Interspersed among the epithelial cells are goblet cells, which secrete mucus (also see fig. 40.5).

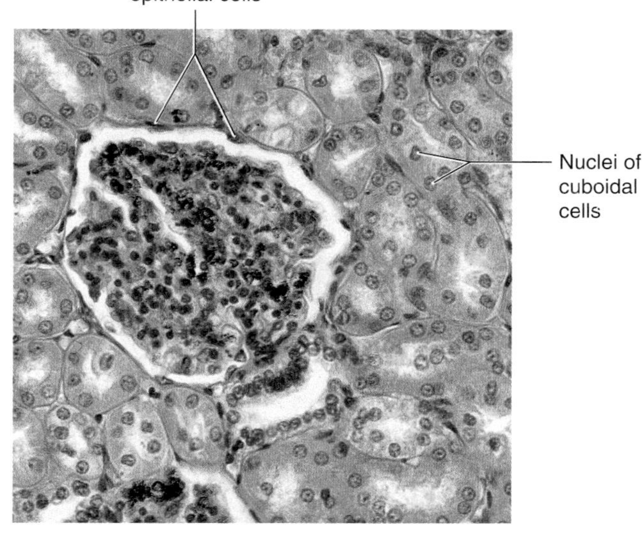

Figure 40.4

Kidney tissue showing simple squamous and cuboidal epithelia. These cuboidal cells form the walls of kidney tubules (see fig. 40.3*b*). The squamous cells are seen on edge because they line a narrow cavity.

a minimal barrier to diffusion. Squamous cells line alveoli of the lungs, the filtration system of the kidneys, and the major cavities of the body. These cells are relatively inactive and are associated with sites of passive movement of water, electrolytes, and other substances.

- **Cuboidal** and **columnar** epithelial cells appear fuller than do squamous cells and are shaped as their names imply. They line the respiratory and intestinal tracts and ducts such as kidney tubules. Cuboidal and columnar epithelial cells often have cilia and secrete fluids.

Let's examine some of your own living epithelial cells.

Procedure 40.1

Examine squamous and cuboidal epithelial cells

1. Gently scrape the inside of your cheek with the tip of a clean toothpick and stir the tip in a small drop of water on a microscope slide.
2. Lay a coverslip over the preparation and place a small drop of methylene blue at the edge of the coverslip. This stain will add color and contrast to the cells.

CAUTION

Be careful not to spill methylene blue—it will stain your skin and clothes!

3. If the stain does not readily diffuse under the coverslip, pull the fluid under by touching a dry paper towel to the opposite edge of the coverslip.
4. Examine your cells under low, then high, magnification.
5. The cells may be clumped, so scan the slide to find cells that have floated free.
6. Sketch a few cells.
7. Examine a prepared slide of a kidney cross section. Simple squamous and cuboidal cells are both common in vertebrate kidneys.
8. Refer to figure 40.4. Simple squamous epithelium surrounds Bowman's capsules, and simple cuboidal cells compose kidney tubules.

Question 1

a. About how many cells did you remove from your cheek?

b. What shape are your cheek cells?

c. What is the approximate diameter (in micrometers) of these epithelial cells? If necessary, refer to the exercise on microscopy (Exercise 2) for instructions on measuring cells.

d. Are nuclei visible in epithelial cells of your cheek?

Procedure 40.2

Examine columnar epithelium

1. Columnar epithelium lines the inner surface of the intestinal tract and trachea. Examine under low magnification a prepared slide of a cross section of small intestine and locate the relatively large, fingerlike villi of the inner intestinal wall.
2. Increase the magnification to medium, then high power, focusing on the single layer of columnar epithelium covering the villi (fig. 40.5).
3. Also examine a prepared slide of a cross section through a trachea (fig. 40.6).

Question 2

a. Are cuboidal cells such as those in a kidney actually shaped like cubes?

b. Are epithelial cells of the trachea similar in size and structure to those lining the intestine?

c. What is the approximate ratio of length to width for epithelial cells of the trachea?

d. Tracheal cells have many small, hairlike projections that move and sweep debris from the surface of the trachea. What are these projections called?

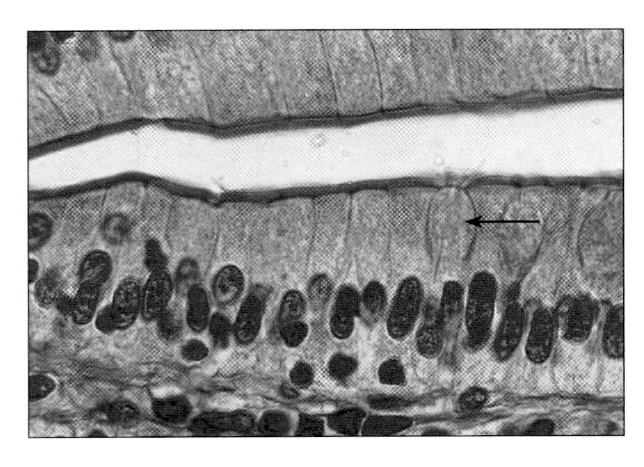

Figure 40.5
Simple columnar epithelium of frog intestine consists of a single layer of elongated cells. The arrow points to a specialized goblet cell that secretes mucus (400×).

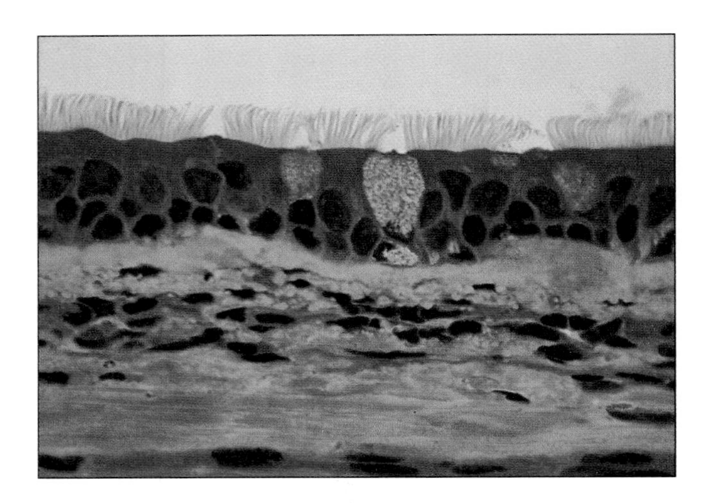

Figure 40.6
Pseudostratified ciliated columnar epithelium taken from the trachea. Notice the tuft of cilia at the top of each cell (500×).

Stratified Epithelium

Examine a slide of a cross section of skin and locate the stratified epithelium (fig. 40.7). **Stratified** tissues are several layers thick. Typically, the upper layer is squamous, the middle one cuboidal, and the basal (bottom) layer columnar. The skin is the most obvious example of stratified epithelium, although "skin" includes tissues other than epithelium.

Question 3

a. Are layers of skin cells distinct, or is there a gradual change in cellular shape from the basal to the surface layers?

b. Skin cells produce **keratin,** a strong fibrous protein found in hair and fingernails. What is the function of keratin?

c. List several functions of skin that relate to the shape and toughness of these epithelial cells.

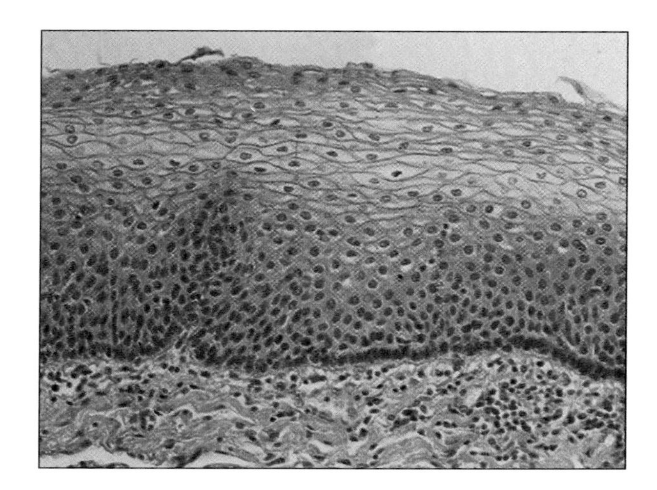

Figure 40.7
Stratified squamous epithelium, such as this cross section of skin, consists of many layers of cells (67×).

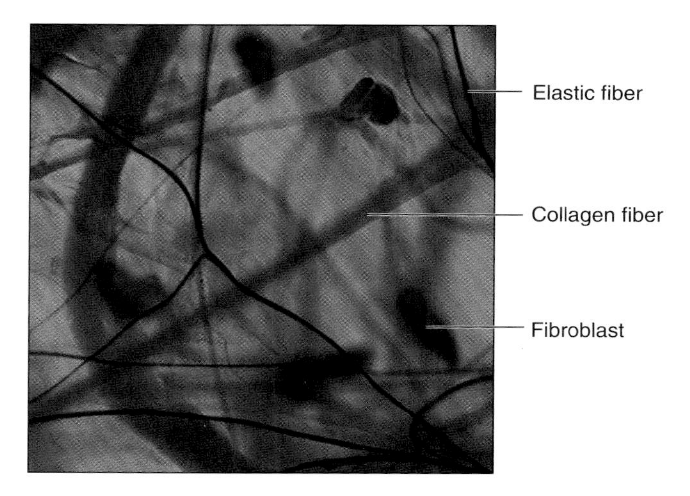

Figure 40.8
Subcutaneous tissue with fibroblasts, which produce collagen and elastic fibers.

Glandular Epithelium

Some glands of the body consist of highly modified epithelial cells that do not function as a protective covering. These cells are more active metabolically than is simple epithelium. For example, **exocrine glands** are derived from tubular invaginations of epithelial layers and include the liver, mammary glands, and pancreas. Cellular secretions of exocrine glands move to the surface and away from the organ via ducts.

Examine a prepared slide of liver. Liver contains many sinuses that carry blood. Note the large nucleus in each **hepatocyte** (liver cell).

Question 4

a. What fluid do hepatocytes secrete and store in the gallbladder?

b. What is the dark spot in the nucleus of a hepatocyte?

CONNECTIVE TISSUE

Connective tissues support and defend the body, and store food. These cells are not tightly packed (as are epithelial cells) and are typically suspended in an extracellular matrix of fibers. Some connective tissues are dispersed and flow in the circulatory system. Classification of connective tissue cells is based as much on function and the nature of the extracellular matrix as on cellular morphology. Connective tissues are divided into two major classes: **connective tissue proper,** which is further divided into loose and dense connective tissues with an abundance of fibers, and **special connective tissues** which include blood, cartilage, and bone, each having a characteristic extracellular matrix.

Connective Tissue Proper

Loose connective tissue consists of cells scattered within an amorphous mass of proteins that form a **ground substance.** Examine a slide of subcutaneous tissue, sometimes called areolar tissue, and note the irregular arrangement of **fibroblasts** and **fibers** (fig. 40.8). Fibroblasts, which are widely dispersed in vertebrate bodies, are irregular branching cells that secrete an extracellular matrix of strong fibrous proteins. The most commonly secreted protein is **collagen,** which represents 25% of all vertebrate protein (fig. 40.9). If all components of the body except collagen were removed, a ghostly mesh of fibers would remain as the framework of the body and its organ systems. Collagen is not the only fiber produced by fibroblasts. **Elastin** fibers are proteins with longer cross-links than those of collagen. This makes elastin fibers elastic.

Examine a slide of tissue having reticulin fibers taken from a lymph gland (fig. 40.10). **Reticulin** is a thin, branching fiber that supports glands such as the spleen and lymph nodes. Reticulin also composes junctions between several other kinds of tissues.

Loose connective tissue includes other cells called **macrophages,** the immune system's first defense against

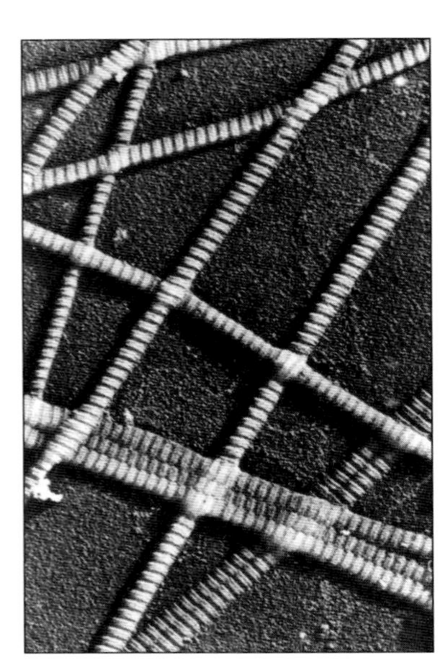

Figure 40.9
Scanning electron micrograph of collagen fibers. Each fiber of this structural connective tissue consists of many individual collagen strands and is very strong.

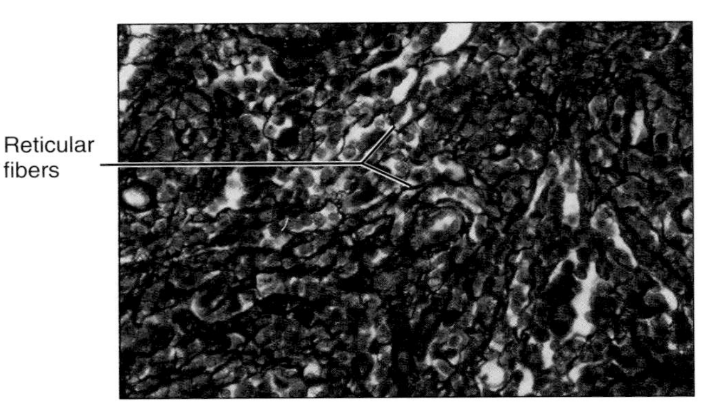

Figure 40.10
Reticular connective tissue in a lymph node.

invading organisms. Macrophage connective tissue consists of many relatively small, round cells. Macrophages are defensive cells that engulf and digest cellular debris, invading bacteria, and foreign particles (fig. 40.11). Macrophages may move individually in the circulating fluids of the body or remain fixed in an organ such as the liver or spleen. Examine a prepared slide of liver tissue. Macrophages of this tissue probably contain small black particles of India ink that they engulfed before the tissue was fixed and preserved.

Adipose cells are found in loose connective tissue and comprise **adipose tissue** (fig. 40.12). Examine a prepared slide of adipose tissue. Each adipose cell contains a droplet of fat (triglycerides). To generate energy, the adipose cell hydrolyzes its stored triglyceride and secretes fatty acids into the blood for oxidation by cells of the muscles, liver, and other organs. The number of adipose cells in an adult is generally fixed. When a person gains weight, the cells become larger, and when weight is lost, the cells shrink.

Question 5

a. Why do adipose cells appear empty?

b. Of what use is the reserve of oil in adipose cells?

Dense connective tissue contains tightly packed collagen fibers; these fibers make dense connective tissue stronger than loose connective tissue. Examine a prepared slide of longitudinal section through a tendon (fig. 40.13). Examine a prepared slide of a ligament if one is available. Tendons connect muscle to bone and derive their strength from this regular, longitudinal arrangement of bundles of collagen fibers. Ligaments bind bone to bone and are similar in structure to tendons.

Question 6

a. How can a fibroblast produce a fiber many times its own length?

b. In what areas of the body would the elasticity of elastin fibers be advantageous?

c. Which tissues in the body require the greatest strength? Explain your answer.

d. Are all the fibers in a tendon oriented in the same direction? Of what importance is this?

Special Connective Tissues

Blood cells and their extracellular fluid matrix called **plasma** perform a variety of tasks in the vertebrate body, including maintaining proper pH and transporting oxygen and carbon dioxide. Practically every kind of substance used by cells is dissolved in plasma. Blood cells are classified as **erythrocytes** (red blood cells), **leukocytes** (white blood cells), or **platelets** (enucleated fragments of large bone-marrow cells).

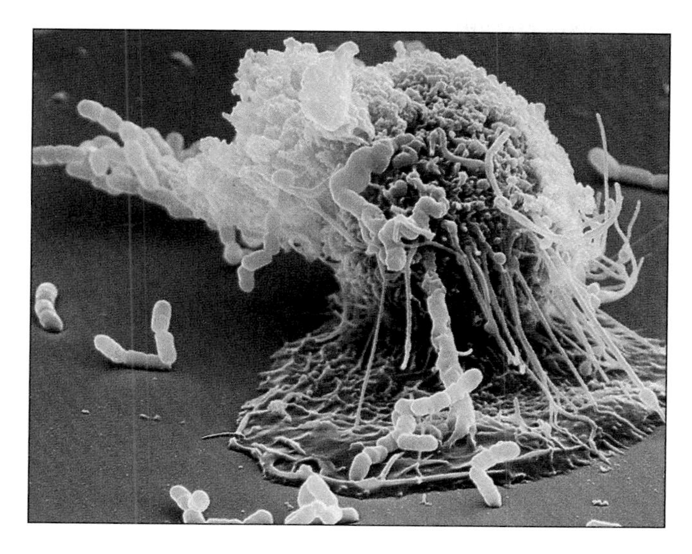

Figure 40.11
A scanning electron micrograph of a macrophage ingesting cells (1800×).

Procedure 40.3

Examine blood components

1. Examine a prepared slide of a human blood smear (fig. 40.14).
2. The most numerous blood cells are erythrocytes, or red blood cells. They sequester **hemoglobin,** which binds and transports oxygen. Notice the uniform shape of erythrocytes; these cells have lost their nuclei and become packets of hemoglobin.
3. Locate some leukocytes (white blood cells), which are much larger than red blood cells and function in defense. Some biologists classify leukocytes as defensive connective tissue.
4. Platelets, which appear as small dark fragments, sequester chemicals and enzymes essential for clotting blood. Locate platelets in your blood smear.
5. If one is available, examine a smear of frog blood cells and contrast their structure with human blood cells. Draw some of these frog cells below.

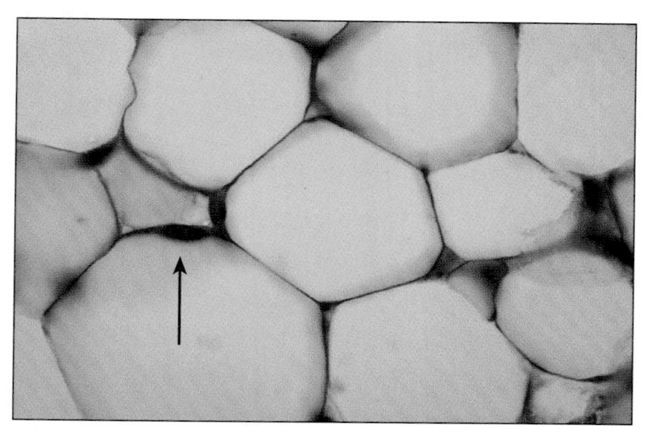

Figure 40.12
Adipose tissue cells contain large fat droplets that push the nuclei close to the plasma membranes. The arrow points to a nucleus (250×).

Question 7

a. Is a nucleus visible in each blood cell? Are nuclei apparent in white blood cells?

b. The shape of the nucleus is often used to subclassify a leukocyte. What variation do you see in the shape of nuclei in leukocytes?

c. What is the ratio of leukocytes to erythrocytes?

d. Do frog blood cells have nuclei?

Cartilage is found in skeletal joints and derives its resilience and support from an extracellular gelatinous matrix of **chondrin.** Chondrin of cartilage may be impregnated with fibers of collagen. This matrix is secreted by cells called **chondrocytes.** As in most connective tissue, cells of cartilage are rather isolated within the extracellular matrix. Chondrocytes in cartilage reside in cavities called **lacunae.**

Examine a slide of hyaline cartilage, which cushions bone surfaces between joints (fig. 40.15). Also examine a slide of elastic cartilage, which commonly occurs in the external ear and voice box (larynx) (fig. 40.16). Elastic cartilage is more flexible than is hyaline cartilage and is rich in fibers.

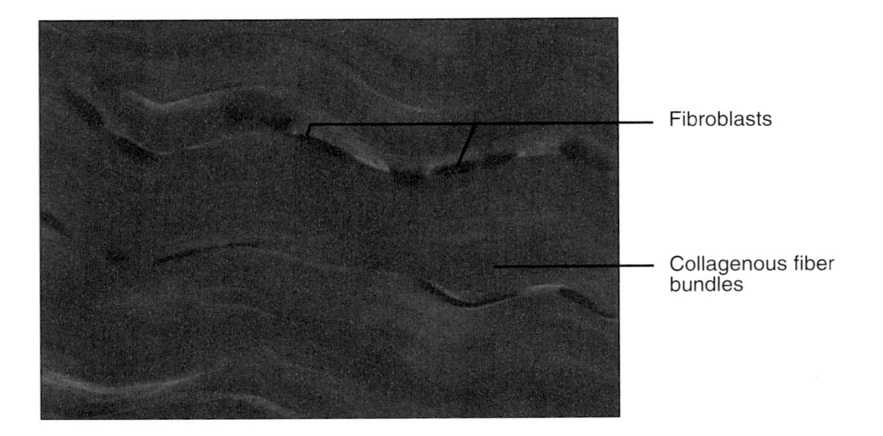

Figure 40.13
Tendon. Dense fibrous connective tissue is a strong tissue that forms tendons, which attach muscle to bone. Bundles of collagen fibers are oriented in the same direction to increase strength.

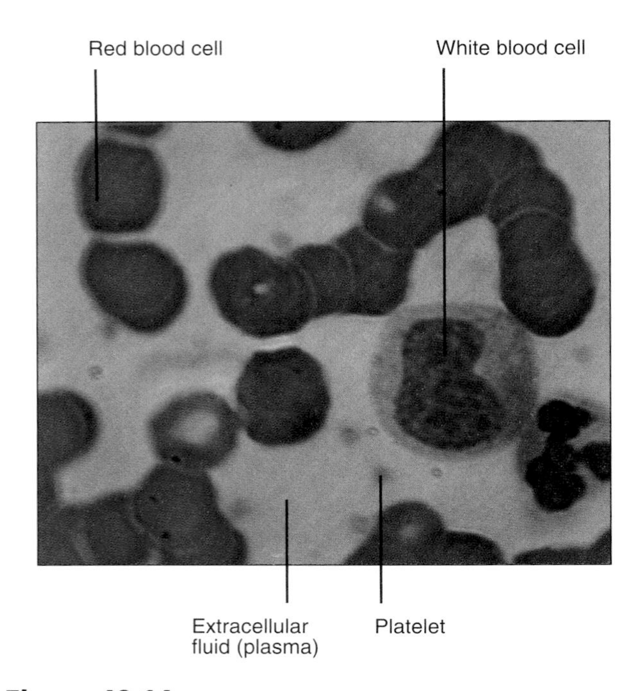

Figure 40.14
Blood is a type of connective tissue. It consists of an extracellular matrix (plasma) in which red blood cells, white blood cells, and platelets are suspended (640×).

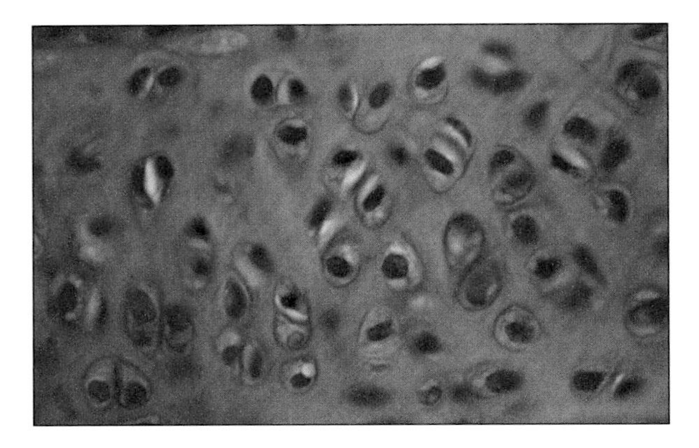

Figure 40.15
Hyaline cartilage cells are located in lacunae surrounded by intercellular material containing chondrin and fine collagenous fibers (250×).

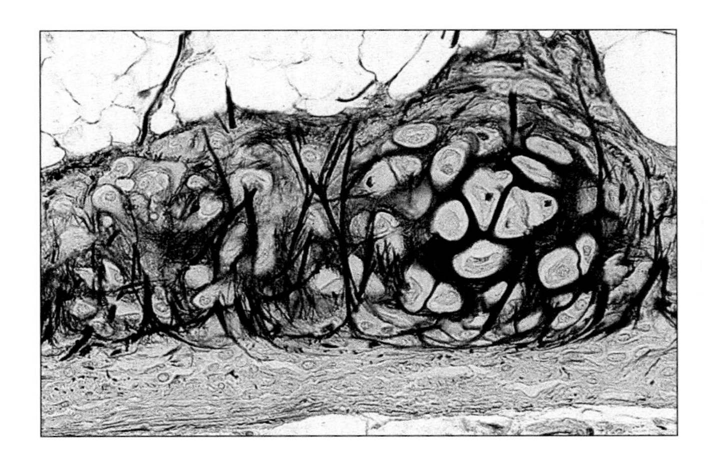

Figure 40.16
Elastic cartilage contains fine collagenous fibers and many dark elastic fibers in its intercellular material (100×).

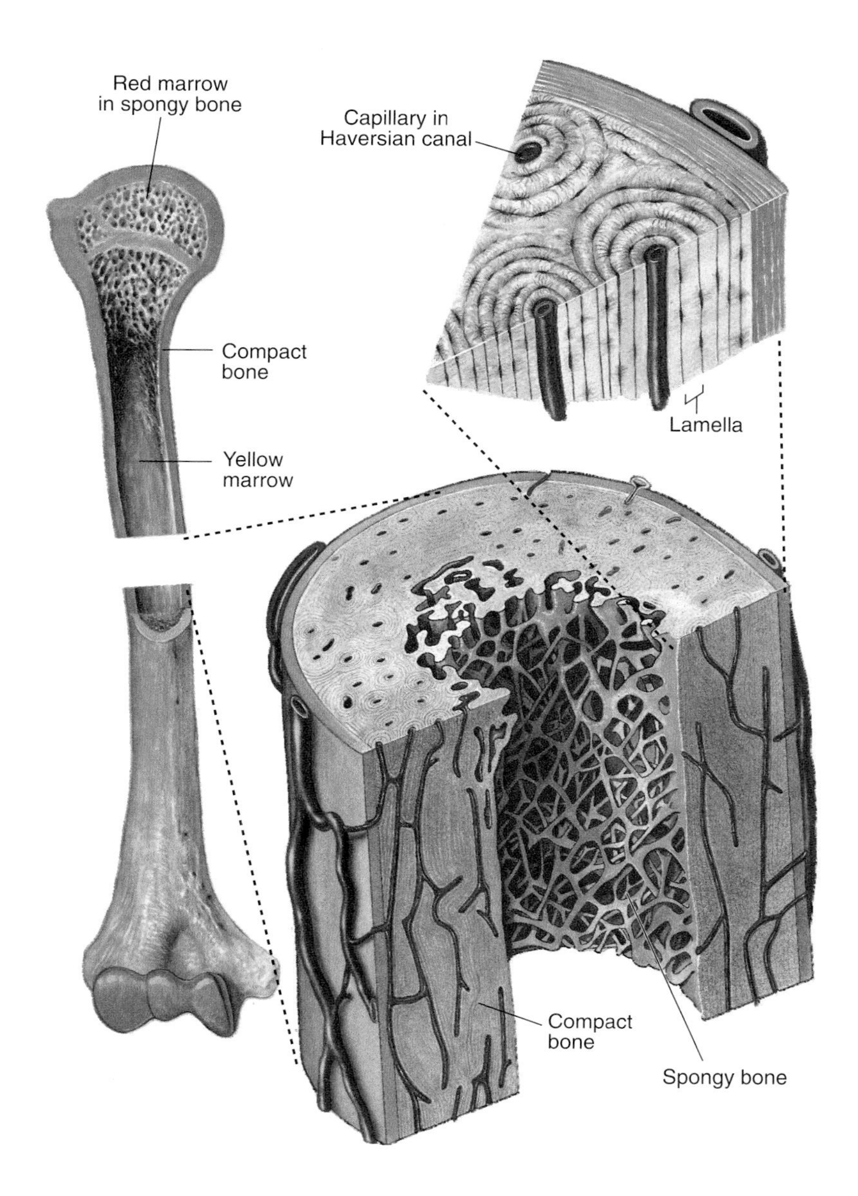

Figure 40.17

Bone shown at three levels of detail. Some parts of bones are dense and compact, providing great strength. Other parts, such as marrow, are spongy and have a more open lattice. Most red blood cells are formed in marrow. New bone is formed by cells called osteoblasts, which secrete collagen fibers as sites for deposition of hard calcium-phosphate crystals. Bone is deposited in thin, concentric layers called lamellae. Lamellae form a series of tubes around narrow channels called Haversian canals, which run parallel to the length of the bone. Haversian canals are interconnected and contain nerves and blood vessels.

Question 8

What are some general characteristics of chondrin that make it adaptive to the function of cartilage?

Bone and its properties are also derived from a strong extracellular matrix with fibers (figs. 40.17, 40.18). Collagen fibers of bone are also surrounded by hard crystals of calcium salts rather than the flexible matrix of chondrin in cartilage. This fibrous and crystalline matrix is maintained by bone cells called **osteocytes.**

Examine a prepared slide of bone. Bone forms in thin concentric layers called **lamellae.** Lamellae form a series of tubes around narrow channels called Haversian canals, which align parallel to the long axis of the bone. **Haversian canals** surround blood vessels and nerve cells throughout bone, and communicate with bone cells in lacunae through **canaliculi.**

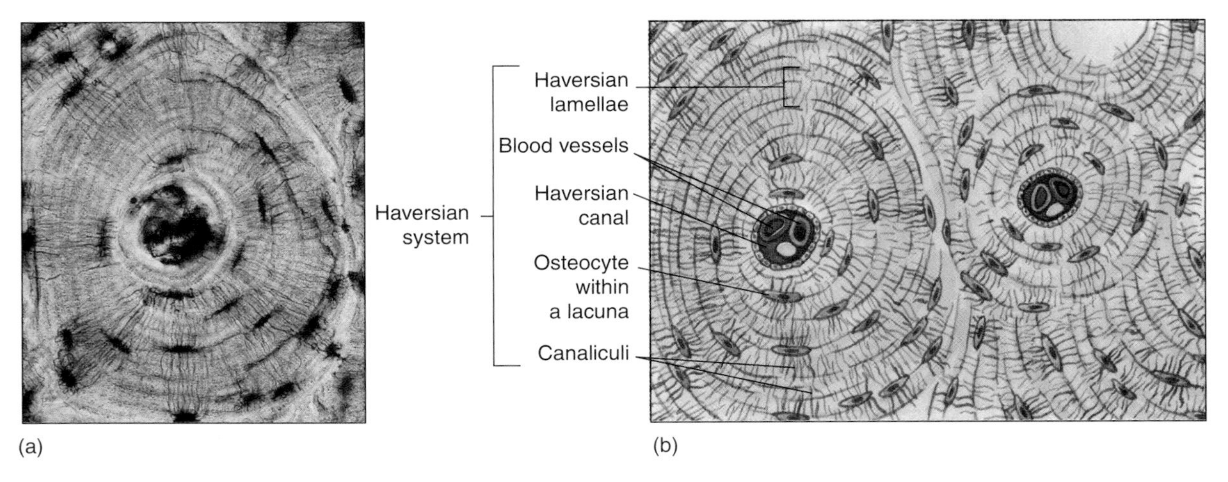

Figure 40.18

Bone, a type of connective tissue. (*a*) Photomicrograph of bone. (*b*) Diagram of a cross section of bone.

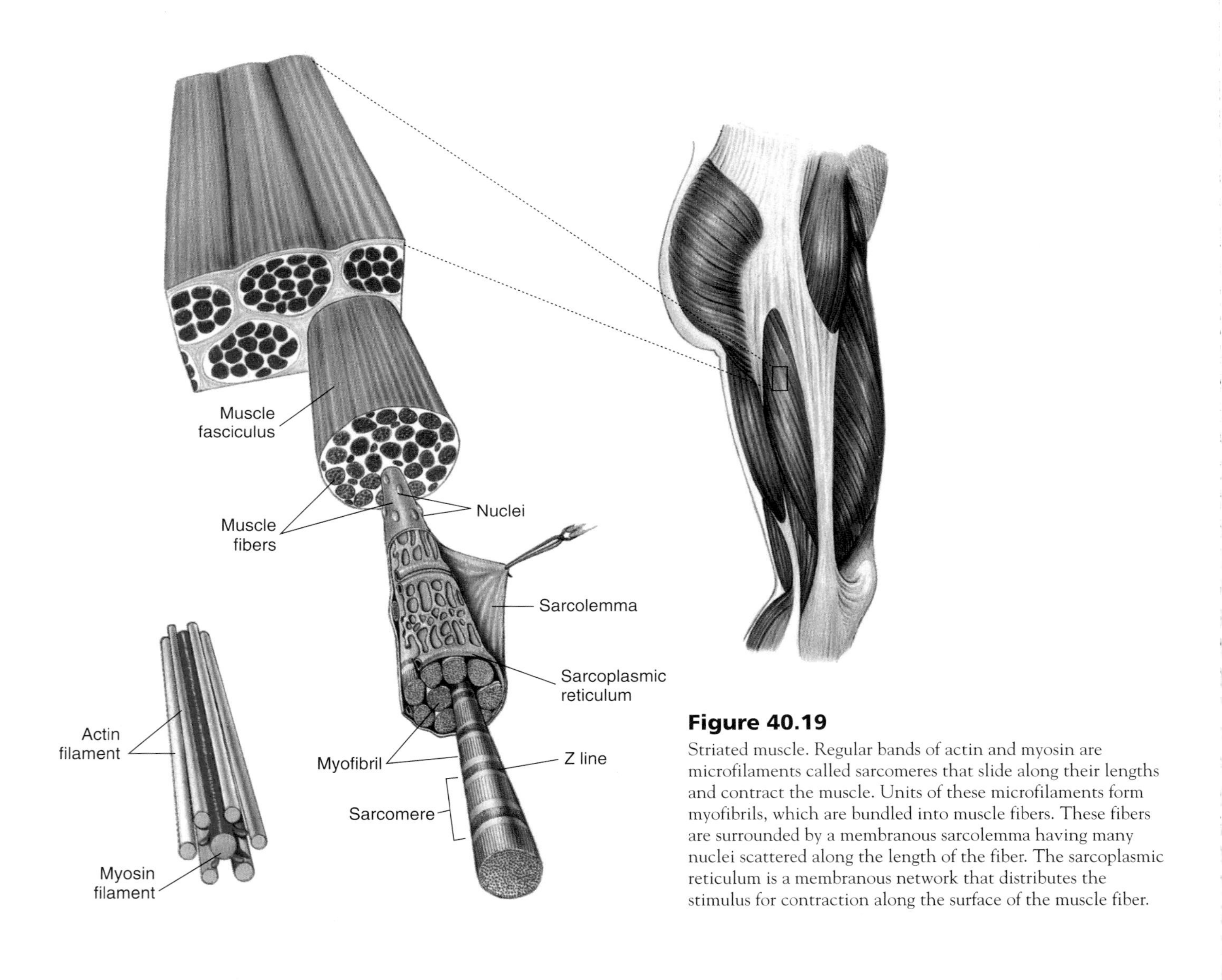

Figure 40.19

Striated muscle. Regular bands of actin and myosin are microfilaments called sarcomeres that slide along their lengths and contract the muscle. Units of these microfilaments form myofibrils, which are bundled into muscle fibers. These fibers are surrounded by a membranous sarcolemma having many nuclei scattered along the length of the fiber. The sarcoplasmic reticulum is a membranous network that distributes the stimulus for contraction along the surface of the muscle fiber.

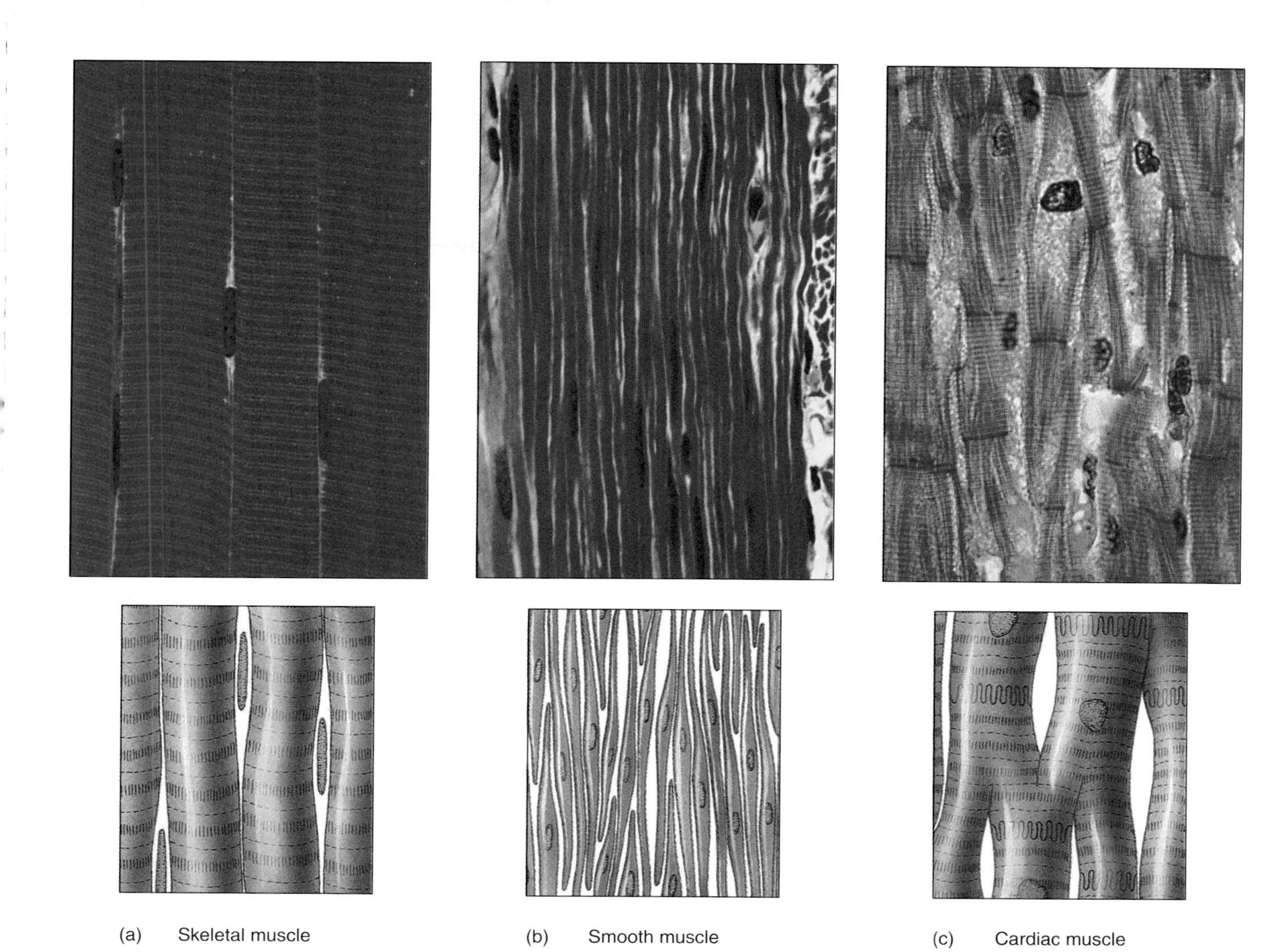

Figure 40.20
Three types of muscle. (*a*) Skeletal. (*b*) Smooth. (*c*) Cardiac. For cardiac muscle, the larger wave lines indicate the intercalated disks.

Question 9
Why is an elaborate system of canals needed in bone more than in cartilage?

MUSCLE TISSUE

The distinctive feature of muscle is its ability to contract, which results from the interaction of **actin** and **myosin** filaments. These proteins occur in other eukaryotic cells but not in such abundance and uniform orientation. Bundles of these contractile filaments, called **myofibrils,** occur within a single muscle cell, and their uniform contraction produces considerable force and movement (fig. 40.19). Vertebrates have three kinds of muscle: **skeletal, smooth,** and **cardiac** (fig. 40.20).

Skeletal Muscle

Review in your textbook how skeletal muscle contracts. Then examine a prepared slide of skeletal muscle (fig. 40.20*a*). Skeletal (striated) muscles are attached to the skeleton and are controlled voluntarily. A skeletal muscle "cell" is a long fiber of regularly arranged contractile units with many nuclei scattered at the periphery of the fiber. The strength and speed of contraction is enhanced by having the contents of many cells coalesced into a fiber rather than functioning as individual cells. The stacked array of actin and myosin filaments within the fibers gives striated muscle its banded (i.e., striated) appearance (fig. 40.20*a*).

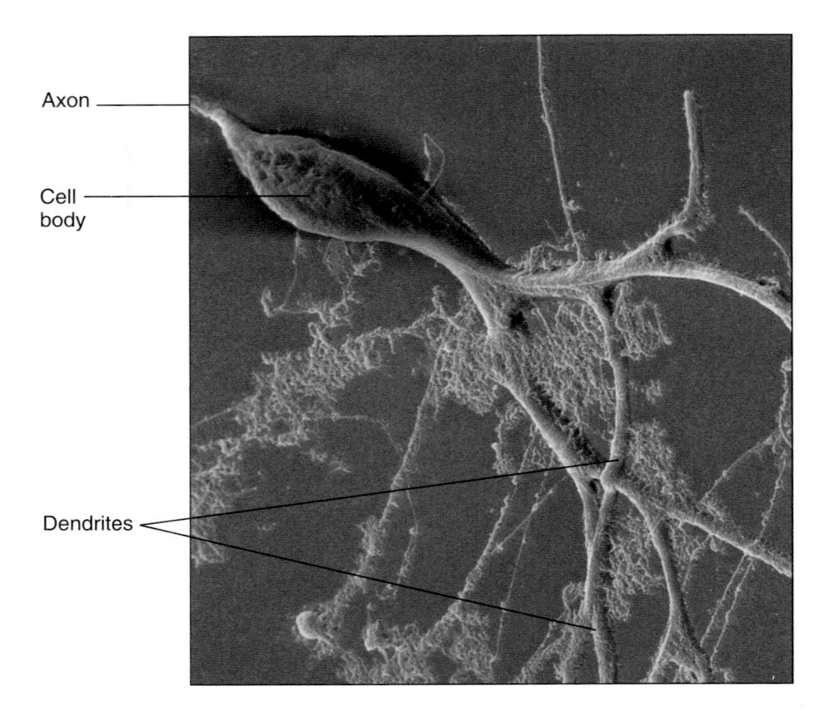

Figure 40.21
Anatomy of a neuron. Neurons are specialized to transmit nerve impulses. A neuron axon receives nerve impulses from other neurons, whereas a dendrite transmits the nerve impulse to subsequent neurons.

Smooth Muscle

Examine a prepared slide of smooth muscle (fig. 40.20*b*). Smooth-muscle cells are long and spindle-shaped, and have a single nucleus. Smooth muscles line the walls of the gut and blood vessels, and their contraction is controlled involuntarily. Smooth muscle is organized into sheets of cells that contract slowly and rhythmically. The uterus and intestine are examples of organs with smooth muscle.

Cardiac Muscle

Examine a prepared slide of cardiac muscle (fig. 40.20*c*). Cardiac (heart) muscle is striated and composed of chains of single, uninucleate cells. However, these cells have specialized junctions called **intercalated disks** between cells that organize them into rather continuous functional fibers similar to those of skeletal muscle. Thus, cardiac cells depolarize and contract more as a unit than do sheets of loosely associated cells of smooth muscle.

Question 10

a. Can you distinguish the small striations perpendicular to the axis of a muscle cell?

b. Of the three types of muscle, which contracts most swiftly?

c. Of the three types of muscle, which contracts without voluntary thought?

NERVOUS TISSUE

Examine a prepared slide of a smear of nervous tissue (fig. 40.21, 40.22). The fourth major class of vertebrate tissue is nervous tissue. Nervous tissue consists of (1) **neurons,** which are cells specialized for transmitting nerve impulses, and (2) **supporting cells,** such as **Schwann cells,** which help propagate the nerve impulse and provide nutrients to neurons.

Neurons consist of (1) a **cell body** containing a nucleus, and (2) cytoplasmic extensions that conduct nerve impulses. **Dendrites** are short extensions of a neuron that usually carry impulses toward the cell body from other cells or sensory systems. **Axons** are long extensions that usually carry impulses away from the cell body. An axon may carry an impulse to a muscle to make it contract or to the dendrites of another neuron. Because cell bodies occur only in

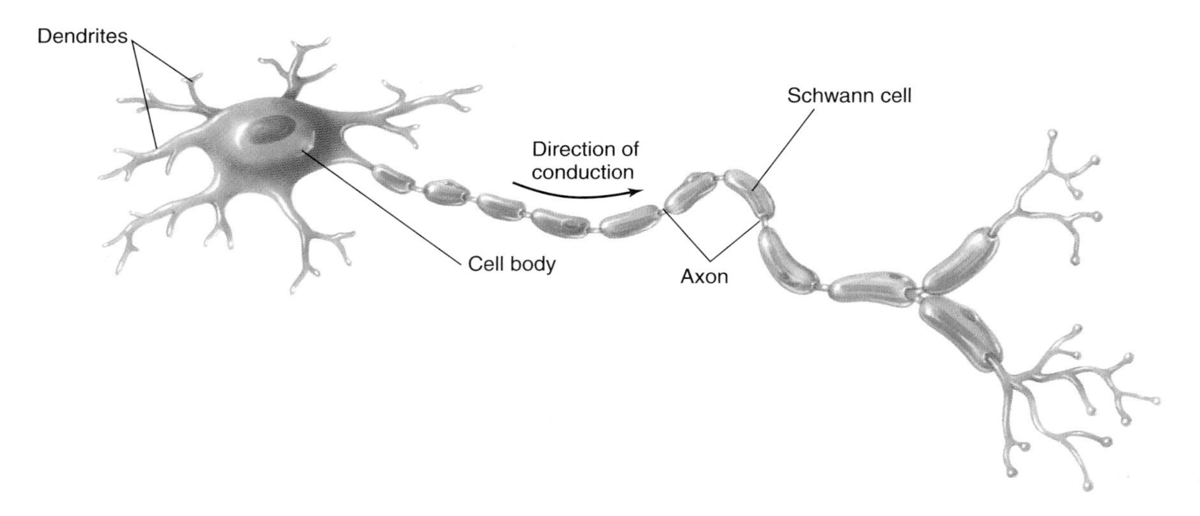

Figure 40.22
A neuron. Dendrites conduct impulses to the cell body. Axons conduct impulses away from the cell body toward effectors (muscles, glands) or toward other neurons.

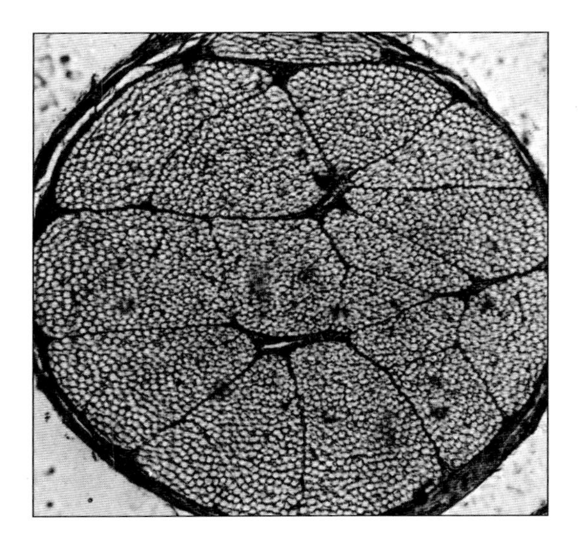

Figure 40.23
Cross section of a nerve. Nerves are a bundle of axons bound together by connective tissue. In this micrograph of a nerve cross section, many myelinated neuron axons are visible, each looking somewhat like a severed hose.

the brain and spinal cord, some axons and dendrites must be a meter long to reach distant parts of the body. Axons and dendrites extending from neurons are often bundled as nerves that look like thin spaghetti among the muscles and organs of the vertebrate body (fig. 40.23).

To review all of the available examples of animal tissues, briefly reexamine each prepared slide. As you view each type of tissue record in table 40.1 the tissue's location and function.

Question 11
What is the difference between a nerve, such as that found in an arm or leg, and a neuron?

TABLE 40.1

A COMPARISON AND ORGANIZATION OF VERTEBRATE ANIMAL TISSUES

Tissue Examined	Location in the Vertebrate Body	Function	Tissue Examined	Location in the Vertebrate Body	Function

Questions for Further Thought and Study

1. Why are vertebrate animal tissues difficult to classify into a single consistent system?

2. If a compound microscope produces an image in only two dimensions (length and width), how could you determine the three-dimensional shape of a cell?

3. How many animal tissues and cell types might be in a typical hamburger?

4. All living cells maintain a polarized membrane. Considering the function of neurons, what is the distinguishing feature of their polarized membrane?

5. How does the structure of bone tissue resemble woody dicots in form and function?

WRITING TO LEARN BIOLOGY

Does the presence of nuclei in frog blood cells indicate that the lack of nuclei in mammalian blood cells is an "advanced" characteristic? Why or why not?

41

Human Biology
The Human Skeletal System

Objectives

By the end of this exercise you should be able to:

1. Identify the major bones of the human skeleton.
2. Understand how bones are held together.
3. Understand what bones form structures such as the elbow and knuckles.

The skeletal system, along with the muscular system (see Exercise 42), determines the shape of an organism, supports other organs, and produces movement. Bones are the main structural element of the skeletal system. Bones are held together by **ligaments,** which are made of dense connective tissue and are slightly elastic.

In this exercise you will study bones of the human body.

Procedure 41.1

What bones compose the human body?

1. Examine the articulated skeleton in the lab. The human skeleton consists of the axial skeleton (skull, vertebrae, sternum, and ribs) and appendicular skeleton (shoulder, arm, hip, and legs).
2. Identify the bones of these parts of the skeleton. Note the geometry of the skeleton, paying particular attention to the shapes of the bones, the textures of the bones, and the planes in which the joints can move (fig. 41.1).
3. Identify and label the bones in figure 41.2. Also identify these bones in your body.

THE APPENDICULAR SKELETON

Shoulder

Clavicle—collarbone

Scapula—shoulder blade

Arm

Humerus—upper arm; is the longest and largest bone of the upper limb

Ulna—longer of the two bones in the forearm; is on the side of the little finger

Radius—shorter of the two bones of the forearm; is on the side of the thumb

Carpals—eight bones in the wrist bound by strong connective tissue

Metacarpals—five main bones in the hand

Phalanges—bones of the fingers

Procedure 41.2

Identifying bones in your body

1. Clench your fist.

Question 1

What bones form the raised knobs of your knuckles?

2. Flex your arm.

Question 2

a. What bones form the elbow?

b. What is a "funny bone"?

3. Use your left hand to hold your right forearm near the elbow. Now rotate your right wrist from palm up to palm down.

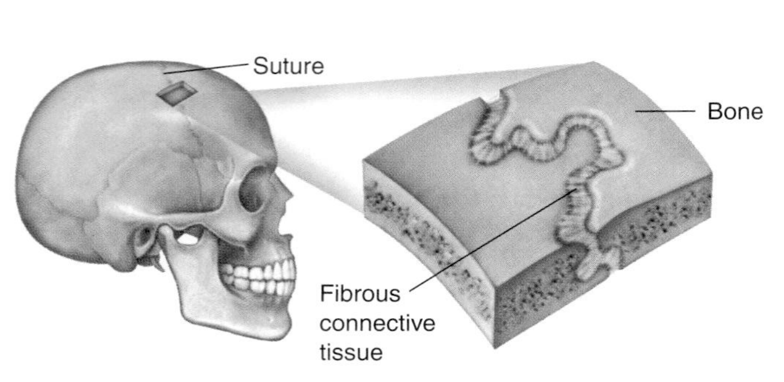

(a) Immovable joint

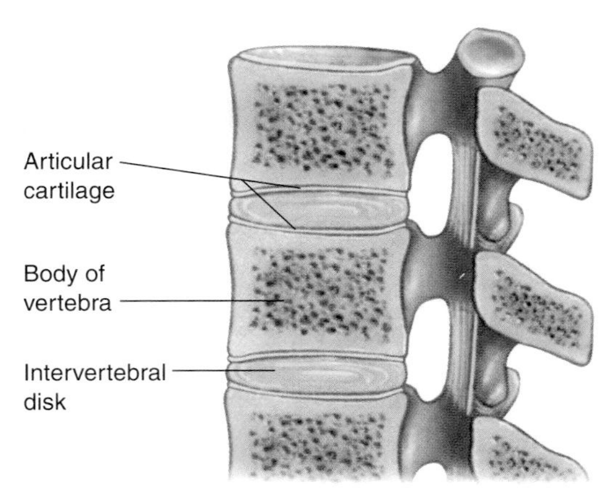

(b) Slightly movable joints

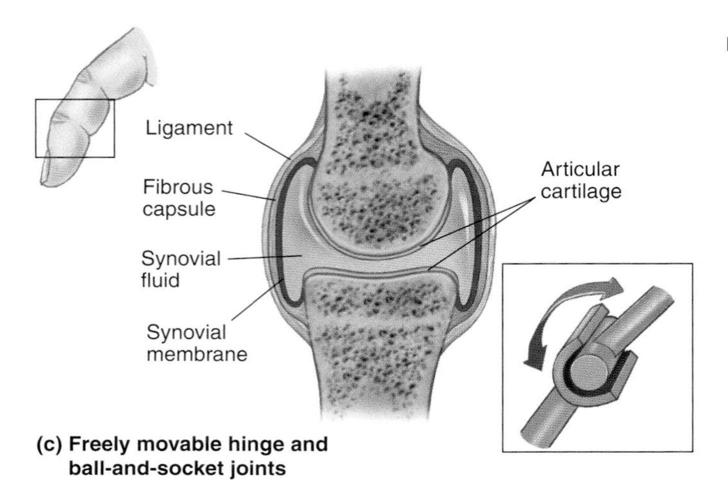

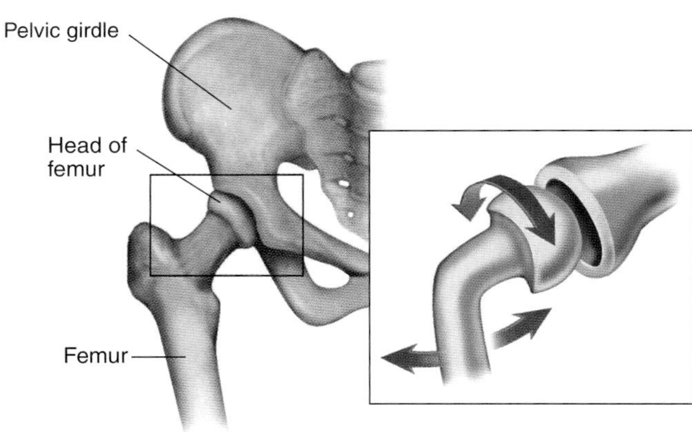

(c) Freely movable hinge and ball-and-socket joints

Figure 41.1

Three types of joints. (*a*) Sutures are immovable joints held together by fibrous connective tissue. (*b*) Cartilaginous joints contain articular cartilage that cushions the two bones of the joint. (*c*) In freely movable joints, a synovial membrane filled with synovial fluid provides lubrication.

Question 3
What bone is stationary?

Leg

Femur—thigh bone; above the knee

Fibula—smaller, more slender of the two leg bones below the knee

Tibia—shin bone; larger of the two leg bones below the knee

Patella—kneecap

Tarsals—seven bones of the ankle and heel

Metatarsals—five long bones of the feet

Phalanges—bones of the toes; two in the big toe and three in each of the other toes

4. Bend your leg at the knee and feel your patella. Then feel the lump just below your patella.

Question 4
What bone forms that lump?

AXIAL SKELETON

Sternum—breastbone

Ribs—normally 24 bones; increase in length from the first through seventh ribs, then decrease in length to the twelfth rib.

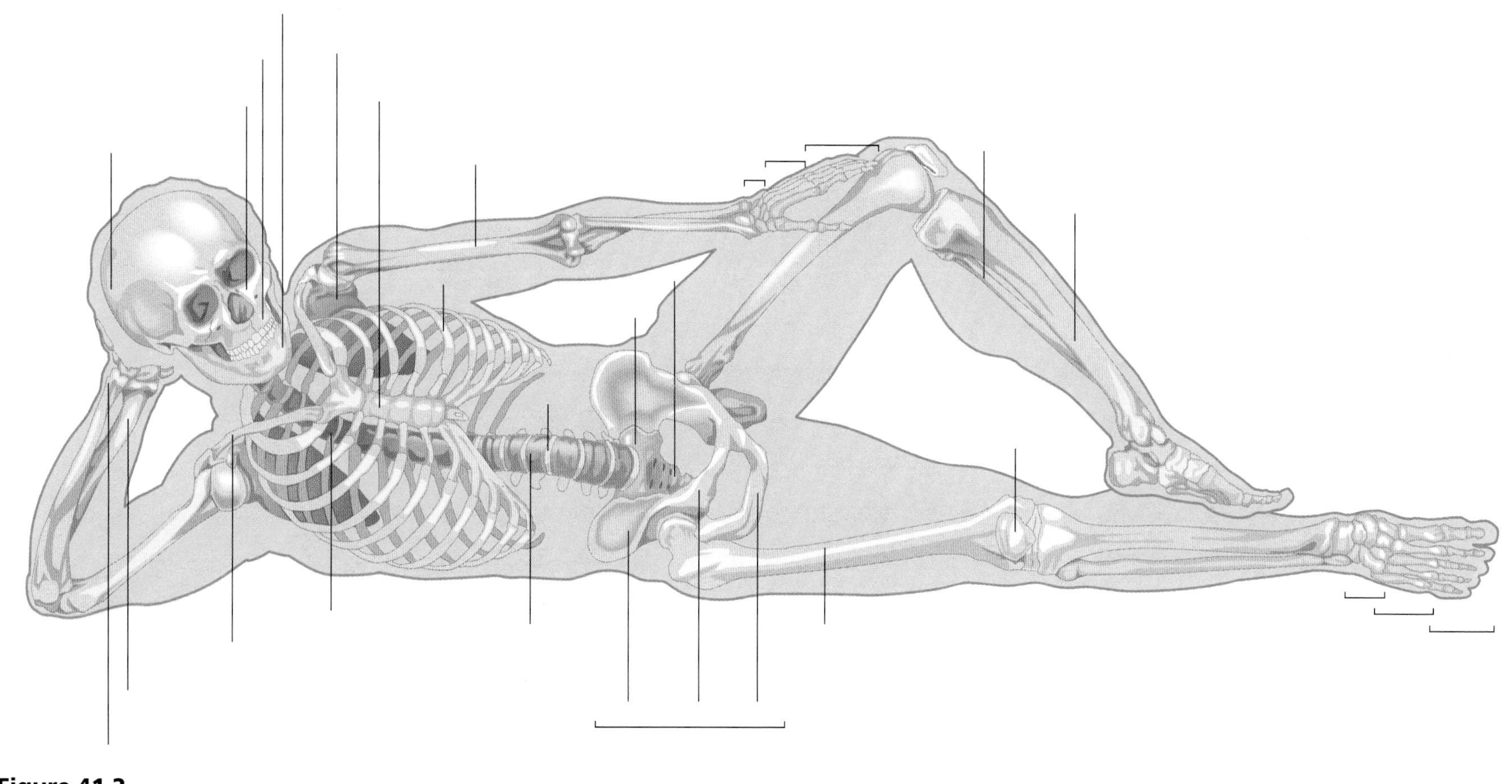

Figure 41.2
Ventral view of the human skeleton.

Vertebrae—26 bones, including the sacrum (forms part of the hip) and the coccyx (tailbone) (sacrum and coccyx are made of fused vertebrae)

Skull—28 bones, including the inner-ear bones; most are fused with immovable joints called sutures that appear as wavy lines

Question 5
How many joints are in the skull?

BONE STRUCTURE AND REMODELING

Re-examine the basic structure of bone (figs. 40.17 and 40.18). Parts of bones are dense and strong, whereas other parts, such as marrow, are spongy. Most red blood cells are formed by bone marrow. Bones are built by cells called **osteoblasts** which secrete collagen fibers as sites for the deposition of hard calcium-salt crystals. These salts are deposited in thin, concentric layers called lamellae (fig. 40.18). When muscles are developed by exercise, the bones they pull against also become thicker and stronger. This is why exercises such as weight lifting increase the mass of bone as well as muscle.

Strong, healthy bone is continually maintained by **bone remodeling,** which is the ongoing replacement of old bone tissue by new bone tissue. As osteoblasts produce collagen and other organic components they become trapped in these secretions. Soon they mature and are called **osteocytes.** Bone is broken down by large cells called **osteoclasts** in a process called **bone resorption.** In humans, bone remodeling replaces bones as many as 10 times during an average lifetime.

Bone remodeling is a balance between bone deposition and bone resorption. If too much mineral is deposited in the bone, the surplus bone tissue often forms thick bumps called bone spurs that can interfere with movement of joints. If too much bone is resorbed, the bones become weak and overly susceptible to fracture.

Question 6

a. What causes osteoporosis?

b. What factors increase a person's chances of getting osteoporosis?

c. What can help prevent the development of osteoporosis?

Questions for Further Thought and Study

1. How many bones are in your body?

2. What is the difference between a ligament and a tendon?

3. What is the longest bone in your body?

4. What bones form your ankle? Your neck? Your chest?

5. Aside from mineral storage, what are the functions of bones?

6. Osteoporosis affects 30 million elderly and middle-aged people in the United States. About 80% of these people are women. Why do older women suffer more from osteoporosis than men?

7. What are some examples of bone disease?

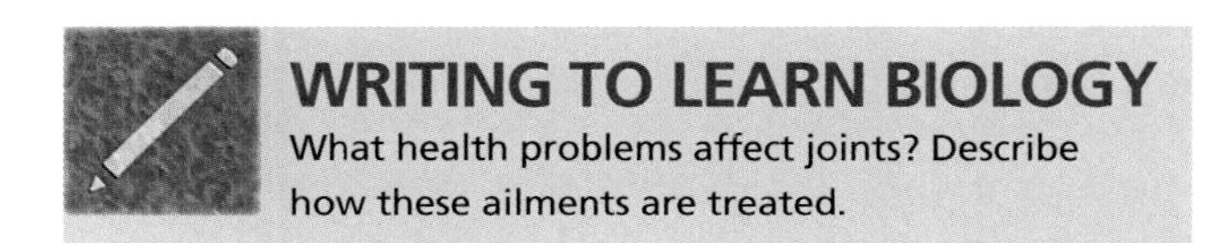

42

Human Biology
Muscles and Muscle Contraction

Objectives

By the end of this exercise you should be able to:
1. Identify the major muscles of your body.
2. Describe how muscles can flex or extend a joint.
3. Describe how the use of a muscle causes fatigue.

Muscles are structures specialized for contraction. Contrary to what many people think, muscles cannot actively lengthen; they can only contract (shorten). Some other force (e.g., gravity or the contraction of another muscle) is necessary to return the muscle to its original (uncontracted) length.

The force generated by a contracted muscle is called **muscle tension.** To move an object, the tension produced by a muscle must exceed the force exerted on the muscle by the object (i.e., the muscle's load). If this does not occur, the muscle cannot move the object.

There are two primary types of muscle contractions: isotonic and isometric. **Isotonic contractions** shorten the muscle, but the tension remains constant. You use this type of contraction when you lift weights. **Isometric contractions** increase the tension generated by the muscle without shortening the muscle. The power of contraction originates with the movement of contractile molecules in the fibers of muscle tissue (fig. 42.1).

Question 1

a. What common activities involve isotonic contractions? Isometric contractions?

b. Which type of contraction develops more muscle tension: isometric or isotonic?

In both types of contractions, the amount of tension that is generated by the muscle is proportional to the number of muscle fibers that contract; the more fibers that contract, the greater the tension. Muscle tone is a sustained contraction of skeletal muscles that produces posture.

Question 2

Why is muscle tone important?

Most joints are movable; they will move in one, two, or three planes, depending on the joint. Movement results from contraction of a skeletal muscle that connects a nonmoving bone (i.e., the origin) to a moving bone (i.e., the insertion) across a joint.

In this exercise you will study how muscles flex or extend joints. The extension of a joint, such as when you straighten your arm, increases the angle between two bones. Flexing a joint, such as when you bend your arm, decreases the angle between two bones. More specifically, the contraction of your biceps muscle flexes your arm, whereas the contraction of your triceps muscle extends your arm. In this example, as is true throughout most of your body, skeletal muscles are arranged in antagonistic pairs: when one contracts, the other relaxes.

In this lab you will also study the major groups of skeletal muscles and how the major movable joints function. You'll also study the mechanics and physiology of skeletal muscle contraction. Before doing this lab, review the introduction to muscle cell biology in Exercise 40.

MAJOR GROUPS OF MUSCLES

The following terms help biologists describe the structure and function of muscles:

Extensor—muscle that straightens a joint

Flexor—muscle that bends a joint

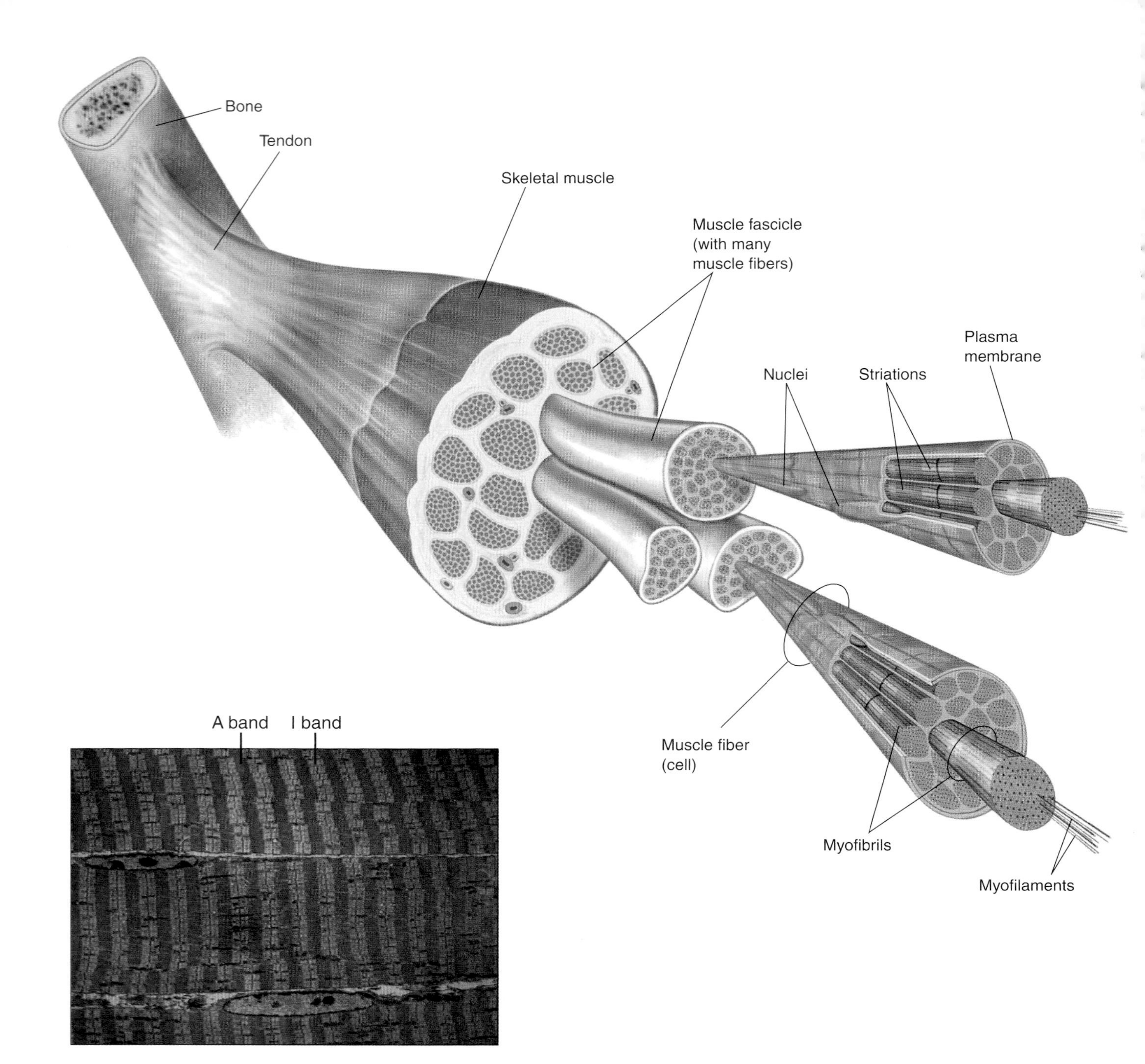

Figure 42.1

Arrangement of myofilaments in a muscle fiber. This diagram shows a large-to-small view of the muscle fiber.

Insertion—where a muscle attaches at its more movable end

Origin—where a muscle attaches to a relatively fixed position

Use your textbook and other materials available in the lab to identify the following groups of muscles. After locating these muscles on yourself or your partner, label figures 42.2 and 42.3.

Shoulder and Trunk

Deltoid—inserts on humerus; originates on clavicle. When arm is at rest, the deltoid is the outer muscle along the upper third of the humerus. When the arm is raised, the deltoid is the hard mass of muscle above the shoulder joint. The short, thick deltoid raises the arm to horizontal or slightly higher.

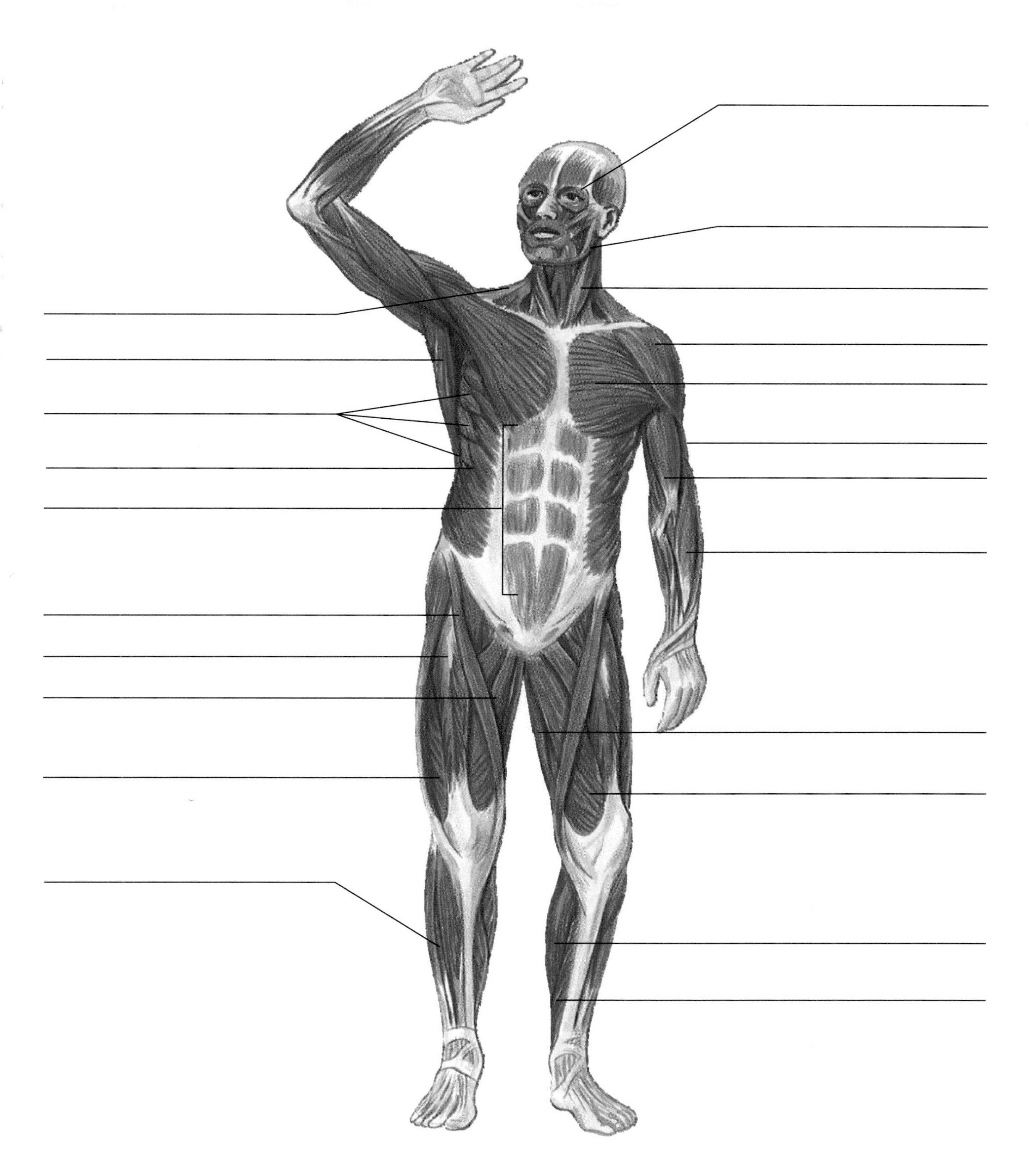

Figure 42.2
Ventral view of superficial muscles of the human body.

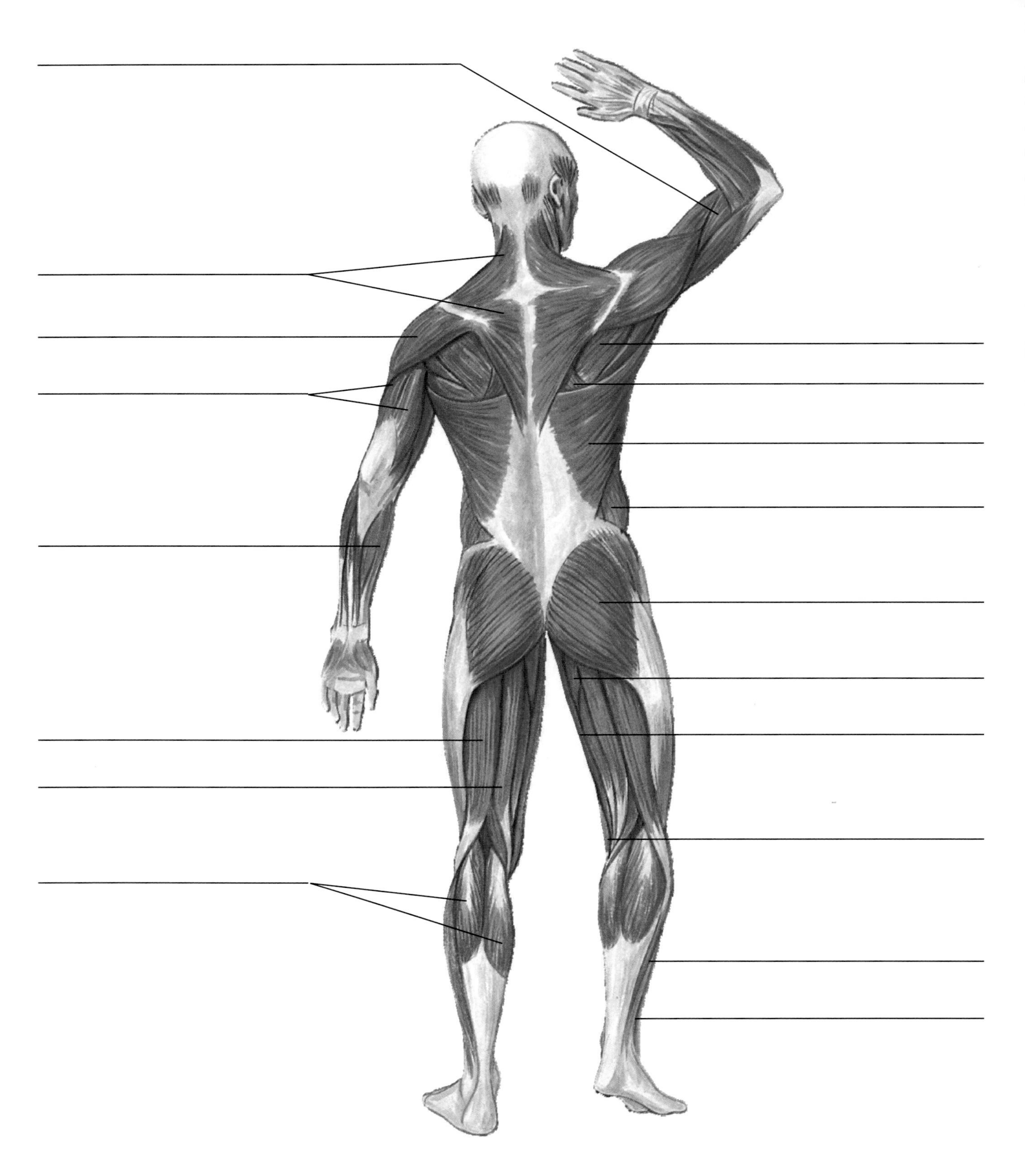

Figure 42.3

Dorsal view of superficial muscles of the human body.

Pectorals—large, triangular muscle covering the upper part of the chest. Inserts on the humerus; originates from the clavicle, upper ribs, and sternum. If your arm is fixed, such as during climbing, the pectoral helps pull the chest upward. The pectorals are the "breast" of poultry and are the main flight muscles of birds.

Trapezius—inserts on the clavicle and along the scapula; originates along the upper dorsal midline. Aids in lifting with the arms or carrying loads on shoulders; braces and shrugs the shoulders.

Latissimus—large sheet of muscle in back. Inserts on the upper part of the humerus, and originates along the mid-dorsal line. Moves the arm downward. The latissimus is a primary muscle used in a swimming stroke or in bringing the arm forcibly downward.

Arm

Triceps—inserts on ulna and originates from scapula and humerus. The triceps extends the arm at the elbow and is the primary muscle for doing a pushup.

Biceps—inserts on the radius; originates on the scapula. The biceps flexes the arm at the elbow and is the primary muscle for doing a pullup.

Procedure 42.1

What muscles flex and extend the forearm?

1. Find a partner.
2. Feel the muscles of your partner's upper arm as it is extended and flexed.
3. Repeat this exercise as your partner holds a weight and then with his or her elbow pointed at the ceiling.

Question 3

a. What muscle flexes the forearm?

b. What is its origin? Its insertion?

c. Which muscle extends the forearm?

d. What is its origin? Its insertion?

Wrist extensors—muscles on the upper side of the lower arm that raise the wrist upward.

Wrist flexors—muscles on the lower side of the lower arm which bend the hand at the wrist. With your palm up, you can see the tendons of the wrist flexors, especially if you are lifting something heavy.

Finger flexors and extensors—similar to those of the wrist. Identify these muscles and the long tendons that attach to them by clenching and extending your fingers. Note that the tendons are crossed from extensor muscles to the middle and ring fingers.

Procedure 42.2

How does the structure of your hand affect its movement?

1. With your fingertips resting on the table, raise only your ring finger as high as it will go.
2. Now raise your middle finger also.

Question 4

Did your ring finger also go higher? Why or why not?

Skull

Masseter—the main muscle that clenches the jaw. You can feel this muscle at your temples and on either side of your cheekbone.

Leg

Hamstring—a set of three muscles on the back of the thigh that bend the leg at the knee. Originate on the coxal bone and femur; insert on the fibula. You can feel the hamstring's tendons at the back of your knee joint when you bend your knee while standing on the other leg. Hamstrings are so named because butchers use these tendons to hang up hams.

Quadriceps—large muscle on the anterior part of the thigh that originates on the coxal bone and inserts on the tibia. The quadriceps extends the knee and enables you to stand from a squatting position. Also provides much of the power for kicking a ball.

Gastrocnemius—the calf muscle; originates on the femur and inserts (by the Achilles tendon) on the heel bone (one of the metatarsals). The gastrocnemius enables you to stand on tiptoe and extend your foot.

Procedure 42.3

What muscles flex and extend the lower leg and foot?

1. Feel the muscles of your partner's thigh as he/she flexes and extends the lower leg against an externally applied force.

Question 5

a. What muscle extends the lower leg?

b. What are its origins? Its insertions?

c. What muscle flexes the lower leg?

d. What are its origins? Its insertions?

2. Repeat the previous observation for your partner's lower leg.

Question 6

a. What muscle group extends the foot?

b. What are its origins? Its insertions?

c. What muscle flexes the foot?

d. What are its origins? Its insertions?

e. What common activities involve contraction of the gastrocnemius?

Toe flexors and extensors—several muscles in the lower leg that curl or extend the toes. Tendons from the extensors are visible atop your foot when you raise your toes.

Procedure 42.4

Can any tendons be manipulated manually?

Grab your ankle with your thumb on the Achilles tendon and squeeze hard. What happens? Explain your answer.

Hip

Gluteus—large, powerful muscle in the posterior pelvic region. Inserts on the femur and originates from the coxal bone. The gluteus supports the pelvis and trunk on the femur (you can show this by standing on one leg and feeling the muscle). Used in climbing, cycling, jumping, and regaining an erect position after bending forward.

Abdominals—set of muscles below the chest that flatten and compress the abdomen. Abdominals bend the body forward and from side to side; also used to urinate and defecate.

CAUTION

Do not do the following exercises if you have heart or lung problems. Stop immediately if you feel faint.

Procedure 42.5

How fast do muscles fatigue?

Work with two lab partners as you do this procedure.

1. Squeeze a tennis ball as rapidly as possible with your hand. While you squeeze, a partner will call out "time" every 15 sec for 3 min. Your other lab partner will count and record your number of contractions during each 15-sec period.

0–15 sec ________ contractions
15–30 sec ________ contractions
30–45 sec ________ contractions
45–60 sec ________ contractions
60–75 sec ________ contractions
75–90 sec ________ contractions
90–105 sec ________ contractions
105–120 sec ________ contractions
120–135 sec ________ contractions
135–150 sec ________ contractions
150–165 sec ________ contractions
165–180 sec ________ contractions

2. Wait one minute and repeat the experiment. Record your results below:

0–15 sec __________ contractions
15–30 sec __________ contractions
30–45 sec __________ contractions
45–60 sec __________ contractions
60–75 sec __________ contractions
75–90 sec __________ contractions
90–105 sec __________ contractions
105–120 sec __________ contractions
120–135 sec __________ contractions
135–150 sec __________ contractions
150–165 sec __________ contractions
165–180 sec __________ contractions

3. Wait another minute and repeat the experiment. Record your results below:

0–15 sec __________ contractions
15–30 sec __________ contractions
30–45 sec __________ contractions
45–60 sec __________ contractions
60–75 sec __________ contractions
75–90 sec __________ contractions
90–105 sec __________ contractions
105–120 sec __________ contractions
120–135 sec __________ contractions
135–150 sec __________ contractions
150–165 sec __________ contractions
165–180 sec __________ contractions

Question 7

a. Did you produce the same number of contractions in each of the three trials?

b. Why or why not?

4. On figure 42.4, graph the number of contractions versus time for each trial. Connect the data points with a straight line for each trial.

Question 8

a. Does the number of contractions increase or decrease during the experiment? Explain your answer.

b. Is the slope of the line constant in the experiments? What do you conclude from this?

c. What is the physiological basis for muscle fatigue?

d. What are the sources of fatigue?

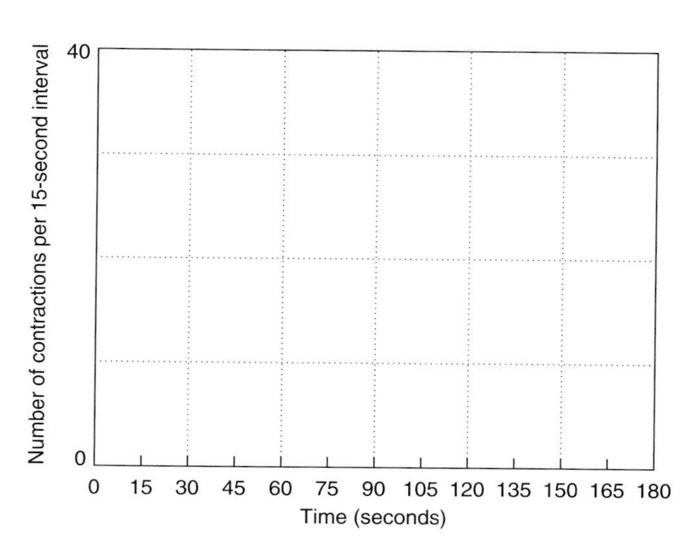

Figure 42.4
Decline in the number of contractions as a muscle fatigues.

INVESTIGATION

Determine Minimum Recovery Time

If given enough rest, muscles will recover from fatigue.

a. Decide how you would determine the minimum recovery time for a simple exercise.

b. Form a hypothesis about the result you expect. Your instructor will advise you about how to write a testable hypothesis. Write your hypothesis here:

c. Decide how you will test your hypothesis. Describe your experimental design here:

d. Do your experiment. What do you conclude? Do your data support your hypothesis?

Questions for Further Thought and Study

1. What would happen if both muscles of an antagonistic pair contracted simultaneously?

2. What is a "pulled" muscle?

3. What other kinds of muscles besides skeletal muscles are there?

4. What is an example of a muscle that is not part of an antagonistic pair?

5. Many people take dietary supplements such as creatine sulfate to improve their strength and endurance. Do these supplements "work"? What is the evidence?

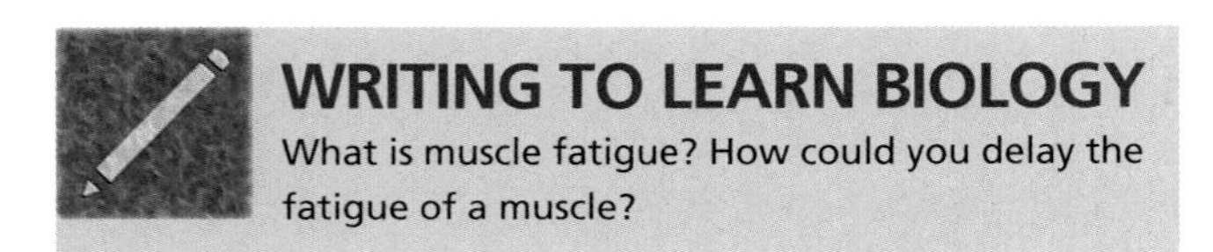

43

Human Biology
Breathing

Objectives

By the end of this exercise you should be able to:

1. Describe the mechanics of breathing.
2. Measure your vital capacity, tidal volume, inspiratory reserve volume, and expiratory reserve volume, and understand how these volumes relate to one another.
3. Measure how exercise and hyperventilation affect your breathing.

In Exercise 11 you studied how cells oxidize sugars to release energy for their activities. This process is called cellular respiration, and in humans and most other organisms it requires oxygen. To get this oxygen for respiration (and to get rid of respiratory waste products such as carbon dioxide), humans have a pulmonary respiratory system in which ventilation and gas exchange occur in specialized organs called lungs (fig. 43.1). We promote gas exchange by forcing air into and out of our lungs as we breathe. The importance of breathing can't be overestimated—every day we breathe about 25,000 times. If our rate of breathing slows too much, we either faint or suffocate.

Air moves in response to pressure gradients. This is true for large masses of air, such as cold and warm fronts that sweep into town, and the relatively small masses of air that move into and out of our lungs. Air always moves from areas of high pressure to areas of lower pressure. For example, when we prick a balloon, air quickly moves from inside the balloon, where pressure is high, to the outside of the balloon, where pressure is lower. Air moving into and out of our bodies as we breathe also moves in response to pressure gradients. As you'll learn in this exercise, we expend much energy to create pressure gradients that help us breathe.

Our lungs are in our thoracic (chest) cavity. We inhale by expanding our lungs, which creates a negative pressure (i.e., suction) in our lungs. This suction pulls air into our lungs. Expanding our lungs is more complicated than it seems, because lung tissue lacks skeletal muscle. To expand our lungs and create the negative pressure needed to inhale, we use our diaphragm and intercostal muscles.

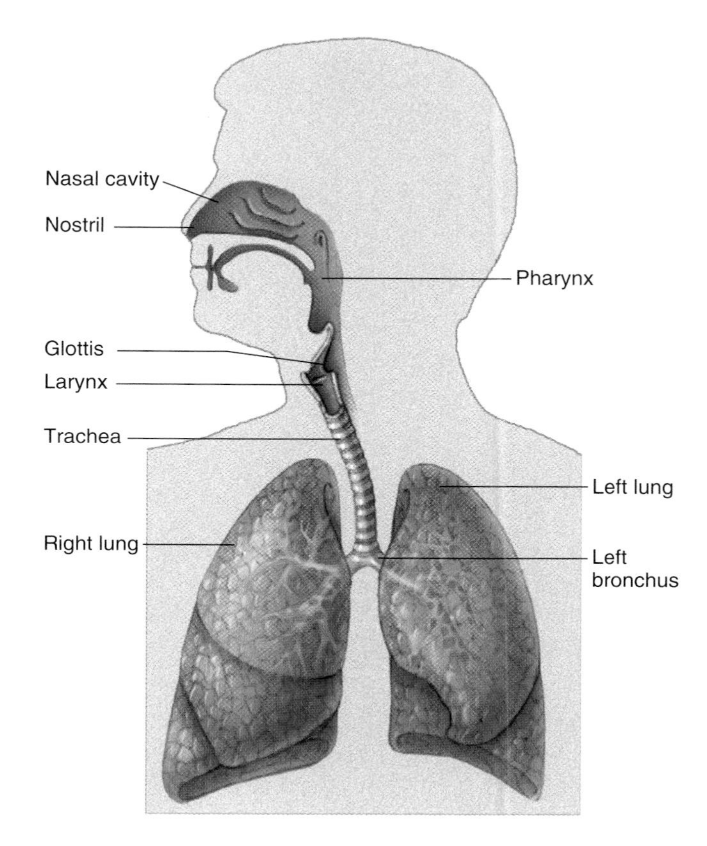

Figure 43.1
The human respiratory system. The human respiratory system contains openings for the intake of air (nostrils and oral cavity), a passageway for the air to the lungs (nasal cavity, larynx, trachea, and bronchi), and an organ that facilitates gas exchange (lungs). Within the lungs, but not shown here, are alveoli (air sacs) and bronchioles (small passages that connect the alveoli).

- The **diaphragm** is a sheetlike muscle separating the abdomen from the chest cavity. It is the primary muscle used in breathing; when it contracts, the diaphram flattens and expands the chest cavity. This expansion of our chest creates a negative pressure (i.e., a partial vacuum), thereby pulling air into our nostrils, mouth, and lungs.

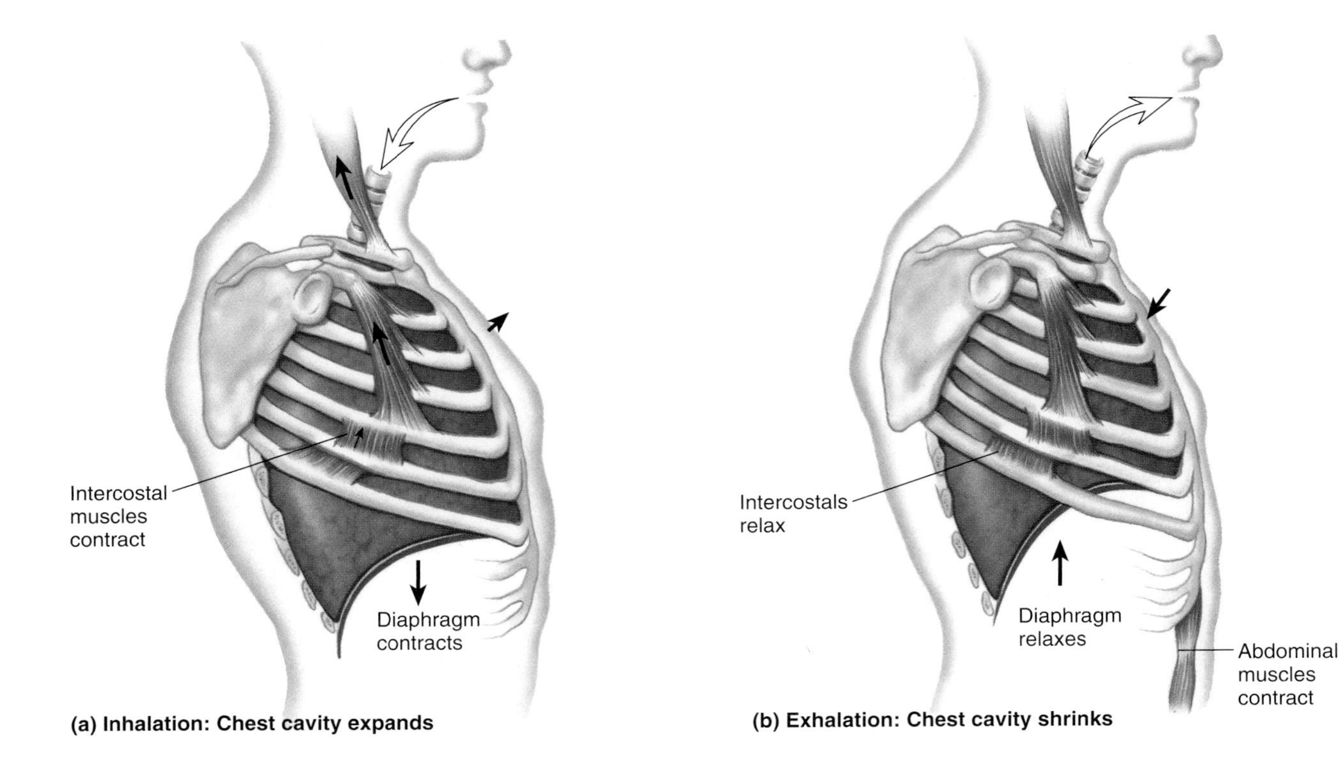

Figure 43.2
Movement of the rib cage and diaphragm during inhalation and exhalation. (*a*) During inhalation, the diaphragm contracts and moves downward, and the chest cavity expands, increasing its volume. As a result of the larger volume, air is sucked in through the trachea. (*b*) During exhalation, the diaphragm and rib cage return to their original positions, reducing the volume of the chest cavity and forcing air outward through the trachea.

- **Intercostal muscles** are located between ribs (fig. 43.2). When these muscles contract, they expand the chest cavity and suck air into our lungs. When other people rub your intercostal muscles, you probably squirm and giggle.

Contracting your diaphragm and intercostal muscles expands your chest cavity. This decreases the pressure in the chest cavity and, because your lungs adhere to the lining of your chest cavity, your lungs expand. This expansion decreases the pressure in the lungs, causing air to move into the lungs. Relaxing the diaphragm and intercostal muscles shrinks the chest cavity, thereby increasing the pressure in the lungs. This forces air out of the lungs because pressure there exceeds that in the atmosphere.

To better understand this, examine the lung model, a simple device that helps demonstrate the principles underlying inhalation and exhalation. The model consists of a glass tube, balloons, enclosed space, and a rubber sheet (fig. 43.3). The glass tube of the model represents air passages, the balloons represent the lungs, the space represents the thoracic cavity, and the rubber sheet at the base of the model represents the diaphragm. Rhythmically pull and push the diaphragm to simulate breathing.

Question 1

a. What happens to the lungs? Why?

b. What would happen if the seal at the base of the jar were broken?

c. What causes a collapsed lung?

d. Is a collapsed lung functional? Why or why not?

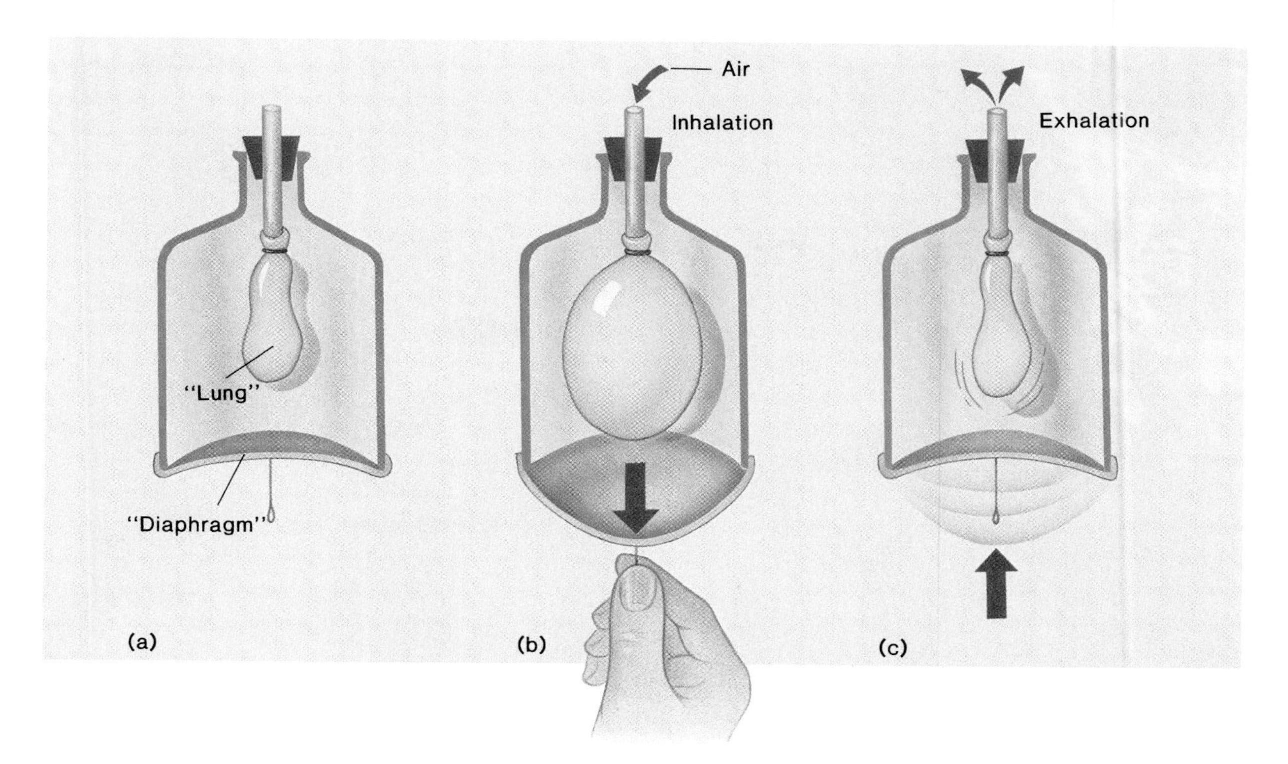

Figure 43.3

A simple experiment that shows how we breathe. In the jar is a balloon (*a*). When the diaphragm is pulled down, as shown in (*b*), the balloon expands; when it is relaxed (*c*), the balloon contracts. In the same way, air is taken into the lungs when the diaphragm moves down, expanding the volume of the lung cavity. When the diaphragm pushes back, the volume decreases and air is expelled.

CAUTION

Don't do the following exercises if you have heart or lung problems. Work with a partner in each of the following procedures, and stop immediately if you feel faint.

Procedure 43.1

Measure differences in chest diameter during breathing

1. Wrap a tape measure around your chest. Record your chest diameter: ________ cm
2. Now take a deep breath, hold it, and measure your chest diameter: ________ cm

Question 2

a. How much did your chest enlarge?

b. What caused your chest to enlarge?

c. What is the significance of this change?

3. Place your hand at the bottom of your sternum (chest bone) and take a deep breath.

Question 3

a. What direction did your hand move when you inhaled?

b. When you exhaled?

4. Place your hands on your abdomen and take five deep breaths. Describe what you feel.
5. Repeat this with your hands on your chest. Describe what you feel.

Question 4

Can you take a deep breath without expanding both your chest and abdomen? Why or why not?

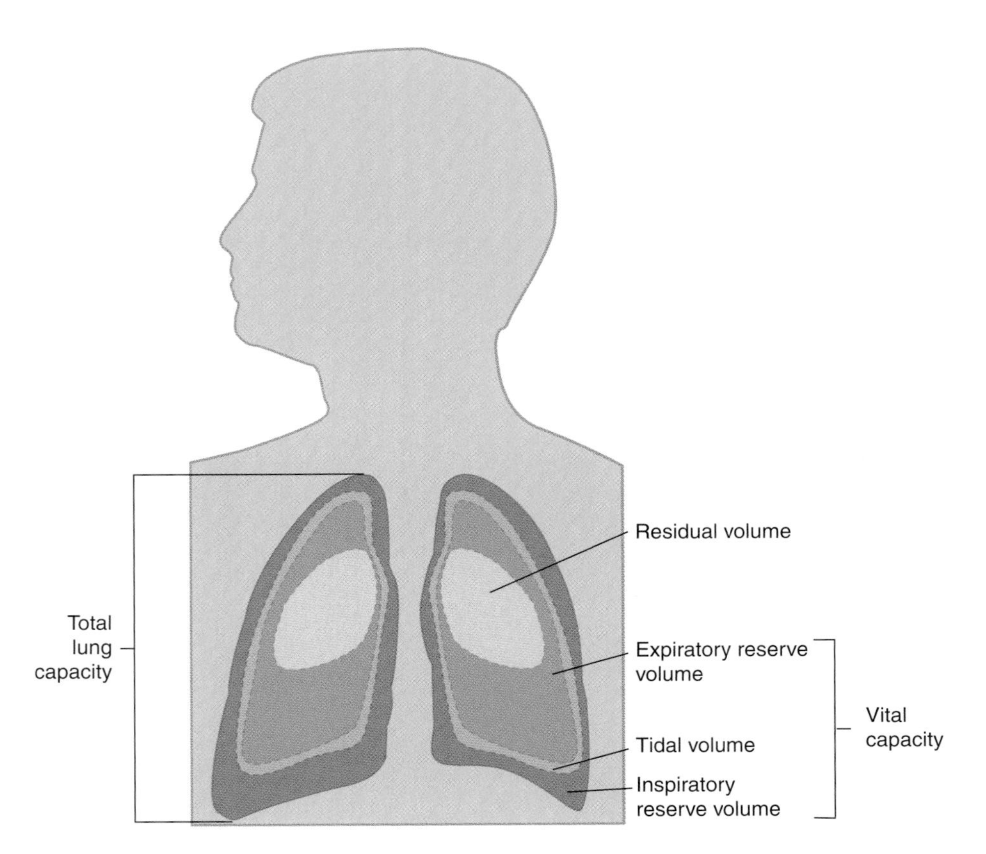

Figure 43.4
Pulmonary volumes and capacities.

6. Exhalation occurs by relaxing your diaphragm and intercostal muscles. To force more air from your lungs you must use your abdominal muscles. Place your hand near your belly button and force as much air out of your lungs as possible.

Question 5

a. What happened to your abdominal muscles when you exhaled?

b. What happened to your intercostal muscles when you exhaled?

LUNG CAPACITY

Your lungs hold several liters of air. This volume of air has several subvolumes (fig. 43.4).

- **Tidal volume** (TV) is the volume of air inhaled or exhaled during a single breath and is the amount of air necessary to maintain the oxygen supply to your tissues. A typical tidal volume is about 500 mL.
- **Expiratory reserve volume** (ERV) is the amount of air that can be exhaled after a normal, quiet exhalation. Typical expiratory reserve volumes range from 800 to 1300 mL.
- **Inspiratory reserve volume** (IRV) is the amount of air that can be inhaled after a person takes a normal breath. Typical inspiratory residual volumes range from 2500 to 3500 mL.
- **Residual volume** (RV) is the air that cannot be exhaled from the lungs. A typical residual volume is about 1200 mL.
- The total of tidal volume plus inspiratory and expiratory reserve volumes is the lung's **vital capacity,** the maximum amount of air that can be inhaled after maximum exhalation. Vital capacity is sometimes used to indicate pulmonary function. A significant decrease in vital capacity is often associated with emphysema, pneumonia, and other lung diseases.

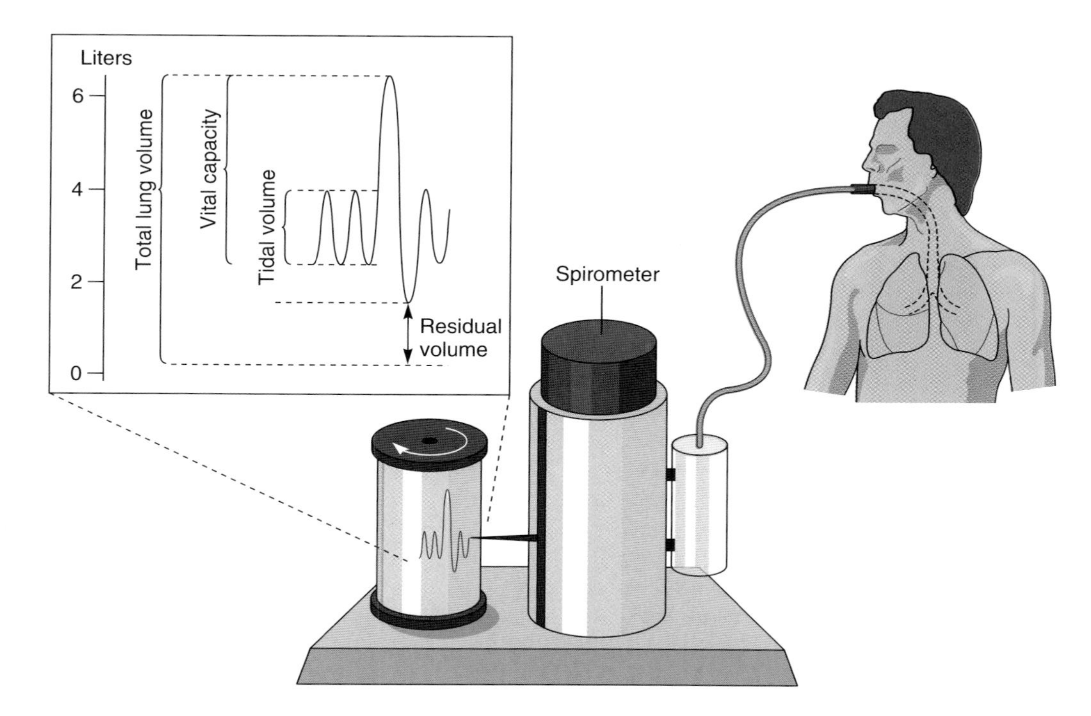

Figure 43.5
Measurement of lung volumes by spirometry.

Measuring Lung Capacity

In this exercise you'll use a spirometer to measure several features of your breathing (fig. 43.5). In the following procedures, record all of your results in table 43.1.

Procedure 43.2

Measure your tidal volume (TV)

1. Set the dial of the spirometer at zero. Place a sterile mouthpiece over the stem of the spirometer. Insert the mouthpiece into your mouth with the spirometer's dial facing upward.
2. Inhale through your nose and exhale through your mouth for five normal breathing cycles.
3. Observe the dial reading. Divide the reading by five to determine your tidal volume for one breath.
4. Record this volume as the first of three tidal volume measurements in table 43.1.
5. Repeat steps 2–3 twice and record these values in table 43.1.
6. Calculate the average of the resulting three values and record this average value in table 43.1.

Procedure 43.3

Measure your expiratory reserve volume (ERV)

1. Reset the dial of the spirometer to zero.
2. Place your mouth on the tube. Inhale normally through your nose, then exhale maximally into the mouthpiece. Observe the reading for this maximal exhalation volume.
3. From this maximal exhalation volume subtract your average tidal volume (determined in procedure 43.2).
4. Record this value in table 43.1 as the first of three ERV values.
5. Repeat steps 2–4 twice and record these ERV values in table 43.1.
6. Calculate your average ERV and record it in table 43.1.

Procedure 43.4

Measure your vital capacity (VC)

1. Reset the dial of the spirometer to zero.
2. Take several deep breaths and exhale completely after each. Then take as deep a breath as possible and exhale slowly and evenly through the spirometer.
3. Record this value in table 43.1 as the first of three VC values.
4. Repeat steps 2–3 twice, being sure to reset the dial after each exhalation. Record these volumes in table 43.1.
5. Calculate your average VC and record it in table 43.1.

TABLE 43.1

MEASUREMENTS OF LUNG VOLUMES

Tidal Volume (TV)	Expiratory Reserve Volume (ERV)	Vital Capacity (VC)	Inspiratory Reserve Volume (IRV)
1st ________	1st ________	1st ________	
2nd ________	2nd ________	2nd ________	
3rd ________	3rd ________	3rd ________	
Average ________	Average ________	Average ________	Calculated value = ________

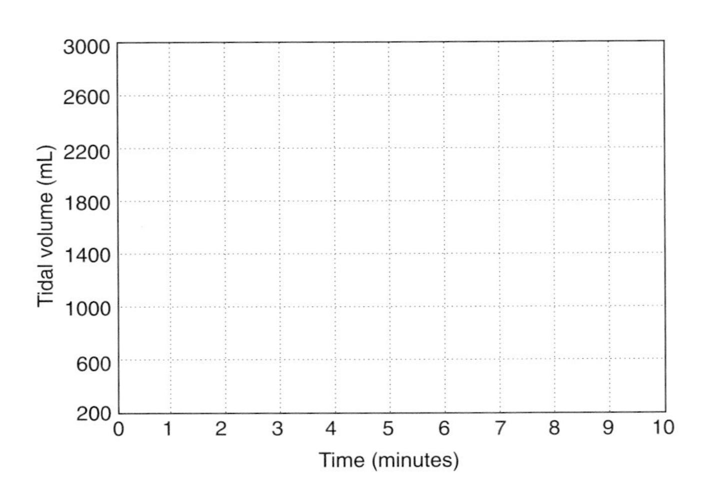

Figure 43.6
Change in tidal volume during recovery from exercise.

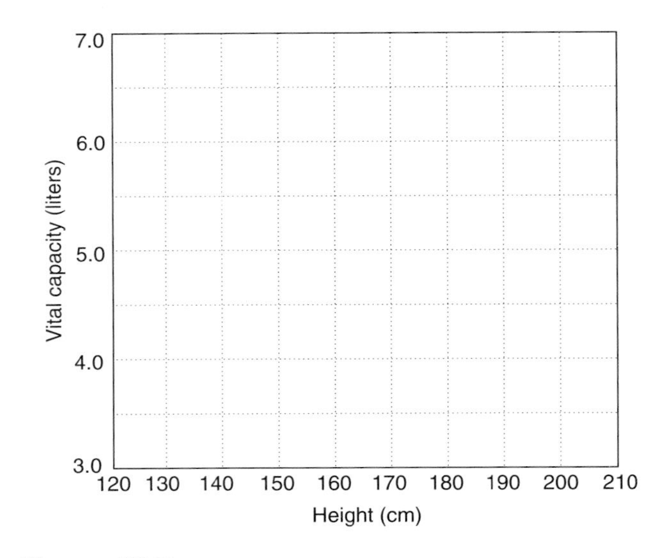

Figure 43.7
The relationship between lung vital capacity and a person's height.

Procedure 43.5

Measure your inspiratory reserve volume (IRV)

Your instrument is not set up to measure your inspiratory reserve volume directly. However, you can determine it by using the following formula:

Inspiratory reserve volume (mL) = **Vital capacity** − **Tidal volume** − **Expiratory reserve volume** = ___ **mL**

People's vital capacity varies with their age, sex, and height. However, a typical vital capacity for men is about 5200 mL, whereas that for women is about 4000 mL. Tidal volume is affected strongly by exercise.

Procedure 43.6

Measure the effect of exercise on your tidal volume

1. Exercise vigorously for 3 min by stepping onto and off of a small stool. Then measure your tidal volume according to procedure 43.2.

   Tidal volume = ________ **mL**

Question 6

a. How do these values compare with your "at rest" values (table 43.1)?

b. How do you explain this difference?

2. Stop exercising and continue to measure your tidal volume every 30 sec for 10 min. Graph your results (fig. 43.6).

Question 7
What is your recovery time?

3. When you've finished the experiment, place the disposable mouthpiece in the collecting bag.

BODY SIZE AND VITAL CAPACITY

A person's vital capacity usually varies with the size of their body. To show this, use a meterstick to measure your height: _________ cm

Write your height, gender, and vital capacity on the chalkboard in the lab. Graph all of the data for your class on figure 43.7. Your instructor may ask you to graph the data for males separately from that of females.

Question 8

a. What do you conclude about the relationship of height and vital capacity?

b. Is this what you would have predicted? Why or why not?

BREATHING RATE

Your rate of breathing is controlled by many factors, the most important of which is the concentration of carbon dioxide in your blood. Slowing your breathing rate slows the release of carbon dioxide into the lungs, thereby increasing the amount of carbon dioxide in the blood.

Procedure 43.7

Measure your breathing rate

1. Determine your breathing rate at rest.
2. Breathe deeply and rapidly for 12 breaths (i.e., hyperventilate).

Question 9
Does deep breathing get easier or harder as time goes by? Why?

INVESTIGATION

Which Muscles Do the Work?

a. Decide how you would determine the relative contributions of the diaphragm and intercostal muscles to your inspiratory reserve volume and/or tidal volume.

b. Form a hypothesis about the result you expect. Your instructor will advise you about how to write a testable hypothesis. Write your hypothesis here:

c. Decide how you will test your hypothesis. Describe your experimental design here:

d. Do your experiment. What do you conclude? Do your data support your hypothesis?

3. After hyperventilating, hold your breath for as long as you can. Record how long you held your breath: _________ seconds
4. Exhale, take a breath, and hold your breath again as long as you can. Again record how long you held your breath: _________ sec
5. Repeat this exercise three more times.

Question 10

a. What pattern do you see in the results?

b. How do you explain these data? That is, how is this response adaptive under these conditions?

6. Rest until your respiratory rate returns to normal. Then run vigorously in place for 3 min. Stop and hold your breath for as long as possible. Record your time: _________ sec

Question 11

a. How does this time compare with others you recorded? Explain your results.

b. What do you think causes the breathing rate to increase: the increase of CO_2 in the blood or the depletion of O_2? How could you test your answer?

Questions for Further Thought and Study

1. What is the clinical significance of vital capacity?

2. Can a collapsed lung be repaired? If so, how?

3. How does smoking affect the various aspects of lung capacity?

4. How would you measure the effects of exercise on vital capacity?

WRITING TO LEARN BIOLOGY

Who do you think would have a shorter recovery time after exercising, a well-conditioned athlete or an out-of-shape professor? Explain your answer.

44

Human Biology
Circulation and Blood Pressure

Objectives

By the end of this exercise you should be able to:
1. Describe the path of blood flow in a four-chambered heart.
2. Describe what causes the sounds made by a beating heart.
3. Describe the structure and function of red blood cells, white blood cells, capillaries, veins, and arteries.
4. Describe how exercise affects blood pressure and pulse rate.

The cells of multicellular organisms are linked by an elaborate circulatory system. In humans, this circulatory system is based on a fast-flowing river of blood that delivers materials, food, and oxygen to cells (see fig. 48.7). Our circulatory system also removes waste products such as carbon dioxide from cells. The circulatory system in humans and other vertebrates is a closed system consisting of a pumping heart, blood, and blood vessels.

In this exercise you'll examine the structure and function of your circulatory system. You'll also make some diagnostic measurements of your circulatory system, including your pulse rate and blood pressure.

HEART

Your heart is a muscular organ that beats about 100,000 times per day. It has four chambers: a left and right atrium, and a left and right ventricle. These chambers are separated by one-way valves that help control blood flow. Study the path of blood flow through the mammalian heart (see figs. 48.8, 48.9).

The right atrium pumps blood to the right ventricle, which then contracts and pumps blood to the lungs. Blood returning from the lungs enters the left atrium, which contracts and pushes blood into the left ventricle. Contraction of the left ventricle forces blood throughout the body, after which the blood completes the cycle by returning to the right atrium.

Examine a heart from a cow. Use your fingers to trace the path of blood flow.

Question 1

a. Which of the chambers has the thickest wall?

b. How does this relate to the function of that chamber?

Each beat of the heart produces a characteristic sound. The first sound is a low pitched "lub" made by the mitral and tricuspid valves closing when the ventricles start to contract. Soon thereafter you hear a "dub" made by the pulmonary and aortic valves closing after the ventricles have contracted. In both cases, blood falling back on the cuplike valves snaps these valves closed, much like wind snaps open a parachute. If any of the valves do not close completely, there is a turbulence in the heart that can be heard as a **heart murmur.** Heart murmurs often sound like sloshing liquid.

Obtain a stethoscope. To best hear the "dub" sound of your heart, press the stethoscope against the fifth or sixth rib, slightly left of center. To hear the "lub" sound, press the stethoscope against the second rib.

Question 2

a. About how fast is your heart beating?

b. Why can't you hear these sounds when you press the stethoscope against your neck or leg?

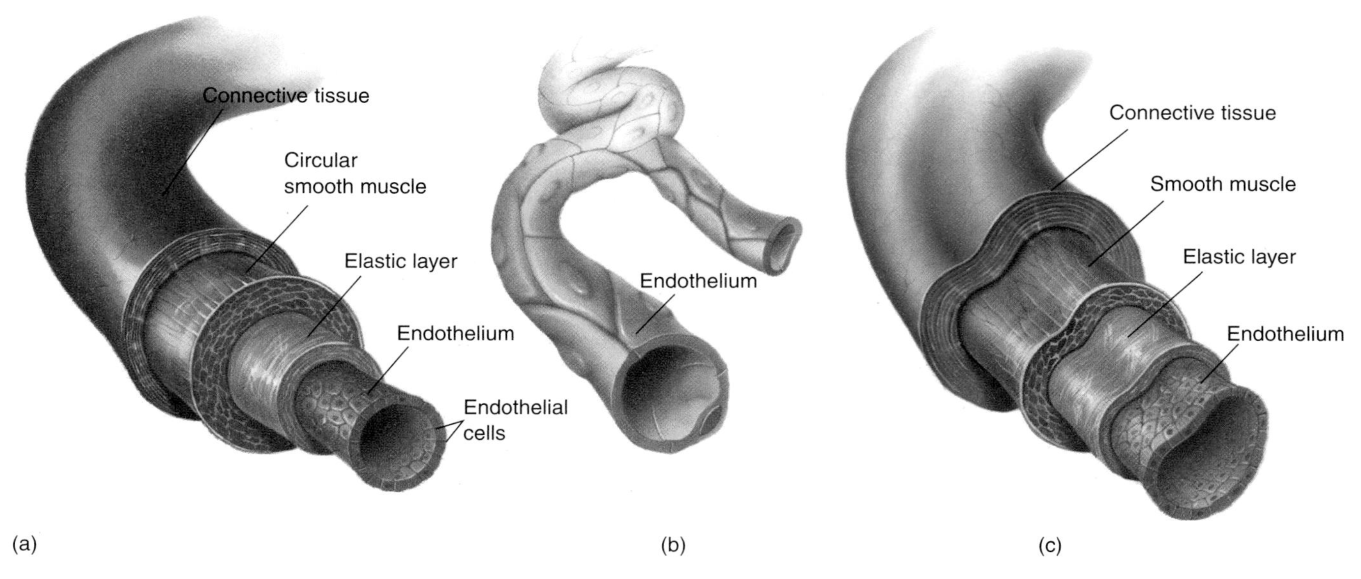

Figure 44.1

Structure of important blood vessels. (*a*) Artery. (*b*) Capillary. (*c*) Vein. Note that an artery has a thicker layer of smooth muscle than a vein.
From Peter H. Raven and George B. Johnson, *Understanding Biology*, 3d edition.

BLOOD

Blood is a type of connective tissue (see fig. 40.12). About 55% of blood is a straw-colored fluid called **plasma.** Suspended in plasma are cells, the most abundant of which are **red blood cells,** or **erythrocytes.** There are about 250,000 erythrocytes in a drop of blood.

Use the low, then high, magnification lenses of your microscope to examine a prepared slide of human blood. The pink cells lacking nuclei are red blood cells. The larger cells stained bluish-purple are **white blood cells,** or **leukocytes.**

Question 3

a. What's in plasma?

b. What is the function of plasma?

c. What is the shape of each kind of blood cell?

d. What is the function of each kind of blood cell?

e. What part of each cell is stained most intensely?

f. What is the approximate ratio of red blood cells to white blood cells in human blood?

g. What are some diseases of blood? What are the symptoms of these diseases?

Blood Vessels

Blood vessels include arteries, capillaries, and veins (fig. 44.1). **Arteries** carry blood away from the heart. Arteries consist of three concentric layers: an outer layer of connective tissue, a middle layer of smooth muscle, and an inner layer of epithelial cells. Examine a prepared slide of an artery.

Question 4

a. Which of the layers is thickest?

b. What does this tell you?

c. How would the deposition of plaque in an artery affect the function of the artery?

Veins carry blood to the heart. Veins have the same three layers that arteries have. Examine a prepared slide of a vein.

Question 5

a. Which of the layers is thickest?

b. How is this different from an artery?

c. What does this tell you about the functioning of veins?

Capillaries connect arteries and veins, and have a diameter slightly larger than that of a single red blood cell. Examine a prepared slide of an artery, vein, and capillary.

Question 6

a. Which has the largest diameter?

b. Which has the smallest diameter?

c. Which has the thinnest wall?

d. What does this tell you about where gas exchange occurs?

Blood Circulation in Goldfish

Use a net to catch a goldfish from the aquarium in the lab. Gently wrap the goldfish in cotton soaked in water from the aquarium. Let the tail protrude from the cotton. Place the fish in a petri dish and examine the tail with a dissecting microscope. Note the moving blood. Gently return the goldfish to the aquarium.

Question 7

a. Is the blood moving at a steady rate and in the same direction?

b. What does this tell you?

c. Are you looking at an artery, vein, or capillary?

d. How do you know?

HUMAN PHYSIOLOGY

CAUTION

Do not do the following exercises if you have heart or lung problems. Stop immediately if you feel faint.

Pulse

When your heart contracts it forces blood into arteries. This surge of blood stretches the artery coming from the heart. Surge after surge of blood from your beating heart produces waves of blood that pulse through your arteries. These pressure waves of blood are known as **pulse.** The pulse rate indicates the number of heart contractions per minute. Typical pulse rates usually range from 65 to 80 contractions per minute, but well-conditioned athletes may have rates as low as 40 contractions per minute.

Many arteries are well positioned for measuring your pulse (fig. 44.2).

Procedure 44.1

Measure the effect of exercise on pulse rate

1. Find your pulse by placing your second and third fingers on the thumb side of your inner wrist (this is where the radial artery passes into the hand). Press down slightly. Count your pulse for 15 sec: ________ beats in 15 sec
2. Multiply this number by 4 to convert this to beats per minute:
   Beats in 15 sec × 4 = ________ beats per min
3. Repeat this measurement three times and average your results. Record your average resting pulse rate: ________ beats per minute
4. Measure your pulse at your common carotid artery.

Question 8

How does this pulse rate compare with that measured at your wrist?

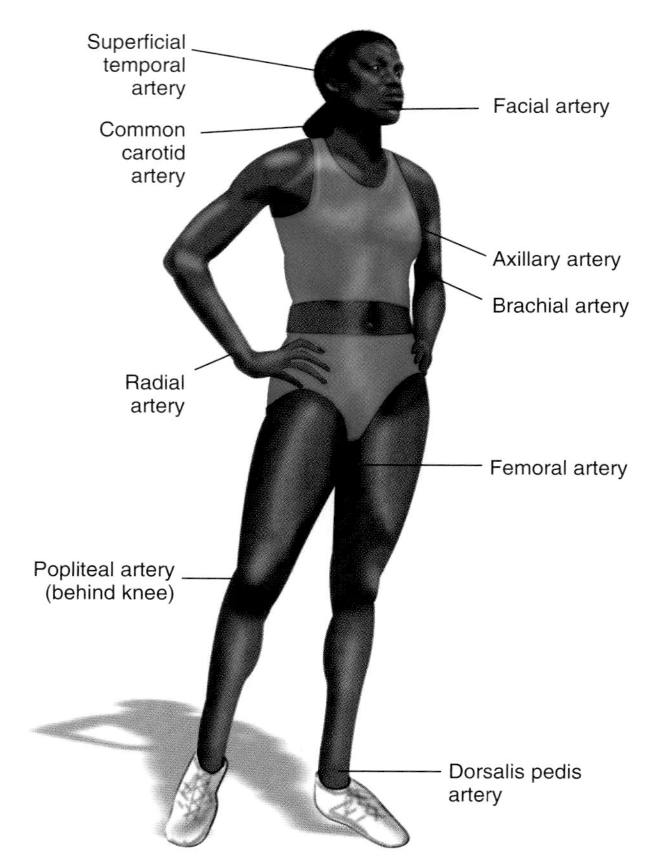

Figure 44.2
Many arteries in our bodies are well-positioned for measuring a pulse.

5. Hold your breath for 15 sec. Then measure your pulse for another 15 sec while still holding your breath.

Question 9

a. How does holding your breath affect your pulse rate?

b. How do you explain this?

CAUTION

Do not do the following exercises if you have heart or lung problems. Stop immediately if you feel faint.

6. Run vigorously in place for 5 min. Then sit down and immediately measure your pulse rate.

Question 10

a. How does exercising affect your pulse rate?

b. How do you explain this?

Blood Pressure

Blood pressure is the pressure exerted on the surface of blood vessels by blood. This pressure circulates blood through arteries, veins, and capillaries. The increased pressure that results from blood leaving the heart is the **systolic pressure.** When the heart relaxes, the arteries return to their original diameter and, in the process, squeeze blood forward through the cardiovascular system. This pressure when the heart relaxes is the **diastolic pressure.**

Question 11

a. Is blood pressure the same throughout the circulatory system?

b. How do you know?

Blood pressure, like barometric pressure, is measured in units called millimeters of mercury (mm Hg). This unit is based on a measuring device called a *manometer,* which is an inverted tube of liquid mercury. Pressure against the mercury reservoir at the base of the tube raises the column of mercury—the more pressure, the higher the column. Thus, the greater the number of millimeters of mercury, the greater the pressure. For example, a pressure of 100 mm Hg would raise a column of mercury 100 mm, whereas a pressure of 160 mm Hg would raise a column of mercury 160 mm.

Blood pressure is usually measured in the brachial artery just above the elbow (fig. 44.2). Systolic pressure there typically ranges from 100 to 140 mm Hg, with the average being near 120. Diastolic pressure typically ranges from 70 to 85 mm Hg, with an average of about 80. Blood pressure is reported as systolic pressure/diastolic pressure. Thus, a typical blood pressure is "120 over 80" (expressed as 120/80).

The difference between systolic and diastolic pressure is **pulse pressure**—this is what you feel in arteries when you touch your skin. Blood pressure is affected by many factors, including a person's intake of salt, the volume of blood, age, and the elasticity of blood vessels (fig. 44.3).

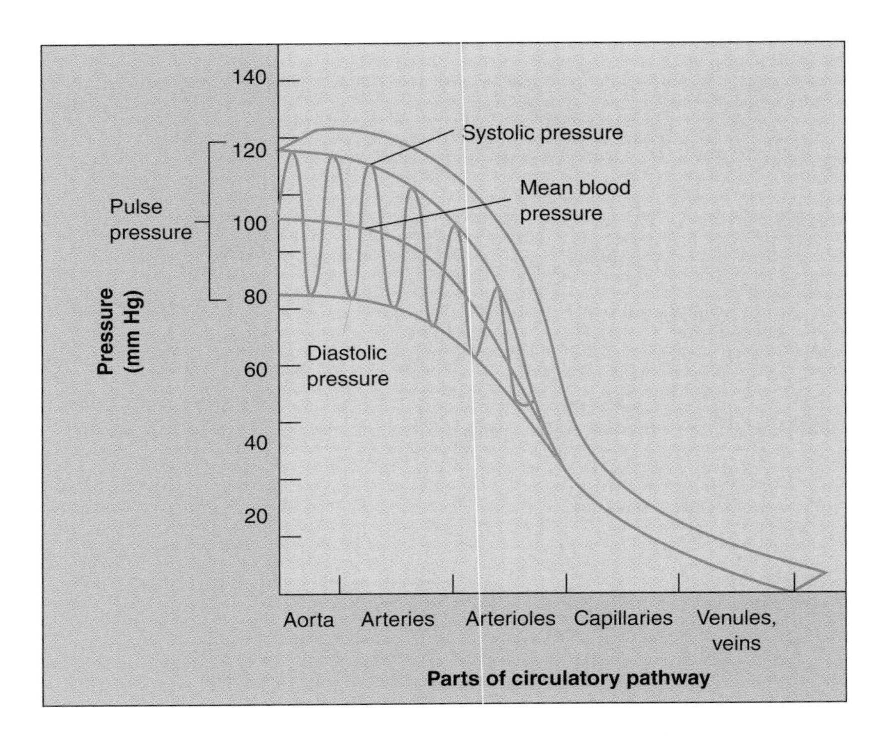

Figure 44.3
The changing pressures associated with the pulse wave in the human circulatory system.

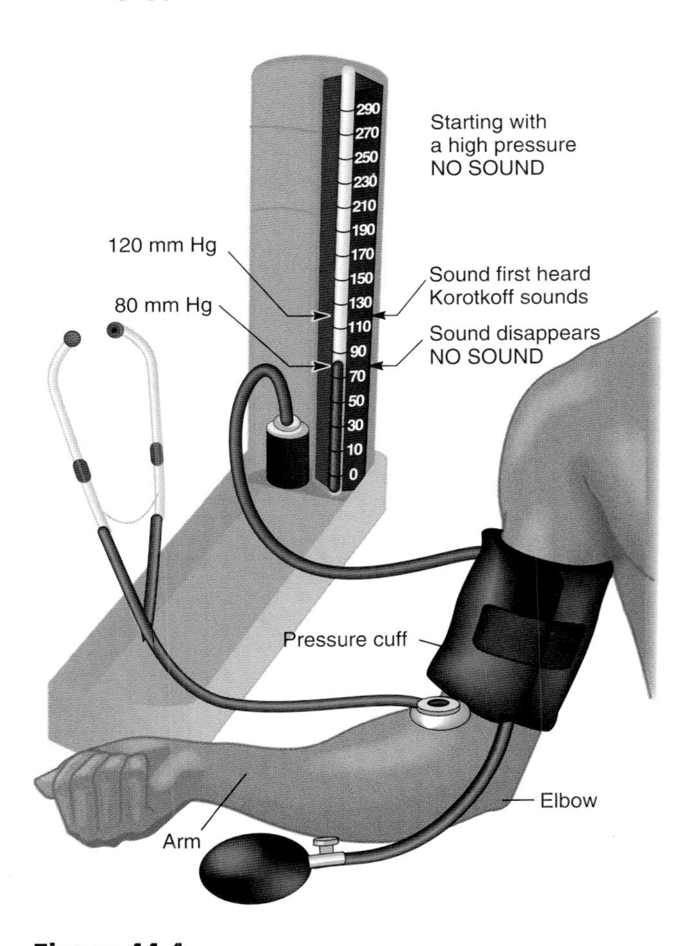

Figure 44.4
Measurement of blood pressure with a sphygmomanometer.

Blood pressure is measured with a stethoscope and a sphygmomanometer, which is a calibrated mercury manometer attached to a hollow, inflatable cuff. The principle is simple: You first tighten the cuff until blood flow through the artery stops. Then you use a stethoscope to determine when blood flow resumes through the artery (fig. 44.4).

Procedure 44.2

Measure the effect of exercise on blood pressure

1. Do all of the following experiments in pairs, alternately serving as subject and experimenter. Have your lab partner lie down and relax for 2 min. Attach the cuff around his/her arm above the elbow. Tuck the flap of the bag under the fold.
2. Inflate the cuff to about 200 mm Hg. Because this pressure exceeds the subject's systolic pressure, the brachial artery in the arm collapses, thus stopping blood flow through the artery. You'll feel no pulse in your partner's wrist when pressure in the cuff is 200 mm Hg.
3. Place the bell of the stethoscope under the cuff and over the brachial artery just above the elbow. Again inflate the cuff to a pressure of about 200 mm Hg. Slowly release pressure in the cuff. When the pressure falls below the systolic pressure, blood spurts through the artery. This flow of blood occurs quickly and produces vibrations and turbulence that can be heard with a stethoscope as loud, tapping sounds.

TABLE 44.1

THE EFFECT OF POSTURE ON BLOOD PRESSURE

Posture	Blood Pressure (mm Hg) (systolic pressure/diastolic pressure)	Pulse Rate (beats/sec)
Prone	______	______
Standing (10 sec after rising)	______	______
Standing (5 min after rising)	______	______
Standing (7 min after rising)	______	______
Standing (9 min after rising)	______	______

The pressure at which you hear these so-called Korotkoff sounds is the systolic pressure.

4. Continue to slowly release pressure from the cuff. As the pressure drops, the sounds become louder and more distinct as more blood flows through the artery. At this point, the flow of blood is continuous but still turbulent. When the cuff pressure reaches the diastolic pressure, blood flow is normal (i.e., nonturbulent) and the sounds disappear. The pressure at which the sound disappears is the diastolic pressure.
5. Repeat this pressure until you obtain consistent measurements. However, do not keep the cuff inflated around your partner's arm for more than a minute or so at a time.
6. Record the average blood pressure: ________ mm Hg
7. When you stand up, gravity causes arterial pressure to decrease in the upper parts of your body and increase in the lower parts of your body. Indeed, standing up from a prone position has an effect on brachial artery blood pressure that is equivalent to losing about 500 mL of blood. Measure your partner's pulse and blood pressure while he or she is reclined.
8. Have your partner stand up. Measure his or her pulse rate and blood pressure immediately and 5, 7, and 9 min after rising. Record your results in table 44.1.

Question 12

How did your partner's blood pressure change? Why?

9. Have your partner do bench-step exercises for 3 min at a rate of 60 steps per min. Have your partner lie down again and continue to measure your partner's blood pressure at 2-min intervals until it returns to normal. Plot these blood pressures on figure 44.5.

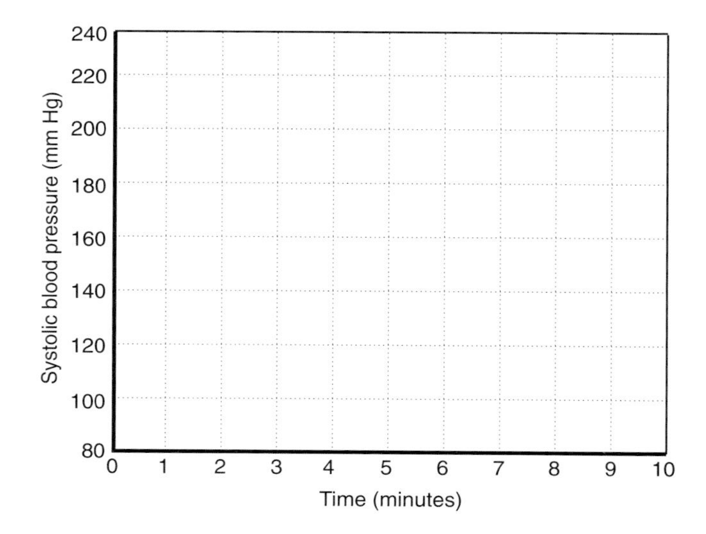

Figure 44.5

Graph of systolic blood pressure on the *y*-axis and time on the *x*-axis. The data you plot on this graph will describe the recovery of systolic blood pressure after exercise.

Question 13

a. How does exercise affect blood pressure?

b. Which is affected most, systolic pressure or diastolic pressure?

c. How do you explain this?

d. What was your partner's recovery time? That is, how long did it take for your partner's blood pressure to return to normal?

e. What other physiological changes occurred as your partner recovered?

10. Temperature also affects blood pressure. To show this, have your partner put his or her hand in cold (5°C) water. Then remeasure the blood pressure.

Question 14

a. How does cold affect blood pressure?

b. Are systolic and diastolic pressures affected similarly?

c. Compare your data with those of your classmates. What is the average blood pressure for your class?

Venous Blood Pressure

Blood moves slowly and at low pressure through capillaries. After exiting the capillaries, there is no mechanism (e.g., heart) to pump and increase the blood pressure so it remains low, and blood continues to move slowly through the veins. One-way valves prevent the blood under such low pressure from flowing backwards.

Procedure 44.3

Blood flow in veins

1. Hang your hands down at your side. Note the veins on the back sides of your hands.
2. Raise your hands above your head.

Question 15

What happens to the veins? Why?

3. Have your partner hold his or her hand out to heart-level and next to a meterstick taped to the wall. Slowly raise your hand; the height where the veins disappear is the venous pressure measured in centimeters (cm) of water. Convert this measurement to mm Hg using the following formula:

mm Hg = cm water × 0.73 = ________

Question 16

a. How does blood pressure in veins compare with that in the brachial artery?

b. Why would you expect such a difference?

Procedure 44.4

Locate valves in veins

1. Compress the vessels near your right elbow until the veins stand out.
2. Lay the index finger of your left hand on a vein near your wrist.
3. Move your thumb along and on top of the vein toward your elbow (i.e., toward your heart).
4. Lift your thumb and note what happens to blood in the vein. If blood refills all of the vein, repeat the experiment by placing your finger where the thumb reached. Continue until you reach a point at which the blood does not return toward the finger when the thumb is lifted.

Question 17

What blocks the backflow of blood in veins?

AN ANALYSIS OF YOUR RISK OF CARDIOVASCULAR DISEASE

A number of factors are suspected or proven to influence your risk of developing heart disease. Factors such as exercise, diet, family history, and smoking are clearly associated with the probability of suffering from cardiovascular disease later in life, even though no single factor is a guaranteed predictor. To assess your risk, complete the following questionnaire provided by the Arizona Heart Institute. To complete the questionnaire, simply record the number of points assigned to each level of each risk factor. Compare your total points with the ranges associated with high, medium, and low risk. Although this is not a definitive test, it will heighten your self-awareness of the consequences of your lifestyle.

Arizona Heart Institute's Heart Test for Men*

Age

- 51 and over 10
- 35–50 6
- 34 and under 1

Family History

If you have parents, brothers, or sisters who have had a heart attack, stroke, or heart bypass surgery at:

- Age 55 or before 5
- Age 56 or after 3
- None or don't know 0

Personal History

Have you had:

- A heart attack 20
- Angina, heart bypass surgery, angioplasty, stroke or blood vessel surgery 10
- None of the above 0

Smoking

Current smoker: How many cigarettes per day?

- 5 or more 20
- 4 or fewer 10

or

Previous smoker who quit less than 2 years ago: How many cigarettes per day did you smoke?

- 5 or more 10
- 4 or fewer 5

or

Never smoked or quit more than 2 years ago 0

Blood Pressure

If you have had your blood pressure taken in the last year, was it:

- Elevated or high (either or both readings above 160/95 mm Hg) 6
- Borderline (between 140/90 and 160/95 mm Hg) 3
- Normal (below 140/90 mm Hg) or don't know 0

Exercise

Do you engage in any aerobic activity, such as brisk walking, jogging, bicycling, or swimming for more than 20 minutes:

- Less than once a week 6
- 1 or 2 times a week 3
- 3 or more times a week 0

Diabetes

If you have diabetes (blood sugar level above 140 mg/dL), your age when you found out:

- 40 or before 3
- 41 or older 2
- Do not have diabetes 0

Blood Fats

If you have had your cholesterol and blood fat levels checked in the last year, score your risk here:

- Over 240 mg/dL 6
- 200–240 mg/dL 3
- Cholesterol under 200 mg/dL 0
- If your HDLs are lower than 35 . . . . add 1

or

If you know your cholesterol to HDL ratio, use this section to score your risk:

- 7.1 and above 6
- 3.6–7.0 3
- 3.5 or below 0

or

If you do not know your blood fat levels, use this section to score your risk: Which of the following best describes your eating pattern?

- High fat: red meat, "fast" foods, and/or fried foods daily; more than 7 eggs per week; regular consumption of butter, whole milk, and cheese 6

*© *Arizona Heart Institute, Phoenix, AZ. Reprinted by permission.*

- Moderate fat: red meat, "fast" foods, and/or fried foods 4–6 times per week; 4–7 eggs weekly; regular use of margarine, vegetable oils, and/or low-fat dairy products 3
- Low fat: poultry, fish, and little or no red meat, "fast" foods, fried foods, or saturated fats; fewer than 3 eggs per week; minimal margarine and vegetable oils; primarily nonfat dairy products . . . . 0 ☐

Use a score from only one section above.

Body Mass

Calculate your body mass index (BMI) as follows:

Weight (pounds): ________ × 0.45 = ________ (W)
Height (inches): ________ × 0.025 = ________ (H)

The BMI equals W divided by the square of H. That is,

$$BMI = W \div (H \times H)$$

Example: a man is 170 pounds and 5 feet 10 inches (70 inches) tall:

$$W = 170 \times 0.45 = 76.5 \quad H = 70 \times 0.025 = 1.75$$
$$BMI = W \div (H \times H) = 76.5 \div (1.75 \times 1.75)$$
$$= 76.5 \div 3.06$$
$$= 25.0$$

- If your BMI is 27 or greater 2
- If your BMI is below 27 . . . . 0 ☐

Now measure your waist and hips and divide your waist measurement by your hip girth:
Example: your waist is 30 and your hips are 34:
$30 \div 34 = 0.88$

(waist) ________ ÷ **(hips)** ________ = ________

- If your waist-to-hip ratio is 0.96 or greater 1
- If your ratio is 0.95 or less . . . . 0 ☐

Stress

Are you easily angered and frustrated:

- Most of the time 6
- Some of the time 3
- Rarely. . . . . 0 ☐

Total Score ☐

What Your Risk Factor Score Means

15 Points or Below: Low Risk

Congratulations! Maintain your heart-healthy status by watching your weight, blood pressure, and blood fat (cholesterol and HDL) levels; get regular check-ups and don't smoke. Retake this test every year to monitor your heart-health risk profile.

16–32 Points: Medium Risk

Our experience indicates that your medium risk level warrants attention. Personal factors or lifestyle habits may be increasing your vulnerability to heart disease. We strongly recommend you schedule an appointment with your doctor for an evaluation, and take this test with you to get advice on how you can improve your heart-health status.

33 Points or Above: High Risk

Your potential for experiencing a heart attack or stroke is significant. You must take action NOW. If you are not already being treated for heart disease, we urgently advise that you see your doctor immediately and take this test with you. You must seek ways to reduce your risk!

Arizona Heart Institute's Heart Test for Women*

Age

- 51 and over 5
- 35–50. 2
- 34 and under . . . . 0 ☐

Family History

If you have parents, brothers, or sisters who have had a heart attack, stroke, or heart bypass surgery at:

- Age 55 or before 5
- Age 56 or after. 3
- None or don't know. . . . . 0 ☐

Personal History

Have you had:

- A heart attack 20
- Angina, heart bypass surgery, angioplasty, stroke, or blood vessel surgery 10
- None of the above . . . . 0 ☐

*© *Arizona Heart Institute, Phoenix, AZ. Reprinted by permission.*

Smoking

Current smoker: How many cigarettes per day?

- 5 or more 20
- 4 or fewer 10

If you are a smoker currently taking oral contraceptives and are:

- Under 35 years old . . . . add 2
- 35 years old and over . . . . add 5

or

Previous smoker who quit less than 2 years ago: How many cigarettes per day did you smoke?

- 5 or more 10
- 4 or fewer 5

or

Never smoked or quit more than 2 years ago 0

Blood Pressure

If you have had your blood pressure taken in the last year, was it:

- Elevated or high (either or both readings above 160/95 mm Hg) 6
- Borderline (between 140/90 and 160/95 mm Hg) 3
- Normal (below 140/90 mm Hg) or don't know 0

Hormone Status

If you have undergone natural menopause, your age at its start:

- 41 or older 1
- 40 or younger 2

If you have had a total hysterectomy, your age when it was done:

- 41 or older 1
- 40 or younger 3

If you take an oral estrogen supplement . . . . subtract 2
If you are still menstruating . . . . subtract 1

Exercise

Do you engage in any aerobic activity, such as brisk walking, jogging, bicycling or swimming for more than 20 minutes:

- Less than once a week 6
- 1 or 2 times a week 3
- 3 or more times a week 0

Blood Fats

If you have had your cholesterol and blood fat levels checked in the last year, score your risk here:

- Over 240 mg/dL 6
- 200–240 mg/dL 3
- Cholesterol under 200 mg/dL 0
- If your HDLs are lower than 45 . . . . add 3

or

If you know your cholesterol to HDL ratio, use this section to score your risk:

- 7.1 and above 6
- 3.6–7.0 3
- 3.5 or below 0

or

If you do not know your blood fat levels, use this section to score your risk: Which of the following best describes your eating pattern?

- High fat: red meat, "fast" foods, and/or fried foods daily; more than 7 eggs per week; regular consumption of butter, whole milk, and cheese 6
- Moderate fat: red meat, "fast" foods, and/or fried foods 4–6 times per week; 4–7 eggs weekly; regular use of margarine, vegetable oils, and/or low-fat dairy products 3
- Low fat: poultry, fish, and little or no red meat, "fast" foods, fried foods, or saturated fats; fewer than 3 eggs per week; minimal margarine and vegetable oils; primarily nonfat dairy products 0

Use a score from only one section above.

Diabetes

If you have diabetes (blood sugar level above 140 mg/dL), your age when you found out:

- 40 or before 6
- 41 or older 4
- Do not have diabetes. 0

☐

Body Mass

Calculate your body mass index (BMI) as follows:

Weight (pounds): ________ × 0.45 = ________ (W)
Height (inches): ________ × 0.025 = ________ (H)

The BMI equals W divided by the square of H. That is,

$$BMI = W \div (H \times H)$$

Example: A woman is 120 pounds and 5 feet 6 inches (66 inches) tall:

$$W = 120 \times 0.45 = 54 \quad H = 66 \times 0.025 = 1.65$$

$$BMI = W \div (H \times H) = 54 \div (1.65 \times 1.65) = 54 \div 2.72 = 19.8$$

- If your BMI is 27 or greater 2
- If your BMI is below 27 0

☐

Now measure your waist and hips and divide your waist measurement by your hip girth:
Example: Your waist is 26 and your hips are 36: 26 ÷ 36 = 0.7

(waist) ________ ÷ **(hips)** ________ = ________

- If your waist-to-hip ratio is 0.86 or greater 2
- If your ratio is 0.8–0.85 1
- If your ratio is 0.79 or less 0

☐

Stress

Are you easily angered and frustrated:

- Most of the time 6
- Some of the time 3
- Rarely. 0

☐

Total Score

☐

What Your Risk Factor Score Means

15 Points or Below: Low Risk

Congratulations! Maintain your heart-healthy status by watching your weight, blood pressure, and blood fat (cholesterol and HDL) levels; get regular check-ups and don't smoke. Retake this test every year to monitor your heart-health risk profile.

16–32 Points: Medium Risk

Our experience indicates that your medium risk level warrants attention. Personal factors or lifestyle habits may be increasing your vulnerability to heart disease. We strongly recommend you schedule an appointment with your doctor for an evaluation, and take this test with you to get advice on how you can improve your heart-health status.

33 Points or Above: High Risk

Your potential for experiencing a heart attack or stroke is significant. You must take action NOW. If you are not already being treated for heart disease, we urgently advise that you see your doctor immediately and take this test with you. You must seek ways to reduce your risk!

Questions for Further Thought and Study

1. What is the value of increased blood pressure and pulse rate during exercise?

2. A well-conditioned athlete shows fewer changes in his or her circulation and breathing in response to exercise than does someone who is in poor condition. Why?

3. How would "hardening of the arteries" (arteriosclerosis) affect blood pressure? Why?

4. What is congestive heart failure?

5. Why is high blood pressure dangerous?

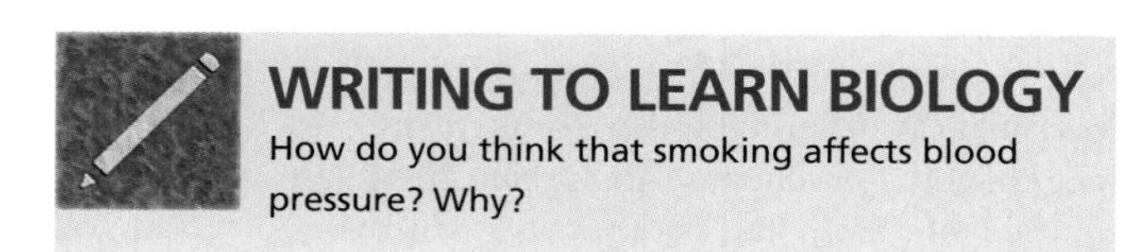

45

Human Biology
Sensory Perception

Objectives

By the end of this exercise you should be able to:

1. Identify the location and basis for your blind spot.
2. Describe afterimages, the distribution of touch and taste receptors, your type of eye dominance, adaptation to stimuli, and the effect of stimulus intensity on perception.
3. Measure your visual acuity, near point, astigmatism, and peripheral vision.
4. Distinguish between nerve deafness and conduction deafness.

Like most animals, you have a complex nervous system that informs your brain of your body's condition and provides the brain with detailed information about the environment. The nervous system also coordinates movements and perceives, translates, and responds to environmental stimuli such as light, touch, and temperature. A deficiency in any part of your sensory systems would be dangerous because you could no longer respond to your ever-changing external and internal environment.

The purpose of this exercise is to increase your awareness of the senses that you use daily. Do all of these experiments with a partner and alternate with him or her as subject and experimenter.

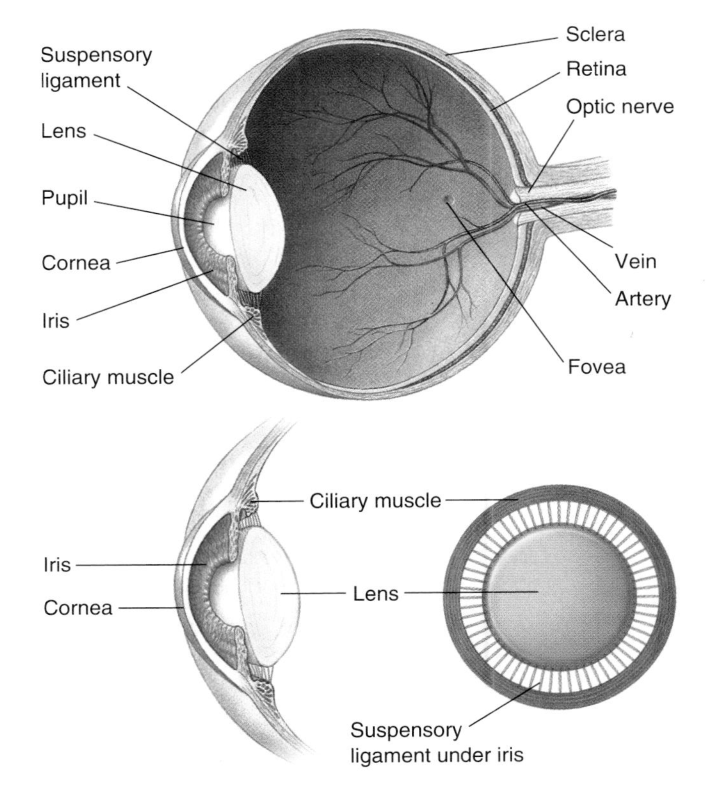

Figure 45.1

Structure of the human eye. The transparent cornea and lens focus light onto the retina at the back of the eye, which contains the rods and cones. The center of each eye's visual field is focused on the fovea. Focusing is accomplished by contraction and relaxation of the ciliary muscle, which adjusts the curvature of the lens.

BLIND SPOT

The **retina** is a layer of photoreceptors on the back inner surface of the eye (fig. 45.1). Most of this surface is covered by photoreceptor cells called rods and cones, which are modified epithelial cells (fig. 45.2). The central fovea of the retina is the region used for color vision and the region giving greatest visual acuity (sharpness of image). Other parts of the retina are important for peripheral vision, but images focused there are not in sharp focus.

The **optic disc** is the region of the retina where blood vessels and the optic nerves enter or leave the retina. This region lacks photoreceptors and is therefore a "blind spot" of the retina. We usually don't notice the blind spot because our brain fills in the blank area for us. However, it's still there.

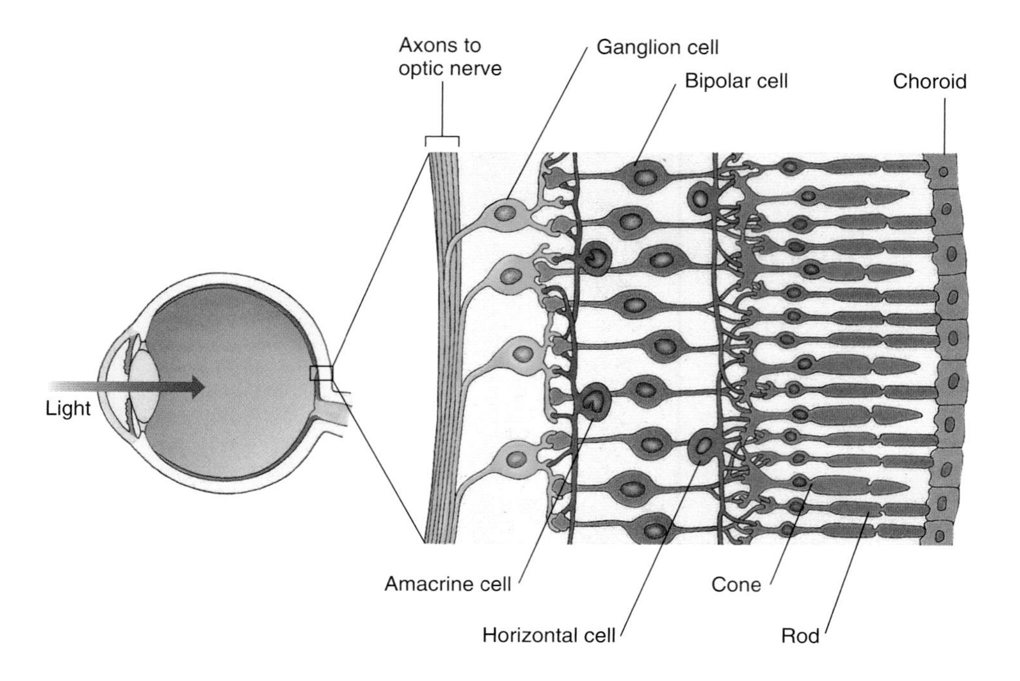

Figure 45.2
Structure of the retina. Note that the rods and cones are at the rear of the retina, not the front. Light passes through several layers of ganglion and bipolar cells before it reaches the rods and cones.

Procedure 45.1

Discover your blind spot

1. Cover your left eye. Hold figure 45.3 about 50 cm (20 in) from your face and directly in front of your right eye.
2. Stare at the cross in figure 45.3. You can also see the circle.
3. Continue to stare at the cross as you slowly bring the figure closer to your eye.
4. At one point the circle will seem to disappear because its image has fallen on your blind spot. Have your partner record that distance.
5. Continue to move the figure closer to your face.

Question 1

a. Does the circle reappear?

b. What does this mean?

6. Test your other eye in a similar manner, but focus on the circle and watch for the cross to disappear.

Question 2

Are the blind spots of your right and left eyes at the same distance?

The fovea is sometimes invaded by blood vessels. When this occurs, the fovea also becomes a blind spot. This condition characterizes a disease called macular degeneration, the most common cause of legal blindness in people older than 65.

EYE DOMINANCE

We're all familiar with preferences for using a particular hand for jobs such as writing and throwing. Our eyes also exhibit right-left dominance.

Procedure 45.2

Determine eye dominance

1. With both eyes open, carefully focus on an object a few feet away.
2. Close one eye, then reopen it.
3. Close the other eye, then reopen it.

Which eye seems more directly in line with the object? If it's the right eye, you are right-eye dominant; if it's the left eye, you are left-eye dominant. If the object is in the middle of both eyes, you are central-eye dominant.

Figure 45.3
Images for detecting a blind spot.

Question 3

a. Do you have right, left, or central dominance?

b. When you look at a distant object, why would you move your head and eyes to an off-center position when one eye is dominant?

Eye dominance is important for how we see and react to our world. For example, right-handed hitters in baseball have their right hand in the upper control position when they hit. Similarly, 65% of baseball players are right-eye and right-hand dominant. Only about 17% are crossed dominant (right hand–left eye or left hand–right eye), whereas another 18% have no eye dominance. These players see the world from a point halfway between both eyes. Interestingly, the best hitters (as judged by batting average) are either crossed dominant or lack dominance. The best pitchers (judged by their earned-run averages) have central-eye dominance.

Question 4

How could eye dominance affect one's ability to hit a baseball?

NEAR POINT

The shortest distance at which an object is in sharp focus is called the **near point:** the closer the distance, the greater your eye's ability to accommodate for changes in distance. This distance gradually increases as we get older. By age 60, this distance is very large, a condition called presbyopia (table 45.1).

Procedure 45.3

Determine your near point

Here's how to determine your near point. If you wear glasses or contacts, keep them on.

1. Hold this page in front of you at arm's length. Close one eye and focus on a word on this page.

TABLE 45.1

TYPICAL VALUES FOR DISTANCE TO NEAR POINT BY AGE GROUPS

Age	Distance to Near Point (cm)
10	9
20	10
30	13
40	18
50	53
60	83
70	100

2. Slowly move the page toward your face until the image of the word is blurred. Then move the page away until the image is sharp. Have your partner record the distance between your eye and the page. That distance is your near point.

Question 5

How does our near point affect what we do and see?

AFTERIMAGES

Images sensed by your eyes do not disappear immediately after you close your eyes or shift your glance. Rather, an image lingers as an afterimage.

Procedure 45.4

Demonstrate afterimage

1. Place a large sheet of white paper and another sheet of black paper on your laboratory bench a short distance apart.
2. Place a bright blue index card on the black paper. Stare at the card intently for 30 sec.
3. Quickly shift your gaze to the white paper. Record your observations.
4. Repeat this procedure with a yellow card and record your observations.

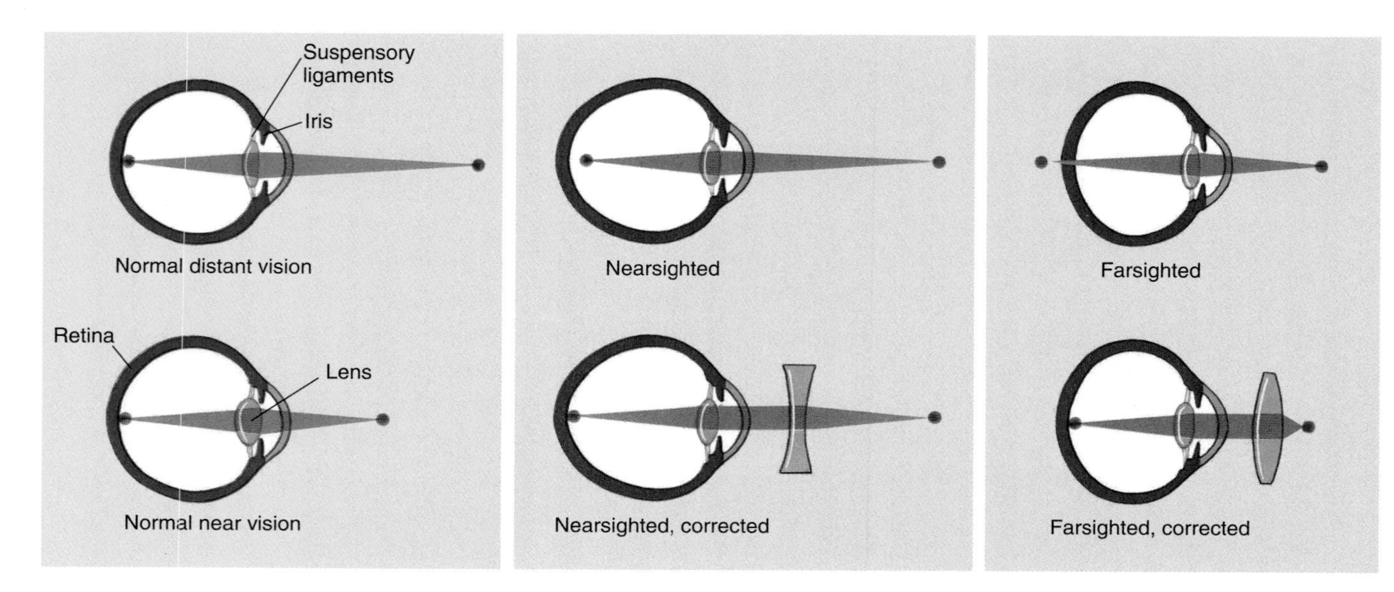

Figure 45.4

Focusing the human eye. (*a*) In people with normal vision, the image remains focused on the retina in both near and far vision because of changes produced in the curvature of the lens. When a person with normal vision stands 20 ft or more from an object, the lens is in its least convex form and the image is focused on the retina. (*b*) In nearsighted people, the image comes to a focus in front of the retina and the image thus appears blurred. (*c*) In farsighted people, the focus of the image would be behind the retina because the distance from the lens to the retina is too short.

5. Repeat the experiment, this time placing the card initially on the white paper and shifting your stare to the black paper. Record and provide an explanation for your results below.

VISUAL ACUITY

Visual acuity is the sharpness of a visual image and is usually measured with a Snellen eye chart. The size of letters on the chart are such that you should be able to read the first line (i.e., the letter E) of the chart from 200 ft away. Conversely, line 8 of the chart is tall enough so that it should be read from 20 ft away. Thus, if a person can read line 8 from 20 ft away, his or her visual acuity is designated as 20/20, or 1. If you can see line 8 from 10 feet away but not farther, your acuity is 10/20, or 1/2 of normal. Snellen charts typically have letter sizes for assessing acuity of 20/15, which is better than normal acuity, to 20/200, which is very poor acuity.

The eyes of farsighted people focus the image behind their retina. These people are referred to as farsighted because they see distant objects clearer than close objects. Conversely, the eyes of nearsighted people focus the image in front of the retina; these people see close objects better than they do distant objects. Both conditions can be corrected with glasses or contact lenses provided that the fovea remains functional (fig. 45.4).

Procedure 45.5

Determine visual acuity

1. Stand 20 ft (6.1 m) from a Snellen eye chart, facing the chart.
2. Cover one eye and read the letters that your partner points to on the chart. Begin at the top of the chart and work your way down.
3. Note the lowest row of letters that you can read accurately. Record the number printed next to that row: ________. That number is the farthest distance (measured in ft) that a person with normal vision can read the letters in that row. For example, if the number is 40, then the person has 20/40 vision, meaning that the person can see at 20 ft what a person with normal vision can see at 40 ft.
4. Test both eyes. If you wear glasses or contacts, test your eyes with and without your lenses.

Question 6

What is your visual acuity?

Right eye: ________

Left eye: ________

ASTIGMATISM

Astigmatism usually results from abnormal curvature of the cornea. Test for astigmatism by facing an astigmatism test chart 10 ft (3 m) away. Cover one eye and stare at the circle in the center of the chart. Test both eyes. If all lines radiating from the circle are straight, no astigmatism is present. If the lines appear crooked or wavy, you probably have astigmatism.

PERIPHERAL VISION

Dim light is best perceived by rods, whereas bright light and colors are best perceived by cones. Most cones are in an area called the central fovea, which is a small part of the retina we use when we focus directly on an object and look at its color and detail. In this area of the retina, almost all of the receptors are cones. Rods dominate the periphery of the visual field. This demonstration will show the monochrome nature of peripheral vision.

Procedure 45.6

Test peripheral vision

1. Have your lab partner stare forward.
2. Slowly bring a piece of colored paper into your partner's visual field from behind his or her head.
3. Stop when your partner tells you that the paper has just entered his or her visual field.
4. Ask what color the paper is. If you did the test properly and if your partner didn't cheat, he or she will probably be unable to determine the color of the paper. Why?

COLOR BLINDNESS

Color blindness is a color vision deficiency that usually is inherited. The most common type of color blindness is red-green color blindness that results from a deficiency of red- and green-sensitive cones. People with this deficiency have difficulty distinguishing shades of red and green. A totally color-blind person sees everything as a shade of gray.

Test for color blindness with the Ichikawa Test Book available in the lab. Hold the plates about 60 cm from your partner's face in bright, natural light. Give your partner about 5 sec to view each plate. Without touching the figure, ask your partner to name the number in each mosaic.

HEARING LOSS

There are two kinds of hearing loss: nerve deafness and conduction deafness. Nerve deafness results from damage to the sound receptors or neurons that transmit impulses to the brain (fig. 45.5). Such damage is usually caused by exposure to loud sound and is not correctable. Conduction deafness results from damage that prevents sound vibrations from reaching the inner ear. Conduction deafness is usually correctable by surgery or hearing aids.

You can use a tuning fork to distinguish between nerve and conduction deafness.

Procedure 45.7

Testing for hearing loss

1. Have your lab partner sit down and plug one ear with cotton.
2. Strike the tuning fork against the heel of your hand (do not strike the fork against a hard object).
3. Hold the fork about 25 cm (10 in) away from your partner's ear, with the edge of the fork pointing toward the ear (fig. 45.6). If your partner can hear the fork, he or she has normal hearing or only minimal hearing loss.
4. As the sound fades, have your partner tell you when he or she can no longer hear the fork. Then place the base of the fork against the temporal bone behind the ear. If a slight hearing loss exists and the sound reappears, some conduction deafness is present. You may want to simulate a conduction impairment by having your partner wear an earplug.

Someone with severe hearing loss will not hear the vibrating fork (or will hear it only briefly). That person suffers conduction deafness if the sound reappears when the end of the fork is placed against his or her temporal bone. If the sound does not reappear, the person is nerve deaf.

DISTRIBUTION OF TOUCH RECEPTORS

To perceive two stimuli as separate sensations, the stimuli must be far enough apart to stimulate two touch receptors. Thus, the distribution of touch receptors affects our ability to distinguish among differing numbers of stimuli.

Procedure 45.8

Test for distribution of touch receptors

1. Have your lab partner close his or her eyes.
2. Touch his or her skin with one or two points of a scissors spread far apart. Ask your partner to report the sensation as "one" or "two." Randomly alternate the two-point touches with one-point touches to keep your partner honest.

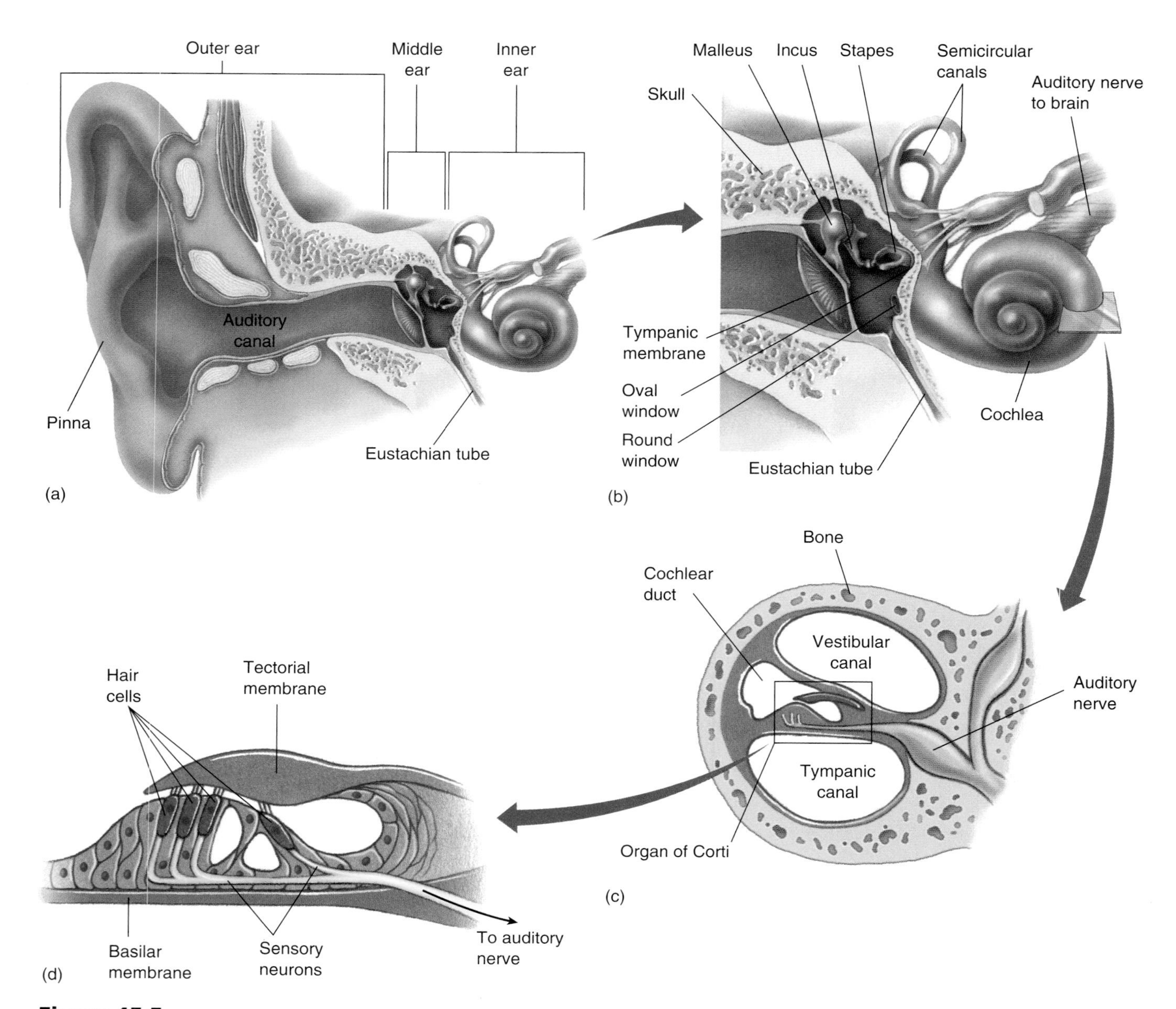

Figure 45.5

Structure of the human ear. (*a*) Sound waves passing through the ear canal produce vibrations of the tympanic membrane, which cause movement of the (*b*) middle ear ossicles (the malleus, incus, and stapes) against an inner membrane called the oval window. Vibration of the oval window sets up pressure waves that (*c* and *d*) travel through the fluid in the vestibular and tympanic canals of the cochlea.

3. Decrease the distance between the two points until your partner reports a one-point stimulus about 75% of the time. Measure this distance as the minimum distance evoking a two-point sensation.
4. Repeat this procedure on the following parts of your partner's body: (a) inside the forearm, (b) back of the neck, (c) palm of the hand, and (d) tip of the index finger.

Question 7

a. Which parts of your body are most sensitive to touch?

b. Which parts are least sensitive?

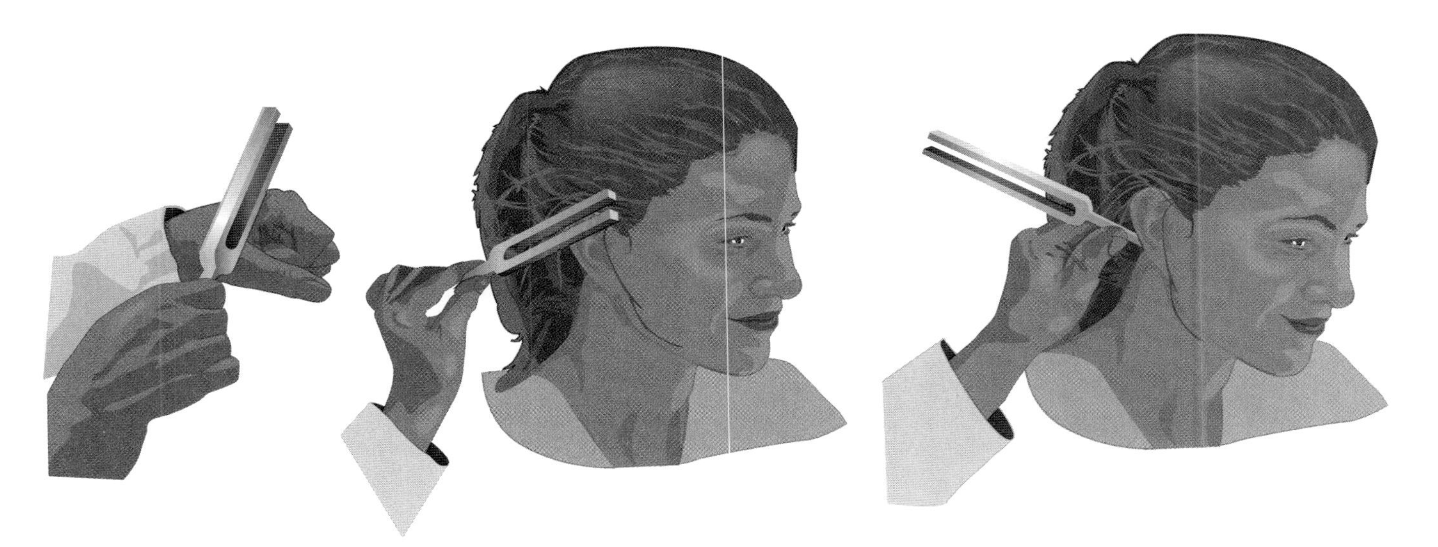

Figure 45.6
Using a tuning fork to test for hearing loss.

c. What is the significance of these differences in sensitivity?

d. How would someone's sensitivity to touch by their fingertips affect their ability to read Braille?

TASTE SENSATION

Taste is a poorly understood sense because our final perception of flavor arises from a combination of other sensory inputs. The texture, smell, and appearance of foods strongly influence our sense of taste. Strictly speaking, the sense of taste refers to sensations originating from taste cells in the mouth, and has traditionally been described as a combination of sensations of salty, bitter, sour, sweet, and possibly other sensations. To cause a taste sensation, food dissolves in saliva, moves through taste pores in taste buds (fig. 45.7), and interacts with neural receptor cells. When food interacts with receptor proteins or with ion channel proteins on the surface of these cells, an electrical impulse is generated and sent to the brain.

Biologists once thought that each of the four areas of the tongue were exclusively responsible for a particular sensation (sweet, sour, salty, bitter). However, current research shows that neurons in all areas of the tongue are widely responsive to a variety of tastes. The final perception of taste is an integration of signals from receptors and is more complex than was first thought. An integration of impulses from taste receptors combined with other senses (vision, touch, and smell) all combine to produce flavor.

Procedure 45.9

Test for taste sensitivity

1. Cut two identical small pieces of a raw onion, raw apple, and raw potato. Have your partner close his or her eyes, clamp the nostrils shut, and open his or her mouth.
2. Put a piece of onion, apple, or potato on your partner's tongue. Ask your partner to identify it without swallowing.
3. Repeat the procedure with the other food.
4. Repeat the procedure with the nostrils open.

ADAPTATION TO STIMULI

Your nervous system can "tune out" stimuli so that you are not constantly bombarded with trivial sensations.

Procedure 45.10

Investigate adaptation to stimuli

1. Have your partner rest his or her forearm on top of a lab bench with the palm facing up.
2. Have your partner close his or her eyes.
3. Place a coin on the inner surface of the forearm. Ask your partner to tell you when the sensation disappears. Record the time between these two events as the adaptation time.
4. Increase the intensity of the stimulus by stacking several coins or by using a heavier object.

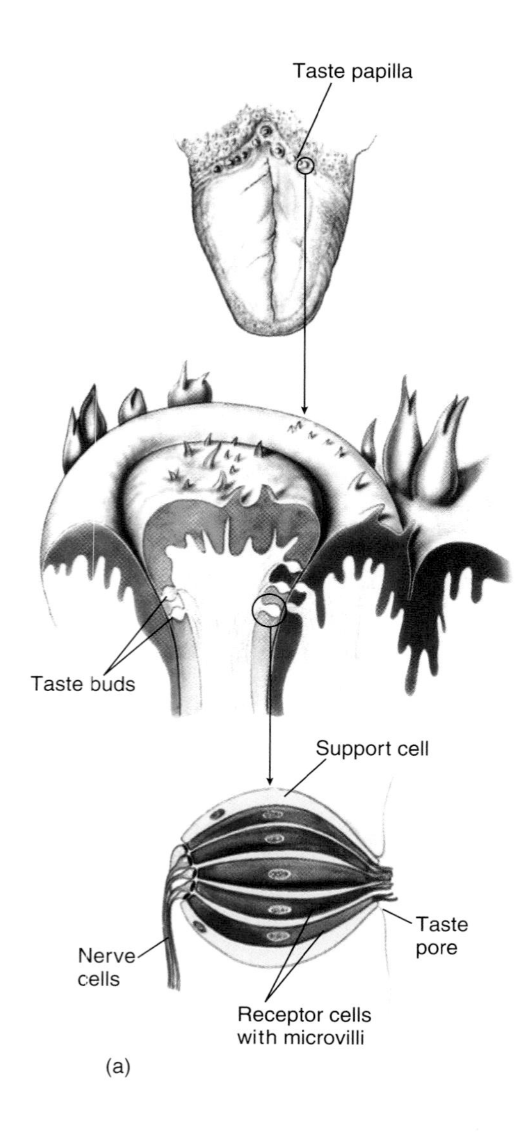

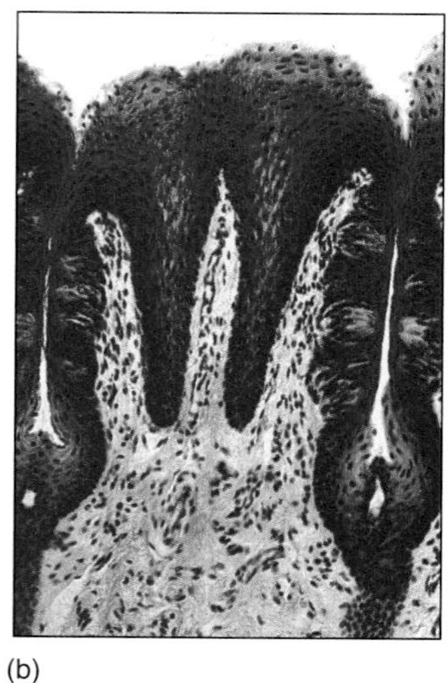

Figure 45.7

Taste. (*a*) Groups of taste buds are typically organized in sensory projections called papillae. (*b*) Individual taste buds are bulb-shaped collections of chemical receptor cells that open into the mouth through a pore.

Question 8
How does the intensity of a stimulus affect adaptation time?

INTENSITY OF SENSATIONS

The intensity of a sensation is usually proportional to the intensity of the stimulus.

Procedure 45.11
Investigate intensity of sensations

1. Fill large pans with ice water, warm water, and water at room temperature.
2. Sequentially place your index finger in the water in each pan. Can you sense the changes in temperature?
3. Repeat the experiment with pans containing water having only slight differences in temperature.

Question 9
What is the smallest difference in temperature that you can detect?

4. Place your left hand in the ice water and your right hand in the warm water (be sure that the warm water has not cooled). Leave your hands in these pans until you “get used to” the temperatures.

Question 10

a. Does the sensation change with time? Why or why not?

b. Would air at the same temperature feel the same? Why or why not?

c. In terms of receptors, what does it mean to “get used to” a sensation?

d. Would your perceptions have changed if the cold water was 0°C and the warm water was 80°C?

e. Why would it be important to be able to perceive the difference between 45°C and 80°C?

f. After at least 1 minute, place both hands in the pan of water at room temperature. Note the sensation. Does it change with time? Why or why not?

g. What do temperature receptors really respond to?

h. Were your receptors giving you useful information in both situations even though the information they gave you changed?

Questions for Further Thought and Study

1. Are organisms aware of all of the activity of their nervous system? Why is this an advantage to an organism?

2. How could you show whether there are specific receptors in your skin that respond to heat and cold?

3. Are all regions of your body equally sensitive to touch and temperature?

4. All animals do not perceive their environment in the same way. List some examples.

5. What extra receptors would help humans survive in today's world?

6. How is our sense of balance related to vision?

WRITING TO LEARN BIOLOGY

What are the advantages and disadvantages of receptor adaptation to an organism?

46

Vertebrate Anatomy
External Features and Skeletal System of the Rat

Objectives

By the end of this exercise you should be able to:

1. Define the terms used to describe the general planes, surfaces, and anatomy of a vertebrate.
2. Identify the external mammalian features of the male and female rat.
3. Identify the bones of a rat skeleton.
4. Compare the skeletal system of a rat with the skeletal system of a human.
5. Discuss the relationship of structure and function for the external features and skeletal system of a rat.

Studying anatomy involves much more than just memorizing and recognizing parts of an organism. It requires understanding the relationship of the function of an anatomical feature to the structure and location of that feature. As you study rat anatomy, remember that your objective is not just to memorize parts of the rat but to understand the overall design of the rat's anatomy and how this structure relates to function.

We have chosen a rat (*Rattus norvegicus*) for dissection because it is a small and familiar vertebrate. In addition, white (albino) rats are bred for laboratory experimentation and are used often as physiological models of vertebrate systems.

Question 1
What features of rats make them a good experimental model to learn about factors that affect humans?

Before you begin dissection, familiarize yourself with the following terms. They are used to describe the location of structures and their relationship to other structures (fig. 46.1).

Dorsal—toward the upper surface (back)

Ventral—toward the lower surface (belly)

Anterior—toward the head

Posterior—toward the tail

Cranial—toward the head

Caudal—toward the tail

Medial—toward the midline of the body

Proximal—toward the end of the appendage nearest the body

Lateral—away from the midline of the body; toward the side

Distal—toward the end of an appendage farthest away from the body

Frontal plane—divides the body into dorsal and ventral halves; two-dimensional plane parallel to the anterior-posterior axis and perpendicular to the dorsal-ventral axis

Transverse plane—a cross section; two-dimensional plane perpendicular to the anterior-posterior axis of the body

Sagittal plane—divides the body into left and right halves; two-dimensional plane parallel to the anterior-posterior axis and parallel to the dorsal-ventral axis

EXTERNAL ANATOMY

Rinse your rat with cold water to remove excess preservative. Then lay it in a small dissecting pan and observe its general characteristics.

The rat's body is divided into six anatomical regions: the **cranial region** or head, the **cervical region** or neck, the **pectoral region** or the area where forelegs attach, the **thorax** or chest region, the **abdomen** or belly, and the **pelvic region** or area where hind legs attach. The terms *right* or *left* refer to the organism's right or left side regardless of how the organism is positioned in the dissecting pan.

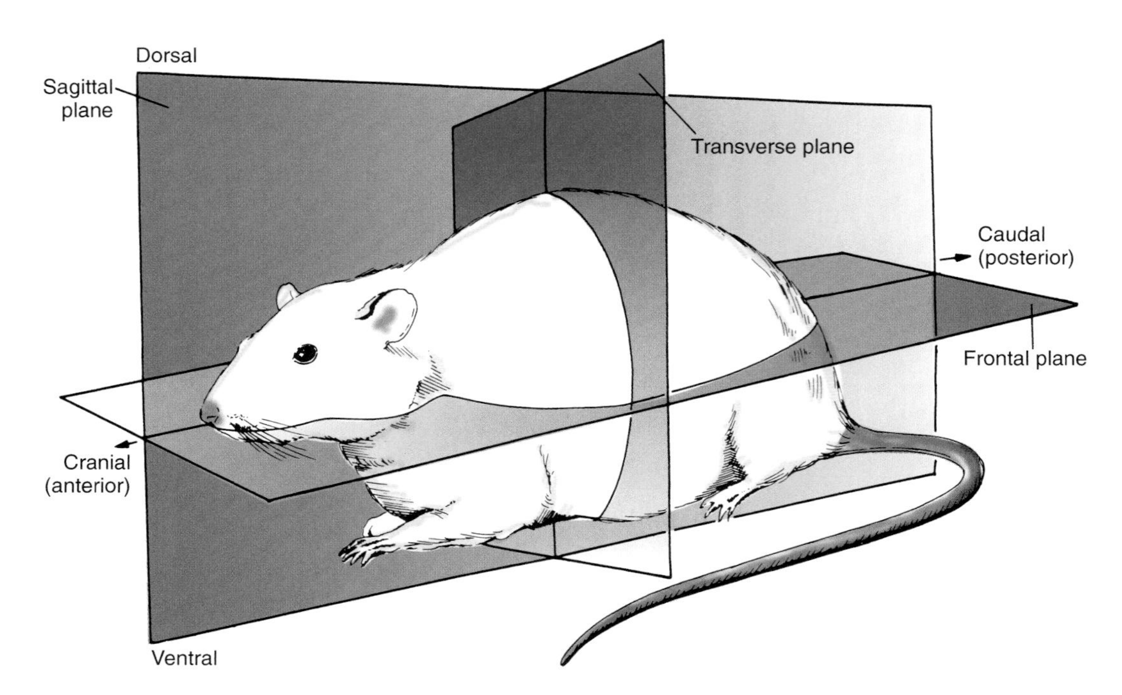

Figure 46.1
Surfaces and planes of section of a rat.

Locate and examine the structures noted in bold print in the following descriptions. A hairy coat called the **pelage** covers most of the rat's body (fig. 46.2). The tactilely (relating to touch) sensitive "whiskers" on the face are called **vibrissae.** The **nares** (nostrils) are at the anterior end of the head. The **subterminal mouth** has a **cleft** in the upper lip which exposes large front teeth called **incisors.** The eyes bulge from the head and enable the rat to see in almost all directions at once. Eyes equipped with large **pupils** are one of many adaptations to nocturnal life. The **nictitating membrane** is located at the inside corner of the **eye.** This translucent membrane may be drawn across the eyeball for protection. Locate the **eyelids**—they are similar to those found in humans. **Ears** are located posteriorly on the head and composed of a **pinna** (external ear) and an **auditory meatus** (ear canal).

On the ventral surface of the rat and just lateral to the midline are six pairs of nipples or **teats.** The **tail** is located posteriorly (caudally) and is covered with hair and scales. Ventral to the base of the tail is the **anus.** Just caudal to the last pair of teats of a female is a small protuberance, the **clitoris,** from which the **urinary aperture** opens. Just caudal to the clitoris is the **vaginal orifice** in a small depression called the **vulva.**

In males, a pair of large **scrotal sacs** lie on each side of the anus and contain the **testes,** which produce sperm cells. The testes may be withdrawn into the abdomen during nonreproductive periods. Just anterior to the scrotal sacs is the **prepuce,** which is a bulge of skin surrounding the **penis.** The penis can be extended from the prepuce through the **preputial orifice.** The end of the penis has a **urogenital orifice** where both urine and sperm cells exit the body. Be sure to examine rats of both sexes.

The rat is **quadruped** (four legs) with four digits and a vestigial thumb on each foot. Both the sole and heel of a rat's foot contact the ground as indicated by the lack of hair from the toes to the heel. This stance is **plantigrade.** Some other animals are **digitigrade** and walk on their toes.

Question 2

a. What external features of rats are common to all mammals? As you continue to study the rat refer to these mammalian features and make additions and deletions.

b. Mammals are very diverse. For example, do all mammals have four legs? Do you recall any mammals with more or fewer than two eyes?

c. Which areas of the rat's body are not covered with hair?

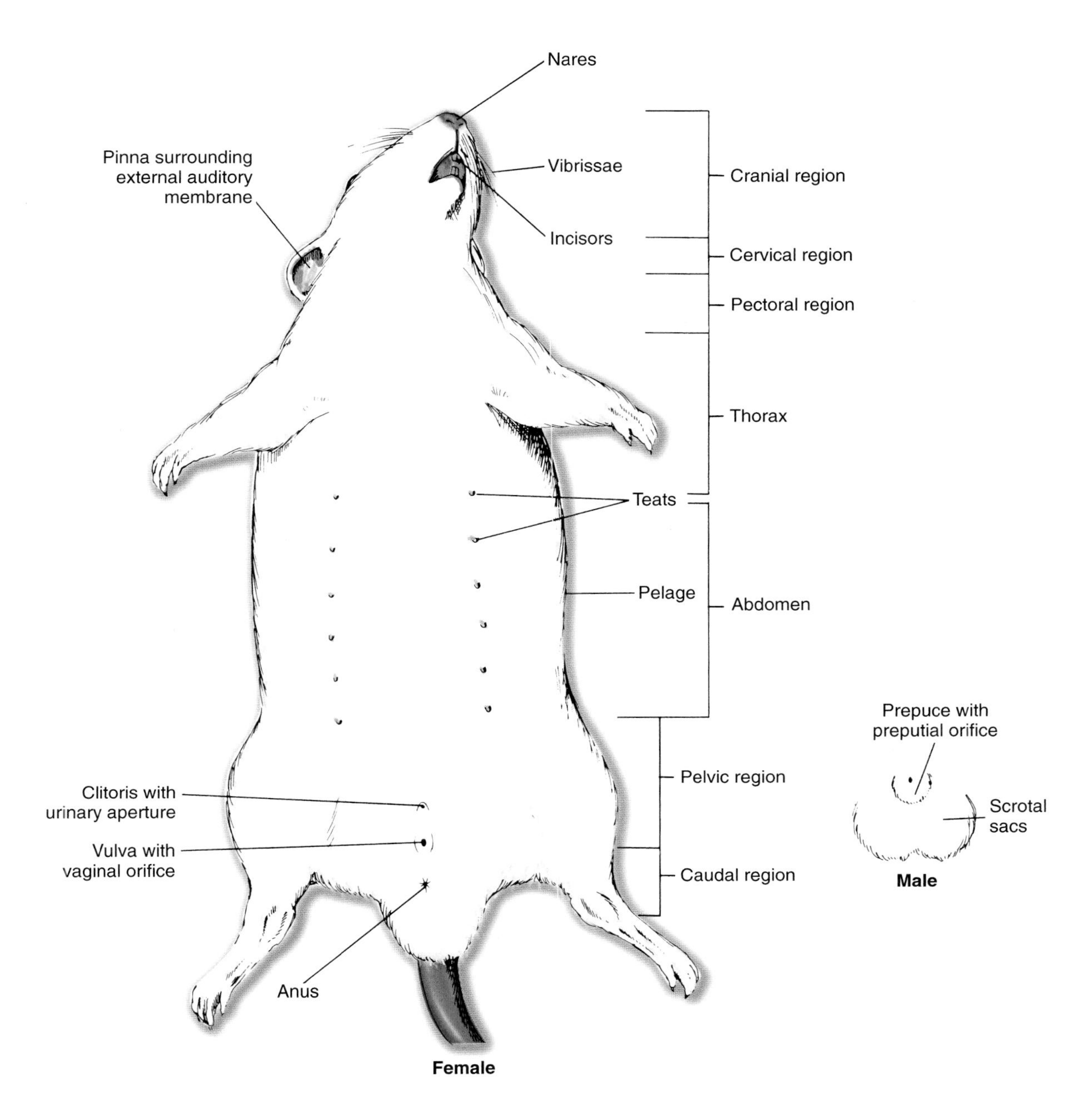

Figure 46.2
External anatomy of male and female rats.

d. Why do these parts of the body have no hair?

e. Can you think of a function of the pelage other than insulation?

f. Can you suggest a function for vibrissae?

g. How long are the vibrissae in relation to the head and in relation to the width of the body?

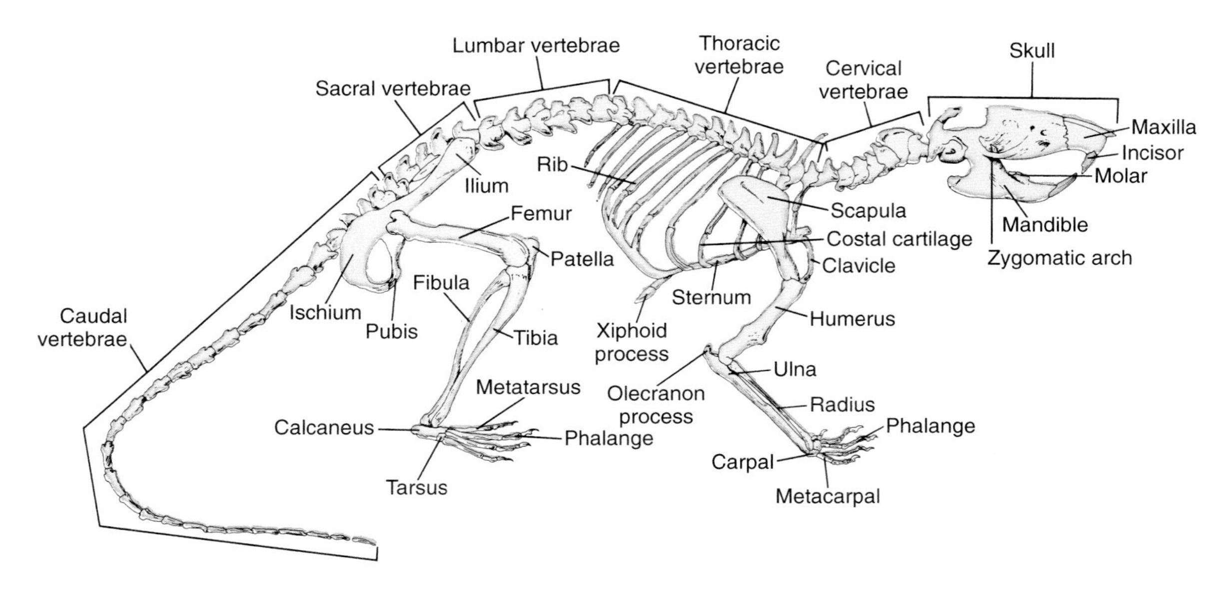

Figure 46.3
The skeleton of a rat.

h. How do the upper incisors mesh with the lower incisors?

i. Are any teeth other than incisors present?

j. Why is the mouth referred to as subterminal?

k. How does the position of eyes on a rat's head differ from that on a human's head?

l. What is the function of a large pinna?

m. Are teats exclusively mammalian structures? Why do you think so?

n. What term similar to quadruped describes animals such as humans with two legs?

o. What animals do you recall that are digitigrade?

p. Are humans plantigrade or digitigrade?

THE SKELETAL SYSTEM

The skeletal system of a rat is studied more easily with a cleaned and articulated skeleton than with your dissected specimen. Examine a rat skeleton and compare its parts with those labeled in figure 46.3. Then compare a rat skeleton with a human skeleton if one is available. Record your observations in table 46.1. In the margin of this page list a few words describing some general differences between these skeletal systems.

Locate and examine the structures noted in bold print in the following descriptions. Prominent features of the **skull** include the **mandible** (lower jaw), **maxilla** (upper jaw), the **zygomatic arch,** and the **cranium** (braincase). The mandible has a broad, flat surface for muscle attachment, and some of the jaw muscles pass under the zygomatic arch and attach to the lateral surface of the cranium. The skull and vertebral column form the **axial skeleton.**

TABLE 46.1

SKELETAL FEATURES OF RATS AND HUMANS

	Rat	Human
Number of bones		
Arrangement of bones		
Presence of unique bones		
Orientation of bones		

Examine the **vertebral column.** It is divided into five groups of vertebrae. From anterior to posterior these groups are the **cervical, thoracic, lumbar, sacral,** and **caudal vertebrae.** Count the number of vertebrae in each group. The shapes of the anterior two cervical vertebrae, the **atlas** and **axis,** differ from the other cervical vertebrae. The atlas and axis form a versatile joint and allow for a variety of head movements. Extending from the thoracic vertebrae are **ribs** that form an enclosure, the **rib cage.** The caudal vertebrae form a long tail.

Compare the vertebral columns and rib cages of rats and humans. List your observations in table 46.2.

Bones of the anterior appendages (forelegs) attach to the **pectoral girdle,** and bones of the posterior appendages (hind legs) attach to the **pelvic girdle.** These bones compose the **appendicular skeleton.** Compare the variation between rat and human bones of the pectoral girdle with the anterior appendages, and the pelvic girdle with the posterior appendages. You will find many similarities between rat and human skeletons.

Question 3

a. How does contraction of muscles passing under the zygomatic arch move the mandible?

b. What are the relative sizes of braincases in humans and rats?

c. Approximately how many vertebrae are in a rat's tail?

TABLE 46.2

FEATURES OF THE VERTEBRAL COLUMNS AND RIB CAGES OF RATS AND HUMANS

	Rat	Human
Number of vertebrae in each group		
Number of ribs		
General shape of vertebrae		
Prominence of spines		
Degree of fusion of caudal and sacral vertebrae		
Attachment of vertebrae to other bones		

d. Does a human have a tail?

e. What does the similarity between the skeletal systems of rats and humans suggest about the ancestry of these two mammals?

Procedure 46.1

Skin your rat

1. Lay your rat on its dorsal surface with the tail toward you. You will need scissors, a scalpel, and patience to remove the skin without damaging the underlying muscles and organs. An incision was already made in the throat to inject colored latex into the circulatory system.
2. Snip anteriorly from this incision toward the lower lip (fig. 46.4). Do not cut deeply. Cut only through the **integument** (skin).
3. Hold the skin up with forceps to prevent cutting muscles or internal organs. Next, snip posteriorly along the ventral midline to the **genitals** (external sex organs).

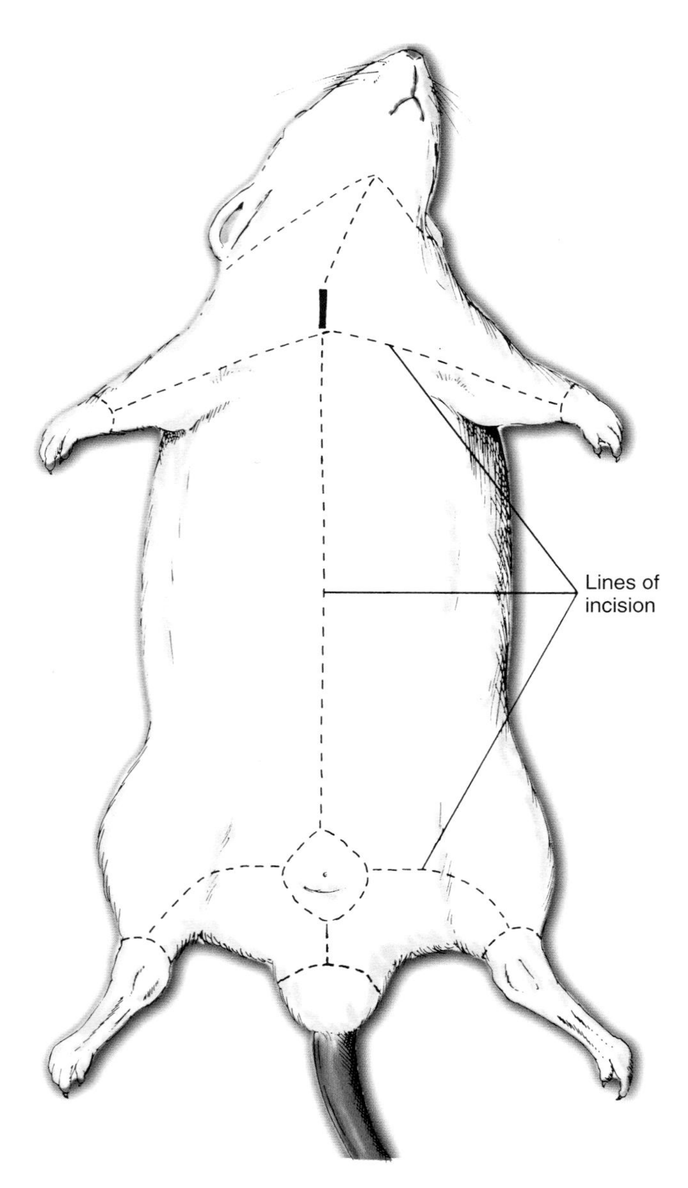

Figure 46.4
Ventral surface of a rat, indicating lines of incision for dissection.

4. Cut through the skin in a circle around the genitals and continue the midventral incision to the tail.
5. Cut a circle around the base of the tail.
6. Make four more incisions on the ventral surface from the midline to the two ankles of the hind legs and the two wrists of the forelegs.
7. Then cut through the skin in a circle around the wrists and ankles.
8. To separate the skin from the body, grasp the flap of skin extending from the chest under a forelimb and force the tip of a blunt probe between the skin and the connective tissue holding the skin to the body. This connective tissue is called **fascia** and is tough yet flexible.
9. Work slowly and gently peel the skin from the body.
10. Remove the skin until you expose the ventral surface of the chest and abdomen.
11. Work toward the dorsal surface. Notice that thin sheets of muscles attach to the skin. These muscles are the **cutaneous trunci,** which arise near the ventral base of the foreleg and spread to the lateral and dorsal surfaces of the chest and abdomen.
12. When you have detached the skin to the dorsal midline, begin to peel the skin anteriorly toward the base of the skull. You may notice a deposit of fat between the shoulder blades. This is **brown fat** and is specialized for heat production if the animal is stressed by low temperatures.
13. Do not remove the skin from the face unless instructed to do so by your instructor.
14. Leave the skin attached at the base of the skull and use it as a protective cape to wrap your specimen when you are not studying it.
15. If available, spray the rat with preservative, then wrap the skin around its body. Place the rat in a plastic bag, seal the bag tightly, and label the bag.

Questions for Further Thought and Study

1. What modifications of the skeletal system would be adaptive for a bipedal vertebrate?

2. What area of the axial skeletal system of a human is weakest? Why?

3. What is the value of a tail to a rat? To a monkey? To a deer?

4. Of what adaptive significance is a ball-and-socket joint at the hip of mammals rather than a hinge joint such as the knee?

WRITING TO LEARN BIOLOGY

What features of the skeletal system diversify movement in humans as compared to movement of a rat?

47

Vertebrate Anatomy
Muscles and Internal Organs of the Rat

Objectives

By the end of this exercise you should be able to:
1. Identify the major muscles of a rat.
2. Define terms describing the location, structure, and movement of muscles.
3. Describe the origin, insertion, and action of selected muscles of a rat.
4. Identify and state the function of major organs of the thorax and abdomen.

Dissection of muscles and internal organs requires patient concentration. Use a blunt probe rather than a scalpel or a needle to separate muscles and to move internal organs. Keep your specimen moist with dilute preservative so tissue won't be damaged as you move muscles and organs to reveal hidden structures.

MUSCULAR SYSTEM

The structure and location of muscles allow efficient performance of complex motions of vertebrates. Muscles are usually arranged on the skeleton in **antagonistic** (opposing) pairs. Contracting one member of the antagonistic pair moves the body, whereas contracting the other muscle restores the body and first muscle to their original positions. For example, after you contract a muscle to bend your arm at the elbow you must contract an opposing muscle to straighten your arm and stretch the first muscle.

Motion produced by muscular contraction also depends on the design of the skeletal system. When muscles contract, they use bones as levers for resistance and motion. As you examine the muscles of a rat you should refer often to a rat skeleton and locate the bones where each muscle attaches.

Muscles are attached to bones by connective tissue called **tendons** that adhere to spines, knobs, ridges, and depressions on bones. The end attached to the bone that does not move during contraction is called the **origin.** The end of the muscle attached to the bone that moves during a contraction is called the **insertion.** Movement caused by the contraction of a muscle is the muscle's **action.** Examine preserved rat and human skeletons and find likely surfaces for attachment of tendons.

Question 1

a. Where on a human skeleton is a likely origin and insertion for a muscle that lifts your knee?

b. Could a muscle arise and insert on the same bone? Why or why not?

The following terms describe muscles that move the body in a particular fashion. Familiarize yourself with these terms and demonstrate each of these terms with movements of your own body and the body of a rat. Also be sure to review the descriptive anatomical terms presented in the introduction of the previous exercise on vertebrate anatomy.

Extensors—increase the angle of a joint

Flexors—decrease the angle of a joint

Abductors—move an appendage away from the midline of the body

Adductors—move an appendage toward the midline of the body

Retractors—move an appendage backward

Protractors—move an appendage forward

Supinators—rotate the palm or bottom of foot upward

Pronators—rotate the palm or bottom of foot downward

Procedure 47.1

Examine and identify muscles

1. Begin examining and identifying the muscles on your skinned rat by picking the stringy fascia away from the surface of the body. The more connective fascia that you remove, the more clearly you will see the muscles.
2. Use a blunt probe to gently separate the various muscles. Work cautiously until you develop a sense of where two muscles overlap. Notice the direction of the fibers within each muscle.
3. Refer to the figures frequently as you separate each muscle.
4. Answer the questions that will guide your observation as you separate muscles in each region of the body.

Figures 47.1, 47.2, and 47.3 illustrate the muscles of a rat in dorsal, lateral, and ventral views. Not all of the muscles of a rat are shown here, but you can easily understand the muscular system by studying those that are illustrated.

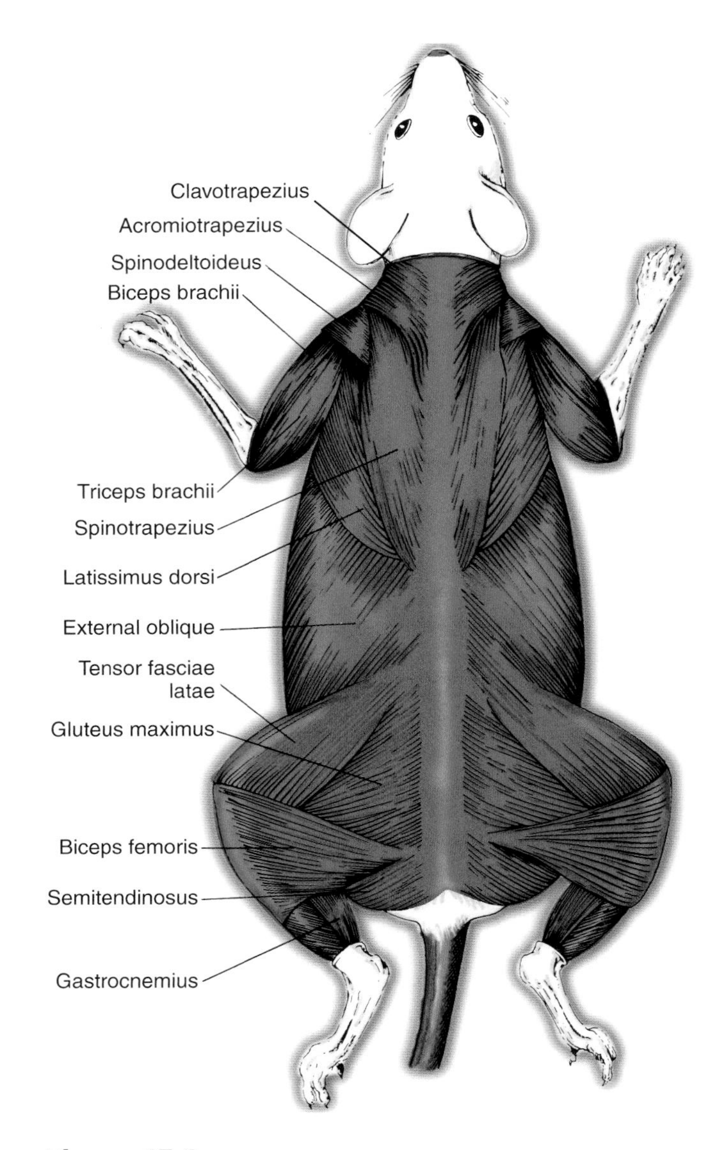

Figure 47.1
Dorsal view of musculature of a rat.

Muscles of the Pectoral Girdle and Appendages

Question 2

a. Which of the muscles on the dorsal surface are shaped like a trapezoid?

b. What bones along the dorsal midline are points of attachment for the **acromiotrapezius, spinotrapezius,** and **latissimus dorsi?**

c. Are the bones named in your previous answer sites of the origin or insertion of these muscles?

d. Which muscle is antagonistic to the **biceps brachii?**

e. Is the biceps brachii an extensor or a flexor?

f. Is the latissimus dorsi a retractor or a protractor?

g. Which dorsal muscle is antagonistic to the **pectoralis major?**

h. Do the pectoralis major and **pectoralis minor** produce different movements?

i. Are the pectoralis major and pectoralis minor antagonistic?

j. What are the origins for the muscles that pull the head down toward the chest?

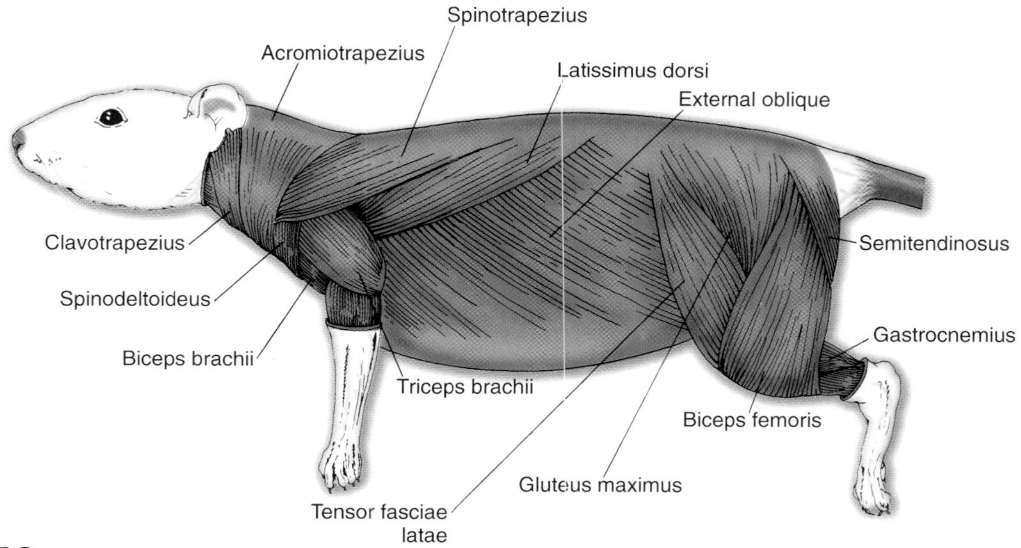

Figure 47.2
Lateral view of musculature of a rat.

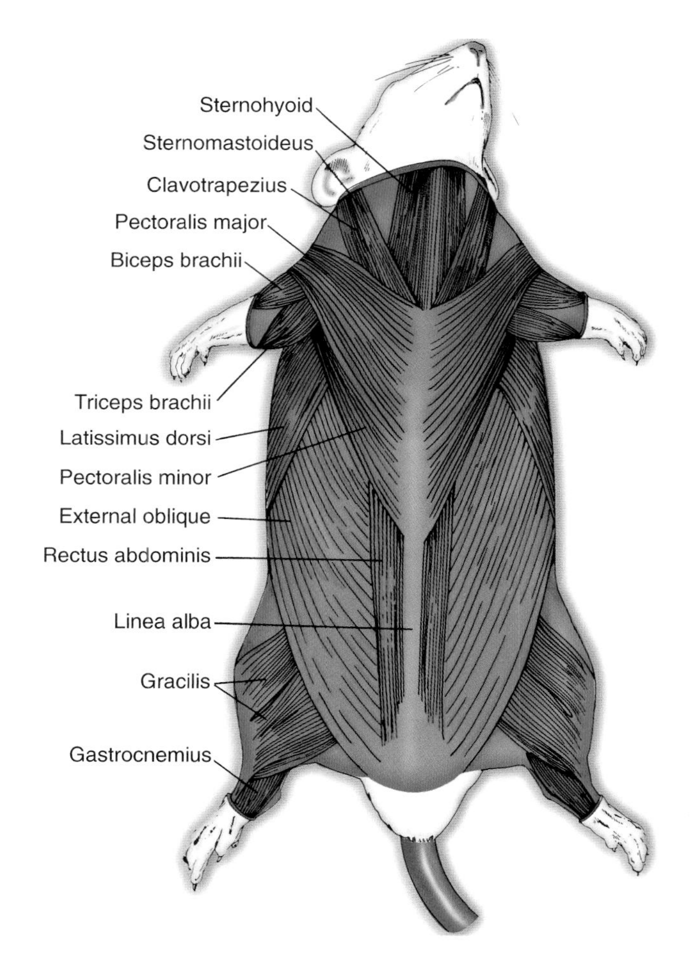

Figure 47.3
Ventral view of musculature of a rat.

Muscles of the Pelvic Girdle and Appendages

Question 3

a. Which of the large lateral muscles are used in power movements of the hind legs?

b. Which of the large lateral muscles extends the knee joint and pulls the leg forward?

c. Which muscle flexes the knee joint and retracts the leg?

d. How is the gastrocnemius used by a running rat?

Origins, Insertions, and Actions of Muscles

The muscular system is functionally associated with the skeletal system. The following is a list of a few selected muscles that illustrate origins, insertions, and actions. Locate the following muscles on your specimen. Then locate their origins and insertions on a rat skeleton and visualize the action of each muscle. If possible, move the bone on your preserved rat or on yourself to show its action.

Biceps brachii—This muscle is located on the anterior surface of the humerus.

Origin: scapula

Insertion: radius

Action: flexes the lower arm

Triceps brachii—This muscle is located on the sides and back of the upper arm.

Origin: humerus and scapula

Insertion: elbow (olecranon)

Action: extends the lower arm

Spinotrapezius—This muscle lays across the dorsal thoracic region of the rat.

Origin: thoracic vertebrae

Insertion: spine of the scapula

Action: moves the scapula upward and backward

Acromiotrapezius—This is the central muscle of the trapezius group.

Origin: cervical vertebrae

Insertion: scapula

Action: moves the scapula dorsally

Latissimus dorsi—This muscle is posterior to and partially covered by the spinotrapezius.

Origin: thoracic and lumbar vertebrae

Insertion: medial side of the humerus

Action: moves the humerus dorsocaudad

External oblique—This muscle covers the sides of the abdominal cavity from the hip to the rib cage. Its fibers are arranged at right angles to the fibers of the latissimus dorsi.

Origin: lumbodorsal fascia and posterior ribs

Insertion: linea alba and pelvis

Action: compresses and retains the viscera and acts with the rectus abdominis to form a flexible ventral and lateral body wall

Cutaneous trunci—This muscle attaches to the skin on the lateral and dorsal sides of the body.

Origin: under the upper front leg

Insertion: skin

Action: moves the skin

Biceps femoris—This large muscle is in two bundles; it is located easily on the side of the thigh.

Origin: ischium

Insertion: distal portion of femur and proximal portion of tibia

Action: abducts the thigh and flexes the shank

Gastrocnemius—This muscle is easy to locate because it forms the bulk of the calf muscle in the lower leg.

Origin: distal end of femur

Insertion: heel (calcaneus) by means of the tendon of Achilles

Action: extends the foot

INTERNAL ANATOMY

Organs of the Head and Neck

Locate and examine the structures noted in bold print in the following descriptions. To locate the **salivary glands (sublingual, parotid,** and **submaxillary glands)** first locate the medial sternohyoid muscle. The salivary glands lie to the side of the anterior part of this muscle. Salivary glands are rather soft, spongy tissue. They secrete **saliva** that lubricates food and contains **amylase** that degrades starch to maltose. Do not mistake lymph glands for salivary glands; the lymph glands are darker, more compact, and circular glands that lie anterior to the salivary glands and are pressed against the powerful jaw muscles. The submaxillary glands are large, roughly oval, and elongate. The sublingual glands touch the submaxillary glands on their anterolateral surface. Parotid glands are best seen in lateral view. Parotid glands extend toward and behind the ear; they are beneath the ear and between the jaw and points of the shoulders. Infected parotid glands in humans produce swellings commonly associated with mumps.

Procedure 47.2

Examine organs of the head and neck

1. After you have located the submaxillary salivary glands, remove them to see structures deeper in the neck.
2. The **thyroid gland** is a gray or brown swelling on either side of the **trachea.** A thin band (isthmus) transverses the trachea. To locate the thyroid gland, expose the trachea by cutting through the **sternohyoid** muscles that cover the ventral surface of the throat.

The two lobes of the thyroid gland lie on each side of the trachea just ventral to the **larynx.** The major secretion of the thyroid gland, **thyroxin,** regulates the metabolic rate of the body. The two lobes are connected by a narrow band of tissue known as the **isthmus of the thyroid.**

The larynx is a rigid, cartilaginous chamber containing the vocal cords. Just caudal to the larynx is the trachea, which conducts air to and from the lungs. The rigid trachea is composed of a series of incomplete rings of cartilage forming a tube.

Question 4
Why is the trachea reinforced with cartilage?

Procedure 47.3

Open the body cavity

1. Make a small incision with your scalpel just caudal to the posterior end of the sternum. Cut no more than 4–5 mm into the body cavity.
2. With your scissors, snip posteriorly for about 1–2 cm through the abdominal wall (toward the pelvis) until you encounter the transversely attached **diaphragm** (fig. 47.4). Try to preserve the attachment of the diaphragm as you follow the next few steps.
3. Do not cut through the diaphragm. Instead, cut laterally in both directions (fig. 47.4) slightly anterior to the diaphragm, and along the line of attachment of the diaphragm to the body wall.
4. To open the thoracic cavity, use scissors to cut the ribs one at a time from the posterior to anterior. This will eventually open the chest cavity.
5. To begin opening the abdominal cavity, make a second abdominal, midline incision just posterior to the diaphragm. Cut laterally in both directions (fig. 47.4) slightly posterior to the diaphragm, and along its attachment in a similar manner to step 3 above.
6. From the second incision cut posteriorly slightly to the right of the white **linea alba** on the midventral surface of the abdomen until you reach the genitals.
7. Cut from the genitals laterally to the junction of each hind limb with the body.
8. After you have opened the animal, take it to the sink and rinse the two body cavities. Be sure that a colander covers the sink drain. After you have rinsed your rat, remove and discard all waste material from the sink.

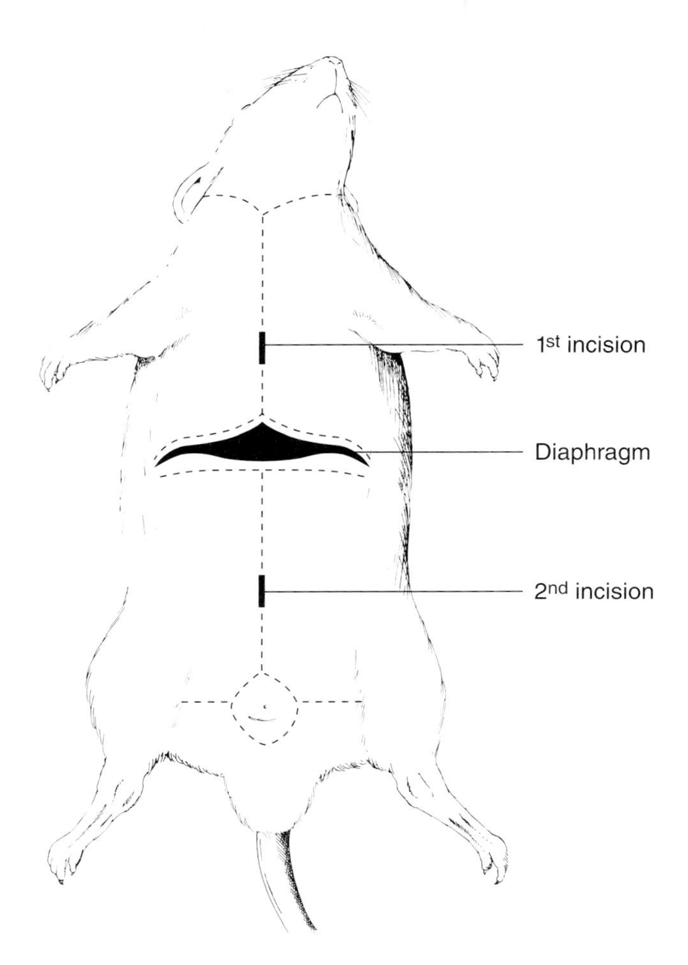

Figure 47.4
Lines of incision to open thoracic and abdominal cavities.

The Thoracic Organs

Locate and examine the structures noted in bold print in the following descriptions. The diaphragm, a thin layer of muscle, separates the **thoracic cavity** from the **abdominal cavity.** The **heart** is centrally located in the thoracic cavity (fig. 47.5). The two dark-colored chambers at the top of the heart are the **atria** (sing., *atrium*) that receive blood from veins. The light brown muscular areas below the atria are the **ventricles,** which are the pumping chambers of the heart. The heart is covered by a thin membrane known as the **pericardium.** Section the heart to see its chambers.

The **thymus gland** lies directly over the upper part of the heart. The thymus functions in the developing immune system and is larger in young rats than in old rats. Compare your rat with other specimens.

Air enters the lungs from the **bronchial tubes** branching from the trachea, which terminate into thin-walled air sacs, the **alveoli** (sing., *alveolus*), where gas exchange occurs. The bronchial tubes and alveoli are embedded in the lung tissue and are not visible. Alveoli are very small. Posterior to the lungs and heart lies the diaphragm. Cramps of the muscular diaphragm sometimes occur when you run; these cramps can cause a sharp pain in your side when you try to breathe rapidly. The diaphragm can also develop a twitch, which may occur every 10 or 20 sec and be rather embarrassing.

Question 5

a. What is the name of this common condition of a twitching diaphragm?

b. Is your rat young or old? How can you tell?

c. The lungs surround the heart on both sides. Are the lungs the same size? Are they both lobed?

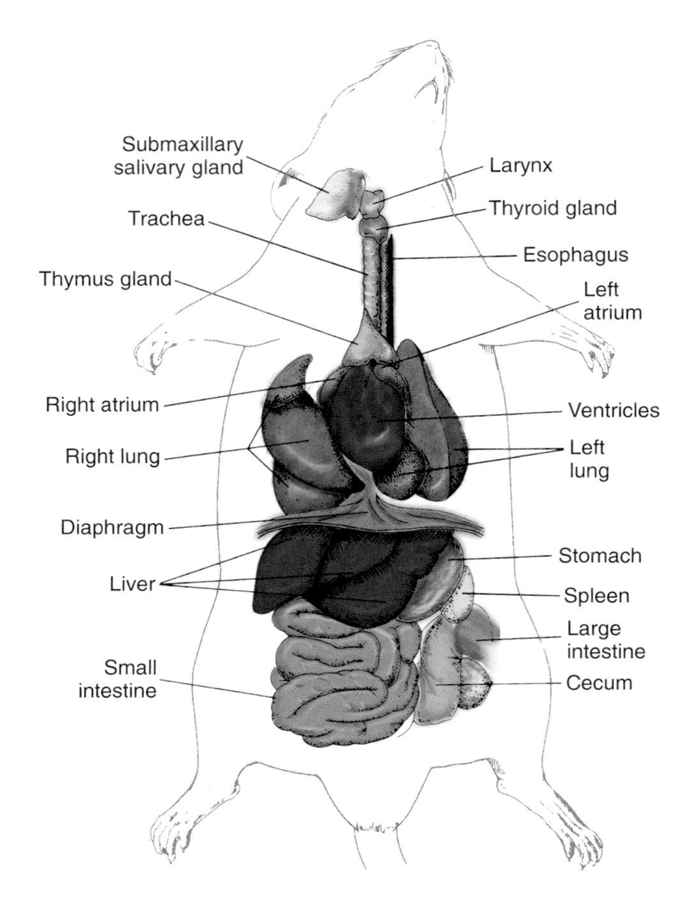

Figure 47.5

Internal organs of the rat. Move the intestines to the right or left to see the dispersed pancreas tissue. Note that muscles cover the esophagus, thyroid, and larynx.

d. Observe the texture of the lungs. Are they muscular?

e. If the lungs are not muscular, how is air brought into the lungs?

f. Is the diaphragm a flat sheet of muscle or does it curve?

g. How does the diaphragm function in breathing?

The Abdominal Organs

Locate and examine the structures noted in bold print in the following descriptions. The **coelom** is the body cavity within which the **viscera** (internal organs) are suspended. The abdominal cavity and viscera are covered by a membranous tissue called the **peritoneum,** which is formed from mesoderm. Recall that the coelomic cavity of evolutionarily advanced invertebrates and vertebrates is always lined by mesodermal tissue. The peritoneum is extensive and occurs in four regions:

- **Parietal peritoneum** covers the walls of the abdominal cavity.
- **Visceral peritoneum** covers the internal organs.
- **Mesenteries** are thin membranes continuous between the parietal and viscera peritoneums. They attach the internal organs to the dorsal body wall.
- **Omentia** (sing., *omentum*) are thin membranes extending from the visceral peritoneums and connect organ to organ.

Note the continuous nature of the peritoneum as you examine the abdominal organs. Locate the **liver,** which is the dark-colored organ suspended just under the diaphragm. The many functions of the liver include producing **bile** to help digest fat, storing glycogen (a high-energy molecule), transforming nitrogenous wastes into less harmful substances, and transforming digested nutrients into various molecules needed by the body. Rats do not have a gallbladder. For these functions the liver is strategically located near the digestive system to receive a rich supply of blood directly from the intestinal tract. The liver has four parts:

- **Median** or **cystic lobe** is located atop the organ. There is a cleft in its central part.
- **Left lateral lobe** is large and partially covered by the stomach.
- **Right lateral lobe** is partially divided into an anterior and posterior lobule, is smaller than the left lateral lobe, and is hidden from view by the median lobe.
- **Caudate lobe** is small, folds around the esophagus and stomach, and appears to be in two small sections seen most easily when the liver is raised.

The **esophagus** pierces the diaphragm and conducts food from the mouth to a muscular enlargement of the digestive tract called the **stomach.** Locate the stomach on the left side just under the diaphragm. The functions of the stomach include food storage, physical breakdown of food, and initial enzymatic digestion of protein. The entrance of the esophagus to the stomach is guarded by the **cardiac sphincter.** Sphincters are circular muscles that control the flow of enclosed fluids by contracting and closing the opening between cavities. The outer margin of the curved stomach is known as the **greater curvature;** the inner margin is the **lesser curvature.** Slit open the stomach and notice ridges, called rugae, which line the stomach. The stomach is attached to the intestine via the **pyloric sphincter.**

The **spleen** is similar in color to the liver and is attached to the greater curvature of the stomach. It is associated with the circulatory system and functions in the formation, storage, and destruction of blood cells.

The **pancreas** is the brownish, flattened gland suspended in membranous tissue near the junction of the stomach and small intestine. This gland has exocrine and endocrine functions. Review these terms in your textbook. The pancreas produces digestive enzymes that are conducted to the intestinal tract through small ducts. The pancreas also secretes and releases **insulin** into the bloodstream to regulate glucose metabolism. The **greater omentum** is the membranous curtain of tissue that hangs from the stomach and contains lymph nodes, blood vessels, and fat.

The **small intestine** is the slender coiled tube that receives partially digested food from the stomach. The small intestine continues digestion and nutrient absorption. It consists of three sections: duodenum, jejunum, and ileum. These sections are distinguishable histologically but may be hard to differentiate anatomically. The **duodenum** receives enzymes from the digestive glands; it begins at the **pyloric sphincter** of the stomach and continues under the stomach. The duodenum has the small, fingerlike folds of intestinal wall called **villi.**

Question 6

a. Is the membrane attaching the spleen to the stomach a mesentery or omentum?

b. What is the function of villi?

Procedure 47.4

Examine abdominal organs

1. Locate the liver.
2. Locate the stomach, slit it open, and notice the rough folds called rugae.
3. Slit open the duodenum and examine the villi.

Locate and examine the structures noted in bold print in the following descriptions. The **jejunum** is the central portion of the small intestine. The **ileum** is the posterior portion of the small intestine that empties into the **colon** (large intestine). The **ileocecal valve** controls flow of food into the colon.

The colon is the large greenish tube extending from the small intestine to the **anus.** The colon is where the final stages of digestion and water absorption occur. It contains a rich flora of bacteria that help digestion. The colon also secretes a lubricant of mucus for the feces. The colon consists of five continuous sections which can be identified by their appearance and location.

- **Cecum** is a large, flattened sac in the lower third of the abdominal cavity. It is a blind-ending pouch located caudal the junction of the ileum and is often green. The cecum is homologous to the appendix in humans and is especially rich in microorganisms that aid in digestion.
- **Ascending colon** leads toward the right lateral lobe of the liver on the right side of the body cavity.
- **Transverse colon** crosses the body cavity under the liver and stomach.
- **Descending colon** descends the left side of the body cavity. You may need to move the small intestine aside to see it clearly.
- **Rectum** is the short, terminal section of the colon between the descending colon and the anus. The rectum temporarily stores feces before they are evacuated from the body.

Review the internal organs by listing all of the structures and areas of the digestive system through which food passes. Refer to your textbook and add a phrase describing the nature of digestion occurring in each area.

Questions for Further Thought and Study

1. Which muscles of a quadruped are most modified to achieve an upright, bipedal stance?

2. Where on the human body are muscles attached to the skin most developed? What is the function of the movement of these muscles?

3. Why might a human need a gallbladder, but a rat does not?

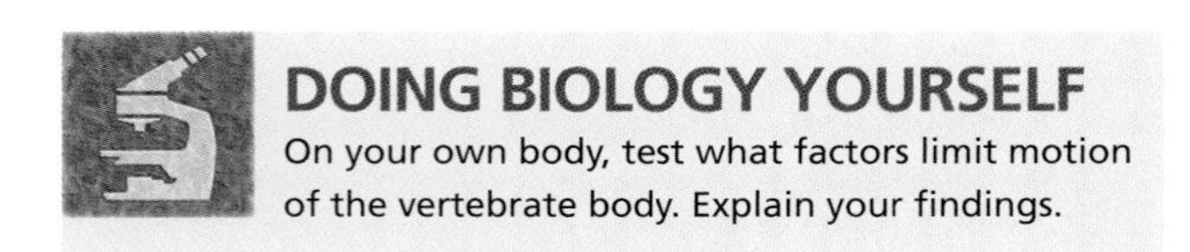

DOING BIOLOGY YOURSELF
On your own body, test what factors limit motion of the vertebrate body. Explain your findings.

48

Vertebrate Anatomy
Urogenital and Circulatory Systems of the Rat

Objectives

By the end of this exercise you should be able to:

1. Describe the structural relationship of the excretory and reproductive systems of a rat.
2. Identify the major features of the male and female reproductive systems.
3. Identify the major arteries and veins of a rat.
4. Trace the flow of blood through the heart, arteries, and veins.
5. Describe and list the parts of the "pulmonary heart" and "systemic heart."
6. Describe the function and structure of the hepatic portal system.
7. Trace the pathway of oxygen and carbon dioxide through the blood vessels of the respiratory system.

The excretory, reproductive, and circulatory systems distribute fluids and cells throughout the vertebrate body. As you examine the various structures of these systems, note the sequence of the vessels that cells and fluids pass through as they exit or move through the body.

UROGENITAL SYSTEM

The excretory and reproductive systems of vertebrates are closely integrated and are usually studied together as the urogenital system. However, they have different functions. The excretory system removes toxic wastes. The reproductive system produces gametes, an environment for the developing embryo, and hormones associated with sexual development. Before you begin dissecting the rat, review figures 48.1, 48.2, 48.3, and 48.4, which show various aspects of the human urogenital systems. As you study the rat's urogenital system compare its anatomy with that of a human.

Procedure 48.1

Observe the excretory and reproductive organs of a rat

1. Locate and observe the organs as described.
2. Do not remove organs; just move them to see other structures.

Excretory Organs

Locate and examine the structures noted in bold print in the following descriptions. The primary organs of the excretory system are the **kidneys.** Examine a model or demonstration dissection of a vertebrate kidney available in the lab (fig. 48.4). A sagittal section through the kidney reveals the **renal artery,** which transports blood to the kidney for filtering, and a **renal vein,** which transports filtered blood away from the kidney. The **ureter** is a tube that conducts urine from the kidney to the **urinary bladder.** The base of the ureter is expanded on the kidney to form a small chamber called the **renal pelvis,** which collects urine produced by the many tubules of the kidney. The outer area of the kidney is the **cortex** and the inner area is the **medulla.** Review in your textbook the fine structure and function of the tubules that form the bulk of the kidney. If you pull the intestines to either side, you can see a pair of kidneys on the dorsal abdominal wall (figs. 48.5, 48.6). Locate the delicate ureters by wiggling the kidneys and watching for movement on the body wall posterior to the kidneys. Remove one kidney and dissect it. Observe its internal structure; compare it to the model (fig. 48.4).

The **urethra** carries urine from the bladder to the **urethral orifice** where urine is expelled. The small, yellowish glands embedded in fat atop the kidneys are the **adrenal glands.**

Question 1
What hormone do the adrenal glands produce? Why have two kidneys?

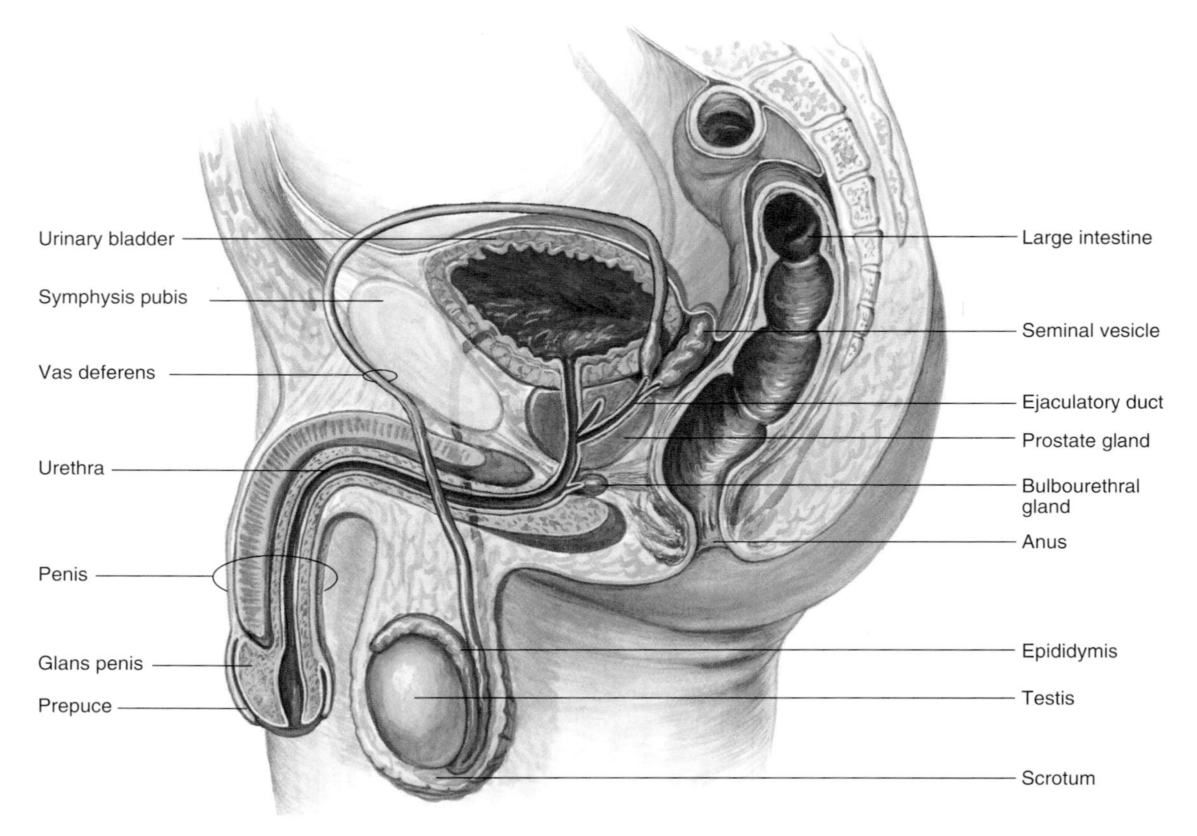

Figure 48.1
The human male reproductive system consists of two testes that produce sperm, ducts that carry the sperm, and various glands. Muscular contractions propel the sperm through the vas deferens past the seminal vesicles, prostate gland, and bulbourethral gland, where most of the liquid of the semen is added. The semen passes through the urethra of the penis to the outside of the body.

Reproductive Organs of the Male Rat

Locate and examine the structures noted in bold print in the following descriptions. The major reproductive organs of the male are the **testes,** which are in the **scrotal sac** (fig. 48.5). Cut through the sac to expose a testis. On the surface of the testis is a coiled tube, the **epididymis,** which collects and stores maturing sperm cells produced within the testis. The tubular **vas deferens** conducts sperm from the epididymis to the **urethra,** which carries sperm through the penis and out of the body. Review gamete production in Exercise 14.

Question 2
Where in the body do sperm and urine share the same passages? What male structures serve only for excretion, which only for reproduction, and which for both?

The lumpy, brown glands located to the right and left of the urinary bladder are the **seminal vesicles.** The white gland below the bladder is the **prostate gland.** The seminal vesicles and prostate gland secrete materials forming much of the **seminal fluid** (semen) that activates and transports sperm cells.

Reproductive Organs of the Female Rat

Locate and examine the structures noted in bold print in the following descriptions. The short gray tube lying dorsal to the urinary bladder is the **vagina** (fig. 48.6). The vagina divides into two **uterine horns** extending toward the kidneys against the dorsal body wall. This **duplex uterus** (two uterine horns) is common in some groups of nonprimate mammals. It will accommodate more than one offspring (a litter) during a reproductive cycle. In contrast, a **simplex uterus** common in primates such as humans has a single, medial chamber for development of an embryo. At the tips of the uterine horns are small, lumpy glands called **ovaries.** The tiny tubes between the ovaries and the uterine horns are the **oviducts.** Oviducts capture eggs produced by the ovaries and conduct the eggs to the uterine horns. You may need a dissecting microscope to help locate the oviducts.

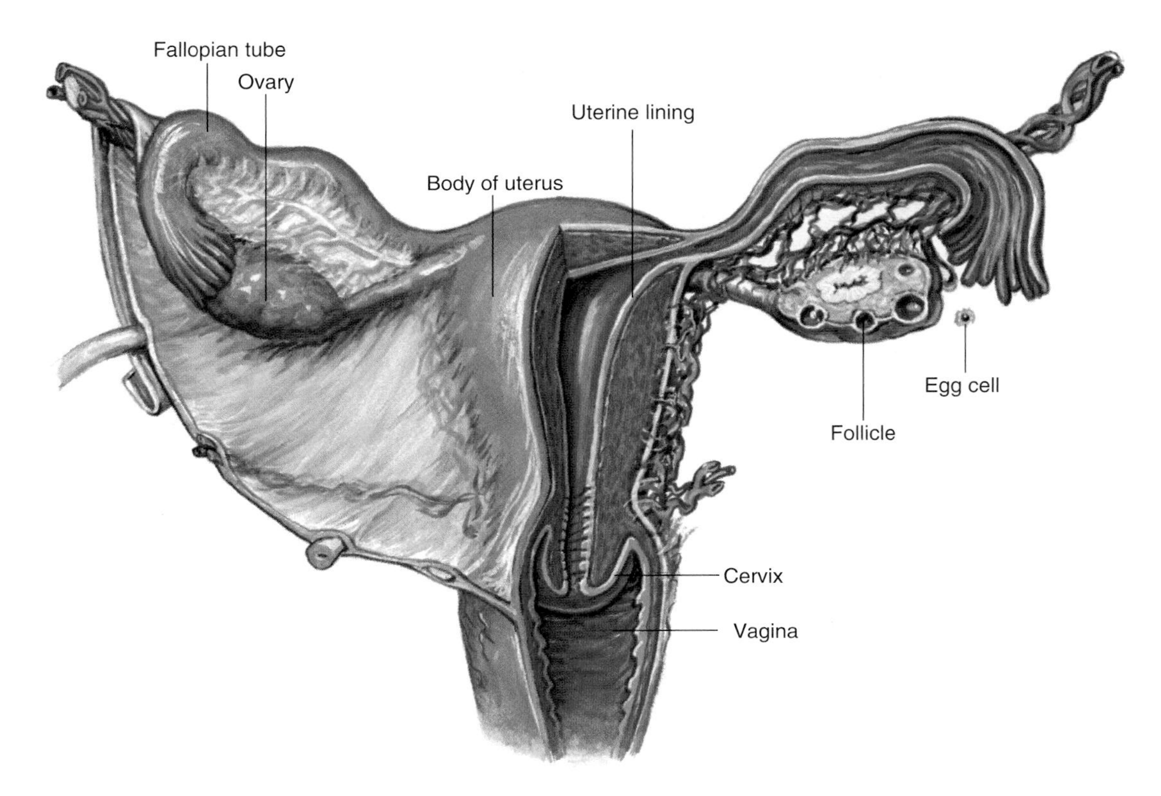

Figure 48.2

The female reproductive system includes two ovaries that produce eggs. After ovulation, an egg travels down the oviduct (uterine tube) to the uterus. If the egg is not fertilized, it is shed when the uterine lining is lost during menstruation.

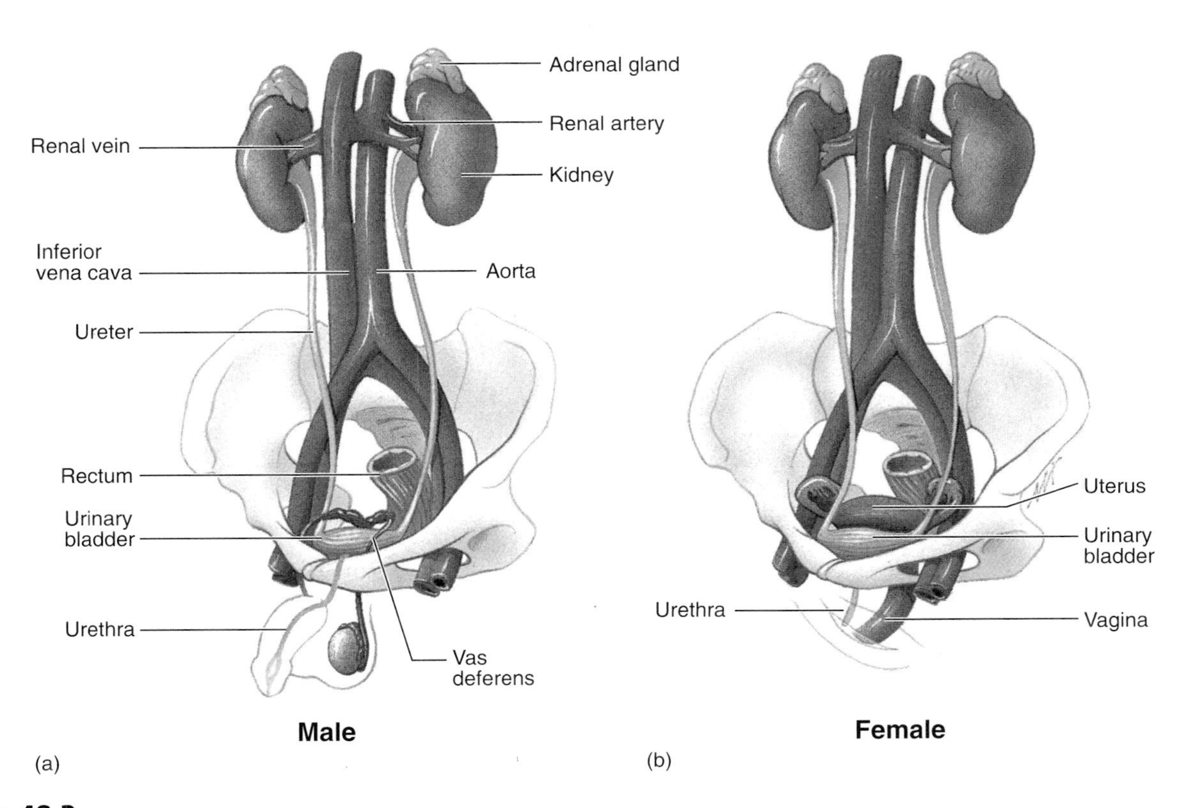

Figure 48.3

(*a*) Male and (*b*) female urinary systems shown in relation to the pelvis.

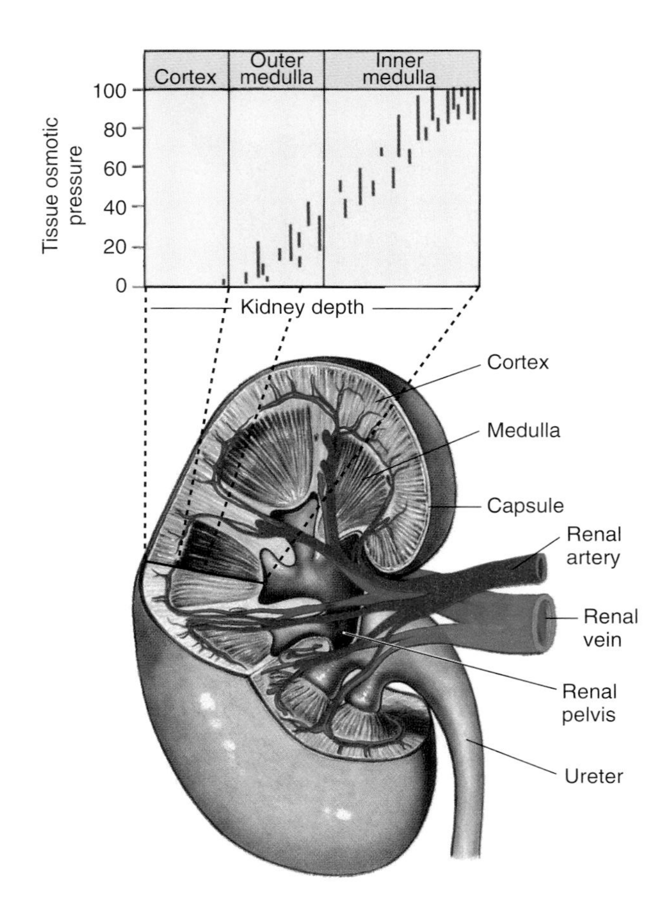

Figure 48.4

Structure of the human kidney. By varying the concentration of solutes around capillaries in different regions, the kidney extracts fluid from the blood. As dilute filtrate derived from blood circulates through tubules looping through the medulla, water is reabsorbed in response to the surrounding osmotic concentrations, and waste products are routed (as urine) to the renal pelvis and ureter.

Question 3

What female structures serve only for excretion, which only for reproduction, and which for both?

CIRCULATORY SYSTEM

Locate and examine the structures noted in bold print in the following descriptions. The general structure of the circulatory system of the rat is almost identical to that of humans. Blood circulates in two general systems. **Pulmonary circulation** carries blood through the lungs for oxygenation and back to the heart. **Systemic circulation** moves blood through the body after it has left the heart. As you examine the blood vessels notice that the arterial system, especially

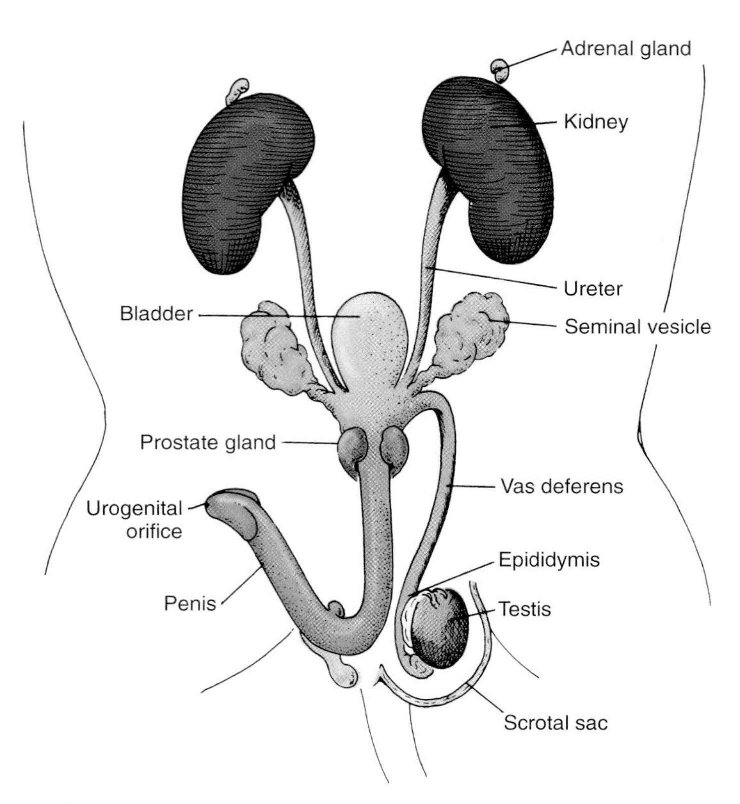

Figure 48.5

Urogenital system of a male rat.

near the heart, is distinctly asymmetrical. In future courses on comparative anatomy you will learn the derivation of this asymmetry.

Examine figures 44.1, 44.2, 48.7, 48.8, 48.9, and 48.10, which show aspects of the human circulatory system. As you study the rat's circulatory system, compare its anatomy with that of a human.

Procedure 48.2

Observe the circulatory system of a rat

1. Locate organs of circulation.
2. Follow the path of circulation as described.

Begin your study of the circulatory system at the heart. The arteries and veins have been injected with colored latex so that you can see them more clearly. Do not cut through any arteries or veins as you dissect muscle tissue from around these vessels.

Circulation of Blood Through the Heart and Pulmonary System

You have already located the heart in an earlier lab. Examine it again; also examine the vertebrate heart shown in figures 48.9 and 48.10.

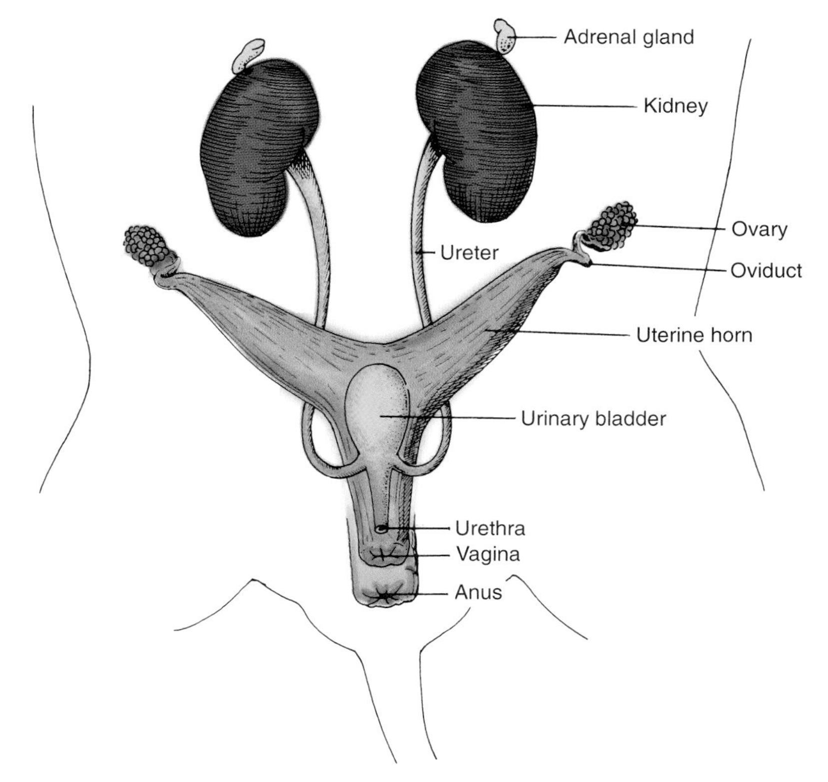

Figure 48.6
Urogenital system of a female rat.

Blood from the posterior portion of the body enters the right atrium of the heart through the **inferior vena cava.** The terms *superior* and *inferior* are often used to describe the vena cavae in humans, while cranial and caudal are sometimes used for other mammals such as rats. Blood from the anterior portion of the body enters through the **right** and **left superior venae cavae.** Blood flows from the right atrium, through the **tricuspid valve** to the **right ventricle.** It is then pumped through the pulmonary **semilunar valve** and into the **pulmonary trunk.** The pulmonary trunk divides dorsally into the right and left **pulmonary arteries;** these are the only arteries in the body that carry **deoxygenated** blood. Blood then flows through the pulmonary arteries to the lungs where it is **oxygenated.** Blood returns from the lungs to the **left atrium** of the heart via four **pulmonary veins** and flows through the **bicuspid (**or **mitral) valve** to the **left ventricle.** The pulmonary veins are best traced by beginning at the left atrium and proceeding to the lungs.

Circulation Through the Heart and Systemic Arteries

Blood leaves the **left ventricle** of the heart through the **aortic semilunar valve** and flows into the **aorta** (fig. 48.9). The aorta has four general areas. Label the first three of these areas in figure 48.9. The (1) **ascending aorta** begins at the semilunar valve of the left ventricle and passes outside and over the left and right atria. The aorta then bends to the left and passes dorsally to the heart. This bend is the (2) **aortic arch.** The aorta continues as the (3) **descending aorta** and passes along the mid-dorsal wall from the thorax through the diaphragm to the abdomen. Caudal to the diaphragm the aorta continues as the (4) **abdominal aorta.** You must move the stomach, liver, and intestines to the side to see the abdominal aorta as it passes along the vertebral column (fig. 48.11).

Question 4

a. How do the thicknesses of the atrial and ventricular walls differ? How does wall thickness relate to function?

b. Which structures compose the pulmonary heart? Which compose the systemic heart?

Branches of the Aortic Arch and Descending Aorta

Coronary arteries serve the heart muscle but are not easily seen because they branch from the aorta within the heart.

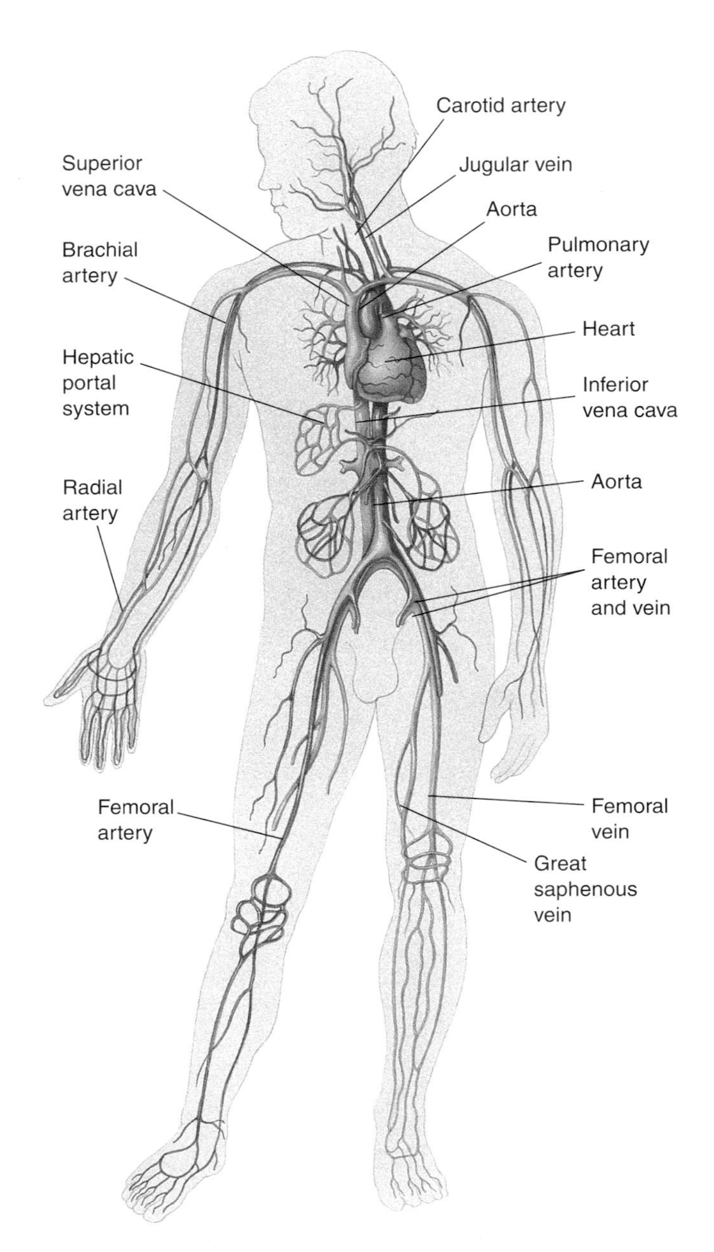

Figure 48.7

Human circulatory system. A muscular heart pumps blood through arteries, veins, and capillaries. Veins carry blood to the heart, and arteries carry blood away from the heart to all parts of the body. Blood passes from arteries into small, thin-walled capillaries where nutrients, oxygen, and wastes can diffuse into and out of tissues. Capillaries coalesce to form veins, which route blood to the liver and kidney for filtration and back to the heart. The circulatory systems of all vertebrates are similar.

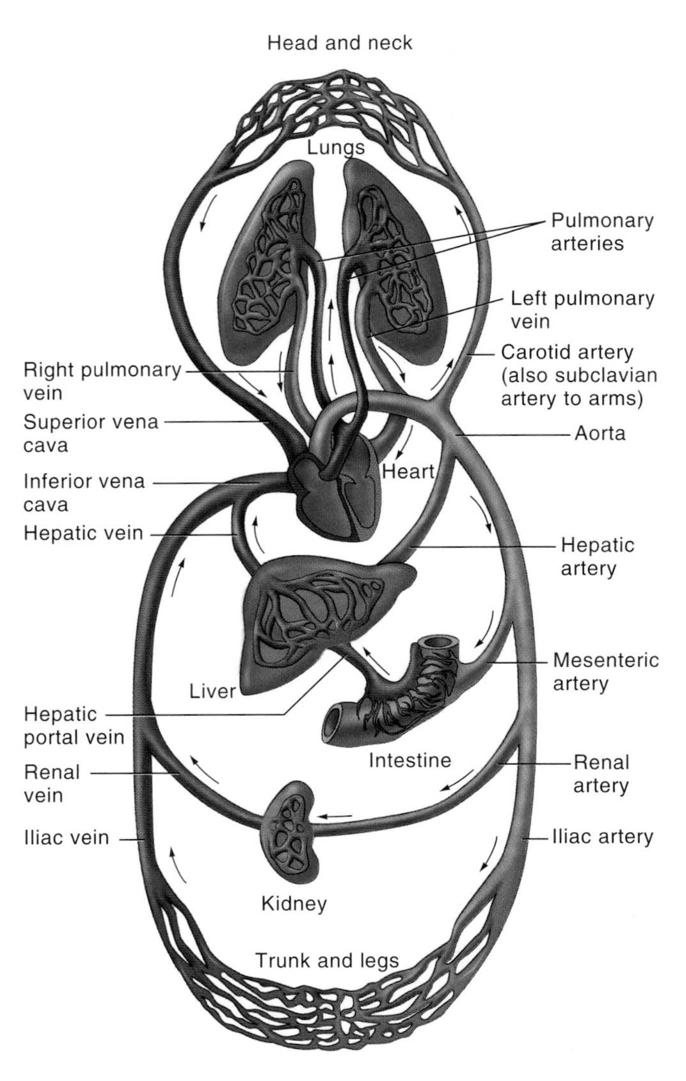

Figure 48.8

Pulmonary and systemic circulation—the journey of blood.

The first visible artery branching from the aorta is the **brachiocephalic artery** (fig. 48.11), which originates from the aortic arch. The brachiocephalic artery is short and divides into the **right common carotid artery,** which supplies the right side of the neck, and the **right subclavian artery,** which supplies the right shoulder and arm. At the most anterior part of the bend in the aortic arch you should find the **left common carotid artery,** which conducts blood up the left side of the neck. Immediately to the left of the left common carotid artery is the left **subclavian artery.** This artery supplies blood to the left shoulder and arm.

Branches of the Abdominal Aorta

Push the abdominal viscera to the left to locate the abdominal arteries. The first arterial branch from the abdominal aorta below the diaphragm is the **celiac artery.** This artery delivers blood toward the stomach, liver, spleen, and pancreas. The second artery arising from the abdominal artery is the **superior mesenteric artery,** which is larger than the celiac artery. Notice that the superior mesenteric artery delivers blood directly to the small intestine and intestinal

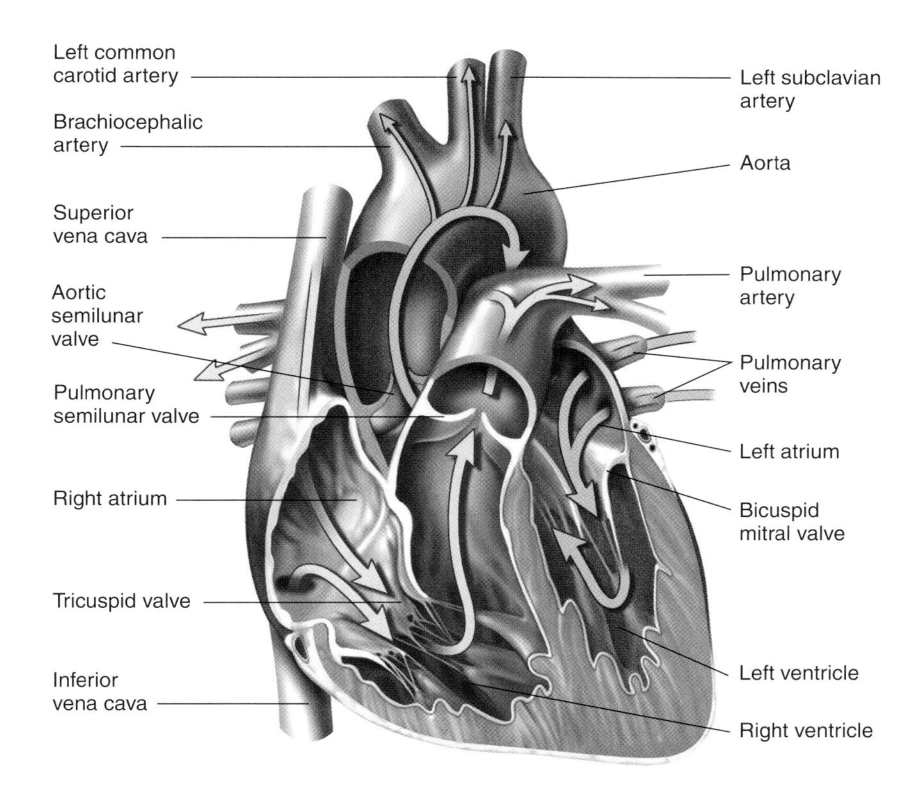

Figure 48.9

Path of blood through the human heart. The mammalian heart is divided into four chambers. Blood enters the right atrium from the superior vena cava and passes into the right ventricle through a valve that prevents backflow of blood. From the right ventricle, blood moves through the pulmonary artery and the lungs. Oxygenated blood from the lungs returns to the heart via the pulmonary veins. This blood enters the left atrium and then the left ventricle, from which it enters the general circulatory system of the body through the aorta.

mesenteries. The **renal arteries** are short and lead directly to the kidneys. Just posterior to the renal arteries are the **genital arteries** leading to the testes of a male and the ovaries of a female. Farther along the posterior end of the abdominal aorta you will find a pair of **iliolumbar arteries** leading to the dorsal muscles of the back. Next, the **inferior mesenteric artery** leads to intestinal mesenteries. The abdominal aorta also gives rise to the **caudal artery,** which passes into the dorsal body wall and supplies blood to the tail. Gently pull the posterior end of the abdominal aorta slightly to the side to see the caudal artery. The abdominal aorta finally divides posteriorly to form the **iliac arteries,** which deliver blood to the pelvic area and hind extremities.

Question 5

a. Which abdominal arteries are paired? Which are singular?

b. Which artery is a significantly different length in mature males and females? Why?

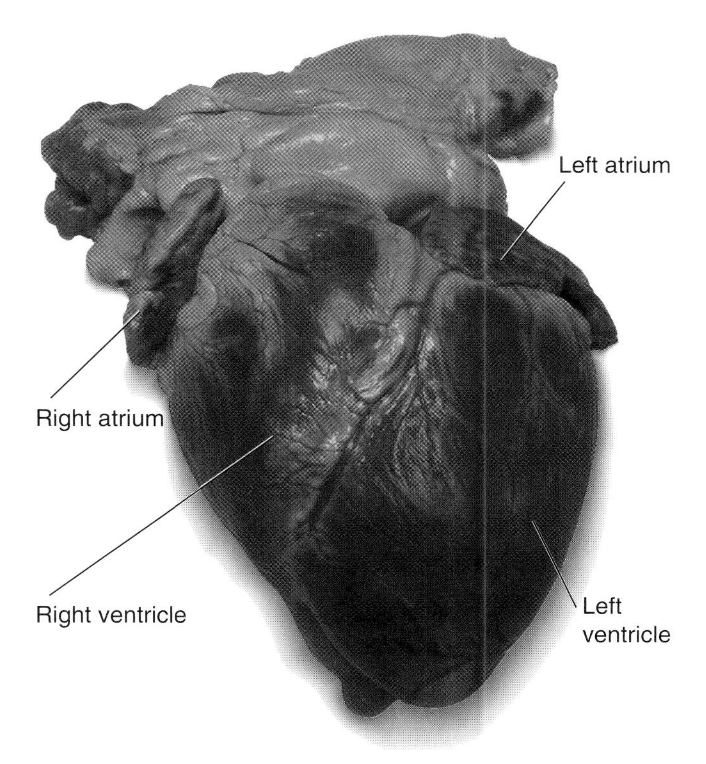

Figure 48.10

The human heart.

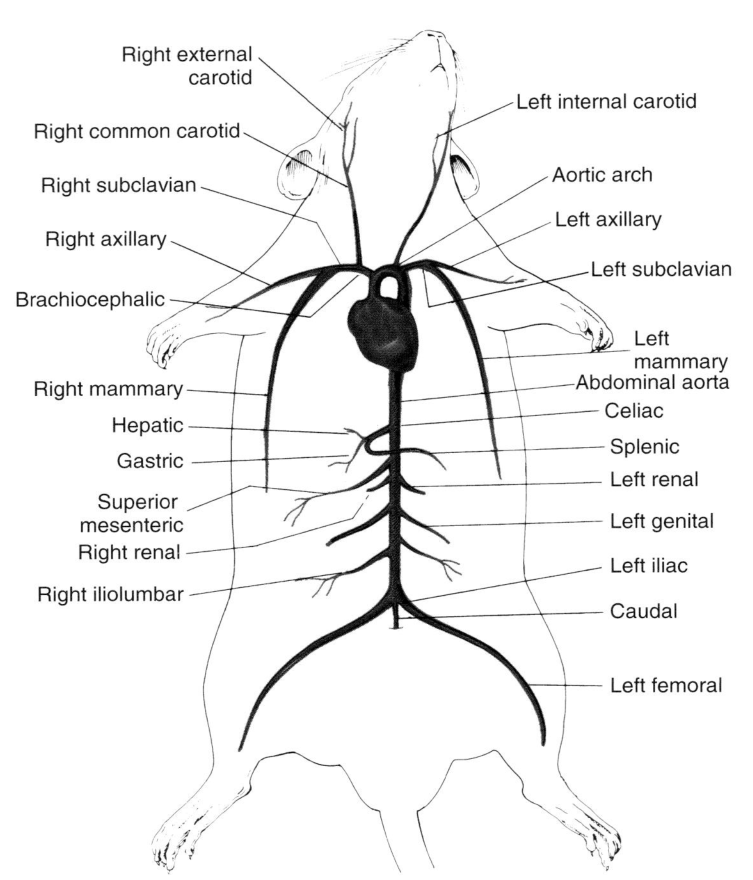

Figure 48.11

Arterial system of a rat.

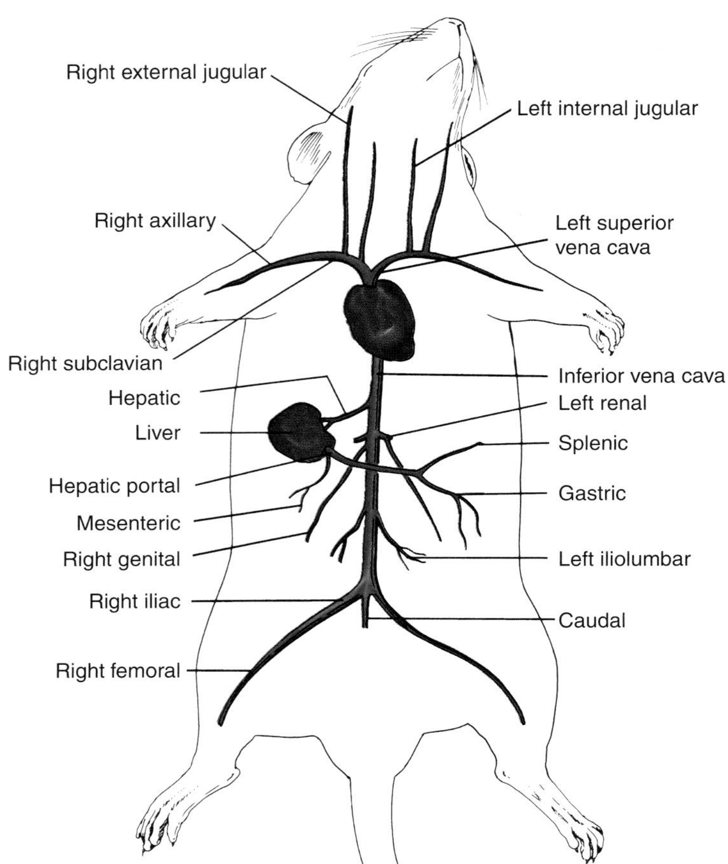

Figure 48.12

Venous system of a rat.

Secondary Branches of the Thoracic Arteries

The **right subclavian artery** branches from the brachiocephalic artery. The right subclavian artery is short, passes under the clavicle, and branches into the **right internal mammary artery** (look along the inside of the ventral chest wall), and the **right axillary artery** leading toward the armpit. The **right common carotid** passes anteriorly along the neck after branching from the brachiocephalic artery and gives rise to the **right external carotid artery,** servicing the face, and the **right internal carotid artery,** servicing the inner areas of the head. The left common carotid artery arises from the aortic arch and carries blood along the left side of the neck. This artery gives rise to the **left external** and **internal carotid arteries.** The left subclavian artery branches from the aortic arch and divides to form the arterial branches similar to those on the right side.

Secondary Branches of the Abdominal Arteries

The celiac artery gives rise to (1) the **hepatic artery,** which supplies blood to the liver, (2) the **gastric artery,** which leads to the stomach, and (3) the **splenic artery,** which leads to the spleen and pancreas. Carefully trace the celiac artery from the abdominal aorta until you find these branches. If the arteries were cut during dissection, you will not be able to positively identify these branches. The **iliac arteries** give rise to many branches and continue on the inner thighs as the left and right **femoral arteries.**

The Systemic Veins

The **left** and **right superior venae cavae** conduct blood from the upper part of the body into the right atrium (fig. 48.12). Trace these veins from the atrium. As you follow a superior vena cava anteriorly, it produces a small **internal jugular vein** and continues as the **subclavian vein.** The subclavian vein then divides anteriorly into the **external jugular vein** and an **axillary vein.** The jugular veins drain blood from the head, and the axillary veins bring blood from the shoulders and arms.

The **inferior vena cava** carries blood from the lower body to the right atrium. The **hepatic vein** drains the liver and enters the inferior vena cava near the diaphragm. **Renal veins** drain the kidneys. **Genital veins** lead from the gonads

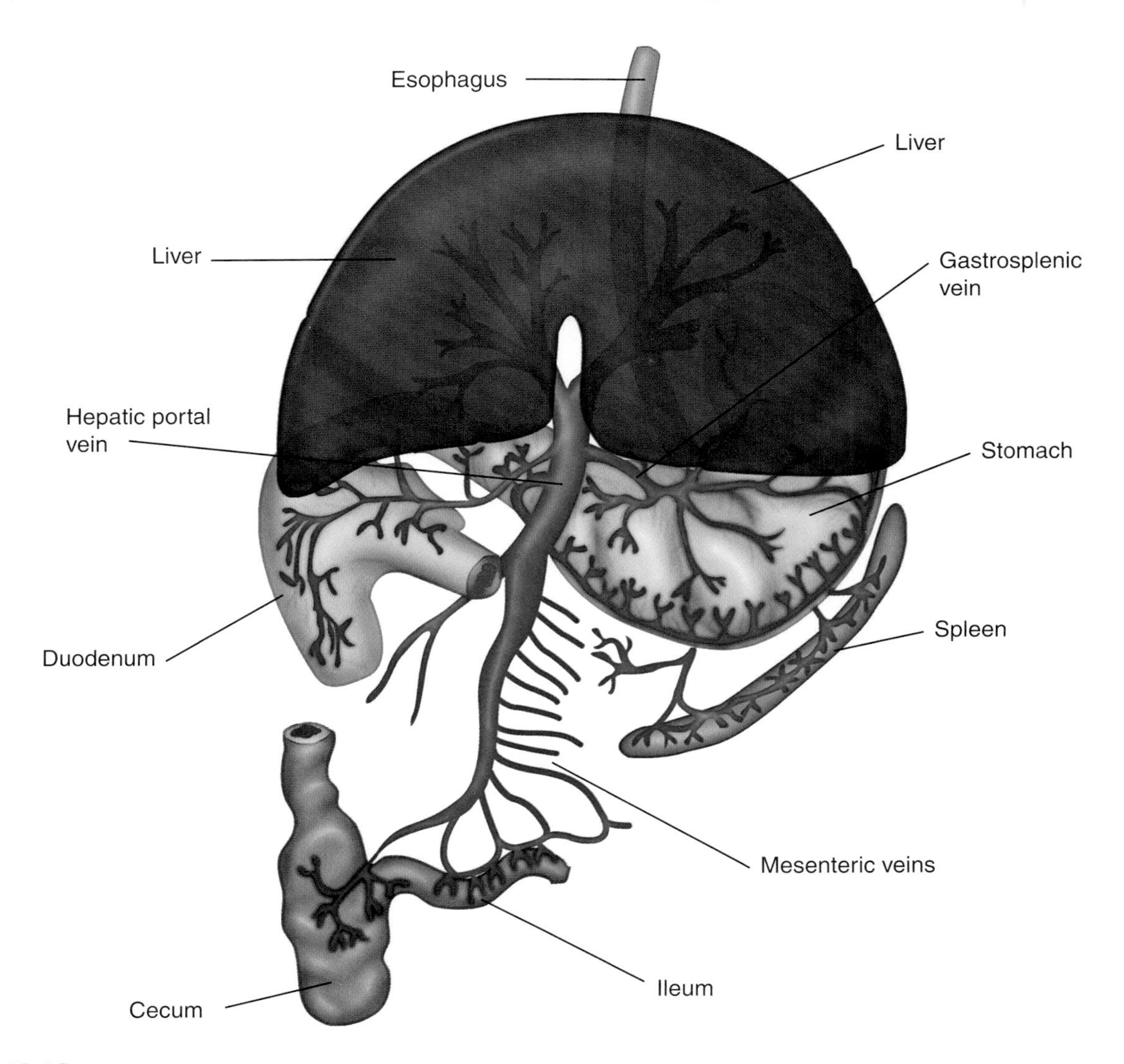

Figure 48.13
Hepatic portal system.

either to the inferior vena cava or to the renal veins. The **iliac** and **femoral veins** drain the legs, and the **caudal vein** drains the tail.

Question 6

a. Which veins are paired? Which are singular?

b. Why are veins in appendages buried deeper than arteries?

The Hepatic Portal System

A **portal system** is a system of veins that carries blood from one bed of capillaries to another bed of capillaries (fig. 48.13). The **hepatic portal system** carries blood from capillaries in the mesenteries, small intestine, spleen, stomach, and pancreas to the liver. Specifically, the **gastric, splenic,** and **mesenteric veins** drain the digestive system and unite to form the **hepatic portal vein,** which carries blood to the liver. In the small intestine, digestive enzymes break down food, and the products pass across the intestinal wall into the bloodstream. Thus, the liver is strategically located to receive blood after nutrients have been absorbed in the intestinal tract. The liver cells can easily modify these nutrients and secrete materials into the circulatory system. In the liver, blood from the hepatic portal vein passes through another capillary system, which ultimately coalesces to form the hepatic vein leading to the inferior vena cava. In some preserved rats the hepatic portal system has been injected with yellow latex. Ask your instructor to help you locate the portal system if your rat was not injected.

After you have located all of the arteries and veins on your specimen, test your knowledge by tracing the flow of blood from the kidney to the liver. Name all of the vessels and organs through which a blood cell might pass. You will be surprised at the number of vessels and structures you mention.

Question 7

Does the hepatic portal vein carry oxygenated or deoxygenated blood? Why should blood pass through the liver after leaving the digestive system and before entering the other organs?

Questions for Further Thought and Study

1. Trace the path of a blood cell as it moves from the capillaries of the brain to the small intestine.

2. Do you recall any vertebrates with external reproductive organs on their dorsal surface? What advantages or disadvantages would this provide?

3. Body fluids include urine, blood, food, and reproductive fluids. Trace the path of each as it moves through the body.

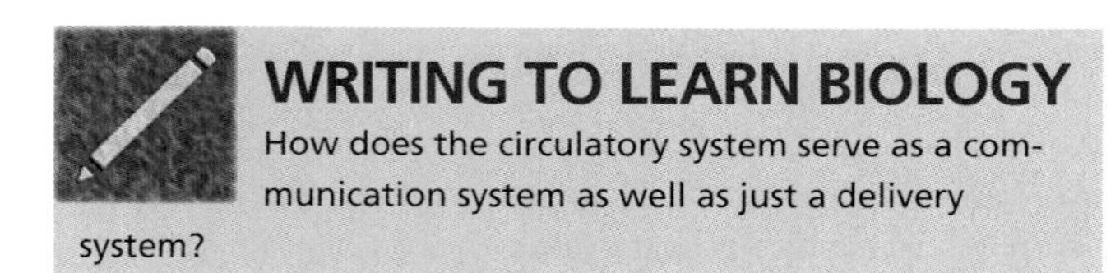

49

Embryology

Comparative Morphologies and Strategies of Development

Objectives

By the end of this exercise you should be able to:

1. Describe the early stages of embryological development common to advanced invertebrates and vertebrates.
2. Understand the formation of a three-layered embryo with ectoderm, endoderm, mesoderm, and a presumptive digestive cavity.
3. Relate the major structures of early embryos to the environment in which they develop.

All sexually reproducing multicellular organisms begin life as single-celled zygotes and progress through stages of **growth, differentiation,** and **morphogenesis.** Growth is the irreversible increase in size of an organism. Differentiation is the structural and functional specialization of groups of cells. Morphogenesis is the development of pattern, shape, and form. **Embryology** is the study of these processes in the early development of an organism.

Early stages of embryology are similar in advanced invertebrates and vertebrates, and include formation of body cavities and multiple cell layers. During these stages, embryos must overcome fundamental problems. For example, development always requires energy (i.e., food), protection, as well as a waste disposal system. The specific morphologies and structures involved in the development of these functions vary among different groups of animals. That is, different organisms have different adaptations to cope with the same problems.

During this exercise you will compare features and developmental strategies of the three earliest stages of development of the sea star, frog, and chick. These developmental stages are:

- **Fertilization**—Male and female gametes fuse to form a zygote.
- **Cleavage**—Zygote divides into a larger and larger number of smaller and smaller cells that eventually form a hollow sphere of many cells. Cleavage ends when groups of cells begin to differentiate.
- **Gastrulation**—Cells of the sphere formed during cleavage continue to divide and move inward to form three cellular layers.

SEA STAR DEVELOPMENT

Examine either a preserved or living *Asterias* on display and review in your textbook the general characteristics of echinoderms. *Asterias* is a common sea star belonging to the invertebrate phylum Echinodermata (see Exercise 39).

Fertilization

If models are available in addition to living and prepared material, use them to examine sea star development. Examine a prepared slide of an unfertilized sea star egg and compare it with a slide of a zygote (fig. 49.1*a*). Eggs and sperm of a sea star are extruded from the body into seawater, and fertilization occurs externally. If living organisms are available, your instructor will prepare the organisms and stimulate them to shed eggs and sperm. These small eggs have only a small amount of yolk that is distributed evenly in the cytoplasm. Sea star eggs are **isolecithal,** meaning that their yolk is distributed evenly. This is important because the amount and distribution of yolk in an organism's egg strongly affects its pattern of development.

Question 1
Are both the egg and zygote of a sea star the same size?

Cleavage

Examine prepared slides of the two- or four-celled cleavage stage (fig. 49.1*b*) and of the blastula of a sea star (fig. 49.1*c*).

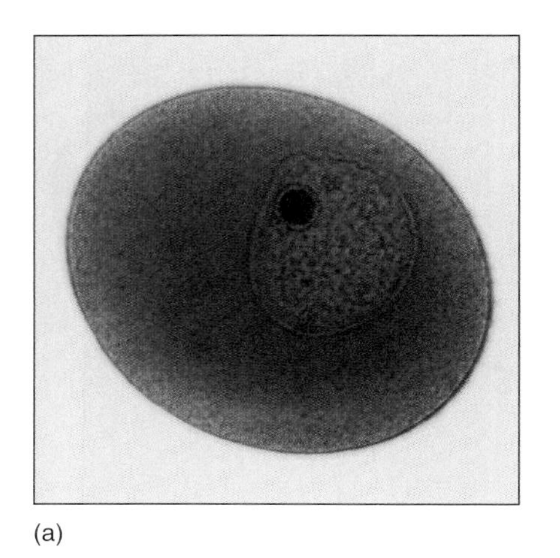

(a)

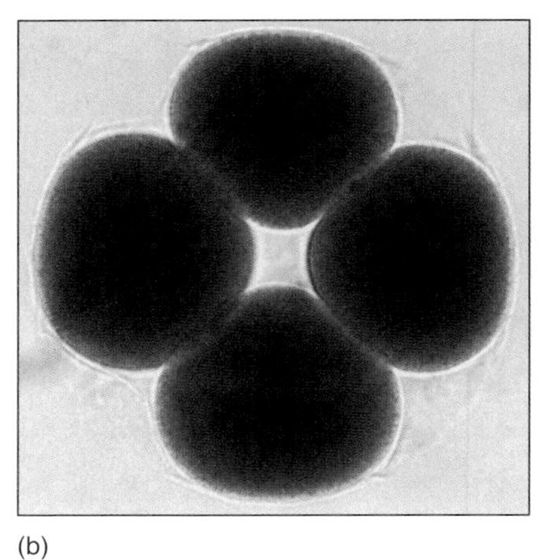

(b)

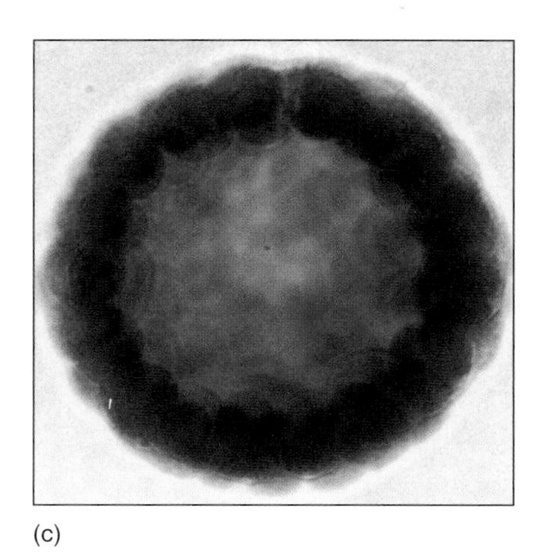

(c)

Figure 49.1
Early stages in sea star embryology. (*a*) Zygote. (*b*) Early cleavage. (*c*) Blastula.

Approximately 1 h after fertilization, the zygote cleaves (divides) to form two cells. Cleavage of the entire zygote is termed **holoblastic,** meaning that only a small amount of yolk is present and easily divides. Continued cleavage forms a ball of 16–32 cells called a **morula,** and later a hollow sphere of many cells called a **blastula.** The fluid-filled cavity within the blastula is the **blastocoel,** and the cells are called **blastomeres.**

Question 2

a. Are the cells of the two-celled stage of a sea star similar in size?

b. How does the size of an embryo during early cleavage compare with that of the zygote? Is the sea star embryo growing during cleavage?

c. How many distinct layers of cells do developing sea stars have at the blastula stage?

d. The blastomeres of a mature blastula eventually develop cilia and the entire blastula can rotate and move. How is movement important to a developing sea star?

Gastrulation

Examine prepared slides of gastrulation stages and early feeding stages of sea star development (fig. 49.2). Gastrulation occurs after the blastula has matured and consists of many cells. During early gastrulation, cells invaginate into the blastocoel and form a **gastrula** and a new cavity, the **archenteron** (fig. 49.2*a, b*). After invagination the gastrula has two germ layers of cells, the endoderm and ectoderm. The blastocoel soon disappears, and the archenteron (sometimes called the gastrocoel) becomes the presumptive digestive tract.

The outer layer of gastrula cells is the **ectoderm** and will form the skin and nervous system of the mature organism. The inner layer is the **endoderm** and will form the digestive tract. Refer to your textbook for a list of other organs formed by these two germinal layers. The third and final germ layer is the **mesoderm** that forms between the ectoderm and endoderm. The mesoderm forms from cells that disassociate from the endoderm. The mesoderm produces muscular tissue and parts of the reproductive and circulatory systems.

The development of organs is called **organogenesis.** In sea stars, gastrulation is followed by organogenesis and the formation of a motile and feeding larval stage, the **bipinnaria larva** (fig. 49.2*c*). This stage is followed by the **brachiolaria larva** (fig. 49.2*d*). These larvae attach to the substrate and undergo considerable metamorphosis. Soon thereafter, sea star arms grow from the body with no relation to the original bilateral symmetry. Eventually, the maturing organism detaches from the substrate as a mobile, **juvenile** sea star (fig. 49.2*e*).

Question 3

Why are the ectoderm and endoderm called "germ" (from the word *germinate*) layers?

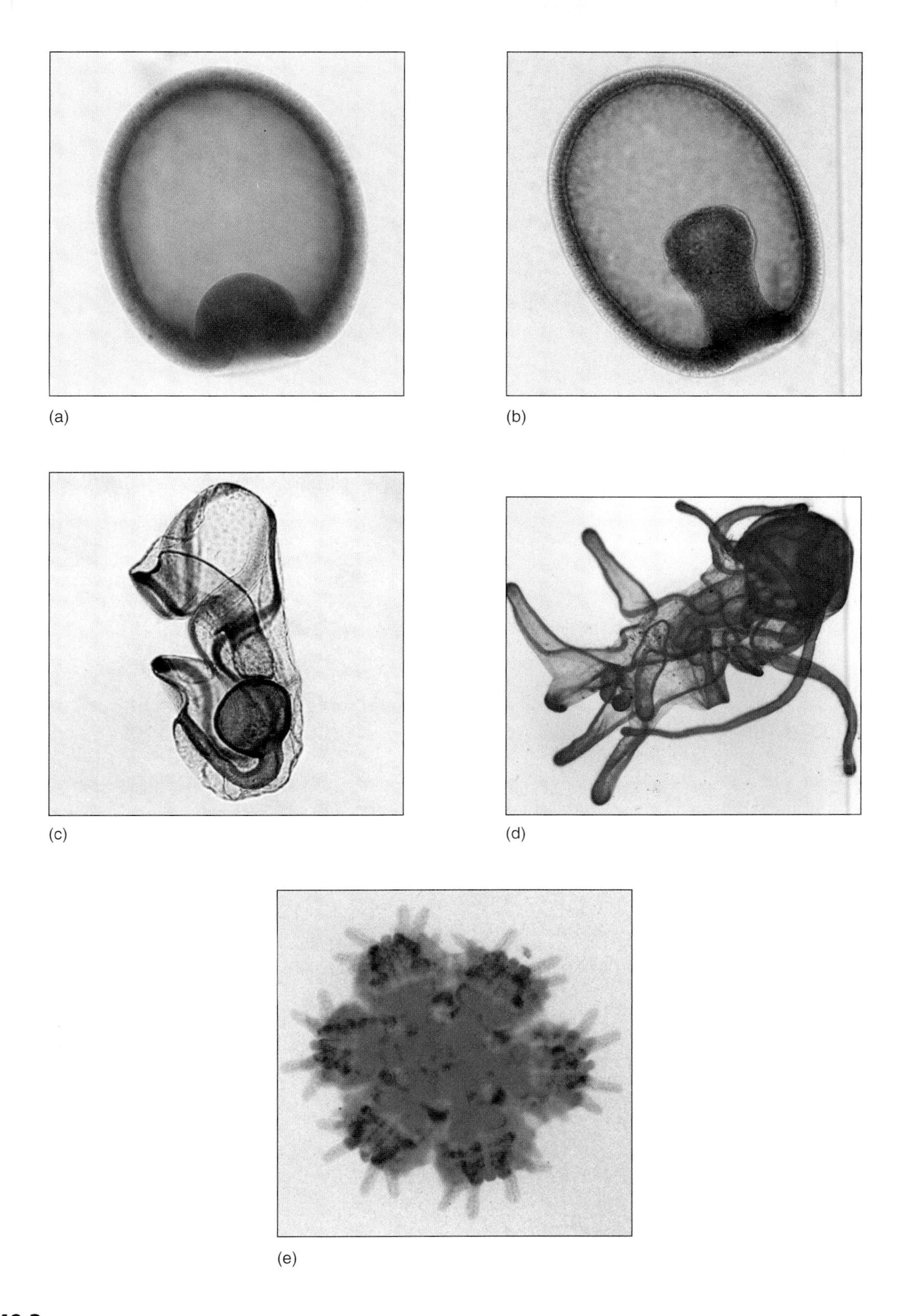

Figure 49.2
Developmental stages of a sea star. (*a*) Early gastrula. (*b*) Late gastrula. (*c*) Bipinnaria larva. (*d*) Brachiolaria larva. (*e*) Juvenile sea star.

Figure 49.3
External fertilization. When frogs mate, as these two are doing, the clasp of the male induces the female to release a large mass of mature eggs, over which the male discharges his sperm. A gelatinous mass of eggs is visible in the background.

FROG DEVELOPMENT

Review in your textbook the general characteristics of chordates, vertebrates, and amphibians. Frogs are vertebrates in class Amphibia of phylum Chordata (Exercise 39). Although frogs often live on land, they require freshwater to lay eggs and reproduce.

Fertilization

Examine prepared slides of a frog egg and zygote. If models are available, use them to examine frog development.

Fertilization is external for frogs, but it coincides with contact between a male and female frog during **amplexus** (fig. 49.3). In this process a male clasps a female between his front legs and applies pressure to stimulate her to release eggs. The male then releases sperm cells into the water immediately surrounding the eggs. Fertilization occurs externally.

Penetration of an egg by a sperm cell activates the cytoplasm to become specialized in certain regions. A darkly pigmented hemisphere, the **animal pole,** and a lighter hemisphere, the **vegetal pole,** appear opposite each other. The vegetal pole contains a significant amount of yolk; thus, frog eggs are **telolecithal** (fig. 49.4).

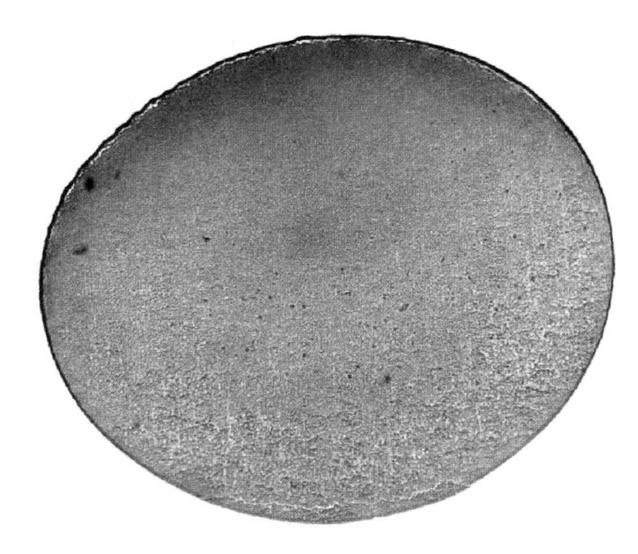

Figure 49.4
Frog zygote.

Question 4

a. How large is a frog zygote compared to that of a sea star?

b. Are any areas of the frog zygote more darkly pigmented than other areas?

Cleavage

Examine slides of early and late cleavage of a developing frog embryo (fig. 49.5) and a slide of a cross section of a frog blastula (fig. 49.6).

Cleavage of frog zygotes is holoblastic, but the resulting cells are not equal in size. As they continue to divide a morula develops. Fluid then collects in the center of the morula and the hollow mass of cells forms a blastula. Vegetal pole cells of the blastula are laden with yolk and divide more slowly than do the animal pole cells. This produces an asymmetrical blastula.

Question 5

a. Which are larger, animal pole cells or vegetal pole cells?

b. Is the developing blastocoel visible in late cleavage?

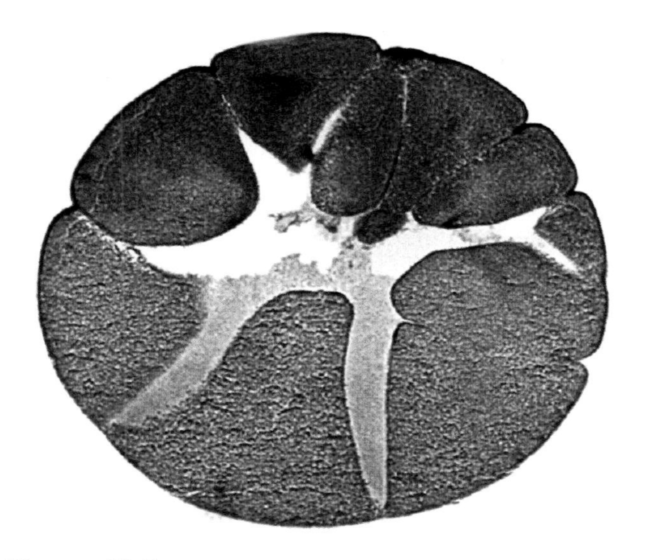

Figure 49.5
Frog late cleavage, cross section.

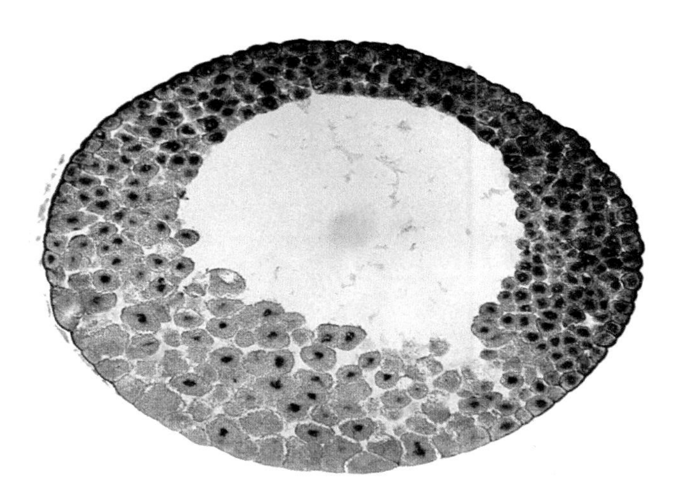

Figure 49.6
Frog blastula, cross section.

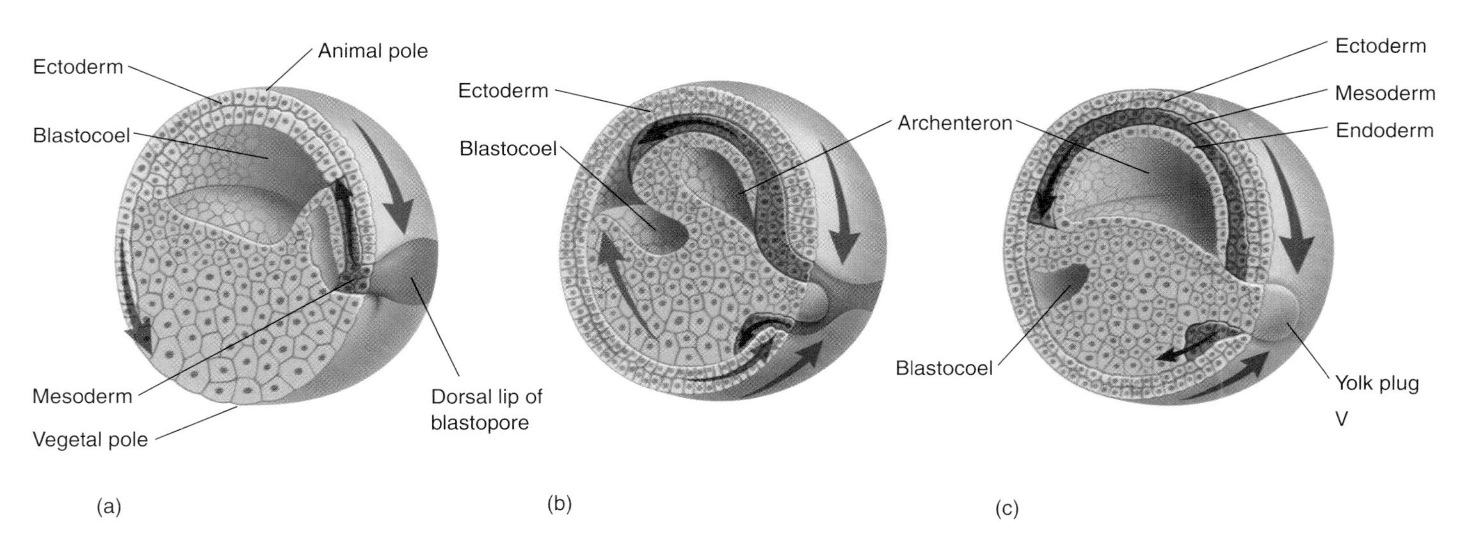

Figure 49.7
Frog gastrulation. (*a*) A layer of cells from the animal pole folds down over the yolk cells, forming the dorsal lip. (*b*) The dorsal lip zone then invaginates into the hollow interior, or blastocoel, eventually pressing against the far wall. The three principal tissues (ectoderm, endoderm, and mesoderm) become distinguished here. (*c*) The inward movement of the dorsal lip creates a new internal cavity, the archenteron, which opens to the outside through the yolk plug remaining at the point of invagination. The gastrula is a three-layered system with a cavity destined to become the digestive tract. In phyla such as protostomes, the blastopore will give rise to the mouth; in other phyla (deuterostomes), it gives rise to the anus. Humans and frogs (phylum Chordata) are deuterostomes.

c. How does orientation of the blastocoel within the blastula compare with that of a sea star?

Gastrulation

Examine a cross section of a frog gastrula (fig. 49.7). Because yolk-laden cells of the vegetal pole of a frog blastula are larger than cells of the animal pole, the layer of blastomeres does not simply bend inward and invaginate as it does in sea stars. Instead, the layer of animal-pole cells moves, grows down over the vegetal cells, and moves inward at a depression called the **blastopore.** This growth and movement is called **involution.**

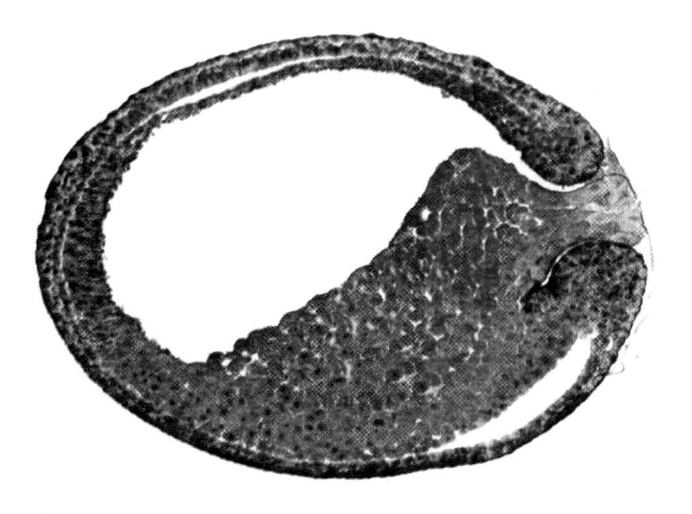

Figure 49.8
Frog gastrula, cross section.

The yolk cells are enveloped during gastrulation. However, some of them protrude as a **yolk plug** through the **blastopore** (fig. 49.8). The mesodermal layer develops between the endoderm and ectoderm from cells proliferating within the dorsal lip of the blastopore.

Gastrulation in vertebrates is followed by formation of the **neural tube (neurulation)** and other organs (organogenesis), thereby producing a functional larva.

Question 6

a. Is the degenerating blastocoel visible in the frog gastrula?

b. What is the relative size of the frog gastrula compared to the blastula and zygote?

c. How much yolk protrudes through the blastopore of a frog gastrula?

d. What is the major function of the larval stage in the overall life history of an organism?

e. Are all of the organ systems fully developed in a free-swimming larva? Which organ systems are unlikely to be fully developed?

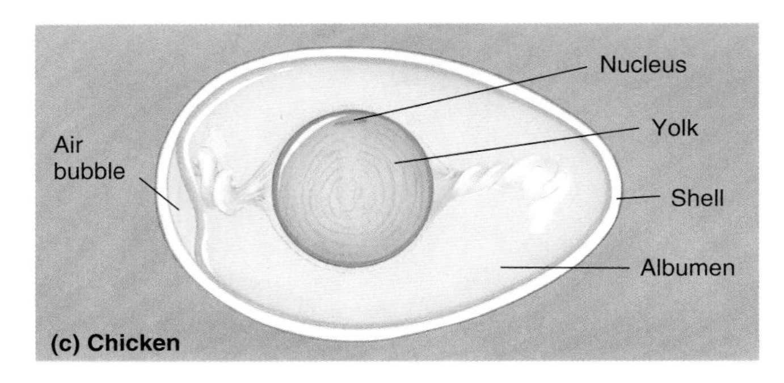

Figure 49.9
Chicken egg.

f. How has the developing frog embryo dissipated its metabolic wastes?

CHICK DEVELOPMENT

Review in your textbook the characteristics of bird eggs. Birds are vertebrates and have developmental structures and strategies adapted for a totally terrestrial existence. Specifically, birds (and reptiles) have eggs with protective shells, a large amount of yolk, and an intricate membrane system for various functions.

Fertilization

Crack open and examine an unfertilized chicken egg (fig. 49.9). If models are available in addition to living and prepared material, use them to examine chick development.

Fertilization occurs internally before a hard shell is produced. The egg cell is a large yellow mass of yolk with a small area of cytoplasm on its surface called a **germinal disc,** which contains the egg nucleus. Cells formed from the germinal disc later form the embryo. Surrounding the egg cell is clear, watery **albumen** that supplies the embryo with water and food in addition to the yolk. Albumen is surrounded by two shell membranes and a hard shell that is porous to gases.

Question 7

a. Is all of the albumen of a chicken egg the same consistency?

b. If you pull a probe through the albumen it seems to cling to the probe and to itself. What macromolecules are abundant in albumen and make it viscous?

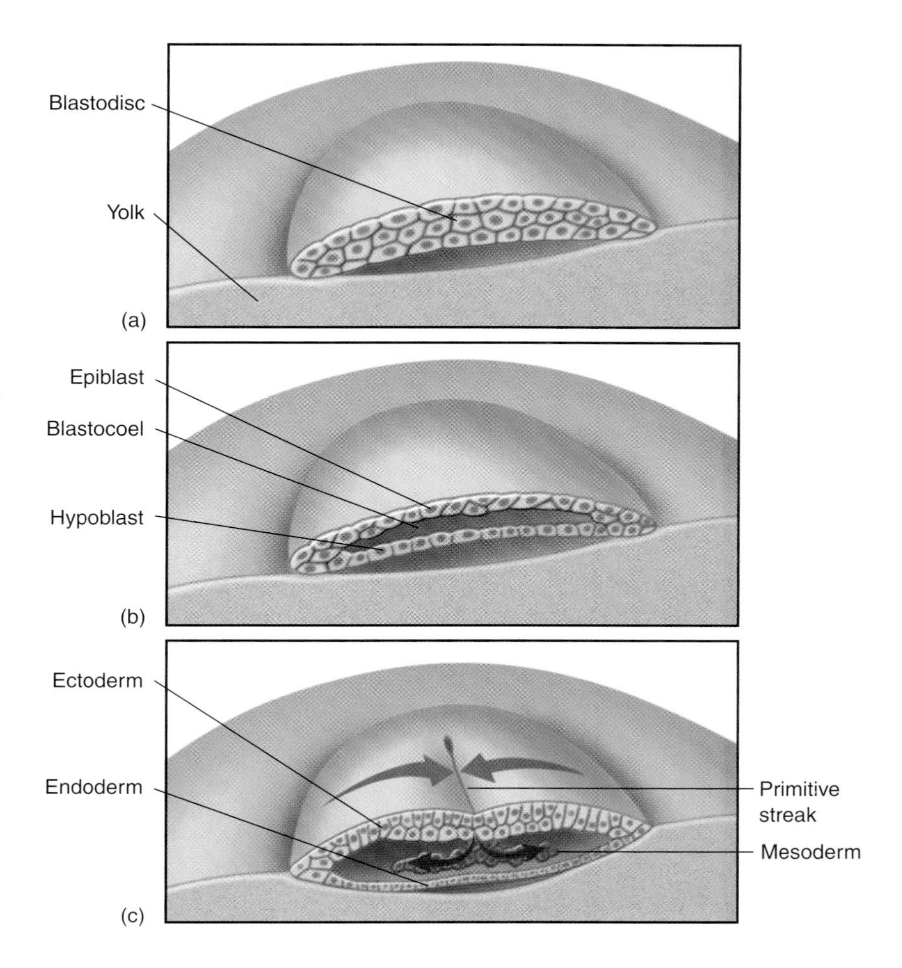

Figure 49.10

Gastrulation of the chick blastodisc. (*a*) The two sides of the chick blastodisc are not separated by yolk. (*b*) The upper layer of the blastodisc differentiates into ectoderm, the lower layer into endoderm. (*c*) Among the cells that migrate into the interior through the dorsal primitive streak are future mesodermal cells.

c. Why must an eggshell be porous?

d. What functions does the shell have?

Cleavage

Yolk is so abundant in bird eggs that the entire zygote cannot divide; instead, cleavage is confined to the germinal disc. Such partial cleavage is termed **meroblastic** and produces a flat blastodisc of developing cells lying atop the bulky yolk. As the **blastodisc** develops, it separates from the underlying yolk and then forms two layers: the **epiblast** and **hypoblast** separated by a blastocoel (fig. 49.10a, b). The epiblast will become the ectoderm.

Gastrulation

As in the sea star and frog, gastrulation involves movement of cells into the hollow blastula to establish a multilayered system. However, the opening of the blastodisc is not a round blastopore; rather, it is a linear furrow called the **primitive streak** (fig. 49.10c). Cells migrate across the surface of the blastodisc and into the primitive streak. Most of the migrating cells become the mesoderm, some of whose cells merge with the hypoblast to form the endoderm. The primitive streak becomes the midline of the developing chick.

Surrounding the embryo are networks of arteries and veins leading to extraembryonic membranes. These membranes in eggs developing in a terrestrial environment are important because they assume functions performed by the aquatic environment for other organisms such as frogs and sea stars (fig. 49.11).

The **amnion** surrounds the embryo, the **yolk sac** surrounds the yolk, and the **allantois** forms a waste disposal sac.

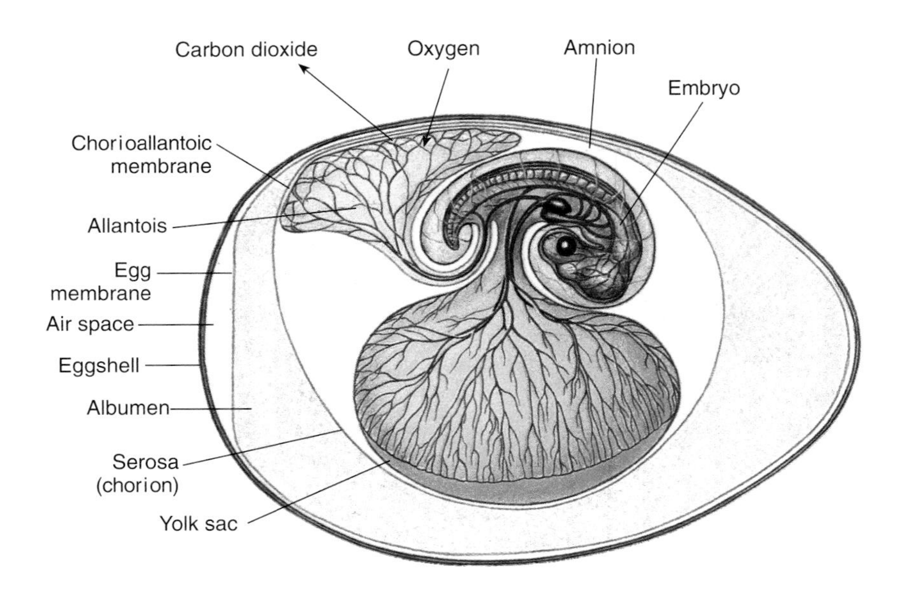

Figure 49.11
The amniotic egg is perhaps the most important feature that allows some vertebrates to live in a variety of terrestrial habitats. An egg protects the embryo from drying out, nourishes it, and enables it to develop outside of water. An egg contains a large nourishing yolk and abundant albumen with nutrients and water. Nitrogenous wastes are excreted into the saclike allantois for storage. The amnion surrounds the developing embryo, and the chorion helps control diffusion of substances into and out of the embryo. The shell offers protection but is porous for diffusion of oxygen and carbon dioxide.

TABLE 49.1

FUNCTIONS OF THE MAJOR MEMBRANES OF A DEVELOPING CHICK

Membrane	Function
Amnion	
Allantois	
Chorion	
Yolk sac	

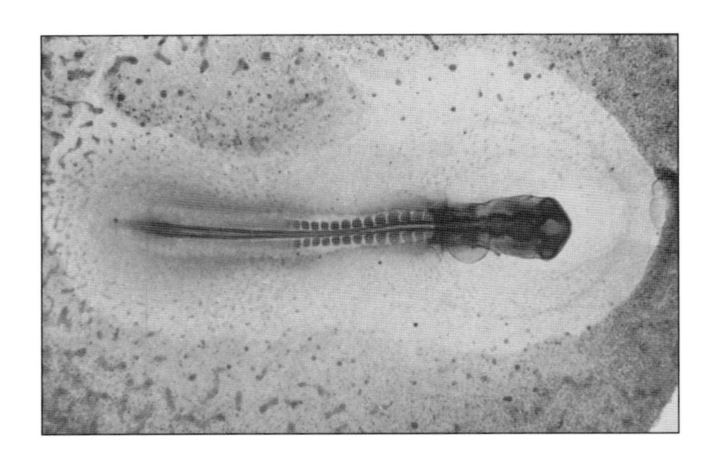

Figure 49.12
Chick embryo at 33 hours, whole mount.

The **chorion** surrounds the entire embryo and yolk. Review in your textbook the function of each of these membranes and complete table 49.1.

Procedure 49.1
Examine chick embryos

1. Observe the demonstration specimens of living 33- and 72-h chick embryos.
2. Examine plastic mounts of 24-, 33-, and 72-h chick embryos, and follow the formation of the brain, heart, and eye.
3. Compare the living and mounted specimens with figures 49.12 and 49.13.

After 33 h of development the blastodisc is about 20 mm in diameter. The newly developed heart can be seen on the right side. Two veins bring blood to the heart, and a single ventral aorta drains blood from the heart. The developing brain continues as a neural tube the length of the body. After 72 h of development the anterior part of the embryo lies on its left side. The central nervous system and eye have developed considerably.

Question 8

a. Has organogenesis begun for 33- and 72-h chick embryos?

b. Is the heart beating in 33- and 72-h chick embryos?

c. Sea star and frog embryos do not have membranes similar to those of a chick embryo. How do sea star and frog embryos accomplish the functions of these membranes?

d. Chickens lack a larval stage. How do they compensate for this?

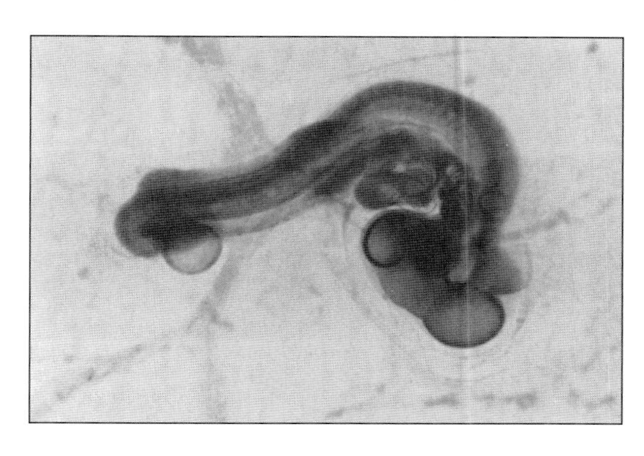

Figure 49.13
Chick embryo at 72 hours, whole mount.

REVIEW

To review the comparative features of early development of the sea star, frog, and chick, complete tables 49.2 and 49.3 with appropriate descriptions and associated structures.

TABLE 49.2

A COMPARISON OF EMBRYOLOGICAL FEATURES OF A DEVELOPING SEA STAR, FROG, AND CHICK

Feature	Sea Star	Frog	Chick
Relative egg size			
Amount of yolk			
Distribution of yolk			
Blastulation			
Gastrulation			
Larval stage			

TABLE 49.3

A COMPARISON OF ECOLOGICAL ASPECTS OF EARLY DEVELOPMENT OF SEA STARS, FROGS, AND CHICKS

Feature	Sea Star	Frog	Chick
Environment of development			
Fertilization			
Mechanism of waste disposal			
Physical protection			
Parental care			

Questions for Further Thought and Study

1. Where does a sea star embryo get its nutrition during prelarval stages of development?

2. More recently evolved organisms have modified the stages of embryological development. Would you expect early or late stages to be modified the most? Why?

3. The basic stages of embryological development are remarkably similar for a wide range of organisms. How would you explain such consistency?

4. As cells invaginate and move about, do they "know" where they are? If not, how is movement controlled? If so, how do they perceive their position?

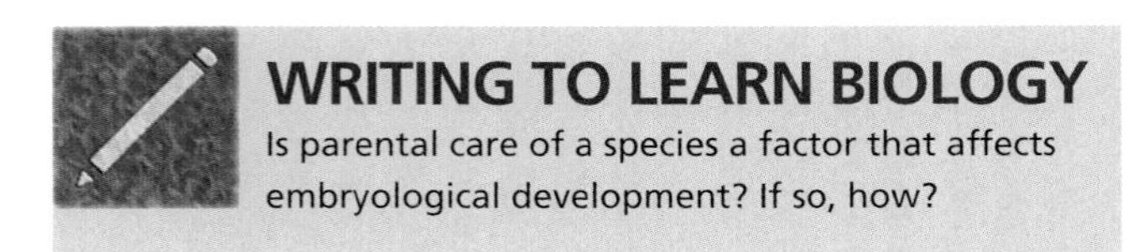

WRITING TO LEARN BIOLOGY

Is parental care of a species a factor that affects embryological development? If so, how?

50

Animal Behavior
Taxis, Kinesis, and Agonistic Behavior

Objectives

By the end of this exercise you should be able to:
1. Test how various stimuli influence the behavior of brine shrimp and pill bugs.
2. Classify a behavior as an example of an agonistic behavior, kinesis, or taxis.

Biologists explain animal behavior by observing animals and by studying the physiology of behavior. Explanations of animal behavior are often complex. To minimize this complexity, biologists often study animals that have a limited range of behaviors and that have ecology, evolution, and senses we understand. Complex behaviors are usually understood only after understanding simple and isolated behaviors.

Orientations are behaviors that position an animal in its most favorable environment. There are two types of orientations: taxis and kinesis. A **taxis** is a movement toward or away from a stimulus. Prefixes such as *photo-* and *chemo-* are usually added to describe the nature of the stimulus. For example, a fruit fly that flies toward light is positively **phototactic.** A **kinesis** is a random movement that is not oriented to the direction of the stimulus; that is, the stimulus initiates, but does not necessarily orient, the movement. The speed of the kinesis is determined by the intensity of the stimulus.

Some animals also exhibit **agonistic behaviors.** These behaviors usually occur when the animal is in a confrontation where there may be an attack or withdrawal. An agonistic behavior that results in retreat or avoidance is a **submission,** whereas one that produces a more forceful response is an **aggression.** Agonistic behaviors seldom lead to death; rather, they usually help an animal maintain its territory. In many animals, agonistic behaviors make the animal look larger or more threatening.

In this exercise, you'll study three examples of animal behavior: agonistic behavior in Siamese fighting fish, kinesis in pill bugs, and taxis in brine shrimp. You'll also design and do your own experiments to better understand these behaviors. Throughout this exercise, think of the adaptive significance of each behavior that you study.

AGONISTIC BEHAVIOR IN SIAMESE FIGHTING FISH

In male Siamese fighting fish (*Betta splendens*), the sight of another male triggers an innate, intraspecific, ritualized series of responses toward the intruder (figs. 50.1, 50.2). These agonistic behaviors include broadside movements and facing movements. During broadside movements, the *Betta* turns its side toward the opponent. Typical aspects of broadside movements include tail flashing (closing and reopening the caudal fin), tail beating (sudden movement with its caudal fin toward an opponent), and pelvic fin flickering (moving its pelvic fin vertically opposite an opponent).

After these broadside movements, a fish may perform a facing movement, at a right angle to the head of its opponent, while holding its fins erect. Simultaneously, the fish usually erects its gill covers (opercula); the branchiostegal membranes, which are under the gill covers, may protrude during the display. The fish may also perform other agonistic behaviors, such as darkening its skin, slightly arching its back, raising its dorsal fin, lowering its head, darting, biting, or nipping an opponent.

In this exercise, you will determine (1) what stimulus initiates agonistic behavior in *Betta,* and (2) what parts of the stimulus are most important to trigger the response. Before you do this exercise, familiarize yourself with the anatomy of a Siamese fighting fish; be sure you can identify the dorsal fin, pectoral fin, ventral fin, gill cover, and tail (fig. 50.1). Also know how to recognize the agonistic behaviors of these fish. Figures 50.2*a* and *b* show *Betta* in nonaggressive poses, whereas Figures 50.2*c* and *d* show *Betta* in aggressive poses.

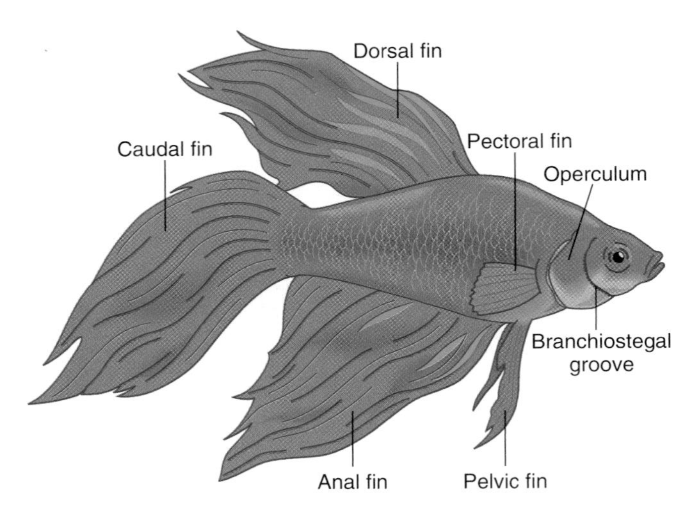

Figure 50.1
Male Siamese fighting fish.

Procedure 50.1

Understand agonistic behavior in Siamese fighting fish

1. The *Betta* you will study have been kept in separate aquaria such that the fish cannot see each other or their own reflections. Agonistic behaviors occur most frequently when the water temperature is 28°C.
2. Because some fish are more aggressive than others, you'll have to do a preliminary test to determine which fish are most likely to respond to the stimuli. Do this by holding a small mirror against the side of each aquarium so that each fish can see its reflection. Observe the fish's reaction.
3. Select the four or five most aggressive males for your experiment. Assign an identification number to each fish.
4. Decide which stimulus you will test (e.g., color of the opposing fish).
5. Form a hypothesis about the result you expect. Your instructor will advise you about how to write a testable hypothesis. Write your hypothesis here.
6. Decide how you will test your hypothesis. Describe your experimental design here.

In your experimental design, use photocopies of figure 50.1 to discover what parts of the stimulus are sufficient to trigger agonistic behavior. For example, if your group decides to test whether color causes the agonistic response, color each of the images of the same shape a different color. Then attach each image to a stick or straw and present the images to the fish one at a time.

7. While doing your experiment, do not present the image to a fish for longer than 5 min. Also, be sure that you study only one variable at a time. For example, do not color the fish with different colors; if you do, you'll not be able to isolate which variable (i.e., shape or color) causes the response.
8. Count and record in table 50.1 the number of times each display occurs during a test.

Question 1

a. Are the displays "all or none"? That is, do fish ever exhibit partial responses?

b. What is the simplest stimulus that initiates a response?

c. Which is more important for triggering agonistic behavior: color, size, or movement?

Nonaggressive postures **Aggressive postures**

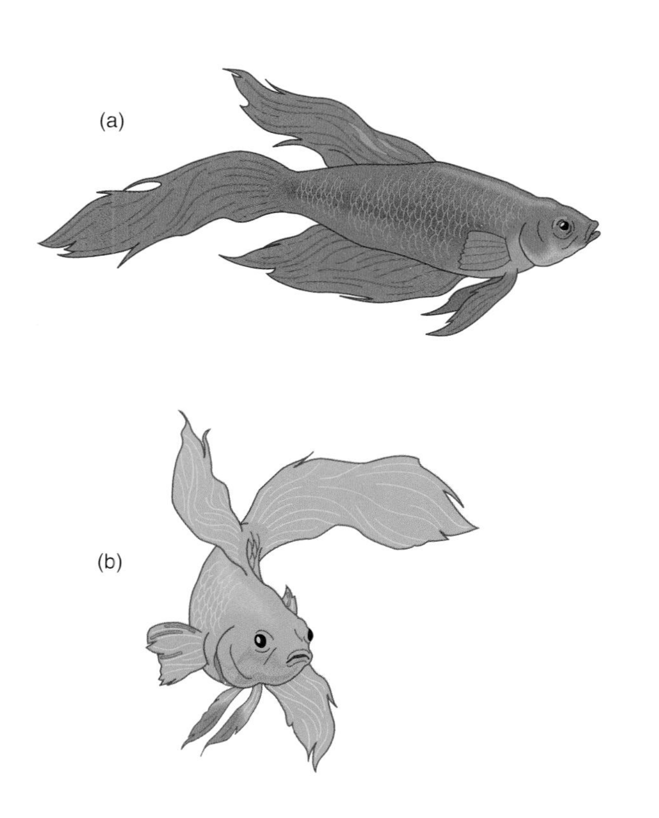

Figure 50.2
Male Siamese fighting fish in (*a*, *b*) nonaggressive postures and (*c*, *d*) aggressive postures.

d. After repeated identical stimuli, does a fish become "conditioned"? That is, does the duration of the behavior change or stop?

e. Will another species of fish trigger the behavior?

f. Will a female *Betta* trigger agonistic behavior in a male? If so, how does the response compare with that elicited by a male?

g. What is the adaptive significance of agonistic behavior?

KINESIS IN PILL BUGS

Pill bugs (also called sow bugs and rolypolies) are terrestrial crustaceans that spend much of their time avoiding dry environments (fig. 50.3). Pill bugs are easily collected in warm weather under logs, flower pots, and in leaf litter. As their name suggests, rolypolies often respond to mechanical stimuli by rolling into a ball.

In this exercise, you will study pill bug behavior in wet and dry environments.

TABLE 50.1

AGONISTIC BEHAVIORS IN SIAMESE FIGHTING FISH

Behavior	Test 1	Test 2	Test 3	Test 4	Test 5
Pelvic fin flickering					
Tail flickering					
Broadside movements					
Tail beating					
Facing movements					
Dorsal fin erect					
Gill cover erect					
Back arching					
Head lowered					
Skin darkening					
Biting and nipping					

Figure 50.3
Pill bugs are terrestrial crustaceans whose behavior is strongly influenced by moisture.

Question 2
How do you think that pill bugs will respond when they're placed in an environment whose surface is moist? In one that is dry?

Procedure 50.2
Study kinesis in pill bugs

1. Place five pill bugs in each of two large petri dishes, one containing wet filter paper and the other containing dry filter paper. Place the dishes in a dark drawer.
2. After 5 min, observe the pill bugs in the dishes. Before you open the drawer, assign each of the following tasks to a member of your group:
   - Count the number of pill bugs moving in each dish.
   - Count the number of turns (changes in direction) per minute for a single pill bug in each dish.

TABLE 50.2

OBSERVATIONS OF KINESIS IN PILL BUGS

Dish	Number of Pill Bugs Moving	Rate of Movement	Number of Turns per Minute
Wet			
Dry			

- Choose a moving pill bug in each dish. Determine its rate of movement by counting the number of times that the animal circles the dish.

3. Record your observation in table 50.2.

Question 3

a. What other stimuli might affect the behavior of pill bugs?

b. What is the adaptive significance of a kinesis?

Figure 50.4

A young brine shrimp. This crustacean has segmented appendages for swimming and pigmented eyespots for detecting light.

TAXIS IN BRINE SHRIMP (*ARTEMIA SALINA*)

Brine shrimp (*Artemia*) are small crustaceans that sense stimuli with two large compound eyes and two pairs of antennae (fig. 50.4). *Artemia* live in salt lakes and can be bought at most pet stores. In this exercise, you will study how various stimuli affect behavior of these shrimp.

Procedure 50.3

Understand taxis in brine shrimp

1. Place brine shrimp in a test tube of saltwater provided by your instructor. Then place the tubes against a black background to see the shrimp. Do not feed or disturb the shrimp.
2. Note their behavior in this relatively stimulus-free environment.

Question 4

a. Are the shrimp near the top or bottom of the tube? Are they moving? Are they in groups or are they solitary?

b. What stimuli (e.g., light, food, acids) might initiate taxis in *Artemia?*

c. How do you predict the shrimp will respond to each of these stimuli?

Procedure 50.4

Test how stimuli influence behavior of brine shrimp

1. Choose one of the stimuli listed in your answer to Question 4.
2. Form a hypothesis to determine how this stimulus might influence the behavior of the brine shrimp. Write your hypothesis here.

3. Decide how you will test your hypothesis. Describe your experimental design here.

4. Do your experiment. What do you conclude? Do your data support your hypothesis?

Question 5

a. What is the adaptive significance of each response listed in Question 4?

b. Did the responses of the shrimp match your predicted responses?

INVESTIGATION

How do other stimuli influence the behavior of pill bugs?

a. Choose one of the stimuli listed in your answer to Question 3.

b. Form a hypothesis to determine how this stimulus affects behavior. Write your hypothesis here.

c. Decide how you will test your hypothesis. Describe your experimental design here.

d. Do your experiment. What do you conclude? Do your data support your hypothesis?

Questions for Further Thought and Study

1. Killdeer are ground-nesting birds common throughout North America. When a predator approaches, they often feign broken wings. What is the selective advantage of this type of innate behavior?

2. Mating (reproductive behavior) often involves a complex sequence of events that helps an animal find, court, and mate with a member of the same species. What are some examples of these behaviors? Do humans exhibit mating behaviors? If so, what are they?

3. What are the adaptive advantages of agonistic behaviors that are not followed by damaging fights? What animals have strong displays that are not followed by a damaging fight? What animals do engage in damaging fights?

4. How could each of the behaviors that you studied today have arisen by natural selection?

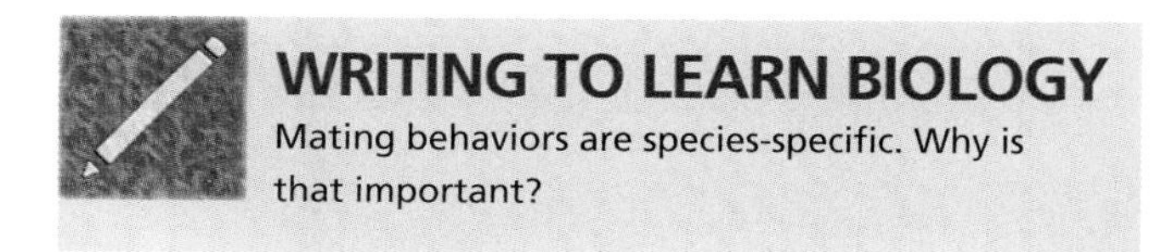

Dissection of a Fetal Pig

I

Objectives

By the end of this exercise you should be able to:
1. Perform a whole-body dissection of a vertebrate animal.
2. Identify the major anatomical features of the vertebrate body in a dissected specimen.

Fetal pig anatomy provides an excellent model of general mammalian anatomy. Although there are some significant differences, the body plan is the same, and the functional relationships within and between the anatomical systems model that of humans as well as other commonly studied mammals.

MATERIALS NEEDED

- Preserved (plain or double-injected) fetal pig
- Dissection tools and trays
- Mounted fetal pig skeleton (optional)
- Storage container (if specimen is to be reused)

CAUTION

Read the directions and safety tips for this exercise carefully before starting any procedure. In a short course, a well-preserved specimen can be used from time to time throughout your studies. If you will be looking at your dissected specimen from time to time during the next several months, make sure that it is kept in the appropriate container under conditions suggested by your instructor.

THE EXTERNAL ANATOMY

Observe the usual precautions when working with a preserved or fresh specimen. Heed the safety advice accompanying preservatives used with your specimen. Use protective gloves when you handle your specimen. Avoid injury with dissection tools and dispose of your specimen as instructed.

Procedure A.1

Examine the external characteristics of your specimen

1. Determine the anatomical orientation of the specimen. Which direction is anterior? Posterior? Which direction is ventral? Dorsal? Identify sagittal, transverse, and frontal planes in your specimen.
2. Identify these externally visible features:
   pinna (auricle)
   external nares (nostrils)
   umbilicus (umbilical cord)
   forelimbs
   hind limbs
   thoracic region
   abdominal region
   nipples
   anus
   tail
3. Determine the sex of your specimen by examining the external genitals:
   Female—Immediately anterior to the anus, on the ventral surface, is the vulva with an opening to the vagina and the urethra. The vulva is also called the urogenital opening.
   Male—In the male fetal pig, there is a rather loose area of skin immediately posterior to the anus, perhaps even hiding the anus from view. Around the time of birth, each testis will descend from its position inside the body into the space under this skin. The skin will pouch out to form the outer wall of the scrotum. Just posterior to the umbilical cord is the distal end of the penis with its prepuce, or skin-fold covering. Locate the opening of the urethra in the penis.

SKIN, BONES, AND MUSCLES

Procedure A.2

Remove the skin from the specimen

1. Place the animal in a tray; make sure the animal's ventral surface is facing you.
2. Pull up on the skin over the neck and puncture it with the tip of a scissors. Slide the bottom tip of the scissors into the subcutaneous area under the skin.
3. Begin cutting along the lines indicated in figure A.1*a*. Notice that the posterior cuts are different for male and female specimens.
4. Be careful not to cut into the skeletal muscles under the skin or through the base of the umbilical cord.
5. With your forceps, pull the two flaps of skin over the neck away from the animal's body. Notice the areolar tissue under the skin that is pulled apart as you remove the skin. Sometimes it helps if you scrape at the loose connective tissue under the skin with your scalpel as you peel the skin away. The skinning process is difficult unless you have patience and proceed slowly. Pull the flaps of skin over the abdomen and over the groin area away in a similar fashion. If you have a male pig, leave the skin posterior to the umbilical cord in place for now.
6. Pin the flaps of skin to the floor of the dissection tray or cut them away from the body entirely. If you plan to study the musculature of the dorsum, you must similarly remove the skin from the animal's back.
7. Now examine the skin, identifying these features of the integument:
   - Dermis—the thick inner layer of the skin
   - Epidermis—the thinner outer layer of the skin
   - Hypodermis—the subcutaneous tissue under the skin proper, made of areolar and adipose tissue

Explore the shape of the skinned fetal pig body. How many bones of the fetal pig's skeleton can you see or feel? If you have a mounted fetal pig skeleton available, identify as many of the bones of the skeleton as you can. If you become stumped, refer to figure A.1*a,b*. Notice the similarity between the fetal pig's skeletal plan and that of a human. One difference that is easy to see is in the vertebral column: The pig has a different number of each type of vertebral bone compared with a human and has numerous caudal vertebrae instead of a single coccyx. Also, because the fetal pig's skeleton is just beginning its development, many of the bones are cartilaginous rather than bony.

Observe the fetal pig's musculature. Some of the external muscles of the torso can be separated from each other for easier viewing. Slide a probe into the loose connective tissue joining adjacent muscles and run the probe along their margins. Using figure A.2 as a guide, try to identify the major skeletal muscles of the fetal pig's body. Because the animal was not yet active when the specimen was prepared, the muscles appear underdeveloped and make muscle identification difficult.

CARDIOVASCULAR STRUCTURES

Procedure A.3

Examine cardiovascular structures

1. Cut flaps in the neck, abdomen, and groin areas as you did with the skin. Be careful not to damage any visceral organs with your scissors as you cut. Likewise, avoid injuring the internal structures associated with the umbilicus. Fold back the flaps and anchor them with pins or remove them.
2. Open the ventral body cavity by cutting into its muscular wall in a manner similar to your earlier cut into the skin over the abdomen.
3. Locate the heart near the middle of the thoracic cavity, in the mediastinum. Can you identify the four chambers? After identifying the major vessels (see steps 4 and 5), you may want to remove the heart and dissect it.
4. Locate the aorta, which is the large artery leaving the heart and arching posteriorly. If you have a double-injected preserved specimen, the systemic arteries are filled with red latex and the systemic veins are filled with blue latex. If not, the arteries usually can be distinguished from veins because they are stiffer and lighter in color than veins. Trace the major branches of the aorta, naming them if you can. Use figure A.3 if you need help.
5. Locate the anterior vena cava and note where it drains into the heart. It is sometimes called the precava and is analogous to the superior vena cava in a human. Follow its tributary veins and identify them with the help of figure A.3. Locate the posterior vena cava (postcava) and trace its tributaries. Once you have cut into the ventral body cavity, you may be tempted to cut and remove organs. It is important that you keep everything as intact as possible. You may pull organs to the side to view deeper structures, but avoid making cuts.

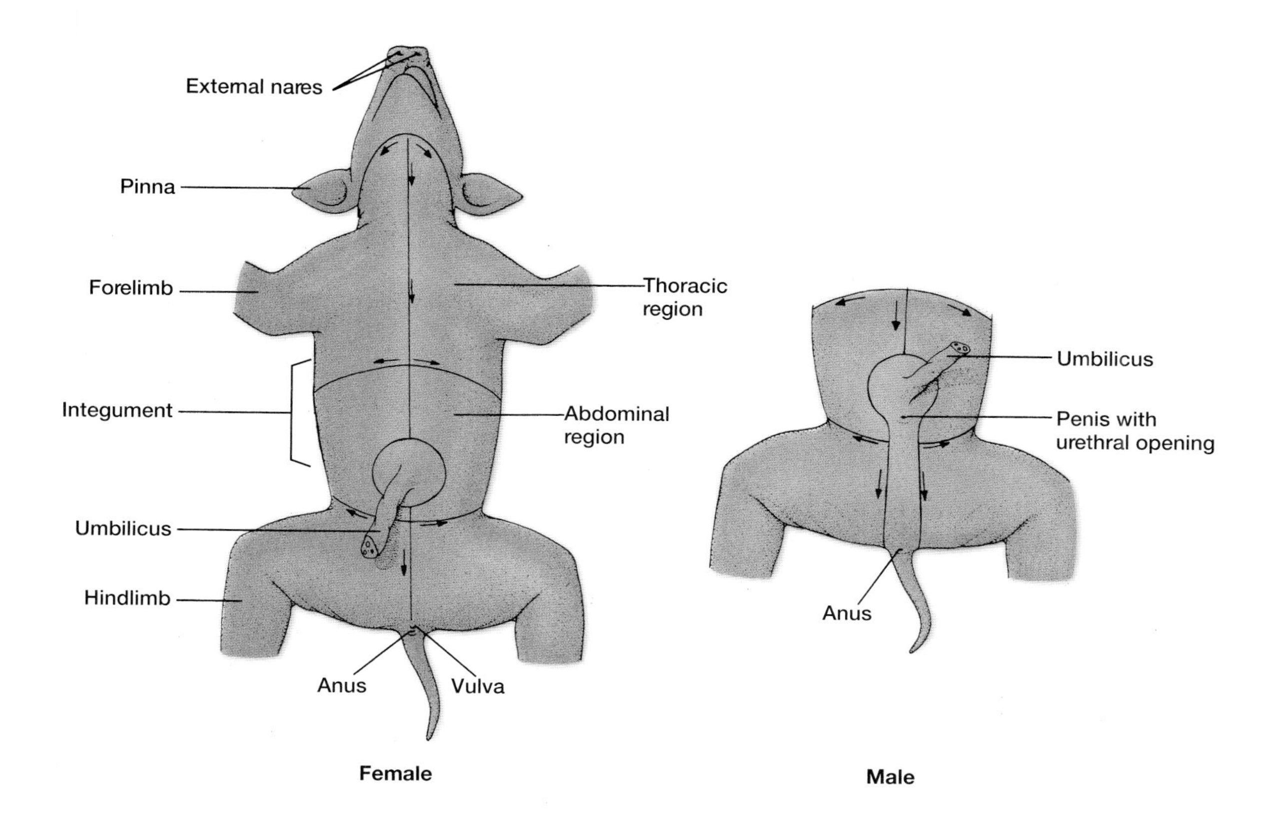

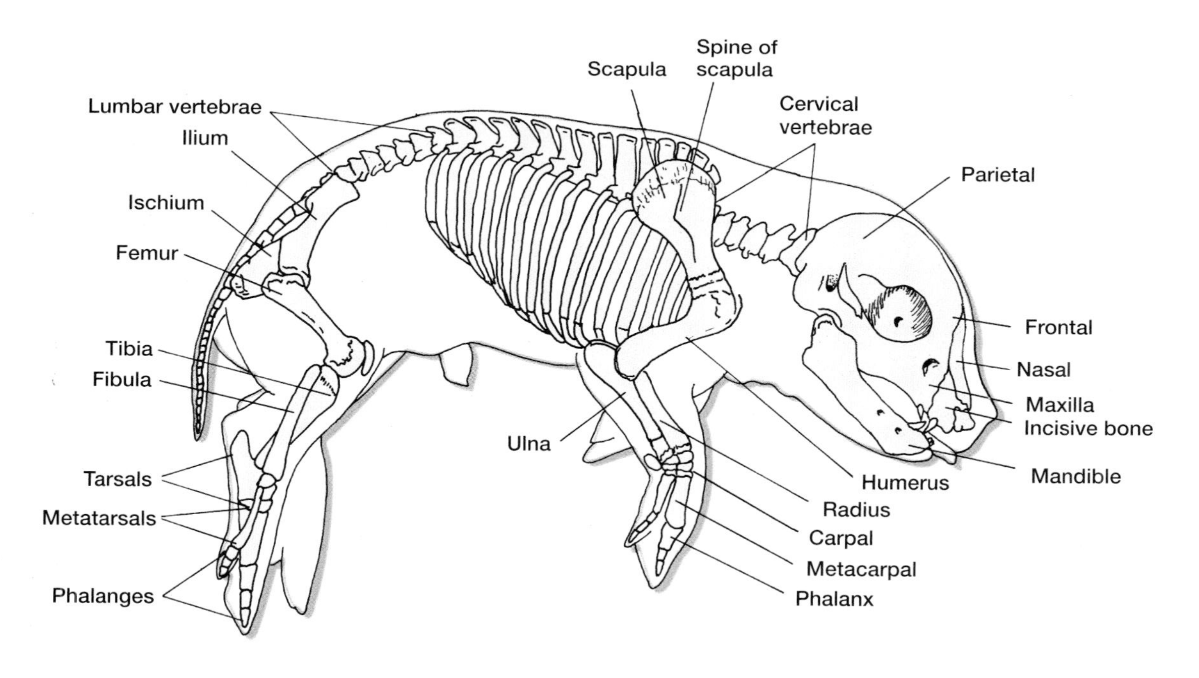

Figure A.1
(*a*) External aspect of a female (above left) and male (above right) pig. (*b*) Fetal pig skeleton.

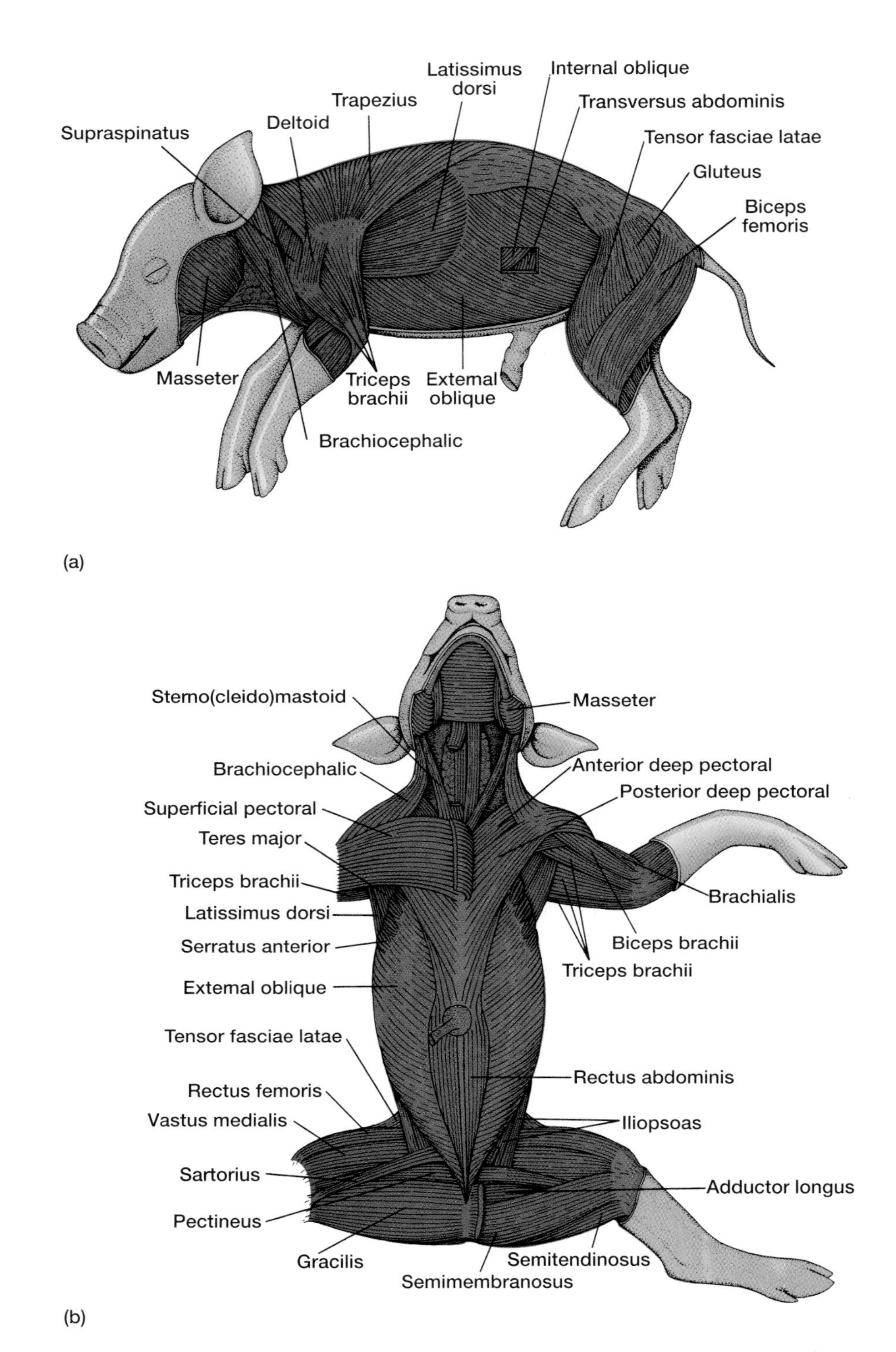

Figure A.2

(*a*) Lateral view of fetal pig musculature. (*b*) Ventral view of fetal pig musculature.

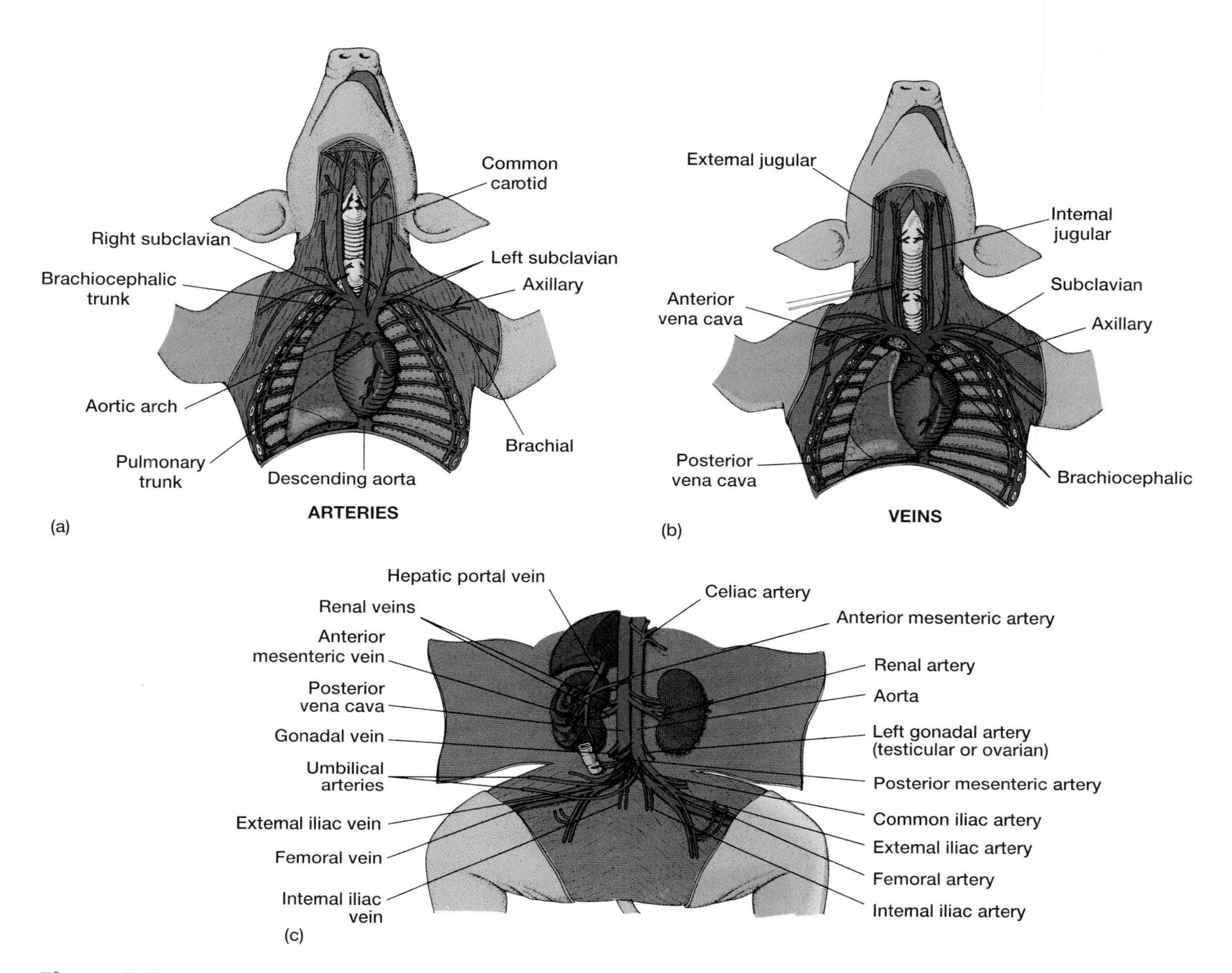

Figure A.3

(*a*) Arteries and (*b*) veins of the anterior body. (*c*) Major vessels of the posterior body.

THE VISCERA

The viscera, or major internal organs, can be seen within the ventral body cavity. Use figures A.4 and A.5 to guide you in locating the following:

1. Locate some of these features of the lower respiratory system:
   larynx
   trachea
   primary bronchi
   lungs (Can you distinguish the parietal and visceral pleurae?)
   diaphragm
2. Locate these structures of the digestive system:
   esophagus
   stomach
   liver (four separate lobes)
   gallbladder
   pancreas
   small intestine
   mesentery
   large intestine (spiral colon)
3. Locate these lymphatic organs:
   spleen
   thymus
4. Locate these features of the urinary system:
   kidney
   renal cortex
   renal pyramid
   renal pelvis
   renal calyx
   ureter
   urinary bladder
   urethra
5. Try to locate these endocrine glands in your specimen:
   thyroid gland
   thymus gland
   pancreas
   adrenal glands
   testes
   ovaries
6. Identify these structures associated with the male reproductive system:
   testes
   epididymis
   ductus deferens
   seminal vesicle
   penis
7. Find these female reproductive system structures:
   ovaries
   oviducts (fallopian tubes)
   uterus (Notice that the fetal pig uterus has a Y shape, with right and left uterine horns.)
   vagina
8. Carefully examine the umbilical cord, noting these structures:
   umbilical vein
   umbilical arteries
9. In the space below, sketch a cross-section of the umbilical cord and label all identifiable structures. Unless your lab group has both a male and a female specimen, you may want to temporarily trade specimens with a group that has a pig of the opposite gender from yours. By doing so, you will be able to find the features of both reproductive systems.

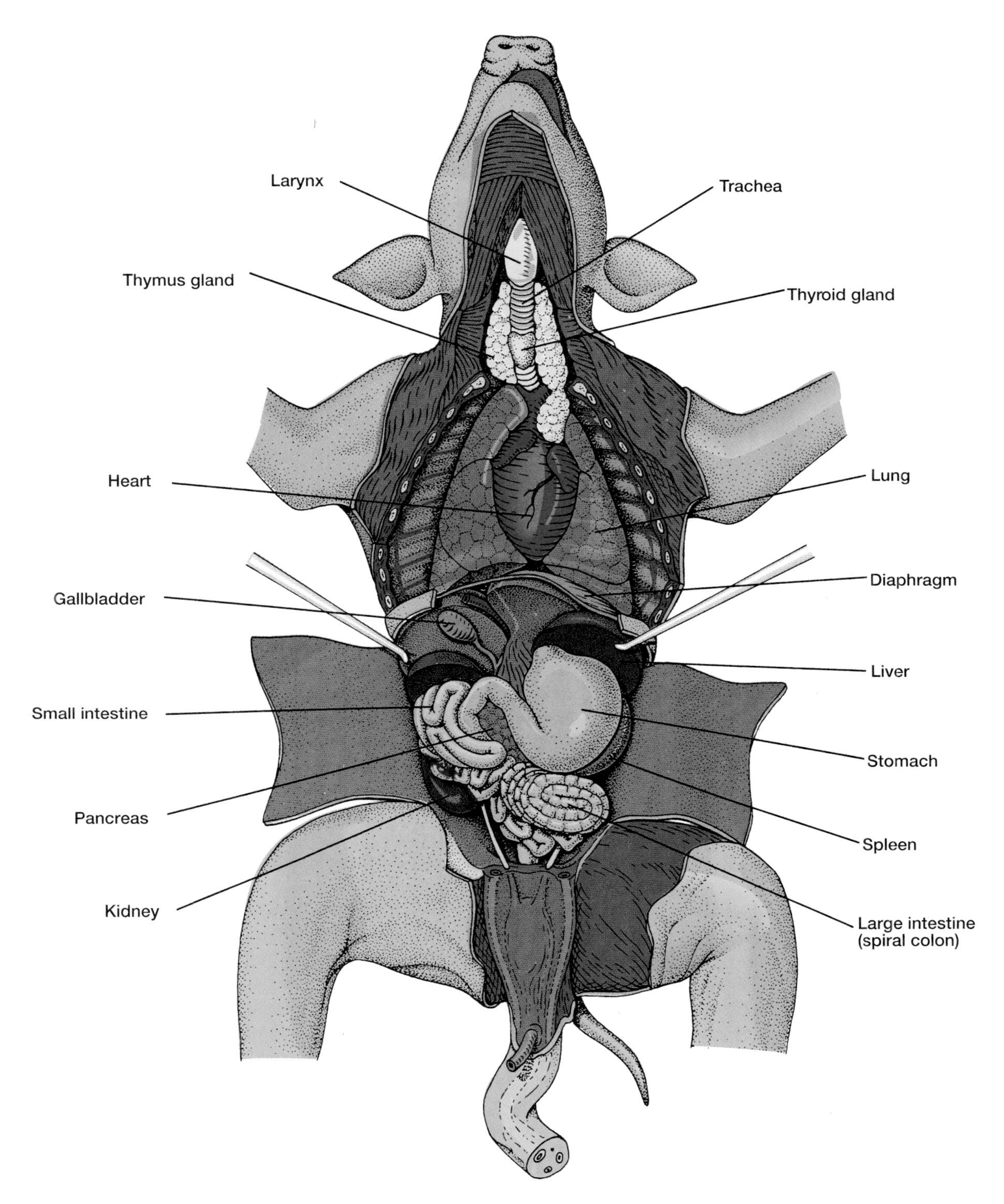

Figure A.4
Major visceral organs of a fetal pig.

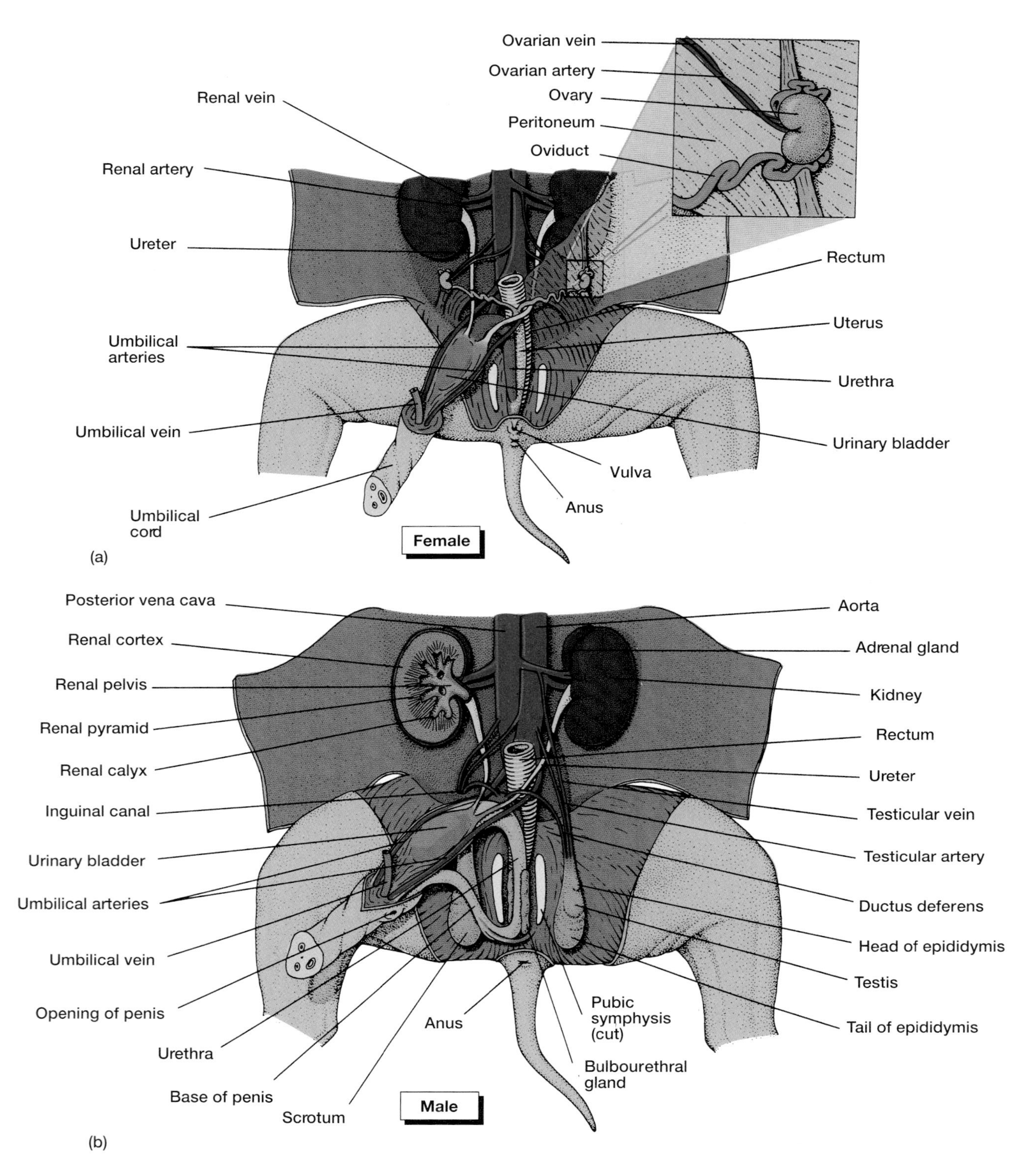

Figure A.5

Urinary and reproductive organs of (*a*) a female and (*b*) a male fetal pig.

II

How to Write a Scientific Paper or Laboratory Report

Your instructor may occasionally ask you to submit written reports describing the work you did in the lab. Although these reports will probably not be published in scientific magazines or journals, they are important because they will help you learn to write a scientific paper. A scientific paper is a written description of how the scientific method was used to study a problem.

Understanding how to write a scientific paper (such as a lab report) is important for several reasons. Scientists become known (or remain unknown) by their publications in books, magazines, and scientific journals. Regardless of the presumed importance of a scientist's discoveries, poor writing delays or prohibits publication because it makes it difficult to understand what the scientist did or the importance of the work. Poor writing usually indicates an inability or unwillingness of a scientist to think clearly.

Scientific papers are the vehicle for the transmission of scientific knowledge; they are available for others to read, test, refute, and build on. Few skills are more important to a scientist than learning how to write a scientific paper.

Before you finish reading this appendix, go to the library and browse through a few biological journals such as *American Journal of Botany, Ecology, Journal of Mammalogy*, or *Journal of Cell Biology*. Make photocopies of one or two of the articles that interest you. As you'll see, scientific papers follow a standard format that reflects the scientific method.

PARTS OF A SCIENTIFIC PAPER

Almost all scientific papers have these parts:

- Title
- List of authors
- Abstract
- Introduction
- Materials and methods
- Results
- Discussion
- References

Understanding this format eases the burden of writing a scientific paper, because writing is an exercise in organization. Refer to the journal articles you photocopied in the library as you read this appendix.

Title

The title of a paper is a short label (usually fewer than 10 words) that helps readers quickly determine their interest in the paper. The title should reflect the paper's content and contain the fewest number of words that adequately express the paper's content. The title should never contain abbreviations or jargon (jargon is overly specialized or technical language).

List of Authors

Only those people who actively contributed to the design, execution, or analysis of the experiment should be listed as authors.

Abstract

The abstract is a short paragraph (usually fewer than 250 words) that summarizes (1) the objectives and scope of the problem, (2) methodology, (3) data, and (4) conclusions. The abstract contains no references.

Introduction

The introduction concisely states why you did the work. Avoid exhaustive reviews of what has already been published; rather, limit the introduction to just enough pertinent information to orient the reader to your study.

The introduction of a scientific paper has two primary parts. The first part is a description of the nature and background of the problem. For example, what do we already know (or not know) about the problem? This description is developed by citing other scientists' work, to give a history of the study of the problem, and by pointing out gaps in our knowledge. The second part of the introduction states the objectives of the study.

Materials and Methods

The materials and methods section describes how, when, where, and what you did. It should contain enough detail to allow another scientist to repeat your experiment, but it should not be overwhelming.

Materials include items such as growth conditions, organisms, and the chemicals used in the experiment. Avoid trade names of chemicals and describe organisms with their scientific names (e.g., *Zea mays* rather than "corn"). Also describe growth conditions, diet, lighting, temperature, and so on.

Methods are usually presented chronologically, and this discussion is often subdivided with headings. Examples of methods include sampling techniques, types of microscopy, and statistical analyses. If possible, use references to describe methods.

Experiments described in a scientific paper must be reproducible. Thus, the quality of materials and methods is judged by the reader's ability to repeat the experiment. If a colleague can't repeat your experiment, the materials and methods section is probably poorly written.

For most lab reports, do not copy the experimental procedures word for word from the lab manual. Rather, summarize what you did in several sentences.

Results

The results section is the heart of a scientific paper. It should clearly summarize your findings and leave no doubt about the outcome of your study. For example, state that "All animals died 29 hours after eating cyanide" or "Table 1 shows the influence of 2,4-D on leaf growth." Keep it simple and to the point.

Tables and graphs are excellent ways to present results but shouldn't completely replace a written summary of results. Tables are ideal for presenting large amounts of numerical data, and graphs are an excellent way to summarize data and show relationships between independent and dependent variables. The variable that the scientist established and controlled during the experiment is the **independent variable.** It is presented on the *x*-axis of the graph. Protein content of a diet might be an independent variable in an experiment measuring weight gain by an animal (fig. A.6). Similarly, time and temperature are often independent variables.

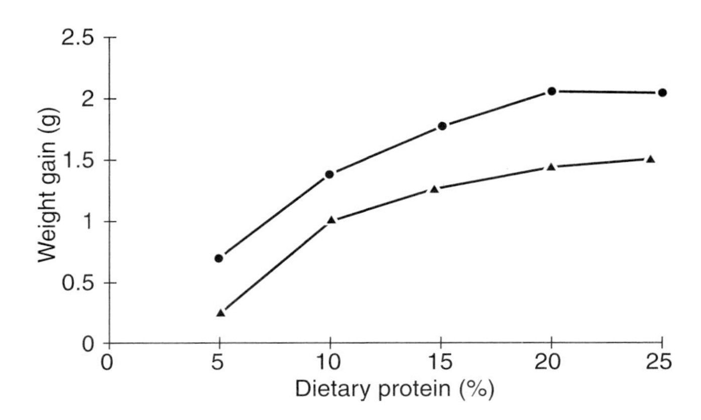

Figure A.6
Sample graph from a scientific paper.

The **dependent variable** changes in response to changes in the independent variable and is presented on the *y*-axis of the graph. Weight and growth rate are examples of dependent variables that may change in response to light, temperature, diet, and so on. Graphs must also have a title (e.g., "Influence of Temperature on Root Elongation"), labeled axes (e.g., "Temperature," "Root Elongation"), and scaled units along each axis appropriate to each variable (e.g., °C, mm h^{-1}). Place tables and graphs on separate pages from the text.

Discussion

It's not enough to simply report your findings; you must also discuss what they mean and why they're important. This is the purpose of the discussion section of a scientific paper. This section should interpret your results relative to the objectives you described in the introduction and answer the question "So what?" or "What does it mean?" A good discussion section should do the following:

- Discuss your findings; that is, present relationships, principles, and generalizations. Point out exceptions and lack of correlations.
- Don't conceal anomalous results; rather, describe unsettled points. State how your results relate to existing knowledge.
- State the significance and implications of your data. What do your results mean? If your data are strong, don't hesitate to use statements beginning with "I conclude that..."

References

Scientists rely heavily on information presented in papers written by their colleagues. Indeed, the Introduction, Materials and Methods, and Discussion sections of a paper often contain citations of other publications. The format for these citations varies in different biological journals. The following citation for an article is in the format recommended by the Council of Biology Editors:*

White, H.B., III. Coenzymes as fossils of an earlier molecular state. J. Mol. Evol. 7:101–104; 1976.

A FEW SIMPLE RULES FOR WRITING EFFECTIVELY

Informative sentences and well-organized paragraphs are the foundation of a good scientific paper. Listed below are a few rules to help you write effectively. Following these rules won't necessarily make you a Hemingway, but it will probably improve your writing.

- *Write clearly and simply.* For example, "the biota exhibited a 100% mortality response" is a wordy and pretentious way of saying "all of the organisms died." Remember, keep it simple and straightforward.
- *Keep related words together.* Consider the following sentence taken from a scientific publication: "Lying on top of the intestine, you perhaps make out a small transparent thread." Do we really have to lie on top of the intestine to see the thread? The author meant that "a small transparent thread lies atop the intestine."
- *Use active voice.* Write "Good writers avoid passive voice," not "The passive voice is avoided by good writers." Here are some other examples of passive voice:

  Poor: My first lab report will always be remembered by me. (passive)

  Better: I'll always remember my first lab report. (active)

  Poor: Examination of patients was accomplished by me. (passive)

  Better: I examined patients. (active)
- *Write positively.* For example, write "The rats were always sick" instead of "The rats were never healthy." Use definite and specific sentences. For example, write "It rained every day for a week" instead of "A period of unfavorable growth conditions set in."
- *Be sure of the meaning of every word that you use, and write exactly what you mean.* Refer to a dictionary and thesaurus to ensure clarity and proper word usage. For example, you allude to a book, and elude a pursuer.
- *Delete unnecessary words.* For example:

Replace	*With*
The question as to whether	Whether
Advance notice	Notice
At this point in time	Now
Be that as it may	But
In the event that	If
General consensus	Consensus
Young juvenile	Juvenile
Student body	Students
Due to the fact that	Because
Chemotherapeutic agent	Drug

- *Use metric measurements* (see Exercise 1).
- *Be sure that each paragraph conveys a single major idea and has a topic sentence.* The topic sentence should state the main idea of the paragraph.
- *Have a friend or colleague read a draft of your writing and suggest improvements.*
- *Don't plagiarize.* Learn to summarize and be sure to cite all references from which you extracted information.

A neat and typed presentation is a must. If you use a word processor, remember to use the spell checker. Carefully proofread to catch mistakes. Put your work aside for at least 24 hours before you proofread. If you're interested in learning more about improving your writing, read *The Elements of Style* (3rd ed.), by W. Strunk and E. B. White (New York: Macmillan, 1979).

* *Council of Biology Editors style manual: A guide for authors, editors, and publishers in the biological sciences. 5th ed. Council of Biology Editors; 1983.*

Conversion of Metric Units to English Units

Units of Length

The meter (m) is the basic unit of length.

1 m = 39.4 inches (in)	
= 1.1 yard (yd)	1 in = 2.54 cm
1 km = 1000 m = 10^3 m	
= 0.62 miles (mi)	1 ft = 30.5 cm
1 cm = 0.01 m = 10^{-2} m	
= 0.39 in = 10 mm	1 yd = 0.91 m
1 nm = 10−9 m = 10^{-6} mm	
= 10 angstroms (Å)	1 mi = 1.61 km

Units of area are squared (two-dimensional) units of length.

1 m^2 = 1.20 yd^2 = 1550 in^2 = 1.550×10^3 in^2

1 hectare = 10,000 square meters (m^2) = 2.47 acres

Measurements of area and volume can use the same units.

1 m^3 = 35.314 ft^3 = 1.31 yd^3

1 cm^3 (cc) = 0.000001 m^3 = 0.061 in^3

Units of Mass

The gram (g) is the basic unit of mass.

1 g = mass of 1 cm^3 of water at 4°C = 0.035 oz

1 kg = 1000 g = 10^3 g = 2.2 lb

Units of Volume

The liter (L) is the basic unit of volume. Units of volume are cubed (three-dimensional) units of length.

1 liter = 1000 cm^3	
1 liter = 2.1 pints = 1.06 qt	1 cup = 240 mL
1 liter = 0.26 gal = 1 dm^3	
1 mL = 0.035 fl oz	

Units of Temperature

5 × degrees Fahrenheit = (9 × degrees Celcius) + 160

For example:

40°C = 104°F (a hot summer day)

75°C = 167°F (hot coffee)

−5°C = 23°F (coldest area of freezer)

37°C = 98.6°F (human body temperature)

Credits

LINE ART

Exercise 39

Figure 39.22: From Wallace/Mahan, *Introduction to Ornithology*. Copyright © 1975. Reprinted by permission of Prentice Hall, Upper Saddle River, New Jersey; **39.25:** From Milton Hildebrand, *Analysis of Vertebrate Structure*, 4th edition. Copyright © 1995 John Wiley & Sons, Inc. Reprinted by permission of John Wiley & Sons, Inc.

Exercise 44

Text: © Arizona Heart Institute, Phoenix, AZ. Reprinted by permission.

PHOTOS

Exercise 1

Figures 1.1-1.3b: Biologyimages.com.

Exercise 2

Figures 2.3-2.7b: Biologyimages.com; **2.8:** Dr. E.R. Degginger; **2.9:** Courtesy of Leica, Inc., Deerfield, Illinois.

Exercise 3

Figure 3.1: © David M. Phillips /Visuals Unlimited; **3.2:** Courtesy T.D. Pugh and E.H. Newcomb, University of Wisconsin Botany Dept.; **3.3a,b:** Biologyimages.com; **3.4:** © K.G. Murti/Visuals Unlimited; **3.5:** Courtesy of Jean M. Shatley; **3.6a:** © Dwight Kuhn; **3.7:** © J.D. Litvay/ Visuals Unlimited; **3.8:** Courtesy of Kenneth Miller, Brown University; **3.9:** John D. Cunningham/Visuals Unlimited; **3.11:** © Don W. Fawcett/Visuals Unlimited; **3.12:** © John D. Cunningham/Visuals Unlimited; **3.13b & 3.14b:** © M. Abbey/Visuals Unlimited.

Exercise 5

Figure 5.3: © BioPhoto Associates/Photo Researchers, Inc.; **5.5a:** © Manfred Kage/Peter Arnold, Inc.; **5.5b:** © Michael Pasdzior /The Image Bank /Getty Images; **5.5c:** © George Bernard /Animals Animals /Earth Scenes; **5.5d:** © Oxford Scientific Films/Animals Animals/Earth Scenes; **5.5e:** © Scott Blackman/ Tom Stack & Associates.

Exercise 6

Figures 6.2-6.11: EDVOTEK; **6.12e:** George Kantor.

Exercise 7

Figure 7.2: Biologyimages.com.

Exercise 8

Figures 8.1 & 8.10a: © Runk/ Schoenberger /Grant Heilman Photography; **8.10b:** © Alfred Owczarzak/Biological Photo Service.

Exercise 9

Figure 9.2b: From J. David Robertson, *Medical Cell Biology*, 1979 by Charles Flickinger. © W.B. Saunders, Co.

Exercise 11

Figure 11.1: Biologyimages.com.

Exercise 12

Figure 12.1: © R. Calentine /Visuals Unlimited; **12.7 & 12.8:** Biologyimages.com; **12.10a:** Courtesy of Kenneth Miller, Brown University; **12.10b:** Biologyimages.com.

Exercise 13

Figure 13.2b: © BioPhoto Associates/ Photo Researchers, Inc.; **13.4:** © David M. Phillips/Visuals Unlimited; **13.5a-d:** © Ed Reschke; **13.6a-d:** Biologyimages.com.

Exercise 14

Figure 14.6: © Ed Reschke.

Exercise 15

Figure 15.1: Courtesy of Ulrich K. Laemmli; **15.3:** Courtesy of Dr. Stanley N. Cohen.

Exercise 16

Figure 16.1b: © Frank B. Sloop, Jr., M.D.; **16.2(both) & 16.4a, b:** © McGraw-Hill Companies/Bob Coyle, photographer.

Exercise 17

Figure 17.1: Biologyimages.com; **17.2:** © Mary Evans Picture Library /Photo Researchers, Inc.; **17.4:** © Michael Fogden; **17.7:** © Eric Grave/Phototake; **17.8:** © M.I. Walker/Photo Researchers, Inc.; **17.9:** © Philip Sze/Visuals Unlimited; **17.10:** © BioPhoto Associates/Photo Researchers, Inc.; **17.11:** © Carolina Biological Supply/Visuals Unlimited.

Exercise18

Figure 18.3(skull): © John Reader/SPL/Photo Researchers, Inc.; **18.3(jaw):** The Natural History Museum, London; **18.4:** © McGraw-Hill Companies/Bob Coyle, photographer.

Exercise 19

Figure 19.1: © Brian Parker/Tom Stack & Associates.

Exercise 21

Figure 21.2: Courtesy National Museum of Natural History, © 2001 Smithsonian Institution.

Exercise 22

Figure 22.1: © John D. Cunningham/Visuals Unlimited; **22.2:** © Fred Ward 1997.

Exercise 23

Page 232(top): © Abraham & Beachey/BPS/Tom Stack & Associates; **(bottom):** Friedrich Widdel/Visuals Unlimited; **23.1:** © David M. Phillips/Visuals Unlimited; **23.2a-d:** © Dr. Tony Brain/David Parker/SPL/Photo Researchers, Inc.; **23.3a-c:** © David M. Phillips/Visuals Unlimited; **23.4:** © G. Musil/Visuals Unlimited; **23.A:** © T.J. Beveridge/Tom Stack & Associates; **23.5:** © Leon J. LeBeau/Biological Photo Service; **23.8:** © Cabisco/Visuals Unlimited; **23.9:** © Fred Marsik/Visuals Unlimited; **23.10a:** © Sinclair Stammers/Photo Researchers, Inc.; **23.10b:** © E.C.S. Chan/Visuals Unlimited; **23.10c:** © Runk/Schoenberger/Grant Heilman Photography; **23.10d:** © Ron Dengler/Visuals Unlimited.

Exercise 24

Figure 24.1: © M.I. Walker/Photo Researchers, Inc.; **24.3a-d:** © Carolina Biological Supply/Phototake; **24.4:** © Philip Sze; **24.6:** © John D. Cunningham/Visuals Unlimited; **24.7:** © Kingsley Stern; **24.9a:** © Heather Angel; **24.9b:** © E.C.S. Chan/Visuals Unlimited; **24.10a:** © Eric Grave/Photo Researchers, Inc.; **24.10b:** © Dr. D.P. Wilson/Photo Researchers, Inc.; **24.11:** Courtesy J.D. Pickett Heaps; **24.13a:** © Michael Abbey/Visuals Unlimited.

Exercise 25

Figure 25.1a: © M. Abbey / Visuals Unlimited; **25.3:** © Dwight Kuhn; **25.5a:** © Ed Reschke; **25.5b:** © Edward S. Ross; **25.7a:** © Brain Parker/Tom Stack & Associates; **25.7b:** © Manfred Kage/Peter Arnold, Inc.; **25.8:** © John D. Cunningham/Visuals Unlimited; **25.10:** © Ed Reschke.

Exercise 26

Figure 26.4: © Dr. Jeremy Burgess/SPL/Photo Researchers, Inc.; **26.6:** © Richard H. Gross; **26.7a:** © Ed Pembleton; **26.7b:** © Kjell Sandved/Butterfly Alphabet, Inc.; **26.8a:** © David Phillips/Visuals Unlimited; **26.8b:** © BioPhoto Associates/Photo Researchers, Inc.; **26.9:** © J. Michael Eichelberger/Visuals Unlimited; **26.10:** © George B. Chapman/Visuals Unlimited; **26.11:** © Ed Reschke; **26.13a:** Biologyimages.com; **26.13b:** © Hans Reinhard/Bruce Coleman, Inc.; **26.13c:** Biologyimages.com; **26.15:** © Bruce Iverson; **26.16:** © Carolina Biological Supply/Phototake; **26.17a:** © Runk/Schoenberger/Grant Heilman Photography; **26.17b:** © L. West/Photo Researchers, Inc.; **26.17c:** © Ed Reschke / Peter Arnold, Inc.

Exercise 27

Figure 27.3a,b: Ed Reschke/Peter Arnold Inc.; **27.4:** © Triarch/Visuals Unlimited; **27.5a:** © John D. Cunningham/Visuals Unlimited; **27.5b:** © Robert & Linda Mitchell; **27.6a:** © Ed Reschke; **27.6b:** © Bruce Iverson; **27.8b:** © Edward S. Ross; **27.9 & 27.10:** Courtesy of G.S. Ellmore; **27.11:** © William E. Ferguson.

Exercise 28

Figure 28.2a: © Runk/Schoenberger/Grant Heilman Photography; **28.2b,c:** © Kjell B. Sandved/Butterfly Alphabet; **28.3:** Jack M. Bostrack/Visuals Unlimited; **28.5:** © Kingsley Stern; **28.6:** © Wm. Ormerod/Visuals Unlimited; **28.7:** Biologyimages.com; **28.8:** © Edward S. Ross; **28.9a,b:** Biologyimages.com; **28.10:** © Ed Reschke; **28.11:** Biologyimages.com; **28.12:** © Kingsley Stern.

Exercise 29

Figure 29.2a,b: © Kingsley Stern; **29.3:** © Runk/Schoenberger/Grant Heilman Photography; **29.4a,b:** © Robert & Linda Mitchell; **29.5:** © William E. Ferguson; **29.7:** © Phil Gates/Biological Photo Service; **29.8:** Biologyimages.com; **29.9:** © George J. Wilder/Visuals Unlimited; **29.10:** Biologyimages.com; **29.11a:** © Robert & Linda Mitchell; **29.11b:** © Science VU/Visuals Unlimited; **29.11c:** Walter H. Hodge/Peter Arnold Inc.

Exercise 30

Figure 30.1: © Edward S. Ross; **30.2:** © Runk/Schoenberger/Grant Heilman Photography; **30.6:** © Carolina Biological Supply/Phototake; **30.7:** Biologyimages.com; **30.8a-d & 30.11a-c:** © Biodisc, Inc.; **30.15:** © E.R. Degginger/Bruce Coleman, Inc.

Exercise 31

Figure 31.1a: © Lynwood M. Chace/Photo Researchers, Inc.; **31.1b:** © John D. Cunningham/Visuals Unlimited; **31.2b, 31.3:** © BioPhot; **31.4a-31.5:** © Ed Reschke; **31.6d:** © BioPhot; **31.7:** Biologyimages.com; **31.9:** © Ed Reschke; **31.10:** © BioPhot; **31.11:** © Ed Reschke; **31.12:** © Triarch / Visuals Unlimited; **31.13-31.15:** © BioPhot.

Exercise 32

Figure 32.1: © Patrick W. Grace/Photo Researchers, Inc.; **32.3:** © Terry Ashley/Tom Stack & Associates.

Exercise 33

Figure 33.1: © Larry Simpson/Photo Researchers, Inc. **33.2:** © Kingsley Stern; **33.3a,b:** © BioPhot; **33.4(both):** © Professor Malcolm B. Wilkins, Botany Dept., Glasgow University; **33.8:** Sylvan Wittwer/Visuals Unlimited, Inc.

Exercise 35

Figure 35.1a: © Nancy Sefton/Photo Researchers, Inc. **35.1b:** © Daniel W. Gotschall/Visuals Unlimited; **35.2:** © Carolina Biological Supply/Phototake; **35.4:** © Robert & Linda Mitchell; **35.5:** © Carolina Biological Supply/Visuals Unlimited, Inc.; **35.6:** © Robert & Linda Mitchell; **35.9a:** © Ed Reschke; **35.11:** © A. Bannister/N.H.P.A.; **35.12a:** © Carolina Biological Supply/Visuals Unlimited; **35.13:** © N.G. Daniel/Tom Stack & Associates; **35.15:** © Runk/Schoenberger/Grant Heilman Photography; **35.17, 35.18:** Biologyimages.com.

Exercise 36

Figure 36.1a: © T.E. Adams/Visuals Unlimited; **36.1b:** © Stan Elems/Visuals Unlimited; **36.2b:** Biologyimages.com; **36.3:** © Carolina Biological Supply/Phototake; **36.4:** © E.J. Cable/Tom Stack & Associates; **36.5:** © Carolina Biological Supply/Phototake; **36.6a:** From

A Pictorial Presentation of Parasites Edited by H. Zaiman, M.D.; **36.9:** © Carolina Biological Supply/ Visuals Unlimited; **36.10:** © Larry Jensen/Visuals Unlimited; **36.11:** E.L. Schiller/Armed Forces Institute of Pathology; **36.12:** © Science VUFIP/Visuals Unlimited; **36.14 (both):** © Stan Elems/Visuals Unlimited.

Exercise 37

Figure 37.2a: © Milton Rand/Tom Stack & Associates; **37.2b:** © Kjell Sandved/Butterfly Alphabet, Inc.; **37.2c:** © Marty Snyderman; **37.3:** © Walter E. Harvey/Photo Researchers, Inc.; **37.4:** © Science VU - Polaroid/Visuals Unlimited; **37.5:** © F. McConnaughey/Photo Researchers, Inc.; **37.7a:** © Fred Bavendam; **37.7b:** © James Watt/Animals Animals/Earth Scenes; **37.7c:** © Edward S. Ross; **37.8a:** © Robert De Goursey/ Visuals Unlimited; **37.13:** © Dwight Kuhn.

Exercise 38

Figure 38.1: © Tom McHugh / Photo Researchers, Inc.; **38.4, 38.5:** © Carolina Biological Supply/ Phototake; **38.9a:** © Adrian Wenner/Visuals Unlimited; **38.10:** © Runk/Schoenberger/Grant Heilman Photography.

Exercise 39

Figure 39.1a: © Alex Kerstitch; **39.1b,c:** © Carl Roessler/Tom Stack & Associates; **39.1d:** © William Ober, M.D.; **39.1e:** © Carolina Biological Supply/Phototake; **39.2b:** © Kjell Sandved/Butterfly Alphabet, Inc.; **39.5 & 39.6b:** Biologyimages.com; **39.11b:** © David Hall/Photo Researchers, Inc.; **39.12a:** © Peter Scoones/Woodfin Camp & Associates; **39.13:** © John D. Cunningham/Visuals Unlimited; **39.15a:** © Russ Kinne/Photo Researchers, Inc.; **39.16b:** Jeff Rotman; **39.18:** Courtesy of C.P. Hickman; **39.19:** © Tom McHugh/ Photo Researchers, Inc.; **39.20:** Courtesy of C.P. Hickman; **39.21a:** © Karl H. Switzk/Photo Researchers, Inc.; **39.23a:** U.S. Fish and Wildlife Service; **39.23b:** © Tom McHugh/Photo Researchers, Inc.; **39.24:** © G. Ronald Austing/ Photo Researchers, Inc.

Exercise 40

Figure 40.3a,b: © Ed Reschke; **40.3c:** © Visuals Unlimited; **40.4:** Biologyimages.com; **40.5:** © Manfred Kage/Peter Arnold, Inc.; **40.6:** © Ed Reschke; **40.7:** © Fred Hossler / Visuals Unlimited; **40.8:** Courtesy of Dr. Jerome Gross, Massachusetts General Hospital; **40.9:** © Ed Reschke; **40.10:** © R. Calentine/Visuals Unlimited; **40.11:** © Manfred Kage/Peter Arnold; **40.12-40.15:** © Ed Reschke; **40.16:** Biologyimages .com; **40.18a:** © Ed Reschke; **40.20a-c:** © John D. Cunningham/ Visuals Unlimited; **40.21:** © Lennart Nilsson, *Behold Man*, Little Brown and Company, Bonier Fatka; **40.23:** © E.R. Lewis/ BPS/Tom Stack & Associates.

Exercise 42

Figure 42.1: © D.W. Fawcett and J. Venable/Visuals Unlimited.

Exercise 45

Figure 45.7b: © Ed Reschke.

Exercise 48

Figure 48.10: © A. & F. Michler/Peter Arnold, Inc.

Exercise 49

Figure 49.1a-49.2e: Biologyimages.com; **49.3:** © Hans Pfletschinger/Peter Arnold, Inc.; **49.4-49.8:** Biologyimages.com; **49.12 & 49.13:** © Carolina Biological Supply/Phototake.

Exercise 50

Figure 50.3: © Richard Walters/Visuals Unlimited; **50.4:** © Carolina Biological Supply/Visuals Unlimited.

Notes

Notes

Notes

Notes

Notes

Notes

Notes

Notes

Notes

Notes